2nd International Conference on Science and Innovated Engineering

IOP Conference Series: Materials Science and Engineering
Volume 854

Malacca, Malaysia
9 - 10 November 2019

Editors:

Usman
Nelly Safitri
Sulaiman Wadi Harun
Anas Abdul Latiff
Mohd Fauzi Ab Rahman
Muhammad Taufiq Ahmad
Muhammad Farid Mohd Rusdi

ISBN: 978-1-7138-4734-2
ISSN: 1757-8981

Printed from e-media with permission by:

Curran Associates, Inc.
57 Morehouse Lane
Red Hook, NY 12571

Some format issues inherent in the e-media version may also appear in this print version.

Institute of Physics
Dirac House, Temple Back
Bristol BS1 6BE UK

Phone: 44 1 17 929 7481
Fax: 44 1 17 920 0979

techtracking@iop.org

Additional copies of this publication are available from:

Curran Associates, Inc.
57 Morehouse Lane
Red Hook, NY 12571 USA
Phone: 845-758-0400
Fax: 845-758-2633
Email: curran@proceedings.com
Web: www.proceedings.com

TABLE OF CONTENTS

IOP Conf. Series: Materials Science and Engineering **854** (2020) 011001 doi:10.1088/1757-899X/854/1/011001

THE 2ND INTERNATIONAL CONFERENCE ON SCIENCE AND INNOVATED ENGINEERING
9 - 10 November 2019, Melaka, Malaysia

Banner

Preface

The 2nd International Conference on Innovative Technology, Engineering, and Sciences is an inaugural conference organized by the Lhoksomawe State Polytechnique (PNL) and Optical Society of Malaysia (OSM). The main goal of i-COSINE is to bring together researchers, scientists and experts in universities, companies, institutions, communities, agencies, associations and societies to provide them a unique platform for sharing worldwide ideas as well as the recent developments on science and engineering technology. This conference has received nearly 90 participants from various field of technology, engineering, and sciences. It has attracted researchers and academicians from ASEAN region to gather in UNESCO World Heritage City of Melaka, Malaysia. A compilation of articles that represent the significant results of the conference has been selected to publish in this volume.

Two core research topics have been compiled and listed as follows:

1. Engineering and Technology
    - Chemical Engineering
    - Civil Engineering
    - Electrical Engineering
    - Environmental Engineering
    - Mechanical Engineering
    - Food Technology

2. Science
    - Chemistry
    - Physics
    - Mathematics
    - Green and Social Innovation

IOP Conf. Series: Materials Science and Engineering **854** (2020) 011001 doi:10.1088/1757-899X/854/1/011001

Editors
Usman
Nelly Safitri
Sulaiman Wadi Harun
Anas Abdul Latiff
Mohd Fauzi Ab Rahman
Muhammad Taufiq Ahmad
Muhammad Farid Mohd Rusdi

Event Pictures

IOP Conf. Series: Materials Science and Engineering **854** (2020) 011001 doi:10.1088/1757-899X/854/1/011001

List of Committee

Advisor
Sulaiman Wadi Harun (University of Malaya, Malaysia)
Waleed S. Mohammed, Thailand (Bangkok University, Thailand)
Marwan (Syiah Kuala University, Indonesia)
Kaharudin Dimyati (University of Malaya, Malaysia)
Zahriladha Zakaria (Universiti Teknikal Malaysia Melaka, Malaysia)

Conference Chair
Usman
Hazli Rafis Abdul Rahim

Secretary
Ahmad Haziq Aiman Rosol
Zuraidah Harith
Alfian Putra
Busra

Committee Member
Aminah Ahmad
Arni Munira Markom
Mohd Erfy Ismail
Muhammad Fahmi Hisamudin
Muhammad Taufiq Ahmad
Muhammad Farid Mohd Rusdi

IOP Conf. Series: Materials Science and Engineering **854** (2020) 011001 doi:10.1088/1757-899X/854/1/011001

Ahmad Razif Muhammad
Anas Abdul Latiff
Mohd Fauzi Ab Rahman
Afiq Arif Aminuddin Jafry
Nabilah Kasim
Haziezol Helmi Mohd Yusof
Mohd Adzimnuddin Mohd Nor Azami
Malathy Batumalay
Mursyidah
Suryati
Azwar
Nelly Safitri
Musbar
Elwina
Hamdani
Zulkarnaini

IOP Conf. Series: Materials Science and Engineering **854** (2020) 011002 doi:10.1088/1757-899X/854/1/011002

Name	Institution
Noraini Marsi	Universiti Tun Hussein Onn Malaysia
Z Erwanto	Politeknik Negeri Banyuwangi
E Kurniasih	Lhokseumawe State of Polytechnic
Husaini	Laboratory of Computational Mechanics, Department of Mechanical Engine
Suryati	Politeknik Negeri Lhokseumawe
Muryanto muryanto	Research Center for Chemistry – LIPI
Nazaruddin	Politeknik Negeri Lhokseumawe
Ratna Sari	Lhokseumawe State Polytechnic
Salahuddin	Politeknik Negeri Lhokseumawe
Lim JW	INTI International University
Syarwan	Politeknik Negeri Lhokseumawe
Mulizar	Politeknik Negeri Lhokseumawe
Tomi Erfando	Universitas Islam Riau
WN Wari	Politeknik Negeri Banyuwangi
S D Said	Chemical Engineering Dept. Syiah Kuala University, Indonesia
B Nizamani	University of Malaya, Malaysia
M K Afdhol	Universitas Islam Riau
N Ahmed	UiTM Terengganu Kampus Dungun
S Diana	Politeknik Negeri Lhokseumawe
Yanty Maryanty	Politeknik Negeri Malang
S Omar	UiTM Terengganu
Faisal Rizal	Politeknik Negeri Lhokseumawe
Omer A Alawi	University of Malaya
N Lusi	Politeknik Negeri Banyuwangi
Faisal Mahmud	University of Malaya
Meri Andriani	Samudra University
Syafruddin	Politeknik Negeri Lhokseumawe
Abdul Muhyi	Politeknik negeri lhokseumawe
Victoria Kristina Ananingsil	Soegijapranata Catholic University
Marlina Marzuki	Politeknik Negeri Lhokseumawe
Zulkarnaini	Lhokseumawe State Polytechnic
Muhammad Nasir	Politeknik Negeri Lhokseumawe
Yanni Sudiyani	Research Center for Chemistry, Indonesian Institute of Sciences
Salmyah	Lhokseumawe State Polytechnic
Milawarni	Politeknik Negeri Lhokseumawe
Dasmi Husin	Politeknik Negeri Lhokseumawe
Arni Munira Markom	Universiti Teknologi MARA, Cawangan Johor
Salahuddin	Politeknik Negeri Lhokseumawe
Abdolreza Toudehdehghar	INTI International University
Mustika Amri	Politeknik Negeri Lhokseumawe
T Handhika	Gunadarma University
A F Pratiwi	State Polytechnic of Cilacap
Abdolreza Toudehdehghar	INTI International University
R A M Yusoff	Universiti Teknologi Malaysia
A A A Jafry	Universiti Teknologi Malaysia
Yayat Ruhiat	Universitas Sultan Ageng Tirtayasa
N Tri Suswanto Saptadi	Universitas Atma Jaya Makassar
D Andrian	University of Indonesia
T Karima	Universitas Indonesia
Nur Farah Syuhada Mohan	Universiti Teknikal Malaysia Melaka

K N Yudyawati	Universitas Indonesia
Nurul Fadillah	Samudra University
N Novianda	Samudra University
T Rihayat	Politeknik Negeri Lhokseumawe
T Rihayat	Politeknik Negeri Lhokseumawe
T Rihayat	Politeknik Negeri Lhokseumawe
Suryani	Politeknik Negeri Lhokseumawe
N Safitri	Politeknik Negeri Lhokseumawe
Ahmad Ihsan	Samudra University
M S Rosid	Physics Department, Universitas Indonesia
R Widyarta	Universitas Indonesia
M A Naufal	Universitas Indonesia
M F Khoir	Universitas Indonesia
Murni	Gunadarma University
Noor Hindryawati	Mulawarman University
C Huda	Universitas Airlangga Surabaya, Universitas Kanjuruhan Malang
D P Lestari	Gunadarma University
Murni	Gunadarma University
A Singh	INTI International University
E S Nugroho	Universitas Indonesia
L Lindawati	Politechnic of South Aceh
Mohd Sufian bin Abdul Hai	Universiti Teknologi Malaysia Kuala Lumpur
Zulfikar	Politeknik Negeri Lhokseumawe
Busra	Politeknik Negeri Lhokseumawe
Atthariq	Politelnik Negeri Lhokseumawe
M S Rosid	Universitas Indonesia
Yanty Maryanty	politeknik negeri malang
Amri	Politeknik Negeri Lhokseumawe
Fauziah Che Mat	Politeknik Port Dickson
M F A Rahman	Universiti Teknikal Malaysia Melaka
Nelly Safitri	Politeknik Negeri Lhokseumawe
E Elwina	Politeknik Negeri Lhokseumawe
N Rosli	Universiti Sains Malaysia
M A M Johari	University of Malaya
N Nahar	Politeknik Negeri Lhokseumawe

IOP Conf. Series: Materials Science and Engineering **854** (2020) 011002 doi:10.1088/1757-899X/854/1/011002

Email address
norainimarsi@gmail.com
zulis.erwanto@poliwangi.ac.id
ekakurniasih@pnl.ac.id
husainiftm@unsyiah.ac.id
suryati@pnl.ac.id
moeryanto.mrt@gmail.com
nazar_aw@yahoo.com
ratnasari@pnl.ac.id
salahuddin.mt@pnl.ac.id
jinwei.lim@newinti.edu.my
kikiyolanda956@gmail.com
fliah360@gmail.com
tomierfando@eng.uir.ac.id
naris@poliwangi.ac.id
syahiddin@che.unsyiah.ac.id
nizamanibilal@yahoo.com
afdhol@eng.uir.ac.id
norizan.ahmed@gmail.com
selviediana@pnl.ac.id
yanty.maryanty@polinema.ac.id
suzianaomar@gmail.com
ariputrap97@gmail.com
omer.adnan85@gmail.com
nurainilusi@poliwangi.ac.id
nashrul@um.edu.my
meri_tind@unsam.ac.id
irsyafruddin@yahoo.co.id
miftahulhamzijaisal@gmail.com
kristina@unika.ac.id
marlina.marzuki78@gmail.com
zulkarnaini@pnl.ac.id
nasiraceh76@gmail.com
sudiyani@gmail.com
salmyah@pnl.ac.id
milawarnimila@gmail.com
dasmihusin@pnl.ac.id
arnimunira@uitm.edu.my
salahuddintik@pnl.ac.id
abdolreza.toudehdehghan@newinti.edu.my
mustika.amr@gmail.com
handhika.tri@gmail.com
art_dhita@yahoo.com
abdolreza.toudehdehghan@newinti.edu.my
r.addayusoff92@gmail.com
r.addayusoff92@gmail.com
guntaray@untirta.ac.id
ntsaptadi@yahoo.com
davidandrian92@yahoo.co.id
tanisakarima@gmail.com
nursyuhada0907@gmail.com

syamsu.rosid@ui.ac.id
nurulfadillah@unsam.ac.id
novianda_tif@unsam.ac.id
teukurihayat@yahoo.com
teukurihayat@yahoo.com
teukurihayat@yahoo.com
suryanisalim@yahoo.com
nellysafitri@pnl.ac.id
ahmadihsan@unsam.ac.id
syamsu.rosid@ui.ac.id
syamsu.rosid@ui.ac.id
ahganiya@gmail.com
Khoirfarkhan25@gmail.com
murnipskm@staff.gunadarma.ac.id
hindryawati@gmail.com
choirul@unikama.ac.id
dewi_putrie@staff.gunadarma.ac.id
achmad_fahrurozi@staff.gunadarma.ac.id
arwinders.jigiris@newinti.edu.my
ezranugroho@Gmail.com
lindawati203@gmail.com
msufianahamid@gmail.com
zul_elka@pnl.ac.id
muhammad.arhami@gmail.com
atthariq.huzaifah@pnl.ac.id
syamsu.rosid@ui.ac.id
yanty.maryanty@polinema.ac.id
amri@pnl.ac.id
mbhisyam@um.edu.my
mfauziar@utem.edu.my
nellysafitri@pnl.ac.id
elwina@pnl.ac.id
nurizatirosli@student.usm.my
swharun@um.edu.my
nahar@gmail.com

IOP Conf. Series: Materials Science and Engineering **854** (2020) 011003 doi:10.1088/1757-899X/854/1/011003

Peer review statement

All papers published in this volume of *IOP Conference Series: Materials Science and Engineering* have been peer reviewed through processes administered by the proceedings Editors. Reviews were conducted by expert referees to the professional and scientific standards expected of a proceedings journal published by IOP Publishing.

IOP Conf. Series: Materials Science and Engineering **854** (2020) 012001 doi:10.1088/1757-899X/854/1/012001

The Mechanical Performance of Pipe Based on Fiberglass Reinforced with Plastic Waste (FRPW) Composites

Noraini Marsi[1,2], Lingeshavaran Subramaniam[1], Anika Zafiah Mohd Rus[2], Nurulsaidatulsyida Sulong[2], Muhammad Haikal Mohd Fodzi[1], Nur Athirah Huzaisham[1], Nik Alnur Auli Nik Yusuf[3], Salwa Mahmood[1], Muhammad Farid Shaari[1] and Hafizuddin Hakim Shariff[4]

[1]Innovative Manufacturing Technology (IMT), Faculty of Engineering Technology, Universiti Tun Hussein Onn Malaysia (UTHM), Hub Pendidikan Pagoh, KM 1, Jalan Panchor, 84600 Pagoh, Muar, Johor.
[2]Sustainable Polymer Engineering, Advanced Manufacturing and Materials Center (SPEN-AMMC), Universiti Tun Hussein Onn Malaysia (UTHM), 86400 Parit Raja, Batu Pahat, Johor.
[3]Faculty of Bioengineering and Technology, Universiti Malaysia Kelantan Jeli Campus, Locked Bag No.100, 17600 Jeli, Kelantan, Malaysia.
[4]Kim Hoe Thye Industries Sdn. Bhd. No.99, Lot 143, Jalan Air Manis, Bukit Mor, Muar, Johor.

Email: mnoraini@uthm.edu.my

Abstract. The project research present the mechanical performance of pipe based on fiberglass reinforced with plastic waste (FRPW) in plant application system. The use of FRPW able to reduce corrosion problem faced by oil and gas industry. In this study involved four types of combination ratio of fiberglass reinforced with plastic waste (FRPW) of 1.0: 0.5; 1.0:1.0; 1.0:1.5; and 1.0:2.0. The fabrication process started with grinding process of plastic waste into small size in the range of 0.1 mm. Fiberglass then reinforced with plastic waste by mixed with resin and hardener with ratio of 2: 1 and poured into the cylinder mould. There is a possibility fiberglass from 10-40% by weight result in substantial increase in elastic modulus, accompanied by an increase in strength with reduced ductility 1.0 of ratio. The tensile test showed clearly exhibited that 1.0 of plastic waste reinforced fiberglass with stand the higher maximum force value of 2.69 kN. For the bending test ratio of 1.0 of plastic waste withstand the higher bending strength at 5.29 kN. Ratio of 1.0 FRPW is more suitable for produced pipe in plant application system due to matrix-reinforcement bonding for each pipe sample after conducting tensile strength. The result obtained that ratio 1.0 of FRPW shown good matrix-reinforcement bonding.

1. Introduction

Plastics have substantial benefits in terms of their low weight, durability and lower cost relative to many other material types. These properties make manufactured polymers astoundingly helpful, and since we figured out how to make and control those, polymers have turned into a fundamental piece of our lives. Particularly finished the most recent 50 years plastics have immersed our reality and changed the way that we live [1]. Fiberglass can also be a less expensive alternative to other materials. Composite materials such as fiberglass reinforced plastics (CFRP) exhibit outstanding mechanical performance at low material densities, thus offering the potential of significant weight reduction in structural lightweight components [2]. Fiberglass reinforced with plastic waste, FRPW can enhance mechanical properties. It additionally demonstrates that option of plastic to solid causes some lessening in mechanical properties for example, compressive strength, tensile strength, and flexural strength. Along these lines fiberglass reinforced with plastic waste, FRPW is best method to produce pipe in plant application system and also able to reduce corrosion problem.

The word ''waste'' projects a vision of a material with no value or useful purpose. However, technology is evolving that holds promise for using waste or recycled plastics to make an array of high performance composite products that are themselves potentially recyclable [3]. The plastic waste is now a serious environmental threat to modern civilization. The plastic waste are widely used in most of manufacture industry to refabricate the product that needed and also to reduce cost of production. The following products which is made from plastic waste can be used to reduce pollution. Approximately 50 per cent of plastics are used for single-use disposable applications, such as packaging, agricultural films and disposable consumer items, between 20 and 25% for long-term infrastructure such as pipes, cable coatings and structural materials, and the remainder for durable consumer applications with intermediate lifespan, such as in electronic goods, furniture, vehicles, and more [4]. Plastics are inexpensive, lightweight and durable materials, which can readily be moulded into a variety of products that find use in a wide range of applications. As a consequence, the production of plastics has increased markedly over the last 60 years.

2. Literature Review

The plastic waste is now a serious environmental threat to modern world. 50 percent of the plastic we use, we use just once and throw away. The amount of plastic waste consumed annually has been increase in weight. Studies demonstrate that regular corruption of plastic takes around 100 to 1000 years. The plastic wastes accounts for $10.62 \pm 5.12\%$ of the total stored wastes in the old landfill, among which, 69.13% is plastic bags include white PE plastic bags accounted for 11.34% colored PE plastic bags 29.77% and other plastic bags 28.02% and 30.87% plastics such as PP, PVC, PS and others [5]. Plastic waste is composed of several toxic chemicals, and therefore plastic pollutes soil, air and water. Since plastic is a non-biodegradable material, land-filling using plastic would mean preserving the harmful material forever. Figure 1 shows the source of plastic waste [6].

In spite of developing question, plastics are basic to present day life. Plastics made conceivable the improvement of PCs, mobile phones, and the greater part of the lifesaving advances of present day solution. Lightweight and useful for protection, plastics help spare non-renewable energy sources utilized as a part of warming and in transportation. Maybe most essential, economical plastics raised the way of life and made material wealth all the more promptly accessible [7]. The plastic waste are widely used in most of manufacture industry to refabricate the product that needed. Plastic waste also applications like protection, during their service life, plastic materials spare in excess of 140 times the vitality required for their generation. They are likewise champion with regards to ensuring merchandise and sustenance, subsequently decreasing breakage and waste. Due to the interesting properties of plastic materials, they are utilized as a part of an extensive variety of use areas, for example, packaging, building and construction, automotive and aeronautics, electrical and electronic to help us to live and build a more sustainable world [8].

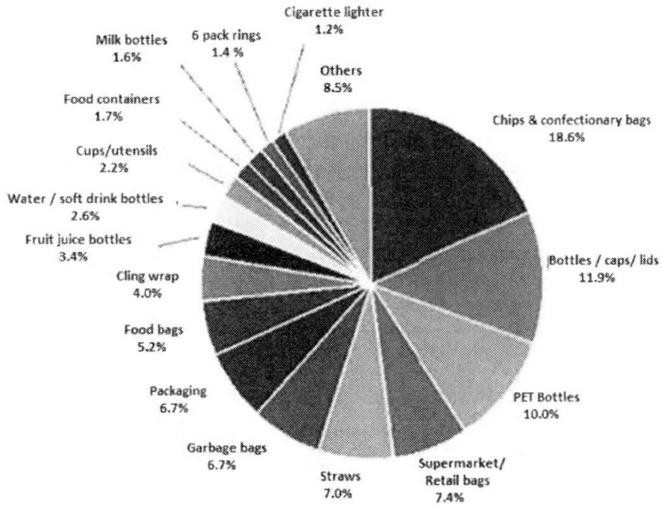

Figure 1: The source of plastic waste [6]

3. Methodology

3.1. The preparation of samples

The preparation of samples was involved grinding process of plastic waste with size in the range of 1 – 3mm. The plastic waste was provided by Kim Hoe Thye Industries Sdn. Bhd. The plastic waste were mixed with the epoxy and hardener within three minutes. The mixture was spread thoroughly on the fiberglass. Next another layer of fiberglass was again spread thoroughly on the existing fiberglass by using paint roller. Four samples were prepared to produce pipe from fiberglass reinforced by different ratio of plastic waste which is the ratio of 1:0.5, 1:1.0, 1:1.5 and 1:2.0 as shown in Table 1 and Figure 2.

Table 1. Samples with different ratio of plastic waste

Samples	Plastic waste	Fiberglass	Ratio of epoxy resin and hardener
A	0.5	1.0	3:1
B	1.0	1.0	3:1
C	1.5	1.0	3:1
D	2.0	1.0	3:1

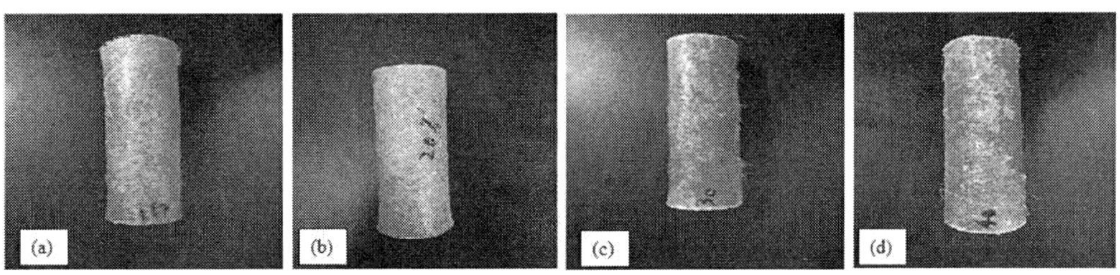

Figure 2: Pipe samples by different ratio of fiberglass reinforced plastic waste (FRPW); (a) 1:0.5, (b) 1:1.0, (c) 1:1.5, (d) 1:2.0

4. Results and discussions

4.1. Tensile strength test

Figure 3 exhibited the graph of force versus time for different ratio of fiberglass reinforced plastic waste (FRPW). It is clearly exhibited that 1.0 of plastic waste reinforced fiberglass withstand the higher maximum force at 2.79 kN, followed by 2.0 of plastic waste reinforced fiberglass at 2.26 kN, the ratio of 1.5 of plastic waste can sustain at 1.20 kN and 0.5 ratio which can hold up to 1.55 kN.

The ratio of 1.0 plastic waste gives higher tensile performance due to the composition of plastic waste is optimum and more consistent compared to other fillers and reinforcements. Generally, the addition of fiberglass from 10-40% by weight result in substantial increase in elastic modulus, accompanied by an increase in strength with reduced ductility [9]. Tensile properties of a composite material are mainly depending on fiber strength, modulus, fillers, fiber length and orientation, fiber-matrix interfacial bonding and fiber content [10]. Previous researcher was observed that the presence of 5% maleic anhydride polypropylene copolymer in fiberglass reinforced polypropylene composites results in an increase of tensile and flexural strength by 41% and 45%, respectively [11].

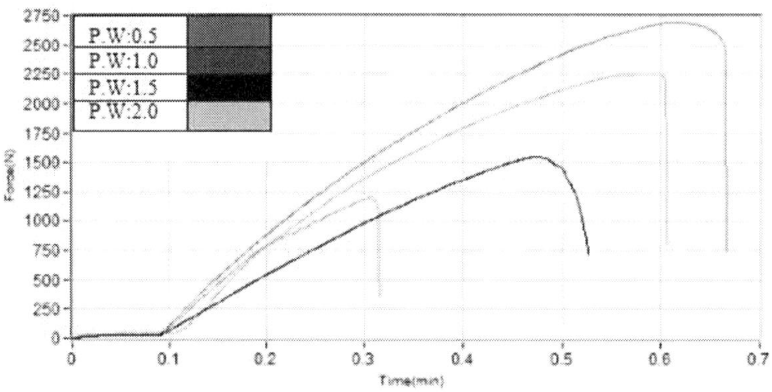

Figure 3: Tensile test results at different ratio of Fiberglass Reinforced Plastic Waste (FRPW)

4.2. Bending Test

The preparation of samples was involved Figure 4 exhibited the bending strength test for different ratio of Fiberglass Reinforced Plastic Waste (FRPW).

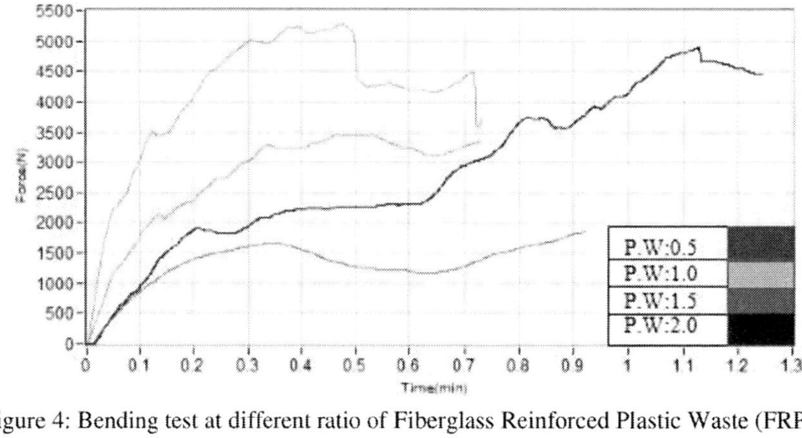

Figure 4: Bending test at different ratio of Fiberglass Reinforced Plastic Waste (FRPW)

It is shows that 1.0 of plastic waste reinforced fiberglass withstand the higher maximum force value of 5.29 kN, followed by 2.0 of waste plastic reinforced fiberglass with value of 4.89 kkN, then 1.5 of waste plastic reinforced fiberglass with value of 3.45 kN and lastly 0.5 of waste plastic reinforced fiberglass which can hold up till value of 1.85 kN. It was expected that fibrous fraction of FRPW would have a significant reinforcing effect and lead to a higher improvement on flexural behavior. Fiber reinforcing has several advantages to enhance the mechanical properties [12]. Plastic waste matrix itself has relatively less strength and hence this research work will focus on the enhancement of mechanical properties preferentially flexural strength and compressive strength through glass fiber reinforcement and binder pitch [13].

Acknowledgement

The author would like to thank the Universiti Tun Hussein Onn Malaysia for supporting this project under GPPS Grant by UTHM Scheme (Phase 2018), vot H340, H341 and Contract Grant, vot H464.

References

[1] Milios, L., Davani, A.E. and Yu, Y. (2018). Sustainability Impact Assessment of Increased Plastic Recycling and Future Pathways of Plastic Waste Management in Sweden. Recycling. MDPI. vol.3(3), pp.33.

[2] Rafiee, M., Nitzsche, F. and Labrosse, M. R. (2019). Fabrication and Experimental Evaluation of Vibration and Damping in Multiscale Graphene/ Fiberglass/ Epoxy Composites. Journal of Composite Materials, vol. 53(15).

[3] Mwanza, B. G. and Mbohwa, C. (2017). Drivers to Sustainable Plastic Solid Waste Recycling: A Review. Procedia Manufacturing, Vol. 8, pp. 649-656.

[4] Lebreton, L. and Andrady, A. (2019). Future Scenarios of Global Plastic Waste Generation and Disposal. Palgrave Communication 5, Article number 6.

[5] Zhou, C., Fang, W., Xu, W., Cao, A. and Wang, R. (2014). Characteristics and the Recovery Potential of Plastic Wastes Obtained from Landfill Mining. Journla of Cleaner Production, Vol. 80, pp. 80-86.

[6] Hahladakis, J.N., Velis, C.A., Weber, R., Iacovidou, E. and Purnell, P. (2018). An Overview of Chemical Additives Present in Plastics: Migration, Release, Fate and Environmental Impact During Their Use, Disposal and Recycling. Journal of Hazardous Materials, Vol. 344, pp. 179-199.

[7] Gielen, D., Boshell, F., Saygin, D., Bazilian, M.D., Wagner, N. and Gorini, R. (2019). The Role of Renewable Energy in the Global Energu Transformation. Energy Strategy Reviews. Elsevier. Vol. 24, pp. 38-50.

[8] Luzi, F., Torre, L., Kenny, J.M. and Puglia, D. (2019). Bio- and Fossil-Based Polymeric Blends and Nanocomposites for Packaging: Structure-Property Relationship. Materials, vol.12(3), pp.471.

[9] Azrin Hani, A.R., Azman, N.S., Ahmad, R., Mariatti, M. and Roslan, M.N. (2016). Ballistic Impact Response of Woven Hybrid Coir/Kevlar Laminated Composites. MATEC Web of Conferences, Volume 78, Article Number 01048.

[10] Marsi, N. and Rus, A.Z.M (2017). The Effects of Biopolymers Composite Based Waste Cooking Oil and Titanium Dioxide Fillers as Sueprhydrophobic Coatings. IOP Conference Series: Materials Science and Engineering, Volume 226(1), Article Number 012161.

[11] Biswal, M., Mohanty, S. and Nayak, S.K. (2011). Mechanical, Thermal and Dynamic-Mechancial Behaviour of Banana Fiber Reinforced Polypropylene Nanocomposites. Polymer Composites, vol.32(8).

[12] Marsi, N., Rus, A.Z.M. and Sulong, N. (2017). Vibration Performance of Lightweight Concrete Coated Biopolymer based on Used Cooking Oil. ARPN Journal of Engineering and Applied Sciences. Vol.12(14), pp.4236-4242.

[13] Patel, K., Gupta, R., Garg, M., Wang, B. and Dave, U. (2019). Development of FRC Materials with Rcycled Glass Fibers Recovered from Industrial GFRP-Acylic Waste. Advances in Materials Science and Engineering. Hindawi. Volume 2019, Article 41497078, pp.1-15

IOP Conf. Series: Materials Science and Engineering **854** (2020) 012002 doi:10.1088/1757-899X/854/1/012002

The Innovation of Interlock Bricks with A Mixture of Bagasse Ash Without Combustion

Z Erwanto[1], D D Pranowo[1], A Holik[2], M S Amin[1], F Darmawan[1]

[1] Department of Civil Engineering, Politeknik Negeri Banyuwangi, Banyuwangi, Indonesia
[2] Study Program of Agribusiness, Politeknik Negeri Banyuwangi, Banyuwangi, Indonesia

E-mail: zulis.erwanto@poliwangi.ac.id

Abstract. Nowadays, the wall building material is dominated by clay brick and the process requires combustion. Therefore, the need for the development of interlock brick to introduce wall material that is easy to install, strong, without combustion and environmentally friendly. The shape design of interlock brick is made with Modified Tanzanian Interlock Brick (MTIB). Interlock brick size is 25.1 x 12.6 x 7.8 cm. The method of brick making with the experiments of various variations of the mixture to know the product specifications of suction, absorption, salt content and compressive strength with a quality target of grade 50 kg/cm^2 that qualified SNI 15-2094-2000. The result was an MTIB that meet the requirements of SNI in a mixture with a ratio of 2 Clay: 3 Sand: 3.5 Portland Cement: 1.5 Bagasse Ash. Specifications of MTIB product obtained results compressive strength of 63.78 kg/cm^2, absorption of 19%, the salt content of 4.14%, volume weight of 1.549 g/cm^3 and a specific gravity of 1.81 and water suction power of 57.34 g/dm^2/minute. MTIB innovations made from the waste of bagasse ash and without combustion process could be utilized for efficient civil building construction, fast, economical, and environmentally friendly compared with other brick materials.

1. Introduction

1.1. Background

The rapid development of buildings, fences, and housing in Indonesia have made the need for wall material increasing. At present, the wall building material is dominated by bricks made from clay and the drying process requires combustion. Generally, blocks need plaster and precision in the field. In line with these needs, materials that are easy to install, strong and environmentally friendly are in desperate need.

Therefore, it is necessary to develop interlock bricks to introduce wall materials that are easy to install, strong and environmentally friendly. In the case of developing interlock bricks, this time avoids the combustion process by adding a mixture of cement. The goal is to become an interlocking brick that is environmentally friendly and maintain good survival for the next generation. Brick making is generally very dependent on clay which must be extracted as raw material from the brick. But gradually these raw materials will run out or be subject to excavation restrictions because they damage the environment.

IOP Conf. Series: Materials Science and Engineering **854** (2020) 012002 doi:10.1088/1757-899X/854/1/012002

Therefore, the need to reduce clay raw materials with other materials to minimize the use of clay in the manufacture of interlock bricks. One of them is by adding silica-containing materials that are like cement if they react with water. Bagasse ash is one of the wastes which consists of inorganic salts and is rich in silica (Si). This waste is challenging to dispose of because the function of bagasse ash is less useful for plantations or agriculture, even by some farmers consider bagasse ash as a destroyer of soil fertility [12].

According to previous research, silica will react with cement and water to form calcium silicate hydrate which functions as an adhesive [19]. Another study related to the study of bricks with additive material of bagasse ash or the like in research [1], [2], [3], [5], [7], and [20]. Based on the above, a study was conducted on the innovation of interlock bricks with a mixture of bagasse ash using clay as a base material mixed with added material in the form of cement to avoid the process of combustion, sand, water and printed using a hydraulic machine, like as the previous research with using interlock brick such as [4], [6], [8], [9], [10] and [13]. By using these materials and tools, it is hoped that interlock bricks are easily installed, environmentally friendly and achieve SNI standard compressive strength.

1.2. Research Objectives

The research objective is to find out the novelty of interlock bricks from bagasse ash without the combustion process.

2. Theoretical Basis

2.1 Brick

Bricks are part of the building used to make a building, and materials are useful for making bricks originating from the ground with or without a mixture of other materials which are then burned at high temperatures so they cannot be destroyed again if soaked in water [5]. The definition of brick according to the Red Brick as Building Material NI-10 [11], is a building element for the construction of buildings and made from the soil with or without a mixture of other materials, burned high enough, so that it cannot be destroyed again if soaked in water. A light red brick is a red brick whose weight is less than 1.2 kg/dm^3. While hollow red brick is red brick with a total cross-sectional area of more than 25% of its cross-sectional area.

According to SNI 03-0349-1989 Concrete Bricks for Couples Walls, concrete bricks are a type of brick-shaped building made from the main ingredients of Portland cement, water, and aggregates used for wall pairs. The original brick is divided into solid concrete bricks and hollow concrete bricks. Solid concrete brick is brick that has a solid cross-section of 75% or more of the total cross-sectional area and has a solid volume of more than 75% of the entire brick volume while perforated concrete brick is a brick which has a cross-sectional area of more than 25% over the cross-sectional area and more hole volume of a 25% total volume limit [14].

2.2 Brick Quality Requirements

In this study using SNI 15-2094-2000 because interlock brick is a transformation of solid or perforated red brick with a maximum volume of 15%, while the size of interlock brick hole with MTIB form is 5.72% [17]. Bricks for wall pairs must meet the following quality requirements:

2.2.1 Visibility. The solid red brick for the wall pair must be a long rectangular prism, have elbow ribs, flat planes and do not show cracks.

2.2.2 Compressive Strength. The magnitude of the average compressive strength and the variation coefficients allowed for solid red bricks for wall pairs are by **Table 1** and can be calculated using the following **Equation 1**:

$$\tau = P / A \tag{1}$$

With, τ is the average compressive strength (kg/cm^2), P is the maximum load (kg), A is the area size press (cm^2)

According to SNI 15-2094-2000 which discusses solid red brick for wall pairs, bricks for wall pairs according to the lowest compressive strength are divided into 3 (three) classes, likes **Table 1.**

Table 1. Average Compressive Strength and Coefficient of Variation in Red Brick (mm Unit) [17]

Classes	Minimum average compressive strength of 30 bricks tested kg/cm² (MPa)	The coefficient of variation of the average compressive strength under test (%)
50	50 (5)	22
100	100 (10)	15
150	150 (15)	14

The coefficient of variation is the statistical parameter that is most widely used to determine the amount of variability in a sample. The magnitude of the figure is estimated through price s; the price of s is not absolute but varies from sample to sample; the coefficient of variation can be determined by **Equation 2** [7]. As for the standard coefficient of variation can be seen in **Table 1**.

$$v = \frac{Sd}{\tau} x\ 100\% \qquad (2)$$

With, v is the coefficient of variation, Sd is the standard deviation of the compressive strength (kg/cm^2), and τ is the average compressive strength (kg/cm^2).

2.2.3 Volume Weight. The volume weight is the weight of the unit volume to determine the weight of the contents [17]. Calculation of volume weight can be calculated using **Equation 3** as follows:

$$VW = \frac{W}{V} \qquad (3)$$

With, VW is the *volume weight* (kg/cm^3), W is the *weight* of brick (kg), and V is the volume of brick (cm^3).

2.2.4 Salinity. The Salinity which is soluble salts can cause structural damage "efflorescence" on the surface of the brick are magnesium sulfate ($MgS0_4$), sodium sulfate ($NaS0_2$), potassium sulfate (KSO_4), with a total salt content maximum of 1.0% [17]. In SNI 10-1978:6 subchapters 6.2.5, the salinity standards for bricks are as follows:

- It is not dangerous if less than 50% of the brick surface is covered by a thin layer of white due to the crystallization of dissolved salts.
- There is a possibility of harm if 50% or more of the brick surface is covered by a thick white layer due to the crystallization of soluble salts, but the parts of the brick surface do not become powder or detach.
- Harm if more than 50% of the brick surface is covered by a thick white layer because the crystallization of soluble salts and the parts of the brick surface become powdered or released.

The salt content test aims to determine the amount of salt contained in bricks. If in the process of installing salt there is a salt content, weathering will occur due to dissolved salts and will result in a bad bond between the brick and the mixture. Testing of salinity can be calculated using **Equation 4** as follows [18]:

$$Salinity = \frac{F\ salt}{F\ brick} x100\% \qquad (4)$$

With, Salinity is a salt content (%), F_{salt} is the extent of salt fields attached to bricks (cm^2), and F_{brick} is the extensive brick (cm^2).

2.2.5 Water Absorption. Water absorption is the ability of bricks to absorb the percentage of water weight. The testing of water absorption is an essential factor because it is one of the properties of bricks which dramatically affects the strength of brickwork. Water absorption on the brick must be controlled to prevent a lot of water loss from the mixture being used and to avoid cracking during plastering. Determining water absorption is used in **Equation 5** [17]. If the absorption of water is less than 20%, the brick is considered good.

$$WA = \frac{B-A}{A} x100\% \quad (5)$$

With, WA is the water absorption (%), A is the dry weight of brick (fixed) in kg, and B is the weight of brick after soaking for 24 hours (kg).

2.2.6 Power of Suction Brick. The testing of suction power to the brick aims to measure the amount of suction power on the interlocking brick. Suction power dramatically influences the strength and quality of the brick pairing work. Different suction power will cause differential voltage and cracks. Therefore, it is important to equalize the suction power before installation. The suction power of bricks can be calculated using **Equation 6** [17].

$$\omega = \frac{B-A}{F} \quad (6)$$

with, ω is the power of suction brick (g/dm^2/minute), A is the dry oven weight of brick (g), B is the weight of the brick after soaking per minute (g), and F is the brick base area which is related to water (dm^2/minute).

2.2.7 Specific Gravity of Brick. Specific gravity is the ratio between the weight of the volume unit of a material to the weight of the contents of the same volume at the specified temperature. The specific gravity of the brick can be calculated using **Equation 7** as follows [15]:

$$\gamma = \frac{W_{sat}}{W_{sat}-W} \quad (7)$$

with, γ is the specific gravity of brick, W_{sat} is the saturated weight of brick (g), W is the saturated weight of brick in water (g).

3. Methodology

3.1. Experimental Techniques

This research is quantitative through an experimental approach, in the form of non-combustion brick testing with the addition of bagasse. The parameters sought in this study were compressive strength, specific gravity, volume weight, salt content, suction power, and absorption. By regarding the effect of bagasse ash as additional material for brick making without combustion [20]. So this research will start from the testing of materials characteristics, design of specimens, manufacture of samples, maintenance of samples to testing. The test object is made in the form of a Modified Tanzanian Interlock Brick (MTIB) brick size of 250 x 125 x 75 mm and with a mixture of bagasse additives.

In making bricks, the clay used must be clean of the 4.75 mm sieve and not too plastic dirt. In order not to cause high dry strength properties that affect the results of brick drying interlock. Therefore, before the process of making interlocking, it is necessary to do a 4.75 mm sieve and atterberg boundary testing of clay. The sand used in making interlock is a local sand brick originating from sand mines, Kalipuro district, Banyuwangi Regency. In making interlock bricks, sand is used to reduce shrinkage and facilitate drying.

Supplementary material in the manufacture of interlock bricks is bagasse ash from waste from the Company of Glenmore Sugar Industry like **Figure 1**. Bagasse ash must be clean of garbage and sieved with no. 200 to obtain a grain size equivalent to cement. With fine particle granules, the hydration will be faster because the hydration starts from the surface of the grain.

Figure 1. Bagasse Ash Waste Form Glenmore Sugar Industry

3.2. Instrument
Manual hydraulic interlock brick molding machine used to print a mixture of materials that have been mixed to form a 250 x 125 x 75 mm interlock brick. In the use of a manual hydraulic interlock brick molding machine, a mold will be made in the form of the expected MTIB mold like **Figure 3**. In **Figure 2** there is a picture of a manual hydraulic interlock brick molding machine.

Figure 2. Manual Hydraulic Molding Interlock Brick Machine

3.3. Mix Design of Interlock Brick
After the tools and materials were prepared, then the next step is to do the mix composition design of the interlock brick design. For the mix composition design, the interlock brick design was measured by using a volume comparison of materials with several trial-and-error alternative mix designs to determine right proportions and achieve quality with the SNI standard compressive strength.

The best composition ever tested with a mixture of 2 Clay: 3 Sand: 3 Portland Cement: 2 Calcium: 2 Bagasse ash produces a quality of 40.49 kg/cm^2. For the composition is assumed total is 12 times the volume of the patch (2 kg) can produce 15-20 interlock bricks. If using dye material to add its artistic material, can use cement dye with a composition of 5 ounces per 24 kg. As for alkaline resin material is quite 3 or sufficiently adjusted to the composition. The alkali function is to make the soil pH not too acidic, so bonded cement to be strong and solid, increasing the color density, and harden. Example comparison for 1 m^3 red brick: 150 kg Cement: 450 kg Fine sand: 900 kg red soil + (2 kg alkali red brick dissolved 50 lt water), all dry ingredients are stirring in a dry state. For my Preliminary testing, if the composition without Ash with ratio 3 Clay: 4 Sand: 2 Cement: 3 Calcium have a quality of 33.73 kg/cm^2, and then without clay with ratio 4 Sand: 3 Cement: 3 Calcium: 2 Bagasse ash has a quality of 53.15 kg/cm^2.

The conclusions of my study, the comparison of the volume of bagasse ash with cement must be more cement to obtain good compressive strength on the interlocking brick. Based on the composition of previous studies regarding interlock bricks, the mix design of brick interlock design with a comparison of 2 clay: 3 sand: 3 Portland Cement: 2 bagasse ash without calcium, this is based on development and research. The specimen to be made in 30 pieces with the form of MTIB.

IOP Conf. Series: Materials Science and Engineering **854** (2020) 012002 doi:10.1088/1757-899X/854/1/012002

3.4. Making Interlock Brick

Simion Hosea Kintingu made an important modification to repair the interlocking brick to fit the requirements of the Tanzanian people. The size of the bricks is 300 x 150 x 100 mm, similar to the size of Thailand and Bamba type. However, it does not cover the lack of brick Tanzanian Interlock Brick system, the stability of the wall during construction in the vertical direction can not be covered properly. So it requires mortar and reinforcement to achieve its maximum strength [10].

The form of interlock brick that will be made is an interlocking brick in the form of Modified Tanzanian Interlock Brick (MTIB) which is a modified form of Thai brick with Tanzanian with a size of 250 x 125 x 75 mm shown in **Figure 3**. The Modified Tanzanian Interlock Brick (MTIB) has the advantage of a pyramid locking system with a hole in the middle. This form was considered better because the force excitation transformation can be distributed evenly in the horizontal direction. In its application in the field, it does not require people who have special skills (anyone can install it correctly and quickly) because the installation is almost similar to the Lego kids game.

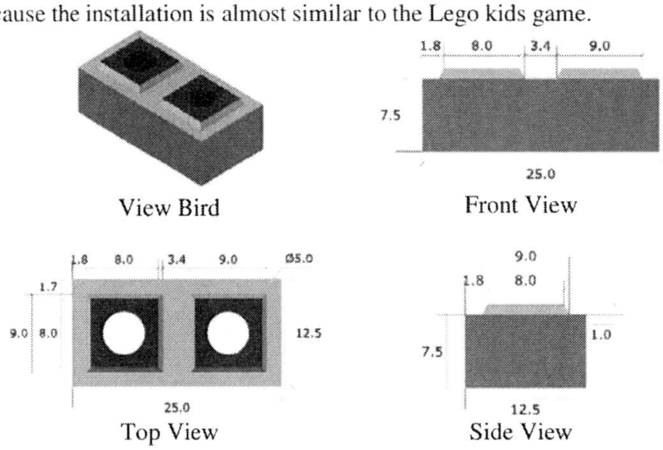

Figure 3. Modified Tanzanian Interlock Brick

4. Results and Discussion

4.1. Testing of Atterberg Limits for Clay Soils

The average moisture content at the liquid limit (LL) is 35.19%, plastic limit (PL) is 29.67%, and for index plasticity (PI) value is the difference in the liquid limit at the plastic limit of 5.52 %. In testing the atteberg boundary which is a soil sample taken from one of the conventional red brick entrepreneurs Banyuwangi Regency can be graphed in **Figure 4.**

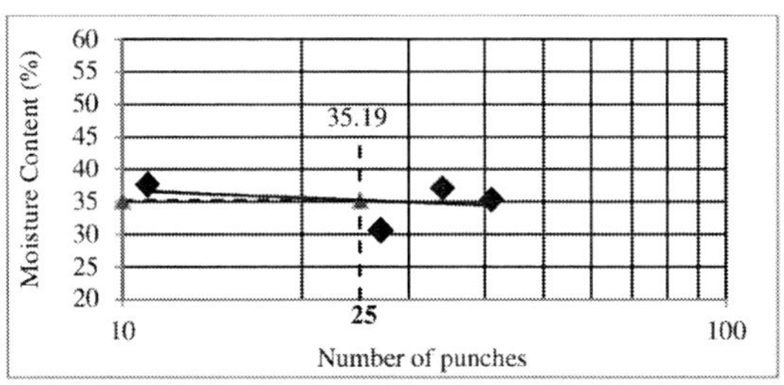

Figure 4. Atterberg Limits for Clay Soils

4.2. Testing of Specific Gravity for Fine Aggregate

From the results of testing the specific gravity for fine aggregate (sand), it was found that the average density of sand was 2.33 grams. This result was not included in standard density according to ASTM

C128-78, which is 2.5-2.7 grams. So that the sand material can not be used in the process of brick making because it can affect the quality of bricks so it should replace with sand material that has better quality.

4.3. Testing of Sieve Analysis for Fine Aggregate

From the results of sand filter analysis, it was found that the sand had a fine aggregate modulus of 3.563%, while the required fineness modulus of the granules is 1.50-3.80%. Based on the current requirements, the test results can be stated for this sand will produce a good mixture if used for interlock brick mixture.

4.4. Making Interlock Brick Test Materials

The test object is made in the form of MTIB brick. To molding bricks in the form of MTIB, a press with a hydraulic press of about 6 tons was used. By the planned mix, in this study used three types of mixtures measured by volume comparisons were as follows:

1. 2 Clay: 3 Sand: 2.5 Portland Cement: 2.5 Bagasse Ash
2. 2 Clay: 3 Sand: 3 Portland Cement: 2 Bagasse Ash
3. 2 Clay: 3 Sand: 3.5 Portland Cement: 1.5 Bagasse Ash

In the process of making interlock brick specimens, the materials needed are clay, sand, portland cement, and bagasse ash. With enough water added until the mixture is saturated and can be grasped (not runny). **Figure 5** shows the test object in the form of MTIB. After the interlocking brick is molded, it is allowed to dry at room temperature. After drying, the interlocking brick is ready to be tested according to the age of the test.

Figure 5. Drying of Interlock Brick Test Objects

4.5. Testing of Interlock Brick Salt Content

Figure 6 shows the highest salt content found in mixture 1, and the lowest is found in mixture 3. It shows that the percentage of salt content in interlock brick specimens is influenced by the amount of bagasse ash added. The more bagasse ash added the higher the content of the salt in the interlocking brick. Similarly, the opposite.

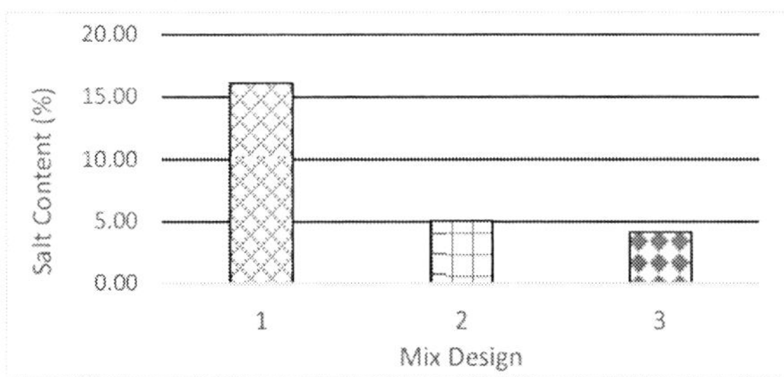

Figure 6. Salt Content of Interlock Brick With Bagasse Ash Mixture

4.6. Testing of Interlock Brick Suction Power

Of the three mixed variations tested, none met the SNI 15-2094-2000 standard because the value produced from each specimen was >20 g /dm²/minute. It could be concluded that the amount of suction power on the interlocking brick is influenced by the amount of addition of bagasse on the brick composition. The more addition of bagasse, the higher the suction power. Vice versa, if the addition of bagasse ash is less, the suction power will be smaller.

4.7. Testing of Interlock Brick Absorption

In **Figure 7**, it can be seen that mixture 1 has the highest absorption and mixture 3 has the lowest absorption of all mixed variations. Therefore, it can be concluded that the amount of water absorption on the interlocking brick is influenced by the addition of bagasse ash to the composition applied. The more bagasse ash added, the higher the absorption capacity produced. Vice versa, the less bagasse ash is added, the smaller the absorption power generated by the interlocking brick will be.

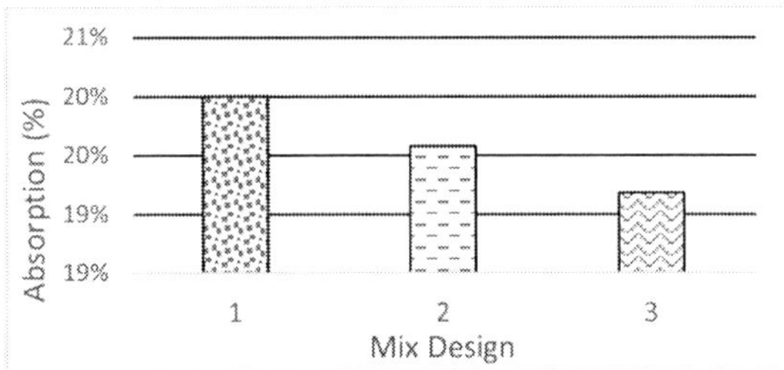

Figure 7. Average Absorption of Interlock Brick

4.8. Testing of Interlock Brick Specific Gravity

In **Figure 8**, it can be seen that the addition of bagasse ash affects the specific gravity produced by interlock bricks. The more bagasse ash added, the smaller the particular severity produced. If the less bagasse ash is combined, the higher the value of the specific gravity produced.

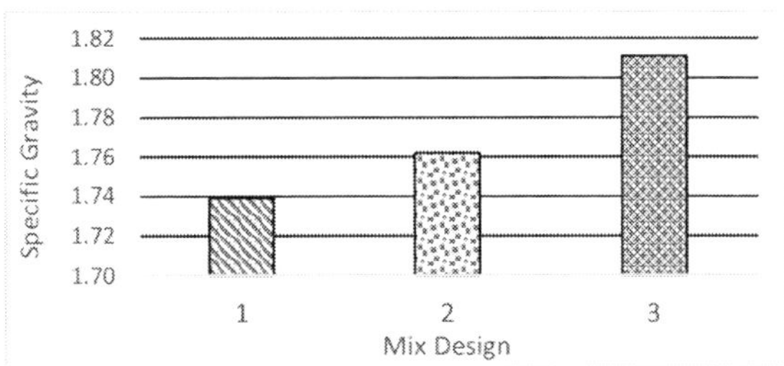

Figure 8. Average Specific Gravity of Interlock Brick

4.9. Testing of Interlock Brick Volume Weight

In **Figure 9**, shows that the volume weight of the interlocking brick is influenced by how much bagasse ash was added to the interlock brick composition. The more addition of bagasse then the weight of the amount produced will be lighter. Vice versa, if the addition of bagasse ash is less, the weight of the resulting volume will be more substantial.

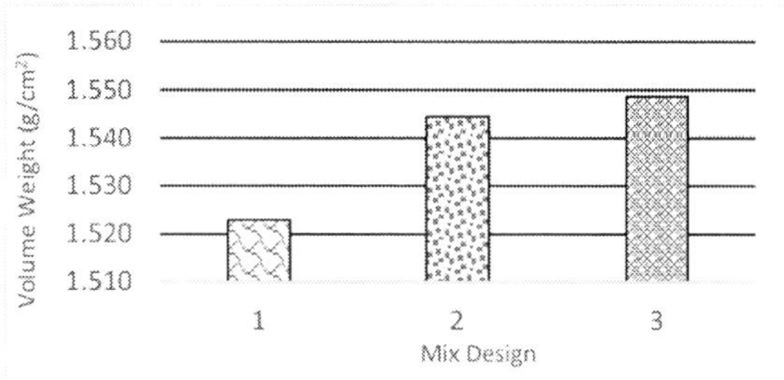

Figure 9. Average Volume Weight of Interlock Brick

4.10. Testing of Interlock Brick Compressive Strength

In **Figure 10**, it can be seen that the addition of bagasse ash is very influential on the compressive strength of the interlocking brick. The more bagasse ash added, the smaller the compressive strength produced by the interlocking brick. Vice versa, the less bagasse ash is added, the higher the compressive strength of the interlocking brick. The results of the compressive strength of interlock bricks with a mixture of bagasse ash have the highest compressive strength in variations of mixture 3 that is 2 Clay: 3 Sand: 3.5 Portland Cement: 1.5 Bagasse Ash, with an average compressive strength of 63,78 kg/cm^2 and entered in class 50 in accordance with SNI 15 - 2094 - 2000.

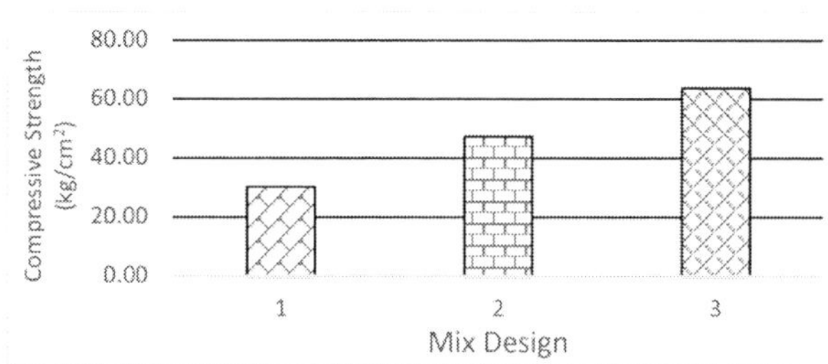

Figure 10. Average Compressive Strength of Interlock Brick

4.11. Recapitulation of Test Results

From the results of all tests carried out on traditional bricks and interlock bricks with bagasse ash mixture with three mixed variations, it can be seen the results of the test analysis in **Table 2.**

Table 2. Recapitulation of Test Results for Conventional Bricks and Interlock Bricks with Bagasse Ash Mixes

No.	Type of Testing	Unit	SNI Standard	Conventional Bricks	Interlock Brick		
					Mix 1	Mix 2	Mix 3
1	Salinity	%	<50	0	16.14	5.05	4.14
2	Suction power	g/dm²/menit	Max. 20	139.06	58.75	57.34	57.34
3	Absorption	%	Max. 20	36	20	20	19
4	Specific gravity	-	-	1.66	1.74	1.76	1.81
5	Volume weight	g/cm³	-	1.11	1.511	1.537	1.543
6	Compressive strength	kg/cm²	Min. 50	12.11	30.26	47.28	63.78

4.12. Advantages of Modified Tanzanian Interlock Brick (MTIB)

The advantages of interlock bricks are the Modified Tanzanian Interlock Brick as follows:

1. The simple placement of connections between interlock brick cavities.

2. The composition of interlock bricks made from a mixture of bagasse ash waste which is environmentally friendly.
3. This interlock brick is environmentally friendly because it does not go through the combustion process.
4. In the implementation of the field, interlock bricks are faster than conventional bricks because they are mutually binding.

Figure 11. *Modified Tanzanian Interlock Brick*

5. Conclusions

In adding bagasse ash to the composition of the interlocking brick without combustion process with the Modified Tanzanian Interlock Brick (MTIB) form, it dramatically affects the quality of the interlocking brick. The best composition is in mixture 3, which is a ratio of 2 Clay: 3 Sand: 3.5 Portland Cement: 1.5 Bagasse Ash. The specifications are the weight until 3332 - 3426 g with a salt content of 4.14% which will continue to increase if bagasse ash is added more and more, water suction power of 57.34 g/dm²/minute, absorption of 19%, specific gravity of 1.81, volume weight of 1.549 g/cm³ and a compressive strength of 63.78 kg/cm² fulfills class 50 quality. If more bagasse ash is added, it will further reduce the compressive strength of the interlocking brick.

Acknowledgments

Thank you for the source of funding PNBP of Politeknik Negeri Banyuwangi year 2019 budget in the scheme of research based on Master's research plan. The appreciation is also given to the research team for their cooperation so that we can complete our research.

References

[1] Abdurrohmansyah, et al. 2015 Studi Kuat Tekan Batu Bata Menggunakan Bahan Additive (Abu Sekam Padi, Abu Ampas Tebu Dan Fly Ash) Berdasarkan Spesifikasi Standar Nasional Indonesia (SNI) (Lampung: JRSSD Edisi September, Vol. 3, No. 3) page 541 – 552.
[2] Alamsyah A A 2017 Pemanfaatan Abu Ampas Tebu (Bagasse Ash Of Sugar Cane) Sebagai Bahan Pengisi (Filler) Dengan Variasi Tumbukan Pada Campuran Aspal Panas ATB (Asphalt Treatd Base) (Malang: Seminar Nasional Teknologi dan Rekayasa (SENTRA) Fakultas Teknik Jurusan Teknik Sipil Universitas Muhammadiyah Malang)
[3] Febrianto E D C 2010 Pengujian Kualitas Bata Merah Yang Menggunakan Abu Ampas Tebu Dengan Berbagai Presentase (Jember: Universitas Jember)
[4] Habsya C, et al. 2014 Lockbrik Modular Beton Untuk Alternatif Bahan Dinding Yang Memenuhi Mutu SNI Dengan Biaya Murah (Surakarta: Sinektika Vol. 14, No. 2)
[5] Handayani S 2010 Kualitas Batu Bata Merah dengan Penambahan Serbuk Gergaji (Semarang: Jurusan Teknik Sipil, Fakultas Teknik, Universitas Negeri Semarang (UNNES))
[6] Haryadi T 2016 Teknis dan Aplikasi Interlocking Compressed Block Sebagai Material Sustainable dan Potensinya Pada Arsitektur (Bandung: Program Studi Arsitektur, Universitas Katolik Parahyangan)
[7] Hidayat N 2016 Karakteristik Batu Bata Dengan Campuran Abu Ampas Tebu dan Gypsum (Banyuwangi: Politeknik Negeri Banyuwangi)
[8] Indra W R and Dachyar E M 2012 Aplikasi Bata Interlocking Untuk Rekonstruksi Pasca Letusan

Gunung Merapi (Jakarta: Jurnal Sains dan Teknologi Indonesia Edisi Desember, Vol. 14, No. 3) Page 8 – 14.
[9] Kafrain I G Y 2018 Dinding Bata Interlock Pulutan (Universitas Katolik De La Salle Manado: Jurnal Realtech Vol. 14, No. 1)
[10] Kintingu S H 2009 Design Of Interlocking Bricks For Enhanced Wall Construction Flexibility, Alignment Accuracy And Load Bearing (The University Of Warwick: School of Engineering)
[11] NI-10 1984 Bata Merah Sebagai Bahan Bangunan (Bandung: Yayasan Lembaga Pendidikan Masalah Bangunan)
[12] Noerwasito T 2004 Abu Tebu Pabrik Gula, Bata Efisien Enerji (Surabaya: Dimensi Teknik Arsitektur Vol. 32, No. 1) Page 57 – 62.
[13] Primayatma 1993 Peranan Semen Portland dan Bahan Lain Terhadap Campuran Tanah Liat Sebagai Bahan Batu Bata Merah Tanpa Pembakaran (Jakarta: IBG)
[14] SNI 03-0349-1989 Bata Beton Untuk Pasangan Dinding (Jakarta: Badan Standarisasi Nasional)
[15] SNI 03-1969-1990 Metode Pengujian Berat Jenis Dan Penyerapan Air Agregat Kasar (Jakarta: Badan Standarisasi Nasional)
[16] SNI 03-1973-1990 Metode Pengujian Berat Besi Beton (Jakarta: Badan Standarisasi Nasional)
[17] SNI 15-2094-2000 Bata Merah Pejal Untuk Pasangan Dinding (Jakarta: Badan Standarisasi Nasional)
[18] SNI-10.1978:6 Standar Nasional Indonesia Kuat Tekan dan Penyimpangan Ukuran Batu Bata (Jakarta: Badan Standarisasi Nasional)
[19] Sudarsana I K 2011 Karakteristik Batu Bata Tanpa Pembakaran Terbuat Dari Abu Sekam Padi Dan Serbuk Batu Tabas (Bali: Jurnal Ilmiah Teknik Sipil Vol. 15, No. 1) Page 93-101
[20] Zulkaiddah N, et al. 2016 Pengaruh (SiO2) Dalam Ampas Tebu Dan Sekam Padi Sebagai Bahan Tambahan Pembuatan Batu Bata Tanpa Pembakaran (Makassar: Jurusan Sipil, Fakultas Teknik, Universitas Hasanuddin)

IOP Conf. Series: Materials Science and Engineering **854** (2020) 012003 doi:10.1088/1757-899X/854/1/012003

Transesterification Condition of Crude Palm Oil to Biodiesel Using Solid Catalyst Zeolite/NaOH

E Kurniasih, P Pardi, Adriana

[1] Department of Chemical Engineering, Study Program of Industrial Chemical Technology, Polytechnic State of Lhokseumawe, Aceh

[2]Department of Chemical Engineering, Study Program of Oil and Gas Production, Polytechnic State of Lhokseumawe, Aceh

Email :ekakurniasih@pnl.ac.id

Abstract. The use of NaOH as a catalyst for biodiesel production has long been used and provides high conversion conversions. Homogeneous NaOH phase during the transesterification reaction causes soap to form at the end of the process. Soap is diseparasi from biodiesel by washing and decantation techniques. The presence of soap can reduce the quality of biodiesel, especially at the flash point of biodiesel. Heterogeneous catalyst zeolite / NaOH is synthesized from natural zeolite and NaOH through the impregnation method. Natural zeolite is contained in natural rock obtained from Ujong Pancu, Aceh Besar. The synthesis of natural zeolite catalyst was initiated by activating zeolite powder with 4 M HCl for 10 hours at 90oC. Activated zeolite was impregnated using 2.5% NaOH at a ratio of 1: 2 (w / v) at 90oC for 2 hours. Application of zeolite / NaOH in transesterification reaction using crude palm oil off grade. The results showed that the use of zeolite / NaOH can prevent the formation of soap with characteristics that have met SNI 04-7182-2006 standards.

1. Introduction

To deal with the fossil fuel crisis, biodiesel has been recognized as an alternative to replace the role of petroleum diesel as a fuel. Fossil fuels have caused excessive emissions resulting in global warming, severe weather changes, which can threaten the future of humanity. Compared to fossil fuels, biodiesel is superior because it can reduce environmental pollution and global warming, making it even more existent among many researchers who want to get biodiesel with high ester content and according to SNI standards.

Biodiesel can be derived from vegetable oils, animal fats, microalgae, and is biodegradable, renewable and non-toxic (Wu et al, 2016). Biodiesel is formed by binding to the alkyl group from alcohol, usually the alkyl used in the form of methyl is the first choice because of its reactive nature.

In the process of biodiesel synthesis, a catalyst is used to accelerate the reaction rate of methyl ester formation. There are 2 types of catalysts, namely homogeneous catalysts and heterogeneous catalysts. According to Pena et al (2013) base catalysts are usually preferred over acid catalysts, because of their ability to catalyze transesterification reactions at relatively low reaction temperatures and high reaction rates which are about 4000 times faster than acidic catalyst reaction rates. High biodiesel yields can also be achieved at short reaction times, and most importantly this catalyst is cheap and easy to obtain (Mansir et all, 2017). One type of homogeneous base catalyst is NaOH.

NaOH as a homogeneous base catalyst itself is widely used because of its reactive nature, but at the end of the process a product separation process is needed. In addition it requires the neutralization and separation of the final reaction mixture which leads to a series of environmental problems related to the use of large amounts of water and energy. For this reason, the role of heterogeneous catalysts as a substitute for homogeneous catalysts is needed, by using heterogeneous catalysts the product separation process is easier to do. In addition, heterogeneous catalysts have several advantages including easier

IOP Conf. Series: Materials Science and Engineering **854** (2020) 012003 doi:10.1088/1757-899X/854/1/012003

catalyst separation by physical methods, such as set-tling and centrifugation, and environmental pollutant reduction (Pena et al, 2013). Examples of heterogeneous catalysts that are commonly used are zeolites.

Zeolites are three-dimensional, crystalline solids, made from tetrahedral that are interrelated with alumina and silica, so they are widely used in various applications such as adsorbents, catalysts, solar energy storage, thermal adsorption storage and in the pharmaceutical industry because of their unique porous properties (Soonmin, 2017).

Zeolite as a catalyst has been recognized for its superiority, as evidenced by the large number of researchers involving zeolite as a catalyst to accelerate the reaction rate, recent studies such as those conducted by Zhang et al (2017) who analyzed the zeolite structure of fatty acid methyl esters (FAME) and Lawan et al (2019) which uses a CaO / Zeolite catalyst in biodiesel production.

Both homogeneous NaOH catalysts and zeolite heterogeneous catalysts, both have advantages as catalysts because of their nature. Therefore, in this study the biodiesel synthesis process was carried out using a zeolite / NaOH catalyst, which was a catalyst synthesized from zeolites impregnated with NaOH. Thus it is expected that the rate of formation of biodiesel can take place quickly due to the presence of zeolite / NaOH catalysts containing reactive NaOH, as well as the formation of soap and facilitate the separation of catalysts at the end of the process, to obtain biodiesel with high ester content and according to SNI 04-7182-2006 standards.

2. Method

2.1 Raw Material

The raw materials used are crude palm oil (CPO) as the triglyceride source, Methanol (p.a), zeolite/NaOH as the heterogeneous catalyst, H_2SO_4 98%, NaOH (p.a), hexane, ethanol. CPO as the main ingredient analyzed free fatty acid to know the initial free fatty acid content.

2.2 Esterification

At first, crude palm oil (CPO) was performed free fatty acid (FFA) analysis to determine the free fatty acid content before esterification reaction was performed. CPO is reacted with methanol and an H_2SO_4 as the acid catalyst. It is added with a ratio of 15% (w/v). The esterification reaction is pre-treatment process for CPO to reduce free fatty acid (FFA). The reaction lasts for 120 minutes, at temperature 65°C and 400 rpm on stirring speed (scale of 1000 rpm). If CPO as trigliceride source contains many FFA, it will interferethe formation of ester bonds with methanol.

2.3 Transesterification

The transesterification reaction is also called alcoholization, is the conversion stage of triglyceride (vegetable oil) to alkyl ester, reacted with the alcohol to produce a by-product of glycerol. The CPO from esterification reaction is referred to crude ester. It is used as a feedstock in the transesterification reaction. crude ester is reacted with methanol with the addition of zeolite/NaOH heterogeneous catalyst, temperature, mole ratio and reaction time according to the Taguchi design Reactan are stirring at speed of 400 rpm (scale 1000 rpm).

The reaction temperature is adjusted to the boiling point of methanol. After the reaction time is reached, the transesterification reaction is stopped. Crude products are separated by separating funnels using the decantation method. Biodiesel is in the upper layer, and glycerol is in the lower layer.

Biodiesel which has been separated from glycerol is purified using aquadest to remove the formed side products. Crude biodiesel is then distilled to take up excess methanol. Distillation is carried out at a temperature of 68oC. Pure biodiesel is then carried out a final ALB analysis to determine the decrease in ALB content.

IOP Conf. Series: Materials Science and Engineering **854** (2020) 012003 doi:10.1088/1757-899X/854/1/012003

3. Characteristic analysis

Biodiesel products with the highest ester content, characteristics were analyzed in the form of density, viscosity, acid number, flash point. The results of product analysis are adjusted to SNI 04-7182-2006 standards.

4. Result and Discussion

Biodiesel production uses a heterogeneous zeolite/NaOH catalyst with a catalyst ratio of 0.5% -2.5% (w/w). The use of zeolite / NaOH allows the recycle of the catalyst because the catalyst is not homogeneous in the mixture so that it is easily separated at the time of purification. From the results of the study (Figure 1 and 2), obtained the highest ester content at a catalyst ratio of 2.5% (w/w) of 63.51%. The use of zeolite/NaOH catalyst also does not result in the formation of byproducts in the form of soap in large quantities, because the core of NaOH has been trapped in the zeolite framework so that it does not form a saponification reaction with free fatty acids in triglycerides (CPO).

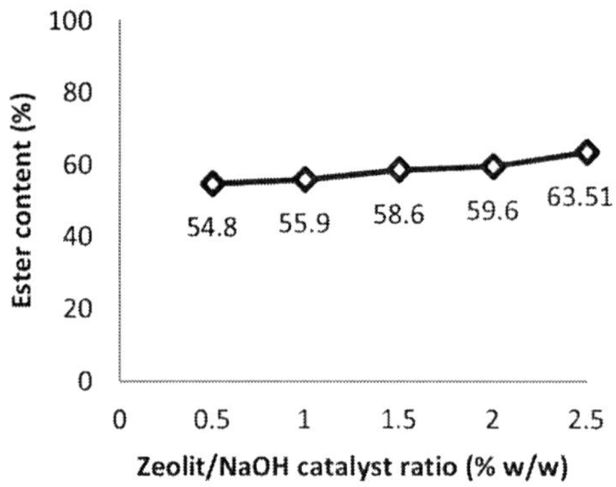

Figure 1. Correlation of Zeolite / NaOH Catalyst Ratio to Esther Content

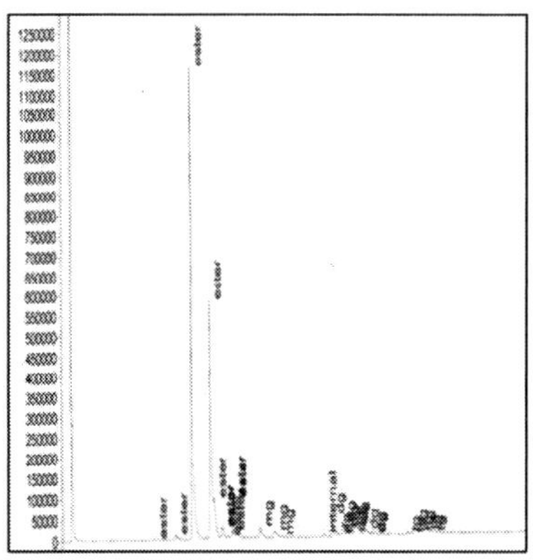

Retention Time	Komponen	%
19,2	Trigliserida	9,89
11,06	Monogliserida	15,2
15,4	Digliserida	11,2
5,14	Ester	63,51

Figure 2. Gas Chromatography Zeolite / NaOH Catalyst Biodiesel Product 2.5% (w/w)

5. Product Characterization

Product characterization aims to determine the extent to which biodiesel products exhibit properties as alternative fuels. Products containing large quantities of esters have properties close to or in accordance with biodiesel standards. The characteristics tested include acid number, density, kinematic viscosity, no flame, water content, total glycerol which is considered to have a major influence to determine the quality of a body cell product. In this research, the product with the highest ester content to be determined for the analysis of the characteristics includes :

Tabel 1. Biodiesel Product Characterization

Parameter	Biodiesel Zeolite/NaOH	Standar SNI 04-7182-2006
Acid Value	0,4	Maks 0,8 mg KOH/gr
Density	0,860	0,850-0,890 kg/liter
Kinematic Viscosity	3,7	2,3-6,0 mm^2/s
Flash Point	110	Min 100°C
Water Content	0,01	Maks 0,02%
Glycerol Total	0,24	Maks 0,24%

The Table 1 shows that the characteristics of zeolite / NaOH biodiesel meet SNI 04-7182-2006 standards.

6. Conclusion and Suggestions

Based on the results of gas chromatography analysis of biodiesel products with zeolite/NaOH catalysts, it is recommended to increase the zeolite/NaOH catalyst ratio above 2.5% (w/w) to improve the alkaline atmosphere in the reaction mixture system.

Acknowledgements

Thank you to the Ministry of Research, Higher Education for funding this research through the Decentralization Research Grant program. Applied Chemistry Laboratory and Process Unit, Pilot Plant Laboratory, Oil and Gas Laboratory which has facilitated this research activity and the Palm Oil Research Center, Medan.

References

[1] Lawan, I., Garba, Z.N., Zhou, W., Zhang, M., Yuan Z. 2019. Synergies between the microwave reactor and CaO/zeolite catalyst in waste lard biodiesel production. Renewable Energy 145 (2019) 2550-2560.

[2] Mansir, N., Teo, S.H., Rashid, U., Saiman, M.I., Tan, Y.P., Alsultan, G.A., Taufiq-Yap, Y.H. 2017. Modified waste egg shell derived bifunctional catalyst for biodiesel production from high FFA waste cooking oil. A review. Renewable and Sustainable Energy Reviews.

[3] Pena, R., Romero, R., Martinez, S.L., Natividad, R., Ramirez, A. 2013. Characterization of KNO3/NaX catalyst for sunflower oil transesterification. Fuel 110 (2013) 63–69.

[4] Soonmin, H. 2017. Zeolite and application: A Review. Ideal Internasional E - Publikasi Pvt. Ltd.

[5] Wu, Z., Chen, C., Guo, Q., Li, B., Que, Y., Wang, L., Wan, H., Guan, G. 2016. Novel approach for preparation of poly (ionic liquid) catalyst with macroporous structure for biodiesel production. Fuel 184 (2016) 128–135.

[6] Zhang, Y., Xia, C., Lin, M., Duan, Q., Zhu, B., Peng, X., Wang, B., Yuan, S., Liu, Y., Shu, X. 2017. Sponge-structured titanosilicate zeolite with high catalytic activity in epoxidation of fatty acid methyl ester. Catalysis Communications.

IOP Conf. Series: Materials Science and Engineering **854** (2020) 012004 doi:10.1088/1757-899X/854/1/012004

Effects of strain rate due to SMAW welding on mechanical properties of AISI 1005 low alloy steel

Husaini[1]*, K Akbary[1], I Hasanuddin[1], T E Putra[1], and H Husin[2]
[1]Laboratory of Computational Mechanics
Department of Mechanical Engineering – Universitas Syiah Kuala
Darussalam – Banda Aceh 23111, Indonesia
[2]Department of Chemical Engineering, Syiah Kuala University, Darussalam - Banda Aceh 23111, Indonesia

*Email: husainiftm@unsyiah.ac.id

Abstract. The welding process can affect the mechanical properties of the metal. This study was aimed at examining the effects of welding on the strain rate of AISI 1005 carbon steel. Shield metal arc welding, utilizing an E6013 electrode with a diameter of 2.5 mm, a double V-notch with an angle of 120°, and a welding position of 1G, was used to weld low carbon steel. A tensile test was performed to determine the mechanical properties of the steel at various strain rates to represent the loads that it would be subjected to. The test was performed at strain rates of 3 mm/min, 9 mm/min, 15 mm/min and 20 mm/min. The results showed that as the strain rate of steel increased to 20 mm/min, the maximum tensile strength of 447.67 MPa and highest yield strength of 362.94 MPa were obtained, while the lowest values were obtained at a strain rate of 3 mm/min. An examination of the fractured surfaces using SEM showed that at the strain rates of 3 mm/min, 15 mm/min and 20 mm/min smaller voids were formed.

1. Introduction

Low carbon steel applications are beginning to flourish, and it is estimated this metal is being used in more than 60% of heavy vehicles, platinum-fuel offshore oil vessels, car frames and in building construction [1]. This is because low carbon steel has a high thermal conductivity resistance, high toughness, good machinability and good weldability [2]. There are many methods available for joining different materials, and one such method is the welding process, which is used more often because it results in a high level of toughness and strength. However, the welding process will bring about changes to the crystal structure of the material such that there will be a decrease in the physical and mechanical properties of the welded material [3, 4, 5]. Welding produces a change in the mechanical behaviour of the base metal, the welding metal and in the heat-affected zones (HAZ), thereby causing complications in the weld under local stress and strain conditions.

Therefore, the purpose of this study was to determine the impact of the strain rate on the mechanical properties of AISI 1005 low alloy steel due to the influence of welding.

2. Methodology

2.1. Materials

The specimen used in this study was AISI 1005 low carbon steel. The chemical composition of AISI 1005 is: C (0.06 wt.% Max), Si (0.12 wt.%), Mn (0.35 wt.%), P (0.004 wt.%), S (0.050 wt.%), and Ni

(0.50 wt.%). The filler metal used an E6013 electrode, which is the standard used by AWS, with an electric welding wire with a tensile strength of 60 ksi for all the welding positions [6]. The chemical composition of E6013 is: C (0.12 wt.%), Mn (0.3-0.6 wt.%), Si (0.35 wt.%), S (0.035 wt.%), and P (0.04 wt.%).

2.2. Preparation of Specimens

First, a groove was made in the specimen before the welding process. The double V-notch with a seam angle of 120° and a diameter of 10 mm was very suitable for use on AISI 1005, as shown in Fig. 1 [6]. After the welding process, the AISI 1005 material was used as a tensile test specimen with the Standard AWS B4.0, as shown in Fig. 2 [7].

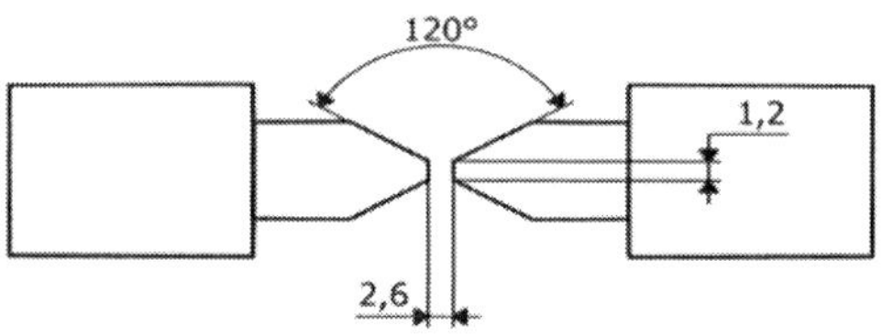

Figure1. Double V welding camp.

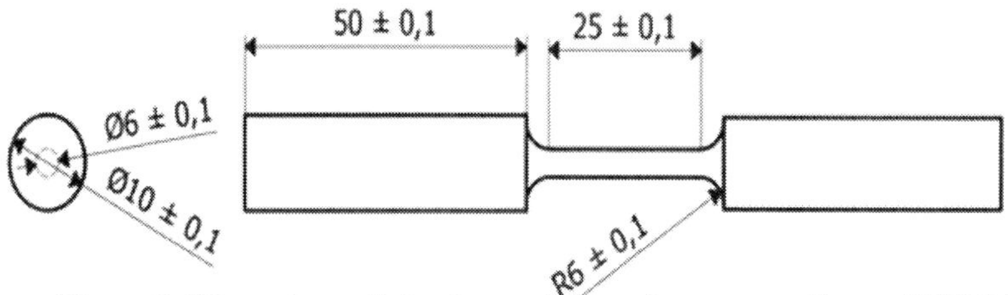

Figure 2. Dimensions of circular tensile specimens according to AWS B4.0 [7]

2.3. Tensile Ttest

A universal testing machine with a maximum capacity of up to 10 kN was used to conduct the tensile test at room temperature. Tensile tests are performed to determine the mechanical properties such as the tensile stress and yield stress of materials [8].

2.4. Microstructural Test

A microstructural test, using scanning electron microscopy (SEM) in combination with computer software for image analysis, was carried out to determine the shape of the visible fault and the cause of the fracture. The specimens for the SEM test were cut according to the required size of the fracture that occurred in the tensile test. The area of the fracture was scraped with various grades of sandpaper, and then etched with 5 ml HCL, before the SEM test was conducted.

3. Results and Discussion

3.1. Tensile Properties

From the tensile test results at various strain rates of the welded AISI 1005 material, it was found that the highest yield strength (0.2% offset) and tensile strength of 362.94 MPa and 447.67 MPa, respectively, occurred at a strain rate of 20 mm/minute, while the lowest yield strength (0.2% offset) and tensile strength of 142.68 MPa and 406.39 MPa, respectively, occurred at a strain rate of 3 mm/min., as shown in Table 1. This showed that the faster strain rate increased the yield strength and tensile strength of AISI 1005. This was a result of strain hardening on the metal that arose as a consequence of the more difficult formation and movement of dislocations. Hardening occurs when there is increased resistance to dislocation movements. Following the symptoms of melt, there was a continuous increase in stress to maintain deformation. In other words, the flow stress of the deformed metal increased with increasing strain [9]. The curve of the tensile test for each strain rate is shown in Fig. 3.

Table 1. Value of mechanical properties of four strain rate variations

Crosshead speed (mm/ min.)	Yield strength (0,2% *offset*) (MPa)	Maximum Tensile Strength (MPa)
3	142,68	406,39
9	328,62	421
15	208,68	438,16
20	362,94	447,67

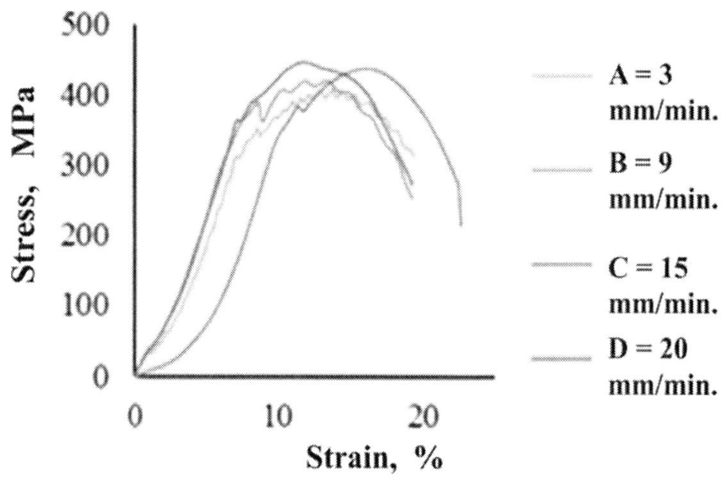

Figure 3. The curve of tensile test results with four strain rate variations.

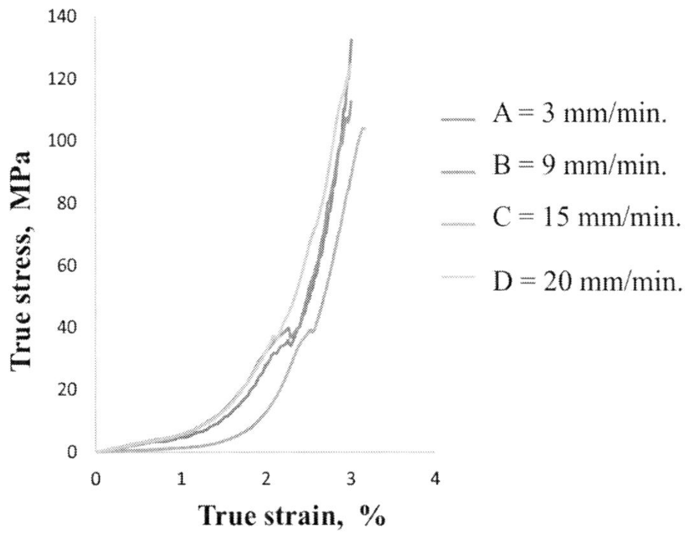

Figure 4. Curve of true stress vs. true strain.

Fig. 4 shows the true stress that resulted in material toughness. The higher the value of the true stress, the higher the decrease in the ductility. The results of the true maximum tensile strength at each strain rate showed that the highest strain rate of 20 mm/min., produced an true maximum tensile strength of 20884.92 MPa, while the lowest strain rate of 3 mm/minute produced an true maximum tensile strength of 5984.31 MPa.

Table 2. The true maximum stress value.

Crosshead speed (mm/min.)	True maximum tensile strength (MPa)
3	5984,31
9	6072,67
15	7493,06
20	20884,92

3.2. Analysis of Fractured Surface of Specimen.

An observation was made of the macro structure on the fractured surface to determine the type of fracture that was experienced after the tensile test. The fracture surface could be observed microscopically or macroscopically.

The fractured surface of the specimen at each strain rate can be seen in Fig. 5. The results of the macro analysis in Fig. 5(a) and (b) shows that the fault was ductile as it had the distinctive characteristics of a ductile fracture, given that the fracture was fibrous and dark (dull), and that it formed an angle of 45° in the direction of the pull. Fig. 5(c) shows a resilient fracture, which was very evident due to the appearance of the cup and cone [10-12]. Fig. 5(d) shows that the fracture was very resilient as it was fibrous and dull, and formed an angle of 45° in the direction of the tensile force.

An examination of a fractured surface using a scanning electron microscope (SEM) can give important information about the type of fracture that has occurred. The broken surface of each specimen at 4 different strain rates can be seen in Fig. 6-9.

Figure 5. The fracture surface of the tensile test specimen with strain rate (a) 3 mm/min. (b). 9 mm/min. (c). 15 mm/min. and (d). 20 mm/min.

Figure 6. (a) Fracture surface of specimens with a strain rate of 3 mm/min. (b) SEM image.

In Fig. 6, the surface fracture was fibrous and dark (dull). The fractured metal surface of the weld was observed using a scanning electron microscope with a magnification of 600x, and it revealed that fracture that was formed had a considerable number of smaller voids.

IOP Conf. Series: Materials Science and Engineering **854** (2020) 012004 doi:10.1088/1757-899X/854/1/012004

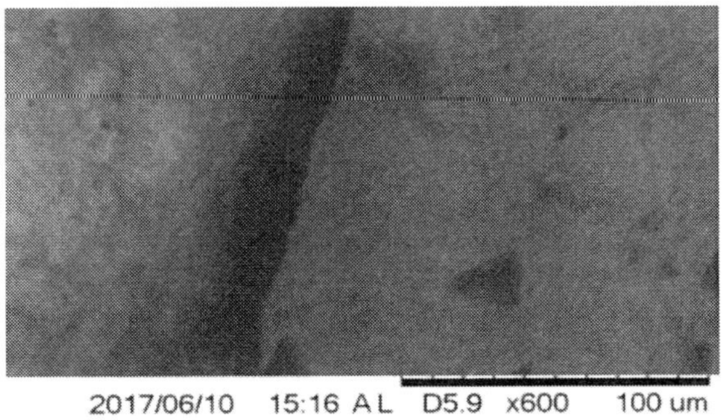

Figure 7. SEM image of fracture surface of specimens with strain rate of 9 mm/min.

Fig. 7 shows that the fracture surface of the specimen was fibrous and dull. The fractured metal surface of the weld was observed using SEM with a magnification of 600x, and revealed a crack propagation with large voids compared to the previous fractured surface, specifically at a strain rate of 3 mm/min.

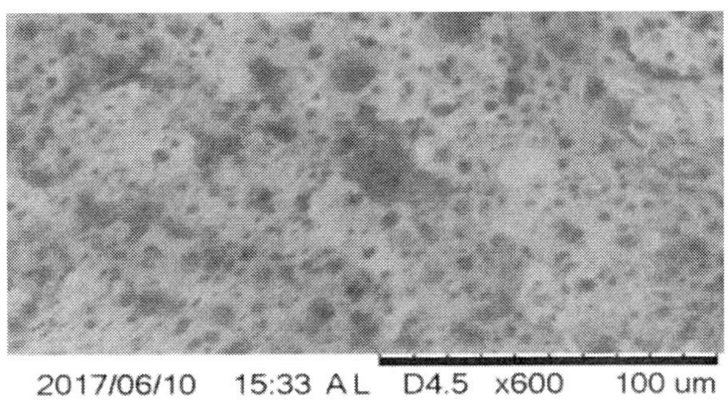

Figure 8. SEM image of fracture surface of specimens with strain rate of 15 mm/min.

Fig. 8 shows the results of the tensile test at a strain rate of 15 mm/min. on a fractured surface such as glass. The fractured metal of the weld was observed using SEM with a magnification of 600x, and revealed a broken surface with the smallest voids.

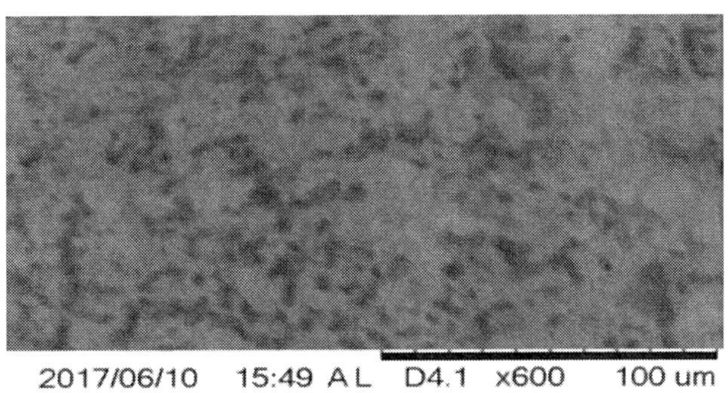

Figure 9. SEM image of fracture surface of specimens with strain rate of 20 mm/min.

Fig. 9 shows the tensile test results at a strain rate of 20 mm/min., where the fractured surface was characterized by considerable plastic deformation, before and during the crack propagation process. Faults were observed in the metal weld using SEM with a magnification of 600x, in which several large voids and small voids were seen in the solid metal.

4. Conclusions

Based on the results of the research the following conclusions could be drawn:

1. The strain rate due to welding caused the mechanical properties of the steel to increase. The highest mechanical properties were obtained at a strain rate of 20 mm/min., where the maximum tensile strength was 447.67 MPa and the maximum yield strength was 362.94 MPa.
2. There was an increase in the maximum tensile strength with each increase in the strain rate. The lowest tensile strength of 406.9 MPa and lowest yield strength of 142.68 MPa were attained at a strain rate of 3 mm/min.
3. Voids appeared on the fractured surface of the specimens, with smaller voids being formed at a strain rate of 20 mm/min., while at a strain rate of 9 mm/min., the specimens had large voids on the fractured surface.

Acknowledgment

The authors would like to express their gratitude to Universitas Syiah Kuala for financial support for research through the Laboratory research grant No.313/UN11.2/PP/PNBP/SP3/2019, and partly financial support from Profesor research grant No. 03/UN11.2PP / PNBP /SP3/2019.

References

[1] S R Nathan V Balasubramanian S Malarvizhi A G Rao 2015 *Journal of Science Direct Defence Technology* **11** 308
[2] ASM International 2005 *ASM Metal Handbook - Printed in the United States of America*
[3] G Magudeeswaran V Balasubramanian G Madhusudhan Reddy 2014 *Journal of ScienceDirect Defence Technology* **10** 47
[4] J Arifin H Purwanto I Syafa'at 2017 (in Indonesian) *J. Mech. Eng. Universitas Wahid Hasyim Semarang* **13** 121
[5] G E Linnert, 1965. *Welding Metallurgy*. AWS. Ohio.
[6] H Wiryosumarto T Okamura 2000 Teknologi Pengelasan Logam (in Indonesian) PT. Pradya Paramita Jakarta.
[7] AWS B4.0 Standard Methods for Mechanical Testing of Welds
[8] Husaini Zuhaimi 2016 Int. Joural of Technology **7** 456
[9] Husaini K Kishimoto M Hanji M Notomi 2016 ARPN *J. of Eng. and Appl. Science* **11** 885
[10] Husaini M Rachman M N Machmud 2018 AIP Conf. Proc. **1983** 050003
[11] Husaini N Ali J K Hamza S E Sofyan 2019 *IOP Conf. Proc. Series: Mater. Sci. Eng.* **523** 012065
[12] Husaini K Kishimoto M Notomi T Shibuya *Fatig. and Fract. of Eng. Mater. and Struct.* **24** 895

IOP Conf. Series: Materials Science and Engineering **854** (2020) 012005 doi:10.1088/1757-899X/854/1/012005

Different Techniques of Solar Rooftop Combo-PV/T System Implementation: Materials and Installations

Suryati, Nelly Safitri, Misriana, Widdha Mellyssa

Electrical Engineering Department, Politeknik Negeri Lhokseumawe

Email: suryati@pnl.ac.id

Abstract. Sun radiation been utilized to produce clean and environmental-friendly solar energy. Beside has been used as heat energy, it also has been used as electricity. Solar photovoltaic and thermal (PV/T) collector is a combo system that utilize solar PV technology and solar heater technology all together in one simply module. At the time PV system is irradiated with solar energy, the cell temperature increases significantly. As the heat that extracted is passing through a heat extracting fluid in such way under the module, then this heat is transferred become hot air and water. This paper aims to reveal several materials that been used to design solar combo-PV/T collector modules and their gains and drawbacks, and the installation onto households' rooftops. Water-based and nanofluids-based PV/T modules are two technologies that being considered in this paper. All the comparisons that have been made turned out to match the assumption that the greater the difference of temperature between ambient temperature and temperature of the cells of the module, the less efficient the electrical efficiency and electrical output of the PV module and also to overcome the problem of low thermal conductivity. The considered water-nanofluids PV/T solar collector modules then put into account for better performance to improve the raised problem of thermal conductivity.

1. Introduction

As sun radiation been widely used over several decades to produce clean and environmental-friendly heat and electricity, the performance of solar thermal and solar photovoltaic (PV) as the collector of sun radiation must put into account. People used to install solar thermal and solar PV separately on their rooftop in order to get hot air and hot water and also electricity. However, these solar energy collector's technologies are less aesthetic, and need more space on the rooftop.

For that reason, the solar PV/T collector technology is developed. This technology is a hybrid sun radiation collector technology that deploy solar PV technology and solar heater technology in one module. When a PV system is irradiated with solar energy, the cell temperature increases significantly. The greater the temperature differentiation between ambient temperature and cell temperature occurred, the less efficient then happened to the electrical efficiency and electrical output of the PV module. In order to increase the electrical efficiency, this excess heat is extracted by passing a heat extracting fluid (air or water) under the module. This is the fundamental of PV/T collectors integrated method [1-3].

Several references have reviewed two PV/T solar collector modules technologies based on water [4-5] and nanofluids [6-7]. In one hand, water-based PV/T mostly utilizes water as coolant in PV/T system. the water flow can be categorized as natural, force and hybrid system flows, respectively [4], yet, this PV/T modules technology has low thermal conductivity problem. On the other hand, the nanofluids PV/T that suspending nanoparticles in water has improved the raised thermal conductivity problem, and consequently, performed a significant positive effect on the heat transfer.

IOP Conf. Series: Materials Science and Engineering **854** (2020) 012005 doi:10.1088/1757-899X/854/1/012005

This paper reviewed the differences of those two techniques of PV/T modules which for further study might be used as a consideration to build a better module that performs water-nanofluids as coolants for PV/T collectors.

2. Performance of PV/T solar collector module

As indicates in [8], the combination of PV cells and solar thermal collector into a single module allows the cooling of PV cell and simultaneously utilizing the extracted heat for domestic use. PV/T solar collector module can reduce the operating temperature and improve system efficiency. It also produces thermal energy and electricity simultaneously. The collector module is consisting PV panel, insulation and frame. Physically, PV/T collector consists of one or more glass sheets or a transparent material placed atop an absorbing plate with air flowing around it. One way to increase the collector's efficiency is by using heat transfer area through finned absorber, corrugated surfaces, and porous media [9-10].

Figure 1 illustrates the fundamental mechanism of solar PV/T collector. Electricity production is provided primary priority in PV/T system applications. For that reason, the module absolutely preferable to keep PV unit operating temperatures maintain the lowest possible in order to get the appropriate efficiency levels.

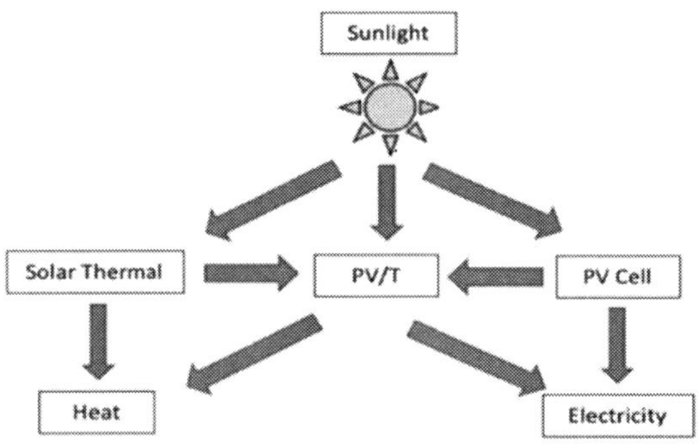

Figure 1. Fundamental work of solar PV/T collector.

PV/T collector can be classified into four types base on heat transfer medium; air-based PVT collector, water-based PVT collector, the combination of water/air-based PVT collector, and nanofluid-based PVT collector [1].

This section discusses the water- PV/T and nanofluids-based solar collector modules, respectively. The comparison then made to describe the benefits and drawbacks of those two modules due to then improving the modules by combining them into the water-nanofluids PV/T solar collector module. The benefits and drawbacks are gained from the overall energy and exergy efficiency of the considered solar PV/T collector modules.

2.1. *Water-based PV/T*

The development of water-based PV/T collectors is interesting and very promising area of research since currently this PV/T collector module is also used as water heater. Despite the cost-effective improvements on solar energy conversion, the uptake rate is still lower than the independent PV and thermal systems [4]. Figure 2 indicates the material structure of water-based PV/T collector module. A sandwiched structure of four layers namely photovoltaic cells lamination beneath the cover with a small air gap/tubes or flowing channels through the absorber which closely adhered to the PV cell layer and a thermally insulated layer located right below the flow channels. All the layers are fixed into a framed module using the adequate clamps or various joining methods and connections [11].

Studies on water-based PV/T system mostly focus on determining suitable water flow rate and temperature, sizing of water tubes for optimization and structural configurations including parts, components, connections etc.

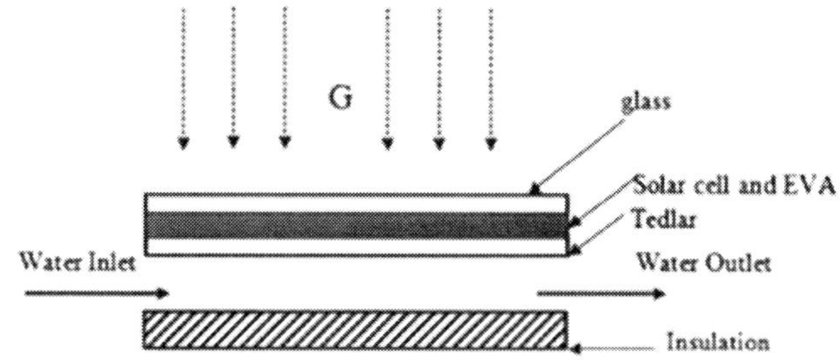

Figure 2. Material structure of water-based PV/T solar collector.

Figure 3 shows the mechanism of water-based PV/T collectors from the cross-sectional perspective point of view. Water-based PV/T collectors has similar structure as the conventional flat plate solar collectors. The absorber consist of tube is attached to the PV panel in order to use for heating purposes. The collector could achieve the enhance cooling effectiveness due to the high thermal mass of the water over the air [8].

In accordance to reference [12], the techniques are used to enhance the performance of water-based solar PV/T collector can be classified in the term of physical parameter and operating condition toward the environment due to seek the thermal efficiency based on six operating factors, namely ambient temperature, irradiance, cell efficiency, wind speed, PV cell covering factor and ratio of mass water to collector area.

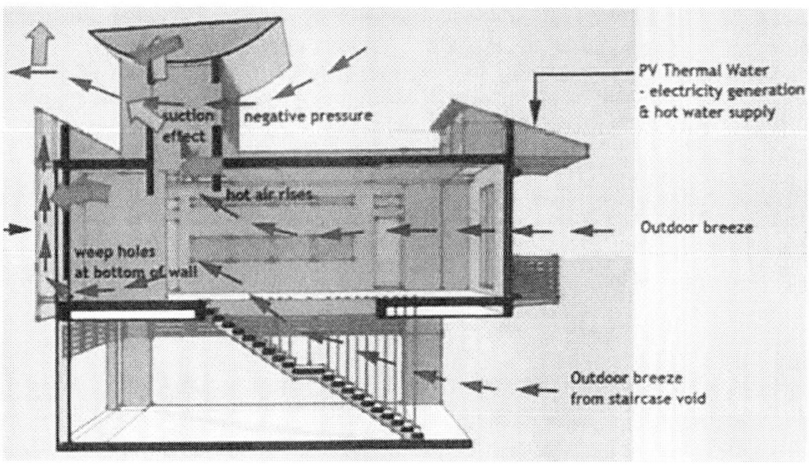

Figure 3. Water-based PV/T fundamental work, cross-sectional perspective point of view.

Mechanisms of absorbing and transferring the residual heat dissipated as a result of exposure of solar cells to solar radiation is designed in different configurations and dimensions due to optimize the module performance. Solar PV/T collector water-based type is effectively recommended for use in low latitude regions since the mains water supply temperatures are below 20°C in a year and function best in all climates except for instances prone to freezing [11].

The overall energy and exergy efficiency of the water-based solar PV/T collector as stated in [8-12]. For the energy efficiency, it is calculated in such way using equations as the overall efficiency of PV/T. Whereas $\eta_{PV/T}$ is the sum of solar collector thermal efficiency, η_{th} and PV's electrical efficiency, η_{PV} used to evaluate the performance of PV/T (1). The thermal efficiency of flat plate water-based solar PV/T collector is a ratio of the useful thermal energy, Q_u to the overall incidence irradiations, I (2). The electrical efficiency of PV modules, η_{PV} which is the ratio of measured output power to the overall incident solar radiation (3).

$$\eta_{PVT} = \eta_{th} + \eta_{PV} \quad (1)$$

$$\eta_{th} = \frac{Q_u}{I} \quad (2)$$

$$\eta_{PV} = \frac{P_o}{IA_c} \quad (3)$$

In addition, the exergy efficiency can be defined as the ratio of the thermal efficiency of an actual system compared to an idealized or reversible version of the system for heat [12]. The exergy efficiency can be calculated using exergies for thermal and PV (electrical). By assuming the insignificant effects of potential and kinetic energy changes, then the exergy efficiency of a water-based solar PV/T collector is defined as ratio of net output exergy to net input exergy [8]. It can be calculated as (4).

$$\eta_{Ex,PVT} = \frac{Ex_o}{Ex_i} = 1 - \frac{Ex_d}{Ex_i} \quad (4)$$

Some benefits of water-based solar PV/T collector can be found in [12], they were in the term of, i) bifunctionality: the way the module produces electrical energy and heat simultaneously; ii) efficiency: as the combination of energy and exergy efficiency resulting high independency systems; iii) improved thermal comfort: reduces air conditioning load; and iv) the facts that water works better as a working fluid than air, thereby increasing electrical energy harvesting. The water-based solar PV/T collector also has limited drawbacks. Once flat plate collectors are not supposed to be tracked, they have great amount of cosine losses, therefore less total energy falls on a rigid surface during the day. In addition, the maintenance cost for this type of solar collector is not cheap [13].

As for the installation, water-based solar PV/T collector is most suitable for areas with limited space. It is also consistently higher efficiency per unit area and can be easily mounted on existing roofs with minimal modifications. When come to the aesthetics point of view, it is more uniform than side-by-side systems.

2.2. Nanofluids-based PV/T

The use of nanofluids in solar PV/T module is given certain research consideration. Nanofluid material can be used as both for heat transfer medium and as optical filter that isolates the thermal toward PV and thermal receiver. According to [6] and [14], to produce nanofluids, the main aspects to consider are the nanoparticle materials and base liquids, although this process presupposes the close control of a set of important parameters in order to obtain functional thermal fluids. Nanofluids as a working fluid in the solar PV/T collector achieved high temperature in the thermal part of the system. it is also improved the performance of the system (electrical and thermal efficiencies) and reduced the required PV/T system area due to high production system per unit area [6].

Extensive work on the flowing nanofluids through optical filter improved the efficiency of the PV/T system. According to [16], various factors such as the concentration, flowrate, irradiance of the nanofluids, size of the nanoparticle, geometry of microchannel, type of base fluid are affected the efficiency of nanofluid-based solar PV/T collector. The most common methods to extracting heat from the PV units via nanofluids by employing heat collector at the rear end of the panel and using nanofluid as a liquid in spectral splitting filter joined on the front surface of PV module. Sometimes, both methods are used simultaneously in order to increase the efficiency. Nanofluid used as the cooling media. The overall system efficiency may be improved with minimal changes to the design of the structure by using a working fluid with improved heat transfer properties.

Figure 4 illustrates the diagram of a nanofluids-based solar PV/T collector. In single pipe system, there are two sections of pipe; primary section and secondary section. Primary section is set underneath the rear surface of the PV module having aluminium sheet in between. Primary section further elongates above the upper surface of it. Nanofluid enters from the inlet of primary section, thus, absorbing heat of the module. Heated nanofluid further passes over the PV's upper surface, in turns filtering the solar radiation. Part of radiation having wavelength equal to silicon bandgap is filtered and rest of the section is absorbed by the nanofluid flowing in the secondary channel which gets out of the secondary pipe at secondary outlet. Air exists between upper surface of PV module and secondary channel section. As the air gets hot, it flows in upward direction and the cool air still remains in contact with PV surface. It is assumed that no convection current is produced in the air.

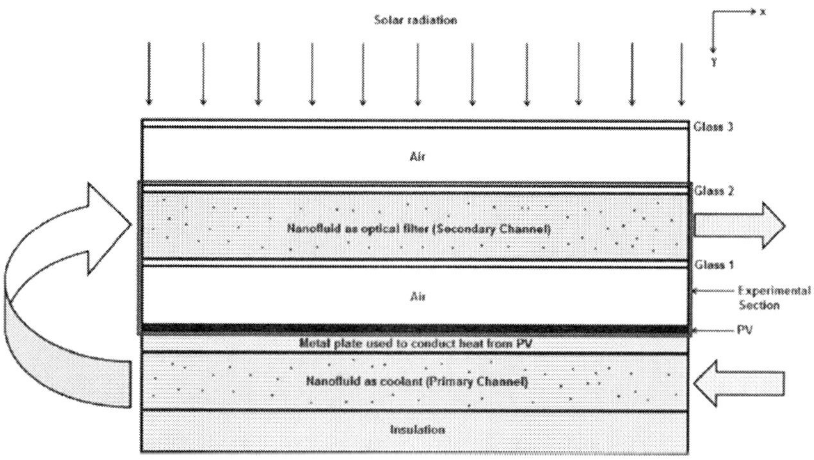

Figure 4. Schematic diagram of nanofluids-based solar PV/T collector [14].

The overall energy efficiency of the nanofluids-based solar PV/T collector as stated in [15-17]. Energy balance of nanofluids-based solar PV/T collector calculated by using (5), then the overall efficiency of the system is found by (6).

$$\dot{E}_{in} = \dot{E}_{el} + \dot{E}_{th} + \dot{E}_{loss} \tag{5}$$

$$\eta_{pvt} \cong \frac{\dot{E}_{el} + \dot{E}_{th}}{\dot{E}_{th}} \Rightarrow \eta_{pvt} = \eta_{pvt} = \eta_{th} + r \times \eta_{el} \tag{6}$$

Some benefits of nanofluids-based solar PV/T collector can be found in [17], a small area of nanofluid-based PV/T system may produce the amount of energy equivalent to a large area of a standard PV-only can produce. The optimized optical and thermal nanofluids-based PV/T hybrid system is a reliable solution to electrify remote off-grid regions at a low cost and to provide a potentially large supply of useful thermal energy. Also stated in [16], by using surfactant in the nanofluid can also surge the performance. The measures that can refine the performance of these systems include: i) glazing can drastically improve the nanofluid-based solar PV/T collector's performance; ii) simultaneously using optical filters over the surface and thermal collector at the rear end can also elevate performance; and iii) applying alternating magnetic field around the flow channel can increase the performance of system if the Ferro-nanoparticles are being used. The limited of nanofluids-based solar PV/T collector stated in [16], while using nanofluids include instability, agglomeration, high pumping power, and erosions. Stability improvement is the most important need of the hour in order to further proceed towards commercial use of nanofluids, as no perfect method of preparation and processing of stable nanofluid has been determined up-to-date. However, it is very possible to improve these limitations.

As for the installation, nanofluids-based solar PV/T collector is most suitable for areas with narrow space. It is also reliably higher efficiency per unit area and can be easily mounted on existing roofs with minimal modifications. When come to the aesthetics point of view, it is more uniform than side-by-side systems.

3. Performance of considered water-nanofluids PV/T collector module

The capability of nanofluids when it is combined with water to enhance the performance of solar PV/T collector has been studied in many different techniques. The effectiveness of utilized water-nanofluids as absorber fluids in solar PV/T collector strongly depends on the type of nanoparticles and base fluid, volume fraction of nanoparticles, radiative properties of nanofluids, temperature of the liquid, size and shape of the nanoparticles, pH values, and stability of the nanofluids [18]. Figure 5 shows the schematic diagram of water-nanofluids solar PV/T collector.

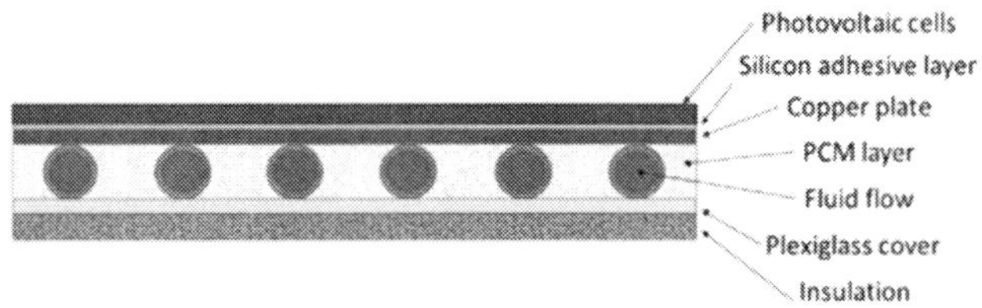

Figure 5. schematic diagram of water-nanofluids solar PV/T collector.

In reference [7], a combination of water and CuO nanofluids was investigated. The influence of CuO nanofluid confected the heat transfer coefficients. Since the excellent thermo-physical properties of the nanofluid comparative to water, then the simultaneous operation, the thermal performance of each fluid is directly associated with its counterpart. The overall energy and exergy efficiency of the water-nanofluids solar PV/T collector are expressed in (6) and (7).

$$\varepsilon_{pvt} \cong \frac{\dot{Ex}_{th} + \dot{Ex}_{el}}{\dot{Ex}_{in}} = \varepsilon_{th} + r\varepsilon_{el} \tag{7}$$

By using the considered water-nanofluids solar PV/T collector, the electrical and thermal efficiencies were found higher than those of the water-based and nanofluid-based PV/T systems. We modelled the eq. (1) – (7) in such way using MATLAB. Figure 6 illustrates the comparison of the parameter namely electrical and thermal output power, and also electrical and thermal efficiencies of water-based, nanofluids-based and water-nanofluids based solar PV/T collectors, respectively. The electrical and thermal outputs of those solar PV/T collectors as the critical parameters are compared showed that the most reliable solar PV/T collector is water-nanofluids solar PV/T collector.

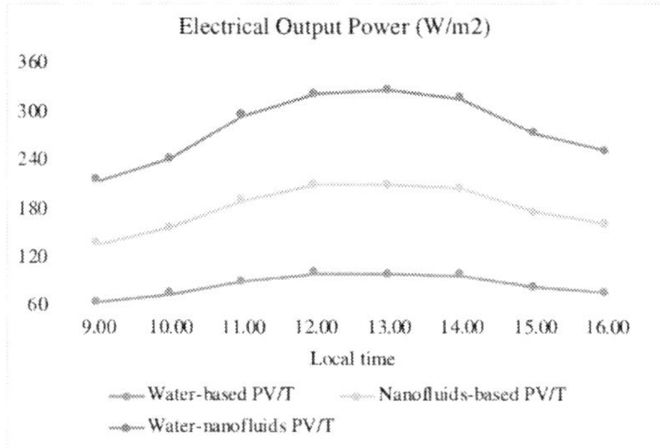

(a). Electrical output power of solar PV/T collectors

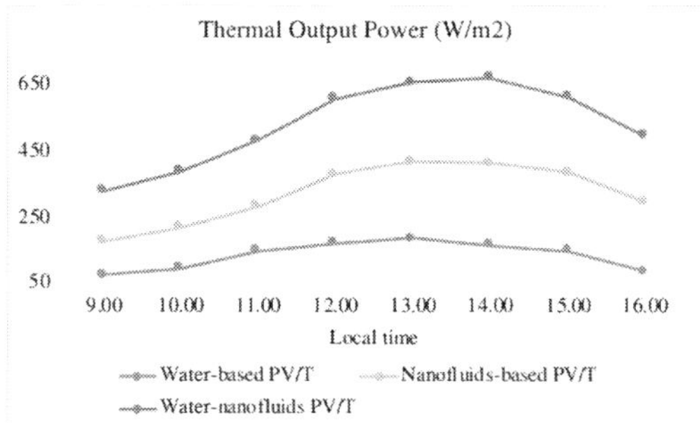

(b). Thermal output power of solar PV/T collectors

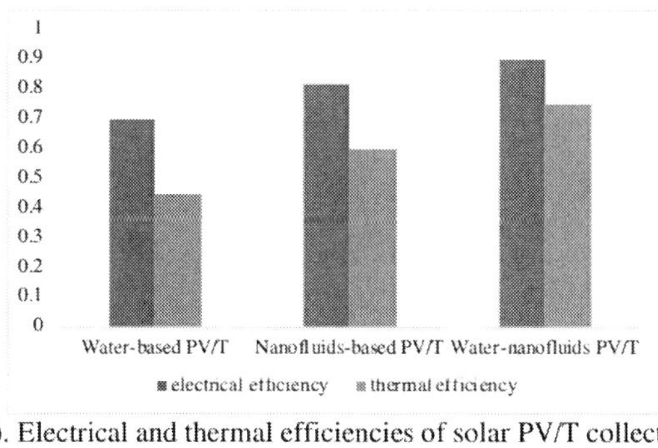

(c). Electrical and thermal efficiencies of solar PV/T collectors

Figure 6. Comparison of types of solar PV/T collectors in the efficiencies point of view.

Some benefits of water-nanofluids solar PV/T collector can be found in [19], namely: i) wide array of commercial applications due to its coreless capabilities; ii) energy and money conserving out of electricity; iii) excellent thermal insulation in a thin layer system which no influence to the weight; iv) no moisture infiltration, which gives the insulation the benefits of corrosion resistance; v) easy installation and has high efficiency. Thus, It was found that the use of nanofluids and combining it with water in the solar collector field can play a crucial role in increasing the efficiency of this device.

4. Conclusion

As can be seen from the reviews of water-based and nanofluids-based solar PV/T collector, and from modelling the considered water-nanofluids solar PV/T collector in MATLAB, the energy and exergy efficiencies of the collectors meet that recently the most efficient one is water-nanofluids solar PV/T collector. In addition, a water-nanofluids solar PV/T collector without any extra energy consumption can be a suitable option to improve the cooling in a PV/T system significantly.

References

[1] A. Fudholi, et al., "R&D of Photovoltaic Thermal (PVT) Systems: An Overview." International Journal of Power Electronics and Drive Systems (IJPEDS), vol. 9(2), pp. 803-10, 2018.

[2] A. Fudholi, et al., "Review of solar photovoltaic/thermal (PV/T) air collector." International Journal of Electrical and Computer Engineering (IJECE), Vol. 9, No. 1, pp. 126~133, ISSN: 2088-8708, DOI: 10.11591/ijece.v9i1.pp126-133, 2019.

[3] W. E. Ewe, A Fudholi, C. H. Yen, N. Asim and K. Sopian, "Performance of Solar Air Collector-Thermoelectric Hybrid System.", International Conference on Sustainable Energy and Green Technology 2018, IOP Conf. Series: Earth and Environmental Science 268 012071 doi:10.1088/1755-1315/268/1/01207, 2019.

[4] N. S. Bt. Rukman, et al., "Energy and exergy efficiency of water-based photovoltaic thermal (PVT) systems: an overview.", International Journal of Power Electronics and Drive System (IJPEDS), Vol. 10, No. 2, pp. 987~994, ISSN: 2088-8694, DOI: 10.11591/ijpeds.v10.i2.987-994, 2019.

[5] N. F. Mohd Razali, et al., "Review of water-nanofluid based photovoltaic/thermal (PV/T) systems.", International Journal of Electrical and Computer Engineering (IJECE), Vol. 9, No. 1, pp. 134~140, ISSN: 2088-8708, DOI: 10.11591/ijece.v9i1.pp134-140, 2019.

[6] F. A. Sachit, "Nanofluids Used in Photovoltaic Thermal (PV/T) Systems: a Review.", International Journal of Engineering & Technology, Vol. 7, No. 3, pp. 599-611, 2018.

[7] M. I. Hussain, J. H Kim and J.T Kim, "Nanofluid-Powered Dual-Fluid Photovoltaic/Thermal (PV/T) System: Comparative Numerical Study.", Energies Journal, Vol. 12, No. 775, pp. 1-19, doi:10.3390/en12050775, 2019.

[8] A. Fudholi, et al., "Overview of Photovoltaic Thermal (PVT) Water Collector.", International Journal of Power Electronics and Drive Systems (IJPEDS), Vol. 9, No. 4, pp. 1891~1898, ISSN: 2088-8694, DOI: 10.11591/ijpeds.v9.i4.pp1891-1898, 2018.

[9] M. Zohri, et al., "Exergy assessment of photovoltaic thermal with v-groove collector using theoretical study. TELKOMNIKA, vol. 16(2), pp. 550-57, 2018.

[10] M. Zohri, et al., "Photovoltaic thermal (PVT) system with and without fins collector: theoretical approach. International Journal of Power Electronics and Drive System (IJPEDS), vol. 8, no. 4, pp. 1756-63, 2017.

[11] P. R. Dongarwar and Arun Nair V, "Review of hybrid photovoltaic thermal(PV/T) water based system.", IJARSE, Vol. 07, Issue 06, pp. 530-539, 2018.

[12] M.A.M. Rosli, et al., "Progress on Flat-Plate Water Based of Photovoltaic Thermal (PV/T) System: A Review.", Iranica Journal of Energy & Environment 5 (4): 407-418, 2014.

[13] K. Moradi, M. A. Ebadian, C-X. Lin, "A review of PV/T technologies: Effects of control parameters.", International Journal of Heat and Mass Transfer, vol. 64 pp. 483-500, 2013.

[14] M. Gupta, N. Arora, R. Kumar, S. Kumar, and N. Dilbaghi, "A comprehensive review of experimental investigations of forced convective heat transfer characteristics for various nanofluids," pp. 1–21, 2014.

[15] S. Saroha, et al., "Theoretical Analysis and Testing of Nanofluids-Based Solar Photovoltaic/Thermal (PV/T) Hybrid Collector.", *Journal of Heat Transfer*, Vol. 137(9), p. 091015, doi: http://dx.doi.org/10.1115/1.4030228), 2015.

[16] H. M. Ali, "Application of Nanofluids for Thermal Management of Application of Nanofluids for Thermal Management of Photovoltaic Modules: A Review.", Microfluidics and Nanofluidics Journal, http://dx.doi.org/10.5772/intechopen.74967, 2018.

[17] S. Hassani et al., "Environmental and exergy benefit of nanofluid-based hybrid PV/T systems.", Energy Conversion and Management, Vol. 123, pp. 431–444, 2016.

[18] N. Bozorgan and M. Shafahi, "Performance evaluation of nanofluids in solar energy: a review of the recent literature.", Micro and Nano Systems Letters – Springer Open Journal, pp. 3-5 DOI 10.1186/s40486-015-0014-2, 2015.

[19] A. K. Hussein, "Applications of nanotechnology to improve the performance of solar collectors – Recent advances and overview.", Renewable and Sustainable Energy Reviews, Vol. 62, pp. 767–792, 2016.

IOP Conf. Series: Materials Science and Engineering **854** (2020) 012006 doi:10.1088/1757-899X/854/1/012006

Black Liquor Pretreatment of Concorb and Enzymatic Hydrolysis for Lignocellulosic Bioethanol

Muryanto muryanto[1*], Eka Mardika Handayani[2] and Yan Irawan[1]

[1]Research Center for Chemistry – LIPI, Kawasan Puspiptek Serpong, Tangerang Selatan 15314, Indonesia
[2]Metrology in Chemistry Laboratory, Centre for Research and Human Resource Development– National Standardization Agency of Indonesia, Building 420, Kawasan PUSPIPTEK Serpong, 15314, Tangerang Selatan, Banten, Indonesia

Email: moeryanto.mrt@gmail.com

Abstract. Lignocellulosic bioethanol from corncob could be a promising alternative fuel because of abundant resources, and it will not interfere with the food supply. Pretreatment is the first step in a lignocellulosic bioethanol process. One of the effective pretreatments is alkaline pretreatment because it gives the cellulose-rich material for enzymatic hydrolysis. However, it produces black liquor as wastewater that provides environmental impacts. The black liquor from Empty Fruit Bunch (EFB) pretreatment was collected and used for pretreatment of corncob to reduce wastewater from bioethanol production. The pretreatment process was conducted in a 5-liter reactor. The pretreated corncob was hydrolysis with cellulase enzyme and converted to ethanol by *Saccharomyces cerevisiae*. The delignification of corncob using EFB-black liquor was 45.82%. The black liquor pretreatment was increasing ethanol production 3.25 times than untreated substrat after saccharification and fermentation process by Saccharomyces cerevisiae (36.39 g/l).

1. Introduction

Corncob is one of the abundant lignocellulosic waste materials in Indonesia. Main components of corncob are cellulose, hemicellulose that can be converted to fermentable sugar for ethanol production. In 2012, corn plantation could produce 20 million tons and generate 1.94 million tons of corncob and estimation of ethanol production from corncob is around 0.38 million tons [1].

Ethanol production from lignocellulosic material consists of three main processes, which are pretreatment, hydrolysis, and fermentation. Pretreatment is needed to produce macroscopic and microscopic biomass structure used for hydrolysis and to break lignin and hemicellulose bonding and crystal structure [2]. Delignification must be done to destroy the lignin structure, reduce lignin content, and give an optimum result in bioconversion lignocelluloses to glucose [3].

The alkaline pretreatment process is widely used in ethanol lignocellulosic production. During alkaline pretreatment, high amount of black liquor is produced. Black liquor contains potentially toxic chlorinated compounds, suspended solids, phenolic, lignin, and spent alkaline [4]. Production 76.46 kg of bioethanol in a pilot plant will produce 3,000 liters of wastewater from a pretreatment process that is dominantly obtained from alkaline solution (sodium hydroxide) [4]. Black liquor wastewater releases black-colored effluent with high chemical oxygen demand (COD) and Total Suspended Solids (TSS) which makes it significantly toxic to the environment. Therefore, the black liquor must be treated before discharge to the environment or can be minimized to utilize or reuse for another purpose. The high pH value in black liquor indicates rich in alkali content of spent alkaline. The high content of NaOH in black liquor has the potential to be used as a solvent in the pretreatment process.

Our previous research successfully used EFB based-black liquor as an alkaline pretreatment solvent for EFB [5,6] and sweet sorghum bagasse [7]. Therefore, in the present study, we use EFB based-black liquor as a solvent in corncob pretreatment to investigate its effect on enzymatic hydrolysis and ethanol production.

2. Material and methods

2.1. Materials

Corncob (Cc) used in this experiment was obtained from Indonesian Center for Agricultural Post Harvest R&D, Ministry of Agriculture. Physical pretreatment was conducted as the first step to reducing the size below 1 cm granule then stored in a plastic bag. The black liquor was collected from alkaline explosion pretreatment of EFB using 10% NaOH solution, 150°C, 4 bar and 30 minutes in Pilot Plant 2nd Generation Bioethanol, Research Center for Chemistry-LIPI. Commercial cellulase enzyme (Cellic® Ctec2 and Cellic® Htec2 supplied from NOVOzyme) were used in all the experiments as sole enzymatic complex. We have analyzed it by measuring its activity as FPU. The enzyme showed an activity of 132 FPU/ml. The organism used for the fermentation process in this study was commercial dry yeast of *Saccharomyces cerevisiae* (Fermipan, French).

2.2. Black Liquor Pretreatment

Black liquor pretreatment was conducted in the chemical explosion reactor [5,6]. 250 g of dry corncob (10 % moisture content) was mixed with 1,250 ml of EFB based-black liquor. The pressure was controlled at 4 bars; the temperature in the reactor was 130°C for 40 minutes. After pretreatment, the pretreated biomass was recovered by filtration, washed with water until neutral and dried in the oven at 50-60°C overnight.

2.3. Saccharification

The saccharification process was conducted in 250 ml Erlenmeyer with 15% (w/v) substrate, and then 70 ml of 0.05 M citrate buffer was added. The samples were closed and sterilized 15 min using an autoclave. After reaching the room temperature, enzyme and buffer citrate addition were added to the sample until the total volume is 100 ml. Two kinds of enzymes, CTec2 and HTec2, were combined with ratio 5:1. The CTec2 loading was 30 FPU g-1 pretreated corncob. The saccharification was conducted in a shaker incubator for 72 hrs at temperature 40°C, and 150 rpm agitation.

2.4. Fermentation

The substrates after saccharification process were added with one percent dry yeast *Saccharomyces cerevisiae*. The fermentation process was conducted in shaking incubator at 32°C with 150 rpm agitation for 72 hours.

2.5. Analysis

Corncob composition was characterized before and after black liquor pretreatment based on a method from National Renewable Energy Laboratory/NREL[8]. Corncob (300 mg, dry weight) was subjected to acid hydrolysis for lignin, cellulose, and hemicellulose content analysis. After hydrolysis, acid-insoluble lignin (AIL) was weighed using Sartorius BS224S, and acid-soluble lignin (ASL) was measured using UV/Vis Spectrophotometer (Optizen 2120 UV) at 205 nm. Total lignin was obtained from the sum of AIL and ASL. On the other hand, after hydrolysis, cellulose and hemicellulose were measured by HPLC (Waters e2695).

The sample (1 ml) was withdrawn from enzymatic hydrolysis and fermentation medium every 24 h. Glucose, xylose, and ethanol were determined using high-performance liquid chromatography/HPLC (Waters e2695).

3. Results and discussion

3.1. Pretreatment

Characteristics, properties, and analysis of the components of untreated and pretreated corncob are shown in Table 1. The main components of untreated corncob are cellulose followed by the lignin and hemicellulose. The total carbohydrate of corncob is 56.08%, and total lignin fraction is 26.47% of the dry biomass. Corncob could potentially be used as raw materials for bioethanol production process due to the cellulose contains.

Pretreatment is the most critical process in converting lignocellulose. Lignin is amorphous heteropolymer which is insoluble in water [9]. Lignin is associated with cellulose and hemicellulose, and it can inhibit enzymatic hydrolysis process. The purpose of the pretreatment is to reduce the amount of lignin in biomass known as delignification. Alkaline pretreatment can improve the ability of cellulase enzyme in the enzymatic hydrolysis process. In this study, pretreatment is conducted using black liquor from alkaline pretreatment process of EFB. Based on Table 1, cellulose content in corncob increased after pretreatment process used black liquor. The cellulose content in pretreated corncob reached 48.24% its mean 1.5 times higher than untreated corncob. Utilization of black liquor as a solvent in the pretreatment process can reduce the use of NaOH in the lignocellulosic bioethanol production process [6].

Table 1. Chemical composition of untreated and pretreated corncob.

Component	Untreated	Pretreated
Cellulose	31.69	48.24
Hemicellulose	24.39	24.67
Lignin	26.47	14.34
Ash	1.05	0.18

Figure 1 shows the effect of EFB-based black liquor pretreatment against delignification, cellulose recovery, and biomass recovery. Alkaline pretreatment process can eliminate lignin, hemicellulose, and other compounds that cause a reduction in the mass of corncob produced. The pretreatment process with black liquor is included in the alkaline pretreatment. Alkaline pretreatment can reduce the lignin content in lignocellulose. Black liquor pretreatment process at 130°C/40 minutes can reduce lignin up to 45.82%. Reduction of lignin, hemicellulose, and compounds other chemicals will cause a reduction the Corncob mass collected. Biomass recovery generated from black liquor pretreatment reached 66.37%. In general, the lignocellulose that has been pretreated with alkaline pretreatment will lose almost half of the initial mass [5]. The temperature and process time of the pretreatment process significantly affect the biomass recovery in alkaline pretreatment because the process temperature has a more dominant role in dissolving the soluble compounds [10]. The alkaline pretreatment process will only dissolve lignin, whereas cellulose is not dissolved, it resulted in the biomass produced with a high cellulose content. Cellulose recovery in this study reached 90%, its mean black liquor pretreatment was useful for the corncob pretreatment process because the only small amount of cellulose was lost during the process.

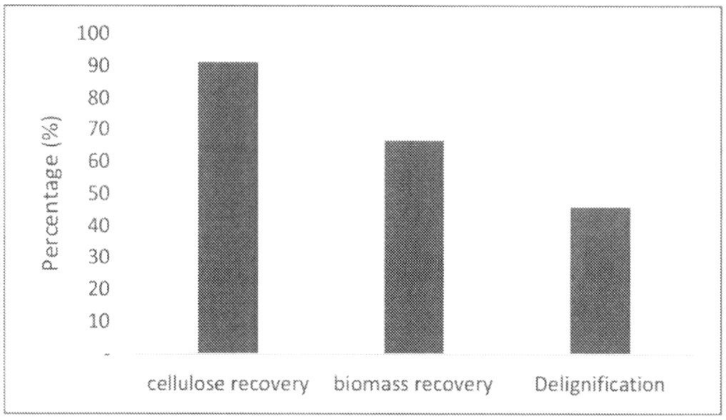

Figure 1. Delignification, cellulose and biomass recovery after pretreatment

The alkaline pretreatment process is suitable for bioconversion of lignocellulose into bioethanol because it give the higher cellulose content in the biomass. Cellulose in lignocellulose will be converted into glucose and fermented into bioethanol. Hence, cellulose content is very important in the process of bioethanol lignocellulose. In black liquor, there is spent alkaline which can be used as a solvent of lignin. Besides to dissolve of the lignin, the black liquor pretreatment process also damages the cellulose structure. NaOH content in the black liquor caused swelling of biomass, and reduce cellulose crystallinity and polymerization [11]. Some of the cellulose will also dissolve and loss in this process. Cellulose recovery in corncob reaches 90.95% its mean only the small amount of cellulose lost due to the black liquor pretreatment process.

Delignification is the main purpose of the pretreatment process. Reduced lignin content can increase the performance of cellulase enzymes in converting cellulose into glucose of the lignocellulose [12]. Sjöström [13] stated that several phases occur in dissolving lignin, namely initial delignification phase, intermediate delignification, and final delignification. The initial delignification that occurs at low temperatures is generally controlled by diffusion. Meanwhile, the next step is controlled by chemical reactions with the temperature above 140°C. In this phase, the reaction rate increases with the rising temperature. Delignification in the black liquor pretreatment process for corncob at 130°C reached 66.37%. The results of our previous studies showed that black liquor cannot be used directly for the same lignocellulose pretreatment due to NaOH content is lower than initial concentration, and the make up NaOH is needed to reach the same result [5],[6]. Smaller NaOH content can be used for pretreatment of lignocellulose which has low lignin content. In the previous study, black liquor may be used for the pretreatment process of bagasse sorghum without the addition of NaOH and the delignification reached 60% [7].

3.2. Saccharification and Fermentation

Saccharification or enzymatic hydrolysis aims to hydrolyze cellulose into monomer sugar (glucose). Corncob cellulose is converted into glucose using a cellulase enzyme. The results of the saccharification process are shown in Figure 2. The process of saccharification is carried out for 72 hours. The glucose content generated by saccharification increases along with the increasing processing time of saccharification. A dose of 30 FPU is an optimum enzyme dose based on our previous studies [14]. Corncob untreated produces the smallest glucose content (24.96 g/l). Meanwhile, pretreated corncob produces glucose with a value of 3.19 times greater than the untreated corncob. The pretreated corncob had higher cellulose and lower lignin content, that made enzyme easier to convert cellulose into glucose. Black liquor pretreatment may increase the concentration of glucose in the enzymatic hydrolysis process for the pretreatment increases the cellulose accessibility by enzymes. One purpose of the pretreatment is to increase the rate of enzyme digestibility [15].

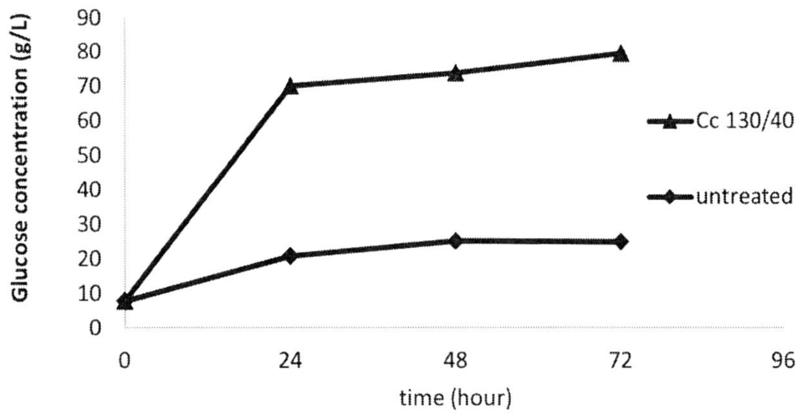

Figure 2. Glucose concentration after enzymatic hydrolysis

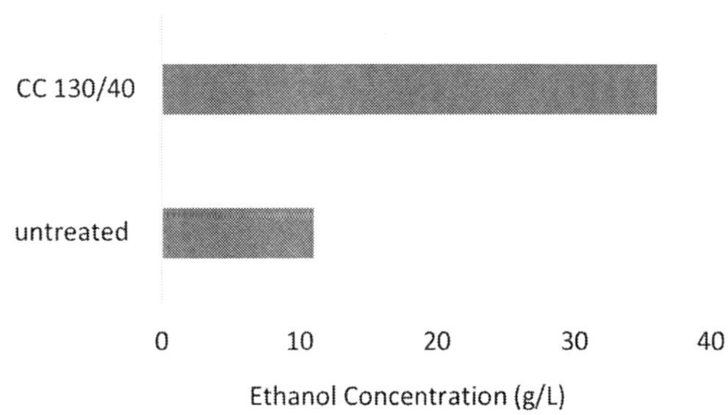

Figure 3. Ethanol concentration after fermentation

Glucose generated by saccharification process is then converted into ethanol using *Saccharomyces cerevisiae* through a fermentation process. *S. cerevisiae* has been known as yeast that can transform hexose sugars (glucose) into ethanol [16]. Figure 3. shows the concentration of ethanol produced by *S. cerevisiae* after the fermentation process for 72 hours. *S. cerevisiae* can convert glucose either from a substrate that has been pretreated or from the untreated substrate. The ethanol concentration obtained from the pretreated substrate (pretreatment at 130°C/40 minutes) equal to 36.39 g/l. Similar to the saccharification process, in the fermentation process, the black liquor pretreatment process gave effect to the concentration of ethanol produced. In the untreated substrate, *S. cerevisiae* only produces ethanol by 11.21 g/l. This small ethanol concentration was due to less glucose contained in the untreated

IOP Conf. Series: Materials Science and Engineering **854** (2020) 012006 doi:10.1088/1757-899X/854/1/012006

substrate for the performance of cellulase enzymes in the saccharification process was hampered by the lignin [17]. It shows that black liquor can be used as a pretreatment process solvent of corncob with good results.

Table 2. Ethanol concentration and yield of ethanol production from corncob

Var	max. ethanol concentration (g/l)	yield ethanol based on pret-CC (g/g pret-CC)[a]	yield ethanol based on untreated-CC (g/g untreated-CC)[b]
untreated	11.21	0.07	0.07
Cc 130/40	36.39	0.24	0.16

[a] yield ethanol based on pretreated corncob = max ethanol/dry weight of pretreated corncob (150 g/l)
[b] yield ethanol based on untreated corncob = max ethanol x rec biomassa/dry weight of corncob (150 g/l)
F = cellulose fraction in biomass

Table 2. shows the ethanol production from the process of saccharification and fermentation of pretreatment substrate with different conditions. The ethanol yield based on pretreated corncob generated by *S. cerevisiae* is reached 0.24 g/g pretreated substrate, its higher than our previous study that used *Rizhopus oryzae* for corncob fermentation [14]. This result showed that black liquor could be used as a pretreatment solvent. When viewed from the ethanol production to the initial substrate of corncob before pretreatment, the ethanol production produced by black liquor pretreatment is higher than the untreated production reached 0.16 g/g untreated substrate(more than two times). This means that the ethanol produced by the black liquor pretreatment is more efficient than untreated process. Pretreatment with the use of black liquor has several advantages, saving the use of NaOH, reduce black liquor waste discharged into the environment and produce the same product with the use of alkaline pretreatment.

4. Conclussion

The use of black liquor as a pretreatment solvent can reduce lignin content in corncob up to 45.82% under the process temperature conditions of 130°C for 40 minutes. The content of cellulose in pretreatment corncob reaches 48.24%, with the results of saccharification using cellulose enzymes which produce glucose concentration of 79.62 g/l. Glucose generated by saccharification at the fermentation using *S. cerevisiae* produces ethanol reached to 36.39 g/l. The parameters generated by the black liquor pretreatment approach the results obtained using alkaline pretreatment. Therefore, the black liquor can potentially be used as a pretreatment solvent of the corncob.

Acknowledgement

The authors are grateful to the Indonesian Center for Agricultural Post Harvest R&D, Ministry of Agriculture for providing corncob.

References

[1] Sudiyani Y, Sembiring K C, Barlianti V, Badri A I and Muryanto M 2015 The opportunity of bioethanol production from lignocellulosic agricultural waste in Indonesia *Energy Science and Technology vol 7 Bionergy* (Studium Press LLC) pp 529–50

[2] Sudiyani Y and Muryanto M 2012 The potential of biomass waste feedstock for bioethanol production *Proceeding of International Conference on Sustainable Energy Engineering and Application* pp 5–10

[3] Taherzadeh M J and Karimi K 2008 *Pretreatment of lignocellulosic wastes to improve ethanol*

and biogas production: A review vol 9
[4] Sari A A, Kurniawan H H, Nurdin M and Abimanyu H 2015 Decolorization of Black Liquor Wastewater Generated from Bioethanol Process by Using Oil Palm Empty Fruit Bunches *Energy Procedia* **68** 254–62
[5] Muryanto M, Triwahyuni E, Abimayu H and Cahyono A 2015 Alkaline Delignification of Oil Palm Empty Fruit Bunch Using Black Liquor from Pretreatment *Procedia Chem.*
[6] Muryanto, Triwahyuni E, Hendarsyah H and Abimanyu H 2015 Reuse Black Liquor of Alkali Pretreatment in Bioethanol Production *Energy Procedia*
[7] Muryanto and Sari A A 2018 Pretreatment of Sweet Sorghum Bagasse Using EFB-Based Black Liquor for Ethanol Production BT - Sustainable Future for Human Security : Environment and Resources ed B McLellan (Singapore: Springer Singapore) pp 85–95
[8] Raymond R and Tina E 1996 *Determination of Carbohydrates in Biomass by High Performance Liquid Chromatography LAP-002* vol LAP-002
[9] Hendriks a T W M and Zeeman G 2009 Pretreatments to enhance the digestibility of lignocellulosic biomass. *Bioresour. Technol.* **100** 10–8
[10] Kim Y, Kreke T, Mosier N S and Ladisch M R 2014 Severity factor coefficients for subcritical liquid hot water pretreatment of hardwood chips. *Biotechnol. Bioeng.* **111** 254–63
[11] Mosier N, Wyman C, Dale B, Elander R, Lee Y Y, Holtzapple M and Ladisch M 2005 Features of promising technologies for pretreatment of lignocellulosic biomass *Bioresour. Technol.* **96** 673–86
[12] Millett M A, Baker A J and Satter L D 1976 Physical and chemical pretreatments for enhancing cellulose saccharification. *Biotechnol. Bioeng. Symp.* 125–53
[13] Sjöström E 1993 *Wood Chemistry - Fundamental and Application (Second Edition)* (San Diego: Academic Press)
[14] Muryanto, Kurniawan H H and Irawan Y 2016 Bioethanol Production from Corncob by Rhizopus oryzae *Proceeding of The National Technology Congress* pp 159–66
[15] Sun Y and Cheng J 2002 Hydrolysis of lignocellulosic materials for ethanol production : a review q *Bioresour. Technol.* **83** 1–11
[16] Claassen P A M, van Lier J B, Lopez Contreras A M, van Niel E W J, Sijtsma L, Stams A J M, de Vries S S and Weusthuis R A 1999 Utilisation of biomass for the supply of energy carriers *Appl. Microbiol. Biotechnol.* **52** 741–55
[17] Vermaas J V, Petridis L, Qi X, Schulz R, Lindner B and Smith J C 2015 Mechanism of lignin inhibition of enzymatic biomass deconstruction *Biotechnol. Biofuels* **8** 217
[18] Muryanto, Sahlan and Sudiyani 2012 Simultaneous Saccharification and Fermentation of Oil Palm Empty Fruit Bunch for Bioethanol Production by Rhizopus oryzae *Int. J. Environ. Bioenergy*

IOP Conf. Series: Materials Science and Engineering **854** (2020) 012007 doi:10.1088/1757-899X/854/1/012007

Reliability Analysis of 20 KV Electric Power Distribution System

Nazaruddin*, Mahalla, Fauzi, Maimun, Subhan, Said Abubakar, Sayed Aiyub

Electrical Engineering Department – Politeknik Negeri Lhokseumawe

Email: nazaruddin@pnl.ac.id

Abstract. The reliability of the distribution system is the ability of a system to supply electrical energy continuously without causing failure to consumers. The reliability of the distribution system can be analyzed based on the SAIDI (System Average Interruption Duration Index), SAIFI (System Average Interruption Frequency Index), and CAIDI (Customer Average Interruption Duration Index) indices. The SAIFI, SAIDI and CAIDI values are determined by simulating using ETAP 12.0.6 software, which is by running reliability assessment analysis. The simulation process was carried out an IEEE 9 bus network distribution system. The simulation results show a SAIFI value of 1.113 frequency/ customer.yr, a SAIDI value of 17.306 hours / customer.yr and a CAIDI value of 15,549 hours / customer interruption. For normal operating conditions, the total demand for loads supplied from secondary substations to the 20 KV distribution network is 12,955 MW and 5,234 MVAR, with losses of 8.06%

Keywords: Reliability, distribution, SAIDI, SAIFI, CAIDI

1. Introduction

Electricity supply for customers requires various components including generating systems, transmission systems and distribution systems. These components are interrelated to one another that forms an electric power system. Electric power system is a collection of electricity centers and substations (load centers) which are connected to one another by a transmission network so that it is an interconnection system.

The distribution system is the part of the power system component that has the most disruptions, so the main problem in operating the distribution system is to overcome interference. Disturbances can occur in the generator system, transmission system and distribution system so that it will have a blackout on the customer [1].

The reliability of an electric distribution system can be calculated by knowing the amount and length of outages experienced by each load. The reliability of the distribution system is defined as the ability of the distribution system to carry out its functions for a specified period of time without failing. To be able to determine the level of reliability of a system, an examination must be made through calculation and analysis of the success rate of performance or operation of the system that is reviewed at a certain period, then compare it with the standards set.

IOP Conf. Series: Materials Science and Engineering **854** (2020) 012007 doi:10.1088/1757-899X/854/1/012007

To find out the reliability of a feeder, a reliability index is established, which is the amount to compare the appearance of a distribution system. Reliability indices that are often used in a distribution system are SAIFI (System Average Interruption Frequency Index), SAIDI (System Average Interruption duration Index). The reliability index is a number or parameter that indicates the level of service or the level of reliability of a consumer electricity supply [4] [5].

This paper will analyze the reliability of the system on a 20 KV primary distribution network. In an electric power system, reliability is the level of service to consumers so that electricity supply can be carried out continuously. The reliability index calculation is done using ETAP software which will be tested on the IEEE-9 bus network distribution system. Network configuration is a radial distribution system that is supplied from one side of the supply end (secondary substation bus).

2. Distribution system modelling

2.1. Primery distribution system configuration

The configuration of the primary distribution system consists of a radial feeder, parallel feeder and ring feeder or loop feeder [2] [3]. A feeder radial is the simplest and commonly used one. It is used the extensively to supply small and medium residential, commercial, industrial and non-critical loads. It derives its name from the fact that the feeder radiates from the secondary substation and branches into sub-feeders and laterals which extend into all parts of the area served. The feeders and sub-feeders are three-phase three-wire (or four wire) circuits. The customers may be of three-phase or single phase. The distribution transformers are connected to the primary feeders, sub-feeders and customers, usually through fused cutouts. The radial feeder is the most economical but the least reliable of the various types since a fault in the feeder means disruption of supply. Figure 1 shows a radial primary feeder.

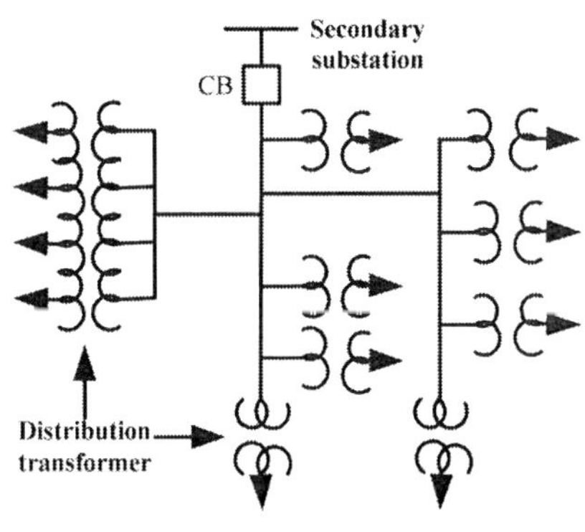

Figure 1. Radial feeder

A system of two or more radial feeders originating from the same or different secondary substations and separately routed through the load areas is known as loop feeder system. If the ends of the two feeders are tied together through normally open switching devices, the resulting arrangement is known as open loop system. If the ends are tied together by means of normally closed switching devices, the result is a ring loop or simply ring feeder. The loop feeder provides a good continuity of service. The feeder and loop component must have sufficient delivery capacity to service loads that might be transferred under emergency condition. This system is the most practical for providing reasonable reliability with nearly full utilization of facilities. Figure 2 shows a loop feeder.

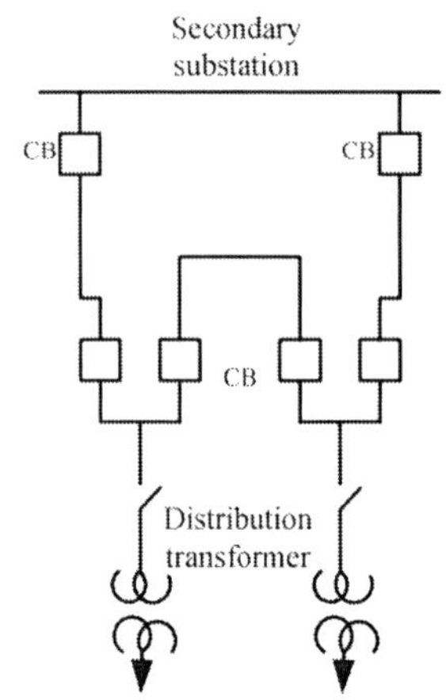

Figure 2. Loop feeder

2.2. Reliability of electric power system

Reliability is the probability of e device performing its function adequately for the intended period of time under specified operating condition [2]. The reliability of an electric power system is the ability of an electric power system to carry out its functions continuously without experiencing interference continuously without experiencing interference.

The reliability of the electric power system is the continuity of supplying electrical energy to customers, for customers who have large capacities such as industries need a better level of service and continuity. If the distribution of electrical energy is interrupted, the production process from the industry will be disrupted.

2.3. Reliability index

To find out the reliability of an electric power system, it is based on a reliability index that is a quantity that describes the performance of an electric power system, the indexes are SAIFI (System Average Interruption Frequency Index), SAIDI (System Average Interruption Duration Index). and CAIDI (Customer Average Interruption Duration Index) [6] [7].

2.3.1. SAIFI (System Average Interruption Frequency Index).

SAIFI describes the average frequency of blackouts for each customer within a year in an area that is evaluated.

$$SAIFI = \frac{Total\ number\ of\ customer\ interruptions}{Total\ number\ of\ costomer\ served}$$

$$SAIFI = \frac{\sum \lambda i\ Ni}{\sum Ni} \quad \text{f./cust.yr} \quad (1)$$

λi :failure rate

Ni: number of customers at the point of load i

IOP Conf. Series: Materials Science and Engineering **854** (2020) 012007 doi:10.1088/1757-899X/854/1/012007

2.3.2. SAIDI (System Average Interruption Duration Index)

SAIDI describes the average duration of blackout for each consumer within a year in an area that is evaluated,

$$SAIDI = \frac{Sum\ of\ costomer\ interruption\ duration}{Total\ number\ of\ customer\ served}$$

$$SAIDI = \frac{\sum Ui\ Ni}{\sum Ni} \quad \text{(hr./cust. yr)} \tag{2}$$

2.3.3. CAIDI (Customer Average Interruption Duration Index).

CAIDI describes the average amount of time (duration) of each outage

$$CAIDI = \frac{Sum\ of\ customer\ interruption\ durations}{Total\ number\ of\ customer\ interruption}$$

$$CAIDI = \frac{\sum Ui\ Ni}{\sum Ni\ \lambda i} \quad \text{(hr./cust. intr)} \tag{3}$$

2.4. Research step

Data processing method is to compute with ETAP 12.0 software can be described as follows:

1) Describe the network configuration to be studied
2) Enter the network parameter values (R and X values) .
3) Entering the voltage value on each bus.
4) Entering power values (P and Q) in the generator
5) Enter the load values (P and Q).
6) Perform computational processes with ETAP software by running load flow analysis to calculate demand for system load requirements for normal operating conditions.
7) Perform reliability assessments to determine system reliability
8) Determine the reliability index.

3. Test performed on system

The system test is carried out on the IEEE-9 bus system research object [8], the research data can be shown in table 1 and table 2:

IOP Conf. Series: Materials Science and Engineering **854** (2020) 012007 doi:10.1088/1757-899X/854/1/012007

Table 1. Load data IEEE-9 bus system

V (KV)	P (KW)	Q (KVAR)	P_G (KW)	Q_G (KVAR)
20	1840	460	0	0
20	980	340	0	0
20	1790	446	0	0
20	1598	1840	0	0
20	1610	600	0	0
20	780	110	0	0
20	1150	60	0	0
20	980	130	0	0
20	1640	20	0	0

Table 2. Line/cable data IEEE-9 bus system

From bus	to bus	R (Ω/KM)	X (Ω/KM)
0	1	0.1233	0.4127
1	2	0.0140	0.6050
2	3	0.7463	1.2050
3	4	0.6984	0.6084
4	5	1.9831	1.7276
5	6	0.9053	0.7886
6	7	2.0552	1.1640
7	8	4.7953	2.7160
8	9	5.3430	3.0264

The configuration of the IEEE-9 bus system distribution network can be seen in Figure 3.

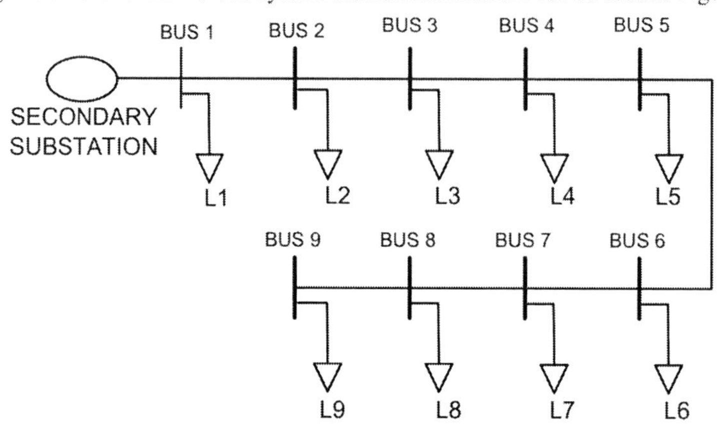

Figure 3. Network configuration of the IEEE 9 bus system

IOP Conf. Series: Materials Science and Engineering **854** (2020) 012007 doi:10.1088/1757-899X/854/1/012007

4. Results and dicussion

System testing conducted in this paper is on the IEEE 9 bus system with a base value of 30 MVA with a voltage of 20 KV and 50 Hz. Network configuration is in the form of a radial distribution system that is supplied from the sending side (secondary substation). The simulation process is carried out using ETAP software, for normal operating conditions the 20 KV radial distribution system is simulated with load flow analysis, which shows the condition of the system at peak load. The system will be analyzed on each bus, with bus 1 (secondary substation bus) considered a slack bus or reference bus. This load flow calculation is done to find out the amount of power supply to the system, the voltage drop and losses in the network. The value of the power demand is 12.955 MW and 5.234 MVAR supplied from the substation, with the largest voltage drop occurring on cable 5 which is 3.37%. The amount of total losses in the IEEE 9 bus system is 1045.1 KW and 1326.5 KVAR.

Then a simulation is performed to calculate the reliability of the system by running a reliability assessment of the ETAP software. The simulation results show the reliability value can be shown in the table 3.

Table. 3 Reliability for the IEEE-9 bus system

Bus number	Average interruption rate (f/yr)	Average outage duration (hour)	Annual outage duration hr/yr
BUS 1	1.033	14.82	15.306
BUS 2	1.053	15.01	15.806
BUS 3	1.073	15.2	16.306
BUS 4	1.093	15.38	16.806
BUS 5	1.113	15.55	17.306
BUS 6	1.133	15.72	17.806
BUS 7	1.153	15.88	18.306
BUS 8	1.173	16.03	18.806
BUS 9	1.193	16.18	19.306

Table 3 shows the reliability of the system which consists of the average number of interruptions and duration of outages of each bus. The average value of the duration of the interruption has increased for the bus farthest away from the power supply side that is bus 9, while the bus closest to the power supply side, the average duration of interruption is very small, namely on bus 1. while buses are on the power supply side, the average duration of disturbance is very small, namely on bus 1.

The duration of outages of each bus has increased for the most distant buses from power supply, the most experienced blackouts are bus 8 and bus 9. The reliability values from table 3 can be made in graphical form, which shows the average amount of interruption and duration of outage as shown in figures 4 and 5.

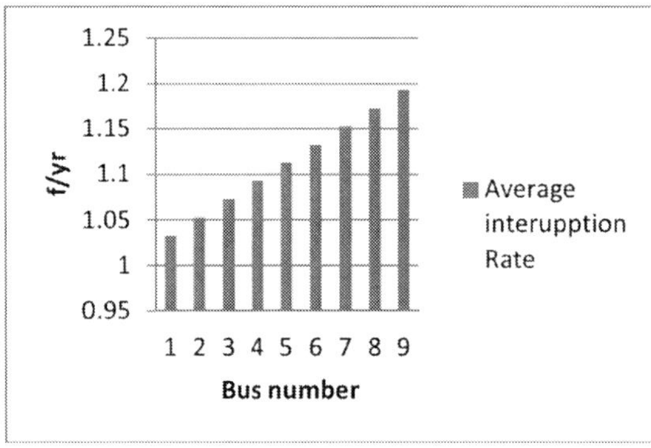

Figure 4. Characterisric of average interruption rate for the IEEE 9 bus system

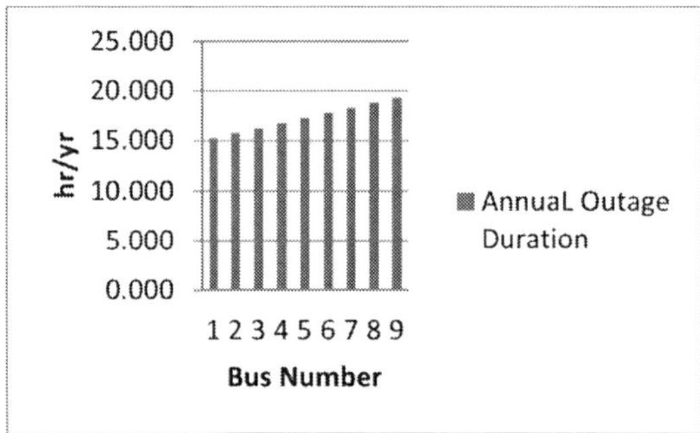

Figure 5. Characteristic of outage duration for the IEEE 9 bus system

The results of the SAIFI, SAIDI and CAIDI reliability index calculations for the IEEE 9 bus system distribution network with ETAP software can be shown in table 4.

Table. 4 Reliability index value for the IEEE 9 bus system.

Type	Index value
SAIFI	1.130 f/cust
SAIDI	17.306 hr/cust.yr
CAIDI	15.549 hr/cust.intr

The results of table 4 show that the SAIFI index value of 1,130 f / cust is still below the IEEE standard std1366-2000 which is 1.45 f / cust [9]. The SAIDI and CAIDI index values from table 4 are 17,306 hr / cust.yr and 15,549 hr / cust.intr respectively, these values exceed the IEEE standard std1366-2000, respectively 2.3 hr / cust.yr and 1.58 hr / cust.intr.

5. Conclusion

From the results of the reliability analysis for the IEEE 9 bus system, the following conclusions can be drawn:

1) Total power demand supplied from substations to the 20 KV distribution network is 12.955 MW and 5.234 MVAR, with losses of 8.06%.

2) The average duration of disturbance has increased for the bus farthest away from the power supply side (bus 9) that is 1.193 times / year.
3) The duration of outages that often appear is on bus 8 and bus 9 respectively 16.03 hours and 16.18 hours.
4) Reliability index value: SAIFI of 1.113 frequency/ customer.yr, SAIDI value is 17.306 hours / customer.yr, and CAIDI value is 15.549 hours / customer interruption

References

[1] Marsudi, D. 1990. "Operasi Sistem Tenaga Listrik". Balai Penerbit & Humas ISTN, Jakarta
[2] Gupta, B.R., 1999, "Power System Analysis and Design", Wheler Publishing, New Delhi.
[3] Gonen, T., 1987, "Electric Power Distribution System Engineering", McGraw-Hill Book Co, Singapore.
[4] Lestari., M.C.W. and Mulyadi, Y., 2017, "Analysis of 20 KV Power Distribution System Reliability Using The Section Technique", ISMEE, IOP Publishing
[5] Ali, K., Wiyagi, R.O. and Syahputra, R., 2017, "Reliability Analysis of Power Distribution System" JET-UMY, Vol. 1. No. 2, pp.67-74., Yogyakarta
[6] T.,Prakash. and Thippeswamy, K., 2017, "Reliability Analysis of Power Distribution System: A Case Study", IJERT, Vol. 6, Issue. 07, pp. 557-561., India
[7] Eminoglui., U. and Uyan., R., 2016, "Reliability Analysis of Electrical Distribution System: A Case Study", IRJES, Vol. 2, Issue 12, pp. 94-102, Turkey.
[8] Babu., G.K. and Samala. R.K., 2015, "Load Flow Analysis Of 9 Bus Radial System Using BFSLF Algorithm", IASET, Vol. 4 Issue 6, pp. 17-22, India.
[9] Sucita., T., Mulyadi, Y and Timotius, C., 2017, "Reliability Evaluation of Power Distribution System with Reliability Index Assessment", ISMEE, IOP Publishing

IOP Conf. Series: Materials Science and Engineering **854** (2020) 012008 doi:10.1088/1757-899X/854/1/012008

Synthesis and Characterization of Pd-Cu-Zr/Alumina Composite Membrane Using Surfactant-Induced Electroless Plating Technique

Ratna Sari[1], Ratni Dewi[1], and Lukman Hakim[2]

[1]Lhokseumawe State Polytechnic, 24301 Lhokseumawe, North Aceh, Indonesia
[2]Malikussaleh University, 141 Reuleut, North Aceh, Indonesia

E-mail: ratnasari@pnl.ac.id

Abstract. The Pd-based membranes consisting of Pd, PdCu and PdCuZr on alumina were prepared by surfactant-induced electroless plating technique. Characterization of these membranes were carried out using X-ray Diffraction (XRD, Shimadzu XRD-7000) for microstructure, Differential Scanning Calorimetry (DSC) and Thermal Gravimetric Analysis (TGA) (Shidmazu DSC/TGA-60) for thermal behavior. Considering TGA result, the PdCuZr/alumina membrane has weight loss from 10 to 9.93 mg (99.3 % residue) at 50-600°C. DSC graph of the PdCuZr/alumina membrane showed that some changes in the number exothemic/endothermic peaks. It endothermic peak was at 399 °C.

1. Introduction

One the most important technologies for high purity hydrogen is membrane separation [1]. Research into membrane properties and the fundamental aspects of displacement events in various membrane operations is important for future membrane technology. The selection of membranes that are suitable for specific purposes is taken into consideration. Much progress has been made in the field of membrane development although some problems still must be solved [1,2].

Currently Pd-based membranes are extensively studied because of their excellent permeability of hydrogen as a result of the solution-diffusion mechanism. Despite its unique ability, the use of pure Pd membrane is hindered. Due to the phase transition, hydrogenation of Pd results in 10% increase in the lattice volume. Consequently, it creates high level of stress leading to the structural distortion. Poisoning or fouling because of sulfur or carbon in operating stream also limits its use [2,3]

Among Pd-alloy membranes, the face-centred cubic (fcc) structure of Pd-Cu membrane have the lower cost and some tolerance to surface poisoning. Nevertheless, the body-centred cubic (bcc) phase of Pd-Cu membrane is highly sensitive to gas impurities. In order to improve the hydrogen permeability and thermal stability, ternary Pd-Cu-M alloys have been studied [4,5].

In term of simplicity and ease of operation, electroless plating (EP) is the most versatile technique for metal ceramic plating. However, this technique suffers with disadvantage of gas bubbles from the oxidation-reduction reaction between Pd-complex and reducing agent. The gas bubbles cause hinder uniform Pd-film deposition due to its adhering to the substrate surface and in the pores.

Surfactant is known to have the ability to reduce interfacial tension between the substrate surface and solution, thereby promoting minimal metal nucleation in the solution and pitting [3]. Till date, only few researchers have been studied effect of adding surfactant to the EP bath. In this study, we try to established relationship between the effect of surfactant and structure/thermal properties of the Pd, Pd-Cu and Pd-Cu-Zr on alumina membrane support.

2. Methodology/ Experimental

2.1. Materials
The material used for membranes support are α-alumina with an outer diameter of 1 cm, 10 cm long, 0.2 cm thick and 100 nm pore size. The chemicals for membranes washing used were sodium hydroxide (Sigma-Aldrich) and isopropanol (Sigma-Aldrich). Stanum chloride (Sigma-Aldrich) and hydrochloric acid (Merck 37%) are chemicals used for the manufacture of sensitizing solutions. The activation solution used consisted of palladium chloride (Aldrich 99.9%) and hydrochloric acid (Merck 37%). The solution used for electroless plating process contained palladium chloride (Aldrich 99.9%), cuprum nitrate (Sigma-Aldrich), zirconium (IV) oxide chloride octahydrate (Merck), cetyltrimetil amonium bromide, CTAB (Sigma-Aldrich), formaldehyde (Sigma-Aldrich), anhydrous sodium hypopospit (Sigma-Aldrich), tetraasetic ethylene diamine acid (Sigma) and ammonium hydroxide (Sigma-Aldrich 28%).

2.2. Preparation of Pd Coated Alumina Ceramic Composite Membrane
All of Pd-based membranes were prepared via a plating method [2] through surfactant-induced electroless plating technique [3]. Synthesis of Pd, Pd-Cu and Pd-Cu-Zr/alumina composite membranes were carried out through several stages, namely cleaning membranes, sol-gel process, sensitizing and activating membranes and coating by surfactant-induced electroless plating method. Membranes cleaning were carried out using sodium hydroxide solution, deionized water and isopropanol in an ultrasonic bath. Then membranes are dried in an oven to remove all cleaning agents in membranes pore. For sensitizing and activating processes, membranes are dipped in a sensitization solution consisting of $SnCl_2$ and HCl, then washed with deionized water. The membranes were then dipped into the activation solution of $PdCl_2$ and HCl. The membranes are washed again with deionized water. These processes are repeated to produce sufficient Pd nucleus on the surface of the membranes support before the electroless plating process is carried out. This process is carried out also for activation of $CuNO_3$ and $ZrOCl_2.8H_2O$. Electroless plating solution used was obtained from mixing $PdCl_2$ with HCl, then dissolved in deionized water. In a separate container, NH_4OH is mixed with EDTA and added with a mixture of $NaPO_2H$ with deionized water, then the mixture is stabilized for several hours. CTAB solution is added into EP solution. Membranes are dipped into the mixture without heating. The surfactant-induced electroless plating process is carried out at room temperature, alkaline pH with constant rotation speed. Membranes are washed with hot deionized water and heated in an oven. The electroless plating process was repeated for cuprum and zirconium plating.

2.3. Characterization
The Pd, Pd-Cu and Pd-Cu-Zr/alumina composite membranes were analysed using X-ray Diffraction (XRD, Shimadzu XRD-7000), Differential Scanning Calorimetry (DSC) and Thermal Gravimetric Analysis (TGA) (Shidmazu DSC/TGA-60).

3. Results and Discussion
Figure 1 presents the XRD patterns of Pd/alumina, PdCu/alumina and PdCuZr/alumina composite membranes fabricated by surfactant-induced electroless plating. The XRD patterns were recorded with a Cu K_α radiation at 40 kV and 30 mA. The X-ray diffractrograms were collected in the range 10-85°. The structural similarity in all composite membranes show that the presence of a small amount of additional elements do not promote the formation of new phase in PdCu/alumina and PdCuZr/alumina composite membrane. The 2θ and d-spacing values corresponding to the five major reflection peaks for Pd/alumina, PdCu/alumina and PdCuZr/alumina for comparison can be seen in Table 1.

IOP Conf. Series: Materials Science and Engineering **854** (2020) 012008 doi:10.1088/1757-899X/854/1/012008

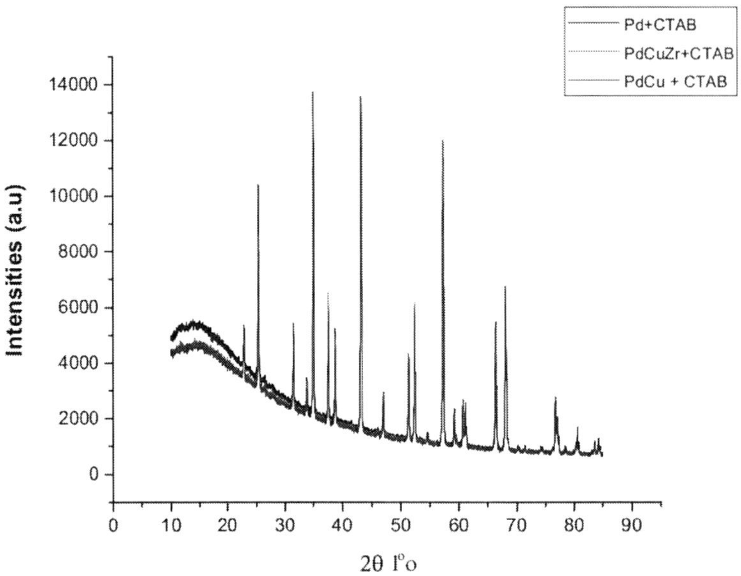

Figure 1. XRD patterns of Pd, PdCu and PdCuZr/alumina composite membrane

Table 1. The 2θ and d-spacings of Pd/alumina, PdCu/alumina and PdCuZr/alumina

Membrane	2θ (°)					d-spacing (A)				
Pd/alumina	25.435	35.009	43.203	57.350	68.060	3.499	2.091	2.0923	1.605	1.376
PdCu/alumina	35.004	43.204	52.392	57.342	68.061	2.561	2.092	1.745	1.605	1.376
PdCuZr/alumina	35.018	43.219	52.405	57.351	68.067	2.560	2.561	1.744	1.605	1.376

The 2θ values indicate that from 57.35° angle XRD reflection peaks, for Pd/alumina. PdCu/alumina and PdCuZr/alumina are in the same position. The peaks of PdCuZr/alumina composite membrane sharper than Pd/alumina and PdCu/alumina. Its indicate the formation of PdCuZr/alumina and PdCu/alumina interstitial alloy. It occur when the atoms of one component are smaller than the other, and the smaller atoms fit into the spaces between the larger atoms [7]. From the diffraction intensity of these membranes, indicating en expansion of the crystall structure of the PdCuZr/alumina membrane.

In the TGA, the samples of Pd/alumina, PdCu/alumina and PdCuZr/alumina composite membranes are heated from 50 °C to 600 °C with 40 °C/min under nitrogen atmosphere. The results of TGA analysis of these membranes are shown in Fig. 2. The thermal curves were a descending curves, which indicate that weight loss of these membrane with increasing temperature have occurred. The sample weight had decreased from 10 to 9.88 mg (98.8% residue), 10 to 9.84 mg (98.4 % residue) and 10 to 9.93 mg (99.3 % residue) respectively for Pd/alumina membrane, PdCu/alumina membrane and PdCuZr/alumina membrane. By comparing the TGA data between these membranes, we have notice that the PdCuZr/alumina membrane can withstand the higher temperature.

IOP Conf. Series: Materials Science and Engineering **854** (2020) 012008 doi:10.1088/1757-899X/854/1/012008

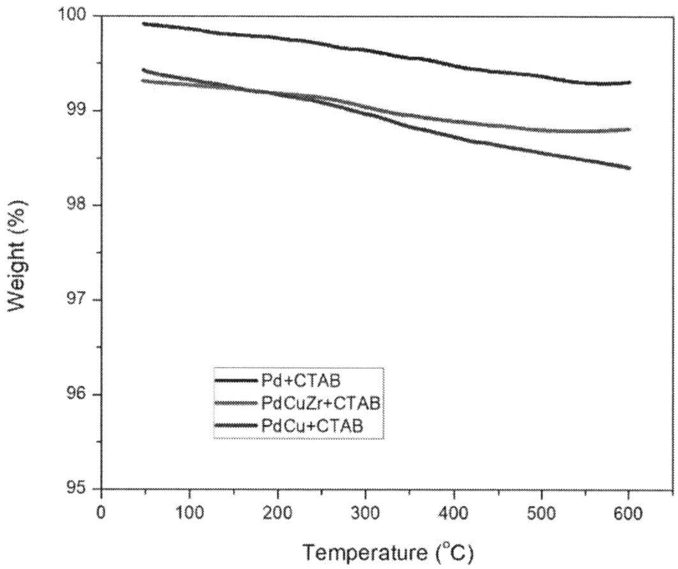

Figure 2. Thermograms of Pd/alumina, PdCu/alumina and PdCuZr/alumina composite membranes

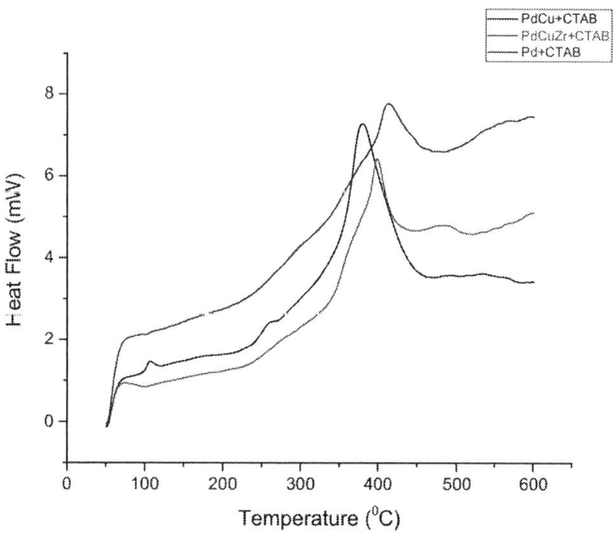

Figure 3. DSC graphs of Pd/alumina, PdCu/alumina and PdCuZr/alumina composite membranes

Figure 3 shows the differential scanning calorimetry (DSC) graphs of of Pd/alumina, PdCu/alumina and PdCuZr/alumina composite membranes. A similar thermal behaviors are observed in the range temperature 50-600 °C. There can be seen some changes in the number exothemic/endothermic peaks and their position. The DSC exothermic decomposition of these membranes in the temperature range from 51 to 53.27 °C, 50 to 52.66 °C, and 50 to 54.21 °C respectively for Pd/alumina membrane, PdCu/alumina membrane and PdCuZr/alumina membrane. The thermograms endothermic of these membrane in the temperature range from 54.74-600 °C (peak at 413 °C), 53.09-600 °C (peak at 380 °C), and 54.2-600 °C (peak at 399 °C) respectively for Pd/alumina membrane, PdCu/alumina membrane and PdCuZr/alumina membrane.

4. Conclusion

This study has explored microstructure and thermal behavior of Pd, PdCu and PdCuZr membranes on alumina fabricated by surfactant-induced electroless plating technique. The PdCuZr/alumina membrane has crystall structure with a maximum heating temperature only decrease 7 % of weight and from DSC thermogram it endothermic peak was at 399 °C.

Acknowledgement

The authors gratefully acknowledge the support for this work by Ministry of Research Technology and Higher Education, Indonesia in Hibah Penelitian Dasar Unggulan Perguruan Tinggi (PDUPT).

References

[1] A. Agarwal, M. Pujari, R. Uppaluri and A. Verma 2015, Surf. Eng. **31**: 221
[2] R. Sari, S. Mulyati, R. Dewi and L. Hakim 2018, IOP Conf. Ser.: Mater. Sci. Eng. **536**: 012046
[3] M.S. Islam, M.M Rahman, and S. Ilias 2012 Int. J. Hydrogen Energy **37**: 3477-3490
[4] Y. Gu, K. Kusakabe, and S. Morooka 2001 Sep Purif Technol **24**: 489-495
[5] S. Nayebossadri, J. Speight and D. Book 2013 J. Name **3**:1-3
[6] S. Nayebossadri, J. Speight and D. Book 2016 ACS Appl Mater Inter 1-38
[7] T. Lai, H. Yin and M. Lind 2015 J. Membrane Sci. **489**: 264-269

Development of a High Performance Remote Terminal Unit (RTU) of Wireless SCADA System for Monitoring Performance of Micro Hydro Power Plant

Salahuddin, Bakhtiar, Yusman, Fadhli

Teknik Elektro Politeknik Negeri Lhokseumawe

Email: salahuddin.mt@pnl.ac.id

Abstract. Remote Terminal Unit (RTU) is one part of SCADA system that functions to collect data from the plant, RTU itself is a unit of minicomputer that is equipped with a standalone system with a smaller physical size. The main objective of this research is to develop a high performance RTU for the wireless SCADA system that is applied to monitoring the performance of microhydro power plants (MHPP). Remote terminal units are built using Arduino Uno as a control center that will regulate analog data traffic from sensors connected to the system. The Remote Terminal Unit processes data from sensors that are converted into digital data and then sent via communication devices to the Main Terminal Unit (MTU) section that is located in the remote control center. The data received from sensor consist of voltage, current, frequency and turbine rotation of a MHPP. The Remote Terminal Unit (RTU) has been able to properly acquire data received by the sensor and then send it to the Main Terminal Unit via an AX.25 protocol based communication device.

Keywords: Remote Terminal Unit, wireless SCADA system, MHPP Parameters, Main Terminal Unit, Protocol AX.25.

1. Introduction

Remote terminal unit (RTU) is a component of SCADA equipment that is designed to monitor substation activity in a microhydro power plant system. Basic information about the generating system is obtained from monitoring the status of the equipment and measuring several important parameters such as voltage, current, frequency and rotation of the turbine in the microhydro power plant. These parameters are very important to observe because they determine the performance of a power plant system[1][2][3]. The information is then processed by RTU and then sent to the Control Center. Conversely, the Control Center can send commands to the RTU. RTU can be connected with one or two Master Stations. In addition to the Master Station, the RTU can also be connected to other RTUs (remote RTUs) via communication lines.

The RTU functions to communicate with the Master Station using a partitioned database and different communication protocols, processing the inputs and outputs that are connected to the RTU through digital and analog input / output modules. RTU also functions to communicate with IEDs (Intelligent Electronic Devices) and can acquire data from IEDs such as smart meters and safety relays. RTU can also be used as an event logger by connecting one or two printers and alphanumeric terminals, so if there is a change in status it can be printed locally. RTU users can configure as needed. This configuration is done using a database system. The database is then loaded into RTU and stored in RAM.[4][6]

In general, RTUs currently use SCADA-based RTUs such as Siemens, ABB, or using PLCs, but are constrained with a high enough price if used to control medium-scale plants. As for the minimum functional requirements that must be fulfilled RTU quite a lot, including being able to receive analog data with a current range of 4-20 mA and voltage of 0-10 Volts, able to receive and process digital data from MTU, able to communicate with MTU, have the ability to start- automatic setup and restoration initialization in case of a power outage, have the ability to download a database from MTU, have a backup battery backup, and others. In addition, it must also have supporting features in providing maximum results on RTU performance, such as supporting the use of analog outputs for standard 4-20mA values, PLC support, etc.[5]

This research will develop a remote terminal unit that has high performance and can acquire data from current, voltage and frequency sensors and turbine rotation sensor on a microhydro power plant. Then the data is sent to the master terminal unit via the ax.25 protocol-based radio communication network and displayed on the master unit using the labview base. This research focuses on the use of hardware that can improve the performance of an RTU, so the data received by the master station is valid data which is sampled directly from the microhydro power plant.[7][8]

2. Literature Review

2.1. Remote Terminal Unit (RTU)

Remote Terminal Unit (RTU) is a unit of microprocessor or microcontroller-based SCADA architecture and is tasked with retrieving data from sensor equipment installed in a plant. The data is then sent to a remote MTU through the communication used. Sending data can be done through a wired communication network or wireless communication network. RTUs are also possible to communicate on a peer-to-peer basis with other RTUs, and can also act as a relay station to other RTUs that cannot be accessed from the MTU. The smallest RTU capacity has less than 10 analog and digital signals to those that have a large capacity above 100 digital signals.[3][6][7].

RTU hardware consist of CPU and memory, analog/digital Input, analog/digital output, counter, digital output interface, power supply and other tools needed to support the work of the RTU. In general the configuration of an RTU is shown in Figure 1.

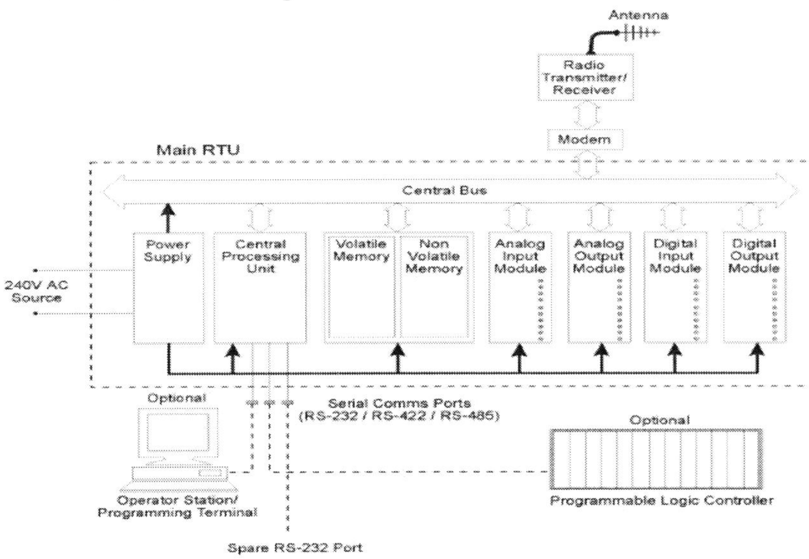

Figure 1. RTU Architecture

IOP Conf. Series: Materials Science and Engineering **854** (2020) 012009 doi:10.1088/1757-899X/854/1/012009

2.2. Microhydro Power Plant (MHPP)

MHPP in principle utilizes the height difference and the amount of water discharge per second that is in the flow of irrigation canals, rivers or waterfalls. This water flow will rotate the turbine shaft to produce mechanical energy[4]. This energy then moves the generator and generates electricity. The MHP illustration can be seen in Figure 2.

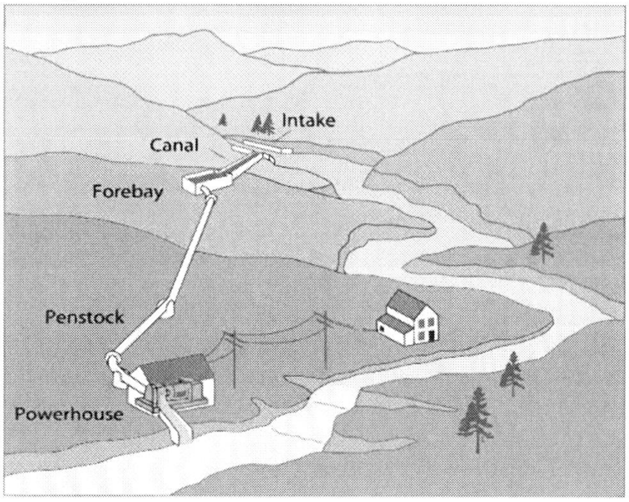

Figure 2. Illustration of Microhydro Power Plant

The construction of the MHPP needs to begin with the construction of a dam to regulate the flow of water that will be used as a propulsion of MHPP[7]. Dams need to be equipped with sluice gates and garbage filters to prevent the entry of dirt or silt. Near the dam built intake building. Then proceed with the manufacture of a conduit that serves to drain water from the intake. At the end of the overflow channel a settling pond was built. This pool functions to settle sand and filter dirt so that the water that enters the turbine is relatively clean. This channel is made with concrete construction and is as close as possible to the turbine house to save fast pipes. Rapid pipes function to drain water before entering the turbine. Usually made of steel pipes that are pulverized, then welded. Flanges are used for connection between pipes.

2.3. SCADA

SCADA (Figure 3) is a system that collects data from various sensors at a factory, plant or in another remote location and then sends this data to a central computer which then manages and controls the data. Initially, SCADA was designed to be on a private network that uses communication lines. Because of the greater scope of its use, the use of this communication line has become impractical, because it introduced an integrated wireless communication system for SCADA[2]. In particular, the SCADA system includes several parts including, operating equipment such as pumps, valves, conveyors and branch breakers that can be controlled by actuators or relays. Instruments in the field or in facilities that are sensitive to conditions such as pH, temperature, pressure, power level and flow rate. Then a close communication network between local processors and instruments and operating equipment. This section includes Programmable Logic Controller (PLC), Remote Terminal Unit (RTU), Intelligent Electronic Devices (IED) and Process Automation Controller (PAC)[1][6].

IOP Conf. Series: Materials Science and Engineering **854** (2020) 012009 doi:10.1088/1757-899X/854/1/012009

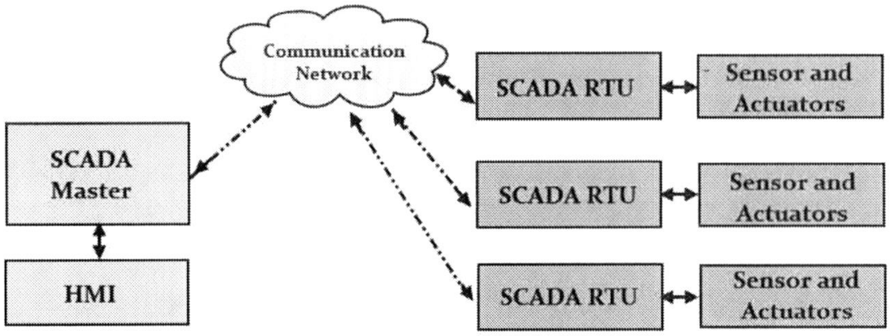

Figure 3. Typical wireless SCADA system

3. Research Methods

Overall the system is designed the same as the general configuration of the SCADA system. Because this RTU was built for the purpose of monitoring the performance of microhydro power plants, the input requirements needed are current sensors, voltage sensors, frequency sensors and turbine speed sensors. The designed RTU is expected to be able to acquire the data with high precision so that the data received by the master station is real data from the microhydro power plant. This RTU was built by using an arduino microcontroller in which there is already an ADC function to convert analog data received from the sensor into digital data.

The system architecture developed consists of two parts. The first part is the RTU circuit consisting of the Arduino Mega 2560 microcontroller, current sensor, voltage and frequency sensor and turbine rotation sensor. The second part is the KYL 1020 radio network communication circuit that functions to send data from the sensor to the MTU section. Changes in parameters that occur in the RTU will be monitored by the Master Terminal Unit (MTU) consisting of personal computers and SCADA Human Machine Interface (HMI) display software.[6]

The design of RTU begins by making a series of simulations using Proteus software, then making software for microcontrollers that will regulate and control the working system as an interface device. The microcontroller then converts the electronic signals received from the sensor device to be processed so that it can be sent using a communication protocol to transmit data via radio communication. Data acquisition devices used are current, voltage and frequency sensors as well as turbine spinning which provide information data from micro hydro power plants that can inform operators about current, voltage and frequency information and turbine rotation. Figure 4 shows the hardware architecture of the proposed systems.[7]

Arduino Mega 2560 Board is an Arduino Board that uses ATmega 2560 Microcontroller IC. This board has relatively many Pin I / O, 54 digital inputs / outputs, 15 of which can be used as PWM output, 16 analog inputs, 4 UART . Arduino Mega 2560 is equipped with 16 Mhz crystals. For relatively simple use, just connect the power from USB to PC / Laptop or via DC Jack using 7-12 V DC adapter.

The data collection section consists of four sensors, namely the current sensor, voltage sensor, frequency sensor and turbine speed sensor. In this study, the ACS712 current sensor was chosen because of its precision and consists of a series of low linear offset Hall effect sensors with copper induction paths located near the mold. When the current enters through the copper conduction path, it will produce a magnetic field which is then felt by the IC Hall effect and converted to proportional voltage. The voltage senor is designed using a parallel resistor that will split the voltage into two parts. Current proportional to the measured voltage must be passed through an external resistor and installed in series with the main transducer circuit. 16x2 LCD is used to display current, voltage, frequency and turbine rotation data and will display all parameters in real time.

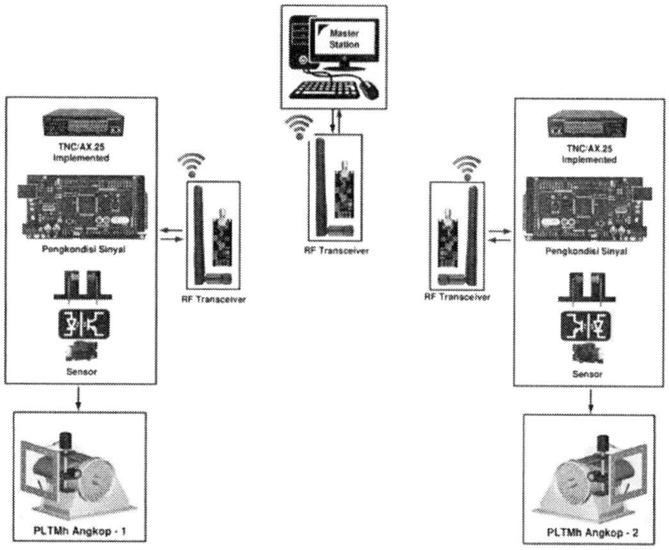

Figure 4. Hardware architecture design

To send data from RTU, the YS-1020UA Wireless Transceiver radio communication module is used. RF data transceiver YS-1020UA is a device that can transmit serial data through aerial media. The device performs the process of laying down digital serial data to a carrier frequency with a higher frequency and then transmitted to the air by the transmitter. At the receiver frequency the carrier containing the data is captured and separated from the data carried. YS-1020UA Wireles Data Transceiver Module can send and receive serial data through air media, with a frequency of 433MHz ISM band and water baud rate of 9600bps. The use of these modules is quite practical because in terms of size they are quite small and directly can be connected to RS232[6]. The module works with a supply of between 3.3 to 5 VDC. In one module can be used as a sender and receiver as well.. Serial data that will be transmitted via RF is fed to the YS-1020 module by a microcontroller in series. Similarly, the data received, will be taken by the microcontroller serially. Figure 5 shows the schematic circuit design using Proteus.

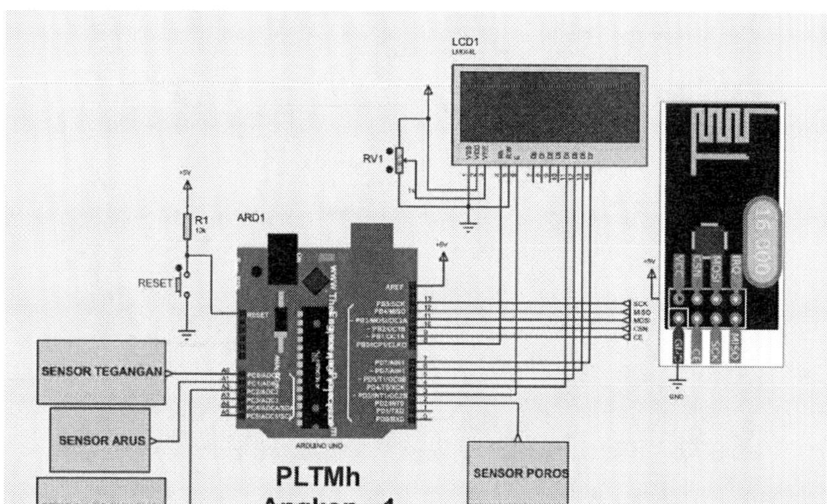

Figure 5. Schematic circuit design using Proteus.

IOP Conf. Series: Materials Science and Engineering **854** (2020) 012009 doi:10.1088/1757-899X/854/1/012009

4. Result and Discusion

Development of RTU for wireless SCADA systems for monitoring the performance of micro hydro power plants via Radio frequency with the ax.25 protocol serves as a monitoring system for several important parameters in micro hydro power plants. This RTU (Figure 6) can work well and can acquire data from sensors and send it to the Master station, so that the control room operator can see and observe the parameters of micro hydro power plants in real time (Figure 7). If an error occurs or interference with one of the parameters, the system will get a warning from the RTU which will then immediately send a signal to the microcontroller. The microcontroller immediately processes the data received and sends a signal to the KYL 1020 radio communication device. So the operator in charge of the control room is alert and immediately calls maintenance or takes other steps to resolve the problem. In general, the overall set of project and testing schemes works well. However, further research will be carried out for continuous improvement.

Figure 6. Remote Terminal Unit (RTU) Complete Hardware System

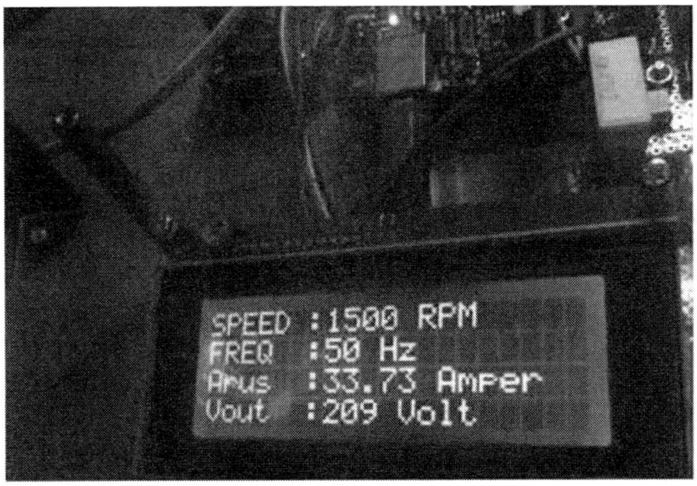

Figure 7. Display of Data from Remote Terminal Unit (RTU)

5. Conclusion

Remote Terminal Unit (RTU) is a small computerized unit device that can be placed at a location far from the monitoring location. RTU functions as a data collection point at the local level and collects parameters from sensors and sends commands for system control. The contribution of this paper is to provide development in the types of RTU and hardware alternatives for the design and development of a series of new models for Remote Terminal Unit (RTU) hardware and its functions. The developed system is designed to monitor the performance of the parameters in the micro hydro power plant and send the data to the monitoring center so that anticipatory steps can be taken if there is a changing pamper and not in accordance with a predetermined arrangement.

References

[1] J. Kudtongngam, P. Tiwatthanont and U. Lewlomphaisarl 2009 *CCAS-SICE* **4** 5653
[2] Salahuddin, Bakhtiar, Yusman and Fadhli 2018 *Proc. of I-COSINE 2018 (Sabang)* Vol. 536, Num. 1 (IOP Conference Series: Materials Science and Engineering) p 012051
[3] Salahuddin S, Bakhtiar B, Yusman Y and Fadhli F 2017 J. Multidisp. Ac. **1** 7
[4] Sobhan N 2016 *Proc. 2nd Int. Conf. on Robotics and AI* (Rawalpindi) Vol 1 p 175
[5] Salahuddin, Bakhtiar, Yusman and Fadhli 2018 *Proc. of MICoMS 2017 (Lhokseumawe)* Vol. 1 (Emerald Publishing Limited) p 525
[6] Badrzadeh B, Bradt M, Castillo N, Janakiraman R, Kennedy R, Klein S, Smith T, and Vargas L 2012 *Proc. IEEE Power and Energy Society General Meeting* (Detroit) p 1
[7] Andreoli AL, Franchin MN and Cagnon JA 2014 *Proc. 11th IEEE/IAS Int. Conf. on Industry Applications* (Juiz de Fora) p 1
[8] Zielinski B 2008 *J. Theor. Appl. Informatics* **20** 199
[9] Tiwari S and Patel RN 2015 Proc. *Conf. on Eng.and Systems (Allahabad)* p 6
[10] Ronan J, Walsh K and Long D 2010 *Proc. 2nd Int. Workshop on Mobile Opportunistic Networking MobiOpp 10 ACM* p 72.
[11] B. Zielinski, (2013), J. *Bull. of the Polish Academy of Sciences Technical Sciences* **61** 639

IOP Conf. Series: Materials Science and Engineering **854** (2020) 012010 doi:10.1088/1757-899X/854/1/012010

A Review on the Air Flow Behaviour in the Intake Pipe

Lim J W*, Narendran N, Chai C E, M I N Ma'arof

Faculty of Engineering and Quantity Surveying (FEQS), INTI International University, Persiaran Perdana BBN, Putra Nilai, 71800 Nilai, Negeri Sembilan Darul Khusus, Malaysia

Email: jinwei.lim@newinti.edu.my

Abstract. The air flow behaviour plays an important role to provide the better ram air effect inside the intake manifold for combustion. The aim of this study is to review various research paper on air flow in the intake manifold. The study is mainly focussed on the geometrical influence in the air flow behaviour in the intake pipe. Most of the studies emphasized on the computational fluid dynamics (CFD) software which is utilized as tool to investigate the air flow behaviour in the pipe. The important variables that taken into consideration are the shape of pipe inlet, intake pipe diameter, and bending angle of the intake pipe. By conducting the review on the papers, the possible alternatives could be obtained to overcome the air restrictions in the intake pipe and increase the performance of the engine.

1. Introduction

Air intake flow is considered as one of the crucial part of study in automotive research and development nowadays. A good designed intake manifold pipe can significantly improve the engine performance. In this paper, the relationship between the intake design and engine performance are discussed in more detailed. According to Y.K Loong et al (2013), the intake manifold design had significantly influences the performance of the engine as the intake of the air could directly increase the combustion. In the engine, the power output and the volumetric efficiency are highly dependable on the amount of air intake into the engine. With high volumetric of air is introduced in the cylinder, more fuel can be added into the combustion chamber, hence, this will result in the increment of peak pressure which will in return will elevate the overall the performance of the engine. Once the CFD simulation is validated against the Experimental Setup, many design parameters were tested using CFD as shown below which include the length and dimension, intake runner opening shape, surface finish and inlet to plenum angle. (Y.K Loong et al. 2013)

2. Design Criteria in Intake Pipe

The geometrical parameters are taken into the considerations in order to improvise the performance of the engine. To achieve the goal of this paper, 3 parameters are focussed which are the diameter, length and angle of bend of the intake pipe. When relating to one of the parameters mentioned which is the diameter of the intake pipe, Mahmoud A. Mashkour developed an analytical solution to estimate the effective diameter for the intake manifold pipe. In the research, the study mentioned the requirement to reduce the knocking effect in the engine. This is due to the increment of the intake pressure increases output pressure in the engine. This might cause the knock tendency to increase. Previously, this problem is encountered when the turbocharger or a supercharger is used to increase the charge of air into the engine. By increasing the pumping work to the engine, it could increase the tendency of the engine to

IOP Conf. Series: Materials Science and Engineering **854** (2020) 012010 doi:10.1088/1757-899X/854/1/012010

knock. Basically, this problem is solved by the efficient air flow velocity in the mean of selection of suitable diameter of the intake pipe (Mahmoud A. Mashkour, 2012). In the other research, B.M Krishna et al.2009 had shown that at lower intake valve opening, profiles are not so smooth and this can be related to the diameter of the intake pipe which takes in the air. In addition, at lower intake valve opening, the air flow velocity decreases due to the air friction caused by fluid and metal wall interaction. Since the inlet valve and ports are the largest flow restrictions in the intake system which results in the large pressure drop across the intake valve. However, these problems are less at higher intake valve opening which results in smooth profiles of velocities. (B.M Krishna et al.2009)

Based on the research studies by Mohd Faisal Hushim et al. (2015) had shown that the air flow behaviour also changes with different intake manifold angles. By the GT Power simulation tool, high angle of the intake manifold will provide better engine performance in term of high brake torque, brake power, brake mean effective pressure and volumetric efficiency. For that, two result characteristics are presented which are the pressure and velocity contour. The simulation is run under steady-state condition on six manifolds which are 30°, 60°, 90°, 120°, 150° and 180°. In the research, the major challenges faced are the air restriction where it tends to reduce the power of the engine. To overcome this issue, the angle of bend acts as the tool to reduce the air friction at the bent part. This proves that the highest angle of bent of the intake manifold pipe able to increase the speed of the air intake thus reducing the possible restrictions. Besides, the studies also prove that the homogeneous charge as one of the key aspects to gain better performance. This suggests that the geometry of the pipe should be designed to improve the air flow and ensure the air-fuel mixed homogeneously so the vaporized fuel can be transported to the combustion chamber to be combusted effectively.

Table 1. Data of the Boundary Conditions

Boundary Conditions	
Parameters	Conditions
Inlet Pressure	101.3 kPa
Outlet Pressure	0 Pa
Intake manifold	No-slip wall and smooth wall roughness
Air Velocity	15 m/s
Air Temperature	300 K

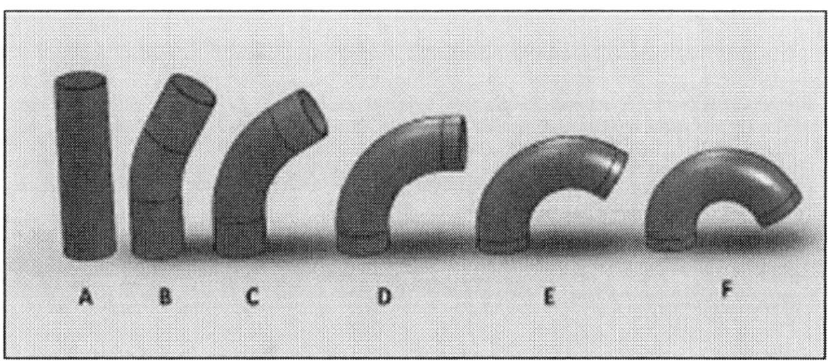

Figure 1. Different design of the intake manifold with varying angle (Source: Mohd Faisal Hushim et al. 2018).

From the figure 4 and 5, it depicts different pressure and velocity contours for each design of intake manifold. For the velocity contour, it shows that the pipe which is a straight pipe flows with the highest homogeneity along the pipe. This homogeneity flow profile was due to no restriction to the air flow from the inlet to the outlet along the pipe. The bending will affect the flow pattern and pressure inside the pipe. From the figure 4 and 5, it was found that air velocity reduces at certain area inside the pipe with the decreases of intake manifold angle, or in other words, high angle of intake manifold will produced higher air velocity.

The velocity contour starts to change when bending was applied to the intake manifold. As can be seen in figure 4 and 5, for the bending intake manifold with its specific angle, high velocity occurred near the inner bending wall while at the outer bending wall low air velocity was observed. This situation happened due to the travel distance of air along the pipe. It can be stated that the bending side work as a shortcut distance for the air to travel inside the pipe. For the inner bending side, the air flow velocity will increase as air flow through this shorter distance. However, due to the longer distance air travelled, it results in lower air velocity. As the bending pipe increases, it reduces the air flow along the pipe, hence flow characteristics will change. Air restriction can reduce power from the engine. Air resistance must be avoided as best as possible. For this case, the bending of the pipe works as a resistance to the air flow. The straight intake manifold is the best manifold that can eliminate air resistance. This proves that the highest angle of intake manifold is the best angle that can produce better performance due to less air flow resistance in it. When gas flows into the intake manifold from the regulator chamber, the vortex is generated in the intersection of the regulator chamber and manifold because a sudden change in flow cross-sectional area. The air distribution of each intake manifold is uneven (Mohd Faisal Hushim et al. 2018). When discussing on the velocity distribution in the intake manifold pipe, S.Aziz et al (2018) also conducted some research on the pipe based on the pipe angle. The studies prove that the velocity of the air flow into the intake manifold is influenced by the bending angle. The bends and junctions will create the turbulent air flow characteristics which will choke the flow in the pipe similar to the venturi effect. (S. Aziz et al. 2018). To prove this, another research done by A.Norizan et al (2017) whereby the venturi tube shape concept on the pipe could improve the air flow. Venturi effect is when the fluid through the shrinking surface area, it will increase the speed of the air and the air pressure is reduced. (A Norizan et al. 2017)

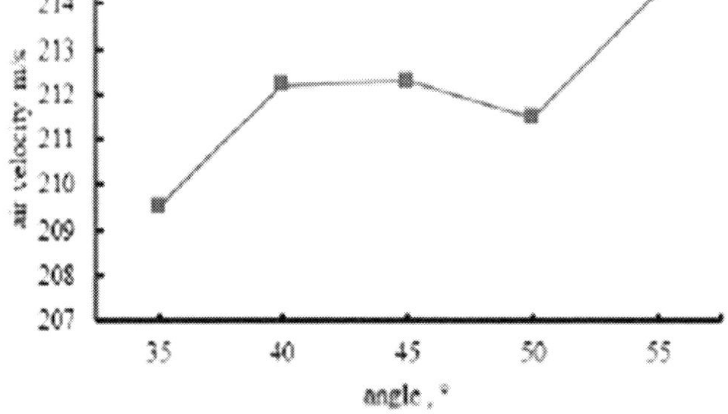

Figure 2. Graphical result for air velocity at different bending angle (Source: S.Aziz et al. 2018)

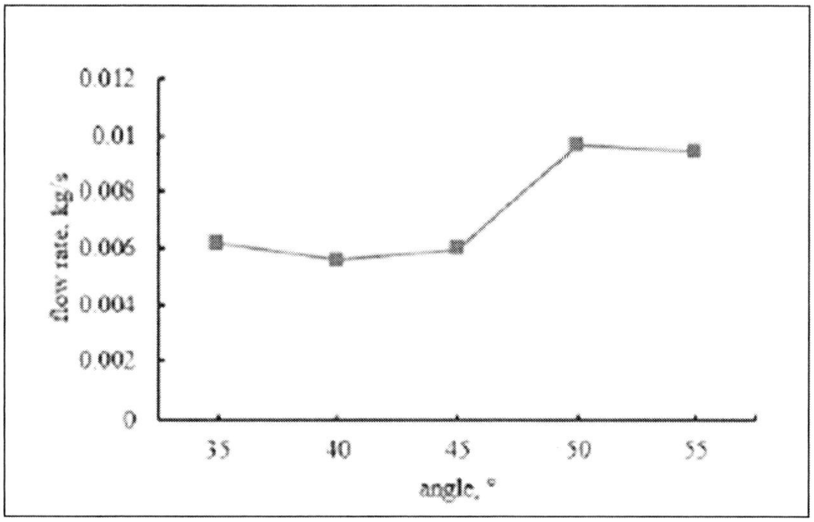

Figure 3. Graphical result for mass flow rate at different bending angle (Source: S.Aziz et al.2018)

For the pressure contour, pressure gradient was observed along pipe for each design. In the research by Mohd Faisal Hushim et al. shows the results where the figure 4 depicts that for manifold with high angle, the pressure contour at the inlet pipe is high before it starts to decline along the pipe. This situation happened due to the velocity of the air at the outlet is higher compared to the inlet. Referring to the pressure contour figures, high pressure region started to develop at the outer bend wall, and low pressure at the inner bend wall. As the angle of manifold decreases, the high pressure region increases and becomes larger. The highest pressure region was observed at the lowest air flow velocity region inside the pipe. It can be seen that lowest pressure region occurred at the bending area due to the highest air flow at this area. (Mohd Faisal Hushim et al. 2018). Referring back to the S.Aziz et al. 2018, the highest pressure loss occurs during the highest bending angle. This is because the air flow collides with the inner wall of the bending angle. This is due to the narrow corner of the pipe that caused the collision to happen. Furthermore, the air travel shorter distance since the length changes with the changes in angle. (S. Aziz et al. 2018)

IOP Conf. Series: Materials Science and Engineering **854** (2020) 012010 doi:10.1088/1757-899X/854/1/012010

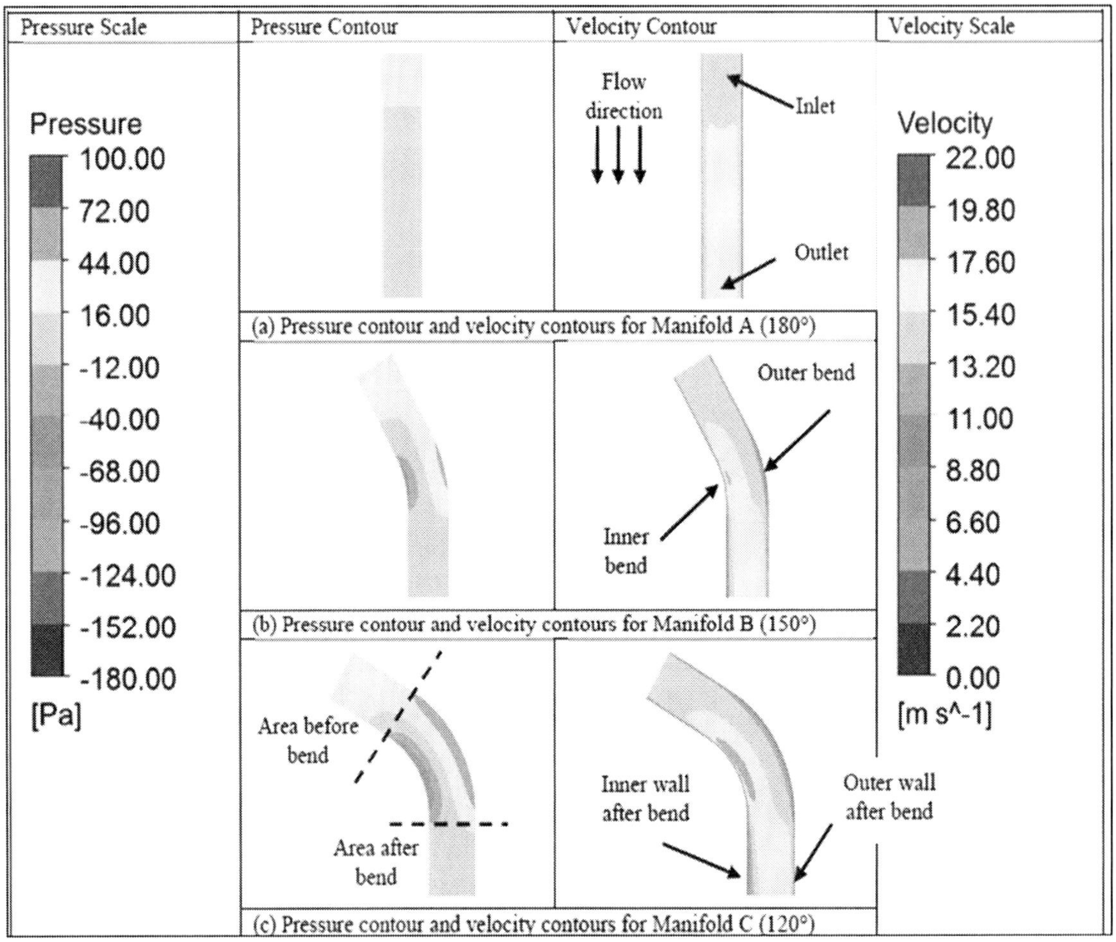

Figure 4. The pressure and velocity contour for 180°, 150° and 120° (Source: Mohd Faisal Hushim et al. 2018).

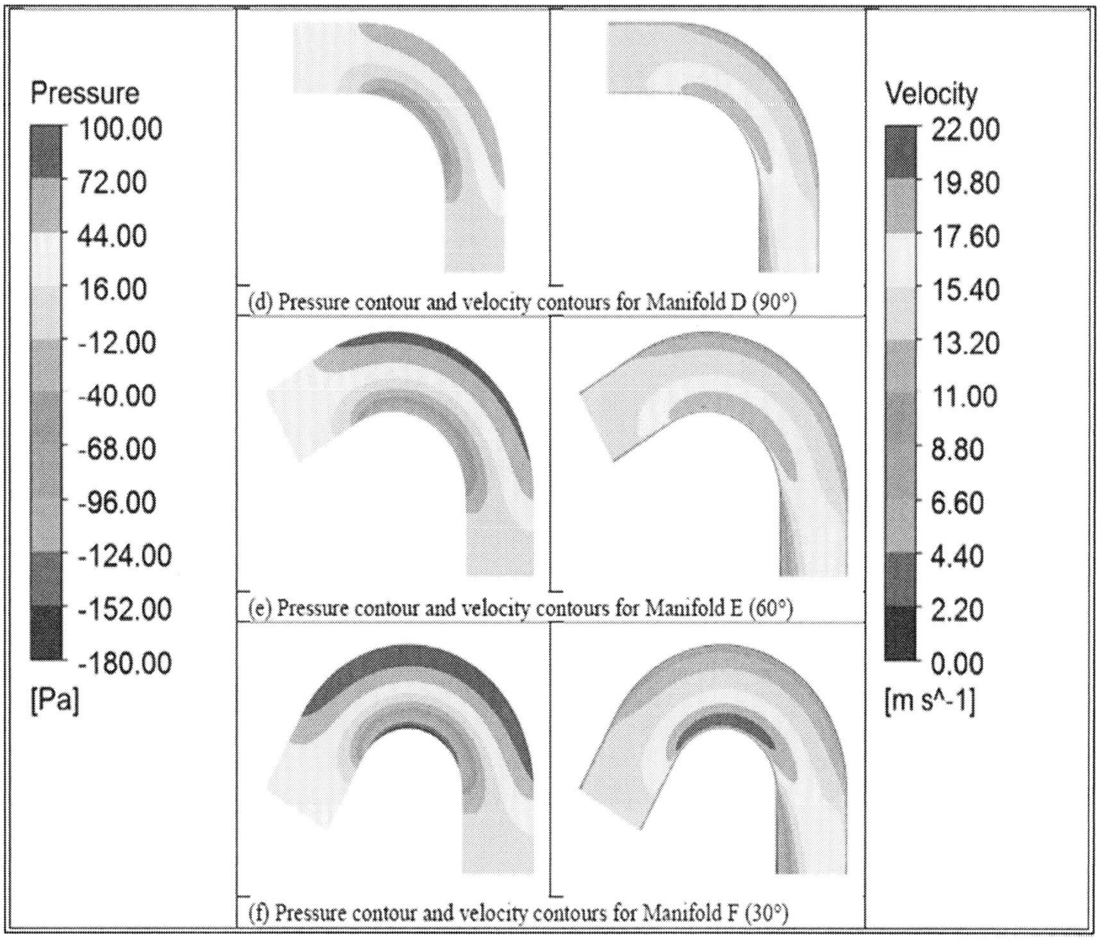

Figure 5. The pressure and velocity contour for 90°, 60° and 30° (Source: Mohd Faisal Hushim et al. 2018).

Figure 6. Intake Pipe re-routed to face front of the motorcycle

In designing the intake manifold, the length of the pipe is focussed by researchers to obtain the possible positive output on the air flow behaviour. An experiment on the experiment and analysis of intake pipe by (Mohd Redzuan bin Hasan, 2013) shows that the length of the pipe has the effect on ensuring the good air flow into the combustion chamber. The analysis shows for the performance and functionality, the engine requires the maximum power at lower range speed. By varying the length of the inlet pipe at constant speed, it results in high indicated mean effective pressure against the pipe length. For the air to flow in the intake pipe, the frequency ratio will be present as well and this is defined as the ratio of the inlet pipe frequency to valve frequency and the frequency ratio is inversely proportional to the intake pipe length and engine speed. In the other perspective, the volumetric efficiency is considered a crucial part in engine combustion and it can be improvised by long pipe. (Mohd Redzuan bin Hasan, 2013). Sachin Singla et al. (2015) had clarified that the improvement of the engine occurs with the manipulation of the intake pipe length. At the low engine speeds, the pipe length needs to be extended whereas at the high engine speeds, the pipe need to be shorten to enable faster propagation of pressure wave and intake of air at high volume flowrate. (Sachin Singla et al. 2015).

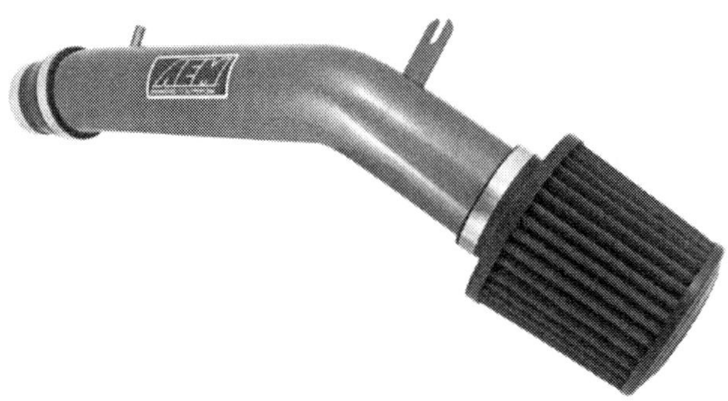

Figure 7. The existing design of the intake pipe

IOP Conf. Series: Materials Science and Engineering **854** (2020) 012010 doi:10.1088/1757-899X/854/1/012010

Figure 8. Modified design applied to current superbike to reduce the tubulence of air in the pipe (Source: Z1 Motorsports M-Spec 350Z Cold Air Intake Kit STREET Version)

2.1. Optimum operating conditions for engine

In considering the design of the pipe in intake manifold, it is crucial to emphasis on the requirement of the engine to operate at maximum performance. The engine depends on the air intake to perform the combustion in the engine. Karlis Barnis conducted a study which shows that if more air enters inside the combustion chamber, the more efficient is the fuel-burning process. This actually increase the engine torque and power. The small changes from in intake pipe design could help the performance by decreasing the flow resistance and modifying the pressure oscillations caused the cyclic intake strokes of the piston. For a suitable operating condition, the goal of these components are aimed to maximize the volumetric efficiency and creates optimum conditions for complete combustion. According to the previous research done in this field by Selamet et al, 2001 they had shown that the propagating pressure waves in fluids will reflect in the cylinder of the piston which focussed on the changes in the cross section of the path. When low pressure is generated near the intake valves by the piston, the intake cycle begins and the fluid in the intake starts to flow into the cylinder creating the momentum and generate a negative pressure wave front. This flowing air-fuel mixture will then transfer the kinetic energy to the pressure generating the positive pressure wave front where it travels back to the intake valve and then to the inlet pipe. This is a type of the oscillations that are considered to be affected by the geometrical changes in the intake manifold pipe and the pipe intake pipe need to be designed in order to ensure the pressure oscillation frequency which aids engine breathing in high engine speed.

By understanding this propagating pressure in the intake manifold pipe, Shuqing Guo et al had conducted an experimental test to study the flow of air in the pipe. By conducting the kinetic experiment, they prove that the flow is smooth as the fluid flow is ordered and does not have influence on each other. Hence, this flow is known as the laminar flow. As the Reynolds number exceed the critical value, the flow become un-ordered state. This flow will automatically affect the variation of velocity changes with time in the pipe under the turbulent flow condition. In their test, it is found that the speed value is not in stable state and the pulsation is high (refer to figure 9). When designing the pipe, the pulsating pressure due to this flow need to be considered as it could increase the load of the air flow energy and the design of the pipe must be able to maintain the flow at optimum point of flow. Basically, the pulsating pressure are due to the combustion that occurs in cylinder which creates the positive and negative wave front in the cylinder. The intake manifold is designed in such a way to control this pressure fluctuations in order to supply enough air flow rate for combustion and prevent other problems caused by this pulsation (Shuqing Guo et al, 2018)

IOP Conf. Series: Materials Science and Engineering **854** (2020) 012010 doi:10.1088/1757-899X/854/1/012010

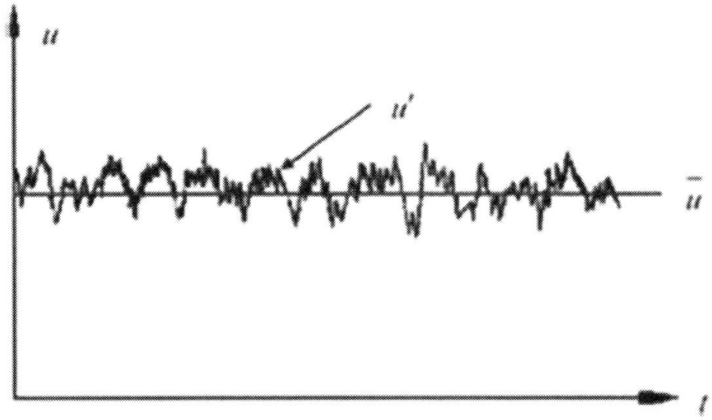

Figure 9. Measured speed of turbulence at a certain point in pipe with change of time (Source: Shuqing Guo et al 2018)

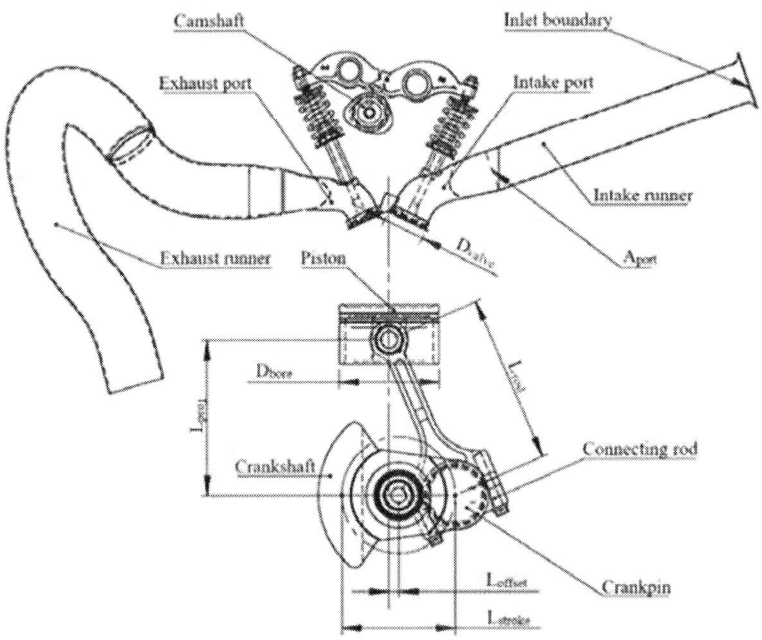

Figure 10. Main Component of Internal Combustion Engine (Source: Latvia University of Life Sciences and Technologies, Latvia)

3. Application of Bernoulli Principles and Reynolds Number

According to the Engineers Edge, the calculation regarding the Bernoulli condition is proved that the considerations need to be taken on the conservation of efficiency of the air flow rate. This condition will also provide the assumptions on the dynamic efficiency of the air flow in the pipe. By modification of the pipe geometry, the air flow behaviour could be manipulated in term of its pressure and velocity according to the pipe design. By testing the design in the simulation, the required amount of the air could be allowed to enter the engine. The air flow behaviour could be calculated using the given equation which applies in the different conditions.

Table 2. The Bernoulli's Equation in Different Conditions

Description	Equation
Bernoulli's Equation (head form) Recommend form for liquids	$\left(\frac{p_1}{\gamma}+\frac{V_1^2}{2g}+z_1\right)=\left(\frac{p_2}{\gamma}+\frac{V_2^2}{2g}+z_2\right)$
Bernoulli's Equation (pressure form) Recommend form for gases	$\left(p_1+\frac{\rho V_1^2}{2}+\rho g z_1\right)=\left(p_2+\frac{\rho V_2^2}{2}+\rho g z_2\right)$

In internal flow in intake pipe, Reynolds numbers have consuming the high percentage to the engine performance. The Reynolds numbers influence the performance because, from Reynolds numbers, the flow profile can be determined. Reynolds number can be effect by the diameter of the pipe and the pipe length. Flow profile is divided by 3 classes:

1. Reynolds numbers $\leq$ 2300 (laminar flow)
2. 2300 $\leq$ Reynolds numbers $\leq$ 4000 (transitional flow)
3. Reynolds numbers $\geq$ 4000 (turbulent flow)

The transition from laminar to turbulent flow is relating with geometry, surface roughness, flow velocity, and type of fluid. The equations related to determine the Reynolds numbers are:

Reynolds numbers=Inertial forces/Viscous forces
Reynolds numbers= $\rho V_{avg} D/\mu$
ρ= density of air (kg/m^3)
V_{avg} = velocity of air in pipe (m/s) D = diameter of pipe (m)
μ = dynamic viscosity (kg/m.s)

The Bernoulli's number is basically used to determine the pressure and velocity distribution along the pipe. Due to the operating condition of the engine, the pressure wave will vary by time to time. This will also influence the intake air velocity as the pulsating pressure will disrupt the air flow. In addition, the Reynold's number will be crucial to investigate the type of the air flow in the pipe. In studies by S.Maharudrayya et al. 2004 shows that the pressure losses in flow distributor plate of the fuel cell depend on the Reynolds number and geometric parameters of the small flow channels or also known as pipe. Based on the experiment conducted, the contour plots of the velocity in the symmetry plane (z = 0) for a sharp 90° bend at different Re. For Re<50, there will be no flow separation in term of the formation of eddies and vortices. Separation at the inner corner appears at Re of 100, while significant recirculation at the outer corner appears at Re of about 200. (S.Maharudrayya et al. 2004).

4. CFD Simulation Results on Intake Manifold Pipe

The calculation results are processed by using FLUENT's own post-processing function to obtain desired results such as pressure, absolute pressure, turbulent kinetic energy, and turbulent energy dissipation rate maps which are used to analyze the air flow. The criterias that are analyzed based on the velocity and pressure of intake manifold to study the air resistance, and also factors that produces air resistance, and the method of reducing the resistance. In addition, the performance of the engine can be improved by conducting the appropriate simulation. Therefore, the required aspects can be analyzed according to the analysis results of the intake manifold. P.Ragupathi et al. (2018) had carried out an experiment of the intake manifold using the CFD tool. In the studies, the CFD able to provide the useful informations and data regarding the flow pattern which has the potential to reduce the development time of the intake system of an engine. The turbulent model is used in the simulation tool to analyze the physical phenomena involved in the kinetic energy. (P. Ragupathi et al. 2018).

In conducting the Ansys simulation, Priyadarsini, I.(2016) had followed each steps need to be done in correct steps to obtain good results. The steps involved are:-

1. Meshing
   - this technique the flow domain is converted or split into various subdomain primitives like hexahedral and tetrahedral.
2. Boundary Conditions
   - It is required to use inflation layer meshing to accurately capture the boundary layer region for any wall-bounded turbulent flows
3. CFD Post Processing
4. -After running the simulation, the expected outcomes are plotted in the form of contours or streamlines (Priyadarsini, I. 2016)

5. Conclusion

The review of all the papers shows that the geometrical parameters had significantly influence the air flow behaviour. The design could be the tool to maximise the optimum intake air into the engine for better air-fuel mixture which will increase the performance of the engine. The good understanding of these parameters and air flow behaviour could produce an efficient intake pipe.

References

[1] Loong, Y.K. and Salim, S.M., 2013. Experimentation and Simulation on the Design of Intake Manifold Port on Engine Performance.

[2] Mashkour, M.A., 2012. Naturally Aspirated Internal Combustion Engine. Journal of Applied Sciences, 12(2), pp.161-167.

[3] Banis, K., 2018. Computational fluid dynamics pressure wave and flow rate analysis of intake runner design in internal combustion engine. RESEARCH FOR RURAL DEVELOPMENT, 1.

[4] Hamilton, L.J., Cowart, J. and Rozich, J., 2009. The Effects of Intake Geometry on SI Engine Performance, 2009. *SAE Technical Paper*.

[5] Selamet, A., Kothamasu, V. and Novak, J.M., 2001. Insertion loss of a Helmholtz resonator in the intake system of internal combustion engines: an experimental and computational investigation. *Applied Acoustics*, *62*(4), pp.381-409.

[6] Maharudrayya, S., Jayanti, S. and Deshpande, A.P., 2004. Pressure losses in laminar flow through serpentine channels in fuel cell stacks. Journal of Power Sources, 138(1-2), pp.1-13.

[7] Hasan, M.R., 2013. Experiment and Analysis of Intake Pipe for Single Cylinder Engine (Doctoral dissertation, UMP).

[8] Hushim, M.F., Alimin, A.J., Razali, M.A., Mohammed, A.N., Sapit, A. and Carvajal, J.C.M.,

2015. Air flow behaviour on different intake manifold angles for small 4-stroke PFI retrofit kit system. ARPN Journal of Engineering and Applied Sciences, pp.1-7.

[9] Guo, S., Huang, S. and Chi, M., 2018. Optimized design of engine intake manifold based on 3D scanner of reverse engineering. EURASIP Journal on Image and Video Processing, 2018(1), p.70.

[10] Norizan, A., Rahman, M.T.A., Amin, N.A.M., Basha, M.H., Ismail, M.H.N. and Hamid, A.F.A., 2017, October. Study of intake manifold for Universiti Malaysia Perlis automotive racing team formula student race car. In Journal of Physics: Conference Series (Vol. 908, No. 1, p. 012069). IOP Publishing.

[11] Aziz, S., Amin, N.A.M., Rahman, M.T.A., Rahman, A., Syayuthi, A.R.A., Majid, M.S.A. and Suhaimi, S., 2018, September. Design and Analysis of an Operative Inlet. In IOP Conference Series: Materials Science and Engineering (Vol. 429, No. 1, p. 012075). IOP Publishing.

[12] Krishna, B.M. and Mallikarjuna, J.M., 2010. Characterization of flow through the intake valve of a single cylinder engine using particle image velocimetry.

[13] Engineer Edge. Pressure Drop Along Pipe Length of Uniform Diameter

[14] Ragupathi, P., Rajkumar, S. and Hussain, R.S., DESIGN AND ANALYSIS OF DIESEL ENGINE INTAKE MANIFOLD USING CFD SIMULATION.

[15] Priyadarsini, I., 2016. Flow analysis of intake manifold using computational fluid dynamics. *International Journal of Engineering and Advanced Research Technology*, *2*(1), pp.1-5.

[16] Singla, S., Sharma, M.S. and Gangacharyulu, D., 2015. Study of design improvement of intake manifold of internal combustion engine. *International Journal of Engineering Technology, Management and Applied Sciences*, *3*, pp.234-242.

[17] Hasan, M.R., 2013. *Experiment and Analysis of Intake Pipe for Single Cylinder Engine* (Doctoral dissertation, UMP).

IOP Conf. Series: Materials Science and Engineering **854** (2020) 012011 doi:10.1088/1757-899X/854/1/012011

Characterization of Fresh Sidoarjo Mud as Material on Geopolymer System

Syarwan[1], Kiki Yolanda Putri[1,2], Ruhana[1], Miswar[1], Amir Fauzi[1,2*]

[1] Civil Engineering Department, Politeknik Negeri Lhokseumawe, Acheh, Indonesia
[2] Geopolymer and Green Technology Research Center, Acheh, Indonesia

Email: syarwan@pnl.ac.id

Abstract. Sidoarjo mud is the alternative cementitious material which derived from a volcano mud eruption in Porong, Sidoarjo, East Java, Indonesia. The calcination of fresh sidoarjo mud in high temperature was known rich in the content of Si and Al that has similar chemical content with the material of OPC. However, the calcination process had caused the consumption of energy, time duration and cost in the production. To solve the problem, this study was required to investigate the fresh sidoarjo mud without the calcination process. However, material of FSM was analyzed to understand its feasibility as a material in geopolymer referring to the material of fly ash. The results showed that the material of FA consisted of quartz (SiO_2), mullite ($3Al_2O_3$ $2SiO_2$ or $2Al_2O_3$ SiO_2) and hematite. While, the material of FSM consisted of quartz (SiO2), mullite ($3Al_2O_3$ $2SiO_2$ or $2Al_2O_3$ SiO_2), magnetite (Fe_3O_4), hematite (Fe_2O_3), and rutile (TiO_2). The particle of FA was amorphous and sphere with the misty surface that caused the absorption of alkaline solution and reduced the workability on fresh geopolymer mortar. While the particle of FSM was the agglomerate irregular shape with the rough surface that also caused the absorption of alkaline solution and reduced the workability on fresh geopolymer mortar. The analysis of FTIR identified that the material FA and FSM had the stretching in the gel of Si-O-Al and Si-O-Si. The compressive strength reached optimum at the replacement of 5% FSM.

1. Introduction

Ordinary Portland Cement (OPC) is a famous material that is often used in the construction as the binder. The production of OPC was known to have increased pollution in the environment. There were more million tons of OPC production in the world that played an important role in the increase of CO_2 emission, exploitation the natural resources and consumed the energy. It was required to solve this problem with the use of alternative material that was from the waste material such as fly ash (FA), silica fume (SF), ground granulated blast furnace slag (GGBFS), metakaolin, rice hush dust.

Geopolymer is a green technology formed by polymerization of silicon, alumina, and oxygen in the triangle structure consisted of Si-O-Al and Si-O-Si which is one of the most stable covalent bonds is the nature. The constituent of geopolymer is the cementitious material from waste material that the rich in the content of Si and Al and alkaline solution derived from soluble alkali metal that is the combination of NaOH and Na_2SiO_3 solution with the certain ratio.

Sidoarjo mud (SM) is a volcano mud eruption caused by drilling gas exploration in Porong, Sidoarjo, East Java, Indonesia. The eruption had released the mud as 18x104 m^3 per day in 2007 and it is considered to continue for the next 25 to 30 years. To date, the volume had impacted the environment and society. The previous research was investigated with the calcination of SM at a high temperature. The results showed that the material of SM was categorized as a cementitious material due to rich in Si and Al content, but it was lower in the content of Ca. The material of SM was an amorphous phase with a crystalline peak of quartz and silicon dioxide (SiO_2). The analysis of FTIR showed that the material of SM formed the gel of Si-O-Si and Si-O-Al on geopolymer binder. Whereas, the particle of SM has an irregular agglomerate shape with the rough surface.

Furthermore, it was mentioned that the finer particle size of SM caused the low flowability that caused the absorption of alkaline solution and reduction of viscosity in geopolymer mixture. The low Ca content in the source material of SM brings to prolong setting time in the geopolymer binder. It was due to the low Ca content generated N-A-S-H gel retarding the reaction of geopolymer which it debilitated the mechanical properties at an early age. Finally, the calcination of SM at the temperature of 800^0C provided the highest compressing strength in geopolymer mortar.

However, the calcination process of sidoarjo mud caused the consumption of energy, time duration and cost in the production. To solve the problem, this study was required to investigate the fresh sidoarjo mud without the calcination process. So that, the characterization of fresh sidoarjo mud as material on geopolymer system is investigated in this study.

2. Material and experimental program

This study used the material of FSM that was from Porong, Sidoarjo, East Java, Indonesia whereas the material of fly ash was from power plantation in Nagan Raya, Aceh, Indonesia. Furthermore, the alkaline solution was generated by a combination of NaOH and Na_2SiO_3 solution with a certain ratio which was obtained from CV. Rudang Jaya company in Medan, Indonesia. The material was evaluated by XRD to analyze the particle structure of the material, FTIR to indicate the geopolymer binder gel and SEM to indicate the shape and surface of the particle. The mix proportion was designed based on the constituent material of geopolymer, whereas the mixing procedure followed the conventional method which referred to the standard of ASTM C305. The fresh and hardened properties of geopolymer mortar were evaluated by workability, setting time and compressive strength.

3. Results and discussion

3.1. Characterization of material

3.1.1. Chemical composition

Chemical composition is one of analysis in the characterization of material linking to the condition of fresh and hardened properties on geopolymer system. Table 1 shows the chemical composition of FA and FSM that is known rich in content of Si and Al. Yet, the contents of Ca in the material of FSM and FA are categorized as low Ca and moderate Ca content. The content of Si and Al formed a gel of Si-O-Si and Si-O-Al that contributed to the mechanical properties of geopolymer mortar. The low Ca content formed a gel of N-A-S-H that caused prolong the setting time in fresh geopolymer mortar whereas the medium Ca content formed a gel of C-A-S-H that bring the short setting time in fresh geopolymer mortar. Furthermore, the contents ratio of Si and Al in the material of FSM and FA is 2.93 and 2.11, respectively. It means that the content ratio of Si and Al of 2,93 caused the worse chemical stability in the air compared to the content ratio of Si and Al of 2.11 that indicated the presence of efflorescence on the surface due to the higher residual free K^+.

Table 1. Chemical composition of fly ash and FSM

IOP Conf. Series: Materials Science and Engineering **854** (2020) 012011 doi:10.1088/1757-899X/854/1/012011

Chemical composition	Analysis, % by weight	
	Fly ash (FA)	FSM
SiO_2	37.16	32.5
Fe_2O_3	18.79	26
Al_2O_3	17.61	11.1
CaO	8.72	6.71
MgO	6.43	5.37
K_2O	0.79	4.25
SO_3	1.96	4.24
LOI	8.02	9.83

3.1.2. Crystalline analysis of material.

The crystalline identification of FA and FSM particle was investigated by X-ray diffraction test (XRD) as seen in Figure 1. The spectrum in the particle of FA and FSM follows the similar peak of dispersion refers to the database of XRD. The trend of the graph indicates the similarity of dispersion peak in the material of FA and FSM. It means that both of material consists of quartz (SiO_2), mullite ($3Al_2O_3\ 2SiO_2$ or $2Al_2O_3SiO_2$) and hematite (Fe_2O_3). However, the maximum dispersion peak is at 20°-30° that indicates the amorphous material.

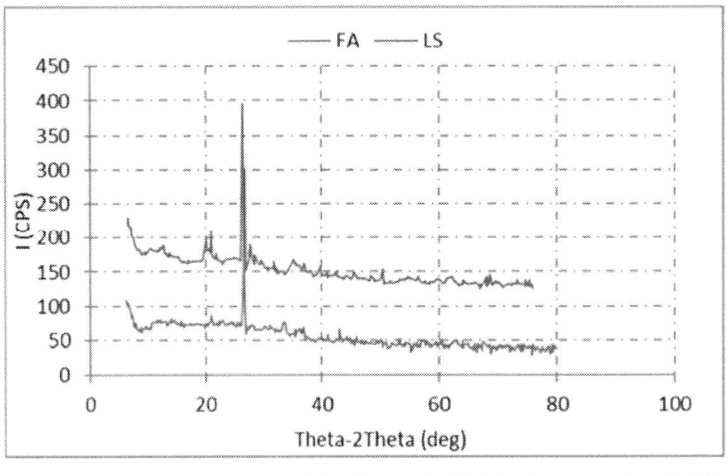

Figure 1. Identification material of FA and FSM by the analysis of XRD

3.1.3. Surface image of material

Figure 2 shows the scanning electron microscope (SEM) image to analyze the particle image of the material. Figure 2a shows the spherical shape of FA particle with the misty surface. The surface of FA in this study was in contrast with the surface of common FA that is a spherical shape with a sleek surface. The misty surface was considered absorbing the alkaline solution on fresh geopolymer mortar so that caused the reduction of workability. Also, the spherical shape of FA was indicated to be able to fill the concavity of geopolymer binder.

Furthermore, figure 2b shows irregular agglomerate shape of FSM particle with the rough surface. Irregular agglomerate shape of FSM caused the particle to be unable to fill the concavity on the geopolymer binder as good as the particle of FA. It brought up the small concavity on the geopolymer binder. Nevertheless, the big concavity was reduced and replaced with the small concavity. Furthermore, the rough surface of FSM absorbed the alkaline solution from the fresh mixture of geopolymer mortar that caused the reduction of workability on fresh geopolymer mortar.

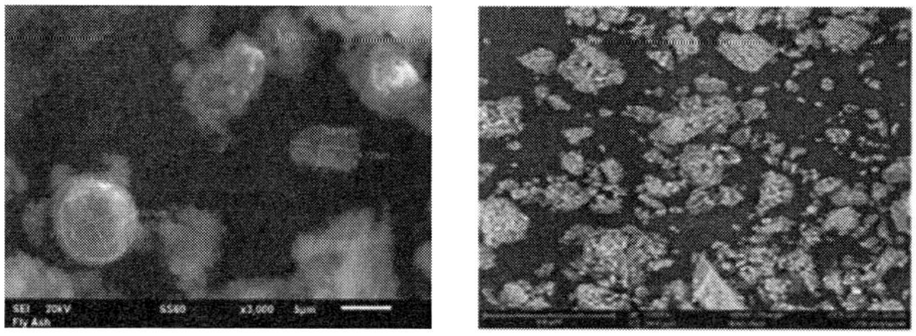

2a. The image of FA particle 2b. The image of FSM particle

Figure 2. The surface image of FA and FSM particle

4. Chemical bonding identification of material

Figure 3 shows the FTIR result of FA and FSM particle that identifies the chemical bonding structure of a material. Peak band at a wavenumber of 3600 cm^{-1} is presented by the material of FSM that indicated stretching and bending of H-O-H. It means that the material of FSM contained more water in its material. It is attributed to the calcination process that was not applied to fresh geopolymer mortar so that it was water buildup in the material. Peak band at a wavenumber of 3400 cm^{-1} is presented by the material of FA and FSM that indicates the presence of Ca $(OH)_2$ in its material. However, material of FSM had deeper peak band compared to the material of FA. It indicates that the presence of more efflorescence on the surface of FSM based geopolymer mortar compared to FA based geopolymer mortar. Peak band in the wavenumber range of 3440 cm^{-1}-1380 cm^{-1} is presented by the material of FSM that indicated the stretching and bending of water band and $CaCO_3$ (carbonate) in its material. Peak band at wavenumber range of 880-1140 cm^{-1} is presented by the material of FA and FSM that indicated the stretching of Si-O-Si and Al-O-Si in the crystalline of quartz and mullite. This wavenumber area determines what the material can be as geopolymer material or not. Furthermore, there was not found the presence of alkali content in the material of FSM. Yet, it was a presence in the material of FA. It is seen through peak band in the wavenumber of 680 cm^{-1}-800 cm^{-1}. Finally, peak band in the wavenumber below 680 cm^{-1} is presented by the material of FA. It indicates that the bending of Si-O was only presented by the material of FA. It means that the gel of Si-O was found in the material of FA that contributed to a high compressive strength of geopolymer mortar.

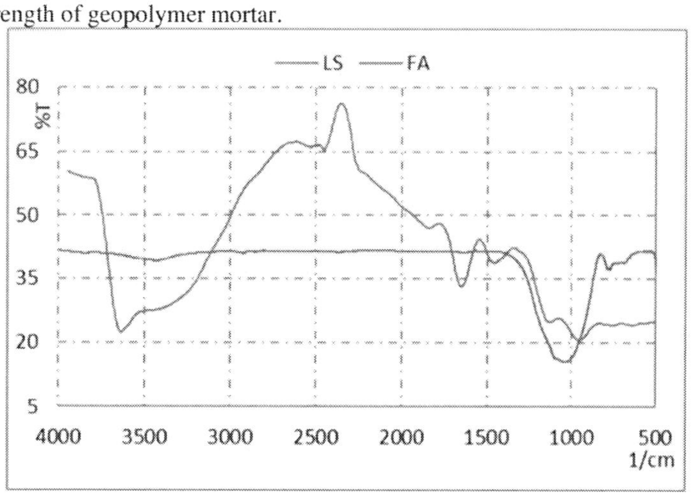

Figure 3. the FTIR analysis of FA and FSM material.

5. Workability

Rheology properties in fresh geopolymer mixture were measured by the workability and setting time. The workability of the fresh mixture was conducted by the flow table apparatus with measuring the diameter flow after a certain number of cycling. The workability was affected by the surface of the material that contacted with the water-cement in the conventional concrete or alkaline solution in the geopolymer concrete. It was known that the glassy surface of material did not absorb the water-cement or alkaline solution in a short time. Whereas the rough surface of material absorbed them by a direct time that caused the loss of workability on the fresh mixture of geopolymer mortar. It can be seen in Figure 4 which the increasing replacement of FSM in the FA based geopolymer mixture caused the reduction of workability. It is attributed to the surface of FSM as mentioned in the SEM section which in the agglomerate and rough surface.

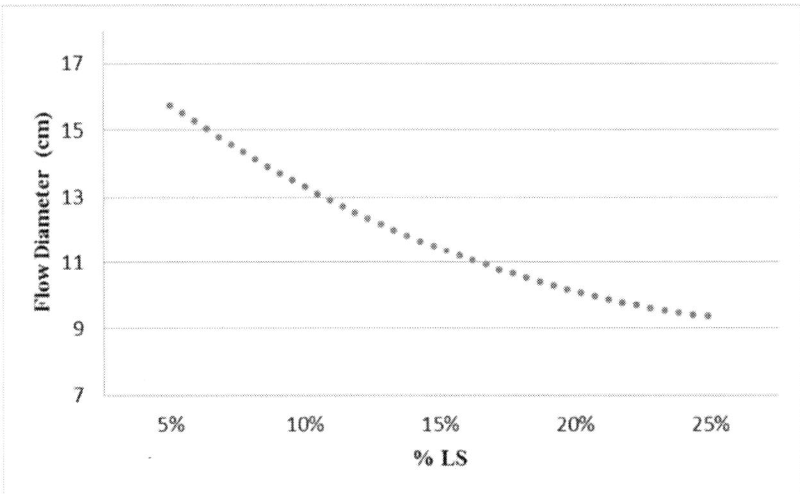

Figure 4. Workability of FSM on FA mortar geopolymer

The particle shape of FSM was an irregular agglomerate shape with the rough surface. Irregular agglomerate shape of FSM was considered not able to fill the concavity as good as the particle of FA. However, the rough surface of the FSM particle caused the absorption of the alkaline solution on the fresh mixture of geopolymer mortar that impacted on the reduction of workability. Figure 6 shows that the increase of FSM replacement causes the reduction of workability. It was attributed with the increase of FSM particle developed the rough surface of FSM that absorbed more alkaline solution on fresh geopolymer mortar.

6. Setting Time

Figure 5 shows the initial and final setting time of FSM replacement on the fresh mortar geopolymer. It is noted that the trend of initial and final setting time is consistent to increase. The figure shows that the increase of FSM replacement caused the reduction of FA in the binder. It means that the percentage of Ca content in the binder will be reduced so that caused the increase of setting time. The initial setting time indicated the plastic condition of fresh geopolymer mixture which was 80% reaction of geopolymer. Whereas, the final setting time indicated the static condition which was rest reaction of geopolymer.

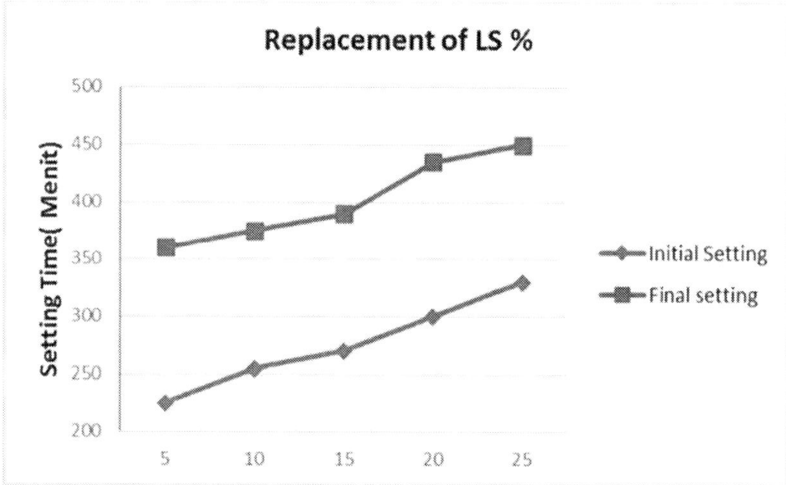

Figure 5. Setting time of FSM on the fresh mortar geopolymer

7. Compressive strength

Compressive strength of FSM replacement on FA based geopolymer is shown in Figure 6. It is seen that the replacement of 5% and 10% FSM provides higher compressive strength compared to the other replacement. However, it was still lower than the compressive strength of FA based geopolymer. It was attributed to the content of Si and Al in the material of FSM. It was known that the chemical content of Si and Al in the material of FSM was lower than the material of FA. The increase of FSM replacement on FA based geopolymer caused the reduction of Si and Al content that impacted on the reduction of gel Si-O-Si and Si-O-Al in the binder. Even, the compressive strength of FSM was lower than the compressive strength of FA. Yet, it was still acceptable to be used as the material in the geopolymer binder. Furthermore, the reduction of compressive strength in the FSM replacement was also caused by retarding of reaction impacting to the prolong setting time. There is a correlation between the compressive strength and setting time which the presence of FSM on the FA based geopolymer mortar retarded the setting time and also reduced the compressive strength.

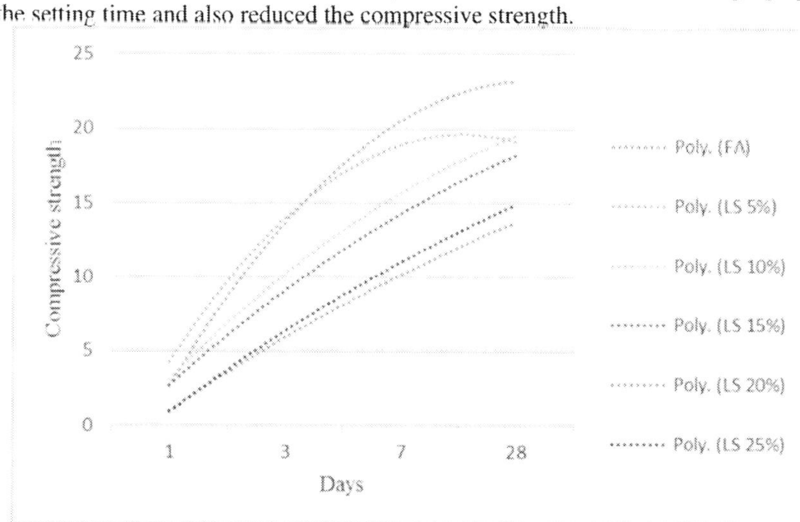

Figure 6. Compressive strength of FSM on the mortar geopolymer.

IOP Conf. Series: Materials Science and Engineering **854** (2020) 012011 doi:10.1088/1757-899X/854/1/012011

8. Conclusion

The study had been investigated with the following conclusion:

1. The material of FSM was rich in the content of Si and Al but poor in the Ca content compared to the material of FA.
2. The analysis of SEM showed that the particle of FSM was irregular agglomerate with the rough surface. While the particle of FA was spherical with the misty surface.
3. The analysis XRD showed that the material of FSM and FA consisted of quartz (SiO_2), mullite ($3Al_2O_3$ $2SiO_2$ or $2Al_2O_3$ SiO_2) and hematite (Fe_2O_3).
4. Peak band at wavenumber range of 880-1140 cm^{-1} was presented by the material of FA and FSM that indicated the stretching of Si-O-Si and Si-O-Al in the crystalline of quartz and mullite.
5. The increase of FSM replacement in the FA based geopolymer mortar caused the reduction of workability, increase the setting time and it did not improve the compressive strength.

References

[1] Aprianti, E. 2017." A huge number of artificial waste material can be supplementary cementitious material (SCM) for concrete production – a review part II". Journal of Cleaner Production, 142: p. 4178-4194.

[2] Shi, Caijun, Jimenez, et al. 2011." New cements for the 21st century: The pursuit of an alternative to Portland cement". Journal of Cement and Concrete Research, 41(7): p. 750-763.

[3] Walkley, Brant, Sani Rackel, et al., 2016. "Phase evolution of C-(N)-A-S-H/N-A-S-H gel blends investigated via alkali-activation of synthetic calcium aluminosilicate precursors", Journal of Cement and Concrete Research vol. 89:p. 120-135.

[4] Safiuddin, Abdus Salam, Jumaat, Mohd Zamin, 2011. "Utilization of palm oil fuel ash in concrete: a review" Journal of Civil Engineering and Management, vol.17(2):p. 234-247.

[5] Bamaga, Hussin, Ismail. 2013." Palm Oil Fuel Ash: Promising supplementary cementing materials". KSCE Journal of Civil Engineering. Vol.17(7):p. 1708-1713.

[6] Islam, Alengaram, Jumaat, Bashar. 2014 "The development of compressive strength of ground granulated blast furnace slag-palm oil fuel ash-fly ash based geopolymer mortar".*Journal of Materials & Design (1980-2015)*.vol.56: p. 833-841.

[7] Fahim Huseien, Mirza, et al. 2017." Geopolymer mortars as sustainable repair material: A comprehensive review". *Journal of Renewable and Sustainable Energy Reviews*. **80**: p. 54-74.

[8] Kroehong, Sinsiri, Jaturapitakkul, et al. 2011. "Effect of palm oil fuel ash fineness on the microstructure of blended cement paste" Construction and Building Materials, vol. 25(11), p. 4095-4104.

IOP Conf. Series: Materials Science and Engineering **854** (2020) 012012 doi:10.1088/1757-899X/854/1/012012

Effect of POFA as a replacement material on fly ash based geopolymer mortar

Mulizar[1], Fazliah[1, 2], Iskandar[1], Aiyub[1], Amir Fauzi[1, 2*]

[1] Civil Engineering Department, PoliteknikNegeriLhokseumawe, Acheh, Indonesia
[2] Geopolymer and Green Technology Research Center, Acheh, Indonesia

Email: mulizar@pnl.ac.id

Abstract. Ordinary Portland Cement (OPC) is widely used as the material for the construction. The production of OPC has caused an issue on the environment, due to the release of CO_2 in the air. The use of by-product or waste material from the industry as an alternative material in the concrete was considered to be a solution to cover this issue. Geopolymer was the advance method that was able to use by-product or waste material to replace OPC as the binder in the construction. However, this material should be rich in content of Si and Al, it was due to the reaction of waste material with the alkaline solution formed the gel of Si-O-Si and Si-O-Al that contributed to the mechanical properties especially compressive strength. Palm oil fuel ash (POFA) is a waste material from the furnace of oil palm that is also rich in the content of Si and Al so that it was considered a new material in geopolymer binder. The method to general geopolymer binder referred to the conventional method. The results showed that the analysis of XRD described the material of POFA in the crystalline with mullite, magnetite, hematite, and rutile. The analysis of FTIR identified that the material POFA has the stretching in the gel of Si-O-Al. The analysis of SEM showed that the particle of POFA has a rough surface that reduced the workability and longer setting time compared to the mixture of FA based geopolymer. Furthermore, the maximum compressive strength is 23 Mpa with the replacement of 5% POFA.

1. Introduction

Ordinary Portland Cement (OPC) is a famous material that is often used in the construction as the binder. The production of OPC was known to have increased pollution in the environment. There were more million tons of OPC production in the world that played an important role in the increase of CO_2 emission, exploitation the natural resources and consumed the energy. The use of waste material from the industry as the alternative material in the construction was believed to be able to control the emission of CO_2 from the environment [1]. Yet, the waste material can be used only in high Si and Al content. The total replacement of OPC in the construction with the waste material was known as a geopolymer system which it was green technology having lower CO_2 compared to OPC. Geopolymer is a material of inorganic polymeric that is formed by an aluminosilicate and calcium content. The binder of geopolymer was generated by the reaction of Si, Al and Ca content with the alkaline solution that formed the gel of Si-O-Si and Si-O-Al contributing on the mechanical properties, whereas the gel of N-A-S-H or C-A-S-H contributed to the setting time [2,3].

Palm oil fuel ash (POFA) is a waste material from the disposal of fuel for palm oil production. Some southeast Asian countries, especially Indonesia, Thailand, and Malaysia were famous for having the production of palm oil in abundant scale, so that appeared a new issue in the environment. Some investigation indicated that the material of POFA contained SiO_2 of 29.9%, Al2O3 of 1.9%, and CaO

IOP Conf. Series: Materials Science and Engineering **854** (2020) 012012 doi:10.1088/1757-899X/854/1/012012

of 26.9%, so that was categorized as the pozzolanic material [4,5]. Furthermore, the increase of POFA replacement in the FA geopolymer mortar reduced the compressive strength so that it was able to be a sustainable material in the construction [6,7].

The feasibility of POFA as a binder in the geopolymer referred to the material of fly ash and metakaolin. It was due to these materials were famous as the binder in the geopolymer system. However, the material of POFA contained the low Ca content that caused the prolonged setting time during the reaction [8]. So that, the previous investigation treated POFA base geopolymer by using the high temperature to accelerate the reaction. This study used the material of POFA as the replacement in fly ash class C base geopolymer to improve the Ca content in the material, so that it did not require the treatment to accelerate the reaction.

2. Material and experimental program

This study used the material of POFA that was from the Syaukat Sejahtera Crude Company Oil in Bireun, Indonesia whereas the material of fly ash was from power plantation in Nagan Raya, Acheh, Indonesia. Furthermore, the alkaline solution was generated by a combination of NaOH and Na_2SiO_3 solution with a certain ratio which was obtained from CV. Rudang Jaya company in Medan, Indonesia. The material was evaluated by XRD to indicate the particle structure of the material, FTIR to indicate the geopolymer binder gel and SEM to indicate the shape and surface of the particle. The mix proportion was designed based on the constituent material of geopolymer, whereas the mixing procedure followed the conventional method which referred to the standard of ASTM C305.The fresh and hardened properties of geopolymer mortar was evaluated by workability, setting time and compressive strength.

3. Results and discussion

3.1. Characterization of material

3.1.1. Chemical composition

The basic investigation to evaluate the characterization of material was started by analysis of chemical composition as seen in Table 1. The oxide content of Si, Al and Ca in the particle of fly ash is 37.16%, 17,61%, and 8.72%, respectively. Furthermore, the particle of POFA has the oxide content of Si, Al, and Ca of 59.3%, 6.32%, and 6.71%, respectively. It is seen that the content of Si in the particle of POFA is significantly higher than the particle of fly ash. Yet, it is in contrast with the content of Al. It provides the Si/Al content of 2.1 in the particle of fly ash, whereas the particle of POFA is 9.38. It means that the higher Si/Al content in the particle of POFA caused worse chemical stability in the air. It was indicated by the presence of efflorescence on the surface which was due to the higher residual free K+ in the particle of POFA. However, it provided a higher mechanical property with the presence of more Si-O-Si bonds and residual silica as reinforcement. Also, the content of Ca in the particle of fly ash is categorized as moderate Ca content contributing on the normal setting time that is similar to the setting time of OPC (5-6 hours), whereas the particle of POFA is categorized as low Ca content causing the prolong setting time.

IOP Conf. Series: Materials Science and Engineering **854** (2020) 012012 doi:10.1088/1757-899X/854/1/012012

Table 1: Chemical composition of fly ash and POFA

Chemical composition	Analysis, % by weight	
	Fly ash	POFA
SiO_2	37.16	59.3
Fe_2O_3	18.79	4.15
Al_2O_3	17.61	6.32
CaO	8.72	6.71
MgO	6.43	3.49
K_2O	0.79	8.52
SO_3	1.96	2.72
Na_2O	0.49	0.35
P_2O_5	3.67	2.32
TiO_2	0.75	0.25
LOI	3.11	5.87

3.1.2. Crystalline analysis of material

The crystalline identification of FA and POFA particle was investigated by X-ray diffraction test. The result of this study can be seen in Figure 1 that shows the spectrum in the particle of FA and POFA follows the similar peak of dispersion refers to the database of XRD. The trend of the graph indicates that the material of FA consists of quartz (SiO_2), mullite ($3Al_2O_3$ $2SiO_2$ or $2Al_2O_3SiO_2$) and hematite (Fe_2O_3) while the material of POFA consists of quartz (SiO_2), mullite ($3Al_2O_3$. $2SiO_2$ or $3Al_2O_3$. SiO_2), magnetite (Fe_3O_4), hematite (Fe_2O_3), dan rutile (TiO_2). The figure also shows that the dispersion peak is at 20°-30°, which indicates that the material was categorized as the amorphous material.

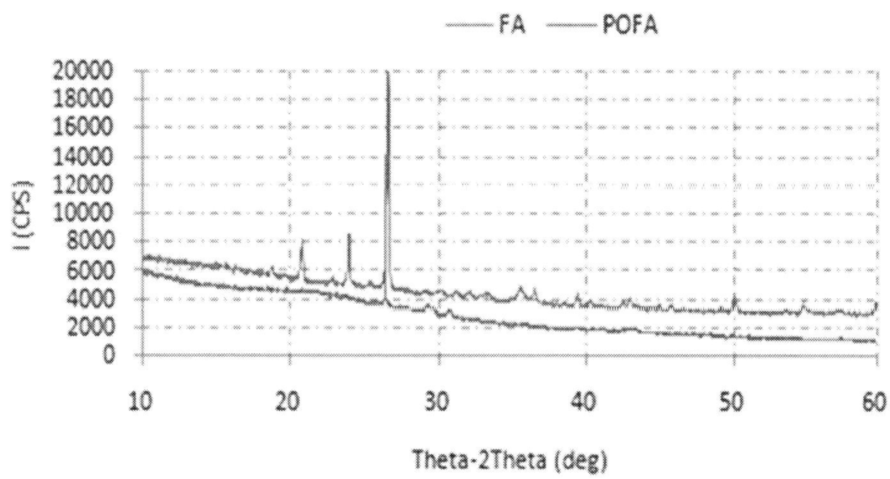

Figure 1. Identification material of FA and POFA by the analysis of XRD

3.1.3. Surface image of material

The particle image of the material was analyzed by scanning electron microscope (SEM). This image in figure 2a shows that the particle of FA has the shape of spherical that is covered by the misty surface. It impacted rheology properties especially the workability of fresh mixture. The misty surface of FA absorbed the alkaline solution in the fresh mixture so that caused the reduction of workability. Also, the spherical shape of FA was indicated to be able to fill the concavity of the binder.

IOP Conf. Series: Materials Science and Engineering **854** (2020) 012012 doi:10.1088/1757-899X/854/1/012012

Furthermore, figure 2b shows that the particle shape of POFA is spherical with the surface of agglomerate and rough. The spherical shape of POFA is similar to the particle shape of FA. It indicates that the particle of POFA contributed as the filler in the geopolymer binder. However, the surface of agglomerate and rough also absorbed the alkaline solution from the fresh mixture of geopolymer so that reduced the workability. Generally, there is no difference between the particle of POFA and FA.

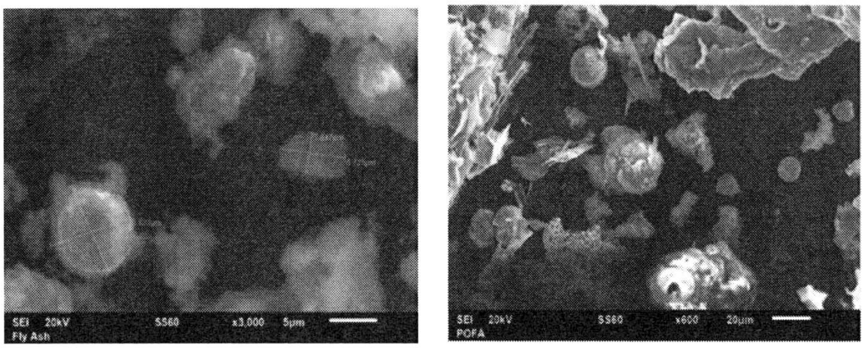

2a. The image of FA particle2b. The image of POFA particle

Figure 2. The surface image of FA and POFA particle

4. Chemical bonding identification of material

Figure 3 shows the FTIR result of FA and POFA particle that identifies the chemical bonding structure. Peak band at a wavenumber of 3400 cm-1 is presented by the material of FA and POFA that indicates the presence of Ca $(OH)_2$ in its material. However, they have the opposite peak in the wavenumber. The material of FA has the bottom peak number, whereas the material of POFA is at top. It indicates that the material of POFA has more $Ca(OH)_2$ compared to the material of FA. It was proven by the presence of efflorescence on the surface of geopolymer mortar. Peak band in the wavenumber range of 3440 cm-1-1380 cm-1 is presented by the material of POFA that indicated the stretching and bending of water band and CaCO3 (carbonate) in its material. Peak band at the wavenumber range of 880-1140 cm-1 is presented by the material of FA and POFA that indicate the stretching of Si-O-Si and Al-O-Si in the crystalline of quartz and mullite. Furthermore, there was not found the presence of alkali content in the material of POFA. But, it was a presence in the material of FA. It is seen through peak band in the wavenumber of 680 cm-1-800 cm-1. Finally, Peak band in the wavenumber below 680 cm-1 is presented by the material of FA. It indicates that the bending of Si-O was only presented by the material of FA. It means that the gel of Si-O was found in the material of FA that provided a high compressive strength on geopolymer mortar.

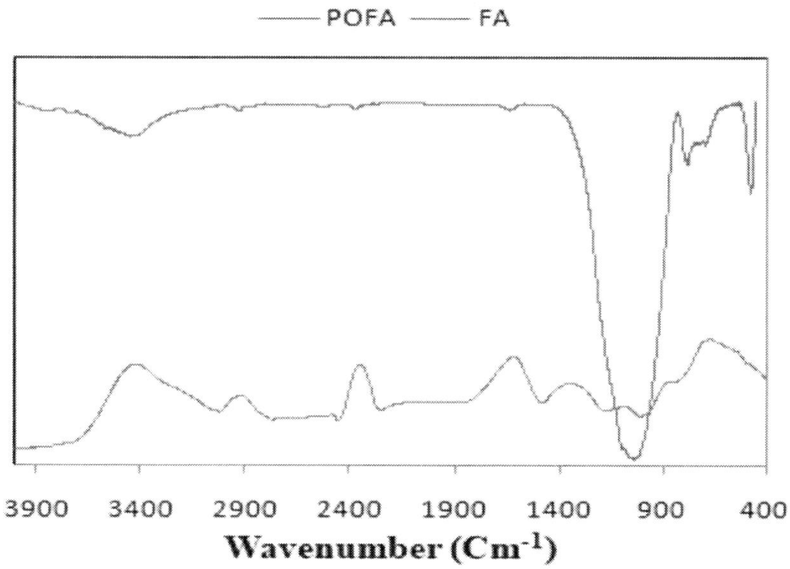

Figure 3. the FTIR analysis of FA and POFA material.

5. Workability

Rheology properties in fresh geopolymer mixture were measured by the workability and setting time. The workability of the fresh mixture was conducted by the flow table apparatus with measuring the diameter flow after a certain number of cycling. The workability was affected by the surface of the material that contacted with the water-cement in the conventional concrete or alkaline solution in the geopolymer concrete. The glassy surface of material did not absorb the water-cement or alkaline solution in a short time, whereas the rough surface of material absorbed them by a direct time that caused the loss of workability on the fresh mixture of geopolymer mortar. It can be seen in Figure 4 which the increasing replacement of POFA in the FA based geopolymer mixture caused the reduction of workability. It is attributed to the surface of POFA as mentioned in the SEM section which in the agglomerate and rough surface.

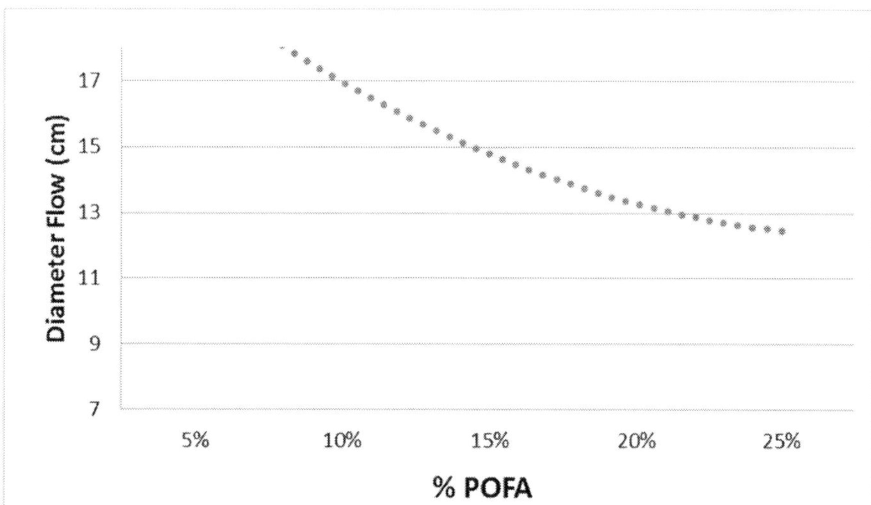

Figure 4. Workability of POFA on FA mortar geopolymer

IOP Conf. Series: Materials Science and Engineering **854** (2020) 012012 doi:10.1088/1757-899X/854/1/012012

The particle shape of POFA is spherical with the surface of agglomerate and rough. The spherical shape of POFA is similar to the particle shape of FA. It indicates that the particle of POFA contributed as the filler in the geopolymer binder. However, the surface of agglomerate and rough also absorbed the alkaline solution from the fresh mixture of geopolymer so that impacted the workability. Figure 4 shows that the increase of POFA replacement causes the reduction of workability. It was attributed with the increase of rough surface of POFA in the fresh mortar geopolymer mixture that absorbed more alkaline solution in the fresh mortar geopolymer mixture. Generally, there is no difference between the particle of POFA and FA.

6. Setting Time

Figure 5 shows the initial and final setting time of POFA replacement on the fresh mortar geopolymer. It is noted that the trend of initial and final setting time is consistent to increase. The figure shows that the increase of POFA replacement caused the reduction of FA in the binder. It means that the percentage of Ca content in the binder will be reduced so that caused the increase of setting time. The initial setting time indicated the plastic condition that described 80% reaction of geopolymer. Whereas, the final setting time indicated the static condition that described rest reaction of geopolymer.

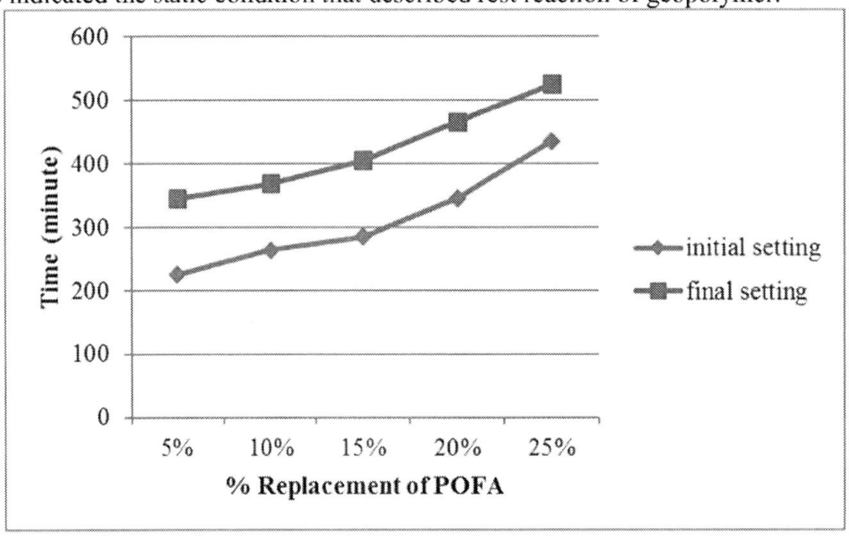

Figure 5. Setting time of POFA on the fresh mortar geopolimer

7. Compressive strength

Figure 6 shows the compressive strength of FOFA replacement on FA based geopolymer. It is seen that the replacement of 5% POFA provides higher compressive strength compared to the other replacement. However, it was lower than the compressive strength of FA based geopolymer. It was attributed to the content of Si and Al in the material. It was known that the chemical content of Si and Al in the material of POFA was lower than the material of FA. The increase of POFA replacement on FA based geopolymer caused the reduction of Si and Al content so that caused the reduction of gel Si-O-Si and Si-O-Al in the binder. Even, the compressive strength of POFA was lower than the compressive strength of FA. Yet, it was still acceptable to be used as the material in the geopolymer binder. Furthermore, the reduction of compressive strength in the POFA replacement was also caused by retarding of reaction causing the prolong setting time. There is a correlation between the compressive strength and setting time which the presence of POFA on the FA based geopolymer mortar retarded the setting time and also reduced the compressive strength.

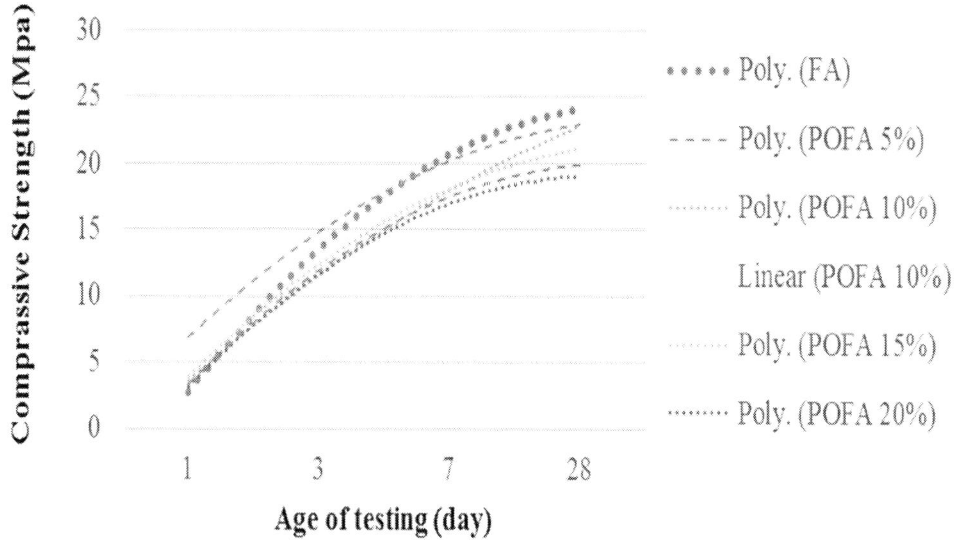

Figure 6. Compressive strength of POFA on the mortar geopolymer.

8. Conclusion

The study had been investigated with the following conclusion:

1. The material of POFA was rich in the content of Si and Al but poor in the Ca content compared to the material of FA.
2. The analysis of SEM showed that the particle of POFA was spherical with the surface of agglomerate and rough. While the particle of FA was spherical that is covered by the misty surface.
3. The analysis XRD showed that the material of FA consisted of quartz (SiO_2), mullite ($3Al_2O_3$ $2SiO_2$ or $2Al_2O_3$ SiO_2) and hematite (Fe_2O_3). While, the material of POFA consisted of quartz (SiO_2), mullite ($3Al_2O_3$. $2SiO_2$ or $2Al_2O_3$. SiO_2), magnetite (Fe_3O_4), hematite (Fe_2O_3), and rutile (TiO_2).
4. Peak band at wavenumber range of 880-1140 cm-1 was presented by the material of FA and POFA that indicated the stretching of Si-O-Si and Al-O-Si in the crystalline of quartz and mullite.

References

[1] Aprianti, E. 2017."A huge number of artificial waste material can be supplementary cementitious material (SCM) for concrete production – a review part II". Journal of Cleaner Production, 142: p. 4178-4194.

[2] Shi, Caijun, Jimenez, et al. 2011."New cements for the 21st century: The pursuit of an alternative to Portland cement". Journal of Cement and Concrete Research, 41(7): p. 750-763.

[3] Walkley, Brant, Sani Rackel, et al., 2016. "Phase evolution of C-(N)-A-S-H/N-A-S-H gel blends investigated via alkali-activation of synthetic calcium aluminosilicate precursors", Journal of Cement and Concrete Research vol. 89:p.120-135.

[4] Safiuddin, Abdus Salam,Jumaat, MohdZamin, 2011. "Utilization of palm oil fuel ash in concrete: a review" Journal of Civil Engineering and Management, vol.17(2):p.234-247.

[5] Bamaga,Hussin, Ismail. 2013."Palm Oil Fuel Ash: Promising supplementary cementing materials".KSCE Journal of Civil Engineering. Vol.17(7):p.1708-1713.

[6] Islam, Alengaram, Jumaat, Bashar. 2014 "The development of compressive strength of ground granulated blast furnace slag-palm oil fuel ash-fly ash based geopolymermortar".*Journal of Materials & Design (1980-2015)*.vol.56: p.833-841.

IOP Conf. Series: Materials Science and Engineering **854** (2020) 012012 doi:10.1088/1757-899X/854/1/012012

[7] FahimHuseien,Mirza, et al.2017."Geopolymer mortars as sustainable repair material: A comprehensive review". *Journal of Renewable and Sustainable Energy Reviews*. **80**: p. 54-74.
[8] Kroehong, Sinsiri, Jaturapitakkul, et al. 2011. "Effect of palm oil fuel ash fineness on the microstructure of blended cement paste" Construction and Building Materials, vol. 25(11), p. 4095-4104.

IOP Conf. Series: Materials Science and Engineering **854** (2020) 012013 doi:10.1088/1757-899X/854/1/012013

Development of Castor Oil as Local Demulsifier to Overcome Water-Oil Emulsion

Tomi Erfando[1*], Devin Theo Handoko[1], Idham Khalid[1]

[1] Department of Petroleum Engineering, Universitas Islam Riau Kaharuddin Nasution No.113, Pekanbaru, Riau, Indonesia

Email: tomierfando@eng.uir.ac.id

Abstract. Crude oil is generally mixed with water at the time of production. The oil and water mixture at the time of production is known as an emulsion. In addressing the problem of emulsions requires the addition of a chemical compound, commonly referred to as demulsifier to be able to separate water and oil. This research aims to make the formulation of demulsifier locally based on castor oil. This castor oil will be processed into a surfactant compound because it has a very high saturated fat content of 33% of the total weight of the seeds. The manufacture of this demulsifier using saponification method is to heat the castor oil and add the KOH dissolved in aquades, as well as the added glycerin to taste. Then, the process of testing demulsification with a sample of the emulsion of crude oil that has been prepared using the bottle test method. From the test results of several temperature conditions, different concentration and time obtained the best test results by demulsifier in the separation of water and oil at a temperature of 80 ^{0}C and concentrations of 5 ml. In sample 1 demulsifier soap and the soap demulsifier added lemon separates the water by as much as 38 ml and 39 ml, meanwhile, the commercial demulsifier separates the water by 32 ml. Thus, it can be concluded that the demulsifier made from castor oil has a better ability than commercial demulsifier.

1. Introduction

In the stage of production and oil processing is often the problem is the water emulsion in the oil is stable. The formation of this emulsion is generally caused by the presence of resins and asphaltenes which act as "natural emulsifiers", as well as wax and solids (Al-Sabagh, Kandile, El-Ghazawy, & Noor El-Din, 2011). In addressing the problem of emulsions requires a requirement to convert the emulsion into a normal return. This requirement is a material that can stabilize the system between surfaces. This ingredient is usually called a stabilizer, or also known as demulsifier (Gavrielatos, Dabirian, Mohan, & Shoham, 2018). Overcoming this emulsion problem is also done to keep the production facility from damage and improve the oil quality itself (Sulaiman et al., 2015).

According to the research of Sulaiman et al., (2015) showed that the demulsifer produced from the local ingredients is very effective in terms of the cheaper price, effective time and solving the crude oil emulsion in a shorter time than on conventional demulsifier. Researchers conducted a test of demulsification using the bottle test method with several parameters, such as temperature, concentration and time. According to Zhou, Dismuke, Lett, & Penny, (2012) to meet the safety standards in the oil and gas world and reduce the impact that occurred on the environment resulting from the use of demulsifier with chemicals, then it takes a new formulation as a demulsifier with develop from local ingredients. Therefore, in this research try to develop a formulation of demulsifier by using local ingredients namely castor oil.

This castor oil will be processed into a surfactant compound because it has a very high saturated fat content of 55% for the core part of the seed or 33% of the total weight of the seed (Sari, Kasih, & Sari, 2010). In the process of making surfactant compounds using the saponification method. Saponification is the process of alkaline hydrolysis to fat from oil by adding KOH (Naomi, Gaol, & Toha, 2013). Furthermore, the local demulsifier that has been made will be conducted by the process of crude oil samples with a demulsification process using the bottle test method. Using this method demulsifier can quickly separate water and oil with the lowest possible moisture content (Manggala, Kasmungin, & Fajarwati, 2017). Then to know the influence of demulsifier with some observed parameters such as temperature, concentration and time in the research, then the analysis is done Regeresi equations using "Minitab Software" (Sungkawa, 2013).

2. Materials And Methods

The method used in this research is the experimental method. In the mention of demulsifier using the saponification method and for testing the effectiveness of demulsifer that has been made using the bottle test method. The materials used in this study are castor oil, local lemon, KOH, glycerin, equadest, commercial finalizing, and X field oil samples.

3. Research Procedures

3.1. Production of Demulsifier

Manufacture of finalizing according to Paramita et al., (2014) using castor oil with saponification method as follows ; heat the oil range 25,79 gr with a temperature of 70 °C for 30 minutes. Then dissolve KOH 8,15 gr with the aquadest as much as 16,3 gr. Gently insert the KOH and Aquades solution into the previously heated range oil. Stir the solution using magnetic stirrer at 800 rpm for ± 100 minutes. Add the glycerin 10 ml into the castor oil solution. Then add 60 ml of the aquadest into the solution.

3.2. Demulsification With Bottle Test Method

Bottle test procedure according to Erfando et al., (2019) and Impian & Praputri, (2014) including the following ; Prepare a demulsifer formulation that will be injected into the crude oil emulsion. Insert the crude oil emulsion samples into the test bottle as much as 50 ml. Inject the demulsifier formula into the test bottles that have been filled with the crude oil emulsion with each concentration of 1 ml, 3 ml, 5 ml. Place the test bottle into the Waterbath using the temperature to be tested that is 40 °C, 60 °C, 80°C. Bottle test time is for 3 hours and observed every 30 minutes.

4. Results And Discussion

This research aims to determine the ability to demulsifier locally sourced castor oil which is reviewed from several variables, such as temperature, concentration, and timing. This research also aims to determine the effectiveness of the ratio of the demulsifier formula to local oil with a commercial demulsifier to the separation process of the water-in-oil emulsion (W/O). Temperature is an important factor that affects the emulsion stability. The higher the temperature, the more unstable and vice versa, the lower the temperature and the more stable the emulsion (Manggala et al., 2017). This study made a formula for demulsifier as many as 3 formulae, which is a surfactant demulsifier (DS) made of local material namely castor oil, demulsifier a lemon surfactant (DSL) that is a mixture of formulation of Surfactant demulsifier From the available lemon and commercial demulsifier.

4.1. Temperature Conditions 40 ℃

The results of this study at a temperature of 40 ^{0}C in Sample 1 can be seen in **Figure 1**. demulsifier surfactant (DS) is high in breaking down the emulsion at 5 ml concentrations of 12,5 ml. In the commercial demulsifier (DK) only able to break down the emulsion at a concentration of 1 ml by 5 ml. This is because the commercial demulsifier is more effective in high temperature, where the combination of heat and the application of chemicals designed can decide the interfacial film (Sulaiman et al., 2015). Meanwhile, lemon surfactant demulsifier can solve the emulsion by 13 ml with 5 ml concentration. According to Kokal, Al-juraid, & Aramco, (1999) several factors affect the performance of demulsifier namely because of the stability properties of the emulsion, the factors influencing it is low temperature and heavy oil type. According to Wylde, Coscio, & Barbu (2008) in this temperature is seen as the result of high water separation influenced by temperature. Known the lower the temperature then the emulsion will stabilize.

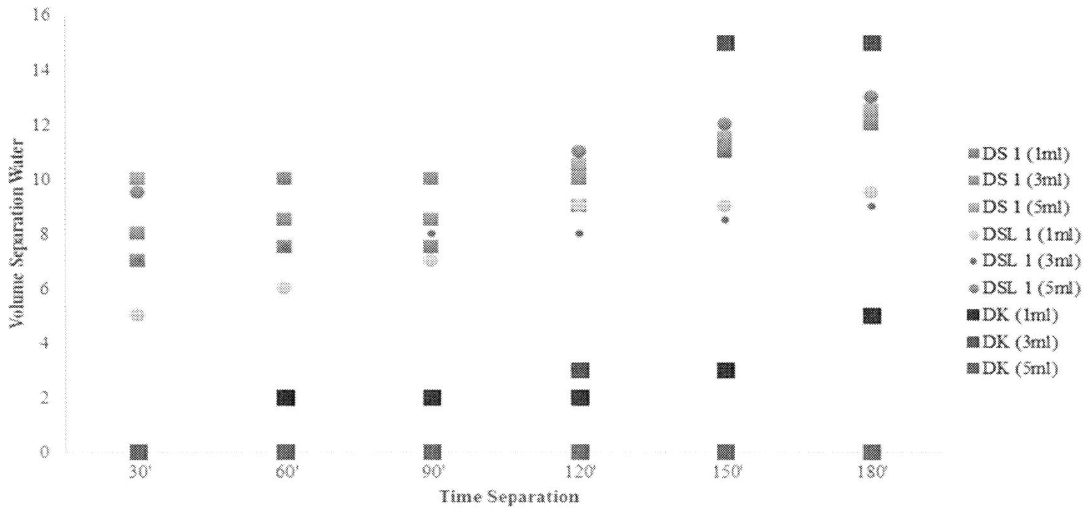

Figure 1. Water separation result of testing time at 40 ^{0}C

4.2. Temperature Conditions 60 ℃

At 60 ^{0}C temperature, the result of water separation from the emulsion has increased compared with the previous temperature, where it is seen that all types of demulsifier have been able to break down the emulsion in all concentrations. From the result in **Figure 2**. It is seen that the surfactant demulsifier (DS) is also able to work as a commercial demulsifier (DK). The result of emulsion breakdown using commercial demulsifier (DK) is 28 ml with 5 ml concentration. Solving the highest emulsions use surfactant demulsifier (DS) of 31,5 ml with a concentration of 5 ml (Sulaiman et al., 2015). Whereas a lemon surfactant demulsifier (DSL) can solve a very high emulsion of 32.5 ml with a concentration of 5 ml since the lemon it self contains citric acid which amounts to 90% of the total organic acids (Sadka, Dahan, Cohen, & Marsh, 2000). Inorganic acids, especially citric acid, has excellent demulsification efficiency in crude oil emulsions because citric acid has more carboxyl groups that can participate in the reaction, resulting in more demulsification efficiency High than other acids (Liu et al., 2018).

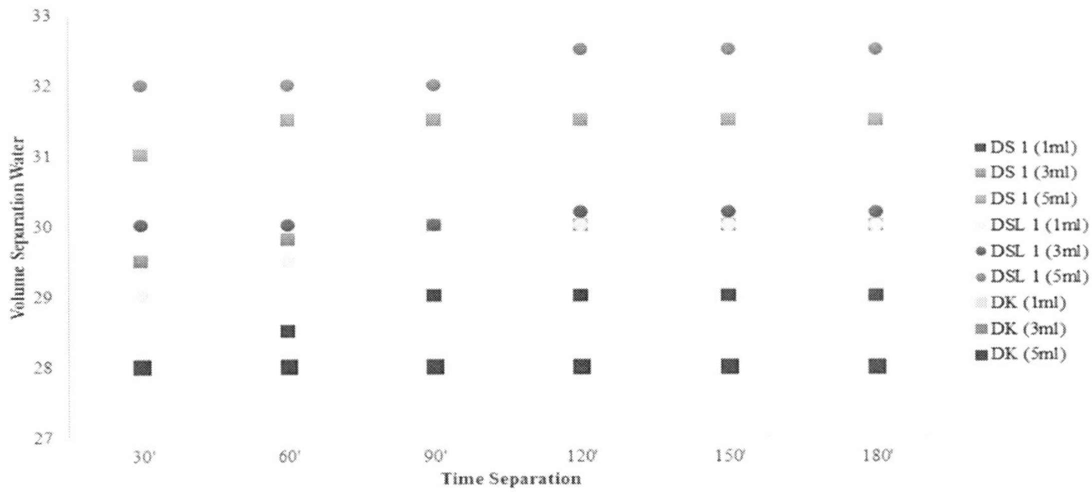

Figure 2. Water separation result to test time in temperature 60 ^{0}C

4.3. Temperature Conditions 80 ℃

At 80 ^{0}C The result of water separation from emulsions in all types of demulsifier both locally and commercially experienced a drastic increase from the previous can be seen in **Figure 3**. It appears that the result of water separation from the emulsion in a commercial demulsifier sample (DK) at a concentration of 5 ml is 39 ml. Where good characteristics of demulsifier one of them are the faster separation (Rita & Hadi, 2017). In other results use surfactant demulsifier (DS) with a concentration of 5 ml can separate water by 38 ml. While a lemon surfactant demulsifier (DSL) with a concentration of 5 ml can separate water by 39 ml. The higher the test temperature then The separation that looks faster and a lot of water is separated. This indicates the high temperature and high concentration of the demulsifier injection greatly affects the demulsification process. The high concentration of demulsifier greatly affects the voltage between fluid loading (Kang, Jing, Zhang, Li, & Wu, 2006).

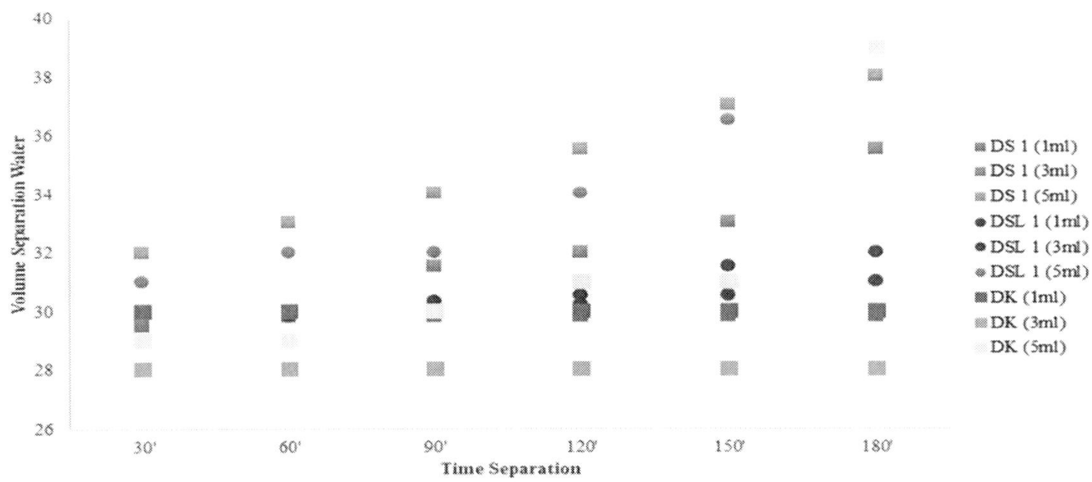

Figure 3. Water separation result to test time in temperature 80^{0}C

4.4. Statistical Analysis Of Regression And Correlation Using Minitab Software

In this study also saw how much influence of temperature and concentration on water separation from oil emulsion with statistical calculations using MINITAB application. Minitab provides various programs for the complete processing of statistical data, such as regression analysis, statistical quality

IOP Conf. Series: Materials Science and Engineering **854** (2020) 012013 doi:10.1088/1757-899X/854/1/012013

control, forecasting with time series analysis and other (Wahyuni, Agoestanto, & Pujiastuti, 2018). Here's the result of statistical calculations with Minitab application.

Table 1. Results of regression and correlation of Software Minitab

Parameters	***P*-Value**	**R-Sq**	**R-Sq (adj)**
Temperature	0	71.0	70.9
Concentration	0.014	1.6	1.3

A comparison of linear regression models determines the influence of X variables on Y (P. Subekti, 2015). If a positive result is obtained, the value of X variables will be equal to the value of the variable Y. From **Table 1**. In regression analysis, the value of P or P-value was obtained at 0.000 which means that the value of P is smaller than the significant criteria of 0.05 so that the free variable has a significant effect on the associated variables. At the output P-value obtained from the temperature is only less than the value of significance ($\alpha = 0.05$) means that the linear regression model meets the criteria of linearity and the changes occurring are very significant (Draper & Smith, 1998). Besides, the value of R-esq (adj) of 70,9% can be seen in **Figure 4**. This means the effect of temperature on water separation in the demulsification process of 70,9% and the remaining 29,1% is the influence of other free variables in addition to temperature (Wahyuni et al., 2018). In this case, it means the correlation of temperature to the separation of the demulsification result is very high because its value is almost close to 1 or 100% and with its positive value indicates that both variables or tested parameters are increased together (Pratomo & Astuti, 2014)

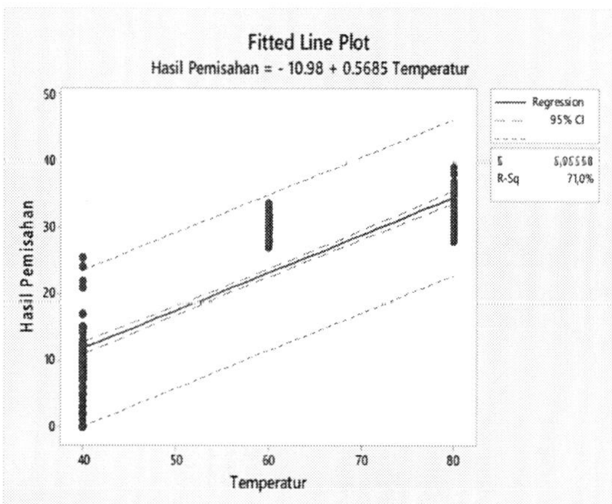

Figure 4. *Fitted Line Plot* temperature vs water separation

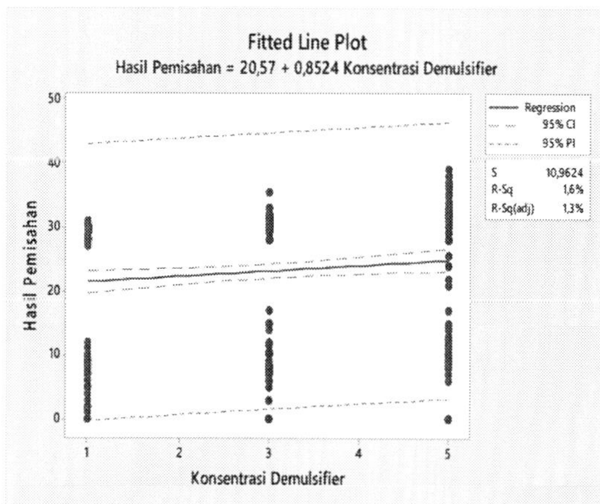

Figure 5. *Fitted Line Plot* concentration vs water separation

In **Figure 5**. The value of the coefficient of determination or R-esq (adj). The value of this coefficient of determination is used to know how large the associated variables are influenced by free variables. From the results of the regression analysis obtained the value of R-esq (adj) by 1,3%, this indicates that the variable on the separation of water is influenced by the free variable (concentration) of 1,3% and the residual amount is 98,7% influenced by variables other than concentration (Wahyuni et al., 2018). In this regard, the higher the centralization of the water separation result from the demulsification process is increasingly higher (Puji Subekti, 2015). Also, the correlation value gained a positive value of 1,3. The result of this value means that the correlation of the concentration on the water separation of demulsification result is low because its value is far from 1 and with positive value indicates that both variables or parameters are tested with (Pratomo & Astuti, 2014). The difference in the efficiency rate of the demulsification of each of the formulas can be due to unequal homogenization of each sample and the durability and quality differences of each formula acting at a certain temperature (Erfando.T, Irma Elfradina, 2019).

5. Conclusion

The retractable conclusion in this study is that local demulsifier samples were able to break down the emulsion throughout the test temperature, the highest separation occurred at 80 ^{0}C with an optimal concentration of 5 ml, namely the surfactant demulsifier (DS) amounted to 38 ml and a lemon surfactant demulsifier (DSL) of 39 ml, while the commercial demulsifier (DK) at a concentration of 5 ml at 40 ^{0}C was not able to break down the emulsion. The effect of temperature and concentration on water separation in the order of 70,9% and 1,3%. In this case that has the greatest influence among some parameters of the regression analysis data is the temperature of 70,9%.

Acknowledgements

Thank you for University Relationship Program (URP) PT. Chevron Pasific Indonesia for the support of financial for the publication of this research.

References

[1] Al-Sabagh, A. M., Kandile, N. G., El-Ghazawy, R. A., & Noor El-Din, M. R. (2011). Synthesis And Evaluation Of Some New Demulsifiers Based On Bisphenols For Treating Water-In-Crude Oil Emulsions. *Egyptian Journal Of Petroleum*, *20*(2), 67–77. Https://Doi.Org/10.1016/J.Ejpe.2011.06.008

[2] Draper, N. R., & Smith, H. (1998). *Applied Regression Analysis* (Vol. 326). John Wiley & Sons.

[3] Erfando.T, Irma Elfradina, N. R. (2019). Effects Of Adding Local Materials On Demulsifier Performance For Oil-Water Emulsions. *International Journal Of Geomate*, *17*(62), 107–112. Https://Doi.Org/10.21660/2019.62.Icee9

[4] Erfando, T., Cahyani, S. R., & Rita, N. (2019). The Utilization Of Citrus Hystrix And Citrus Limon As An Organic Demulsifier Formulation. In *Iop Conference Series: Materials Science And Engineering* (Vol. 509, P. 12145). Iop Publishing.

[5] Gavrielatos, I., Dabirian, R., Mohan, R., & Shoham, O. (2018). Spe-190114-Ms Nanoparticle And Surfactant Oil / Water Emulsions - Is Different Treatment Required? *Spe Western Regional Meeting*.

[6] Impian, D., & Praputri, E. (2014). Optimasi Injeksi Demulsifier Sebagai Respon Terhadap Proses Acidizing, *4*(4).

[7] Kang, W., Jing, G., Zhang, H., Li, M., & Wu, Z. (2006). Influence Of Demulsifier On Interfacial Film Between Oil And Water, *272*, 27–31. Https://Doi.Org/10.1016/J.Colsurfa.2005.07.004

[8] Kokal, S., & Al-Juraid, J. (1999). Quantification Of Various Factors Affecting Emulsion Stability: Watercut, Temperature, Shear, Asphaltene Content, Demulsifier Dosage And Mixing Different Crudes. *Spe Annual Technical Conference And Exhibition*, *56641*. https://doi.org/10.2118/56641-ms

[9] Liu, D., Suo, Y., Zhao, J., Zhu, P., Tan, J., Wang, B., & Lu, H. (2018). Effect Of Demulsification For Crude Oil-In-Water Emulsion: Comparing Co2and Organic Acids. *Energy And Fuels*, *32*(1), 757–764. Https://Doi.Org/10.1021/Acs.Energyfuels.7b03334

[10] Manggala, M. R., Kasmungin, S., & Fajarwati, K. (2017). Studi Pengembangan Demulsifier Pada Skala Laboratorium Untuk Mengatasi Masalah Emulsi Minyak Di Lapangan“ Z”, Sumatera Selatan. In *Prosiding Seminar Nasional Cendekiawan* (Pp. 145–151).

[11] Naomi, P., Gaol, A. M. L., & Toha, M. Y. (2013). Bekas Ditinjau Dari Kinetika Reaksi Kimia, *19*(2), 42–48.

[12] Paramita, N., Fahrurroji, A., & Wijianto, B. (2014). Optimasi Sabun Cair Ekstrak Etanol Rimpang Zingiber Officinale Rosc. Var. Rubrum Dengan Variasi Minyak Jarak Dan Kalium Hidroksida. *Journal Of Tropical Pharmacy And Chemistry*, *2*(5), 272–282.

[13] Pratomo, D. S., & Astuti, E. Z. (2014). Analisis Regresi Dan Korelasi Antara Pengunjung Dan Pembeli Terhadap Nominal Pembelian Di Indomaret Kedungmundu Semarang Dengan Metode Kuadrat Terkecil. *Ilmu Komputer*, (1).

[14] Rita, N., & Hadi, G. (2017). Evaluasi Efisiensi Proses Crude Oil Dehydtation Di Cgs 5 Lapangan X Provinsi Riau. *Jurnal Mineral, Energi Dan Lingkungan*, *1*(1), 16–27.

[15] Sadka, A., Dahan, E., Cohen, L., & Marsh, K. B. (2000). Aconitase Activity And Expression During The Development Of Lemon Fruit. *Physiologia Plantarum*, *108*, 255–262. Https://Doi.Org/10.1034/J.1399-3054.2000.108003255.X

[16] Sari, T. I., Kasih, J. P., & Sari, T. J. N. (2010). Pembuatan Sabun Padat Dan Sabun Cair Dari Minyak Jarak. *Jurnal Teknik Kimia*, *17*(1).

[17] Subekti, P. (2015). Perbandingan Perhitungan Matematis Dan Spss Analisis Regresi. *Seminar Nasional*, *3*(June).

[18] Subekti, P. (2015). Perbandingan Perhitungan Matematis Dan Spss Analisis Regresi Linear Studi Kasus (Pengaruh Iq Mahasiswa Terhadap Ipk). *Prosiding Snatika 2015*, *3*, 70–75.

[19] Sulaiman, A. D. I., Abdulsalam, S., Technology, E., Tafawa, A., Francis, A. O., & Polytechnic, A. (2015). Formulation Of Demulsifiers From Locally Sourced Raw Materials For. *Spe*, *178377–Ms*.

[20] Sungkawa, I. (2013). Penerapan Analisis Regresi Dan Korelasi Dalam Menentukan Arah Hubungan Antara Dua Faktor Kualitatif Pada Tabel Kontingensi. *Jurnal Mat Stat*, *13*(1), 33–41.

[21] Wahyuni, T., Agoestanto, A., & Pujiastuti, E. (2018). Analisis Regresi Logistik Terhadap Keputusan Penerimaan Beasiswa PPA di FMIPA Unnes Menggunakan Software Minitab. In *Prisma, Prosiding Seminar Nasional Matematika* (Vol. 1, Pp. 755–764).

[22] Wylde, J. J., Coscio, S., & Barbu, V. (2008). A Case History Of Heavy Oil Separation In Northern Alberta: A Singular Challenge Of Demulsifier Optimization And Application. *Spe International*, 1–8. Https://Doi.Org/10.2118/117177-Pa

[23] Zhou, H., Dismuke, K. I., Lett, N. L., & Penny, G. S. (2012). Development Of More Environmentally Friendly Demulsifiers. In *Spe International Symposium And Exhibition On Formation Damage Control*. Society Of Petroleum Engineers.

IOP Conf. Series: Materials Science and Engineering **854** (2020) 012014 doi:10.1088/1757-899X/854/1/012014

Aged Reduction Prediction of the Road Plan Due to Overload On the Pancer Road, Pesanggaran District, Banyuwangi

WN Wari, E Suryani, Y Ulfiyati, IW Syafa'at

Department of Civil Engineering, Politeknik Negeri Banyuwangi, Banyuwangi, Indonesia

E-mail: naris@poliwangi.ac.id

Abstract. Jalan Pancer is grade 3 roads with the heaviest axle loads of 8 tons, the age of the road plan is 5 years. But Jalan Pancer has been passed by heavy vehicles with the heaviest axle loads of more than 8 tons which caused premature damage to the road. Therefore this study aims to determine the predicted reduction in the life of the road plan caused by overloading from the vehicles. The research method used is to conduct a traffic survey to get Average Daily Traffic value and identify the heavy vehicles. The data obtained were analyzed to obtain the value of Vehicle Damage Factor (VDF) for each type of vehicle and the value of Cummulative Equivalent Standard Axle-CESA for the life of the road plan. The comparison results among VDF value of the survey and VDF value of the plan conclude that road damage factor greate. Because of the overload conditions the age of the plan is only 1 year 9.7 months, there is a decrease in the life of the road by 3 years 2.3 months in the second year, so it can be concluded that Jalan Pancer suffered early damage before the planned age (5 years).

Keywords: overload, reduced road life plan, road life plan, VDF, Pancer Road, CESA, road damage, Banyuwangi.

1. Introduction

1.1. Backgroud

One of the common problems of road pavement in Indonesia so far is the early damage to pavement. One of the main factors causing road damage is one of them caused by overload. Based on previous observations, that the road pavement in the southern Banyuwangi area often suffered early damage or road damage before the service life period ended. Through the vehicle weight data obtained from the weighbridge, it shows that vehicles passing through roads in the area of southern Banyuwangi exceed the required burden of the Heaviest Axis Load.

One of the roads experiencing these problems is Jalan Pancer, Pesanggaran, Banyuwangi. Pancer Road is a road with class 3 road classification, with the heaviest axle load of 8 tons, with a cross section width is 4 meters. The Pancer Road pavement structure is Hotmix/ACWC, the age of the road plan is 5 years. Jalan Pancer is the main access for the Overlapping Gold Mining Project by PT. Bumi Suksesindo (BSI), also the Pier Development Project on Pancer Beach by PT. Aura Sinar Baru and access to the main attractions of Pulau Merah Beach. This triggers heavy vehicles that pass the permit limit required to cross the Jalan Pancer, including tourism buses, material transport trucks and heavy equipment transport trucks.

IOP Conf. Series: Materials Science and Engineering **854** (2020) 012014 doi:10.1088/1757-899X/854/1/012014

Jalan Pancer was damaged early, the last road improvement project was carried out in year of 2014 with the age of the road plan is 5 years, but in 2016 (only 2 years old) the condition of the road was damaged such as cracks to large holes. Often the occurrence of this overload makes the road damaged earlier, so it is not in accordance with the age of the planned road plan. Then we should be able to take preventive actions on the pavement planning or improvement efforts and the planned age if this overload indication is known. Therefore, research is needed to evaluate the overload on age existing roads in the Pancer Road section using the Vehicle Damage Factor (VDF) calculation formula and the cumulative equivalent standard or Commulative Equivalent Standard Axle (CESA). So that it can be seen decreasing the life of the road plan.

From **Figure 1** below we can see the condition of Jalan Pancer, Pesanggaran subdistrict, Banyuwangi Regency, which is being passed by trucks carrying material to the Pier Development Project on Pancer Beach. This road has a length of ± 2 km and a width of 4 meters of cross section of the road.

Figure 1. Pancer Road Conditions

1.2. Research Objectives
To find out the aged reduction prediction of the road plan that occurs, which is caused by an overload on the Pancer Road, Pesanggaran, Banyuwangi.

2. Theoretical Basis
2.1. Road Pavement Construction.
Pavement is a layer that is located between the basic soil layers and vehicle wheels which functions to provide services to transportation facilities and during the service period is expected to not occur significant damage so that the pavement is in accordance with the expected quality, then knowledge of the nature, procurement and processing of Road pavement making materials are needed [1].
Based on the binding material, road pavement construction can be distinguished including:

a) Flexible pavement construction, which is pavement using asphalt as a binding material. The layers of pavement are carrying and spreading the burden of traffic to the subgrade.
b) Rigid pavement construction, which is pavement using cement as a binder. Boneless concrete plates are placed on the subgrade without a foundation layer. Traffic loads are mostly borne by concrete slabs.
c) Composite pavement construction (composite pavement), i.e. rigid pavement combined with flexible pavement can be either flexible pavement above rigid pavement or rigid pavement over flexible pavement.

2.2. Road classification
The operation of roads must provide roads that provide a sense of security and safety [2]. Operation of the Road will immediately repair the damaged road without seeing who has the biggest impact [3]. The implementation of regional roads can overcome the problems of regional roads by making arrangements, infrastructure, construction and supervision of infrastructure, through various activities such as: [4]
1. Inventory of road service levels and problems
2. Compilation of plans and their implementation programs and determination of the level of road services
3. Planning and optimizing utilization of road sections
4. Geometric improvement of roads and intersections
5. Determination of road class for each section
6. Feasibility test of road functions

Road classification according to road function is divided into 3 types, namely: [5]
a) Arterial Roads, Roads that serve major transportation with characteristics of long distance travel, high average speeds and the number of access roads are efficiently restricted.
b) Collector Roads, Roads that serve transporter / divider transport with characteristics of medium distance travel, average speed is medium and the number of entry roads are limited,
c) Local Roads, Roads that serve local transportation with the characteristics of short distance travel, low average speed and the number of access roads unlimited. *The Age of The Road Plan*

The planned life is the number of years since the road was opened for vehicle traffic until a repair is needed which is structural (until a pavement layer overlay is needed). During the life of the plan, the road maintenance must be carried out such as nonstructural coating which functions as a wear out layer.[1]

2.3. Vehicle Classification
The vehicle classifications are physically distinguished based on dimensions, weight and performance. Vehicle dimensions affect the width of the traffic lane, the width of the hardened road shoulder, the length and width of the parking space[7].

2.4. Vehicle Damage Factor (VDF)
Road damage or better known as Vehicle Damage Factor (VDF), is one of the parameters that can determine pavement thickness significantly and if the weight of the vehicle (especially truck type vehicles) especially with overload load, the VDF value will significantly increase.

The burden of pavement construction has special characteristics in the sense that it has different principles from the burden on other constructions outside of road construction. Understanding the special characteristics of the road pavement construction burden is very important in further understanding, especially relating to pavement construction design, pavement construction capacity, and the process of construction damage concerned.

2.5. Traffic Volume
Traffic volume is the number of vehicles that pass a certain point or line on a cross section of the road. Traffic volume enumeration data is information needed for the planning, design, management and road operation phases.[1]

Traffic volume shows the number of vehicles that crossed one observation point in a unit of time (days, hours, minutes). In relation to determining the number and width of lanes, the unit of traffic volume commonly used is average daily traffic, planning hour volume and capacity.[1]

Traffic volume data can be obtained from routine posts around the location. Vehicle load data retrieval is done by calculating from the axis of the vehicle. The calculation of daily traffic volume is adjusted to the conditions in which departures and departures, for example: school, work, shopping, and evening / evening recreation differ from one another so that they get actual conditions in the field and

IOP Conf. Series: Materials Science and Engineering **854** (2020) 012014 doi:10.1088/1757-899X/854/1/012014

all problems that arise can be summarized well [7]. To search for Average Daily Traffic: can be calculated using the following **Equation 1 and Equation 2**:

$$LHR = (1 + i)^n \text{ x P} \quad (1)$$

$$ADT\,smp = (\text{ ADT }) \text{ x Factor equivalent} \quad (2)$$

Wich is :
ADT = average daily traffic (vehicles/ day)
ADT *smp*= average daily traffic equivalents in a passenger car unit (passenger car unit/day)
i = Traffic growth factor
n = year to n
P = The amount of traffic volume at the beginning of the planned life

2.6. Traffic Loads
The dimensions, weight of the vehicle, and the loaded load will cause compressive force on the vehicle axis. The compressive force is applied to the pavement surface and will contribute to the road surface [8]. Based on this value, the Standard Load configuration is derived for the following axis configurations at **Figure 2** below: [9]

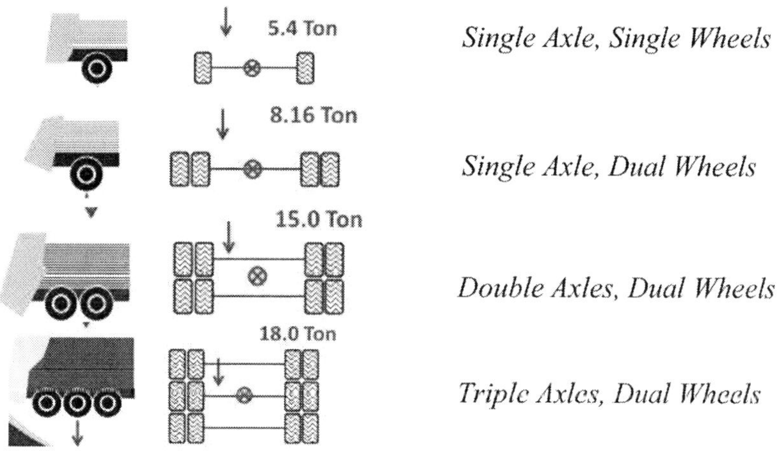

Figure 2. Standard Axis Load Configuration

We can follow the distribution of axis load from various types of vehicles on the empty weight, weight with maximum load and maximum total weight in each type of vehicle where asphalt pavement loads are compressed [10]

The axle load of each vehicle is affected by the location of the vehicle points that vary according to the load of the vehicle, according to the **Equation 3.** below :

$$\text{E empty truck} = \text{E front axis} + \text{E rear axis} \quad (3)$$

2.7. Cumulative Equivalent Numbers
The number of equivalent trajectories over the life of the design can be determined through the cumulative equivalent axle standard (CESA) load axis. To determine the cumulative equivalent standard load axis over the design life, we use **Equation 4**:

$$CESA = LHR \times 365 \times E \times C \times N \quad (4)$$

Which is :
CESA : Load cumulative equivalent standard axis
N : Number of each type of vehicle
365 : Number of days in a year
E : Axis load equivalent
C : Vehicle distribution coefficient
N : Design life relationship factors that are adjusted to traffic development

To find the distribution coefficient (C), it can be seen from **Table 1** according to the criteria of Jalan Pancer, Pesanggaran subdistrict, Banyuwangi. [12]

Table 1. Number of Vehicle Lines and Coefficient (C)

Number of Paths (n)	Light Vehicles *) (<5ton)		Heavy vehicle**) (>5ton)	
	1 way	2 way	1 way	2 way
1 lane	1,00	1,00	1,00	1,00
2 lane	0,60	0,50	0,70	0,50
3 lane	0,40	0,40	0,50	0,475
4 lane	-	0,30	-	0,45
5 lane	-	0,25	-	0,425
6 lane	-	0,20	-	0,40

*) Total weight <5 tons, for example: passenger cars, pick ups, delivery cars
**) Total weight ≥ 5 tons, for example: bus, truck, tractor, semi trailler, trailer.

2.8. Traffic Growth Factors

Traffic growth factor is the number of vehicles that use the road from year to year which is influenced by regional development, increased public welfare and increased ability to buy vehicles. Traffic growth factors are expressed in percent / year. Traffic growth factors can be obtained from the analysis of traffic data, population development, income per capita, regional master plan and others. To calculate traffic growth over the life of a road plan you can use **Equation 5**:

$$F = P\,(1 + i)^n \tag{5}$$

Wich is :
F = Traffic Volume year to n
i = Traffic growth factor
n = year to n
P = The amount of traffic volume at the beginning of the planned life

2.9. Overload

Overload is a condition where a vehicle is loading goods with an amount that exceeds the allowable amount (the heaviest axle load standard)[14]. Overload will cause the Cumulative Equivalent Single Axle Load plan to be reached before the year of the age road plan according to the design. Generally, every road construction has been designed accordingly with applicable design criteria [15]. But so far the overload is still rarely considered in any road design. Because overloading can shorten the service life of the road, this VDF (Vehicle Damage Factor) needs to be included in the pavement design process.

3. Methodology

3.1. Data Collection
There are two kinds of data that needed, among others are primary data and secondary data.

3.1.1. Primary Data
Primary Data is data obtained from direct observation in the field. To get the value of reducing the age of the road plan on Pancer Road, several surveys are needed to get the required data, including:

a. Preliminary survey : preliminary survey to find out the general condition of pancer roads, road classifications, road facilities, and road dimensions. based on preliminary surveys, the classification data of the pancer road is classified as class III road with a cross section width of 4 meters and has the heaviest source load of a maximum of 8 tons. The pancer road is the main access to the tourism and industrial areas, the type of area is a residential area with several shops on the side of the road. This road was repaired in 2014 along 1 km. The pancer road type is 2/2 UD (consisting of 2 lanes, 2 directions, undivided). The dimensions of the road drainage are on the right lane are 0.8 m and 1 m on the left lane, the average right shoulder of the lane is 1.8 m, and 2m for the left lane. Given the width of the road that is not too broad, it will disturb and cause inconvenience for motorists. In addition, this road is often traversed by material transport vehicles, carrying project equipment and other large vehicles.

b. Average daily traffic volume survey: to get traffic volume data on the pancer road. The following **Table 2** is the average traffic volume data survey results on the pancer road:

Table 2. Data from the average daily traffic volume survey

No	Vehicle Type	ADT(Annual Daily Traffict)							ADT (Vehicle/day)	%
		Mon day	Tues day	Wed	Thurs day	Friday	Satur day	Sunday		
1	Light Vehicle	129	132	139	131	114	153	169	138	48,54
2	2 axis truck	242	223	260	234	15	6	7	141	49,55
3	3 axis truck	4	2	6	1	3	2	0	3	0,90
4	Large Bus	0	0	6	2	0	4	8	3	1,00
	Total	375	357	411	368	132	165	184	285	100

3.1.2. Secondary Data. Secondary data is compiled data obtained from related agencies or bodies. Secondary data obtained are ADT data plan, road class data, Aged plan data plan load data vehicle axis data, vehicle type data.

3.2. Techniques of Survey Methods
There are mainly 2 methods by which we can collect data through the survey method.

3.2.1. Preliminary survey. By conducting direct observations in the field, by measuring the dimensions of the road, recording road facilities, and the condition of existing roads

3.2.2. Average Daily Traffic Survey. Conducting of Average Daily Traffic surveys according to conditions where departing and returning schedules, for example: school, work, shopping, and evening/evening recreation differ from one another so that they get actual conditions in the field and all problems that arise can be summarized well, the survey was conducted for 7 days with a 24 hour a day using manual counters tool. The Average Daily Traffic Survey was conducted in the 3rd (three) year from the beginning of the construction of the Pancer road.[7]

IOP Conf. Series: Materials Science and Engineering **854** (2020) 012014 doi:10.1088/1757-899X/854/1/012014

3.2.3. Vehicle Load Surveys. Axis load is influenced by the axle configuration and vehicle load. For example, two of the same vehicles have different axial loads due to differences in load, thus also different in their equivalent numbers. This survey was conducted for 7 days with a 24 hour a day.[1]
In carrying out a vehicle load survey on Jalan Pancer, the type of vehicle for 2-way trucks heading to the Pier Development Project on Pancer Beach obtained vehicle load data from the Project's weighbridge (PT. Aura Sinar Baru). In the type of light vehicle data obtained from the website www.mday.info.com, while Trucks 3 the weight transfer data is obtained from the website www.indotrucker.com and in the type of large Bus vehicles obtained from PT. Abiyan Trans Banyuwangi.

3.3. Data Processing
After primary data and secondary data are obtained, the next step is :
3.3.1. Analyze data using 2 scenarios: Scenario 1 is to find out the value of CESA at the end of the life of the road plan, using ADT data and vehicle weight based on planning data. This scenario is considered a normal condition and is used as a basis for analysis. Scenario 2 is to find out the actual CESA value over the life of the plan, using ADT data from the survey results and vehicle axle loads obtained from the survey results manually

3.3.2. Results and Discussion. After processing the data, it can be seen the effect of overload on the life of the plan and can be known the equivalent value of the VDF (Vehicle Damage Factor) used for planning with VDF survey results which the greater the VDF value, the greater the road damage factor and can find out decrease in road service life from the beginning of construction by interpolation.

4. Results and Discussion

4.1. Comparison of Annual Daily Traffict (ADT) and Traffic Growth
Comparison of planning ADT data with survey ADT data and the magnitude of traffic growth factors can be seen in **Table 3** below.

Table 3. Traffic Growth

No	Vehicle Clasification	ADT First year	ADT Survey 3rd year	traffic growth
		veh/day	Veh/day	
1	Light Vehicle	72	138	0,2426
2	2 Axis Truck	16	141	1,0655
3	3 Axis Truck	0	3	-
4	Large Bus	0	3	-

4.2. Evaluation of Vehicle Loads
Based on the results of the survey of vehicle weight, it was concluded that the weight of the vehicle survey results were much greater than the weight of the planned vehicle based on the standard vehicle axle load, which means that the Pancer Road had overloaded. Vehicle weight data can be seen at **Table 4.**

Table 4. Vehicle Weight Data on the Pancer Road Section

No	Vehicle Clasification	vehicle weight (ton)	
		Design Data	Survey Data
1	Light Vehicle	2	5,1
2	2 Axis Truck	18,2	19,68
3	3 Axis Truck	25	37,72
4	Large Bus	9	13,09

4.3. The comparison of Equivalent Numbers or VDF (Vehicle Damage Factor)

VDF (Vehicle Damage Factor) is one of the parameters that can determine the thickness of the pavement that quite significant and if the vehicle will be heavier (especially overloaded vehicle), the VDF value will significantly increase. According to the type of vehicle Equivalent Number Calculation of vehicles or VDF in scenario 1 which is obtained distribution of Axis Load from Various Types of Vehicles based on Manual Perkerasan Jalan No. 01/MN/BM/83. In the second scenario VDF calculation is calculated by adding up the equivalent number of each axis of each vehicle survey results.

VDF calculation for scenario 2 in light vehicle type with axle configuration (1,1), vehicle weight 5.1 tons with front axle load distribution (As 1) 2.55 tons single axle single wheel and (As 2) 2.55 tons single axis single wheel.

$$VDF = DF_{sb\_depan} + DF_{sb\_belakang}$$
$$= 1\left(\frac{2,55}{8,16}\right)^4 + 1\left(\frac{2,55}{8,16}\right)^4$$
$$= 0,0095 + 0,0095$$
$$= 0,0191$$

VDF calculation for scenario 2 on type 2 axle truck with axle configuration (1,2), vehicle weight 19.68 tons with front axle load distribution (As 1) 6.6912 tons single axle single wheel and (As 2) 12.9888 tons of rear axle double wheels.

$$VDF = DF_{sb\_depan} + DF_{sb\_belakang}$$
$$= 1\left(\frac{6,6912}{8,16}\right)^4 + 1\left(\frac{12,9888}{8,16}\right)^4$$
$$= 0,4521 + 6,4197$$
$$= 6,8718$$

VDF calculation for scenario 2 on 3-axis truck type with axle configuration (1.22), vehicle weight 37.72 tons with front axle load distribution (As 1) 9,4300 tons single-axis single wheel and (As 2) 28,2900 tons of rear axle double wheels.

$$VDF = DF_{sb\_depan} + DF_{sb\_belakang}$$
$$= 1\left(\frac{9,4300}{8,16}\right)^4 + 0,086\left(\frac{28,2900}{8,16}\right)^4$$
$$= 1,7836 + 12,4242$$
$$= 14,2078$$

VDF calculation for scenario 2 on large bus type with axis configuration (1,2), vehicle weight 13.09 tons with front axle load distribution (As 1) 4.4506 tons single axis single wheel and (As 2) 8.6394 tons rear axle double wheel.

$$VDF = DF_{sb\_depan} + DF_{sb\_belakang}$$
$$= 1\left(\frac{4,4506}{8,16}\right)^4 + 1\left(\frac{8,6394}{8,16}\right)^4$$
$$= 0,0885 + 1,2565$$
$$= 1,3450$$

Recapitulation of VDF values for scenario 2 on vehicles crossing Pancer Street, Pesanggaran subdistrict, Banyuwangi, can be seen in **Table 5**.

Table 5. Recapitulation of the VDF value in Scenario 2

No	Vehicle Clasification	Vehicle Weight Survey Data	Load per axis		Angka Ekivalen (VDF)		
			Front Axis	Rear Axis	E Front Axis	E Rear Axis	Amount E
1	Light Vehicle	5,1	2,550	2,550	0,010	0,010	0,0191
2	2 Axis Truck	19,68	6,691	12,989	0,452	6,420	6,8718
3	3 Axis Truck	37,79	9,448	28,343	1,797	12,517	14,3135
4	Large Bus	13,09	4,451	8,639	0,088	1,257	1,3450

The results of the calculation of VDF scenario 2 are compared with the scenario VDF 1. Comparison of the VDF values is the quotient between the VDF values from the survey results with the planned VDF values. This comparison is intended to see the extent of differences in the destruction factors of each type of vehicle, while the results of the comparison can be seen in **Table 6** below.

Table 6. Comparison of VDF (Vehicle Damage Factor) Values

No	Vehicle Clasification	VDF		
		Scenario 1	Scenario 2	Difference
1	Light Vehicle	0,0005	0,0191	3.815%
2	2 Axis Truck	5,0264	6,8718	137%
3	3 Axis Truck	2,7416	14,2078	518%
4	Large Bus	0,3006	1,3450	447%
	Rata-rata	2,0173	7,4577	1.230%

Based on Table 6, there is a comparison of VDF values for each type of vehicle that is> 100% due to the scenario 2 VDF value (survey) not in accordance with the VDF value of scenario 1 (planning), it can be concluded that all types of vehicles are overloaded so that the road damage factor is greater and causes the Pancer Road to experience damage early or before the planned life.

4.4. Cumulative Equivalent Numbers

The pavement design age can be analyzed based on the cumulative Equivalent Standard Axle (CESA) of each scenario. CESA analysis is the analysis of traffic by substituting the LHR value, the vehicle equivalent factor or the coefficient of damage required with ESAL (Equivalent Single Axle Load) units. CESA values are calculated annually from the first year to the end of the planned life. The age of the pavement plan can be analyzed based on the results of the CESA analysis in each scenario. If it is assumed that the value of CESA at the end of the life of scenario 1 plan is the end of the service life period, then the value of CESA at the end of the life of scenario 1 will be achieved in what year in scenario 2

4.4.1. Comparison of CESA values in scenario 1 and scenario 2. The data used in the scenario 1 uses a growth factor of 8.2% per year according to planning data from the Banyuwangi District Transportation Department with a 5-year road planning life and the function of the road is a primary artery with 2 2-way lanes. Based on these data a vehicle distribution coefficient (C) can be obtained by 0.5 while for finding the value of design age factors that have been adjusted for traffic growth (N). In the calculation of the next year, it is only different at n / age of the plan according to what year is calculated. The following **Table 7** shows the differences in the value of CESA in scenario 1 and scenario 2 from the year of 2015 – 2019. CESA values in both scenario 1 and scenario 2 show the value of CESA experienced a significant increase considering the increasing number of LHR and the weight of the vehicle at scenario 2 that was overloaded.

Table 7. CESA Values in Scenario 1 and Scenario 2

No	Year	CESA Values	
		Scenario 1	Scenario 2
1	2015	15.285,69	63.877,19
2	2016	74.516,12	806.462,32
3	2017	188.944,93	6.093.369,97
4	2018	378.738,42	36.432.677,00
5	2019	667.589,80	199.101.861,50

4.4.2. Remaining Life (RL). The CESA value at the end of the design life of scenario 1 is 667,589.80 SAL, while the value of CESA scenario 2 is 199,101,861.50 SAL which is 298.24 times greater than the value of the CESA plan or scenario 1.

$$Remaining\ Life\ \text{(RL) scenario 1} = 100 \times 1 - \left(\frac{CESA\ 2018}{CESA\ 2019}\right)$$

$$= 100 \times 1 - \left(\frac{378.738{,}42}{667.589{,}80}\right)$$

$$= 43{,}27\%$$

$$Remaining\ Life\ \text{(RL) scenario 2} = 100 \times 1 - \left(\frac{CESA\ 2018}{CESA\ 2019}\right)$$

$$= 100 \times 1 - \left(\frac{36.432.677{,}00}{199.101.861{,}50}\right)$$

$$= 81{,}70\%$$

From the results of the calculation of remaining life of the plan, it is known that in scenario 1 and scenario 2 the remaining pavement life for the next 5 years is 43.27% and 81.70% which means the pavement service for 5 years is reduced, this happens because the burden that passes through pavement increases every year.

4.4.3. CESA Values Recapitulation

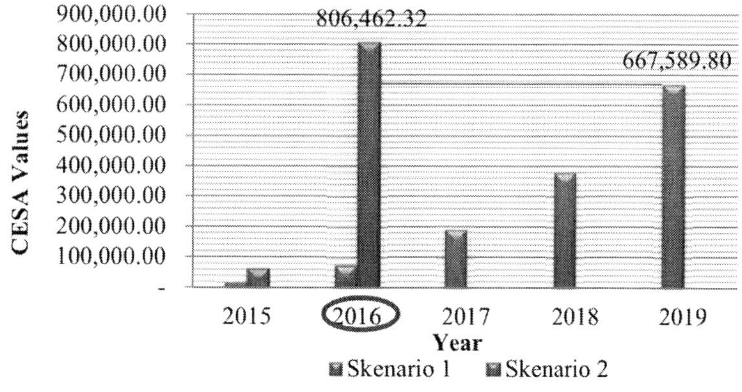

Figure 3. CESA Value Recapitulation

The chart (Figure 3) gives information that from the year of 2016 the results of the CESA value of scenario 2 already exceeded the CESA value of 2019 in scenario 1 (value of planned life). If it is assumed that the value of CESA at the end of the life of scenario plan 1 is the end of service life, then by interpolation we get a reduction in road life (in scenario 2) 3 years and 2.3 months from the planned life, meaning that the pavement age will end at 1 year 9.7 months from when the road was opened.

5. Conclusions

Based on the calculation of the value of VDF, obtained differences in each type of vehicle > 100%. This means that all types of vehicles are overloaded. The results of the average value of scenario 2 VDF is greater than scenario 1, then it can be stated that the road damage factor is greater. Based on CESA analysis, the pavement is planned with a planned age of 5 years and receives a burden of 667,589.80 SAL. If calculated with overload conditions, the planned life is only up to 1 year 9.7 months or there is a decrease in the life of the road plan by 3 years 2.3 months with a load of 806,462.32 SAL in the second year. So it can be concluded that Jalan Pancer suffered early damage before the planned age ended. Therefore, road pavement reconstruction is needed in accordance with the actual load that occurs in the field.

References

[1] Sukirman,S.1999.*Perkerasan Lentur Jalan Raya* (Bandung:Nova)

[2] Pemerintah Indonesia.2009.*Undang-Undang Republik Indonesia Nomor 22, Pasal 23 Tahun 2009 tentang Lalu Lintas dan Angkutan Jalan* (Lembaran RI Tahun 2009 Jakarta : Sekretariat Negara)

[3] Pemerintah Indonesia.2009.*Undang-Undang Republik Indonesia Nomor No 22, Pasal 24 Tahun 2009 tentang Lalu Lintas dan Angkutan Jalan* (Lembaran RI Tahun 2009 Jakarta : Sekretariat Negara)

[4] Pemerintah Indonesia.2009.*Undang-Undang No 22, Pasal 8 Tahun 2009 tentang Lalu Lintas dan Angkutan Jalan* (Lembaran RI Tahun 2009 Jakarta : Sekretariat Negara)

[5] Bina Marga.1997.*Tata Cara Perencanaan Geometrik Jalan Antar Kota* (Jakarta: Departemen Pekerjaan Umum Direktorat Jenderal Bina Marga)

[6] Departemen Pekerjaan Umum, 1997.*Manual Kapasitas Jalan Indonesia.* (Jakarta: Ditjen Bina Marga)

[7] Alamsyah, A. A.2008. *Rekayasa Lalu Lintas*. Malang: Universitas Muhammadiyah Malang

[8] Idris, M., Amelia, S., dan Cahyadi, U. 2009. *Karakteristik Beban Kendaraan Pada Ruas Jalan Nasional Pantura Jawa dan Jalintim Sumatera.* (Bandung : Kolokium Hasil Penelitian Dan Pengembangan Jalan dan Jembatan, Pusat Penelitian Jalan dan Jembatan. Kementrian Pekerjaan Umum)

[9] Kementerian Pekerjaan Umum. 2014. *Spesifikasi Umum 2010 Revisi 3.* (Jakarta : Direktorat Jenderal Bina Marga)

[10] Bina Marga. 1983.*Manual Perkerasan Jalan dengan alat Benkelman beam No. 01/MN/BM/83* (Jakarta: Direktorat Jenderal Bina Marga)

[11] Anonim. 2002. *Majalah Teknik Jalan & Transportasi No. 101*

[12] SNI 03-1732-1989 *Tebal perkerasan lentur jalan raya dengan metode analisa komponen, Petunjuk pelaksanaan* (Jakarta: Badan Standarisasi Nasional)

[13] Departemen Pekerjaan Umum. 2005. *Jumlah Lintasan Ekivalen Selama Umur Rencana Dapat Diketahui Melalui Kumulatif Ekivalen Beban Sumbu Standar / CESA (Cumulative Equivalent Single Axle Load)* (Jakarta : Direktorat Jenderal Bina Marga)

[14] Pemerintah Indonesia.1993.*Peraturan Pemerintah Republik Indonesia No 43 Tahun 1993 tentang Prasarana dan Lalu Lintas Jalan* (Lembaran RI Tahun 2009 Jakarta : Sekretariat Negara)

[15] Martina, R., M.Saleh, S., dan Isya, M. 2015. *Kajian Beban Aktual Kendaraan pada Konstruksi Jalan menggunakan Weigh In Motion.* (Jurusan Teknik Sipil:Universitas Syiah Kuala)

IOP Conf. Series: Materials Science and Engineering **854** (2020) 012015 doi:10.1088/1757-899X/854/1/012015

Inhibition of Free Fatty Acids generation in Crude Palm Oil during storage by using UV-C Light treatment

S D Said[1*], T Maimun[1], T M Asnawi[1] and A Muslim[1]

[1]Bioprocess Engineering Laboratory, Chemical Engineering Department, Faculty of Engineering, Syiah Kuala University, Darussalam-Banda Aceh, 23111, Indonesia

[*]E-mail:syahiddin@che.unsyiah.ac.id

Abstract. During the lengthy storage of crude palm oil (CPO), free fatty acid (FFA) concentration generated in the oil tends to increase which will affect the quality of CPO as the result of lipase enzymatic degradation of triglycerides. Currently, the conventional thermal processes in CPO storage tanks still applied to reduce the FFA formation. However, the thermal method is not very effective in reducing FFA for prolonged heat exposed due to deteriorate the CPO quality and high energy consuming. The alternative UV-C light technique gives some advantages compare with thermal treatment i.e less energy consumption, simple method, cheaper investment and low-cost operation. This study is aimed to evaluate the effects of UV-C light on inhibiting FFA generation in CPO during storage. The evaluation of UV-C light effect on the inhibition of FFA generation was conducted by irradiating CPO samples with various UV-C light exposure time. The FFA level of CPO samples was analyzed by using sodium hydroxide solution before and after treatment. The results show that UV-C light has the ability to deactivate lipase enzyme and consequently inhibit FFA generation in CPO.

1. Introduction

The world palm oil trade is a $60 billions industry with Indonesia and Malaysia produce 85% of the world's supplies [1]. It is predicted in years 2040-2050 CPO production in Indonesia will reach 65 million tons per year [2]. In order to respond the over capacity, the utilization of CPO as a renewable raw material for biorefinery scheme production system tend to increase continuously considering of its environmental benign. Conversion of CPO into biodiesel fuel in Indonesia and in other main CPO producing countries like Malaysia is reaching at a spectacular increase recently. This fact is mostly driven by implementing new paradigm of sustainable development in order not only to reduce the utilization of fossil fuel which produce greenhouse gases (GHGs) emission affecting global warming, but also in term of their national trade and fiscal policy [5].

Quality of CPO is determined by parameters, including free fatty acid (FFA), which needs to be between 2 – 5% [3]. High FFA level may degrade the CPO [4]. The CPO storage tank system of a palm oil mill is very important to maintain the level of FFA in CPO not increasing and being in the desired level of maximum 5% (maximum standard specification) during long storage or transportation time. The greater than 5% FFA level in CPO would decreased its selling price [2]. The average FFA content in CPO products when leaving palm oil mill is 2.3 – 7.2% [3,6]. The increased levels of FFA in CPO are caused by the triglyceride hydrolysis reaction which is catalyzed by lipase enzyme. Lipase enzyme existed in CPO is originated from microbes activities in damage oil-palm fruits during harvesting [4] and the contamination of fungi in CPO [6]. In order to overcome the increase in FFA level, the current enzyme deactivation treatment which has been widely used is by heating the CPO with steam coil and maintaining its temperature in the range of 40–60 ^{0}C [7,8]. Unfortunately, the conventional process has not been able to completely stop the increase in FFA formation. On the other hand, the prolonged heating treatment could destruct the composition of CPO [9].

Recently, UV-C light has been widely utilized to maintaining the quality and content of food ingredients. UV-C light has capability on killing microbes and destructing enzymes structure that acting as destructive agents in food [10]. The use of UV-C light in preventing changes in material quality has advantages compared to the use of heat treatment, i.e the simple system option, cheaper and energy efficient. However, the utilization of UV-C light for destruction of lipase enzyme in CPO is still rarely reported by researchers. The aim of this research is to investigate the effect of UV-C light exposure treatment on CPO in inhibiting the generation of FFA during storage period. In addition, the destruction of lipase enzyme as the effect of UV-C light exposure treatment was also evaluated.

2. Materials and Method

2.1. Experimental set up

Experiments were carried out in a cube shaped cardboard box with an iron frame. The size of the box is 20 (h) x 94 (l) x 30,5 (w) cm. the inside part is painted black. A low pressure UVC lamp of 20 W, wavelength of 253.7 nm, and the lamp length 59 cm (Sankyo Denki Japan) is placed on the top side of the box. The inside temperature of the box when irradiation took place was around 33 °C. Each CPO sample of 100 ml in a beaker glass of 300 ml was placed on the bottom side of the box. The distance between the light source (UV-C lamp) and the surface of the samples is 15 cm.

2.2. UV-C light and lipase enzyme decay

The following steps are arranged to evaluate the effect of UV-C light exposure on lipase enzyme decay:

1. Preparation of acetate buffer and Arabic gums solution (as surfactant)

Acetate buffer solution was prepared by mixing 178 ml of acetic acid solution (0.1 N) with 72 ml of sodium hydroxide solution (0.2 N) in a beaker glass and then stirring until homogeneous. A surfactant solution was prepared by mixing 50 g Arabic gum into 100 ml of distilled water. Then the mixture was stirred while heated until homogeneous [11].

2. CPO hydrolysis with unexposed and exposed enzyme to UV-C light

For CPO hydrolysis using unexposed enzyme, an enzyme solution which is a mixture of 0.5 ml lipase enzyme and 100 ml of acetate buffer was poured into a 300 ml beaker glass, then 5 ml of CPO and 2 ml of Arabic gum solution was added into the solution. Subsequently, the mixture was stirred for 1.5 h and then the FFA level generated in CPO was analyzed. For CPO hydrolysis using exposed enzyme, the enzyme solution was exposed first to UV-C light at desired time prior to adding CPO and Arabic solution [11].

Lipase enzyme decay as an effect of UV-C light exposure was evaluated. The percentage of enzyme decay is calculated by the following equation:

$$\%\ Enzyme\ decay = \frac{Eu-Ee}{Eu}$$

Where: *Ee* = increased level of FFA due to hydrolysis of CPO catalysed by exposed enzyme
Eu = increased level of FFA due to hydrolysis of CPO catalysed by unexposed enzyme

2.3. UV-C light treatment and exposure time

The UV-C light treatment on the inhibition of FFA generation in CPO was conducted in two ways. Firstly, CPO samples were exposed to UV-C light at various time then the samples were placed in a room temperature of 30 °C and kept for 3 weeks. FFA level in CPO were analysed every 1 week (7 days). The experiment was carried out as follows; a 100 ml CPO sample filled in a 300 ml glass beaker then each sample was placed under a UV-C lamp to be exposed for 2, 4, 6, and 8 hours. After the light exposure process, the sample was then placed into a room with a temperature of around 30 °C for 3 weeks and the level of FFA generated in CPO was analysed. For blank test, 100 ml CPO sample in a 300 ml beaker without UV-C light exposure was placed directly in the same room for 3 weeks. Secondly, CPO samples were exposed repeatedly to UV-C light for 2, 4, 6, and 8 hours every day for 3 weeks and then the FFA level generated in CPO was analysed. The effect of the UV-C light treatment on the inhibition of FFA generation was determined as the value of the percentage of FFA inhibition, which is defined as the ratio between the value of the difference of the FFA levels increased in unexposed CPO and that of exposed one to the value of the increase in FFA levels in unexposed CPO. The equation of percentage of FFA inhibition calculation is written as:

$$\%\ FFA\ inhibition = \frac{Yu-Ye}{Yu}$$

where: Yu = increased level of FFA in unexposed CPO
Ye = increased level of FFA in exposed CPO

2.4. Determination of FFA level in CPO
The FFA level was determined by sodium hydroxide titration. Prior to analysis, the CPO stock in 10 L plastic container was poured into a 5 L beaker glass and heated to 60 °C and then stirred until homogeneous. The 5 g of CPO sample from a glass was placed into an Erlenmeyer flask and added with 50 mL of 95% ethanol. The mixture was heated to 40 °C and stirred until all the CPO was completely emulsified. Subsequently, the 1-2 drops phenolphtalein indicator was added into the mixture. The sample was titrated by adding 0.1 N NaOH solution until pink colour developed. The titration volume which indicated the percentage of FFA (w/w) content in the oil was recorded and FFA concentrations was calculated using the following equation [12] :

$$\%FFA = \frac{NM(Va - Vb)}{W}$$

where : N = NaOH normality; Va = the volume of NaOH used in sample titration (mL); Vb = the volume of NaOH used in blank titration (mL); W = sample mass (g); M = molecular mass of the predominant fatty acid (25.6).

3. Results and Discussion

3.1. Effect of UVC light exposure on lipase enzyme deactivation
Lipase enzyme catalyzes hydrolysis of CPO into FFA [4], and UV-C light has an ability to decay enzymes [5]. Prior to evaluating the effect of UV-C light on inhibition of FFA generated in CPO during storage it was needs to characterize the effect of UV-C light on lipase decay. The percentage of lipase decay determined by measuring FFA levels formation in CPO due the addition of pure lipase (control) and UV-C light exposed lipase. Lipase activity test under various UV-C exposed time was conducted to investigate its ability on hydrolysing triglycerides in CPO. Figure 1 revealed that FFA levels increased from its initial value of 7 up to 30%, with an increasing of 23%. Irradiation of UV-C light on lipases decreased its ability in activating CPO hydrolysis. The effect of UV-C light exposure on enzyme deactivation is shown in Figure 1. The longer the exposure time of UV-C light to the enzyme, the lower the increase in FFA levels. The Increased levels of FFA with exposure times of 2, 4, 6, and 8 h was 21.56, 8.33 and 2.24 and 0.56%, respectively. The formation of FFA tends to terminated for exposure time greater than 6 hours. These evidences indicated that UV-C light have decayed the enzyme structure and deactivated enzyme to catalyzing more triglycerides for further hydrolysis. The percentage of enzyme decay depends on exposure time. The longer the exposure time the more enzymes decay. The percentage of enzyme decay with exposure time of 2, 4, 6, and 8 h was 6.52, 63.78, 90.26 and 97.56%, respectively. In comparison, Jeon et al reported the use of intense pulsed light (IPL) irradiation in lipase

IOP Conf. Series: Materials Science and Engineering **854** (2020) 012015 doi:10.1088/1757-899X/854/1/012015

deactivation. Increasing pulse fluence and exposure time decreases activity. IPL induced deactivation causes fragment which then causes changes in tertiary structure of lipase [3].

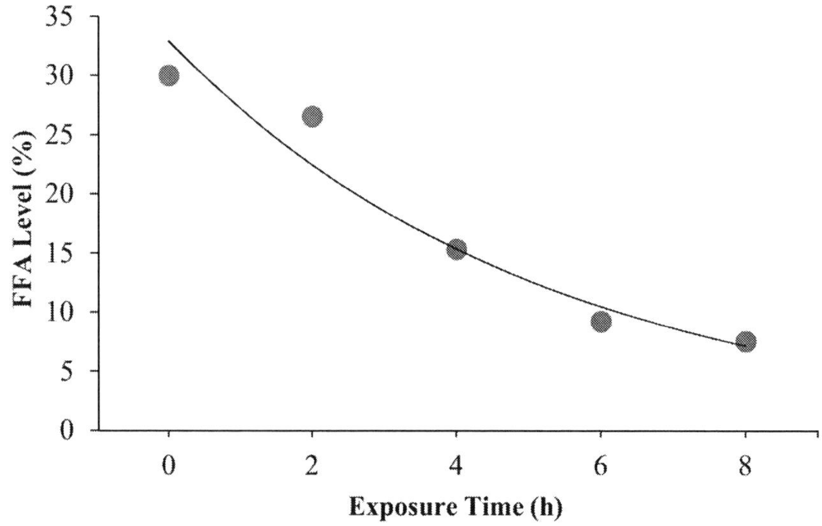

Figure1. Effect of UV-C exposure time on lipase enzyme decay

3.2. Effect of UV-C light irradiation on inhibition of FFA generation

The effect of UV-C light irradiation on the inhibition of FFA generation in CPO was carried out by exposing UV-C light to CPO samples for various times, after that all samples were stored in the room with a temperature of around 30 °C. The FFA generation during storage was monitored by analysing FFA level in each sample every week for 3 weeks of storage. The generation of FFA in CPO increased over time during storage. Figure 2 shows the profile of FFA levels in CPO. The increased FFA levels in CPO that were not exposed to UV-C light after 3 weeks (21 days) of storage time is 1%, this value is higher than those of the exposed one. The longer exposure time of UV-C light to CPO resulted the lower FFA levels. The Increased FFA levels in the exposed samples (6, 12, 18, 24 h) after 3 weeks of storage time was 0.8, 0.7, 0.55 and 0.39%, respectively. The percentage inhibition of FFA generated in CPO as an effect of UV-C light treatment was increased with the increasing UV-C light exposure time. For the exposure time of 6, 12, 18 and 24 h, the percentage FFA inhibition were 10, 22, 38, and 56% respectively.

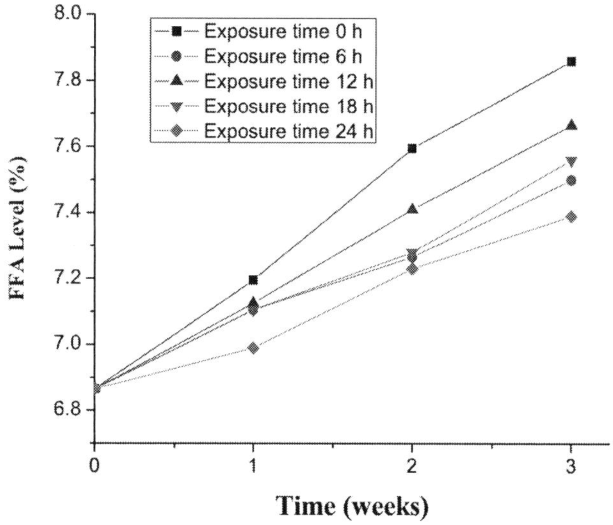

Figure 2. Effect of UV-C exposure time and storage time on FFA level in CPO

3.3. Effect of repeated UV-C light exposure on inhibition of FFA generation in CPO

UV-C light was exposed to each CPO sample repeatedly for; 2, 4, 6 and 8 hours per day for 3 weeks of storage time. The FFA level in each sample was determined by sodium hydroxide titration. During 3 weeks of storage, the not exposed (control samples) FFA content in CPO was increased from 8.12% to 9.2% or an increase of 1.1% from its initial value. Figure 3 shows a decrease in the generation of FFA in CPO as the effect of repeated UV-C light exposure. The values of decreased in FFA levels in CPO depend on time of exposure. Increased exposure time inhibit the increase in FFA generation. The increase in FFA levels in each CPO sample subjected to UV-C exposure time of 2, 4, 6 and 8 hours per day was 0.60, 0.38, 0.23, and 0.13%, respectively. This evidence showed that the UV-C treatment was able to inhibit the increase of FFA level in CPO. Regarding to the inhibition, increasing the exposure time would increase the percentage of inhibition of FFA generation in CPO. The percentage of inhibition of each CPO sample exposed to UV-C light at varied exposure time of; 2, 4, 6, and 8 hours per day was 45.45, 65.45, 79.10, and 87.27%, respectively. The experimental results showed that the inhibition profile of FFA generated in CPO (Fig.3) is in agreement with lipase enzyme deactivation test (Fig.1). These results confirmed that the inhibition of FFA generation in CPO is as a result of lipase enzyme decay by UV-C light exposure.

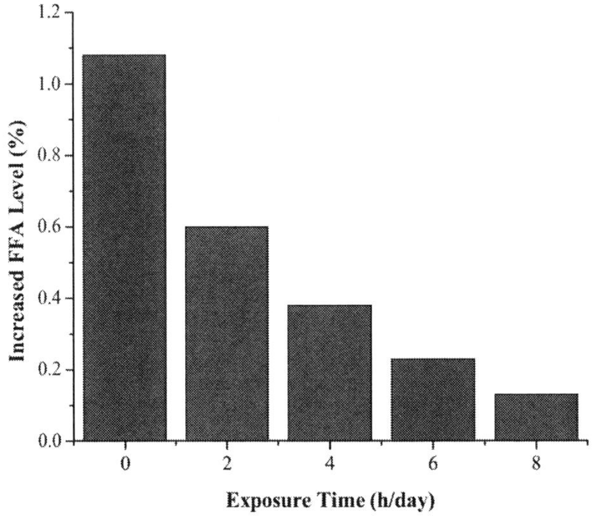

Figure 3. Effect of repeated UV-C exposure time on the increased FFA level in CPO

4. Conclusions

UV-C light can affects lipase enzyme decay to inhibit FFA formation in CPO. The experimental results revealed that the observed variables; UV-C light exposure time significantly contributed on lipase enzyme decay. The percentage of enzyme decay was increased with the increasing UV-C light exposure time. The percentage inhibition of FFA generated in CPO at various UV-C light exposure time of 2, 4, 6 and 8 hours per day for 3 weeks was 45.45, 65.45, 79.10, and 87.27%, respectively. It can be concluded that UV-C light treatment is one of a promising simple method which is able to inhibiting excessive amount of FFA formation in CPO during storage.

Acknowledgements

The authors would like to express our deep gratitude to Mr. M.A Peruzzi and Mr. K. Jodi for their valuable assistance with experimental works at Bioprocess Engineering Laboratory of Chemical Engineering Department, Syiah Kuala University and our thanks are also extended to PT. Karya Tanah Subur, West Aceh, Indonesia for providing CPO sample for this works.

References

[1] https://www.aljazeera.com/ajimpact/malaysia-fighting-billion-dollar-palm-oil-industry-190821144611519.html
[2] C. Gunning-Trant, Y. Sheng, P. Hamshere, T. Gleeson and B. Moir, 2015, What Indonesia wants: Analysis of Indonesia's food demand to 2050, ABARES research report no 15.9, Canberra, November. CC BY 3.0.
[3] M-S. Jeon, K-M. Park, H. Yu, J-Y. Park and P- Chang, 2019, Effect of Intense Pulsed Light on the Deactivation of Lipase: Enzyme-deactivation Kinetics and Tertiary Structural Changes by Fragmentation, *Enzyme and Microbial Technology* ,
[4] O.H. Henry , 2011. Monitoring the free fatty acid level of crude palm oil stored under light of Different Wavelenghts. *American J. of Food Tech., 6: 701-704.*
[5] N.H. Azeman, N.A. Yusof and A.I. Othman, 2015, Detection of free fatty acid in crude palm oil, *Asian J. of Chem*, **27**, No.5, 1569-1573.
[6] L.S Teoh, O. Lasekan, N.M. Adzahan and N. Hashim, 2016, The effect of ultraviolet treatment

on enzymatic activity and total phenolic content of minimally processed potato slices, *J Food Sci Technol*, **53** (7), 3035–3042

[7] A. Khusairi et.al, 2018, Oil palm economic performance and R&D progress in 2019, *J. of Oil Palm Res*. **30** (2), 153-195

[8] D. Cowan, H.C. Holm and H.S. Yee, 2018 Reduction in free fatty acids in crude palm oil by enzymatic remediation, *J.* of *Oil Palm Res*. **24**, 1492-1496

[9] B. Saad et.al., 2006, Determination of free fatty acid in palm oil samples using non-aqueous flow injection titimetric method, *Food Chem*., **102**, 1407-1414

[10] M. Hafiz, R.A. Renjani, A. Haryanto, N. Araswaty and IDM. Subrata, 2016 , Design of temperature and volume control system at crude palm oil (CPO) storage tank, *Proc.of AEASAP, 95–102*

[11] I.M. Noor, M. Hasan and K.B. Ramachandran, 2003, Effect of operating variables on the hydrolysis rate of palm oil by lipase, *Process Biochemistry*, **39**, 13-20

[12] L.S.Teoh, O. Lasekan, N. M. Adzahan, N. Hashim, 2016, The effect of ultraviolet treatment on enzymatic activity and total phenolic content of minimally processed potato slices, *J Food Sci Technol* , **53**(7), 3035–3042

IOP Conf. Series: Materials Science and Engineering **854** (2020) 012016 doi:10.1088/1757-899X/854/1/012016

D-shape Fiber Coated with Indium Tin Oxide for Temperature Sensor Application

B Nizamani[1], M I M Abdul Khudus[2], E Hanafi[1] and S W Harun[1]

[1]Photonics Engineering Laboratory, Department of Electrical Engineering, Faculty of Engineering, University of Malaya, 50603 Kuala Lumpur, Malaysia.
[2]Department of Physics, Faculty of Science, University of Malaya 50603 Kuala Lumpur, Malaysia.

E-mail: nizamanibilal@yahoo.com

Abstract. We designed a temperature sensor by using a D-shape fiber coated with indium tin oxide as a sensing probe. D-shape fiber was fabricated out of conventional single mode fiber by a homemade setup using wheel polishing technique. Insertion loss of D-shape fiber was about 3.5 dB. ITO was deposited over the D-shape fiber by electron-beam deposition technique to have precise deposition of 60nm thin film over the D-shape region of fiber to be utilized as a sensor. It is observed that the transmitted power increased with the increase in temperature with a sensitivity of 0. 0013 dBm/ °C. The proposed sensor also provides a good linearity and stability.

1. Introduction

Temperature sensing has been deployed in many commercial applications over the past few decades such as fire detection, leakage detection, exploration of oil and gas, transmission line monitoring and power cables monitoring. Fiber sensors have recently attracted the researchers due to its predominated ability of being immune to electromagnetic interference. D-shape fibers have been widely used for sensing as they offer benefit of strong evanescent field interaction with metals, semiconductors and metal oxides. Evanescent field interaction varies with the change in surrounding refractive index, thus we coat the evanescent field exposed region of fiber with certain metals, semiconductors or metal oxides. High evanescent field interaction is a vital element for sensing where a good amount of power is needed to interact with the surrounding medium [1].

Indium Tin Oxide (ITO) has transmittance in near-infrared and visible region as high as 90% and electrical resistivity as low as 2×10^{-4} Ωcm [2, 3]. It is widely used in variety of applications such as organic light emitting diode (O-LED) [4], liquid crystal lens [5], solar cells [6] and even the sensors such as ammonia gas sensor [7] and humidity sensor [8]. In contrast to other deposition techniques like vapor deposition [9], sputtering [5], electro spinning [10]. ITO was deposited using electron beam deposition technique to have a control over the accuracy in ITO thickness and to get ease in morphology of deposited structure by getting a uniform thickness of ITO thin film over the D-shape fiber.

2. D-shape fiber preparation and deposition of ITO

We prepared a D-shape fiber out of single mode fiber SMF-28 have core diameter of 9 μm and cladding diameter of 125 μm. The preparation was done using a setup similar to [11], where the fiber buffer was removed first, then it was cleaned and placed over the polishing wheel for fabrication. The polishing wheel was covered with a sandpaper of 1000cw grit size for the side polishing of a fiber. Fiber was then connected to amplified stimulated emission (ASE) for the light source and Optical power meter (OPM) (Thorlabs: PM100D) for the power measurement during the process, as shown in figure 1. In order to obtain a great evanescent field interaction to be useful as a sensor, we measured insertion loss to be 3.5dB while the diameter of fiber was found to reduce at the D-shape region to 71.11 μm from 125 μm which ensured the fiber was side polished and certain amount of cladding was removed as shown in figure 2.

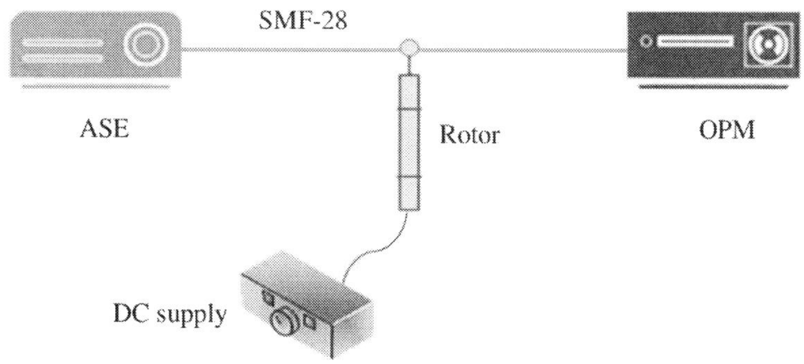

Figure 1. Preparation of D-shape fiber.

Figure 2. 71.11 μm D-shape fiber.

After the preparation of D-shape fiber, it was taken over the glass slide and placed in the electron-beam deposition chamber for the deposition of ITO over the region of side polished fiber. ITO slot was pre-set to be used for deposition and electron beam was applied on ITO to evaporate and form a 60nm layer over the D-shape region. Figure 3 shows the microscopic image of ITO deposited over the D-shape fiber taken with the help of Medilux-12 microscope. Thus, the 60nm thin film of ITO was deposited over a D-shape fiber for the sensing purpose.

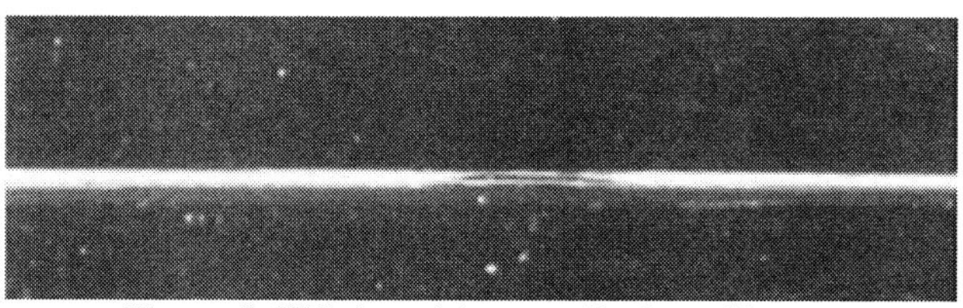

Figure 3. ITO deposition over D-shape fiber

3. Experimental setup

The ITO D-shape fiber was tested for temperature sensor in similar way as of [12] by using a hot plate for temperature measurements. The D-shape fiber coated with ITO was connected to a laser source (ANDO: AQ4321D) at one end and to OPM at other end. Our newly prepared sensor was placed over the hot plate and the multimeter (proskit MT-1860) was connected just beside the sample to have exact temperature readings over the hot plate. Figure 4 illustrates the experimental setup and sensing mechanism of newly prepared D-shape fiber coated with ITO based on evanescent field interaction from the D-shape fiber which is sensitive to the variation of surrounding refractive index. ITO deposited over the D-shape fiber worked as a temperature change detector over the hot plate.

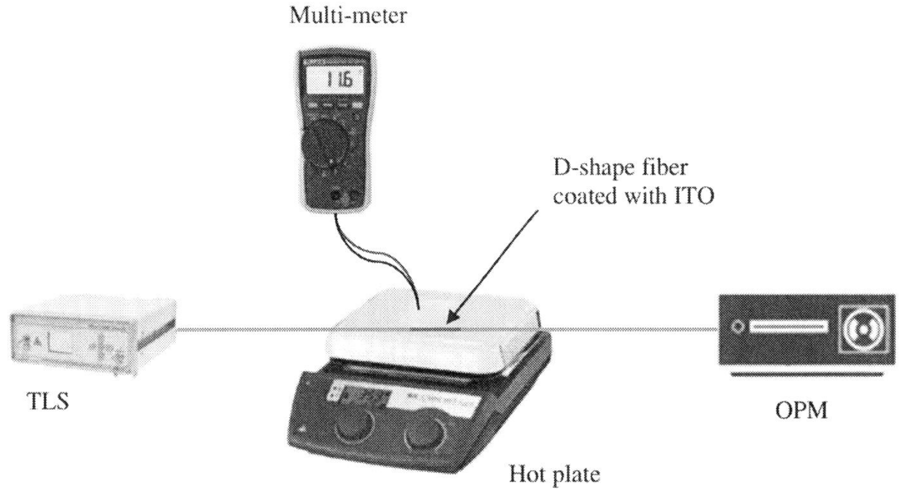

Figure 4. Temperature sensing setup with D-shape fiber coated with ITO.

4. Result and discussions

From the experimental setup in figure 4 the laser source was set to the standard optical communication wavelength of 1550nm. Output power was measured using OPM and a linear increment with increase in temperature was observed in real time. When the temperature was increased from a room temperature of 30 °C to 110 °C, a significant change in output power ensured the sensitivity of our fabricated D-shape fiber. The output power spectrum showed increase in transmittance due to the ITO coating of D-shape fiber. The transmittance trend increased linearly when the temperature raised from 30°C to 110°C, as the output power increased from -5.485 dBm to -5.3688 dBm ensuring the sensitivity to be 0.0013 dBm/ °C with 99% linearity in results as shown in figure 5. The output power readings showed complete stability of power at 80° C for about 300 seconds which proves that our sensor was stable for a long time as shown in figure 6. With the advantage of using a tunable laser source, wavelength sweep was done from 1549.5 nm to 1550.5 nm and stability with wavelength sweep of 1 nm range from 1549.5 nm to 1550.5 was ensured as shown in figure 7. This experiment was repeated for 3 times and similarity of results was obtained with a standard deviation of 0.05. Table 1 sums up the complete response of our newly fabricated sensor based on D-shape polished fiber coated with ITO. However, the fabricated D-shape fiber without coating of ITO did not respond significantly to the temperature sensing.

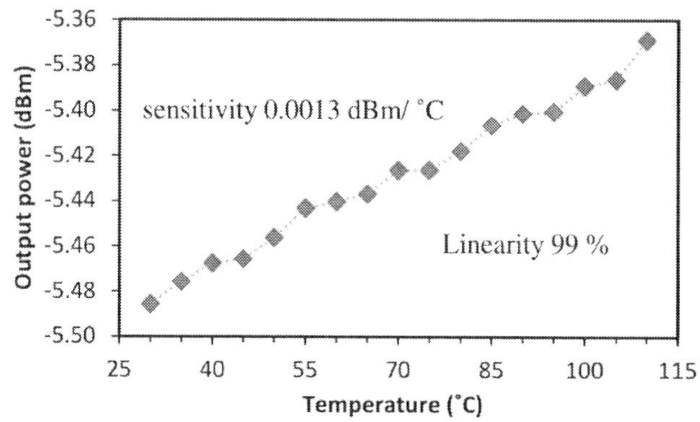

Figure 5. Output power change with the change in temperature.

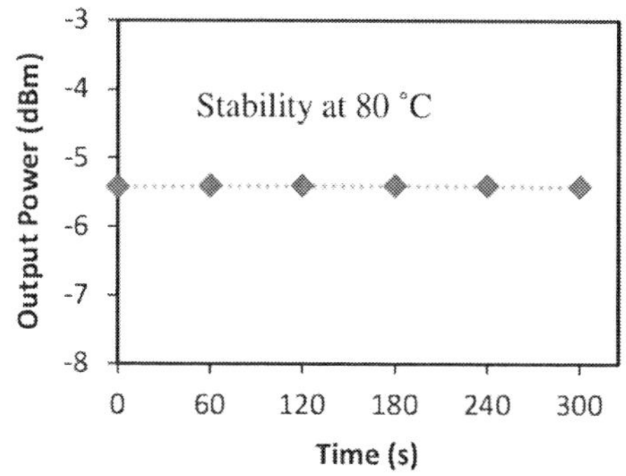

Figure 6. Stability of sensor at 80 °C for 300 seconds at 1550 nm wavelength.

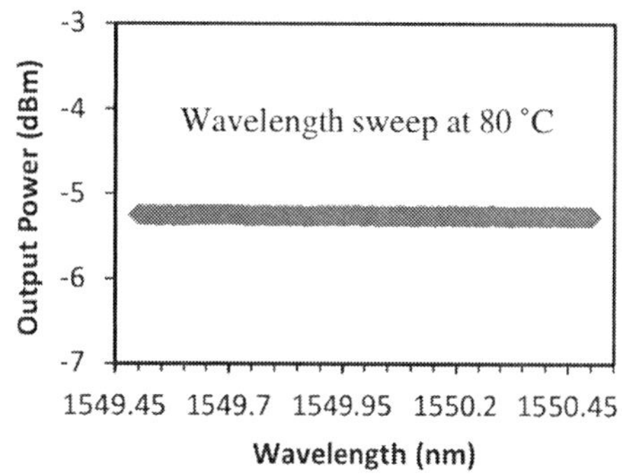

Figure 7. Power stability of sensor at 80 °C in wavelength sweep of 1 nm at the precision of 0.01 nm.

Table 1. Sensing performance parameters of the ITO coated D shape fiber temperature sensor.

Parameters	Values
Linearity	99%
Sensitivity	0. 0013 dBm/ °C
Standard deviation	0.05

5. Conclusion

By fabricating a D-shape fiber out of a single mode silica fiber and coating it with indium tin oxide, we successfully demonstrated a temperature sensor with a highly stable and linear response to the temperature changes recorded from 30°C to 110°C. Thus this simple temperature sensor can be advantageous for variety of temperature sensing applications in highly electromagnetic environments such as industrial applications and transmission line monitoring in the high electromagnetic radiation conditions.

References

[1] Chen GY, Ding M, Newson T and Brambilla G. A review of microfiber and nanofiber based optical sensors. *The Open Optics Journal*. 2013;7(**1**).

[2] Kim H, Gilmore aC, Pique A, Horwitz J, Mattoussi H, Murata H, et al. Electrical, optical, and structural properties of indium–tin–oxide thin films for organic light-emitting devices. *Journal*

of Applied Physics. 1999;86(**11**):6451-61.

[3] Chen C-W, Lin Y-C, Chang C-H, Yu P, Shieh J-M and Pan C-L. Frequency-dependent complex conductivities and dielectric responses of indium tin oxide thin films from the visible to the far-infrared. *IEEE Journal of Quantum Electronics*. 2010;46(**12**):1746-54.

[4] Park M, Chon BH, Kim HS, Jeoung SC, Kim D, Lee J-I, et al. Ultrafast laser ablation of indium tin oxide thin films for organic light-emitting diode application. *Opt Laser Eng*. 2006;44(**2**):138-46.

[5] Hsu CJ, Agrahari K, Selvaraj P, Chiang WF, Huang CY, Manohar R, et al. Application of ultra-thin indium–tin–oxide film in liquid crystal lens. *Optics & Laser Technology*. 2019;119:105603.

[6] Sousa M and da Cunha A. Optimization of low temperature RF-magnetron sputtering of indium tin oxide films for solar cell applications. *Appl Surf Sci*. 2019;484:257-64.

[7] Lin C-W, Chen H-I, Chen T-Y, Huang C-C, Hsu C-S, Liu R-C, et al. On an indium–tin-oxide thin film based ammonia gas sensor. *Sensors and Actuators B: Chemical*. 2011;160(**1**):1481-4.

[8] Babu BM and Vadivel S. High performance humidity sensing properties of indium tin oxide (ITO) thin films by sol–gel spin coating method. *Journal of Materials Science: Materials in Electronics*. 2017;28(**3**):2442-7.

[9] Fung M, Wong K, Chen X, Chan Y, Ng A, Djurišić A, et al. Indium oxide, tin oxide and indium tin oxide nanostructure growth by vapor deposition. *Curr Appl Phys*. 2012;12(**3**):697-706.

[10] Wu X, Wang Y and Yang B. Corn-like indium tin oxide nanostructures: fabrication, characterization and formation mechanism. *Applied Physics A*. 2015;121(**3**):1179-85.

[11] Jafry A, Arof H, Kasim N, Munajat Y, Apsari R and Harun SW. Q-switched Erbium-Doped Fiber Laser Using MoS 2 Deposited Side-polished D-shape Fiber. *Nonlinear Optics, Quantum Optics: Concepts in Modern Optics*. 2018;49.

[12] Jali MH, Rahim HRA, Ashadi MJM, Thokchom S and Harun SW. Applied microfiber evanescent wave on ZnO nanorods coated glass surface towards temperature sensing. *Sensors and Actuators A: Physical*. 2018;277:103-11.

IOP Conf. Series: Materials Science and Engineering **854** (2020) 012017 doi:10.1088/1757-899X/854/1/012017

A Laboratory Scale Synthesis of Ethanol from Agricultural Waste as Bio-based Solvent for Waxy-Paraffinic Crude Oil Mitigation

M K Afdhol, Hidayat F, M Abdurrahman, U Z Husna, N P Sari, R K Wijaya

Department of Petroleum Engineering, Faculty of Engineering, Universitas Islam Riau. Jl. Kaharuddin *Nasution No.113 Km. 11, Pekanbaru, Riau – 28284, Indonesia.*

Email: Afdhol@eng.uir.ac.id

Abstract. Paraffin is a problem commonly faced in petroleum production. The presence of paraffin can inhibit the process of oil flow. Mitigation methods commonly used to overcome this problem is by using chemical injection, which is solvent. In this research, a laboratory experiment was conducted to make bioethanol from agricultural waste that are rice husks and corn cobs. This bioethanol functions as a solvent that is used to reduce the pour point in a waxy-paraffinic crude oil sample. Oil samples were mixed with bioethanol with different mixing ratios, including 5%, 10%, 15%, 20% 50%. The results showed that the alcohol content of the Husk and Corncob rise was 11% and 9% respectively. Then, the best result from mixing of bioethanol with oil reduced the pour point from 43.3 to 41 at mixing ratio of 50%.

1. Introduction

1.1. Wax Paraffin

The problem of paraffin wax is a very crucial problem in the oil field because it can cause oil to not flow when the temperature is below the pouring point (Abdurrahman, Ferizal, Husna, & Pangaribuan, 2018). Paraffin wax is part of hydrocarbons, which cannot be dissolved by crude oil under normal conditions, where the normal conditions of crude oil that have been tested are at a temperature of 70-150 °C and at a pressure of 55-103 Mpa (Wang, Wu, Creek, Shuler, & Tang, 2003).

There are several methods to overcome paraffin wax, one of which is the preventive method of heating and using chemicals, one of the most effective methods is using a hydrocarbon solvent (Afdhol, Lubis, & Siregar, 2019; Khaibullina, 2016). According to Al-yaari & Fahd (2011) for examples of injection schemes using solvents can be seen in Figure 1. The solvent injection device proposed by C-FER Technologies.

Figure 1. Wax (Huang, Wang, Li, Ren, & Zhu, 2017)

1.2. Bioethanol Production

Ethanol is produced through several processes, namely hydrolysis and fermentation using food ingredients that contain starch carbohydrates, such as rice and tubers (Cai, Liu, He, Chai, & Cai, 2018). Biomass containing lignocellulose can be used as a source of reducing sugars such as glucose, xylose and maltose (Mahboubi et al., 2017). In the next step, the alcohol produced must be turned into steam through a distillation process with a temperature of 78.5°C (Salehi, Taghizadeh-alisaraei, Shahidi, & Jahanbakhshi, 2018). According to (Cai et al., 2018) several stages of the biomass process will be carried out to produce ethanol, namely pretreatment, hydrolysis and fermentation. Biomass pretreatment is important because it is to get high ethanol yield. The purpose of pretreatment is to break the structure of lignocellulose so that cellulose is separated from lignin and thus cellulose. Hydrolysis is the process of breaking down polysaccharides in lignocellulosic biomass, which are cellulose and hemicellulose which will be broken down into sugar monomers. Perfect cellulose hydrolysis will produce glucose. Hydrolysis can be done chemically using acids or enzymatically. There are several factors that influence the hydrolysis process, namely carbohydrate content of raw materials, pH of hydrolysis, hydrolysis time, temperature and pressure. Fermentation is microbial activity in food so that the desired product is produced. Common microbes involved in fermentation are bacteria, yeast and mold.

There are important factors that will affect ethanol yield and efficiency. First, under physiological conditions the microbes are added to the media, which depends on the optimal optimum growth conditions of the microbes to be used. Second, the environmental conditions during the fermentation, i.e. pH and temperature. Bioethanol raw materials can be derived from cellulose source materials such as rice straw, bagasse, corncobs and rice husks (Said, Wahid, Diah, & Sabang, 2014). Utilization of biomass has been done a lot such as oil palm shells (Yuliusman, Nasruddin, Afdhol, Amiliana, et al., 2017), coffee grounds and others(Yuliusman, Nasruddin, Afdhol, Haris, et al., 2017). The production of bioethanol from cellulose waste has been developed, rice husk is one of the renewable raw materials used for making bioethanol as it is easily available and very economical (Kouteu Nanssou, Jiokap Nono, & Kapseu, 2016).

1.2.1. Rice Husk

Rice husk (Figure 2) contains several organic compounds namely, lignin, cellulose, hemicellulose, nitrogen compounds, vitamin B and organic acids and contains inorganic compounds in the form of silica (Ebrahimi, Caparanga, Ordono, Villaflores, & Pouriman, 2017). Rice Husk (RH), a broad covering of rice grain which accounts for around 20% of rice, has the potential to produce about 0.4 sugar per 1 gram of dry RH and the sugar process can further be fermented into bioethanol (Ebrahimi et al., 2017).

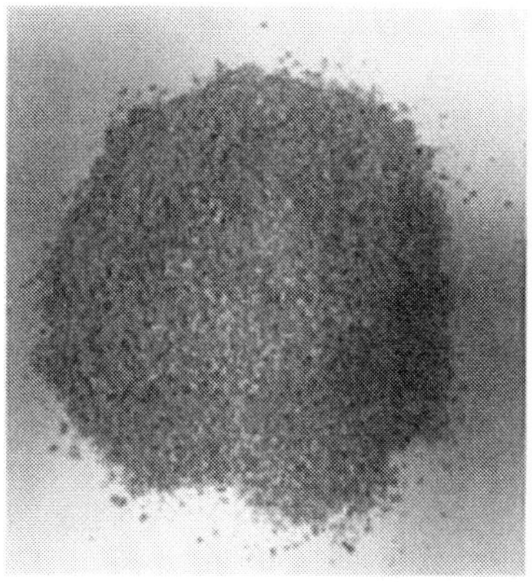

Figure 2. Rice husks (Cai et al., 2018)

Table 1. Composition of Rice Husks

Component	Rice Husks	Rice Bran	Broken Rice
starch	6.9	28.5	68.6
Cellulose	38.4	4.6	1.8
Ash	15.9	8.0	1.5
Hemicellulose	16.1	8.4	3.7
Lignin	16.7	2.8	-
Protein	3.4	12.9	9.9

Source : (Favaro et al., 2017)

It can be seen in Table 1 that the cellulose and hemicellulose content in rice husk is the largest at 38.4% and 16.1% so that it can be used as a source of bioethanol production.

1.2.2. Corn Cobs

In corn plants consist of corn kelobot, corn hair, corn cobs and corn kernels. Corn plants based on systematics (taxonomy) of plants are classified as follows (Sain, 2016).

Table 2. Taxonomic status of maize (Sain, 2016)

Regnum	*Plantae*
Devisio	*Spermatophyte*
Sub Devisio	*Angiospermae*
Classis	*Monocotyphyta*
Ordo	*Graminae*
Familia	*Graminaceae*
Genus	*Zea*
Species	*Zea mays L.*

Corn contains crude fiber which is composed of complex compounds of lignin, hemicellulose and cellulose (lignocellulose), and each one is a compound that can potentially be converted into other compounds biologically. The dry solid content of Corncob by using alkaline, solvent or ammonia based pretreatments produces the greatest amount of ethanol (19-22%), whereas mushroom based pretreatments produce much less (11%) (Zhao, Damgaard, & Christensen, 2018).

Figure 3. Corncob waste (Fachry, Astuti, & Puspitasari, 2013)

The complex compounds of lignin, hemicellulose and cellulose (lignocellulose) each are compounds that can potentially be converted into other compounds biologically and the following table compositions of these compounds from several materials:

Table 3. Composition of Cellulose, Hemicellulose and Lignin from different Lignocellulose (Byadgi & Kalburgi, 2016)

Lignocellulosic Material	***Cellulose*** **(%)**	***Hemicellulose*** **(%)**	***Lignin*** **(%)**
Miscanthus	40	18	25
Coastal Grass	25	35.7	9-18
Corncob	35-40	17-35	7-18
Wheat Straw	30	50	15

From the table above it can be seen that corn is a plant-based material with a compound that is easily converted into alcohol, so this study takes corn as a raw material for making bio-ethanol solvents.

2. Material and Methods

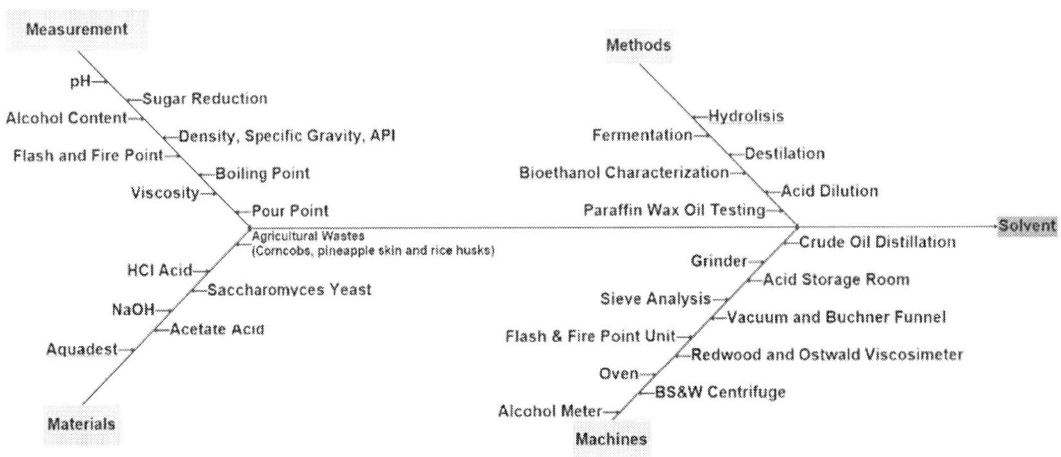

Figure 4. Research's Fishbone

IOP Conf. Series: Materials Science and Engineering **854** (2020) 012017 doi:10.1088/1757-899X/854/1/012017

2.1. Raw Materials
Materials used in this study comprises of rice husks, corncobs, acetic acid, chloric acid, sulfuric acid, NaOH, saccharomyces and distilled water.

2.2. Hydrolysis
Prepare materials that are ready for hydrolysis with sizes 30, 50, 80, 100 and 200 mesh. Enter the ingredients as much as 15 grams into a beaker (Hayuningtyas, Sk, & Sla, 2014). Then add a solution of sulfuric acid and hydrochloric acid as much as 150 ml each with a temperature of 100°C for 2.5-3 hours (Hayuningtyas et al., 2014) at 5% acid concentration. Repeat steps 2 and 3 for acid concentrations of 10%, 15% and 25%. The rice husk from hydrolysis is then filtered, where the hydrolyzed solution is the main product for the fermentation process (Osvaldo, S, & Faizal, 2012)

2.3. Fermentation
The rice husk hydrolysis substrate was ready to be fermented. Prepare anaerobic bacteria, the bacteria used are Saccharomyces Cerevisiae with pH conditions 3-4 (Azizah, N., Al-Baarri, A.N., dan Mulyani, 2012). Entering the results of hydrolysis of rice husk into the fermentation series that has been prepared. Add bread yeast (Utami & Windiyati, 2014) with a mass of 1.5 grams (Jayanti, 2011). Fermented for 5 days at 30°C. (Azizah, N., Al-Baarri, A.N., dan Mulyani, 2012)

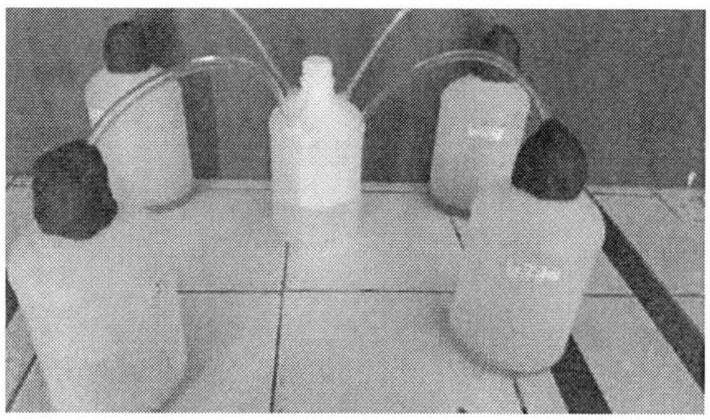

Figure 5. Fermentation Process

2.4. Distillation (ASTM D1160)
The process of distillation (purification) is the stage of separating a substance based on its boiling point. The following stages of distillation: Take a sample of the fermentation that has been done. Weigh flash containing sample + thermometer. Clean and dry the measuring cup, place the measuring cup so that the tip of the outlet goes into the measuring cup of the distillate. Close the measuring cup to prevent the steam from coming out of the tube. Circulate coolant through the condenser. Run to warm up and adjust the temperature slowly so that the distillation liquid drips properly. Stop heating at the end boiling point and allow the distillate liquid to drip on the measuring cup. After cooling, record the volume of the distillate produced.

IOP Conf. Series: Materials Science and Engineering **854** (2020) 012017 doi:10.1088/1757-899X/854/1/012017

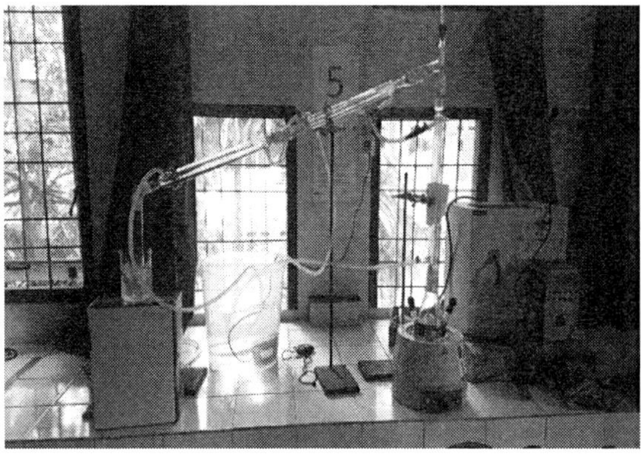

Figure 6. Distillation Tool Series

3. Calculation

3.1. Paraffin Wax Measurement

In this study using crude oil samples taken from field X and the following is the oil properties data in the field:

Table 4. PVT data and crude oil properties in field X

Information	Value	Unit
Oil density	30,8	API
Specific Gravity	0,8713	-
Water phase density	1,00028	gr/cm^3
Oil Viscosity	14,4	cp
Water Viscosity	0,5	cp
Pour Point	100	^{o}F
Flash Point	170	^{o}F
Reservoir Temperature	136	^{o}F
Bottom Hole Pressure	200-400	psig
Reservoir Pressure	532	psig
Bubble Point Pressure	113	psig
Oil formation volume factor	1,049	-
Initial Oil Saturation	63,1	%
Depth Formation	1100-1200	ft

Reservoir fluid composition is also one important parameter that needs to be known in order to be able to choose the type of solvent used and also to predict the formation of paraffin wax. The following is a table of reservoir fluid composition in field X:

IOP Conf. Series: Materials Science and Engineering **854** (2020) 012017 doi:10.1088/1757-899X/854/1/012017

Table 5. Composition of fluids in field X

Component	Percent of Mol	Percent Weight
Hydrogen sulphide	0,00	0,00
Carbon dioxide	0,56	0,08
Nitrogen	0,00	0,00
Methane	0,67	0,04
Ethane	0,67	0,07
Propane	1,51	0,22
Iso-Butane	0,90	0,18
n-Butane	1,57	0,31
Iso-Pentane	1,56	0,38
n-Pentane	1,50	0,36
Hexane	0,35	0,10
Heptane Plus	90,71	98,26
Total	100,00	100,00

The data above becomes a reference for laboratory experiments conducted to find out indications of paraffin wax formation by the methods / tests carried out as described below referred to ASTM D97; Pour the sample into a tube of 10 cc. Insert the tube in the oven and heat it to the test temperature. Remove the tube containing the sample from the oven while the sample is still hot. Observe temperature changes when all samples can be poured

4. Results and Discussion

4.1. Sugar Reduction

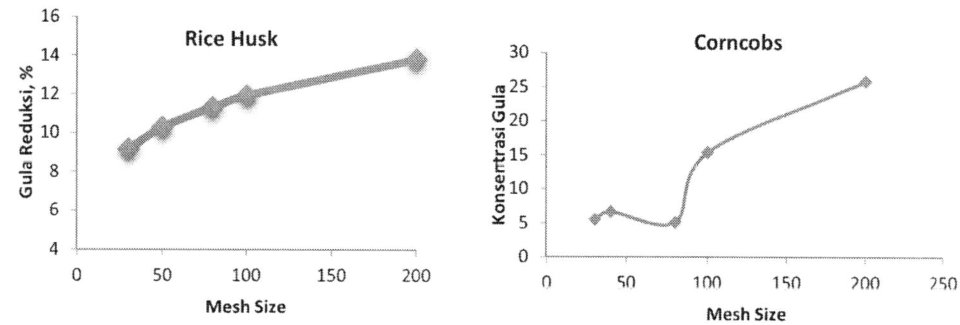

Figure 7. Results of Reducing Sugar Based on Mesh Size Differences

Analysis of the results of reducing sugars on different mesh sizes such as Figure 7 rice husk shows that the smaller the size of the mesh used, the resulting reducing sugars are getting bigger. It was proven that when the 30 mesh size produced reducing sugar 9.17, the mesh size 50 produced the reducing sugar 10.29, the mesh size 80 produced the reducing sugar 11.32, the mesh size 100 produced the reducing sugar 11.92 and the mesh size 200 produced the reducing sugar by 13.8. so it can be concluded that the smaller the mesh used it will produce a large reducing sugar.

Figure 7 corncob shows that the higher the mesh size (the finer) the reducing sugar content will be even higher while the size of 80 mesh will decrease because it can be caused by several factors such as errors during hydrolysis and the sample when tested experiencing changes in properties and after testing as much as 2 times the researcher regards it as an anomaly in the study because viewed from the trendline has shown a graph that increases with the mesh size.

IOP Conf. Series: Materials Science and Engineering 854 (2020) 012017 doi:10.1088/1757-899X/854/1/012017

According to Osvaldo et al., (2012) the size of the raw material used will affect the size of the porosity so that it can maximize contact between the material and the acid to increase the results of cellulose hydrolysis so that the smaller the size of the material used, the greater the yield of reduced sugar.

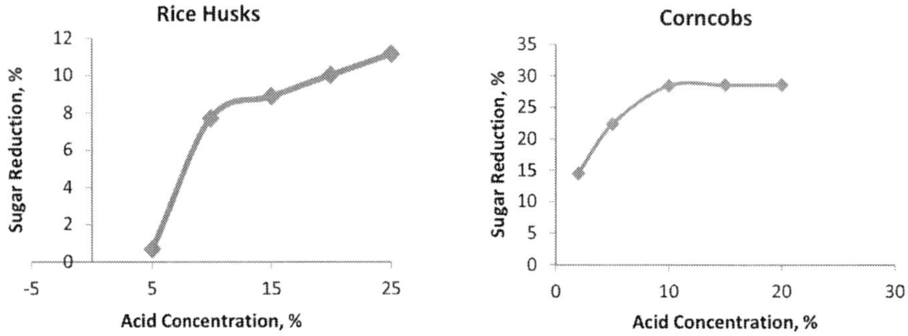

Figure 8. Results of Reducing Sugar Based on Differences in Acid Concentration

Meanwhile, for the observation of Figure 8 rice husk regarding the results of reducing sugars which is influenced by different acid concentrations, it is known that the higher the concentration of sulfuric acid used in the hydrolysis process, the results of the reducing sugars are also greater. Can be seen in the picture when the 5% acid concentration results in reducing sugar by 0.7, 10% acid concentration produces 7.72% reducing sugar, 15% acid concentration produces 8.9% reducing sugar, 20% concentration produces 10.04 reducing sugar and 25% concentration produces reducing sugar equal to 11.17.

From Figure 8 it can be seen that the higher the concentration of acid used, the higher the level of reducing sugar. The higher the reducing sugar levels, the more ethanol will be produced. the highest acid concentration is 20% HCL and also the advantage of using strong acids at low concentrations is that there is no need for acid recovery and the absence of acid ions lost in the process.

4.2. Alcohol Product from Fermentation

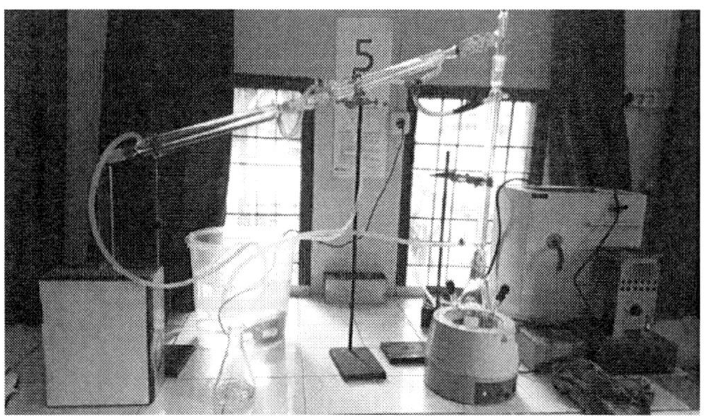

Figure 9. Distillation Circuit

In Figure 9 shows the entire distillation circuit consisting of a core series of sample containers, heaters, thermometers and measuring cups for the distillate reservoir. The boiling point of alcohol is around 78°C (Fachry et al., 2013; Jaya, Arthawan, & Sari, 2012; Salehi et al., 2018). Then the temperature in the distillation device is set to 78°C so that the resulting distillate is a type of bioethanol alcohol.
The percentage of alcohol in the rice husk distillate sample is 8%. In previous studies using rice husk fermented for 5 days produced alcohol by 14.4227% (Utami & Windiyati, 2014)

The bioethanol that has been made is measured by using alcohol meter alcohol meter and after measuring the alcohol content obtained from bioethanol from corncob waste by 11% where this is a fairly high alcohol content which based on research that has been done before the dry solid content of Corncob produce the greatest amount of ethanol (19-22%) (Zhao et al., 2018).

4.3. Alcohol Characterization of Synthetic Solvent and Bio-Solvent

Table 6. Characteristics of synthetic solvents and bioethanol

Properties	***Bioethanol***		**Parasol II**
	Rice Husk	**Corn Cob**	
Density, gr/cc	0.971	0,984	0.8528
SG	0.971	0,984	0.8528
°API	14.13	12.30	34.42
Color	Colorless	Colorless	Colorless
Odor	Aromatic	Aromatic	Aromatic
pH	4	4	9
Viscosity, cp	0.752	0.80	-
Flash Point, °C	48	45	16.1
Fire Point, °C	54	56	-

The advantage of bioethanol from agricultural waste is the value of the sample flash point where the rice husk flash point is 48°C, which means that at room temperature the sample has not been burned so it is safe to use on a laboratory scale. This is the advantage of bioethanol over parasol II, because parasol II has a flash point 16.1 which means below room temperature that is 30°C parasol II will be flammable so bioethanol is more safety than parasol II.

Bioethanol from corncob based on flash point and fire point testing can be said to be more safety for the application of production wells and production pipes, with a flash point of 45°C and a fire point of 56°C. This can be indicated that the bioethanol produced is non-flammable compared to synthetic solvents.

4.4. Paraffin Wax Test Results

Tests carried out to identify the success of solvents in overcoming the problem of paraffin wax in oil, namely testing the pour point (ASTM D97). In this test, two tests were conducted, namely testing the pouring point of oil before mixing the solvent and the pouring point of oil after mixing the solvent. The following table shows the results of tests conducted on three oil samples.

Table 7. Testing of Pour Oil Before Mixing Solvent

Simple Name	**Pour Point**	**Unit**
LGK (11-07-2018)	44	°C
LGK (24-04-2018)	43	°C
LGK (28-06-2018)	43	°C
Average	43,3	°C

In the results table above is the initial pour point temperature of oil containing paraffin wax before the solvent is mixed. Furthermore, to test the feasibility of the solvent from the corncob material that has been made, the solvent is mixed with the oil sample above with the amount of oil sample of 10 ml with a percentage of many solvents added namely 5%, 10%, 15%, 20%, 50% and 100% so we get the test results as in the table below:

Table 8. Testing of the Pour Point of Oil After Solvent Mixing

Sample Name	Pour Point					
	5%	10%	15%	20%	50%	100%
Sample I	44°C	44°C	44°C	43°C	42°C	41°C
Sample II	43°C	43°C	42°C	42°C	41°C	40°C
Sample III	43°C	42°C	42°C	41°C	40°C	39°C
Average	43,3°C	43°C	42,7°C	42°C	41°C	40°C

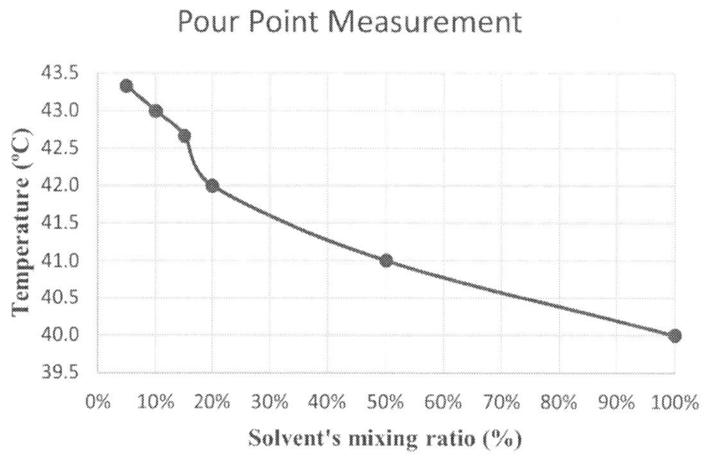

Figure 10. Pour Point Testing After being mixed with Solvent

Seen from the table and figure above, it can be seen that there is a change in value at the pouring point test before and after the solvent is mixed. Changes in the temperature of the pouring point of this oil can indicate the effect of mixing the solvent on the oil containing paraffin wax where the oil in the sample is average with an initial temperature of 43.3°C which means that oil for the first time can flow and by mixing the solvent with the pouring temperature oil into 40°C which shows that after being mixed with oil solvent it is easier to flow at a lower temperature than the initial temperature of the oil so that this can overcome the problem of paraffin wax on oil where with paraffin wax the oil becomes more difficult to flow but with the addition the oil pour point solvent decreases and even at low temperatures the oil can still flow more easily than oil without mixing with the solvent.

5. Conclusion

Based on the results of the analysis in the manufacture of bioethanol and the application of bioethanol to oils with high paraffin levels, the following conclusions are obtained:

1. The characteristics of the bioethanol produced can be seen from the density, SG, viscosity, PH, aroma and physical form that is almost the same as synthetic solvents and corncob itself has a match as bioethanol raw material because it produces an alcohol content of 11% and a risk husk of 9%.
2. Parameters that affect the manufacture of bioethanol such as reducing sugar content, particle size, acid concentration and cellulose content where the highest corncob reducing sugar content is 28.61%, optimal particle size is 200 mesh, optimal acid concentration is 20% HCl and cellulose content is corncob by 30-40%.
3. The average pouring point test results before mixing bioethanol is 43.30C then the oil sample is mixed with bioethanol with a mixing ratio of 5%, 10%, 15%, 20%, and 50% of the sample weight

so that the point temperature has decreased pour at the highest mixing ratio 100% to 40 ^{0}C or a difference of 3.3 ^{0}C.

References

[1] Abdurrahman, M., Ferizal, F. H., Husna, U. Z., & Pangaribuan, L. (2018). *Possibility of Wax Control Techniques in Indonesian Oil Fields*. In *AIP*. doi:10.1063/1.5028059

[2] Afdhol, M. K., Lubis, H. Z., & Siregar, C. P. (2019). Bioethanol Production from Tea Waste as a Basic Ingredient in Renewable Energy Sources. *Journal of Earth Energy Engineering; Vol 8 No 1 (2019): APRIL*. doi:10.25299/jeee.2019.vol8(1).2602

[3] Al-yaari, M., & Fahd, K. (2011). Paraffin Wax Deposition : Mitigation & Removal Techniques. *SPE International Symposium on Oilfield Chemistry 155412*. doi:10.2118/155412-MS

[4] Azizah, N., Al-Baarri, A.N., dan Mulyani, S. (2012). Pengaruh Lama Fermentasi Terhadap Kadar Alkohol, pH, dan Produksi Gas pada Proses Fermentasi Bioetanol dari Whey dengan Substitusi Kulit Nanas. *Jurnal Aplikasi Teknologi Pangan*, *1*(2), 72–77. doi:10.17728/JATP.V1I3.73

[5] Byadgi, S. A., & Kalburgi, P. B. (2016). Production of Bioethanol from Waste Newspaper. *Procedia Environmental Sciences*, *35*, 555–562. doi:10.1016/j.proenv.2016.07.040

[6] Cai, W., Liu, R., He, Y., Chai, M., & Cai, J. (2018). Bio-oil production from fast pyrolysis of rice husk in a commercial-scale plant with a downdraft circulating fluidized bed reactor. *Fuel Processing Technology*, *171*(November 2017), 308–317. doi:10.1016/j.fuproc.2017.12.001

[7] Ebrahimi, M., Caparanga, A. R., Ordono, E. E., Villaflores, O. B., & Pouriman, M. (2017). Effect of ammonium carbonate pretreatment on the enzymatic digestibility, structural characteristics of rice husk and bioethanol production via simultaneous saccharification and fermentation process with Saccharomyces cerevisiae Hansen 2055. *Industrial Crops and Products*, *101*, 84–91. doi:10.1016/j.indcrop.2017.03.006

[8] Fachry, A. R., Astuti, P., & Puspitasari, T. G. (2013). PEMBUATAN BIETANOL DARI LIMBAH TONGKOL JAGUNG DENGAN VARIASI KONSENTRASI ASAM KLORIDA DAN WAKTU FERMENTASI. *Jurnal Teknik Kimia*, *19*(1), 60–69.

[9] Favaro, L., Cagnin, L., Basaglia, M., Pizzocchero, V., van Zyl, W. H., & Casella, S. (2017). Production of bioethanol from multiple waste streams of rice milling. *Bioresource Technology*, *244*(17), 151–159. doi:10.1016/j.biortech.2017.07.108

[10] Hayuningtyas, S. R. I. K., Sk, H., & Sla, S. (2014). Produk bioetanol dari jerami padi (Oryza sativa) melalui hidrolisis asam dan fermentasi dengan Saccharomyces cerevisiae. *Bioteknologi*, *11*(1), 1–4. doi:10.13057/biotek/c110101

[11] Huang, Q., Wang, W., Li, W., Ren, Y., & Zhu, F. (2017). *A pigging model for wax removal in pipes*. In *SPE Production and Operations* (Vol. 32, pp. 469–475). doi:10.2118/181560-pa

[12] Jaya, M. A. S., Arthawan, G. K. A., & Sari, A. N. (2012). Potensi nira kelapa sebagai bahan baku bioetanol. *Jurnal Bumi Lestari*, *12*(1), 85–92.

[13] Jayanti, R. T. (2011). *Pengaruh pH, suhu hidrolisis Enzim A-Amilase Dan Konsentrasi Ragi Roti Untuk Produksi Etanol Menggunakan Pati Bekatul*.

[14] Khaibullina, K. (2016). *SPE-184502-STU Technology to Remove Asphaltene, Resin and Paraffin Deposits in Wells Using Organic Solvents*.

[15] Kouteu Nanssou, P. A., Jiokap Nono, Y., & Kapseu, C. (2016). Pretreatment of cassava stems and peelings by thermohydrolysis to enhance hydrolysis yield of cellulose in bioethanol production process. *Renewable Energy*, *97*, 252–265. doi:10.1016/j.renene.2016.05.050

[16] Mahboubi, A., Ylitervo, P., Doyen, W., De Wever, H., Molenberghs, B., & Taherzadeh, M. J. (2017). Continuous bioethanol fermentation from wheat straw hydrolysate with high suspended solid content using an immersed flat sheet membrane bioreactor. *Bioresource Technology*, *241*(17), 296–308. doi:10.1016/j.biortech.2017.05.125

[17] Osvaldo, Z. S., S, P. P., & Faizal, M. (2012). PENGARUH KONSENTRASI ASAM DAN WAKTU PADA PROSES HIDROLISIS DAN FERMENTASI PEMBUATAN

BIOETHANOL DARI ALANG-ALANG. *Jurnal Teknik Kimia*, *18*(2), 52–62.

[18] Said, M., Wahid, A., Diah, M., & Sabang, S. M. (2014). *Sintesis Bioetnaol Dari Jerami Padi (Oryza Sativa L) Melalui Fermentasi*. In *ISSN 2302-6030* (pp. 178–182).

[19] Sain, A. (2016). *KERAGAMAN GENETIK EMPAT VARIETAS JAGUNG (Zea mays. L) BERSARI BEBAS MENGGUNAKAN MARKA SSRs (Simple Sequence Repeats)*.

[20] Salehi, R., Taghizadeh-alisaraei, A., Shahidi, F., & Jahanbakhshi, A. (2018). Potentiometric of bioethanol production from cantaloupe waste (Magassi Neishabouri Cultivar). *Journal of the Saudi Society of Agricultural Sciences*. doi:10.1016/j.jssas.2018.05.006

[21] Utami, I., & Windiyati, L. (2014). Pembuatan Bioetanol Dari Sekam Padi Menggunakan Kombinasi Soaking in Aqueous Ammonia (Saa) Pretreatment – Acid Pretreatment – Hidrolisis – Fermentasi, *20*(1), 46–53.

[22] Wang, K. S., Wu, C. H., Creek, J. L., Shuler, P. J., & Tang, Y. (2003). *Evaluation of effects of selected wax inhibitors on paraffin deposition*. In *Petroleum Science and Technology* (Vol. 21, pp. 369–379). doi:10.1081/LFT-120018526

[23] Yuliusman, Nasruddin, Afdhol, M. K., Amiliana, R. A., Hanafi, A., & Rachmanda, B. (2017). *Preparation of activated carbon from palm shells using KOH and ZnCl2 as the activating agent*. In *IOP Conference Series: Materials Science and Engineering* (Vol. 180). doi:10.1088/1757-899X/180/1/012282

[24] Yuliusman, Nasruddin, Afdhol, M. K., Haris, F., Amiliana, R. A., Hanafi, A., & Ramadhan, I. T. (2017). Production of activated carbon from coffee grounds using chemical and physical activation method. *Advanced Science Letters*, *23*(6). doi:10.1166/asl.2017.8822

[25] Zhao, Y., Damgaard, A., & Christensen, T. H. (2018). Bioethanol from corn stover À a review and technical assessment of alternative biotechnologies. *Progress in Energy and Combustion Science*, *67*, 275–291. doi:10.1016/j.pecs.2018.03.004

IOP Conf. Series: Materials Science and Engineering **854** (2020) 012018 doi:10.1088/1757-899X/854/1/012018

Generation of Q-switched Erbium-Doped Fiber Laser Using Titanium Dioxide Film Based Saturable Absorber

N Ahmed[1], NH Kamarulzaman[1], NF Zulkipli[3], Z Jusoh[1], HA Rahman[2], SW Harun[3]

[1]Faculty of Electrical Engineering, University Teknologi Mara Cawangan Terengganu Kampus Dungun, 23000 Dungun, Malaysia
[2]Faculty of Electrical Engineering, University Teknologi Mara, Shah Alam 40450, Malaysia
[3]Photonics Engineering Laboratory, Department of Electrical Engineering, University of Malaya, 50603 Kuala Lumpur, Malaysia

Email: noriz207@uitm.edu.my

Abstract. This work demonstrates a generation of a passively Q-switched erbium-doped fiber laser with titanium dioxide (TiO_2) film as saturable absorber. The film is fabricated using a ratio of 2:1 by embedding the rutile nanoparticles (TiO_2) into a polyvinyl alcohol (PVA) host film. Short after placing the SA film on the fiber ferrule, a stable 1563.3 nm pulses appeared at a power pump range from 104.62 mW to 145.83 mW. The maximum output power was recorded at 2.05 mW, inducing a maximum pulse energy of 22.63 nJ. The repetition rate shows an increasing trendline from 81.04 kHz to 90.58 kHz while the pulse width decreases from 5.08 μs to 4.12 μs. The high signal to noise ratio (SNR) of 63 dB indicates that the TiO_2 film has a good performance in generating pulsed laser at 1550 nm wavelength.

1. Introduction

A Q-switched erbium-doped fiber lasers (EDFLs) have a very high demand in applications such as medical treatment, laser cutting technology, optical telecommunication, remote sensing, and a wide range of different applications for industry [1-2]. The excellent beam quality and cost effective of this fiber laser have attracted many researchers in recent years [3-5]. In addition, passively Q-switched technique based on saturable absorber are often preferred as they are more compact in setup and do not require an electronic controller to trigger the pulses, which makes it more favoured compared to active technique [6-7].

In general, saturable absorber can be categorized into two: real SAs and artificial SAs [8]. The artificial SAs however, have difficulties in exploiting optical fiber properties that is homogeneous to saturable absorption, hence struggling to access a better performance [9]. Real SAs, on the other hands, such as Semiconductor saturable absorber (SESAMs), 1D carbon nanotubes (CNTs), graphene, topological insulator (Tis), transition metal dichalcogenides (TMDs) and black phosphorus (BP) have been effectively used to generate pulsed lasers [10-13]. The evolution of saturable absorber technologies has growth tremendously with newly discovered material that provides a strong motivation in fulfilling the needs of photonic applications. However, these SAs have weaknesses that limits their reputation as an effective saturable absorber. SESAMs is a high cost option that have a narrow absorption bandwidth plus having a bulky size [14]. The main drawbacks of CNTs includes low modulation depth and an absorption efficiency that is dependant to their diameter while graphene suffers from a low optical absorption per layer that limits the capability in initiating laser pulses [15]. The exploration of finding a new material continues in providing the laser generation with a low cost, wideband absorption, high damage threshold and a reliability SA. Meanwhile, transition metal oxide (TMOs) have been proposed as an ideal saturable absorber that gained lots of interest in generating pulsed laser [16].

In this paper, Titanium dioxide (TiO_2) has been selected to generate passive Q-switched fiber laser in 1550 nm wavelength. Despite of the fact that the bandgap of TiO_2 is between 3-3.2 ev, this rutile nanoparticles has a remarkable electrical and optical properties such as high refractive index and good transmission in the near-infrared regions [17]. Moreover, with the successful demonstration of passive Q-switched erbium-doped fiber laser, TiO_2 is indeed a suitable candidate that has a good potential as saturable absorber with simplicity and low fabrication cost.

2. Titanium dioxide (TiO_2) film preparation and characterization

The fabrication of taking TiO_2 into thin film was first prepared by mixing the TiO_2 powder and a de-ionized water with a ratio of 2:1 respectively. 30 mg of TiO_2 powder and 15 ml of de-ionized water was blended using a magnetic hot plate stirrer at room temperature for 3 hours. The mixture was then put into the ultrasonic bath for one hour to ensure that the TiO_2 powder is blend well with the de-ionized water. Next step is to mix the TiO_2-de-ionized water solution and the PVA solution with a ratio of 20ml:10ml. The PVA solution was prepared by mixing 1 g of PVA powder with 120 ml distilled water as a solvent. Both solutions were stirred for 1 hour using the magnetic stirrer and again, the mixture was put into the ultrasonic bath for 30 minutes. Finally, the TiO_2 suspension was poured carefully onto a petri dish and left to dry at room temperature for 48 hours to form TiO_2 composite film.

Figure 1(a) shows the image of the TiO_2 thin film on a fiber ferrule. Figure 1(b) compares the percentages of element in the thin film which contain 33.3% of Titanium and 21.61% of oxygen while Figure 1(c) illustrated the field emission scanning electron microscopy (FESEM) of the film. As shown in Figure 1(d), the saturable intensity, modulation depth and non-saturable absorption of TiO_2 film are 0.05kW/cm^2, 13% and 37%, respectively. An absorption of EDFL Q-switched around 13 dB can be seen on Figure 1(e) that shows the linear absorption profile (LAP) of the TiO_2 film.

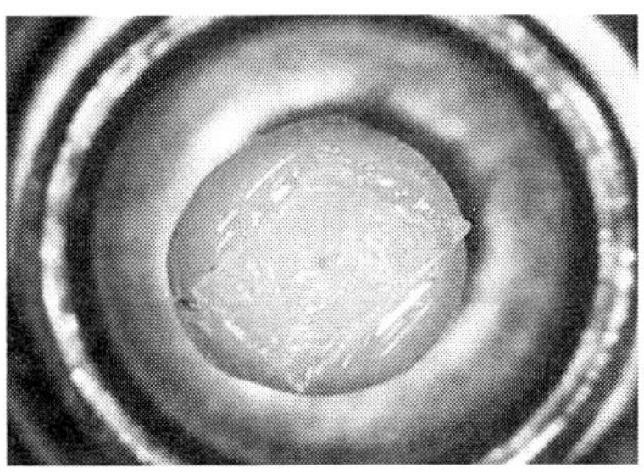

(a)

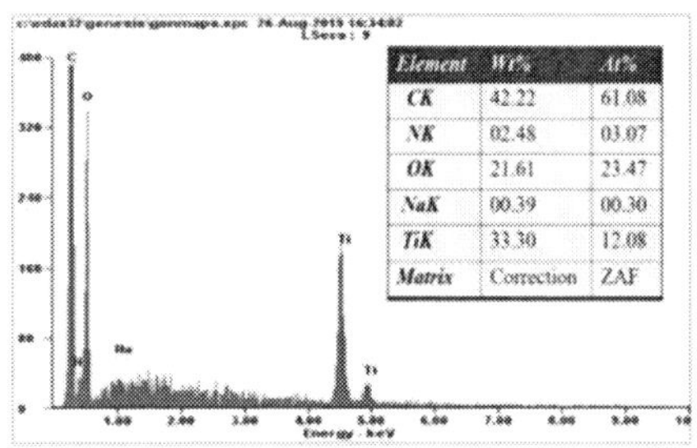

(b)

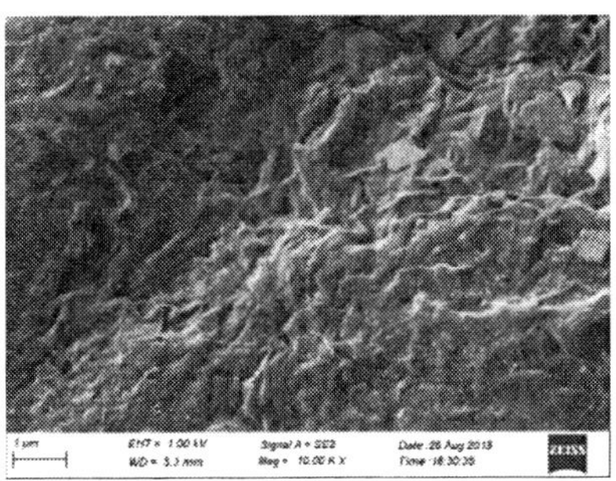

(c)

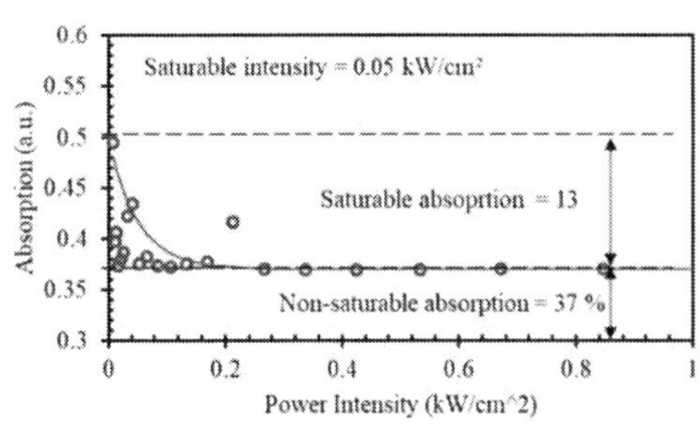

(d)

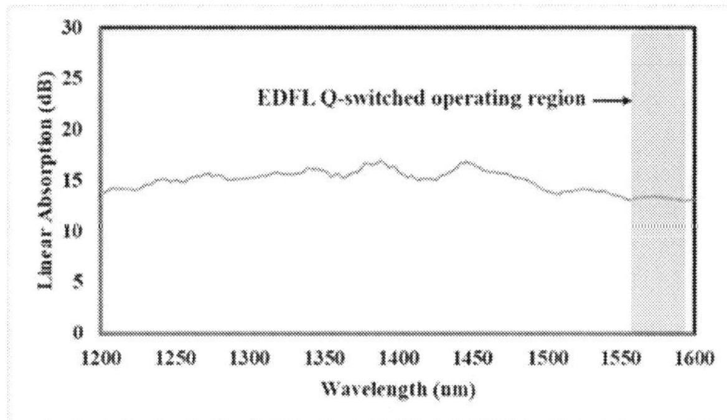

(e)

Figure 1. Characteristics of TiO_2 film (a) the fabricated film attached into an FC/PC fiber ferrule (b) EDS profile (c) FESEM image (d) nonlinear absorption profile (e)) TiO_2 film SA's linear absorption profile

3. EDFL ring cavity setup

The schematic diagram of the proposed laser cavity in generating the Q-switched EDFL is illustrated in Figure 2. The laser cavity consists of 980 nm laser diode, a wavelength division multiplexer (WDM), a 2.8m long erbium-doped fiber (EDF) gain medium, isolator and 90/10 coupler and a film of TiO_2 at the flat end of the fiber ferrule. The 980 nm laser diode pump the erbium-doped fiber (EDF) gain medium based on the forward pumping scheme via the 980/1550 nm port of the wavelength division multiplexer. An optical isolator is used to prevent unwanted feedback of light in the ring cavity. The light then goes into the fiber connecter that has a SA film on it. The film is attached at the flat end of the fiber ferrule by applying index matching gel. After passes the SA, 10% of laser output was tap out by the 3dB coupler and another 90% is retained to circulate inside the ring cavity. The repetition rate and pulse width were measured using a 500-MHz digital oscilloscope while the optical spectrum of the laser is observed by an optical spectrum analyser (OSA) with 50 nm spectral resolution. The output coupler is then connected to a power meter where the average output power is recorded. The repetition rate and pulse stability of the laser is validated using an electrical radio frequency (RF) spectrum signal of a 7.8 GHz RF spectrum analyzer. The total length of ring cavity was about 13 m.

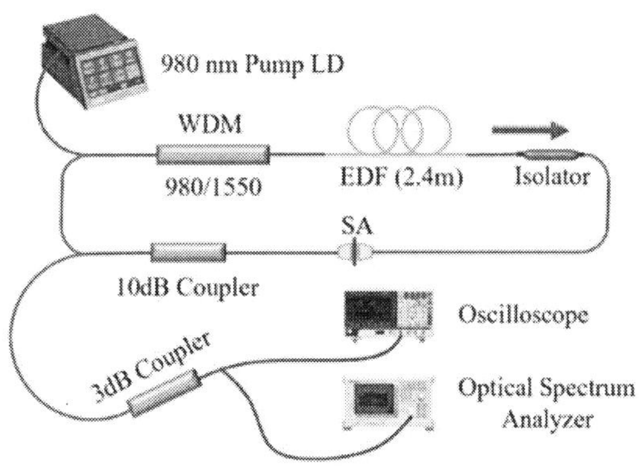

Figure 2. Schematic diagram of Q-switched erbium-doped fiber laser

4. Result and Discussion

Before we start the experiment, the laser cavity has been cleared from any form of impurities especially at the fiber connector in order to make sure that there is no self-Q-switching effect. The oscilloscope observed only continuous wave (CW) signals even when we increase the laser diode pump to the maximum. After inserting the SA into the laser cavity, the CW mode began to disappear. A Q-switching operation started at a threshold input power of 42 mW. As we increase the pump power up to 100mW, a stable pulse train was observed and maintained at a range of 104.62 mW to 145.83 mW. The repetition rate of the pulse is presented in Figure 3(a) and (b).

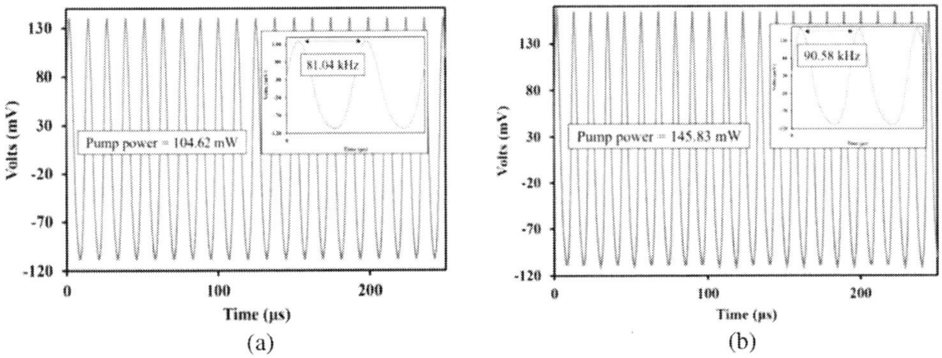

Figure 3. Repetition rate at pump power:(a) 104.62 mW and (b) 145.83 mW

Figure 4(a) depicts the repetition rate and pulse width of Q-switched erbium-doped fiber laser pulses versus pump power. The repetition rate increases from 81.04 kHz to 90.58 kHz with the decreasing pulse width from 5.08 μs to 4.12 μs. The pump power is tuned from 104.62 mW to 145.83 mW. The inversely proportional of the repetition rate and the pulse width is consistent with the typical features of passively Q-switched lasers. Figure 4(b) shows the pulse energy and average output power versus pump power. The average output power rises linearly from 1.58 mW to 2.05 mW, and pulse energy increases from 19.5 nJ to 22.63 nJ.

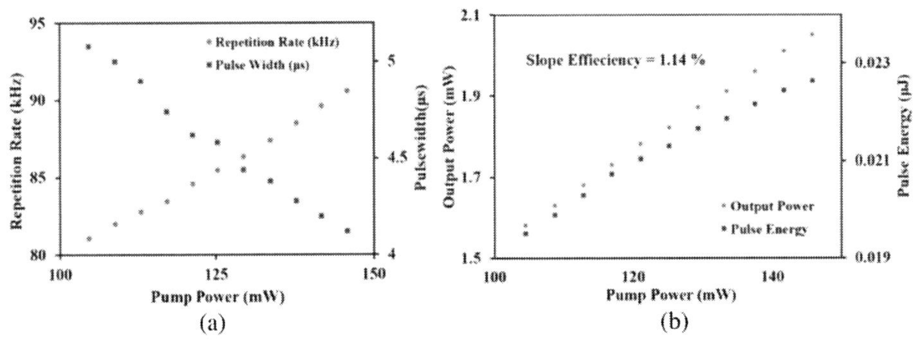

Figure 4. (a) Repetition rate and pulse width versus pump power. (b) Average output power and single pulse energy vary with the increasing pump power.

The output spectrum of the Q-switched laser at a pump power of 104.2 mW, is presented in Figure 5(a), indicates that the Q-switched erbium-doped fiber laser operates at a wavelength of 1563.9 nm. The RF spectrum (Figure 5(b)) shows a signal-to-noise ratio (SNR) of 63 dB for the repetition rate at 90 kHz proving that the laser pulses is very stable.

IOP Conf. Series: Materials Science and Engineering **854** (2020) 012018 doi:10.1088/1757-899X/854/1/012018

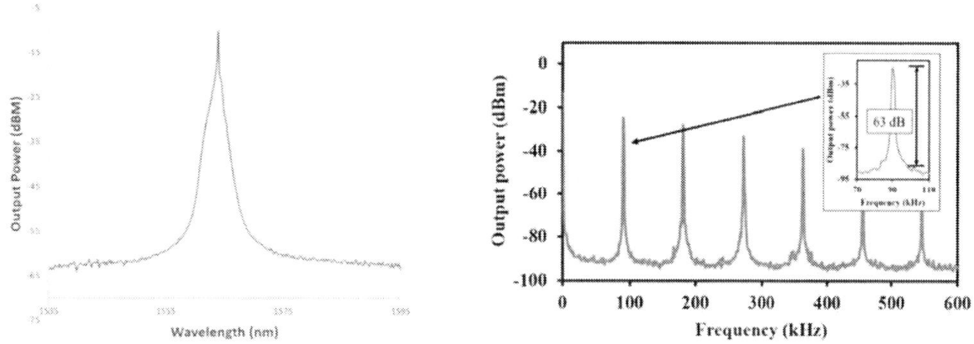

(a) (b)

Figure 5. The characteristics of the q-switching pulses:(a) wavelength and (b) RF spectrum

5. Conclusion

In conclusion, the fabrication ratio of 2:1 of the TiO_2-PVA film as saturable absorber was successfully demonstrated in generating EDFL pulse at 1.55 μm wavelength. A stable repetition rate increases from 81.04 kHz to 90.58 kHz within a range of pump power from 104.62 mW to 145.83 mW. The shortest pulse width was observed at 4.12 μs. At the maximum pump power, a high output power and pulse energy was recorded as 2.05 mW and 22.63 nJ, respectively. A high signal to noise ratio (SNR) of 63 dB validate that the Q-switched laser operation is very stable and confirms that TiO_2 film is an excellent saturable absorber.

References

[1] Guo, H., Feng, M., Song, F., Li, H., Ren, A., Wei, X., … Tian, J. (2016). Q-switched erbium-doped fiber laser based on silver nanoparticles as a saturable absorber. *IEEE Photonics Technology Letters*, *28*(2), 135–138.

[2] Li, H., Xia, H., Lan, C., Li, C., Zhang, X., Li, J., & Liu, Y. (2015). Passively Q-switched erbium-doped fiber laser based on few-layer MoS2 saturable absorber. *IEEE Photonics Technology Letters*, *27*(1), 69–72.

[3] Hattori, H. T., As'Ham, K., Haque, A., Li, Z., & Olbricht, B. (2019). A tale of two tantalum borides as potential saturable absorbers for q-switched fiber lasers. *IEEE Photonics Journal*, *11*(3).

[4] Zulkipli, N. F., Batumalay, M., Samsamnun, F. S. M., Mahyuddin, M. B. H., Hanafi, E., Izam, T. F. T. M. N., … Harun, S. W. (2019). Nanosecond Pulses Generation with Samarium Oxide Film Saturable Absorber. *Chinese Physics Letters*, *36*(7), 074203.

[5] Wu, K., Zhang, X., Wang, J., Li, X., & Chen, J. (2015). WS_2 as a saturable absorber for ultrafast photonic applications of mode-locked and Q-switched lasers. *Optics Express*, *23*(9), 11453.

[6] Baharom, M. F., Rahman, M. F. A., Latiff, A. A., Wang, P., Arof, H., & Harun, S. W. (2019). Lutetium oxide film as a passive saturable absorber for generating Q-switched fiber laser at 1570 nm wavelength. *Optical Fiber Technology*, *50*, 82–86.

[7] Ismail, M. A., Ahmad, F., Harun, S. W., Arof, H., & Ahmad, H. (2013). A Q-switched erbium-doped fiber laser with a graphene saturable absorber. *Laser Physics Letters*, *10*(2).

[8] Woodward, R. I., & Kelleher, E. J. R. (2015). 2D saturable absorbers for fibre lasers. *Applied Sciences (Switzerland)*. MDPI AG.

[9] Rosol, A. H. A., Rahman, H. A., Ismail, E. I., Irawati, N., Jusoh, Z., Latiff, A. A., & Harun, S. W. (2017). Cadmium Selenide Polymer Microfiber Saturable Absorber for Q-Switched Fiber Laser Applications. *Chinese Physics Letters*, *34*(9).

[10] U. Keller, D. A. B. Miller, G. D. Boyd, T. H. Chiu, J. F. Ferguson, and M. T. Asom, "Solid-state low-loss intracavity saturable absorber for Nd:YLF lasers: an antiresonant semiconductor Fabry–Perot saturable absorber," Opt. Lett. 17, 505-507 (1992)

[11] D. Zhou, L. Wei, B. Dong and W. Liu, "Tunable Passively Q-switched Erbium-Doped Fiber Laser With Carbon Nanotubes as a Saturable Absorber," in *IEEE Photonics Technology Letters*, vol. 22, no. 1, pp. 9-11, Jan.1, 2010.

[12] Chen, B., Zhang, X., Wu, K., Wang, H., Wang, J., & Chen, J. (2015). Q-switched fiber laser based on transition metal dichalcogenides MoS_2, MoSe_2, WS_2, and WSe_2. *Optics Express*, *23*(20), 26723. https://doi.org/10.1364/oe.23.026723

[13] J. Y. Liu, F. Y. Zhao, H. S. Wang, W. Zhang, X. H. Hu, X. H. Li, et al., "Generation of dark solitons in erbium-doped fiber laser based on black phosphorus nanoparticles," Optical Materials, 2019, 89: 100-105.

[14] Yuzaile, Y. R., Awang, N. A., Zalkepali, N. U. H. H., Zakaria, Z., Latif, A. A., Azmi, A. N., & Abdul Hadi, F. S. (2019). Pulse compression in Q-switched fiber laser by using platinum as saturable absorber. *Optik*, *179*, 977–985.

[15] Z. Luo *et al.*, "1-, 1.5-, and 2-μm Fiber Lasers Q-Switched by a Broadband Few-Layer MoS_2 Saturable Absorber," in *Journal of Lightwave Technology*, vol. 32, no. 24, pp. 4679-4686, 15 Dec.15, 2014.

[16] P. Harshavardhan Reddy, M.F.A. Rahman, M.C. Paul, A.A. Latiff, A.H.A. Rosol, Shyamal Das, Anirban Dhar, Shyamal Kumar Bhadra, K. Dimyati, S.W. Harun, Titanium dioxide doped fiber as a new saturable absorber for generating mode-locked erbium doped fiber laser, Optik, Volume 158, 2018, Pages 1327-1333, ISSN 0030-4026.

[17] Amirtharajan, Saranya & J. Pandiarajan, Dr & Natarajan, Jeyakumaran & Prithivikumaran, N. (2015). Electrical investigation of TiO2 thin films coated on glass and silicon substrates—effect of UV and visible light illumination. Applied Nanoscience.

IOP Conf. Series: Materials Science and Engineering **854** (2020) 012019 doi:10.1088/1757-899X/854/1/012019

Synthesis and Characterization of Microfiltration Ceramic Membrane Based on Fly Ash and Clay Mixture Using Sintering Method

S Diana, R Fauzan, N Arahman, F Razi, E Elfiana

Chemical Engineering Department, Politeknik Negeri Lhokseumawe, Aceh

E-mail: selviediana@pnl.ac.id

Abstract. Ceramic membranes made from low-cost materials have increasingly been developed. Fly ash is a waste from cement industries, while clay is a natural resource that can be easily found to be used as economical mixed materials in the production of ceramic membrane. The manufacture of microfiltration ceramic membrane from fly ash - clay mixture has been carried out using the sintering method for 4 hours at low sintering temperatures variations at 750 °C (M1), 800 °C (M2), 850 °C (M3) and 900 °C (M4), respectively. The results showed that the ceramic membrane that was sintered at a temperature of 900 °C (M4) had the optimum characteristics with a porosity of 40.82%, a density of 1.0026 gr/cm^3. From scanning elctron microscopy (SEM) analysis showed that the M4 membrane was classified as a microfiltration membrane with an average pore size of 0.64 μm. The energy diepersive x-ray (EDX) analysis showed that the composition of the M4 membrane was dominated by elements O, Si, and Al. Moreover, the x-ray diffraction (XRD) analysis showed that the fraction patterns of the membranes (M1, M2, M3, M4) were dominated by the quartz phase (SiO_2) and illite phase ($(KH_3O)Al_2Si_3AlO_{10}(OH)_2$).

1. Introduction

Currently, ceramic membrane applications for microfiltration processes have been increasingly developed and applied due to the excellent thermal, chemical, and mechanical stability of ceramic membranes and due to high separation efficiency compared to polymer membranes [1-2]. Ceramic membranes are generally made from various expensive, inorganic materials such as silica, alumina, titania, and zirconia. Therefore, a development is made in form of natural ceramic materials with abundant material availability and fabrication techniques with sintering temperatures below 1000 °C to produce ceramic membranes at an economical price [3-4]. Fly ash, which is a waste from cement industry, and clay, which is widely available in nature, can be used as a basic material for making ceramic membranes at an economical price. Furthermore, the characteristics of fly ash and clay which have particle sizes of 1μm - 20μm are also very suitable for microfiltration processes [5].

Research on the manufacture of ceramic membranes made from fly ash and clay using sintering methods have been carried out in recent years. However, those researches were conducted at temperatures above 1000 °C, causing the cost of making ceramic membranes to increase. Jedidi, et.al [6] used fly ash as an economical material for the manufacture of ceramic membranes at a sintering temperature of 1125 °C and produced ceramic membranes with a pore diameter of 4.5 μm and porosity of 51%. Belibi, et.al [7] have succeeded in making a clay-based membrane at a sintering temperature of 1100 °C with a porosity value of 42%. Rekik, et.al [8] produced a ceramic membrane made of clay at a sintering temperature of 1250 °C with a porosity value of 27% and a pore diameter of 0.73 μm. The need for cheap materials and low sintering temperatures pose a new challenge in reducing the cost of making ceramic membranes while still producing the best-quality ceramic membranes. The purpose of this research is to find out about the effect of sintering at low temperatures with variations in temperature of 800 °C, 850 °C, 900 °C, and 950 °C, on the porosity, density, permeability, pore size, and morphology of ceramic membranes made from fly ash and clay mixture.

2. Research Method

2.1. Material

Fly ash and clay with a particle size of 120 μm were used as the main component for making membranes. Fly ash samples which have been produced as waste in cement manufacturing was collected from PT. Solusi Bangun Andalas Indonesia, Banda Aceh, while the clay was purchased from PT. Rudang Jaya, Medan, Indonesia. Polyvinyl Alcohol (PVA) was purchased from Sigma, Aldrich.

2.2. Synthesis of Fly Ash-Clay Microfiltration Ceramic Membranes

A total of 900 grams of fly ash and clay mixture with at a composition of 65% - 35% were put into a beaker glass and then added 4% Polyvinyl Alcohol (PVA). Then, as much as 700 ml of water was added to it gradually, while being stirred until a paste was formed. The membrane was then molded by pouring the paste into a mold. It was needed to make sure that no air was trapped in it by pressing slowly until using a membrane press tool. The membrane was removed from the mold and dried at room temperature for 7 x 24 hours. The membrane was then continued through sintering process at 750 °C, 800 °C, 850 °C and 900 °C for 4 hours.

2.3. Characterization of Fly Ash-Clay Microfiltration Ceramic Membranes

The density of ceramic membranes is obtained by weighing each membrane weight, measuring the diameter and thickness of the membrane, so that the membrane mass and volume of the membrane are obtained. Then the membrane density is calculated based on the following equation:

$$\rho = \frac{M(gr)}{V(cm^3)} \quad (1)$$

where ρ represents the density of membrane (gr/cm3), M represents the mass of membrane (gr), V represents the volume of membrane (cm3). The porosity of each membrane is measured using the Archimedes method as follows:

$$\varepsilon(\%) = \frac{W_3 - W_1}{W_3 - W_2} \quad (2)$$

where W1 represents the weight of dry membrane, W2 represents the weight of the wet membrane measured in water, and W3 represents the weight of the membrane that was filled with water for 48 hours.

The structure and pore size of membrane were analyzed using scanning electron microscope (SEM, ZEISS2074, Center Laboratory, UM, Malaysia). The elements of the membrane were analyzed using energy dispersive x-ray (EDX, ZEISS2074, Center Laboratory, UM, Malaysia). The phase compositions of the membranes were measured using x-ray diffraction (XRD, Analysis Lab, PNL, Indonesia).

The water flux of ceramic membrane was tested using crossflow filtration system with variation operating pressure from 0.25-1.25 bar. The water flux (J, L/m^2.hr) for each pressure variation was calculated using Eq. (3), as follows:

$$J = \frac{V}{At}$$

IOP Conf. Series: Materials Science and Engineering **854** (2020) 012019 doi:10.1088/1757-899X/854/1/012019

(3)

where V represents the volume of the permeate (L), A represents the effective area of the membrane surface (m^2), and t representas the filtration time (h).

3. Result and Discussion

3.1. Porosity and Density of Fly Ash-Clay Ceramic Membrane

Sintering process is a process of compacting powder material by forming bond boundaries between the constituent powders/particles. Bonding between items occurs due to thermal treatment with or without pressure and the sintering temperature is set below the melting point temperature of the constituent particles. The thermal treatment contained in this process causes diffusion between the constituent particles where the particles will come in contact with each other, making an increase in the density of the constituent particles and bettering the contact area between the particles [9].

Figure 1 presents the sintering temperature effects on porosity and density of the membrane. The largest porosity was owned by the membrane which was sintered at 750 °C (48.85%), while the porosity decreased to 40.82% at a sintering temperature of 900 °C. The decrease in porosity of the membrane was caused by the increasing sintering temperature. The increase in temperature caused the contact area between the particles becoming wider and made the structure of the porosity becoming smoother, so that the empty space for porosity was also getting smaller [10].

The porosity of the membrane that is formed also affects the density of the membrane. From figure 2, it can be seen that the membrane density increases with the increasing temperature. The increase in sintering temperature affects the growth and compaction process of the particle where the higher the sintering temperature, the higher the percentage of particle density [11]. Therefore, the largest density is owned by the membrane which is sintered at a temperature of 900 °C, equaling to 1.0026 gr/cm^3.

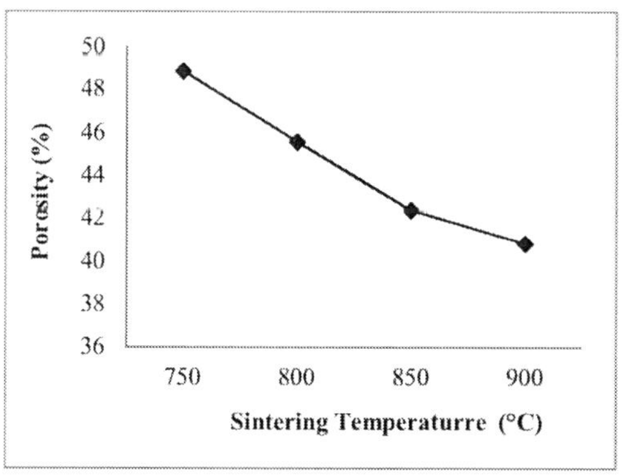

Figure 1. Porosity of Membrane

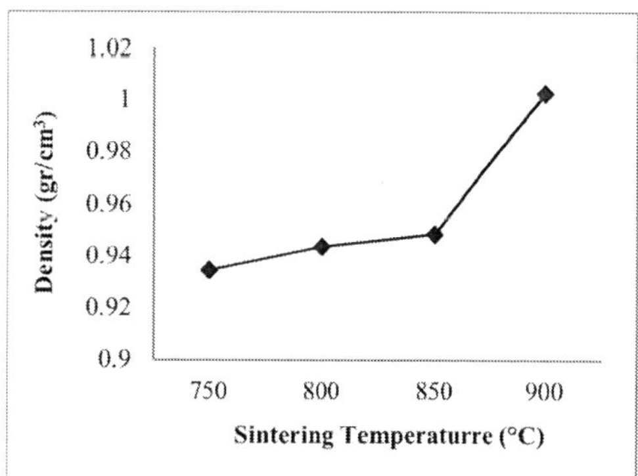

Figure 2. Density of Membrane

3.2. Morphology of Fly Ash-Clay Microfiltration Ceramic Membrane

The structure and pore size of membrane were analyzed using SEM (Scanning Electron Microscope). SEM observations were made on the membrane which was sintered at 750 °C, 800 °C, 850 °C and 900 °C. Figure 3 presents the comparison of the membrane surface structure at different sintering temperatures. At the sintering temperature of 750 °C, the membrane surface was still not homogeneous and the porosity formed still dominated the surface of membrane (figure 3a). The membrane surface

IOP Conf. Series: Materials Science and Engineering **854** (2020) 012019 doi:10.1088/1757-899X/854/1/012019

was increasingly seen to be more homogeneous at a sintering temperature of 900 °C (figure 3d). The increase in sintering temperature at 900 °C caused the fly ash - clay mixture particles to melt, forming the neck of the sintering and bind to each other [12]. Thus, the porosity that was formed was reduced and the pore size of the membrane also shrank by 0.64μm (figure 4).

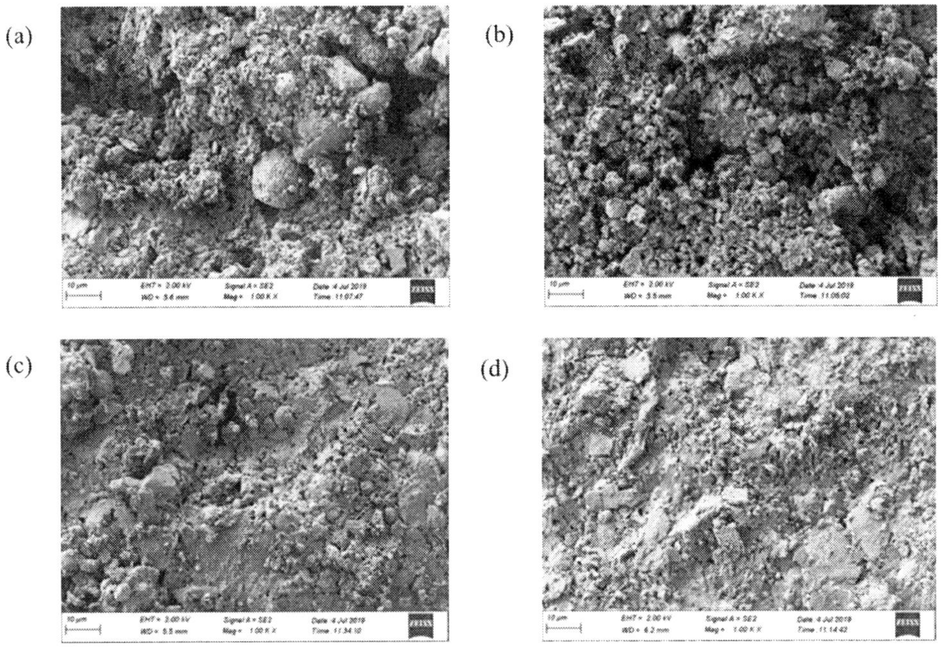

Figure 3. The SEM image of membranes at different sintering temperature: (a) 750°C, (b) 800°C, (c) 850°C, and (d) 900°C

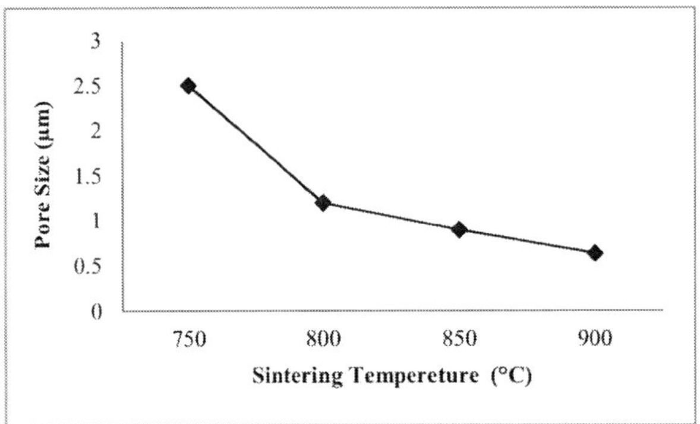

Figure 4. The Pore Size of Membrane

Membrane elements analysis was carried out using Energy Dispersive X-Ray (EDX) to the membrane that had the smallest pore size. It was the M4 membrane which had sintering at 900°C. From figure 5, it can be determined that the main elements of the membrane consisted by O, Si, and Al.

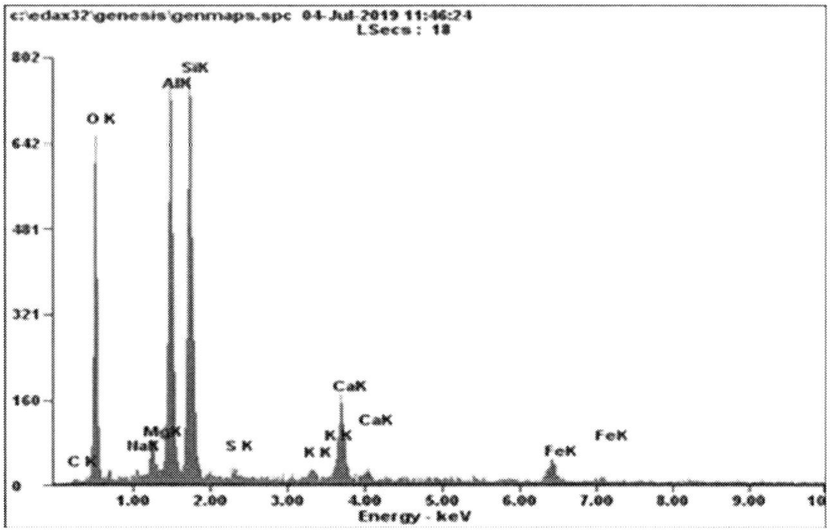

Figure 5. The EDX Analysis

3.3. Effect of Sintering Temperature on Membrane Phase Composition.

The membrane phase composition in this study was analyzed using XRD. Figure 6 presents the XRD patterns of the ceramic membrane with various sintering temperature. It is known that the fraction pattern of all membranes is dominated by quartz (SiO2) and illite ((KH3O) Al2Si3AlO10 (OH) 2) phases. At a sintering temperature of 850 °C, diffraction patterns from the gypsum (CaSO4.2H2O) and graphite (C) phases are formed. At a sintering temperature of 900 °C the graphite phase (C) does not form anymore. The process of mixing materials consisting of fly ash, clay, and PVA causes a deviation in decreasing the melting point, so that at a temperature of 900 °C the graphite phase disappeared.

3.4. The Flux of Fly Ash-Clay Microfiltration Ceramic Membrane

Figure 7 presents the raising of flux value which increases along with the increasing of the pressure and the sintering temperature. The separation occurred because of the driving force caused by the different of input and output pressure of the membrane modul [13]. The highest flux value was gained at the pressure of 1.25 bar and sintering temperature of 750 °C. The increasing the the flux is caused by the increasing of the pressure which resulted to the increasing of the propulsive force given to the solution to pass the ceramic membrane so a big amount of feed solution can pass the ceramic membrane and this process will minimize the effect of concentration polarization. The increase in sintering temperature affects the diffusion process between membrane particles, making it wider, and ultimately affects the structure of the shrinking membrane porosity. Low membrane porosity causes the value of flux and membrane permeability to be low as well [11].

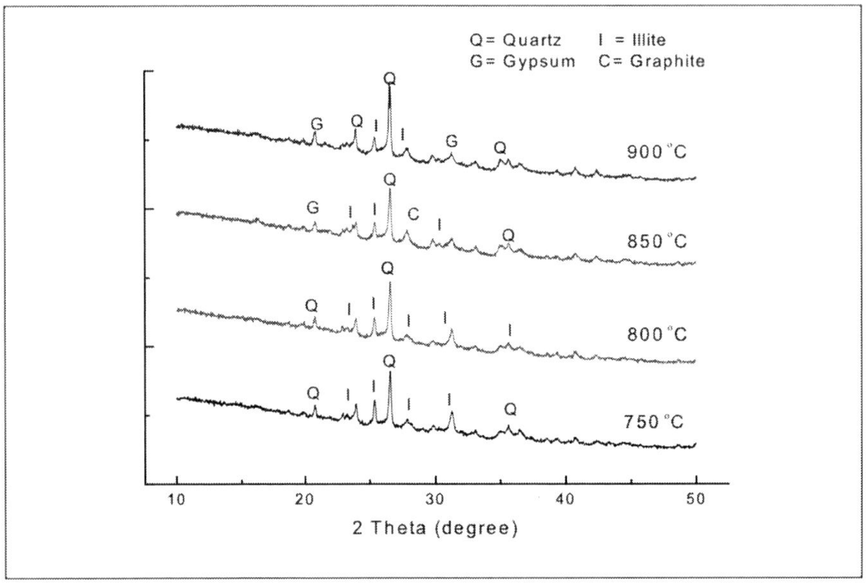

Figure 6. The XRD Patterns of Membranes

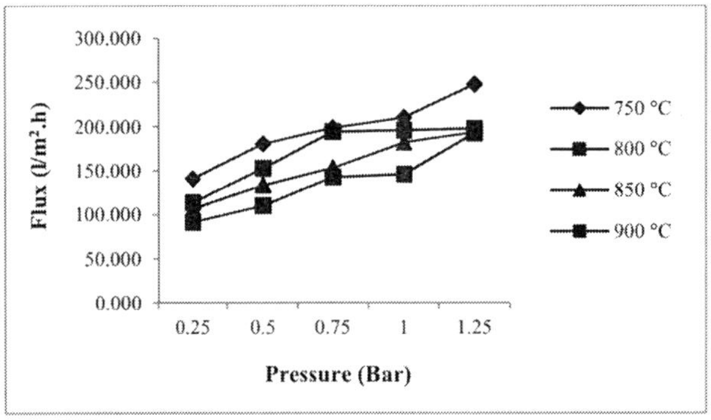

Figure 7. The Water Flux of Membranes

4. Conclusion

Microfiltration ceramic membrane with fly ash - clay mixture material has been successfully made with various sintering temperatures. Ceramic membrane that was sintered at a temperature of 900 °C had a porosity of 40.82%, a density of 1.0026 gr/cm^3, and water flux at a pressure of 1.25 bar of 192.358 Lm^{-2}h^{-1}. Based on the analysis using Scanning Electron Microscopy (SEM), the membranes can be classified as a microfiltration membrane with an average pore size of 0.64 μm.

Acknowledgment

This research was funded by the Ministry of Research, Technology and Higher Education (KEMENRISTEK DIKTI) Republic of Indonesia in 2019. Our appreciation goes to the Director of Lhokseumawe State Polytechnic and Rector of Syiah Kuala University who have supported this research. We also thank Prof. Dr. Nasrul, AR (JTK-Unsyiah) and Dr. Facrul Razi (JTK-Unsyiah) for giving very useful input for this research.

References

[1] Mustafa, K Wyns, S Janssens, A Buekenhoudt and V Meynen 2018 *Sep.Purif.Technology* **193**, 29-37
[2] C Y Huang, C C Ko, L H Chen, C T Huang, K L Tung and Y C Liao 2018 *Sep.Purif. Technology* **198** 79-86
[3] Y Yin, H Wang, D Li, W Jing, Y Fan and W Xing 2016 *Journal Colloid Interface Science* **478** 136
[4] M Yavari, T Le, H Lin 2017 *J. Membr. Sci.* **525** 387.
[5] S Diana, R Fauzan, E Elfiana 2019 *IOP Conf. Ser.: Mater. Sci. Eng* **536** 012089
[6] I Jedidi, S Khemakhem, S Saidi, A Larbot, N Elloumi-Ammar, A Fourati, A Charfi, A B Salah, and R B Amar 2011 *Powder Technology* **208** 427-432
[7] Belibi, Placide & Nguemtchouin, Marie & Rivallin, Matthieu & Nsami, Ndi & Joseph-Marie, Sieliechi & Cerneaux, Sophie & Martin, Ngassoum & Cretin, M 2015 *Ceramics International* **41** 2752-2759
[8] Rekik S, Bouaziz J, Deratani A and Beklouti S 2017 *Periodica Polytechnica Chemical Engineering* **61(3)** 206-215
[9] Taufik Akbar dan Widyastuti 2013 *Jurnal Teknik POMITS* **2** 2301-9271
[10] S M Samaei, S Gato-Trinidad and A Altaee 2018 *Separation and Purification Technology* **200** 198–220
[11] German, Randall 2016 *Materials Science Forum* **835** 50-75
[12] Dong Zoua, Minghui Qiua, Xianfu Chena, Enrico Driolib and Yiqun Fana 2019 *Separation and Purification Technology* **210** 511–520.
[13] E Elfiana, S Diana, A Fuadi and R Fauzan 2019 *IOP Conf. Ser.: Mater. Sci. Eng* **536**

Waste Mud in Oil and Gas CEPU PPSDM Using Bioremediation

Yanty Maryanty, Sigit Hadiantoro, Anang Takwanto, Kamilia, Linga Ardhanariswari

Politeknik Negeri Malang

Email: yanty.maryanty@polinema.ac.id

Abstract. Bioremediation is one method of treating hazardous waste environmentally friendly. This method of waste treatment Utilizes a microbial activity to treat hazardous waste. Oil sludge from the processing of petroleum in PPSDM Migas Cepu is one of the dangerous wastes, so it must be processed first before being discharged into the environment. The bioremediation process of oxygen and nutrients is needed by microbes to decompose harmful compounds in the oil sludge so that the addition of a bulking agent in the form of rice husk and sawdust is needed. Bacteria that help the bioremediation process are heterophilic bacteria, heterophilic bacteria used in this study are aeruginous Pseudomonas, Bacillus cereus and Bacillus spheres. This study aims to determine the effect of the addition of bulking agents on bioremediation and Determine the performance of heterophilic petroleum degrading bacteria in waste. The results of this study indicate the addition of a bulking agent Gives a fairly good result in bioremediation in the reduction in TPH and TCLP values Compared to bioremediation reactors without the addition of a bulking agent. Petrophilic Also bacteria play an important role in the bioremediation process as evidenced in the decline in TPH and TCLP values. The addition of a bulking agent, in a variable one (25; 75) there was a decrease in the value of% TPH namely 56.22%, variable 2 (40:60) there was a decline in the value of% TPH 55.95 and in variable 3 there was a decrease in the value of% TPH by 49, 73%. During the bioremediation process, the increase is in the pH value from 3. 6 to 6.5 indicates the presence of bacterial activity in degrading oil sludge waste. Percent of Percent Nitrogen and Carbon in bioremediation Decreased by 80 and 56%. The success of the bioremediation process was demonstrated by the results of FT-IR analysis, with peaks of 3418.39 cm -1, 2924.65 cm-1, and 1746.22 cm-1.

Keywords: Bioremediation, oil sludge, bacteria heterophilic

1. Introduction

Center for Human Resource Development of Oil and Gas (PPSDM Gas) produce products such as parasol, diesel, and residue. The results of processing these products produce waste in the form of oil sludge. Waste oil sludge is a waste from oil mining. Oil sludge interpreted also as oil sludge which is a heavy metal compound with characteristics that are difficult to remove by water so that it can threaten the health of the environment in the coastal areas and coral reefs. Based on Government Regulation (PP) 101 2014, sludge included in the category of B3 waste (hazardous and toxic) (Munawar et al, 2015).

The negative impact caused by waste oil sludge is waste hydrocarbon contaminated soil would be dangerous because hydrocarbons are toxic and carcinogenic (Zam, 2011). One of the petroleum processing techniques to reduce the negative impact of living beings and the environment is to use a method of bioremediation. Bioremediation can be used in the processing of petroleum sludge and environmentally friendly due to the bacteria used for bioremediation can bind unhealthy ingredients contained in petroleum residual sludge bulking processing is processing waste bioremediation method with additional ingredients used in the bioremediation process to improve the rate of biodegradation in the soil remediation process.

2. Methodology

The research method that will be used that is surveys and experiments in laboratory scale and implemented in PHP Laboratory of Human Resource Development Center for Oil and Gas (PPSDM Migas). The method consists of sampling survey in the field. An experimental method to test heterophilic bacteria to degrade the waste oil sludge.

2.1. Material

Materials used in this research that the oil sludge obtained from the exploration, production to landfill. The bulking agent used is rice husk and sawdust with the same size. The bacteria used in bioremediation as Bacillus spheres, Bacillus cereus, and Pseudomonas aeruginous. Media used in this study is derived from the contaminated soil in PPSDM Migas Cepu.

2.2. Tools

The tools used in this research is spectrophotometry IR, FT-IR instrument, pumpkin Kjeldahl, and a screening tool.

2.3. Start-up Bioremediation

Oil sludge as much as 100 grams placed in containers. The addition of a bulking agent with the variable bulking agent is added to the (25; 75, 40:60 and without an added bulking agent). Addition heterophilic bacteria used in bioremediation, namely with the same concentration.

2.4. Parameter Observation

Soil physical parameters measured were the temperature is measured using a thermometer, water content by gravimetric method (ASTM D2216), the specific gravity and bulk density particles according to the method of immersion in the pycnometer and gravimetric (Black and Hartge 1986). Soil chemical parameters were observed pH is measured using electrometry method (ASTM D4972-01), organic C according to ASTM D 2974-87, N-total by the Kjeldahl method (USEPA Method351.3), and TPH by gravimetric methods (USEPA Method 1664).

3. Results Discussion

In this study, wastewater treatment PPSDM Migas Cepu oil sludge using bioremediation methods. Bioremediation process requires the help of a bulking agent as a bulking agent in the contaminated soil and as additional nutrients for bacteria to degrade the hydrocarbon chain heterophilic. This method has not been done in PPSDM Migas Cepu.

From the physical properties of waste which is processed using bioremediate wastes which originally shaped mud and smelled unpleasant after bioremediation has a shape like normal soil and smell almost like soil. The first variable (25:75) the waste has a crumbly texture compared to other variables. That is because a bulking agent that serves as a bulking agent has a ratio of more than the others. For variable (40:60) outcome is still a waste bioremediation clot because it is saturated by ground and has a stronger odor than the variable one. The smell is caused by the comparison bulking agent less than a variable one.

3.1. Effect of pH before and after bioremediation

According to Hapsari (2014), the optimum pH value is used in the bioremediation method that is 6.7 to 9, while according to Sharma (2012), the optimum pH for growth of bacteria that is 6.5-8. To reduce the amount of nitrogen lost pH should not be more than 8.5. When composting the waste nearly perfect then the pH will be close to 7 in this case because the microbial activity that causes the pH value close to normal (Hapsari, 2014).

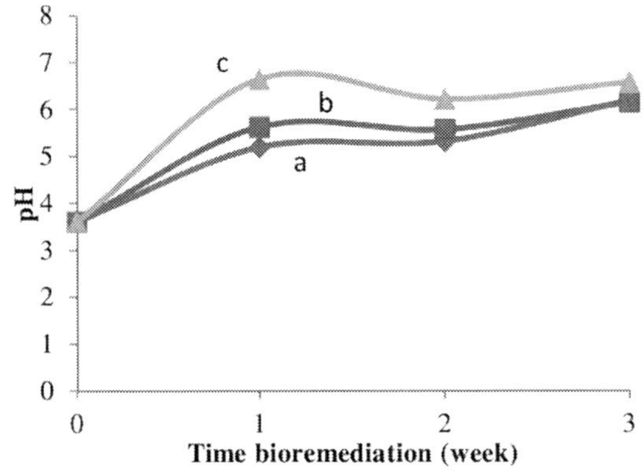

Figure 1. Graph pH Before and After Bioremediation; Description: a = Variable 1 (25; 75), b = Variable 2 (40; 60), c = Variable 3 (without bulking agent)

Original initial pH of 3.6 weekly rise, this happened due to microbial activity degrading conducting enzymatic reactions. pH obtained after bioremediation conforms with the quality standard that after bioremediation pH should be around 5-7, while the pH obtained an average of 6. It can be one piece of evidence that oil-degrading bacteria in waste oil sludge can work with well as the pH produced by the provisions of the optimum pH for microbial degradation of petroleum hydrocarbons in which the range of pH, nutrients for microorganisms soluble and oxygenase enzyme produced by microorganisms degrade hydrocarbons can be optimized.

3.2. Water levels on Bioremediation

The test results showed that the moisture content decreased the percentage of water content. This can be caused by aeration or stirring process that is done every day during the study that led to evaporation. Physical environmental conditions such as temperature and mechanical factors such as aeration and agitation also affect the percentage of degradation. Hydrocarbons spilled in nature will be degraded naturally because of environmental factors, although the rate of degradation was slow. The degradation process includes evaporation, emulsified in water, adsorbed on the solid particles, immersed in water and biodegradable by microbes users hydrocarbons, (Piedat et al., 2008; Ojo, 2006).

For bioremediation of oil sludge process with the addition of a bulking agent affects water levels in optimizing hydrocarbon degradation process. Optimum water content bioremediation process to transfer gas ranges between 25-35%, (Cookson, 1995). In other research humidity ideal for bacterial growth at 25% -28%, while the optimal moisture to degrade hydrocarbons that is 30% -90%.

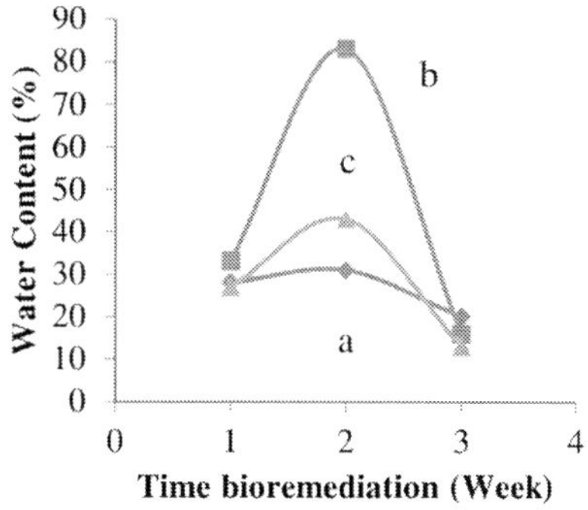

Figure 2. The graph% water content in Bioremediation; Description: a = Variable 1 (25; 75), b = Variable 2 (40; 60), c = Variable 3 (without bulking agent)

3.3. Test Results TPH (Total Petroleum Hydrocarbon)

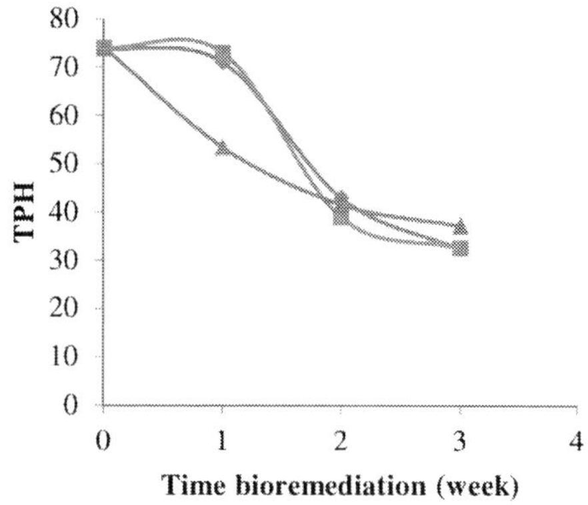

Figure 3. Test Results TPH (Total Petroleum Hydrocarbon) Bioremediation; Information: ■ = Variable 1 (25; 75), ■ = Variable 2 (40; 60), ■ = Variable 3 (without bulking agent)

Testing TPH (Total Petroleum Hydrocarbon) conducted to determine the decrease in hydrocarbon concentration in the reactor. impairment TPH also occurs due to volatilization of short-chain hydrocarbon content. The possibility is reinforced by the data that is to add compost reactor. The results obtained showed a decrease in the concentration of hydrocarbons within 3 weeks as the initial sample which showed a decrease in TPH this can be caused by several factors. These factors, among others, a bulking agent, moisture, pH and agitation (aeration). In the three variables, TPH value also decreased, but the decline is smaller than a variable one and two, it was because the oil sludge The indigenous bacteria can degrade TPH.

Variable 2 has a decreasing value of TPH is best compared to other variables it can be due to the amount of land on bioremediation reactor at most so that the probable number of bacteria that degrade hydrocarbons heterophilic more. The composition of the bulking agent on two variables also influences due to bulking provide nutrients for bacteria to degrade hydrocarbons. While the first variable has a value smaller decline compared variable TPH 2, it is caused by the comparison between the media and the bulking agent.

There are two decisive component degradation, namely the hydrocarbon component as a degradable material acts as a substrate and microbes as degradation act as enzymes in the enzymatic reaction mechanisms. Degradation products are degradable material undergone renovation produces molecules simpler compounds. Overhaul of hydrocarbon molecules takes place continuously during degradation (enzyme) supplied with sufficient nutrients O2 and serves as an external acceptor. Hydrocarbons are degraded in a row by some enzymes, in the presence of oxygen will liberate CO2 gas, water, and simple organic molecules in the form of degradation of organic compounds are acidic, causing the pH change in the media degradation. PH change media degradation is the limiting factor hydrocarbon component degradation process. Based on research that has been done, bulking effect on the decline in value of TPH it is evidenced in the results of TPH value that has been done. Bulking agents can affect the required oxygen transfer microbes to degrade hydrocarbons (Juliani et al, 2011). TPH reduction values may continue to decrease as the time used for bioremediation longer so that microbes can with up to degrade hydrocarbons.

For analysis of total petroleum hydrocarbons decrease supporter, FT-IR analysis to determine the functional groups contained in the oil sludge.

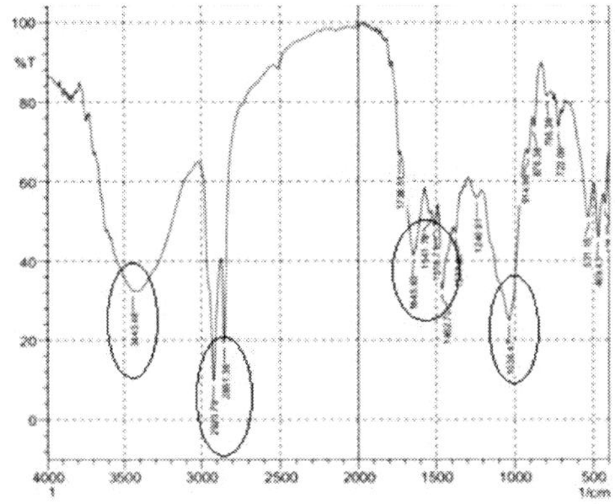

Figure 4. The results of FT-IR Sludge

The results of FT-IR analysis *sludge* who has done the most content that is shown in the peak Phenol 3418.39 cm-1 CH aliphatic alkanes that at peak 2920.79 cm-1, C = C peak alkenes at 1645.93 cm-1, CH alkanes in the peak 1462.70 cm-1 and CF alkyl halide 1036.47 cm-1.

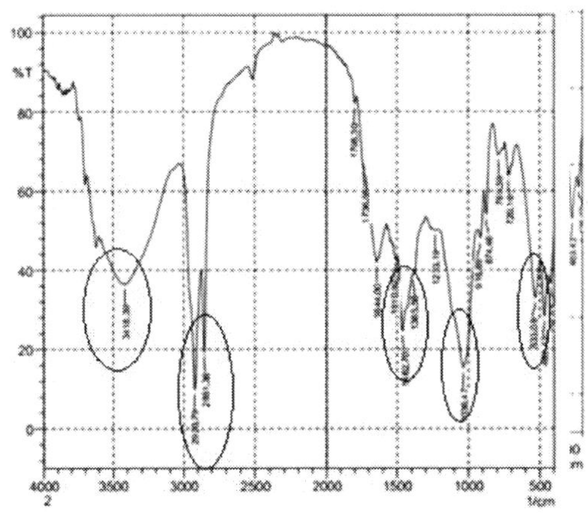

Figure 5. The results of FT-IR Variable 1

IOP Conf. Series: Materials Science and Engineering **854** (2020) 012020 doi:10.1088/1757-899X/854/1/012020

The results of FT-IR analysis of sludge that has done the most phenol content shown on peak 3418.39 cm-1 is CH aliphatic alkanes that at peak 2920.79 cm-1, CF alkyl halides in peak 1036.47 cm-1, CH alkanes in the peak 1462.70 cm-1 and C = C alkenes at peak 1644 cm-1.

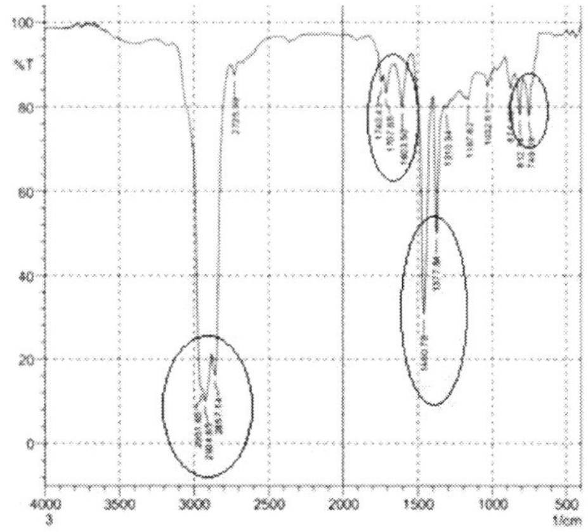

Figure 6. The results of FT-IR Variable 2

The results of two variables on an FT-IR analysis of the highest hydrocarbon content in oil sludge contained in the peak 2924.65 cm-1, which is the functional group of aliphatic alkane CH, CH alkanes in the peak 1460.78 cm -1 and 1377.84 cm-1. The third most functional groups are ketones shown in peak 1707.65 cm-1 and the last most functional groups are aldehyde shown at peak 1740.43 cm-1.

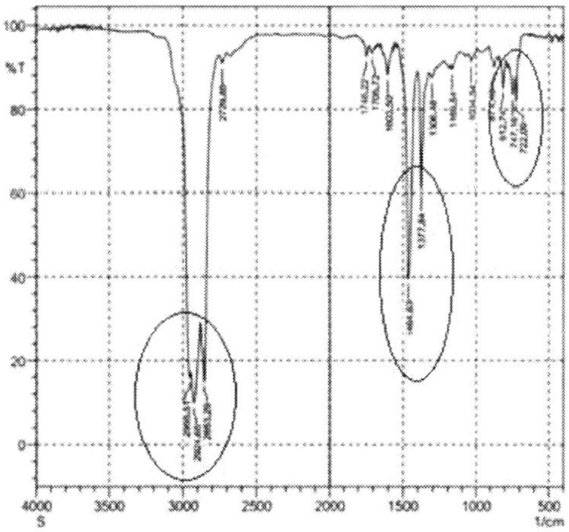

Figure 7. The results of FT-IR Variable 3

FT-IR results of which have been performed on three variables, namely the highest functional groups aliphatic CH alkane, alkane CH, C-Cl Akil halide and alkyl ketones methylene shown in peak 2924.65cm-1, 1464.63 cm-1, 812.74 cm -1, 1746.22 cm-1 and 722.09 cm-1. Based on the theory of the largest hydrocarbon compounds contained in crude oil are alkanes are saturated aliphatic hydrocarbons so that the research results with the same theory that supported their all alkane compounds contained in the sludge, one variable, two variables, and three variables.

3.4. C / N ratio

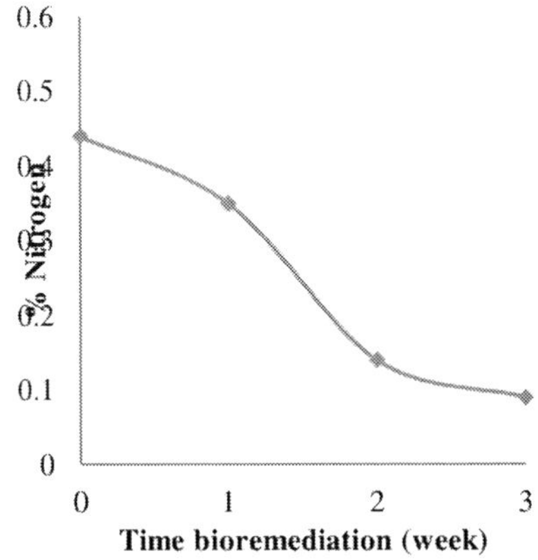

Figure 8. Graph% Nitrogen Bioremediation; Information: ■=% Nitrogen

Analysis of nitrogen in the reactor was conducted by Kjeldahl bioremediation. Samples were analyzed, namely early and sample soil after the bioremediation. Graph of analysis% N on bioremediation show a decline, it based on research shows that bacteria heterophilic the process of degradation of hydrocarbons in sludge
refinery gases PPSDM Migas Cepu (Zam, 2011). Decreased% nitrogen also showed that the bacteria heterophilic need nutrients such as nitrogen in the process of bioremediation to degrade hydrocarbons.

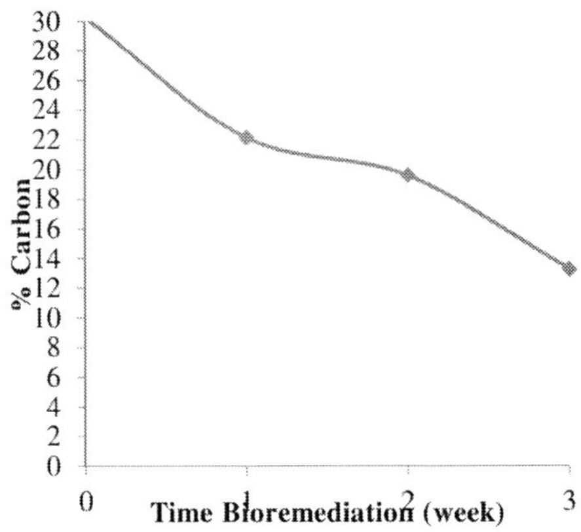

Figure 9. Graph% Carbon Bioremediation; Information: ■ =% Carbon

In addition to nitrogen, carbon heterophilic also required for bacteria to degrade hydrocarbons. Thus, in figure 3.4% carbon reduction also indicates that a process of degradation of hydrocarbons by bacteria heterophilic. Carbon is used as a nutrient heterophilic bacteria. Analysis of carbon content was also analyzed soil samples beginning and end after bioremediation using Walkley & Black titration method.

3.5. Biodegradation Mechanism on Oil Waste

Hydrocarbon components in a material such as petroleum sludge are generally divided into 2 groups, ie, aliphatic and aromatic compounds. Degradation of aliphatic components (alkanes) in contrast to the degradation of aromatic components. The reaction mechanism aliphatic compounds (alkanes) involves the formation of alcohols, aldehydes, and fatty acids. Fatty acids are broken down, the CO2 is released and form a new fatty acid which is a 2-carbon unit shorter than the parent molecule, a process known as beta (β) oxidation. (Nugroho, 2009; Nababan, 2008).

The following is a reaction mechanism of aerobic degradation of alkanes by a microbe using alkane monooxygenases enzyme to convert hydrocarbons into alcohol:

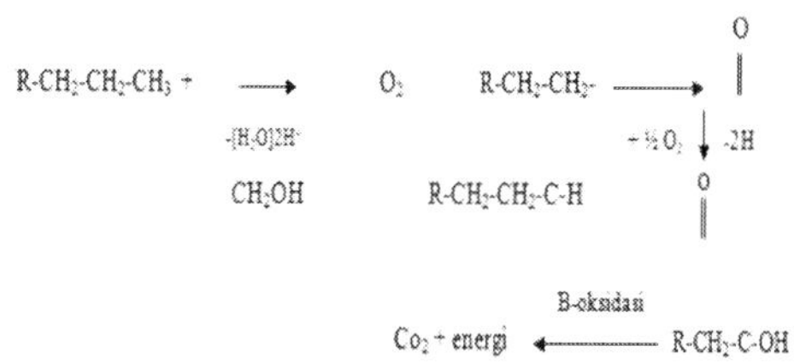

Enzymes involved in the degradation process of hydrocarbons are oxygenase enzyme

The hydrocarbon decomposition process can be accelerated by Pseudomonas aeruginous the required enzyme lactase. The mechanism by adding the enzyme lactase, then the bonding strength of the polymer will be reduced thereby aeruginous Pseudomonas bacteria were able to digest the hydrocarbon component faster (Xuanzhen Li et al., 2012; Acevedo et al., 2011).

Oxygen plays an important role in the metabolism by microbes, oxygen required by microbes to cut hydrocarbon group aliphatic (long-chain) through metabolic processes of hydrocarbons. Petrofilik metabolic processes in bacteria occur oxidation of primary alcohols methyl become. Petrofilik bacterial metabolism also requires dehydrogenase enzyme that functions as oxidizing primary alcohols to aldehydes and henceforth into fatty acids. Hydrocarbons which has to be fatty acids, then digested by bacteria heterophilic through the beta-oxidation process to get the final result of H2O and ATP.

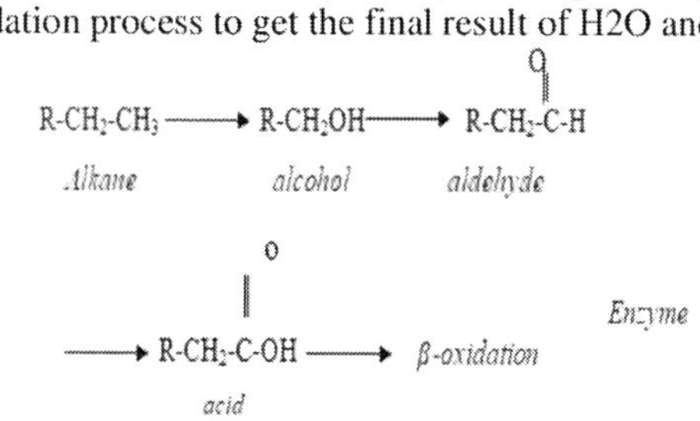

A decisive factor for the degradation process is the contact time (interaction) bacteria with hydrocarbon component. Overview biochemical oxygenase enzyme that is active for the adequacy of nutrient and oxygen supply would improve the performance of the bacteria. Production of the enzyme has a high correlation with the ability to break down hydrocarbons. Overview of organic chemicals from the aspects of the reaction mechanism, ie, aliphatic components undergo oxidation reactions produce alcohol in the presence of oxygen, wherein the carbon atom oxidation number of -3 becomes -1, and if further oxidized will produce aldehyde compounds (oxidation number of carbon atoms changed from -1 to +1) and partly to ketones (carbon atom oxidation number of -1 to +2). Aldehyde product may undergo further oxidation reaction produces a carboxylic acid (carbon atom oxidation number of +1 to +3), then the resulting carboxylic included in the β-oxidation cycle, so that the compound decomposes producing energy and CO2. Increasing the number of bacterial inoculum is known to raise the level of degradation and cause degradation of aliphatic compounds, or aromatic compounds. The other thing is because the enzyme produced more varied in type and degree of decomposition as well as the amount of enzyme that is more than a single culture so that decomposition is faster (Nababan, 2008). Increasing the number of bacterial inoculum is known to raise the level of degradation and cause degradation of aliphatic compounds, or aromatic compounds. The other thing is because the enzyme produced more varied in type and degree of decomposition as well as the amount of enzyme that is more than a single culture so that decomposition is faster (Nababan, 2008). Increasing the number of bacterial inoculum is known to raise the level of degradation and cause degradation of aliphatic compounds, or aromatic compounds. The other thing is because the enzyme produced more varied in type and degree of decomposition as well as the amount of enzyme that is more than a single culture so that decomposition is faster (Nababan, 2008). The hydrocarbon decomposition process can be accelerated by Pseudomonas aeruginous the required enzyme lactase.

3.6. TCLP Test Results (Toxicity Characteristic Leaching Procedure)

Table 1. Analysis results of TCLP

Variables	Cu	Fe	Pb
sludge	33.12	168.29	24,54
1	17.12	133.53	14.57
2	19.83	108.39	18.1
3	29.19	138.11	19.57

TCLP analysis is done by using the AAS method is based on Table 3.1 shows the TCLP analysis also decreased from its initial samples of sludge that has not undergone the bioremediation process, but the results are still far above the TCLP analysis of the quality standards set in this case because the time used for bioremediation less long and nutrients that are used for fewer bacteria. In the analysis of the variables Cu 3 has the greatest level of 29.19 is compared variable 1 and 2 it is due to the variable 3 there is an additional bulking agent so that bacteria heterophilic less nourishment to react with heavy metals and toxic compounds.

Fe and Pb analysis on three variables also have the highest levels of variable 1 and 2, it is also due to variable 3 not given a bulking agent that serves as additional nutrients in bioremediation so that bacterial bioremediation heterophilic at less than the maximum.

4. Conclusion

1. Impairment% TPH in bioremediation process in the variable 1 (25:75) ie 56.22%, variable 2 (40:60) 55.92% and a variable 3 (without bulking agent) 49.73
2. Performance heterophilic bacteria in degrading hydrocarbons in oil sludge can reduce the pH value, TPH, and TCLP. The final pH value is 6 appropriate quality standards. The decline in the value of TPH reached 40% and decreased levels of heavy metals Fe by 60%, Cu 16% and 10% Pb.

5. Suggestions

1. Bioremediation process needs to be done with a longer time, about 60 days
2. It takes the addition of nitrogen as a nutrient for bacteria in the bioremediation process

3. Indigen bacterial isolation needs to be done from the ground mining before bioremediation, and perform treatment according to the characteristics of bacteria used in bioremediation

References

[1] Cooperband, L. (2002), The Art and Science of Composting, A Resource for Farmers and Compost Producers, Center for Integrated Agriculture systems, the University of Wisconsin-Madison.

[2] Elfita and Munawar.2015. "Biodiversity of Bacteria indigen and Contribution in Polluted Environment Management: A Case Study of Some Areas in Indonesia". National Seminar process Biodeversitas Indonesia.1,6 Society, 1359-1363.

[3] Estuningsih, Sri Pertiwi., Napoleon, Duke., Sabarudin., Said, M., Yudhoyono, Bambang.2013. "Indigen Degrading Bacteria Exploration Oil Waste in the Region PT Pertamina UBEP Limes Muara Enim ". Proceedings Semirata Lampung.Lampung.127-133 State University.

[4] Hapsari, Patricia Permana Jati.2014. Oily Lumpur Processing Method of Co-Composting. Final Report of Environmental Engineering Institute of Technology. Surabaya.

[5] Juliani, Any and Rahman, Fudhola.2011. "Bioremediation Lumpur Oil (Oil Sludge) with the addition of compost as a bulking agent and Nutrient Sources Supplement. Journal of Science and Technology Lingkungan.3,1, 001-018.

[6] Mulyana, Nana and Retno, Tri.2013. Bioremediation of Soil Contaminated Waste Oil Lumpur Bulking Agents Using Mixed Microbial consortia Enriched Compost Based Irradiation. A Scientific Journal for The Applications of Isotopes and Radiation.9,2, 139-150.

[7] MunawardanZaidan.2013. "Bioremediation of Oil Waste Technique Klamono Field Biopile in Papua" .Jurnal Science and Mathematics. 1,2,41-46.

[8] Nababan, Bungaria. 2008. Isolation and Test Potential of Solar Oil Degrading Bacteria from Belawan Sea. Dissertation / Thesis Graduate: University of North Sumatra.

[9] Nugroho, A., 2009. The gas production of petroleum biodegradation results: Initial studies applications in microbial enhanced oil recovery (MEOR), Makara Journal of Science, vol. 13 (2): 111-116

[10] Okoh, AI 2003. Biodegradation Of Bonny Light Crude Oil In Soil Microcosm By Some Bacterial Strains Isolated From Crude Oil Flow Stations In Nigeria Saver Pits. African Journal of Biotechnology, vol. 2 (5): 104-108

[11] Okoh, AL, 2006. Biodegradation alternative in the cleanup of petroleum hydrocarbons Pollutants. Journal Biotechnol. And Molecular Biology Review, vol. 1 (2): 38-50

[12] Pikoli, MR, P. Aditiawati, and DI Astuti. 2000. Isolation and identification of bacterial isolates gradually degrading thermophilic petroleum from wells bangko. Department of Biology, ITB, Bandung.

[13] Rossiana N, 2007. Bioremediation of oil sludge with microorganisms and plants of zeolite and testing with sengon (Paraserianthes falcataria L. Nelsen). Dissertation. Padajajran.Bandung University Graduate Program.

[14] Ikhsan Syukria Zam.2011. "Bioremediation of Petroleum Refinery Waste by Using Bacteria indigen In Invitro (Invitro Bioremediation of Oil Refinery Waste by Indigenous Bacteria)". Proceedings of the National Seminar on Chemical Unesa.Surabaya.374-380.

[15] Wu Yucheng, Teng Ying, Li Zhengao, Xuewei Liao, Luo Yongming, 2008. Potential role of polycyclic aromatic hydrocarbons (PAHs) oxidation by fungal laccase in the remediation of an aged Contaminated soil. Journal Soil Biology &

Biochemistry, vol. 40: 789-796

IOP Conf. Series: Materials Science and Engineering **854** (2020) 012021 doi:10.1088/1757-899X/854/1/012021

Passively Q-switched Erbium-doped Fiber Laser using Tungsten Disulfide deposited D-shaped Fiber as Saturable Absorber

S Omar[1], B Musa[1], Z Jusoh[1], MIM Ali[2], AAA Jafry[3], and SW Harun[4]

[1]Faculty of Electrical Engineering, Universiti Teknologi MARA, 23000 Dungun, Terengganu, Malaysia

[2]Faculty of Electrical Engineering, Universiti Teknologi MARA, 40450 Shah Alam, Selangor, Malaysia

[3]Department of Physics, Faculty of Science, Universiti Teknologi Malaysia, 81310 Skudai, Johor, Malaysia

[4]Photonics Engineering Laboratory, Department of Electrical Engineering, University of Malaya, 50603 Kuala Lumpur, Malaysia

Email: suzianaomar@gmail.com

Abstract. This paper generates a stable Q-switching output wavelength 1560 nm by utilizing tungsten disulfide (WS_2) as saturable absorber on D-shaped fiber in erbium-doped fiber laser. D-shaped fiber was prepared by using polishing wheel technique with rough and fine polishing is conducted to ensure efficient light-matter interaction inside D-shaped optical fibre. Proposed Q-switched erbium-doped fiber laser generates shortest pulse duration of 3.25 μs with repetition rate of 108.90 kHz. Q-switched generated was stable under pump power of 84.02 mW to 182.92 mW. Obtaining signal-to-noise ratio of 40.58 dB, Q-switched induced contain pulse energy of 14.60 nJ with output power of 0.64 mW to 1.59 mW. Therefore, proposed D-shaped fiber-WS_2 as saturable absorber was stable and efficient for making a portable laser source based on Q-switched.

1. Introduction

The avalanche development in the short-pulse laser had attracted tremendous research interest in recent years. Thanks to their usefulness in numerous industrial applications, including corrective eye surgery [1], material cutting technology [2], and micromachining of material [3], vast research efforts were conducted to make them faster and efficient. That attempt includes the development of a well-aligned laser configuration to reduced power loss and the incorporation of a material as a pulse initiator to induce a short pulse laser. The ability of the first method requires a few complicated steps involving lens arrangement, careful laser beam confinement, and harmful to eyes as well. In contrast, the latter which manipulates nonlinear optical properties of a material is an eye-safe approach with ease of preparation, which makes it favorable to emit short-pulse laser for various applications. This method configured an all-fiberized laser cavity with a low signal loss, which makes a bunch of photons confined in an optical fiber before they were absorbed by a saturable absorber (SA). Further, the SA modulate the loss and release the light to trigger a train of short pulse in the microsecond to nanosecond pulse duration.

The modulation of loss (Q-factor) triggered by a SA was called a Q-switched. The Q-factor of a laser modulator was express via the ability of a SA to realized pulsed generation in a broad optical bandwidth and a fast response time. In 2004, carbon nanotube (CNT) was utilized as a SA in an all-fiberized laser cavity [4]. It seems promising with excellent optical and physical properties such as ultrafast recovery time, high optical damage threshold, and polarization-insensitive SA. However, chirality and shape-dependent bandgap have always been a limitation for CNT. Several geometry adjustments were needed to expand its operational wavelength in a near-infrared spectrum. Later, Bao *et al.* demonstrated a pulsed laser via the incorporation of graphene as a SA [5]. Graphene owns an ultrafast relaxation time (~200 fs) and bandgap independent operation as it is a zero-bandgap semi metal. Despite its excellencies, graphene possesses a low modulation depth, which leads to weak second-order susceptibility during pulsed generation [6].

Hence, other 2-dimensional materials were introduced to replaces graphene such as topological insulator, black phosphorus, and transition-metal dichalcogenides (TMDs). TMDs comprise of one transition metal element sandwiched between two chalcogenide ions. They were profoundly incredible with fast response time, and a strong second harmonic generation denotes excellent nonlinear absorption properties in the laser cavity. Tungsten disulfide (WS_2) is one of TMDs which was frequently investigated to perform as a suitable SA for wide near-infrared window (1-,1.55- and 2-μm) [7]. It mimics the physical properties of graphene, such as high charge-carrier density, and strong orbit coupling with the ability to undergo frequency doubling effectively, therefore it was better than graphene.

Here, we demonstrated WS_2 coated D-shaped fiber as a SA in an erbium-doped fiber laser (EDFL) cavity. Unlike thin film as SA, D-shaped fiber revealed a long nonlinear optical response between light and material. Thus, it introduced a SA device with high saturable absorption ability and optical damage threshold. The mechanical and thermal stability of the proposed SA was also excellent as we eliminate the parasitic reaction, which was usually induced by a fiber-ferrule [8]. The absence of fiber-ferrule produced an all-fiberized laser cavity without any optical fiber adapter and pigtail, thus decrease splice loss within the cavity. In this work, we obtained a Q-switched EDFL using WS_2 coated D-shaped fiber with a pulse width of 3.25 μs corresponds to a repetition rate of 108.90 kHz. A step-by-step fabrication of the D-shaped fiber was also elaborated to established excellent evanescent field interaction between light and material.

2. Experimental setup

To fabricate the D-shaped fibers, a two-stage polishing process is employed which is adopted from the report by Ahmad *et al.* [9]. While they were using different polishing depth to indicate the amount of core removed, our technique further improves the process by incorporating a more efficient two-stage polishing method using rough and fine sandpapers. The first stage involves polishing with a rough 600-grid sandpaper followed by a fine sandpaper of 3000-grid. Figure 1 shows the set-up of the polishing wheel technique used to fabricate D-shaped fibers.

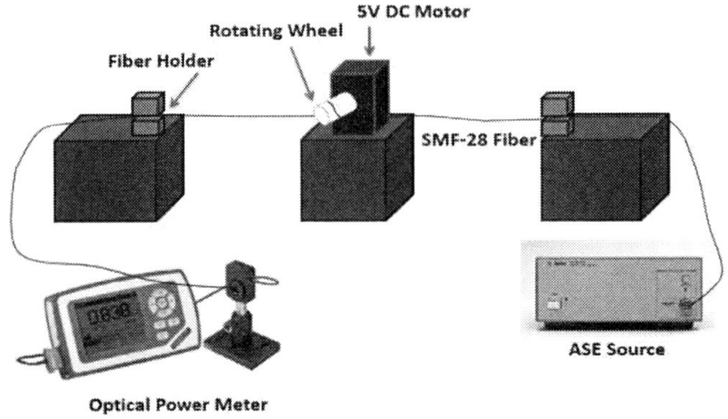

Figure 1. Experimental setup of polishing wheel technique to produce D-Shaped fiber.

IOP Conf. Series: Materials Science and Engineering **854** (2020) 012021 doi:10.1088/1757-899X/854/1/012021

In this set-up, a standard single-mode optical fiber (Corning SMF-28) with core and cladding diameters of 9 and 125 μm, respectively are used to produce D-shaped fibers. The protective jacket of the fiber is removed for approximately 3 mm and then fixed with two fiber holders. The fiber's vibration is kept at minimum during the polishing by maintaining a good tension between the fiber holders. Insertion loss during the polishing process is monitored using the optical power meter (THORLABS, PM 100D. Initially, the stripped part is polished with 600-grid sandpaper until desired insertion loss is observed on the optical power meter. The process normally takes about three minutes to complete. Then, the rotating wheel is adjusted so that the polished region is in contact with 3000-grid fine sandpaper and the fine polishing is conducted for five minutes. Note that, in producing the D-shaped fibers, a careful observation needs to be made as to ensure the interaction between light and material are maximized which in turn will results in a good lasing performance of the generated Q-switched pulses. The fine polishing stage is very crucial to establish efficient light-matter interaction via evanescent field inside D-shaped fiber during lasing. A typical D-shaped fiber fabricated using this method has a lenght of about 1500 mm with the diameter of around 97.33 μm. The insertion loss recorded for these fibers is about 3 dB.

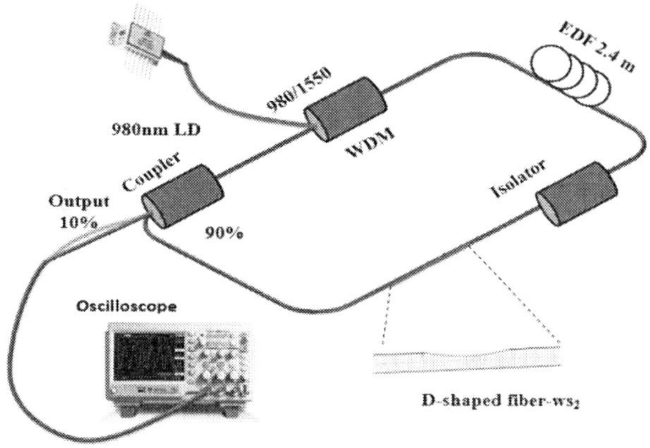

Figure 2. Cavity arrangement for the generation of Q-switched erbium-doped fiber laser.

The laser cavity of the proposed Q-switched fiber laser is depicted in Figure 2. A 980 nm laser diode is used to pump the cavity through a 980/1550 nm wavelength division multiplexer (WDM). A 2.4m erbium-doped fiber (EDF) is introduced in the cavity as a gain medium to generate 1.55 μm region pulse fiber laser. The EDF used has a core diameter of 4 μm, an absorption coefficient of 23 dB/m at 980 nm and a numerical aperture (NA) of 0.16. An isolator is included in the cavity to ensure unidirectional light propagation in which the interactions with D-shaped fiber coated WS_2 to produce Q-switched pulse laser. The fabricated saturable absorber (SA) is optimized by repeatedly depositing 3 μL WS_2 solution by means of micropipette on D-shaped fiber until stable pulses are generated on the digital oscilloscope. WS_2 solution was bought from Sigma Aldrich Malaysia. A 90:10 optical coupler is used to couple out 10% of the light for analysis while another 90% of the light is allowed back to the cavity. Total length of developed erbium-doped fiber laser cavity is 13.8 m. The output characteristics of the fiber laser are monitored by a digital power meter and an optical spectrum analyzer (YOKOGAWA, AQ6370D) with resolution of 0.02 nm. Temporal laser characteristics are measured by using a 350 MHz digital oscilloscope (GWINSTEK, GDS-3352) and 7.8 GHz RF spectrum analyser (ANRITSU, MS2683A) connected via a 7-GHz InGaAS photodetector.

IOP Conf. Series: Materials Science and Engineering **854** (2020) 012021 doi:10.1088/1757-899X/854/1/012021

3. Result and discussion

Figure 3(a) shows a typical pulse train displayed on the oscilloscope at pump power of 51 mW. A stable pulse train is obtained within the pump power region of 84.02 mW to 182.92 mW. The enlarged two pulses enveloped of the pulse train indicates the pulse width of 5.25 μs with a pulse duration of 9.2μs. It is also observed that the Q-switched output is stable with no observable amplitude modulation in the pulse train, indicating the absence of self-mode-locking during the Q-switching operation. The output spectrum of the Q-switched laser at the pump power of 182.92 mW was shown in Figure 3(b), indicates that the laser operates at a wavelength of 1560 nm. The repetition rate of the fundamental pulses is measured as 182.92 kHz with a signal to noise ratio (SNR) of about 38.89 measured using an RF spectrum analyzer with the 10MHz span.

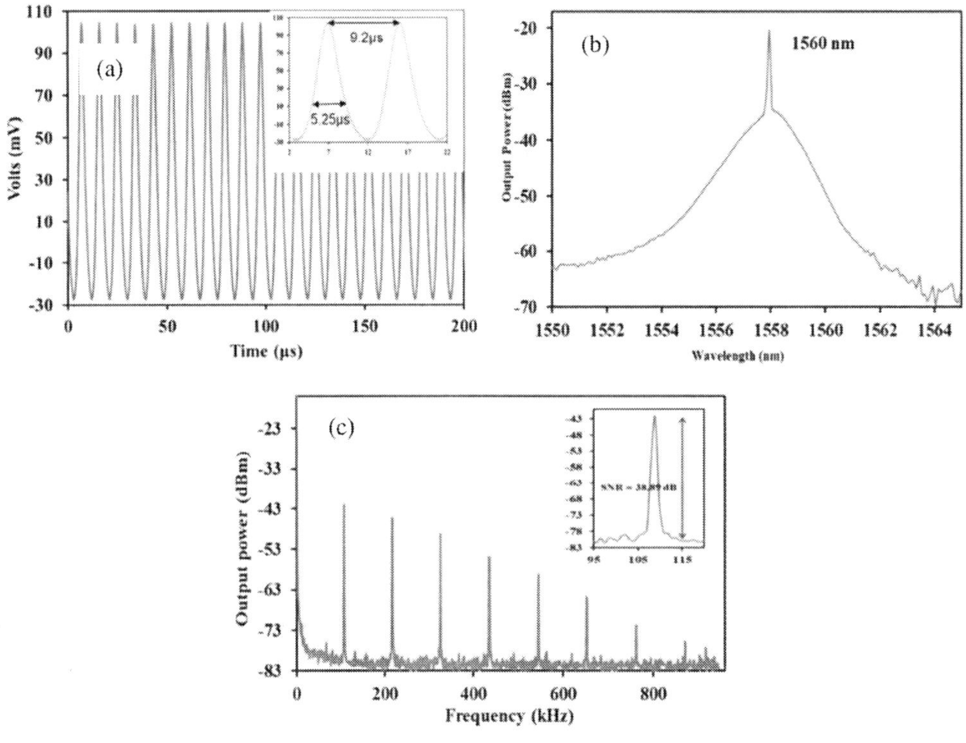

Figure 3. Temporal and spectral characteristics of the WS_2 based Q-switched EDFL; (a) typical pulse train, (b) output spectrum, and (c) RF spectrum with a 10 MHz span.

Lasing performance of Q-switched induced by D-shaped fiber-WS_2 implemented in EDFL is investigated over a variation of input pump power. As depicted in Figure 4 (a), by tuning the pump power within 84.02 mW to 182.92 mW, the pulse duration of 5.25 μs to 3.25 μs captured, while the repetition rate increases. The behavior of the repetition rate and the pulse width indicates typical features of passively Q-switched laser. Within the same pump power, Q-switched pulses record repetition rate of 77.16 kHz to 108.90 kHz. As shown in Figure 4 (b), output power recorded by our ring cavity shows promising performance within the attainable pump power which is from 0.64 mW to 1.59 mW. Those values exhibit maximum pulse energy of 14.60 nJ within the same range of pump power which is also plotted in Figure 4 (b). By referring to the graph of output power as a function of pump power, 1 % of slope efficiency is obtained from generated Q-switched using D-shaped fiber-WS_2 as SA. This is attributed to the excellence configuration of the ring cavity and the effectiveness of D-shaped fiber-WS_2

as a SA. Finally, peak power obtained by varying the pump power from 84.02 mW to 182.92 mW is comparable to other passively Q-switched EDFL which is from 1.53mW to 4.56mW as shown in Figure 4 (c).

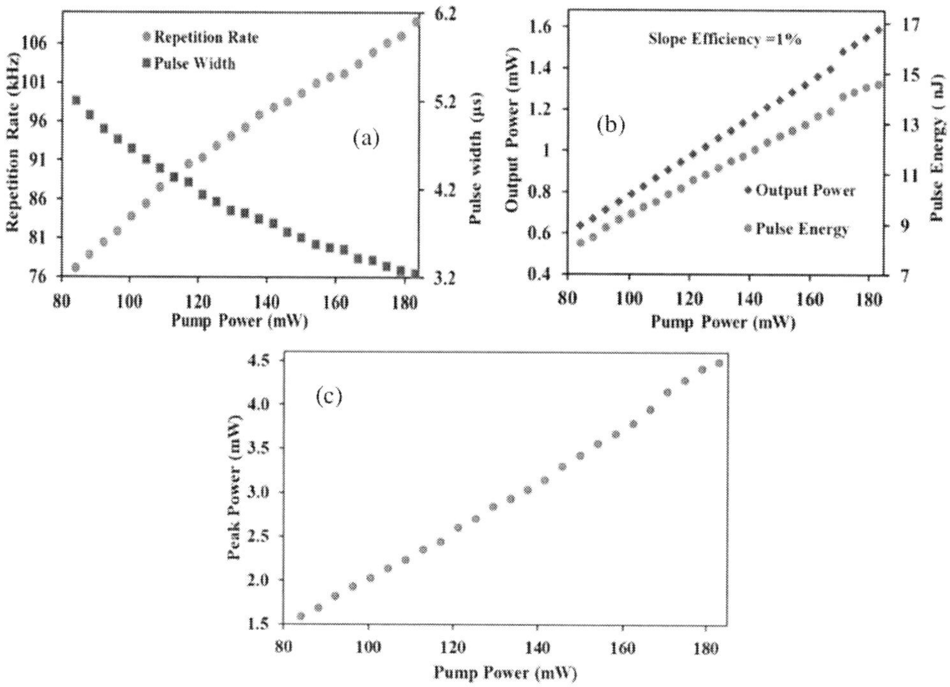

Figure 4. Q-switching performances (a) Repetition rate and pulse width against pump power (b) output power and pulse energy as a function of pump power, and (c) peak power as a function of pump power.

4. Conclusion

Generating Q-switched EDFL was successfully demonstrated using a reliable D-shaped fiber-WS_2 as saturable absorber. Side-polished technique with rough and fine polishing is conducted to ensure efficient light-matter interaction inside D-shaped optical fibre and course efficient light-WS_2 interaction via evanescent field to ensure good propagation of light inside ring cavity. Stable pulse train is observed within pump power of 84.02 mW to 182.92 mW with maximum repetition rate of 108.90 kHz corresponds to shortest pulse duration of 3.25 µs. With maximum pulse energy of 14.60 nJ, Q-switched pulses exhibit maximum output power of 1.59 mW. Stability of Q-switched EDFL is proven with SNR of 38.89 dB. D-shaped fiber with insertion loss of 3 dB has obtained diameter of 97.33 µm which is suitable for making a portable, robust and stable Q-switched laser source. This result showed Q-switched lasers can be realized by utilising the WS_2 based SA inappropriate cavity design.

References

[1] S. Masket, S.E. Masket, Simple regression formula for intraocular lens power adjustment in eyes requiring cataract surgery after excimer laser photoablation, J. Cataract. Refract. Surg. 32(3) (2006) 430-434.

[2] Y. Okamoto, R. Kitada, Y. Uno, H. Doi, Cutting of Solid Type Molded Composite Materials by Q-switched Fiber Laser with High-Performance Nozzle, J. Adv. Mech. Des. Syst. Manuf. 2(4)

(2008) 651-660.

[3] S. Graf, G. Staupendahl, A. Kramer, F.A. Muller, High precision materials processing using a novel Q-switched CO2 laser, Opt. Lasers Eng. 66 (2015) 152-157.

[4] S. Yamashita, Y. Inoue, S. Maruyama, Y. Murakami, H. Yaguchi, M. Jablonski, S.Y. Set, Saturable absorbers incorporating carbon nanotubes directly synthesized onto substrates and fibers and their application to mode-locked fiber lasers, Opt. Lett. 29(14) (2004) 1581-1583.

[5] Q.L. Bao, H. Zhang, Y. Wang, Z.H. Ni, Y.L. Yan, Z.X. Shen, K.P. Loh, D.Y. Tang, Atomic-Layer Graphene as a Saturable Absorber for Ultrafast Pulsed Lasers, Adv. Funct. Mater. 19(19) (2009) 3077-3083.

[6] J. Du, Q.K. Wang, G.B. Jiang, C.W. Xu, C.J. Zhao, Y.J. Xiang, Y. Chen, S.C. Wen, H. Zhang, Ytterbium-doped fiber laser passively mode locked by few-layer Molybdenum Disulfide (MoS2) saturable absorber functioned with evanescent field interaction, Scientific Reports 4 (2014) 7.

[7] M. Zhang, G.H. Hu, G.Q. Hu, R.C.T. Howe, L. Chen, Z. Zheng, T. Hasan, Yb- and Er-doped fiber laser Q-switched with an optically uniform, broadband WS2 saturable absorber, Scientific Reports 5 (2015) 10.

[8] A.A.A. Jafry, N. Kasim, Y. Munajat, R.A.M. Yusoff, M.F.M. Rusdi, M.B.H. Mahyuddin, S.W. Harun, H. Arof, R. Apsari, Passively Q-switched erbium-doped fiber laser utilizing lutetium oxide deposited onto D-shaped fiber as saturable absorber, Optik 193 (2019) 162972.

[9] H. Ahmad, R. Safaei, M. Rezayi, I.S. Amiri, Novel D-shaped fiber fabrication method for saturable absorber application in the generation of ultra-short pulses, Laser Phys. Lett. 14(8) (2017) 7.

IOP Conf. Series: Materials Science and Engineering **854** (2020) 012022 doi:10.1088/1757-899X/854/1/012022

Effect of H_2O_2 as the Foaming Agent on the Geopolymer Mortar using Curing of Room Temperature

Faisal Rizal[1,2], Ari Putra Pratama[1,2], Khamistan[1], Amir Fauzi[1,2]*, Syarwan[1,2], Ahyari Azka[1,2]

[1] Civil Engineering Department, Politeknik Negeri Lhokseumawe, Aceh, Indonesia
[2] Geopolymer and Green Technology Research Center, Aceh, Indonesia

Email: faisalrizal@pnl.ac.id

Abstract. Road construction on the peat has been an issue in the world. Yet, due to the reason of topography and alignment through a peat environment, this work remains to be done. The issue in this study was the bearing capacity of soil due to high water content, creep, deformation high acid, so that was required to find the alternative material to support the base of the road with the light material. This study used fly ash as the alternative material in the foam geopolymer mortar that utilized hydrogen peroxide (H_2O_2) with the percentage of 1%, 2%, and 3% fly ash, respectively, as the blowing foaming agent in fresh geopolymer mortar. The alkaline solution was prepared using the solution of Na_2SiO_3 and NaOH with a ratio of 2.0 and 2.5, respectively. The room temperature was used as the curing for 14 days. The results showed that the optimum compressive strength was obtained about 3.53 MPa with a Na_2SiO_3/NaOH of 2.5 and the percentage of H_2O_2 as 1% fly ash. Furthermore, optimum density was obtained about 0.61 gr/cm^3 with a Na_2SiO_3/NaOH of 2.5 and the percentage of H_2O_2 as 3% fly ash.

1. Introduction

Road construction on peaty soil had been an unavoidable issue. It was due to the topography and alignment of the road going through a peat environment so that caused a decline of low soil capacity, high water content, creeping, deformation and a high degree of acidity [1]. The use of material with a low density (1 gr/cm3) and better compressive strength was an alternative material to replace the aggregate which has been used as a foundation material for road pavement. However, the compressive strength of low-density material has to be maintained or better than the compressive strength of normal material [2].

Geopolymer is a material formed by activating alkali and base material that is rich in silica-alumina content. Alkaline activator was a combination of the solution of sodium or potassium [3,4]. Meanwhile, a base material was a by-product material from industry such as fly ash, rice husk ash, metakaolin, and silica fume. Fly ash is a waste material from the combustion residue of coal in the power plants that were rich in content of Si and Al so that was categorized as geopolymer material [5].

Fly ash had been used as an alternative material on lightweight mortar geopolymer. Some investigations had been conducted with the use of H_2O_2 as a foaming agent on geopolymer mortar [6,7]. Described that the use of Hydrogen Peroxide (H_2O_2) on foam geopolymer depend on the alkaline activator/binder ratio and the Na_2SiO_3/ NaOH ratio. The lowest density was reached by the mixture with an activator/binder ratio of 0.6 and 3% hydrogen peroxide content. The ratio of Na_2SiO_3 to NaOH of 4.5 resulted in a density of 734 kg/m, whereas the ratio of Na_2SiO_3 to NaOH of 2.5 resulted in a density of

IOP Conf. Series: Materials Science and Engineering **854** (2020) 012022 doi:10.1088/1757-899X/854/1/012022

749 kg/m3. The addition of aluminum powder in the range of 0.07 to 0.2% showed the density of 0.64-0.74 gr/cm3. Furthermore, the use of H_2O_2 in the aluminum mixture provided the density of 0.61-1 gr/cm3. Finally, the use of low density with the better compressive strength was still an issue in the previous study so that it is expected to be solved in the current study.

2. Methodology

2.1. Material

This study used the material of fly ash Nagan Raya that was derived from nagan raya, Acheh, Indonesia. The solution of NaOH and Na_2SiO_3 was bought from the market in Medan, Indonesia. The liquid of hydrogen peroxide (H_2O_2) and calcium stearate were bought from the chemical online market.

2.2. Mix proportion

The maximum compressive strength was obtained from the trial experimentation of foam geopolymer mortar. The alkaline activator was a combination of NaOH and Na_2SiO_3 solution with a ratio of 2,0 and 2.5. The solution of NaOH was used about 12M which was produced from 400 gr NaOH in 1000 ml H_2O. Furthermore, the use of the foaming agent in this study was 1.0%, 2.0%, dan 3.0% of fly ash Nagan raya as seen in Table 1. The mix procedure of the geopolymer mortar is illustrated in Figure 1.

Table 1. Mix Proportion of 12M NaOH.

The Ratio of Alcaline Activator	Foaming Agent of H_2O_2	Calcium stearate	Tel is of test			
			Compressive strength and density			
			1 hari	3 hari	7 hari	14 hari
2,0	1.0%	0.1%	3	3	3	3
	2.0%	0.2%	3	3	3	3
	3.0%	0.3%	3	3	3	3
2,5	1.0%	0.1%	3	3	3	3
	2.0%	0.2%	3	3	3	3
	3.0%	0.3%	3	3	3	3

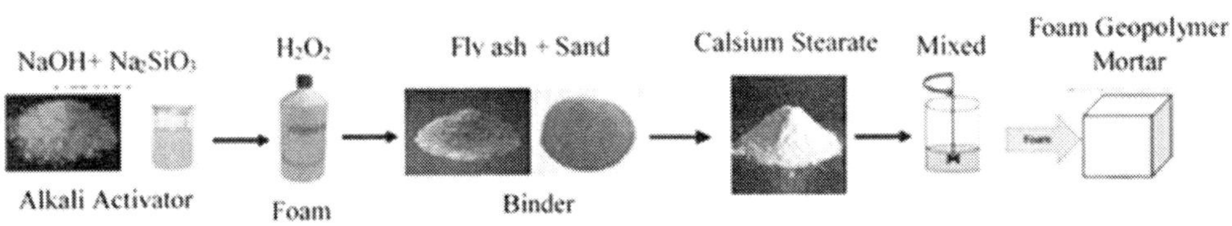

Figure 1. Mix procedure of geopolymer mortar

3. Results and discussion

3.1. X-ray fluorescence

The characteristic material was investigated by XRF analysis that provides the indication of chemical content percentage in the material. Table 2 shows that the material of fly ash nagan raya has the chemical content of SiO_2 and Al_2O_3 of 37.16% and 17.61%, respectively. It means that the material of fly ash nagan raya formed the gel of Si-O-Si and Si-O-Al that contributed to the mechanical properties of geopolymer, especially compressive strength. Furthermore, total chemical content of SiO_2 + Al_2O_3 + Fe_2O_3 is 73.52%. It means that the material of fly ash Nagan raya was categorized as class F fly ash.

IOP Conf. Series: Materials Science and Engineering **854** (2020) 012022 doi:10.1088/1757-899X/854/1/012022

The Ca content of fly ash is 8.72% that is categorized as moderate Ca content. This category provides a similar final setting time with OPC mixture (5-6 hours)

Table 2. Chemical composition of fly ash nagan raya

SiO_2	Al_2O_3	Fe_2O_3	CaO	Na_2O	K_2O	TiO_2	MgO	P_2O_5	SO_3
37,16	17,61	18,75	8,72	0,468	0,788	0,747	6,43	0,139	1,96

3.2. X-ray diffraction

The measurement of fly ash nagan raya by XRD showed in Figure 2. The spectrum in the particle of fly ash nagan raya refers to the database of XRD. The spectrum in figure 2 indicates that the material consists of quartz (SiO_2), mullite ($3Al_2O_3$ $2SiO_2$ or $2Al_2O_3SiO_2$) and hematite (Fe_2O_3). Whereas, the maximum dispersion peak at 20°-30° indicates that the material of fly ash nagan raya was categorized as the amorphous material.

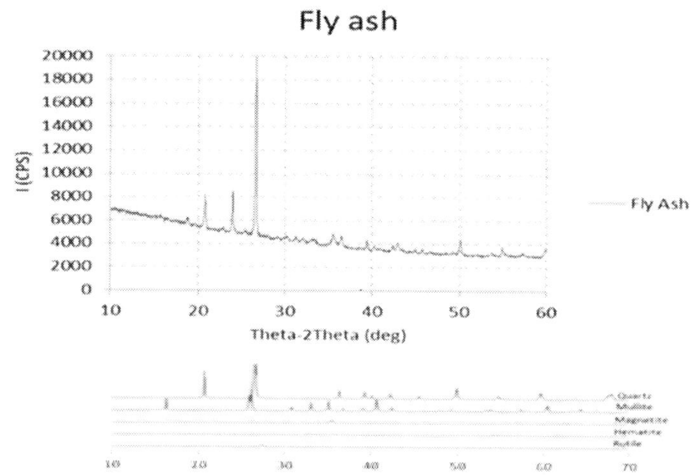

Figure 2. XRD analysis of fly ash nagan raya

3.3. Scanning Electron Microscopy

Figure 3 shows the spherical shape of FA particle with the misty surface. The surface of FA in this study was in contrast with the surface of common FA that is a spherical shape with a sleek surface. The misty surface was considered absorbing the alkaline solution on fresh geopolymer mortar so that caused the reduction of workability. Also, the spherical shape of FA indicated that it was able to fill the concavity of geopolymer binder.

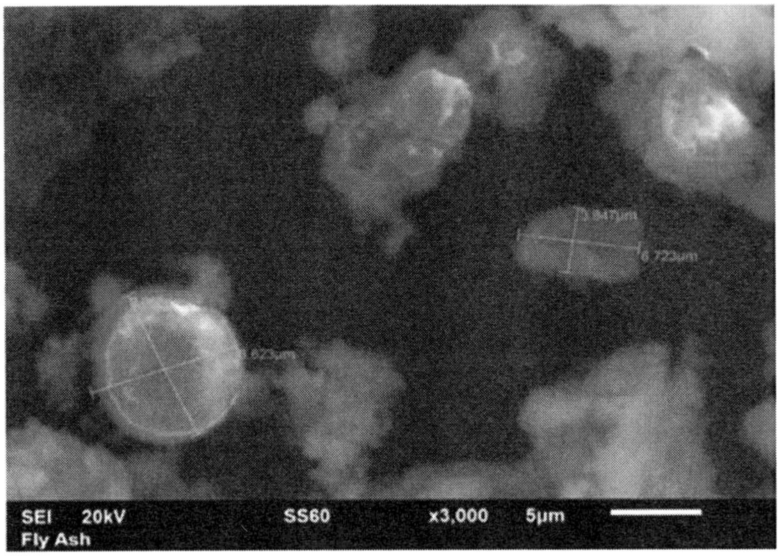

Figure 3. The surface image of Fly Ash particle

3.4. Hardened properties

3.4.1. Compressive strength and density of foaming geopolymer mortar

The specimen for compressive strength used a 50mm x 50mm x 50mm cubes. The test of compressive strength was investigated at 1, 3, 7, and 14 days. Figure 4 shows the difference of Na_2SiO_3 to NaOH ratio of 2.0 and 2.5 on the compressive strength of foaming geopolymer mortar. The figure describes the decrease of compressive strength due to the rise of H_2O_2 percentage in fresh geopolymer mortar. It is attributed to the increase of H_2O_2 percentage caused the increase of the air bubble in fresh geopolymer mortar and finally, it is pore in the geopolymer mortar. Both figures show that the percentage of 1% H_2O_2 in the weight of fly ash provides the optimum compressive strength. However, the ratio of Na_2SiO_3 to NaOH of 2.5 provides better compressive strength than the ratio of Na_2SiO_3 to NaOH of 2.0.

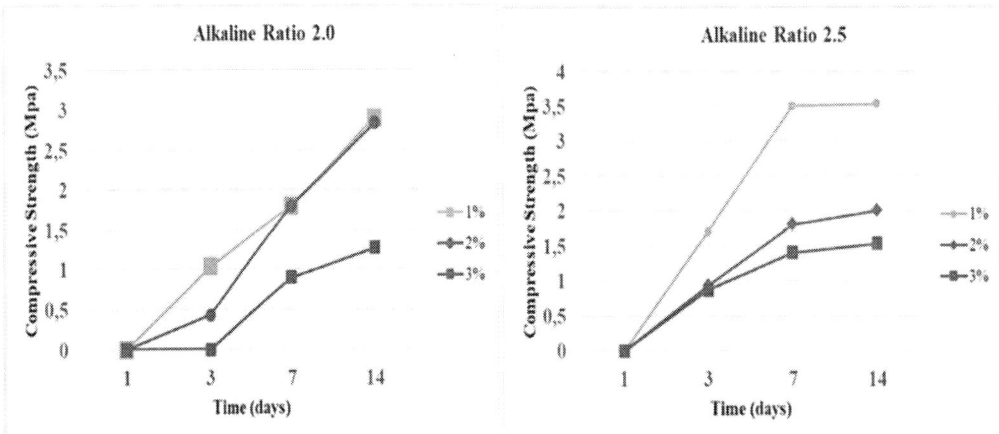

(a) (b)

Figure 4. Compressive strength of foaming geopolymer mortar with (a) the ratio of Na2SiO3 to NaOH of 2.0 (b) the ratio of Na_2SiO_3 to NaOH of 2.5.

IOP Conf. Series: Materials Science and Engineering **854** (2020) 012022 doi:10.1088/1757-899X/854/1/012022

Figure 5 shows the relation between the ratio of Na_2SiO_3 to NaOH on the weight of foaming geopolymer mortar. It indicates that the increase of H_2O_2 percentage in fresh geopolymer mortar reduces the density of geopolymer mortar. It is attributed to the presence of an air bubble that was formed by the reaction of H_2O_2 and binder. Both figures show that the use of different Na_2SiO_3 to NaOH ratio does not impact the density of foaming geopolymer mortar.

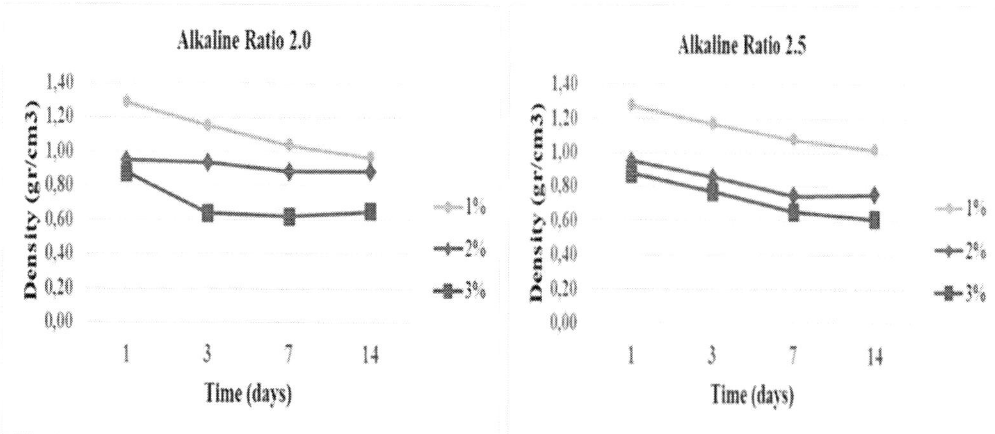

(a) (b)

Figure 5. Density of foaming geopolymer mortar with (a) the ratio of Na_2SiO_3 to NaOH of 2.0 (b) the ratio of Na2SiO3 to NaOH of 2.5

3.5. Relation of density and compressive strength

Figure 6 shows that the relation of density and compressive strength is linear which the decrease of density provides the reduction of compressive strength. The low density was caused by the presence of more pore in foaming geopolymer mortar so that impacted to the compressive strength. It is consistent with the concept of porous material that mentioned high density provides high compressive strength and vice versa. It similar to studies results of Michal lach et. al.

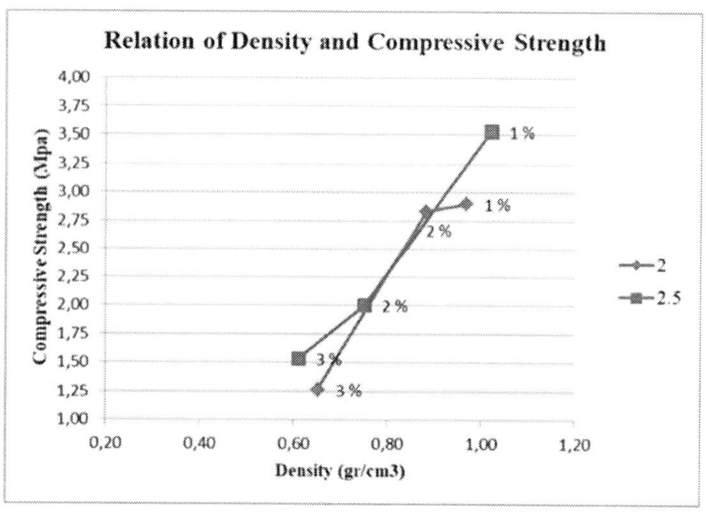

Figure 6. Relation of Density And Compressive Strength

4. Conclusion

This study concluded that the maximum compressive strength was 3.53 MPa that was obtained with the use of Na_2SiO_3 to NaOH ratio of 2.5 and 1% H_2O_2. Whereas, the density was not influenced by the use of Na_2SiO_3 to NaOH ratio. It means that the use of hydrogen peroxide (H_2O_2) caused foaming geopolymer mortar had more pore so that impact on the compressive strength density of foaming geopolymer mortar. This research concluded that fly ash nagan raya was suitable to be used as base material on foaming geopolymer mortar. Furthermore, it was able to be used as the application on direct casting the site.

References

[1] Leith, Peat. et al. 2014. Analysis of operating environments: A diagnostic model for linking science, society and policy for sustainability. Environmental Science & Policy. Vol. 39: P. 162-171.

[2] Sengul, Ozkan. et al. 2011. Effect of expanded perlite on the mechanical properties and thermal conductivity of lightweight concrete. Energy and Buildings. Vol. 43(2): P. 671-676.

[3] Shi, Caijun. et al. 2011. New cements for the 21st century: The pursuit of an alternative to Portland cement. Cement and Concrete Research. Vol. 41(7): P. 750-763.

[4] Zhang, Zuhua. et al. 2011. Geopolymer foam concrete: An emerging material for sustainable construction. Construction and Building Materials. Vol. 56: P. 113-127.

[5] Abdulkareem, Omar A. et al. 2014. Effects of elevated temperatures on the thermal behavior and mechanical performance of fly ash geopolymer paste, mortar and lightweight concrete. Construction and Building Materials. Vol. 50: P. 377-387.

[6] Abdollahnejad, Z.. et al. 2015. Durability Performance of Fly Ash Based One-Part Geopolymer Mortars. Key Engineering Materials. Vol. 634: P. 113-120.

[7] Abdollahnejad, Z.. et al. 2015. Development of Foam One-Part Geopolymers with Enhanced Thermal Insulation Performance and Low Carbon Dioxide Emissions. Advanced Materials Research. Vol. 1129: P. 565-572.

IOP Conf. Series: Materials Science and Engineering **854** (2020) 012023 doi:10.1088/1757-899X/854/1/012023

Thermal Transport Feasibility of (Water + Ethylene Glycol)-Based Nanofluids Containing Metallic Oxides: Mathematical Approach

*Omer A Alawi[1], A R Mallah[2], S N Kazi[2], M N M Zubir[2,1], C S Oon[2]

[1] Department of Thermofluids, School of Mechanical Engineering, Universiti Teknologi Malaysia, 81310 UTM Skudai, Johor Bahru, Malaysia
[2] Department of Mechanical Engineering, University of Malaya, 50603 Kuala Lumpur, Malaysia

Email: omer.adnan85@gmail.com

Abstract. A mathematical approach was used to predict the heat transfer performance of CuO nanoparticles with three different filling ratios [DW], [DW70:30EG] and [DW50:50EG]. There was a discussion on the thermophysical properties. Thermal diffusivity, Prandtl member and pumping power were predicted to find the optimal filling ratio. The effect of nanoparticle shapes on the thermal conductivity and dynamic viscosity was examined. The base fluid of [DW50:50EG] showed significant improvement while DW was optimal for thermal conductivity. Higher thermal conductivity was achieved by DW-based spherical nanofluids, while [DW50:50EG]-based platelets nanofluid showed the best improvement in dynamic viscosity. By increasing the temperature, thermal diffusivity increased while the Prandtl number decreased. The loading of nanofluids did not require more power consumption, and the relative pumping power was less than 1.

Keywords: Metallic oxides nanofluids; Ethylene glycol + water; Thermal transport; Nanoparticle shape; Pumping power

1. Nomenclature

Al_2O_3	Aluminium oxide	$\dot{m}$	Mass flow rate, kg s^{-1}
CuO	Copper (II) oxide	Pr	Prandtl number
DSC	Differential scanning calorimetry	SiO_2	Silicon dioxide
DW	Distilled water	TiO_2	Titanium dioxide
EG	Ethylene glycol	ZnO	Zinc oxide
Fe_3O_4	Iron oxide	ΔP	Pressure drop, Pa m^{-1}
k	Thermal conductivity, W m^{-1} K^{-1}		

Greek symbols

α	Thermal diffusivity, m^2 s^{-1}	ρ	Density, kg m^{-3}
μ	Dynamic viscosity, mPa.s	φ	Volume fraction of nanoparticles
ν	Momentum diffusivity, m^2 s^{-1}		

Subscripts

bf	Basefluid	*eff*	Effective
nf	Nanofluid	*np*	Nanoparticle
r	Ratio		

2. Introduction

In most engineering areas, heat transfer techniques have merged, and much attention has been paid in recent decades to developing more efficient and small heat exchangers. After extensive use of various approaches, such as modifying materials, using extended surfaces or improving process standards, many research activities are now focused on improving the low heat transfer capacities of conventional liquids such as water (DW), ethylene glycol (EG) or engine oils [1]. In this sense, the improvement of heat transfer fluid thermal transport capabilities by dispersing high-conductivity nanoparticles, also known as nanofluids, has become prominent research tools [2].

Nanofluids, including metallic oxides, organic and inorganic materials, were produced using a wide variety of nano-additives. Several studies have shown that nano coolants can be an operational technique for enhancing the performance of heat transport schemes utilizing a combination of EG/DW as a base fluid. [3]. For instance, the experimental results of Vajjha and Das [4] investigated that the thermal conductivity of the [EG60:40DW] combination could be increased by about 12.3% by mixing ZnO solid particles (29 nm) with a volume fraction of 2%. An experimental investigation of the effectiveness of Fe_3O_4 nanoparticles (13nm) was performed on three various forms of ethylene glycol/water mixture with weights 20:80, 40:60, and 60:40 [10]. The researchers realized that the thermal conductivity improvements of the three EG/water combinations stated directly above were 46%, 42% and 33% respectively, with a temperature of 60°C and a volume fraction of 2%.

Thermal conductivity (k) of nanofluids has received much consideration in recent decades due to the high impact of its characteristic on heat transfer performance of thermal installations [5], [6], [7]. However, other thermophysical properties, such as dynamic viscosity (μ), density (ρ) or even specific heat capacity (cp), should also be considered to determine whether the replacement of the conventional liquid with the novel nanofluid would be useful to conduct technical calculations for thermal installations.

Thermophysical properties are very significant factors affecting the nanofluids heat and mass transfer efficiency [8]. Since dynamic viscosity presented a significant impact on internal flow resistance, the Reynolds number, and low pressure, many investigations have been performed on dynamic viscosity of different nanofluids types. As Azmi *et al.* [9] reported by dispersing TiO_2 nanoparticles, the dynamic viscosity of the [EG40:60DW] (volume ratio) mixture could be augmented. For example, if the volume fraction of nanoparticles increased from 0.5% to 1.5%, the dynamic viscosity increase was about 12%. The viscosity variations of (EG/water)-based Fe_3O_4 nanofluid with different fractions of nanoparticle and preparation temperatures were investigated by Sundar *et al.* [10].

The researchers exhibited that the dynamic viscosity of (EG/water)-based nanofluids could be improved by increasing the concentration of nanoparticles and lowering the production temperature. The dynamic viscosity of the base fluid could be improved by 2.9 times by adding a 1% volume fraction of nanoparticles. The dynamic viscosity of various (EG/water)-based nanofluids with the influences of different factors were measured by many studies [6], [11]. Based on their experimental results, the suspension of nanoparticles could increase the viscosity of the base fluid in different amounts. Furthermore, the characteristics of temperature, base fluid, and nanoparticles, including the concentration, particle size, particle type, and particle shape, were the significant factors affecting the viscosity of nanofluids.

The use of water-EG mixture provided a wider range of operating temperatures in the heat transfer study due to the combined properties of water and EG. The current analysis objective is to predict heat transfer performance with three water-EG mixture ratios using CuO nanoparticles. The thermophysical properties and effects of nanoparticles shapes have been investigated. The rate of heat transfer and electricity consumption were predicted.

IOP Conf. Series: Materials Science and Engineering **854** (2020) 012023 doi:10.1088/1757-899X/854/1/012023

3. The Study Methodology

Heat transfer characteristics of (Water + Ethylene Glycol)-based nanofluids containing CuO with a volume fraction of 4% have been investigated using several mathematical correlations. In this research, three different filling ratios [DW, (DW70:30EG), (DW50:50EG)] were discussed. The preparation temperature of nanofluids ranged from 25-50°C and nanoparticle shapes (blades, platelets, cylindrical, bricks, and spherical) to estimate the impact of effective thermal performance of nanofluids. The significance of Prandtl number, thermal diffusivity, the property of enhancement and pumping power have been explained. Base fluids and nanoparticles properties have been listed in Tables 1-4 (data were taken from the National Institute of Standards and Technology).

Table 1. Thermophysical properties of distilled water.

Temp.	ρ *(kg/m³)*	Cp *(J/kg·K)*	k *(W/m-K)*	μ *(mPa-s)*	Pr	α *(m²/s)*
25	997	4073	0.612	0.868	5.8	1.51E-07
30	995	4070	0.619	0.777	5.1	1.53E-07
35	994	4069	0.626	0.701	4.6	1.55E-07
40	992	4067	0.632	0.635	4.1	1.57E-07
45	990	4067	0.638	0.580	3.7	1.59E-07
50	988	4066	0.644	0.532	3.4	1.60E-07

Table 2. Thermophysical properties of (DW70:30EG).

Temp.	ρ *(kg/m³)*	Cp *(J/kg·K)*	k *(W/m-K)*	μ *(mPa-s)*	Pr	α *(m²/s)*
25	1035	3730	0.487	1.864	14.3	1.26E-07
30	1032	3746	0.490	1.623	12.4	1.27E-07
35	1030	3761	0.493	1.428	10.9	1.27E-07
40	1027	3776	0.495	1.269	9.7	1.28E-07
45	1024	3791	0.497	1.137	8.7	1.28E-07
50	1021	3806	0.499	1.026	7.8	1.29E-07

Table 3. Thermophysical properties of (DW50:50EG).

Temp.	ρ *(kg/m³)*	Cp *(J/kg·K)*	k *(W/m-K)*	μ *(mPa-s)*	Pr	α *(m²/s)*
25	1053	3298	0.430	3.367	25.8	1.24E-07
30	1050	3308	0.434	2.936	22.4	1.25E-07
35	1048	3319	0.438	2.565	19.4	1.26E-07
40	1045	3329	0.442	2.244	16.9	1.27E-07
45	1042	3340	0.445	1.968	14.8	1.28E-07
50	1039	3351	0.449	1.730	12.9	1.29E-07

Table 4. Thermophysical properties of CuO nanoparticles [12].

Thermophysical properties	CuO
Density, ρ *(kg/m³)*	6500
Dynamic viscosity, μ *(mPa-s)*	-
Thermal conductivity, k *(W/m-K)*	20
Specific heat, cp *(J/kg-K)*	535.6

3.1. Thermophysical Properties

The thermal conductivity enhancement ratio in the literature was defined as the thermal conductivity ratio of the nanofluid to the thermal conductivity of the base fluid (K_{eff}/K_{bf}). To estimate the thermal conductivity measurements, Maxwell model [13] has been used in current research to estimate thermal conductivity measurements.

$$\frac{k_{eff}}{k_{bf}} = \frac{k_{np} + 2k_{bf} + 2\varphi(k_{bf} - k_{np})}{k_{np} + 2k_{bf} - \varphi(k_{bf} - k_{np})} \quad (1)$$

Nanofluid viscosity, like thermal conductivity, is a vital transport property. Compared to thermal conductivity, nanofluid viscosity literature is still difficult to find. The majority of formulations were extended to express viscosity in accordance with the concentration of the nanoparticles. Temperature is also a significant factor in viscosity, and some correlations have therefore been designed to analyze the effects of temperature on viscosity. Einstein's model [14] was used to estimate the effective viscosity in the present study.

$$\mu_{eff} = \mu_{bf}(1 + 2.5\varphi), \varphi < 0.05 \quad (2)$$

The nanofluid density is usually given to be a mixed density property of base fluid ρ_{bf} and nanoparticles ρ_{np}. Various density meters were employed to evaluate the nanofluid density [15],[16]. The measurement results showed the density of nanofluids using Eq. (3) the mixing theory can be estimated very precisely. Specific heat capacity can simply be described as the heat required to raise 1 g of a substance's temperature by 1°C. The specific heat of solids is normally lower than that of liquids. So, when solid nanoparticles are added to the base fluid, the specific heat is reduced. However, to remove more heat from an ideal coolant, a higher value of specific heat is preferred [17]. For the calculation of the particular nanofluid heat, Eq. (4) is a commonly accepted correlation [18].

$$\rho_{nf} = (1 - \varphi)\rho_{bf} + \varphi\rho_{np} \quad (3)$$

$$Cp_{nf} = \frac{(1 - \varphi)(\rho Cp)_{bf} + \varphi(\rho Cp)_{np}}{(1 - \varphi)\rho_{bf} + \varphi\rho_{np}} \quad (4)$$

3.2. Different Nanoparticle Shapes

This paper discusses five different shapes of nanoparticles (sphere, plates, blades, cylindrical and bricks). To study the influences of these nanoparticle shapes on the thermophysical characteristics of nanofluid, Timofeeva et al. [19] presented the following equations:

$$\frac{k_{eff}}{k_{bf}} = 1 + (C_k^{Shape} + C_k^{Surface})\varphi = 1 + C_k^{\varphi} \quad (5)$$

The effective thermal conductivity can be achieved by utilizing the data from Table 5.

$$\mu_{eff} = \mu_{bf}(1 + A_1\varphi + A_2\varphi^2) \quad (6)$$

Where A1 and A2 are the constants exhibited in Table 6.

Table 5. Effect of the particle shape and the thermal conductivity resistance of nanoparticles [20].

Type	Aspect Ratio	C_K	C_K^{Shape}	$C_K^{Surface}$
Platelets	1:1/8	2.61	5.72	-3.11
Blades	1:6:1/12	2.74	8.26	-5.52
Cylindrical	1:8	3.95	4.82	-0.87
Bricks	1:1:1	3.37	3.72	-0.35

Table 6. Viscosity enhancement coefficients [20].

Coefficient	Platelets	Blades	Cylindrical	Bricks
A1	37.1	14.6	13.5	1.9
A2	612.6	123.3	904.4	471.4

3.3. Prandtl Number and Thermal Diffusivity

The Prandtl number is a non-dimensional number that describes the difference between momentum and thermal diffusivity. Thermal diffusivity is the rate of heat transport from the hot side of a material to the cold side- a measure of how fast a material can absorb heat from its environment. It can be calculated by taking thermal conductivity at constant pressure divided by density and specific thermal capacity.

$$Pr = \frac{\text{Momentum diffusivity}}{\text{Heat diffusivity}} = \frac{\nu}{\alpha} = \frac{\mu/\rho}{\kappa/\rho cp} = \frac{cp\mu}{\kappa} \quad (7)$$

3.4. Pumping Power

The design of heat exchangers for efficient heat transfer and minimum pumping power is important in terms of energy savings and could cause significant errors in the assessment of nanofluid performance (pumping power and heat transfer) in different thermal applications. The pumping power or work required for the circulation of coolant can be calculated using Eq. 8.

$$P_{Pump} = \Delta P \frac{\dot{m}}{\rho} \quad (8)$$

where ΔP is the pressure drop, $\dot{m}$ is the mass flow rate and ρ is the fluid density. In the case of a fully developed condition and a turbulent region in a circular tube with a uniform heat flow on the wall, Eq. 9 could be used to express the pumping power [21].

$$\left(\frac{\dot{W}}{W_{bf}}\right) = \left(\frac{\mu}{\mu_{bf}}\right)^{0.25} \left(\frac{\rho_{bf}}{\rho}\right)^2 \quad (9)$$

4. Results and Discussion

Thermophysical properties of CuO nanoparticles for different temperatures and different base fluids were shown in Figure 1. The values of thermal conductivity increased by increasing the temperature of preparation (25-50°C). CuO/DW nanofluids show the higher thermal conductivity values followed by CuO/(DW70:30EG) and CuO/(DW50:50EG), respectively. By increasing the temperature from 25 to 50°C, the thermal conductivities increased by 5.14%, 4.38% and 2.58% for CuO/DW, CuO/(DW50:50EG) and CuO/(DW70:30EG), respectively. The filling ratio of (DW50:50EG) based nanofluid showed the highest increment of effective viscosity, while distilled water-based nanofluid showed the lowest. Nanofluids containing solid nanoparticles have a higher viscosity than common working fluids; hence, measuring the viscosity is necessary for designing thermal systems and estimating the required pumping power. In engineering applications, it is interesting to investigate the effect of both concentration and temperature on thermophysical property. The density of CuO nanoparticles with a volume fraction of 4%, and the use of three different filling ratios [DW, (DW70:30EG), (DW50:50EG)] is shown in Figure 1. The values of density decreased by increasing the temperature of preparation from 25 to 50°C. CuO/DW based nanofluid showed the highest increase of 22.12% followed by (DW70:30EG) and (DW50:50EG) with 21.16% and 20.67%, respectively at 25°C.

IOP Conf. Series: Materials Science and Engineering **854** (2020) 012023 doi:10.1088/1757-899X/854/1/012023

Most researchers paid greater attention to the thermal conductivity of the nanofluids. Some others emphasized the viscosity of nanofluid. However, specific heat (*Cp*) is also extremely important for nanofluids. The most demanding properties for analyzing energy and exergy are the specific heat of nanofluid. The specific heat capacity of CuO nanoparticle with a volume fraction of 4%, and three different filling ratios [DW, (DW70:30EG), (DW50:50EG)] is displayed in Figure 1. Specific heat of (DW70:30EG) and (DW50:50EG) based nanofluids increased by increasing the temperature while distilled water behaved quite differently.

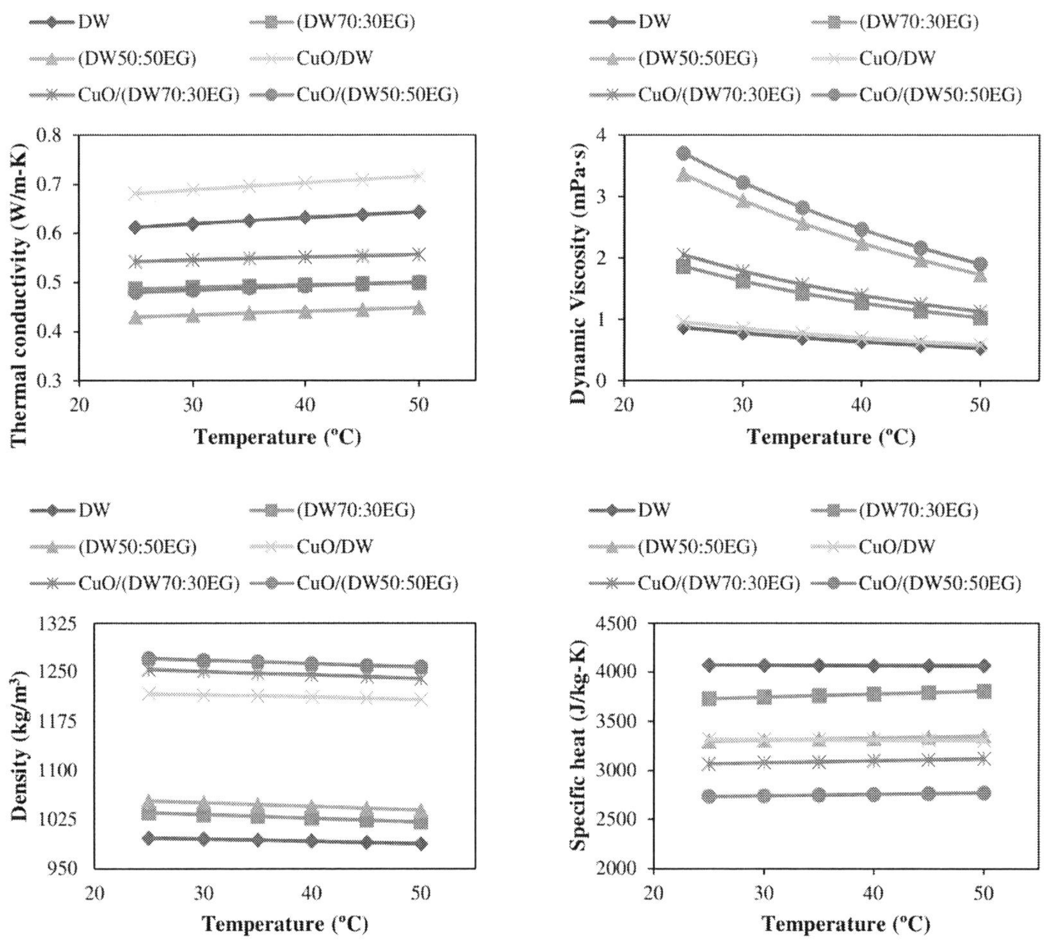

Figure 1. Thermophysical properties of CuO nanoparticles for different temperatures and different basefluids.

The effect of nanoparticle shapes on thermal conductivity and dynamic viscosity of CuO nanofluids was presented in Figure 2. Water-based CuO spherical nanoparticles showed the highest thermal conductivity enhancement of 11.2% while (DW50:50EG)-based CuO bricks nanoparticles showed the lowest at 5%. (DW50:50EG) based platelets nanofluids displayed the highest increment of effective viscosity followed permanently by (DW50:50EG) based cylindrical nanofluids. While the lowest increment was achieved by distilled water-based spherical nanofluids.

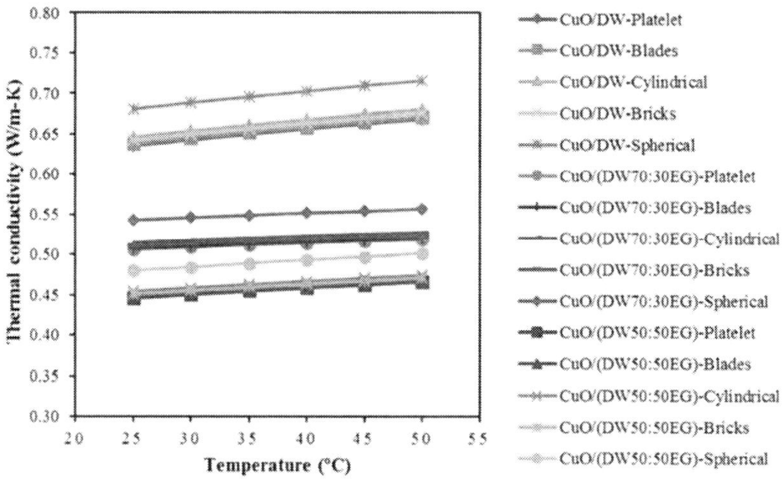

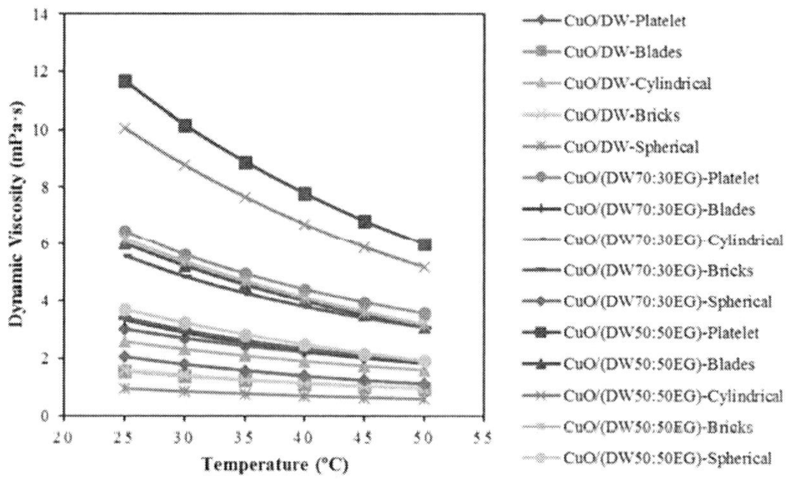

Figure 2. Effect of nanoparticles shapes on thermal conductivity and dynamic viscosity for different temperatures.

As shown in Figure 3, the thermal diffusivity of CuO nanofluids increased by increasing the temperature. The combination of distilled water-based nanofluids achieved the highest thermal diffusivity followed by (DW70:30EG) and (DW50:50EG) based nanofluids, respectively. As shown in Figure 3, Prandtl number decreased by increasing the temperature. CuO/(DW50:50EG) showed the highest values followed by CuO/(DW70:30EG) and CuO/DW.

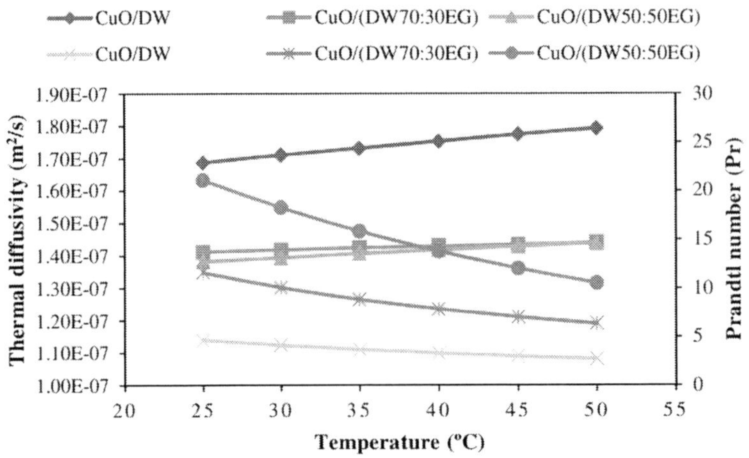

Figure 3. Thermal diffusivity and Prandtl number of CuO nanofluid types at different filling ratios.

Energy consumption and pumping power characteristics are serious limits in terms of economy and energy saving. Pumping power can be considered as an economic performance indicator in the thermal system for evaluating the operability test of fluid and performance of the power plant. Besides, the design of heat exchangers for effective heat transfer and minimum pumping power is important in terms of energy savings and could cause considerable errors when evaluating the performance of nanofluids (pumping power and heat transfer) in various thermal applications. The pumping power of different nanofluid types with three different filling ratios and at various temperatures is compared in Figure 4. Figure 4 showed that there was no increase in the pumping power with the nanofluids loading, and the effect of temperature variation is negligible [22].

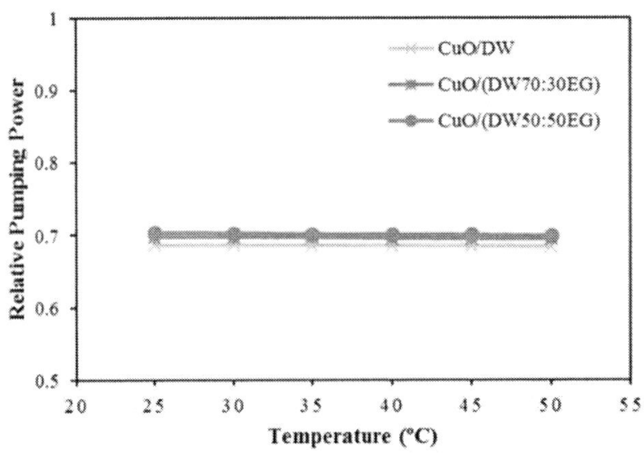

Figure 4. Relative pumping power of CuO nanofluid at different filling ratios and different temperatures.

5. Conclusions

The present paper has mathematically discussed the feasibility of heat transfer characterizes using CuO nanoparticles with three different base fluid mixture ratios [DW], [DW70:30EG] and [DW50:50EG].

Many parameters have been tested during the investigations; the following conclusions have been drawn from the present study.

1. Density decreased by decreasing the temperature. [DW50:EG50] and [DW70:EG30]-based nanofluids exhibited a higher density than water-based nanofluids.
2. Specific heat of [DW50:EG50] and [DW70:EG30]-based nanofluids increased by increasing temperature whereas, the specific heat of water-based nanofluids decreased.
3. Water-based nanofluids showed the highest enhancement in thermal conductivity followed by [DW70:EG30] and [DW50:EG50]- based nanofluids.
4. Dynamic viscosity of water, [DW50:EG50] and [DW70:EG30]-based nanofluids decreased when preparation temperature increased.
5. Thermal diffusivity of water, [DW50:EG50] and [DW70:EG30]-based nanofluids increased by increasing temperature while Prandtl number decreased.
6. Relative pumping power was less than 1, this means no increase in the pumping power with the nanofluids loading, and the effect of temperature variation is negligible.
7. Water-based spherical nanofluids showed the highest thermal conductivity enhancement while [DW50:EG50]-based platelets nanofluids demonstrated the maximum viscosity enhancement.

Acknowledgement
This project is funded by Research University (RU) Grant Faculty Program (No- GPF019A-2019, GPF050A-2018), University of Malaya Research (UMRG) Grant (No- RP045C-17AET) and Research University (RU) Grant (RU019E-2017).

References

[1] M. Mehrali *et al.*, "Effect of specific surface area on convective heat transfer of graphene nanoplatelet aqueous nanofluids," *Exp. Therm. Fluid Sci.*, 2015.
[2] D. Cabaleiro, L. Colla, S. Barison, L. Lugo, L. Fedele, and S. Bobbo, "Heat transfer capability of (ethylene glycol+water)-based nanofluids containing graphene nanoplatelets: Design and thermophysical profile," *Nanoscale Res. Lett.*, vol. 12, no. 1, p. 53, 2017.
[3] Y. Gao, H. Wang, A. P. Sasmito, and A. S. Mujumdar, "Measurement and modeling of thermal conductivity of graphene nanoplatelet water and ethylene glycol base nanofluids," *Int. J. Heat Mass Transf.*, vol. 123, pp. 97–109, 2018.
[4] R. S. Vajjha and D. K. Das, "Experimental determination of thermal conductivity of three nanofluids and development of new correlations," *Int. J. Heat Mass Transf.*, 2009.
[5] W. H. Azmi, K. V. Sharma, R. Mamat, G. Najafi, and M. S. Mohamad, "The enhancement of effective thermal conductivity and effective dynamic viscosity of nanofluids - A review," *Renewable and Sustainable Energy Reviews*. 2016.
[6] H. W. Chiam, W. H. Azmi, N. A. Usri, R. Mamat, and N. M. Adam, "Thermal conductivity and viscosity of Al_2O_3 nanofluids for different based ratio of water and ethylene glycol mixture," *Exp. Therm. Fluid Sci.*, vol. 81, pp. 420–429, 2017.
[7] L. S. Sundar, M. H. Farooky, S. N. Sarada, and M. K. Singh, "Experimental thermal conductivity of ethylene glycol and water mixture based low volume concentration of Al2O3and CuO nanofluids," *Int. Commun. Heat Mass Transf.*, 2013.
[8] H. Kaya, K. Arslan, and N. Eltugral, "Experimental investigation of thermal performance of an evacuated U-Tube solar collector with ZnO/Etylene glycol-pure water nanofluids," *Renew. Energy*, vol. 122, pp. 329–338, 2018.
[9] W. H. Azmi, K. Abdul Hamid, R. Mamat, K. V. Sharma, and M. S. Mohamad, "Effects of working temperature on thermo-physical properties and forced convection heat transfer of TiO_2 nanofluids in water-Ethylene glycol mixture," *Appl. Therm. Eng.*, 2016.
[10] L. Syam Sundar, E. Venkata Ramana, M. K. Singh, and A. C. M. De Sousa, "Viscosity of low volume concentrations of magnetic Fe 3O 4 nanoparticles dispersed in ethylene glycol and water mixture," *Chem. Phys. Lett.*, vol. 554, pp. 236–242, 2012.
[11] S. K. Lim, W. H. Azmi, and A. R. Yusoff, "Investigation of thermal conductivity and viscosity of Al_2O_3/water–ethylene glycol mixture nanocoolant for cooling channel of hot-press forming die application," *Int. Commun. Heat Mass Transf.*, vol. 78, pp. 182–189, 2016.
[12] O. A. Alawi and N. A. C. Sidik, "Mathematical correlations on factors affecting the thermal

conductivity and dynamic viscosity of nanorefrigerants," *Int. Commun. Heat Mass Transf.*, vol. 58, pp. 125–131, 2014.
[13] M. James Clerk, *A treatise on electricity and magnetism*. 2010.
[14] A. Einstein, "A new determination of molecular dimensions," *Ann. Phys.*, 1906.
[15] C. J. Ho, W. K. Liu, Y. S. Chang, and C. C. Lin, "Natural convection heat transfer of alumina-water nanofluid in vertical square enclosures: An experimental study," *Int. J. Therm. Sci.*, 2010.
[16] R. S. Vajjha, D. K. Das, and B. M. Mahagaonkar, "Density measurement of different nanofluids and their comparison with theory," *Pet. Sci. Technol.*, vol. 27, no. 6, pp. 612–624, 2009.
[17] E. Sadeghinezhad *et al.*, "A comprehensive review on graphene nanofluids: Recent research, development and applications," *Energy Convers. Manag.*, vol. 111, no. March 2016, pp. 466–487, 2016.
[18] Y. Xuan and W. Roetzel, "Conceptions for heat transfer correlation of nanofluids," *Int. J. Heat Mass Transf.*, vol. 43, no. 19, pp. 3701–3707, 2000.
[19] E. V. Timofeeva, M. R. Moravek, and D. Singh, "Improving the heat transfer efficiency of synthetic oil with silica nanoparticles," *J. Colloid Interface Sci.*, 2011.
[20] S. M. Vanaki, H. A. Mohammed, A. Abdollahi, and M. A. Wahid, "Effect of nanoparticle shapes on the heat transfer enhancement in a wavy channel with different phase shifts," *J. Mol. Liq.*, 2014.
[21] R. Ben Mansour, N. Galanis, and C. T. Nguyen, "Effect of uncertainties in physical properties on forced convection heat transfer with nanofluids," *Appl. Therm. Eng.*, vol. 27, no. 1, pp. 240–249, 2007.
[22] W. Yu, D. M. France, E. V. Timofeeva, D. Singh, and J. L. Routbort, "Comparative review of turbulent heat transfer of nanofluids," *International Journal of Heat and Mass Transfer*, vol. 55, no. 21–22. pp. 5380–5396, 2012.

IOP Conf. Series: Materials Science and Engineering **854** (2020) 012024 doi:10.1088/1757-899X/854/1/012024

Optimization Design of Electrochemical Machining Process of SKD11 Tool Steel Using Weighted Principal Component Analysis (WPCA)

N Lusi, D R Pamuji, A Afandi and G S Prayogo

Mechanical Engineering Department, State Polythecnic of Banyuwangi, Jalan Raya jember Km. 13 Kabat Labanasem Banyuwangi, Indonesia.

E-mail: nurainilusi@poliwangi.ac.id

Abstract. This research was carried out on electrochemical machining (ECM) process using a workpiece material of SKD 11 tool steel and electrode of brass. Three process variables, i.e., voltage, electrolyte concentration and gap width with three levels for each process variables investigated. Based on the number of process variables and its level, an orthogonal array of L9 and two times replications employed in the design of the experiment. Setting a combination of significant machining parameters to maximize the material removal rate and minimize the surface roughness of the workpiece based on the results of optimization using the Taguchi method and weighted principal component analysis is a combination of voltage factors at level 3 of 48 V, electrolyte concentration at level 2 of 150 g/l, and gap width at level 1 of 1 mm. Machining parameter that has the greatest contribution is an electrolyte concentration which is 41.98% , then the contribution of voltage of 32.33%, and the gap width is 5.63%. Based on the results of confirmation experiments reveal that principal component analysis can effectively acquire the optimal combination of cutting parameters.

1. Introduction

ECM is a type of non-conventional machining process where the tool does not come into contact with the workpiece. The working principle of ECM is based on Faraday's law. The process of ECM applies chemical reactions and is accelerated with electrical energy; it is due to the oxidation reaction and reduction during electrolysis. At ECM the reduction and oxidation reactions are indicated by the decay (electron loss) of the work piece, and the addition of mass positive metal ions mass (attachment of positive metal ions to the tool). In the machining process, the ECM tool (electrode) does not come into direct contact with the work piece. In the machining process, the ECM tool (electrode) does not come into direct contact with the work piece. An electrical current passes through an electrolyte solution between a cathode tool and an anode workpiece [1].

The ECM machining process has been widely used in machining hard-to-cut materials that cannot be processed by conventional machinery, such as stainless steel materials, ceramics, fiber-reinforced composites, super alloys, etc. Material due to its high hardness, strength, brittleness, toughness, and compatibility with low machining properties. Also, with the implementation of the ECM process, all products with complex shapes can be produced with high accuracy and precision [2]. Another advantage of this process is that there is no mechanical energy or thermal energy involved and free-stress [3]. ECM applications are very popular and have been widely used in industries such as automotive, the aircraft industry to engine turbine blades, gears, aerospace, biomedical equipment, construction, and tribology [4]–[8].

ECM processes involve several parameters that will affect ECM machining performance. The use of optimal parameters in the ECM process can significantly improve product quality. Therefore, optimization of process parameters to determine the optimal combination of parameter values in ECM machining is very important to do. Electrolytes are a critical part of ECM which influences the machining process. One of the functions of electrolytes is to absorb the oxidation of metals produced in cutting. Determination of the type of electrolyte solution has a significant effect on the product quality of the ECM process. The use of NaCl can provide higher machining current efficiency compared to NaNO3. This is because the removal of metal starts from a low current density.

Liu et al. [9]experimented micro electrochemical machining (Micro ECM) of Stainless Steel 304 by using composite electrolyte containing neutral salt solution and complexing agents. The complexing agents in electrolyte compete for metal ions to dissolve the insoluble electrolytic products. Experimental results show that the material removal rate (MRR) increased 35% and verified to be effective to increase the machining depth of micro-holes.

Sankar et al. [10] investigated the effect of electrolyte concentration on the Abrasive Assisted-Electrochemical Machining on material removal rate, surface roughness and radial overcut. Aluminium 6061 -boron carbide (5-15 wt.%) composites used as work piece. Surface roughness increase with an increase in electrolyte concentration. The material removal rate of composite material decreased with increasing in boron carbide reinforcement as a result of reinforcement particles disturbed the composite's continuous electrical conductivity. Another machining characteristic such as radial overcut of machined work material increased with increases in electrolyte concentration due to excessive dissolution in and around the reinforced particles.

In the ECM process, the material removal rate is not only influenced by the electrolyte liquid concentration. There are several other process parameters such as gap width, voltage, and current. Determination of the combination of the right process parameters achieving optimum response has been the focus research on ECM processes. In this case, selection of method will be directly related to the quality of the product.

Taguchi method is one method used in process optimization. The application of the Taguchi method was initially only to optimize a single response, while for multiresponse cases there are several researchers who develop multiresponse optimization methods. Chakradhar et al. [11] presented the optimization of the electrochemical machining of EN-31 steel by using Taguchi method combined with grey relational analysis (GRA). The optimal process parameters, i.e., material removal rate can be maximized, and the overcut, cylindricity error and surface roughness can be minimized through this method. Soni et al. [12] applied a genetic algorithm (GA) and Taguchi method to maximize the metal removal rate and minimize the surface roughness and overcutting. The result obtained by using GA, electrolyte concentration and Voltage is the most significant machining parameters for affecting the MRR and Overcut & Surface Roughness. Hoda et al. [13] used an artificial neural network (ANN) to predict values of process parameters such as metal removal rate (MRR), and surface roughness (Ra). Based on experiments that have been carried out it can be stated that ANN is an efficient approach because it reduces the time & effort needed to predict MRR & surface roughness if they are found experimentally using trial and error methods. Deepanshu et al. [14] optimize the response of MRR, overcut and surface roughness (Ra) of material mild steel with a diameter of 50 mm by using a copper electrode and solution of salt water as electrolyte. The method used is the orthogonal array Taguchi L9.

Based on evaluations from existing studies, research on determination setting ECM process parameters to optimize multiple responses that can be observed simultaneously still need to be done. In this study a multi-response optimization was carried out on SKD 11 material processed by ECM, the response to be observed was MRR and surface roughness. Experiment design using the L9 orthogonal array. The multiresponse optimization method that will be used is Taguchi combined with weighted principal component analysis (WPCA).

2. Experimental Procedure

Experiments are conducted by using the prototype of electrochemical machining equipment which is shown in Figure 1. The ECM device consists of a machining chamber, control panel, and electrolyte circulation system [15] . The workpiece position is fixed inside the machining chamber, and the cathode (tool) is attached to the main screw which is driven by a stepper motor. A rectangular block of 20 mm X 12 mm and 6mm height made of SKD 11 tool steel and brass as an electrode which is shown in Figure 2 are chosen as the work piece for carrying out the experimentation to optimize the material removal rate. Sodium chloride (NaCl) electrolyte tends to produce a matte finish with alloy steels[16].

Material removal rate (MRR) = Loss of weight /Machining time (g/min) (1)

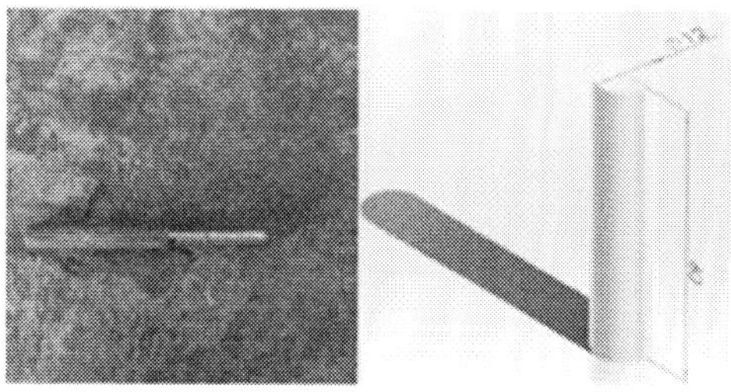

Figure 1. The dimension of The tool

Figure 2. The Electrochemical Machining

Determination of design of experiments is crucial before the experiment began. In these experiments involves three parameters of machining processes, namely voltage, electrolyte concentrations, and the gap width are varied. Each process parameters have three different levels as described in Table 1.

Table 1. Machining Parameters And Their Level

Machining Parameters	**Units**	**Level of Machining Parameters**		
		1	**2**	**3**
Voltage	Volt	24	36	48
Electrolyte concentrations	g/l	100	150	200
Gap width	mm	1	2	3

This research aims to find out and determine the value of the best response from the response of the material removal rate (MRR) and surface roughness (SR), considering the number of parameters included, Taguchi experimental design method was chosen to decide the grouping of investigations. Taguchi method offers some run fewer compared to the full factorial design, with the sequence run as is shown in Table 2.

An L9 orthogonal array has been chosen based on Taguchi quality design concept for experimented. The process parameters interaction effect has been assumed to be imperceptible[17]. The experiment was run randomly and rerun two times to attain the accuracy and validity of the test. Taguchi design method is to identify the machining parameter settings which render the quality of the product or process robust to unavoidable variations in external noise.

Table 2. An L9 Orthogonal Array for Experimentation

Run No.	**Voltage** (v)	**Electrolyte Concentrations** (g/l)	**Gap width** (mm)
1	24	100	1
2	24	150	2
3	24	200	3
4	36	100	2
5	36	150	3
6	36	200	1
7	48	100	3
8	48	150	1
9	48	200	2

3. Weighted Principal Component Analysis (WPCA)

The Weighted Principal Component Analysis (WPCA) method is used to eliminate correlations between responses and to convert correlated responses into uncorrelated response indices called with the principal components (principal components). Then, a value of Combined Quality Loss (CQL) which is defined as the deviation of the MPI (Multi Performance Index) value from the desired ideal value. This CQL functions as a single response function with the aim of reducing the correlation between responses in the process. In this WPCA method, all components are taken into consideration in order to completely explain variation in all responses [18].

3.1. Multiresponse Optimization

The Couple of the Taguchi method and WPCA method following steps to be performed:

1. Converting the experimental data into S/N values.
   The relative quality of a particular parameter design is evaluated using a generic signal to noise (S/N) ratio. Depending on the particular design problem, different S/N ratios are applicable, including larger is better and smaller is better. The S/N ratios for each type of characteristics can be calculated as follows:

   Larger is better (maximize):

$$S/N = -10 \log \left[\sum_{i=1}^{n} \frac{(1/y_i^2)}{n} \right] \quad (2)$$

   and

   Smaller is better (minimize):

$$S/N = -10\log\left[\sum_{i=1}^{n} \frac{y_i^2}{n}\right] \quad (3)$$

2. Normalizing the S/N ratio by using the following equation:

$$X_i^*(k) = \frac{X_i(k) - min\, X_i(k)}{max\, X_i(k) - min\, X_i(k)} \quad (4)$$

3. Calculating pearson of colleration.
   After data normalization a check has to be made whether responses are correlated or not.
4. Calculating the principal component (PCA)
   In order to eliminate response correlation Principal Component Analysis has been applied (Eigen value, Eigen vector, accountability proportion AP and cumulative accountability proportion CAP).
5. Calculating Multi Performance Index (MPI) using the following equation:

$$MPI = \sum_{j=1}^{k} W_j\, Y_i \quad (5)$$

6. Selecting the optimal levels of process parameters.
7. Conducting confirmation experiments and verify the optimal process parameters setting.

4. Experimental Result and Discussion

The experimental result and S/N ratio data for the MRR and surface roughness are depicted in Table 3.

Table 3. Response Data and S/N Ratio

Run No	MRR (g/min)	S/N	Ra (μm)	S/N
1	0.067	-23.479	1.992	-11.097
2	0.08	-21.938	2.367	-7.484
3	0.06	-24.437	3.449	-10.754
4	0.052	-25.680	2.728	-8.717
5	0.07	-23.098	3.562	-11.034
6	0.053	-25.514	4.24	-10.555
7	0.069	-23.223	4.193	-7.952
8	0.153	-16.306	4.271	-12.002
9	0.07	-23.098	4.828	-6.188

Hence the range of data series is too large, and the optimal value of the quality characteristics is too large, and the influence of some parameters might be ignored. The experimental data must be normalized first to eliminate the effect above. A linear normalization of the S/N ratio of experimental result for the experiment for the responses are performed in the range between 0 and 1. There are three different types of data normalization according to whether we require the LB (lower-the-better), the HB (higher-the-better) and NB (nominal-the-best). The S/N ratio obtained from Table 3 was normalized in Table 4.

The next step is converting correlated responses into uncorrelated quality indices called principal components (PC1 and PC2). Principal components are independent (uncorrelated) of each other. Simultaneously, the explained variance of each principal component for the total variance of the responses is also obtained in Table 7. To calculate the value of principal component scores using the following equation[19]:

$$Y_i\,(k) = \sum_{j=1}^{n} X_i^*\,(j)\beta_{kj} \quad (6)$$

IOP Conf. Series: Materials Science and Engineering **854** (2020) 012024 doi:10.1088/1757-899X/854/1/012024

for $j = 1, 2,\ldots k$, the coefficient β_{ij} is called eigenvector.

Table 4. Normalization of Experimental Data

Run No.	Normalized Data	
	MRR	Ra
Ideal Condition	1.0000	1.0000
1	0.2349	0.1557
2	0.3992	0.7771
3	0.1326	0.2147
4	0.0000	0.5651
5	0.2754	0.1665
6	0.0177	0.2489
7	0.2621	0.6967
8	1.0000	0.0000
9	0.2754	1.0000

Table 5. Pearson's correlation coefficient

NO	Correlation between	Pearson's Correlation coefficient (ρ)	Remarks
1	Xi*1 & Xi*2	-0.270	Correlation

Table 6. Results of principal component analysis

	Y_1	Y_2
Eigen value	1.2704	0.7296
Eigen vector	-0.707	-0.707
	0.707	-0.707
AP	0.622	0.365
CAP	0.635	1

In weighted principal component method. all principal components will be used. thus the explained variance can be completely explained in all responses. The principal components are independent to each other. Accountability proportion (W) of individual principal components has been treated as individual priority weights[19]. Finally. multi-response performance index (MPI) has been computed using the following equation:

$$MPI = \sum_{j=1}^{k} W_j\, Y_i \tag{7}$$

Table 7. Principal Components and MPI

No	PC1	PC2	MPI

	Y_1	Y_2	
1	-0.2761	0.3053	-0.0603
2	-0.8316	1.1373	-0.1021
3	-0.2455	0.3253	-0.0339
4	-0.3995	0.6819	0.0004
5	-0.3124	0.3385	-0.0707
6	-0.1884	0.3092	-0.0043
7	-0.6778	0.9718	-0.0669
8	-0.7070	0.4998	-0.2573
9	-0.9017	1.3445	-0.0701

4.1. Analysis of Variance (ANOVA)

Analysis of variance or commonly referred to as ANOVA used to find out which machining parameters significantly affect the responses, in this case, are MRR and surface roughness. ANOVA is a statistical method that can infer some important conclusions by analysis of the experimental data. The method is very useful for revealing the level of significance of the influence of parameters or interaction of parameters on a particular response [15]. Table 8 shown ANOVA of the response quality characteristic.

Table 8. Anova of MPI

Machining Parameters	DF	SS	MS	% Contribution	F_{Value}
Voltage	2	0.0172	0.00863	32.33	6.77
Electrolyte Concentration	2	0.0217	0.01087	41.98	8.52
Gap Width	2	0.0049	0.00250	5.63	1.96
Error	2	0.00255	0.0012	20.06	
Total	8	0.01727	0.00863	100.00	

The percentage of error can be used to evaluate if an experiment possesses feasibility and sufficiency or not, since it is related to the uncertain or uncontrollable factors. The percentage contribution of each parameter also can be seen in the Table 8. The contribution of machining parameters in order are electrolyte concentration, voltage, and, gap width. The Taguchi method response table is used to calculate the average MPI value for each level in the machining parameter. This is done by sorting the MPI value according to the machining parameter level in each orthogonal array column and taking the same level average . The greater the MPI value, the better the corresponding multiple performance characteristic, so that a level that can give the largest average response can be chosen. Based on the response table for the MPI shown in Table 9, the best combination of cutting parameters is the set with a voltage of 48 Volt, electrolyte concentration of 150 g/l, and the gap with of 2 mm.

Table 9. Response Table for The MPI

Machining Parameters	Level 1	Level 2	Level 3	Max-Min
Voltage	0.6547	0.2518	0.1314	0.1062
Electrolyte Concentration	0.4526	0.1434	0.0361	0.1072
Gap Width	0.1073	0.5756	0.5721	0.051

Figure 1 performed ratio plot of MPI. The optimal setting has been evaluated from this plot. In the MPI graph (Figure 3), it is mentioned that the optimal combination of machining parameters for multiple performance characteristics are the third level of voltage, the second level of electrolyte concentration, and Second level of gap width (A_3 B_2 C_2). After the optimal level of machining parameters has been identified, a verification test needs to be carried out in order to check the accuracy of the analysis.

After the optimal level of machining parameters has been identified, the final step is a verification test which needs to be carried out to verify the accuracy of the analysis. The purpose of these tests also to predict and verify the improvement of the performance characteristic using the optimal level of the machining parameters.

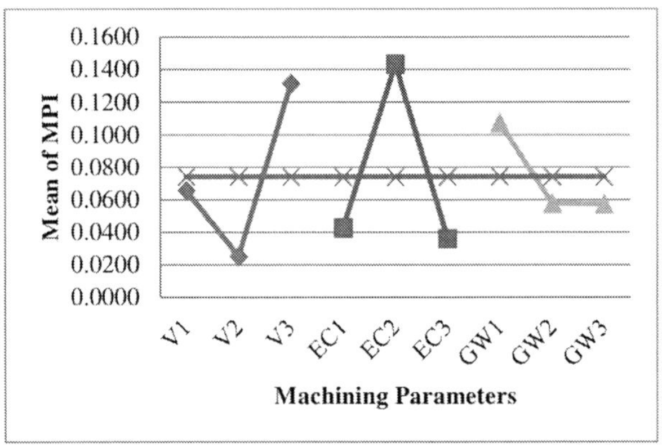

Figure 3. Influence of process parameters on multiple performance statistic

Table 10 compares the result of the confirmation test using the optimal machining parameters $A_3B_2C_2$ obtained by the proposed method and with those of the initial cutting parameters $A_2B_2C_2$. As observed in Table, the material removal rate increases from 0,0768 g/min to 0.1928 g/min and surface roughness was reduced from 4,109 μm to 3,641 μm. Based on the results, it is observed quality characteristics can be greatly improved through this study.

Table 10. Result of Confirmation Tests

	Initial	**Optimal Process Condition**	
		Prediction	**Experiment**
Setting Level of machining parameters	$A_2B_2C_2$	$A_3B_2C_2$	$A_3B_2C_1$
Material Removal Rate (g/min)	0.0768	0.1158	0.1928
Surface Roughness (μm)	4.109	3.188	3.641
MPI	0.5317	0.7616	0.6474

5. Conclusion

This research presents an application of the Taguchi Method coupled with WPCA as a optimization design of Electrochemical Machining for SKD11 tool steel. The results are summarized as follows:

1. Based on analysis of variance, the major controllable parameters significantly affecting the multiple performance characteristics are voltage, gap width and electrolyte contribution with a desired total contribution of 99,96%.
2. The optimal combination of the machining parameters obtained from the proposed method is the set with voltage at level 3, electrolyte concentration at level 2, and gap width at level 2. The corresponding confirmation tests show that metal removal rate increase by 60.17% and surface roughness decrease by 11,49%. The MPI also increases by 17,87%.
3. The proposed algorithm greatly simplifies the optimization design of machining parameters with multiple performance characteristics. Thus, the solutions from this method can bean useful reference for tool manufacturers and operators who are willing to search for an optimal solution of machining conditions.

References

[1] H. Hocheng. Y. H. Sun. S. C. Lin. and P. S. Kao. "A material removal analysis of electrochemical machining using flat-end cathode." *J. Mater. Process. Technol.*. vol. 140. no. 1–3 SPEC.. pp. 264–268. 2003.

[2] H. Sohrabpoor. S. P. Khanghah. S. Shahraki. and R. Teimouri. "Multi-objective optimization of electrochemical machining process." pp. 1683–1692. 2016.

[3] P. Asokan. R. R. Kumar. R. Jeyapaul. and M. Santhi. "Development of multi-objective optimization models for electrochemical machining process." pp. 55–63. 2008.

[4] D. Deconinck. S. Van Damme. C. Albu. L. Hotoiu. and J. Deconinck. "Study of the effects of heat removal on the copying accuracy of the electrochemical machining process." *Electrochim. Acta.* vol. 56. no. 16. pp. 5642–5649. 2011.

[5] R. Mathew and M. M. Sundaram. "Modeling and fabrication of micro tools by pulsed electrochemical machining." *J. Mater. Process. Technol.*. vol. 212. no. 7. pp. 1567–1572. 2012.

[6] K. P. Rajurkar. H. Hadidi. J. Pariti. and G. C. Reddy. "Review of Sustainability Issues in Non-Traditional Machining Processes." *Procedia Manuf.*. vol. 7. pp. 714–720. 2017.

[7] X. Chen. Z. Xu. D. Zhu. Z. Fang. and D. Zhu. "Experimental research on electrochemical machining of titanium alloy Ti60 for a blisk." *Chinese J. Aeronaut.*. vol. 29. no. 1. pp. 274–282. 2016.

[8] Z. Pandilov. "Application of Electro Chemical Machining for materials used in extreme conditions Related content Application of Electro Chemical Machining for materials used in extreme conditions." 2018.

[9] L. Guodong. L. Yong. K. Quancun. and T. Hao. "Selection and Optimization of Electrolyte for Micro Electrochemical Machining on Stainless Steel 304." *Procedia CIRP*. vol. 42. no. Isem Xviii. pp. 412–417. 2016.

[10] M. Sankar. A. Gnanavelbabu. K. Rajkumar. and N. A. Thushal. "Electrolytic concentration effect on the abrasive assisted-electrochemical machining of an aluminum–boron carbide composite." *Mater. Manuf. Process.*. vol. 32. no. 6. pp. 687–692. 2017.

[11] D. Chakradhar and A. V. Gopal. "Multi-Objective Optimization of Electrochemical machining of EN31 steel by Grey Relational Analysis." *Int. J. Model. Optim.*. vol. 1. no. 2. pp. 113–117. 2011.

[12] S. K. Soni and B. Thomas. "A comparative study of electrochemical machining process parameters by using GA and Taguchi method." *IOP Conf. Ser. Mater. Sci. Eng.*. vol. 263. no. 6. pp. 8–15. 2017.

[13] H. H. Abuzied. "Prediction of Electrochemical Machining Process Parameters using Artificial Neural Networks." vol. 4. no. 01. pp. 125–132. 2012.

[14] D. Shrivastava. A. Sharma. and H. Pandey. "Experimental Investigation of Machining Parameter in Electrochemical Machining." *Int. Res. J. Eng. Technol.*. no. V. pp. 2395–56. 2015.

[15] M. K. Das. K. Kumar. T. K. Barman. and P. Sahoo. "Optimization of Surface Roughness and MRR in Electrochemical Machining of EN31 Tool Steel Using Grey-taguchi Approach." *Procedia*

IOP Conf. Series: Materials Science and Engineering **854** (2020) 012024 doi:10.1088/1757-899X/854/1/012024

Mater. Sci.. vol. 6. no. Icmpc. pp. 729–740. 2014.

[16] S. Ayyappan and K. Sivakumar. "Investigation of electrochemical machining characteristics of 20MnCr5 alloy steel using potassium dichromate mixed aqueous NaCl electrolyte and optimization of process parameters." *Proc. Inst. Mech. Eng. Part B J. Eng. Manuf.*. vol. 229. no. 11. pp. 1984–1996. 2015.

[17] N. Lusi. K. Muzaka. and B. O. P. Soepangkat. "Parametric optimization of wire electrical discharge machining process on AISI H13 tool steel using weighted principal component analysis (WPCA) and taguchi method." *ARPN J. Eng. Appl. Sci.*. vol. 11. no. 2. 2016.

[18] M. K. Das and B. Polytechnic. "Effect of Process Parameters on MRR and Surface Roughness in ECM of EN 31 Tool Steel Using WPCA." vol. 4. no. 2. pp. 45–63. 2017.

[19] A. Biswas. S. Bhaumik. G. Majumdar. S. Datta. and S. S. Mahapatra. "Bead geometry optimization of submerged arc weld: exploration of weighted principal component analysis (WPCA)." in *Applied Mechanics and Materials*. 2012. vol. 110. pp. 790–798.

IOP Conf. Series: Materials Science and Engineering **854** (2020) 012025 doi:10.1088/1757-899X/854/1/012025

Development of Web Based Real Time Remote Laboratory for Teaching and Learning

Faisal Mahmud[1], *Mohd Nashrul Mohd Zubir[2], Mohd Yazed Bin Ahmad[3], Zuraidah Binti Abdullah[4], Nahrizul Adib Bin Kadri[3], Unaizah Hanum Binti Obaidellah[5], Tousif MD. Amin Faisal[6]

[1] Department of Computer Science (Applied Computing), Faculty of Computer Science & Technology, University of Malaya
[2] Department of Mechanical Engineering, Faculty of Engineering, University of Malaya
[3] Department of Biomedical Engineering, Faculty of Engineering, University of Malaya
[4] Department of Educational Management, Planning and Policy, Faculty of Education, University of Malaya
[5] Department of Artificial Intelligence, Faculty of Computer Science & Technology, University of Malaya
[6] Department of Software Services, Leads Corporation Ltd, Bangladesh

E-mail: nashrul@um.edu.my

Abstract. A concept has been developed that utilizes the internet to conduct live laboratory exercises under remote condition. The idea about remote laboratory is based on Human-Computer Interaction (HCI) philosophy, controlling system and communication. The integration of these three components plays an important role in making the facility accessible and operational online. In order to realize this concept, an open-source development tool called Node-RED which links the user and the remote facility via a web interface platform has been adopted. Node-RED is a programming tool for connecting hardware, Application Programming Interface (API) and online services in various ways. The node-red dashboard allows users to carry out remote experiments. This strategy would also allow users to leverage internet resources to implement laboratory exercise by eliminating relevant time and space limitation problems. An experimental setup was created to demonstrate the above-mentioned remote laboratory model. Various types of sensors and actuators were used to replace physical interaction between the end-user and the laboratory deployment. A microcontroller was employed to process and relay the data between the host computer and the equipment. It was shown that the communication between the web and remote infrastructure was successfully established and the facility was able to be accessed and operated via the web-based interface using internet protocol. This endeavor serves as an important milestone towards realizing internet-based education particularly in science, technology and engineering streams which primarily involves physical laboratory exercise.

Keywords: Remote laboratory, distant learning, E-Learning, master-slave system, human-computer interface

IOP Conf. Series: Materials Science and Engineering **854** (2020) 012025 doi:10.1088/1757-899X/854/1/012025

1. Introduction

The classical practice of conducting an experiment in the laboratory has been changing radically by the advanced utilization of computers, electronic devices and the internet. This has given birth to a new concept of laboratory practice called 'remote lab'. Remote laboratory conceptual layout and educational approaches are designed to suit specific hardware and software [1]. By implementing the concept, learners are able to experience live experiment using dedicated tools from the website along with the internet connection [2]. As the facility is operated under the internet platform, it can be utilized from all over the world. It enables users to perform distant laboratories that can provide around-the-clock active accessibility via web system [3]. In the fields of engineering and science in particular, where practical laboratory experiments are crucial to improve and complete lessons in the classroom, the implementation of this concept would serve as a new paradigm in the knowledge transfer process [4]. Institutions can overcome financial obligations as well as increasing more learning opportunities [5]. Moreover, remote browser-accessible laboratories concept bare close resemblance to online learning mainly due to the fact that learners are not required to be placed with a specific piece of laboratory equipment [6]. Following this system, distant user can set predefined parameters as their requirements in the user-friendly interface and enable to view in graphical format [7]. Users are able to operate real experimental setup to do various aspects of exercises without space and time constraints since communications and all kinds of interactions between users and systems will be transmitted via web services [8] [9]. This concept is also suitable for class and training operations that encourage interactive involvement on top of the typical instructional approaches [10].

2. Methodology

2.1. Hardware Infrastructure

A simple science-based experiment has been chosen for demonstrating the above remote lab concept. The activity deals with creating and observing the color spectrum spread by a prism subjected to an artificial light source. The traditional way of carrying out this study requires the user to place a prism, light source and display screen at the appropriate angle and distance to allow the light to be refracted and fall on the screen as separate individual colors. To make it possible for the user to conduct this experiment without physical contact with the components, an experimental set-up was built, incorporating modules and functions to allow the system to be operated under remote conditions as shown in figure 1. In addition, a single degree of freedom (SOF) Internet Protocol (IP) camera was deployed to enable the user to view the facility at different angles such that the user would have the feeling of being present in the lab.

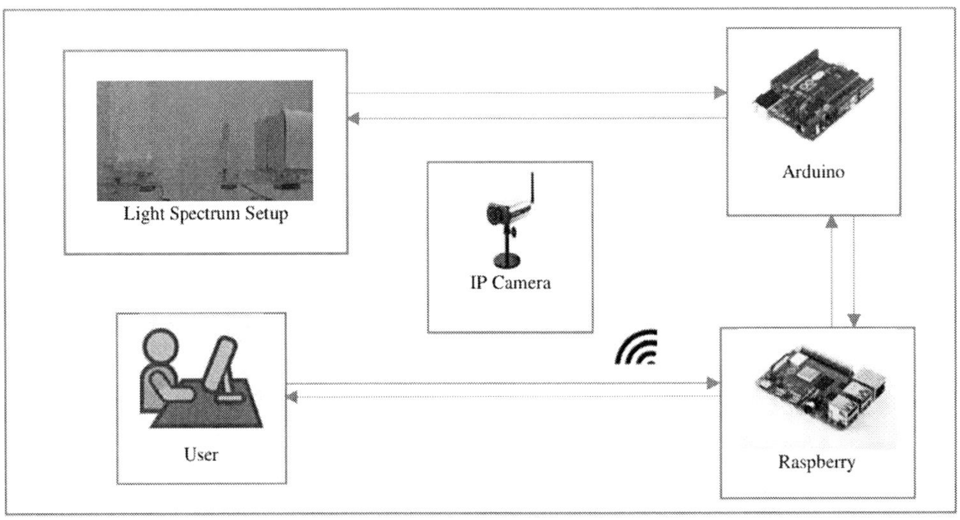

Figure 1. System configuration layout for remote laboratory

The system is built based on four important hardware modules namely server, microcontroller, sensor and actuator as well as the camera. The list of hardware for the current setup is given in Table 1.

Table 1. List of hardware used for the remote lab system

Module	Instrument
Server	Raspberry Pi
Microcontroller	Arduino UNO
Sensor and actuator	Temperature and relative humidity sensors, Stepper motors and DC motors
Camera	TP-Link IP camera model 250

2.2. Specific Web System

An open-source tool, Node-RED was adopted to develop the web system. Node-RED is a software development tool for connecting equipment, APIs and internet services in convenient and interactive ways. It offers a browser-based editor (Figure 2) that makes it simple to connect different modules together using a broad variety of nodes in the palette that can be deployed in a single click at run time.

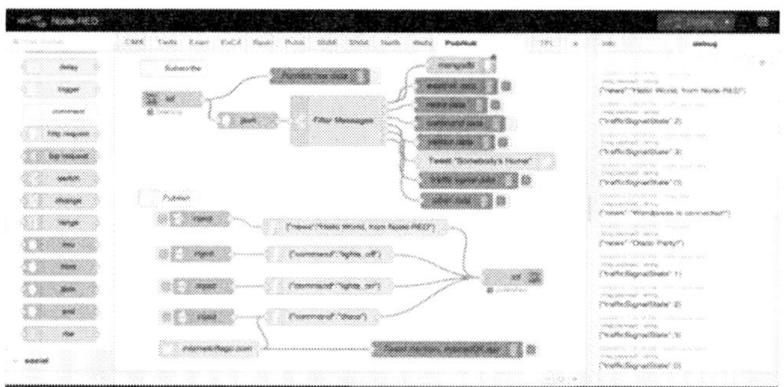

Figure 2. Node-RED web-based editor

Further, this software can also be used to incorporate JavaScript functions for flows of bits and configuring nodes to accomplish signal communication from the web environment to the controller as shown in Figure 3.

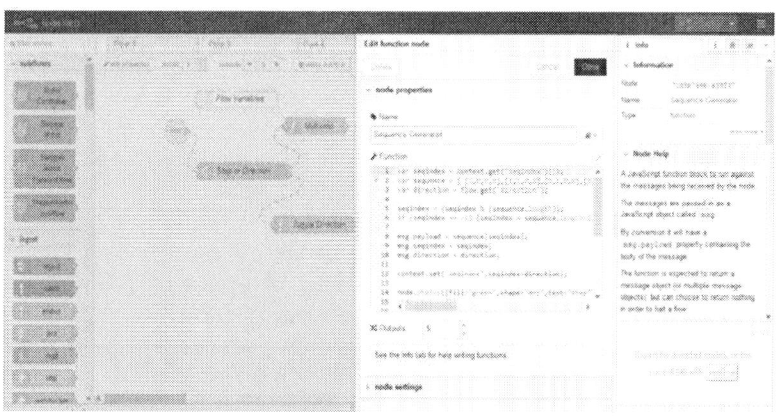

Figure 3. Node-RED JavaScript

The backend programming will be translated into a user-friendly dashboard shown in Figure 4 which allows effective user-hardware interface. After deploying all flows and nodes successfully, all control interfaces such as switches, buttons and toggle-buttons would appear in the dashboard during compilation. Thus, the user would only be interacting via the dashboard from a web browser to take control over modules in the laboratory.

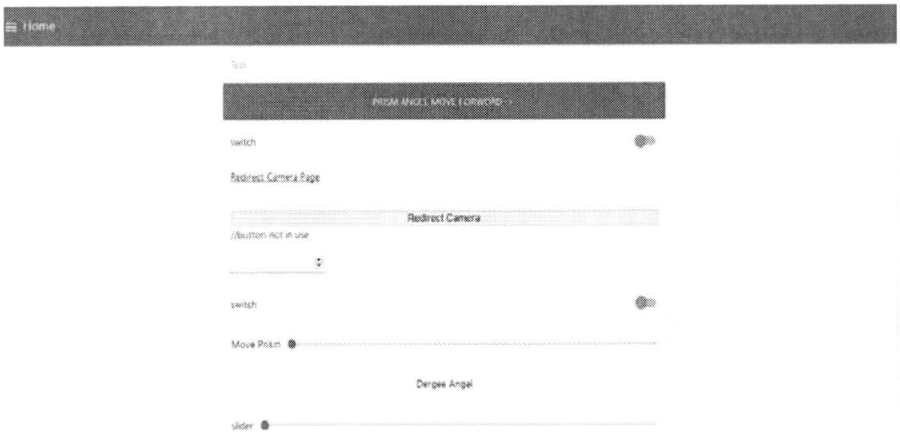

Figure 4. Node-RED dashboard

2.3. User Access Functionality

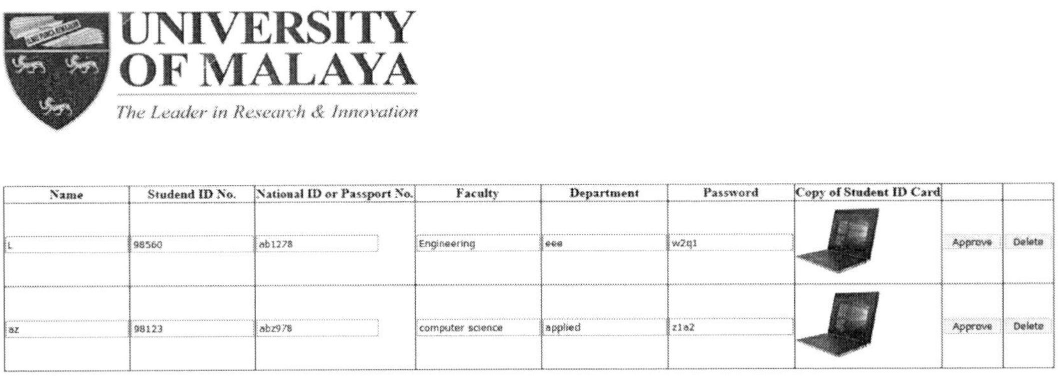

Figure 5. Admin Panel

In order to establish a secure laboratory operation, an access system has to be developed which allows authorized users to control the facility. In the present work, a web-based access module has been developed which prompts the user to register and specify the access details prior to being allowed to communicate with the system. To systematically enable this access functionality, a system called 'client-host' swap mode was introduced. An authentication method was formed within the system to ensure the safety of the server-side from malicious threats. This is accomplished by denying connectivity protocol from unauthorized users. At the heart of the process, two accounts under administrator and user profiles will be launched as the user enters the website. The administrator has the full right to take control over the user profiles. The user can only be granted access upon verification by the admin. Further, under the current access protocol design, the admin has the option to delete user access requests as shown in Figure 5. This would give the admin more control on monitoring the laboratory condition by tracking the user's history.

IOP Conf. Series: Materials Science and Engineering **854** (2020) 012025 doi:10.1088/1757-899X/854/1/012025

To gain access to the remote facility, users must first apply via the registration site as shown in Figure 6. Upon completing the form on the web page which includes specifying their id and password, the users will be directed to a second window layer that prompts them to insert their log in details (Figure 7). All this information will be channeled to the admin portal for verification and approval before they are granted secure access to the facility.

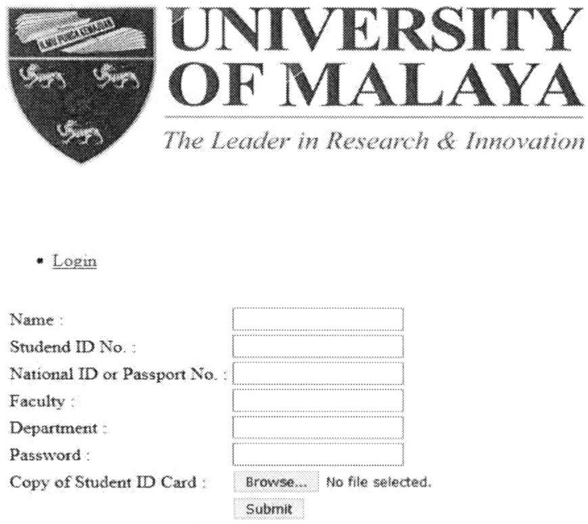

Figure 6. User Registration panel

.

Figure 7. User Login Page

3. Result and Discussion

The complete layout of the overall system encompassing the hardware tools and the communicative interface is illustrated in Figure 8. In detail, the microcontroller is responsible for transmitting output signals to operate the actuation devices as well as receiving information from the sensor modules to be displayed in the web interface. Users send a specific command to interact with the facility by pressing buttons or switches from the Node-RED interface. As the microcontroller used in the current project is a serial device that communicates via a USB serial connection, the Serial IN and OUT nodes are used within the software to synchronize the data packet going in and out from the controller. This phase is critical since any missing information during this data communication would be detrimental to the execution of each task. Upon successful recognition between the software and hardware pins, the triggering signal will be securely transmitted via the Raspberry PI server for performing the command respectively. Modules like motors connected to the controller will move following specific signals. On the other hand, a specific IP address is released by the access point for the IP camera to view the facility online. The flow of communication between the facility and the user is given in Figure 9.

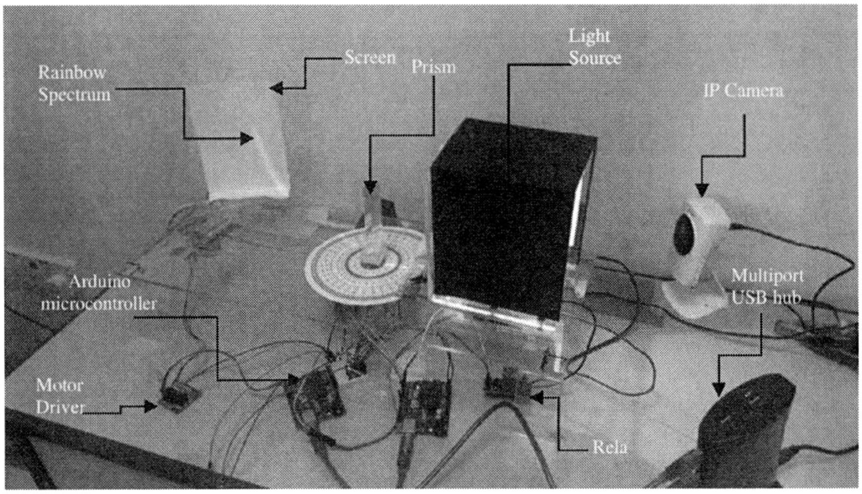

Figure 8. Setup for light spectrum

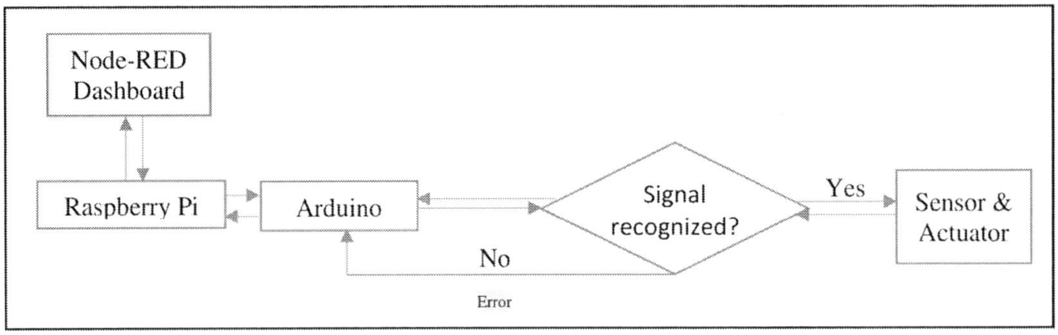

Figure 9. Communication flow between the user and lab facility

In addition, to enable more freedom for the user to view the facility and pinpoint the location of interest during an experiment, a motor has been mounted on the camera which enables 180degree rotation covering the entire lab infrastructure as Shown in Figure 10. The user would benefit from getting more realistic vision mobility that resembles the user presence in the lab.

IOP Conf. Series: Materials Science and Engineering **854** (2020) 012025 doi:10.1088/1757-899X/854/1/012025

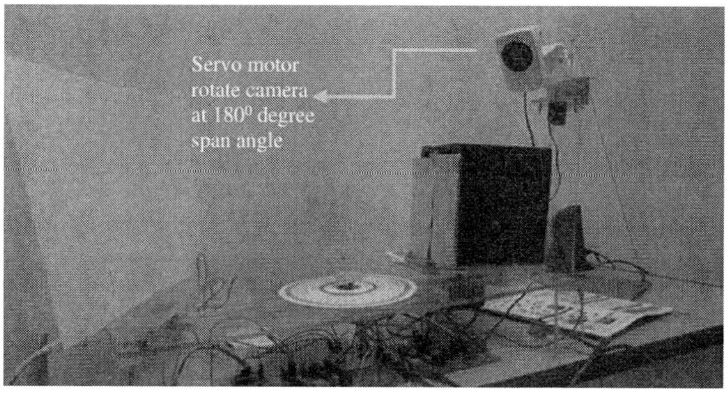

Figure 10. Camera positioning system to view the facility

4. Conclusion

An innovative framework for remote laboratory based on open source development tools has been developed. The main feature of the system is that, it can be accessed by the user from a browser-based system hosted in a special computer that is connected to various hardware modules within the experimental setup to perform laboratory exercises remotely. The framework was developed using open source web based software called "Node-RED" which serves to connect the user and the remote facility under web platform. This tool operates by relaying the command from the user towards the server and controller which translates into a sequence of signals to operate the sensors and actuators to accomplish remote interaction. In addition, a dashboard is provided by Node-RED as a web-interface to establish communication between the client and the hardware. An experimental set-up was established to demonstrate the concept. It was shown that the facility was successfully operated remotely using the above interface from a web browser. In brief, the project established the sustainability of converting the current manual laboratory facility into a remote system that enables the user to conduct experiments under a remote management protocol using internet communication technology.

Acknowledgement

This project is funded by UM-Living Lab grant (No- LL040-18SUS) and Knowledge Transfer Programme (KTP) grant (No- MRUN2019-3C, MRUN2019-3D). The authors wish to thank Faisal Nabil, Mr. Fauzi, and Mr. Hasanul for the technical assistance in building the lab facility

Reference

[1] S. C. Sivakumar, W. Robertson, M. Artimy, and N. Aslam, "A web-based remote interactive laboratory for internetworking education," *IEEE Trans. Educ.*, vol. 48, no. 4, pp. 586–598, 2005.

[2] A. Ferrero, S. Salicone, C. Bonora, and M. Parmigiani, "ReMLab: A Java-based remote, didactic measurement laboratory," *VIMS 2002 - 2002 IEEE Int. Symp. Virtual Intell. Meas. Syst. Distrib. Intell. Sens. Adv. Integr. Virtual Environ.*, vol. 52, no. 3, pp. 117–122, 2002.

[3] P. Sládek, L. Pawera, and J. Válek, "Remote laboratory- New possibility for school experiments," *Procedia - Soc. Behav. Sci.*, vol. 12, pp. 164–167, 2011.

[4] C. S. Tzafestas, N. Palaiologou, and M. Alifragis, "Virtual and remote robotic laboratory: Comparative experimental evaluation," *IEEE Trans. Educ.*, vol. 49, no. 3, pp. 360–369, 2006.

[5] J. Ma and J. V. Nickerson, "Hands-on, simulated, and remote laboratories: A comparative literature review," *ACM Comput. Surv.*, vol. 38, no. 3, p. 1, 2006.

[6] J. E. Corter, J. V. Nickerson, S. K. Esche, C. Chassapis, S. Im, and J. Ma, "Constructing reality: A study of remote, hands-on, and simulated laboratories," *ACM Trans. Comput. Interact.*, vol. 14, no. 2, 2007.

[7] D. Hercog, B. Gergič, S. Uran, and K. Jezernik, "A DSP-based remote control laboratory," *IEEE Trans. Ind. Electron.*, vol. 54, no. 6, pp. 3057–3068, 2007.

[8] Z. Aydogmus and O. Aydogmus, "A web-based remote access laboratory using SCADA," *IEEE Trans. Educ.*, vol. 52, no. 1, pp. 126–132, 2009.

[9] P. Orduña, P. H. Bailey, K. Delong, D. López-De-Ipiña, and J. García-Zubia, "Towards federated interoperable bridges for sharing educational remote laboratories," *Comput. Human Behav.*, vol. 30, pp. 389–395, 2014.

[10] F. Esquembre, "Facilitating the Creation of Virtual and Remote Laboratories for Science and Engineering Education," *IFAC-PapersOnLine*, vol. 48, no. 29, pp. 49–58, 2015.

IOP Conf. Series: Materials Science and Engineering **854** (2020) 012026 doi:10.1088/1757-899X/854/1/012026

The Impact of Anthropometry on Terasi Packaging

Meri Andriani[1], Rita Syntia[2]

[1]Dept. Industry Engineering, Samudra University,
[2]Dept Mechanical Engineering, Samudra University

Email: meri_tind@unsam.ac.id

Abstract The packaging is a protective material for the product to last longer. The purpose of this research is to design an ergonomic packaging using anthropometric methods for terasi products. Terasi is a spice that has special characteristics for Acehnese food. The results obtained in designing packaging with the anthropometric method using the dimensions of the width of the palm. The percentile used is 5% percentile with a mean value of boys and girls of 7.11 centimeters.

Keyword: anthropometry, design, packaging, percentile, terasi.

1. Introduction

Terasi is a fermented product [1], made of small shrimp that are black to brown [2]. Terasi one of the seasonings for various kinds of food, especially in Aceh food. Acehnese foods that use terasi seasoning, for example, are chili lado terasi, grilled fish, chili paste and various Acehnese foods. The increasing number of processed shrimp paste products now makes shrimp paste a best-selling product on the market. Products are complex properties that can be touched, felt, not touched, which are accepted by buyers to satisfy the wants and needs of consumers [3]. The wants and needs of consumers towards the product are seen from the quality, including the quality of terasi. The quality of shrimp paste is seen from the aroma and taste of the shrimp paste itself. But to attract the first time consumer interest in the product is packaging.

The packaging is a container to protect products from dirt and make products last longer. Aside from being a product protector, it can also attract consumers to buy the product [4], besides that, it is also necessary to pay attention to the packaging in its use. The use of convenient packaging can make it easier for humans to do their jobs and vice versa. The terasi packaging in the market is only in the form of a very simple plastic wrap so that the terasi packaging is uncomfortable in its use, this is evidenced by the results of the interview at the time of observation of some consumers. Consumers of terasi complain that when they open the packaging terasi, the terasi often spills over so that it makes the work increase in addition to cooking, namely cleaning the scattered of the terasi. Consumer complaints about terasi are a problem in this study and research is limited to dried terasi products.

Anthropometry is the study of adjusting the size of the tool/machine to the size of the human body [5], in other words anthropometry is used to determine the dimensions that will be used on terasi packaging, while percentiles are used to determine the size of the tool/machine that is adjusted to human size. Human size is used in the design of terasi packaging so that humans are comfortable [6] in using packaging. The size that does not fit between the size of humans and products can cause various complications [7] in its use. Different human sizes one of which is caused by ethnic differences [8], including Indonesia. Therefore, this research focuses on the Acehnese tribe, because the Acehnese like the flavor of terasi cuisine, in accordance with the research conducted, so this research aims to design terasi packaging which is limited to powdered terasi.

2. Methods

The study was conducted with adult objects people in the age from 25 to 55 years who volunteered to become samples. The number of research objects is 100 women and 100 men. The first determination of the anthropometric dimensions of the hands that will be used is the beginning of the research, found in Table 1.

Table 1. Anthropometry for the Design of Terasi Packaging

Picture	Aim
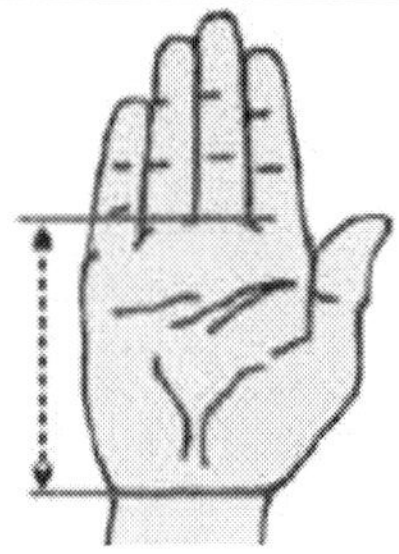	Reference in determining the length and width of the packaging

The anthropometric dimension of the hand used to design the terasi packaging is the length dimension of the palm of the hand. The dimensions of the palm length are used as a reference in determining the length and width of the packaging.

Measurements were taken directly in the field by an ergonomics laboratory assistant who had been trained for ninety days. The martin body instrument is a tool used in measurements for reasons of being more practical in the field, and samples also feel more comfortable using the instrument.

3. Result

Results can be made after knowing the dimensions of the hand used. Statistical tests were performed on the dimensions of the length of the palm. The data uniformity test, including the statistical test performed, is shown in Figures 1 through 4.

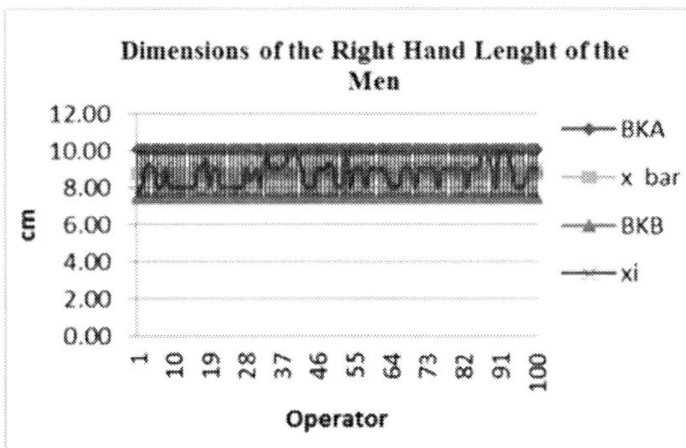

Figure 1. Male Right Hand Anthropometry Data

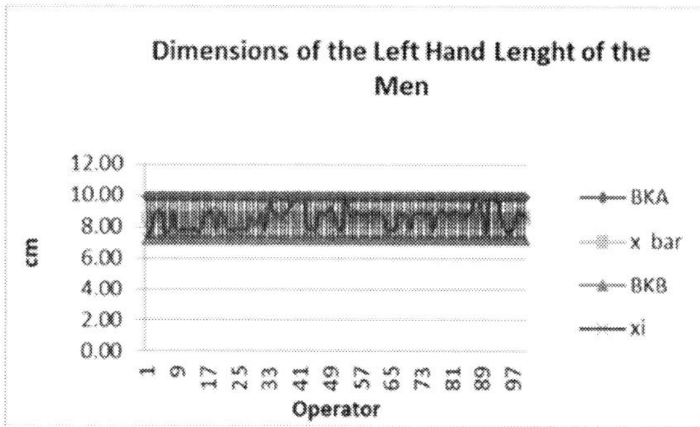

Figure 2. Men Left Hand Anthropometry Data

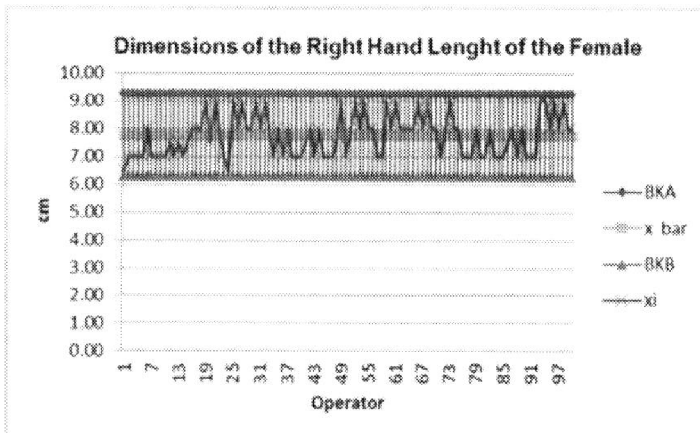

Figure 3. Men Right Hand Anthropometry Data

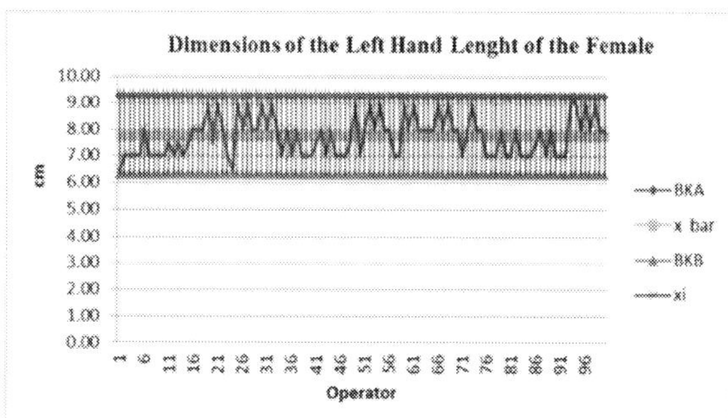

Figure 4. Female Right Hand Anthropometry Data

Data is between control limits so all data is uniform. Data sufficiency test and normality test are carried out, to see the data collected is sufficient and normally distributed, it is found in table 2.

Table 2. Data Validity Test

Men/Female	N	N'		Information
		Right Hand	Left Hand	
Men	100	9	9	Valid
Female	100	15	15	Valid

Table 3. Data Normality Test

Men/Female	D_α	D		Information
		Right Hand	Left Hand	
Men	0,134	0.036	0.047	Normal
Female	0,134	0.019	0.025	Normal

Data obtained from the results of the data adequacy test are all valid and have a normal distribution.

4. Discussion

Designing powdered terasi packaging is the goal of research. The preliminary study in our study showed the inconvenience of using terasi packaging

The design of the terasi packaging is done to be comfortable in its use. Comfort is related to ergonomics. Ergonomics is defined as a human-related study [9] with its activities. Ergonomically designed based on percentiles, are in table 4.

Table 4. Percentile for the Right Hand

No	Men/Females	Mean (cm)	σ (cm)	Percentiles 5 (cm)	Percentiles 50 (cm)	Percentiles 95 (cm)
1	Men	8,73	0,65	7,76	8,73	9,79
2	Females	7,76	0,74	6,54	7,76	8,98

There are statistical differences in the dimensions of the length of the palm of the hand between Men and Females, where the dimensions of Men are bigger than Females

This research also has limitations that need to be considered. First, the low number of samples needed in the study, given the large Acehnese population, so it is necessary to increase the number of samples for subsequent research. Second, the measurement instruments used are still manual, which may have problems related to lack of reliability, so more sophisticated and reliable measurement methods are needed.

Apart from the shortcomings, the design of terasi packaging is a first step in designing packaging using anthropometry and percentiles for the Acehnese tribe. The design of convenient terasi packaging can be seen in figure 5.

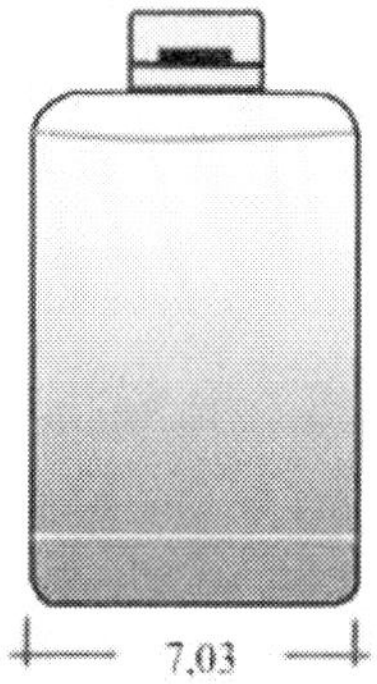

Figure 5. Terasi Packaging Design

The percentile used to design terasi packaging is percentile 5 (P5) with the reason that consumers who have a small dimension of palm length can also use the packaging.

Determination of terasi packaging design is obtained from the first, the average size dimensions of the right-hand length and the length of the left hand, the second the average of the sexes, found in table 5.

Table 5. Determination of Terasi Packaging Size

	Right Hand (cm)	Left Hand (cm)	Average (cm)
Men	7,67	7,50	7,58
Females	6,54	6,40	6,47
	Average		7,03

5. Conclusion

In designing this terasi packaging, only one anthropometric dimension is needed, the dimension of hand length, with the 5th percentile.

Acknowledgment

The authors thank LPPM and PM Samudra University for funding in this study.

References

[1] F. A. Karim, F. Swastawati, and A. D. Anggo, "Pengaruh Perbedaan Bahan Baku Terhadap Kandungan Asam Glutamat Pada Terasi," *J. Pengolah. dan Bioteknol. Has. Perikan.*, vol. 3, pp. 51–58, 2014.

[2] yulian dani Sanjaya, Sumardianto, and P. H. Riyadi, "Pengaruh Penambahan Ekstrak Rosella (Hibiscus sabdariffa Linn.) Terhadap Warna Dan Kualitas Pada Terasi Udang Rebon (Acetes sp.)," *Biomaterials*, vol. 29, no. 34, pp. 4471–4480, 2016.

[3] F. Indriani, "Studi Mengenai Orientasi Inovasi , Pengembangan Produk dan Efektifitas Promosi Sebagai Sebuah Strategi Untuk Meningkatkan Kinerja Produk," vol. 3, no. 2, pp. 82–92, 2006.

[4] T. Susetyarsi, "Kemasan Produk Ditinjau dari bahan Kemasan, Bentuk Kemasan dan Pelabelan Pada Kemasan Pengaruhnya Terhadap Keputusan Pembelian Pada Produk Minuman Mizone di Kota Semarang," vol. 4, no. 3, pp. 19–28, 2012.

[5] M. Andriani, M. T. Hasan, N. Nazaruddin, and N. Ninafahriana, "Application of Anthropometry to Overcome Musculoskeletal Problems," *J. Phys. Conf. Ser.*, vol. 1114, no. 1, 2018.

[6] I. W. Taifa and D. A. Desai, "Anthropometric measurements for ergonomic design of students' furniture in India," *Eng. Sci. Technol. an Int. J.*, vol. 20, no. 1, pp. 232–239, 2017.

[7] S. J. Mirmohammadi, A. H. Mehrparvar, M. Mostaghaci, M. H. Davari, M. Bahaloo, and S.

IOP Conf. Series: Materials Science and Engineering **854** (2020) 012026 doi:10.1088/1757-899X/854/1/012026

Mashtizadeh, "Anthropometric hand dimensions in a population of Iranian male workers in 2012," *Int. J. Occup. Saf. Ergon.*, vol. 22, no. 1, pp. 125–130, 2016.

[8] N. P. Rasco, B. Dytoc, J. L. Guerrero, A. Lopez, and B. Custodio, "A Comparative Analysis of the Ergonomic Anthropometry and Usability of Locally Designed and Foreign Designed Smartphones," *Procedia Manuf.*, vol. 3, no. Ahfe, pp. 5927–5933, 2015.

[9] M. Andriani, M. T. Hasan, R. Akram, and C. I. Erliana, "An Analysis of Workload and Worker ' s Complaint In Tofu Industry," 2019.

IOP Conf. Series: Materials Science and Engineering **854** (2020) 012027 doi:10.1088/1757-899X/854/1/012027

Increased Absorption Capacity Of Organokaolin Adsorbents In Absorb Metals Using Microwave Technology

Syafruddin, Alfian Putra, Suci Ramadhana, Pardi
Teknologi Rekayasa Kimia Industri, Teknik Kimia, Politeknik Negeri Lhokseumawe

Email: irsyafruddin@yahoo.co.id

Abstract. This study about the absorption of iron and magnesium as metal ions in water solutions and kaolin were used as adsorbents. The power of Kaolin as adsorbent still very low so it is necessary to modify kaolin using acyl sulfonate (ABS) surfactant which is an organic compound. The aim is to increase the absorption capacity of kaolin in absorbing metal ions contained in water using modified kaolin to get effective absorption results in absorbing metal ions, especially iron, this research was carried out by varying microwave irradiation time 5, 15, 30, 50 and 80 minutes using surfactant kaolin modification and without surfactant modification, the waste used was artificial waste with an initial concentration of 100 mg / l. The samples were analyzed using Atomic Absorption Spectrophotometers (AAS), while the characteristics used were Fourier Transform Infrared Spectroscopy (FTIR) and Scanning Electron Microscope (SEM). Kaolin modification with microwave use in the adsorption process results in a decrease in artificial Fe waste at 3 minutes irradiation time on modification of efficiency reduction of 72% and Mg waste at 30 minutes decreased by 61%.

Keywords: adsorption, adsorbent, metal ion, kaolin, microwave, surfactant.

1. Introduction

One of the environmental problems that occur in many areas is the occurrence of water pollution caused by the presence of metal ions in the water such as iron and magnesium, metal ions in the water if consumed continuously will adversely affect human health. Recent efforts to control metal ion waste are increasingly developing, leading to efforts to find new methods that are cheap, effective, and efficient. One method for removing metal ions in water using the adsorption process [1]. The adsorption process is more widely used in industry because it has several advantages, which are more economical and also do not cause toxic side effects and are able to remove organic and inorganic materials. Adsorption is a process of waste absorption that occurs above the surface of the adsorbent due to the attraction between the adsorbate molecule and the adsorbent surface [2].

The use of conventional adsorbents requires relatively more expensive operational and regeneration costs. Conventional adsorbents that are often used in the adsorption process are alumina, activated carbon, silica gel, these adsorbents have good adsorption ability but are not economical. In this study using alternative adsorbents derived from nature, in the form of kaolin because in addition to having good adsorption capability, these adsorbents are also more economical [3].

using pure kaolin for the adsorption of lead, zinc and cadmium. From the results of the research it turns out that the adsorption capacity of pure kaolin is still low, when compared to activated carbon, zeolite and bentonite [4].

One of the efforts to improve the absorption capability of kaolin, trying to modify kaolin using acyl sulfonate surfactant (ABS). [5]. Although the absorption of heavy metals in kaolin that has been modified by surfactants has been better than using pure kaolin, the absorption rate of the metal is only 78%. The absorption capacity of kaolin is thought to still be increased by modifying the process using microwave technology. The use of microwaves in the adsorption process will increase electromagnetic energy in the process, the energy can penetrate the particles and interact between the molecular so that the reaction will occur better and the process will be more homogeneous faster than the conventional absorption process [6]. Microwaves can heat the reaction mixture very quickly, uniformly and directly without obstructing the problem of heat transfer through walls as is the case with conventional heating [7]. In addition, the use of microwaves can save energy [8].

Even so, until now information about the use of microwave technology in the adsorption process is still very limited. Therefore, this research will try to apply microwave technology to increase kaolin absorption of Fe and Mg metal ions in water. The results of this study are expected to provide new information as an alternative in increasing the absorption capacity of kaolin adsorbents

2. Method

The materials used in the form of kaolin contained in nature, acyl sulfonate surfactants (ABS) which are positively and negatively charged, are used for the modification process of activating kaolin as an adsorbent. Aquadest, HCl is used in the activation process using acid to remove metal content and organic compounds found in kaolin to increase its absorption [9].

Size of 100 mesh kaolin particles. The equipment used included oven, pH meter, microwave, desiccator, glass beaker, stirring rod, filter paper, Determination of initial and final concentrations for Fe and Mg components using Shimadzu UV-1800 Atomic Absorption Spectrophotometer (AAS), while determining the characteristics for length wave before and after treatment using Fourier Transform Infra Red (FTIR) -IR spectrophotometer and Scanning Electron Microscope (SEM).

2.1. Activation Kaolin

Activation is done using 100 mesh kaolin. Kaolin is heated using an oven at 105 ° C for 3 hours. After being cool mixed with 1N HCL then allowed to stand for 1 hour, then washed with aquadest until neutral (pH = 7) [5]. Drying again using an oven with a temperature of 105 ° C for 2 hours.

2.2. Modifivcation of Kaolin with Anionic Surfactant

Modification is carried out by mixing solid kaolin and anionic surfactants (organocalin). Modification is carried out with a ratio of 45% of the total weight [10]. Precipitation is carried out for 4 hours before being stored in a desiccator before use.

2.3. Microwave Adsorption Process

Adsorption is done by weighing 5 g of organokaolin and then putting it into erlenmeyer. A total of 100 mL of artificial waste samples were then covered with aluminum foil. Then it was put into the microwave with an operating temperature of 55 ° C and analyzed using AAS and adsorbent in the characteristic test using FTIR and SEM.

3. Result and Discussion

Research that has been done shows the performance of organokaolin adsorbents in absorbing artificial waste using microwaves, from the analysis of the ability of kaolin adsorbents modified with surfactants in reducing the concentration of iron and magnesium ions (Fe and Mg)$^{2+}$ in artificial wastewater solutions with an initial concentration of 100 mg / L. To see the performance of the adsorbent before and after the activation process is obtained and the data for the reduction of Fe and Mg metal content is obtained.

IOP Conf. Series: Materials Science and Engineering **854** (2020) 012027 doi:10.1088/1757-899X/854/1/012027

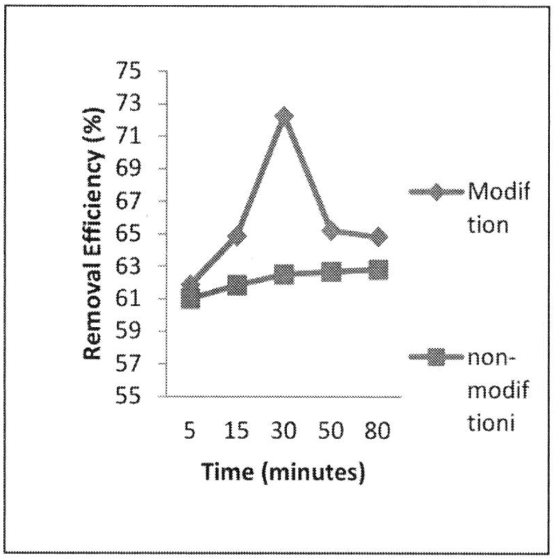

Figure 1. Efficiency of Fe metal removal with microwave irradiation time

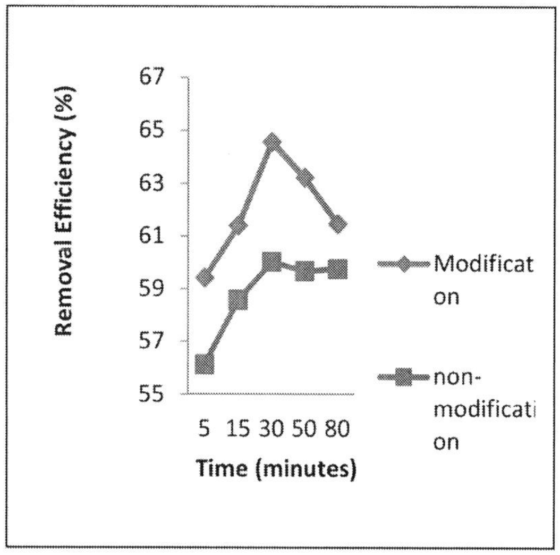

Figure 2. Efficiency of Mg metal removal with microwave irradiation time

Figures 1 and 2 show very significant differences between modification and non-modification adsorbents. The modified adsorbent is more efficient at absorbing Fe metal than non-modified adsorbent. This is because the modified kaolin using anionic surfactant (ABS) is able to reduce Fe metal content by 72.25% in the first minute which is 30 minutes, this is because Fe metal contained in the waste has been absorbed in the organokaolin pores due to increased movement the molecules found in wastewater, so that the interaction between organokaolin and metal ions is more common, the longer the irradiation time decreases the absorption efficiency obtained due to saturation of the adsorbent in absorbing Fe metal. while the absorption efficiency on the non-modification results obtained the best results at 30 minutes irradiation time of 63%.

This shows that in addition to the increase in the performance of adsorbents due to the microwave adsorption process can also accelerate the process of waste absorption because of the presence of microwaves that can heat the reaction mixture very quickly, uniformly and directly without obstructing the problem of heat transfer through the wall so that it captures more ions Fe and Mg metals (Deng, et al, 2015), This is consistent with previous studies which explained that the preparation of synthetic clay samples using microwave irradiation can increase silica levels in clay [11].

3.1. Determination of the characteristics of the adsorbent using Fourier Transform Infrared Spectroscopy (FTIR)

The characterization of functional groups in kaolin is very important, this is due to knowing the functional groups of Si-O-Al bonds. The existence of these functional groups will indicate effectiveness in the process of artificial waste absorption and will indicate effectiveness in sample preparation.

IR spectra of kaolin modification of anionic surfactant before adsorption (Figure 3) shows that the vibration changes of Si-O strain at wavelengths 10496.13 cm-1 and 1126.4 cm-1 became O-H deformation (840.96 cm-1). In the anionic surfactant spectra after Fe metal adsorption, the vibration of Si-O deformation (699.23 cm-1) changes to a vibration of O-H deformation with a wavelength of 894.58 cm-1 and 911.4 cm-1.

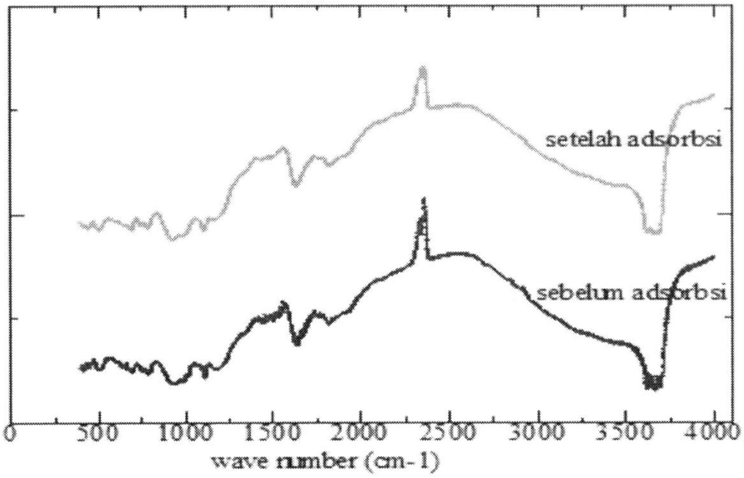

Figure 3. IR Kaolin Modification Spectra

The increase in spectrum area at wave number 1113.94 cm-1 indicates that the appearance of Si-O vibration is strain. The absorption peak experienced a change in absorption intensity which showed the absorption process of Fe metal in the modified kaolin surfactant. The last peak in both spectra shows the octahedral O-H vibration.

3.2. Determination of adsorbent characteristics using Scanning Electron Microscope (SEM)
The results of SEM photo analysis on organokaolin material before the metal adsorption process showed that the organokaolin surface still had many large pores (Figure 4). The results of SEM photo analysis on organokaolin material before the metal adsorption process showed that the organokaolin surface still had many large pores (Figure 5).

Figure 4. Organokaolin surface morphology before metal adsorption

Figure 5. Organokaolin surface morphology after metal adsorption

The results of SEM photo analysis on organokaolin material before the metal adsorption process showed that the organokaolin surface still had many large pores (Figure 4). Meanwhile, in observing organokaolin after the metal adsorption process (figure 5), there are cavities that have coincided and have small pores. These cavities and pores are formed due to the absorption of Fe metal in organokaolin.

4. Conclusion

The results of the study can be concluded Absorption of artificial wastes from Fe and Mg using Organokaolin adsorbents can be increased by using a microwave. At 30 minutes irradiation time, the highest Fe metal adsorption wastes were obtained by 72% while Mg was obtained at 30 minutes irradiation time of 61%.

Acknowledgements

I would like to express my special thanks of gratitude to my team research and my student and which also helped me in doing a lot of Research

References

[1] Bogusz, A., Oleszczuk, P., Dobrowolski, R., 2015. Application of laboratory prepared and commercially available biochars to adsorption of cadmium, copper and zinc ions from water. *Bioresour. Technol.* 196, 540–549.

[2] Du, Zhaolin , Zheng, T., Peng, W., Linlin, H ., Yanxia, W., 2016. Fast microwave assisted preparation of a low-cost and recyclable carboxyl modified lignocellulose-biomass jute fiber for enhanced heavy metal removal from water. Bioresource Technology, vol. 201. hal 41–49

[3] Halim, A. Mohammed,A., Zeinab K. 2013. *Preparation and Formation of Zeolite 5A from Local Kaolin Clay for Drying and Desuphurization of Liquefied Petroleum Gas* Iraqi Journal of Chemical and Petroleum Engineering Vol.14 No.1

[4] Salary, K., Hashemian, S., Baei, M. T. 2017. *Sb(V) removal from copper electrorefining electrolyte: Comparative study by different sorbents*. Transactions of Nonferrous Metals Society of China, Vol. 27. Hal: 440-449.

[5] Putra, A., Novia ,L., Hesti ,M., 2015: *Adsorption of Lead Ion in Water Solutions using Kaolin-Surfactant Modified as Adsorben*, Biopropallindustri, vol- 6: 81-87

[6] Deng, S., Zhang, G., Wang, X., Zheng, T., Wang, P., 2015. Preparation and performance of polyacrylonitrile fiber functionalized with iminodiacetic acid under microwave irradiation for adsorption of Cu(II) and Hg(II). Chem. Eng. J. 276, 349–357.

[7] Poewardi, B. 2017. Synthesis of active natural zeolite adsorbents with the help of microwaves for CO2 adsorption, Journal of Natural Material Engineering and Sustainable Energy, Vol. 1, No. 1. pp. 1-7.

[8] Liao, P., Malik Ismael, Z., Zhang, W., Yuan, S., Tong, M., Wang, K., Bao, J., 2012.Adsorption of dyes from aqueous solutions by microwave modified bamboo charcoal. Chem. Eng. J. 195–196, 339–346.

[9] Balver, C., Munoz,M ,Vicence, A, 2002 chemical activation of a kaolinite under acid and alkaline condition. Chem. Matter. Vol 14 pp. 2033-2043.

[10] Wahyuni, N. 2010, Modification of Kaolin with Benzalkonium Chloride Surfactant and its Characteristics using an Infrared Spectrophotometer. Journal of Science and Applied Chemistry, 1 (4): 1-14.

[11] Ismail, M. A., Eltayeb, M. A., and Abdel, Mage M.A. 2013. Elimination of Heavy Metals From Aqueous Solution Using Zeolite LTA Synthesis From Suandese Clay. Joulnal of Chemical Sciences. International Science Congrass Association. Vol 5. pp 93-98.

IOP Conf. Series: Materials Science and Engineering **854** (2020) 012028 doi:10.1088/1757-899X/854/1/012028

Use of Software Engineering on the Project Implementation Schedule of Lamreung Limpok Bridge

Abdul Muhyi[1], Miftahul Hamzi[2*], Busra[1]

[1] Civil Engineering Department, Politeknik Negeri Lhokseumawe, Aceh, Indonesia
[2] Geopolymer and Green Technology Research Center, Aceh, Indonesia

Email: Aidil.muhyi@gmail.com

Abstract. The development of technology is growing along with the time. It affects on development of construction management which several application programs are offered to facilitate construction management in term of implementing schedule and project data. The application program had facilitated the management to entry project data, manage project activities, project reports and control of project activities on lamreung limpok bridge, Lamgapang, Aceh besar, Aceh, Indonesia. The bridge incurred the expense of Rp.27.179.501.000.00 which it was supported by Aceh government budget 2018. The bridge connected the area of ulee kareng and limpok which the duration of project was 141 working days. This project consisted of 8 parameters in the activity that is the general work, drainage, work relate to land, aggregate pavement, asphalt pavement, structure, finishig, and maintenance. The issue of this study was the reach of project implementation schedule. This study will redesign the project schedule with the use of Microsoft Project 2013, The result showed that the duration of project was 122 working. Furthermore, the Gantt Chartand PDM had referred to the standard.

1. Introduction

The use of computer application programs has facilitated project managers to entry the project data, managing and controlling the project activities that had related to resources on the project. Microsoft project is an application on the processing time of a construction engineering having the ability to plan an activity, organizing, controlling time and cost. This study was conducted using the software application to facilitate the managing of construction project especially the schedule of project implementation on the bridge of Lamreung-Limpok. The length of the bridge is 276.61m with the width of 7m which it cross flood flow of the krueng acheh. The schedule of project implementation was considered as the issue so that the use of microsoft project 2013 on the project management was expected to be able to facilitate this issue.

The construction management aimed that the involvement of resources in the construction projects can be applied by the project manager. The resources of construction projects consisted of manpower, materials, machines, money and methods. The characteristic of construction project had three dimensions that was unique, resources and organization. Whereas, the process of finishing had three constraints consisted of the standard specifications, time schedule and cost[2].

The schedule of project was the determine of project activities schedule from the preparation work until finishing including material, labour and duration of activities. The parameters of in this study were gantt chart, PERT (Project Evaluation and Review Technique), and CPM (Critical Path Method). There were four process of project schedule in the precedence diagram method (PDM) such as description of work, the list of work, duration of work and PDM network [1].

Furthermore, the PDM network should link with the arrangement of plan node that described the number of activities with the relevant duration, finishing of the PDM diagram by completing the attributes and symbols, connecting the nodes with the arrows that described the interdependence between activities, figuring the earliest start (ES), earliest finish (EF), latest start (LS) and latest finish (LF) that identified the critical activities, critical paths, float and project completion duration [3].

The software of microsoft project was used to create and manage the project designs. Due to the facilities with the amazing benefits of this software to entry the project data, so that often caused by the operator. The Microsoft Project was used to manage the schedule of activity so that the construction project was evaluated throughout the stages of project task. Microsoft Project 2010 has advantages such as the ability to handle the activity schedule, organizing, controlling the time and cost that converted data input into an data output [5].

The Microsoft project provided the elements of project management by the combination of facilitated, capability, and flexibility so that was able to manage projects more efficiently and effectively. We got the information, control project work, schedules, financial reports, and control the team work of the project. Microsoft Project is one of the software provided which is a project administration software used to plan, manage, monitor and report data from a project [4].

2. Methodology

The location of project was in Lamreung Lamgapang, Krueng Barona Jaya, Acheh Besar, Acheh. The collection of data was carried out by literature and field study. The secondary data from PT. PRIMA MANDIRI PERAPEN was used as the data in this study. The processing of data was done by using the microsoft project application with the schedule techniques of gantt charts and PDM. However, analysis of work duration data should be calculated by manual. Then, the item of activity and duration data was inputted to Microsoft project 2013

3. Results and discussion

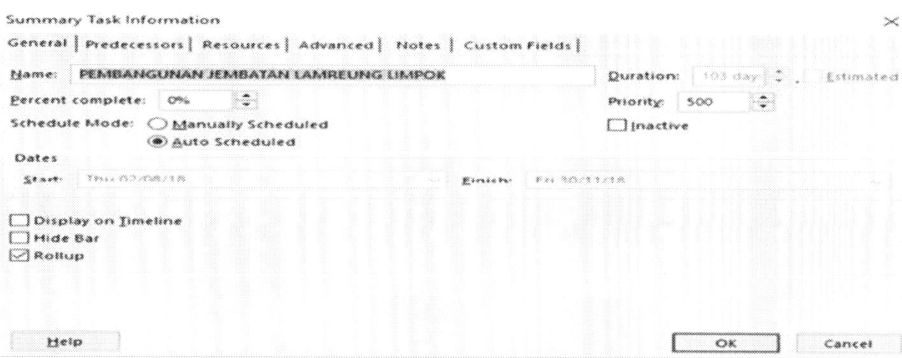

Figure 1. The schedule of project completion from task Information Summary

Figure 2. The schedule of finishing project from project summary task on PDM

IOP Conf. Series: Materials Science and Engineering **854** (2020) 012028 doi:10.1088/1757-899X/854/1/012028

Based on the analysis of microsoft project 2013, the result shows that the bridge project of Lamreung Limpok was finished by 103 working days compared to the secondary data (122 working days). This result was obtained by the display of task information and project tasks summary on PDM as seen in Figure 1 and Figure 2.

This study is calculated by microsoft project 2013 to evaluate the resource of equipment, labour and material as seen in Table 1.

Table 1. The schedule to use resource of equipment in the bridge of Lamreung Limpok

No	Resource Name	Work
1	*asphalt mixing plant*	40 hrs
2	*asphalt finisher*	40 hrs
3	*asphalt sprayer*	16 hrs
5	*compressor 4000-6500 Ν m*	24 hrs
6	*concrete mixer 0.3-0.6 m3*	72 hrs
7	*crane 10-15 ton*	2.184 hrs
8	*dump truck 3,5 ton*	928 hrs
9	*excavator 80-140 hp*	24 hrs
10	*generator set*	40 hrs
11	*motor grader >100 hp*	96 hrs
12	*wheel loader 1.0-1.6 m3*	72 hrs
13	*three wheel roller 6-8 t*	32 hrs
14	*tandem roller 6-8 t.*	40 hrs
15	*tire roller 8-10 t.*	8 hrs
16	*vibratory roller 5-8 t.*	96 hrs
17	*water pump 70-100 mm*	1.272 hrs
18	*water tanker 3000-4500 l.*	1.016 hrs
19	*welding set*	224 hrs
20	*asphalt distributor*	16 hrs
21	*base camp*	128 hrs
22	*truck concrete mixer*	2.544 hrs
23	*batching plant*	1.272 hrs

Table 2. The schedule to use resource of labour in the bridge of Lamreung Limpok

No	Resource Name	Work
1	Pekerja	35.078 hrs
2	Tukang	1.936 hrs
3	Mandor	1.960 hrs
4	Operator	5.720 hrs
5	Pembantu Operator	5.792 hrs
6	Sopir / Driver	3.880 hrs
7	Pembantu Sopir I Driver	3.824 hrs

IOP Conf. Series: Materials Science and Engineering **854** (2020) 012028 doi:10.1088/1757-899X/854/1/012028

Table 3. The schedule to use resource of material in the bridge of Lamreung Limpok

NO	Resource Name	Work
1	Pasir	13.749,77 M3
2	Agregat Kelas A	84,37 M3
3	Agregat Kelas B	112,5 M3
4	Batu Gunung/Kali	211,22 M3
5	Timbunan Tanah	1.056 M3
6	Timbunan Pilihan	105,6 M3
7	As pal	332,35 kg
8	Kerosen I Minyak Tanah	151,52 liter
9	Semen I PC (50kg)	1.092.100,16 kg
10	Laston AC – WC	172,03 ton
11	Laston AC – BC	31,32 ton
12	Baja Tulangan Polos	3.870,35 kg
13	Baja Tulangan Ulir	83.914,03 kg
14	Kawat Seton	83.984,4 kg
15	Pipa Baja Diameter 100 MM	36 M'
16	Pipa PVC	1.116 M'
17	Paku	4.146,51 kg
18	Kayu Perancah	518,31 M3
20	Expansion Joint Asphaltic Plug Tipe Fixed	61 M'
21	Marmer	2 buah
22	Adukan Semen	2
23	Sandaran (Railling)	558 M'
24	Elastomerik Sintetis	64 buah
25	Agregat Kasar	3.255,11 M3
26	Agregat Halus	2.021,07 M3
27	Unit Pracetak Gelagar Tipe I bentang 31,6 meter	36 buah
28	Seton Diafragma	324 M3
29	Panel full deph slab	1.377 buah
30	Cat Marka	614,25 kg
31	Thinner	330,75 liter
32	Blass Bit	141,75 kg
33	Patok Pengarah	50 buah
34	Kerb Pracetak	200 M1
35	Lampu Penerangan Jalan Lengan Tunggal	20 uah

4. Conclusion

Efforts to improve the quality of schedule plan on a project provided the increasing of project performance. The use of Microsoft project 2013 on the management project provided the short time and was able to reschedule the duration if the job has been delayed from the first schedule.

References

[1] Andrian, n. (2017). sistem informasi penjadwalan proyek dan performansi biaya pada pt . kelana buana sulawesi selatan Nurfatwa Andriani Y Dosen Jurusan Teknik Industri. Universitas Teknologi Sulawesi, 2(April), 61–70.

[2] Ervianto, W. I. (2005). Manajemen Proyek Konstruksi. In Andi (p. 268).

[3] Maddeppungeng, A., Suryani, I., & Iskandar, M. (2015). Analisis Pengendalian Penjadwalan Pembangunan Gedung Administrasi Universitas Pendidikan Indonesia (UPI) Kampus Serang Menggunakan Metode "Work Breakdown Structure (WBS)" Dan Kurva-S. *Jurnal Fondasi, 4*(1), 88–98.

[4] Tjakra, J., Arsjad, T., & Malingkas, G. Y. (2013). Perataan Tenaga Kerja Menggunakan Microsoft Project Pada Pekerjaan Peningkatan Jalan, 1(10), 671–677.

[5] Wijaya, G. D., Marsiano, F., & Limanto, S. (2010). studi kasus penjadwalan proyek pada proyek rumah toko x menggunakan microsoft project 2010 Pada studi kasus ini dibantu software Microsoft Project 2010 untuk merencanakan jadwal proyek dan meneliti metode CPM untuk mengetahui durasi proyek diluar dari pek, 1–8.

IOP Conf. Series: Materials Science and Engineering **854** (2020) 012029 doi:10.1088/1757-899X/854/1/012029

Separation of Oleoresin from Nutmeg Using Ultrasound Assisted Extraction and Hexane as Solvent

Victoria Kristina Ananingsih, Bernadeta Soedarini, Emerentiana Karina

Department of Food Technology, Soegijapranata Catholic University, Semarang, Indonesia

Email: kristina@unika.ac.id

Abstract. Indonesia has many herbs and spices that function as ingredients for making herbal medicine or for cooking. Nutmeg (*Myristica fragrans*) is an Indonesia's origin plant that has many health benefits. It produces specific flavors come from its bioactive compounds, namely oleoresin. Oleoresin can be extracted from nutmeg to get the concentrated compounds; hence it can be applied easily as the ingredient of food and beverage products. Ultrasonic assisted extraction (UAE) is applied in this research since it has higher yield compared to other extraction methods. This research is aimed to optimize the process condition of ultrasonic assisted extraction of nutmeg to get the good qualities of oleoresin. Solvent used in this research was hexane with extraction temperatures of 39, 45, 52 °C, and extraction times of 30, 45 and 60 minutes. Ratio between nutmeg and solvent were 5:100, 10:100 and 15:100. Frequency used was 45 kHz. The results showed that the optimum process condition to separate oleoresin from nutmeg were at extraction temperature of 52 °C, extraction time of 60 minutes and ratio between nutmeg and hexane of 5:100. This optimized process condition produced yield 62% of extracted oleoresin. Antioxidant activity and phenol content of this optimized oleoresin were 44,78% and 0,11 ppm respectively.

Keywords: nutmeg, oleoresin, ultrasound, hexane

1. Background

Nutmeg (*Myristica fragrans*) is an original plant from Indonesia that grows well in tropical climates at altitudes below 700 meters above sea level. When the nutmeg is cut into two parts, the nutmeg seeds appear wrapped in a red layer called mace. Nutmeg seeds can be used to give flavor to food products due to the presence of oleoresin and essential oils contained in them. As a country of origin of nutmeg, Indonesia supplies nearly 75% of the world's nutmeg and mace seed needs. The popularity of nutmeg on the global market began in the 16th century since Ternate, Tidore, Ambon and the Banda Islands became the main pathway for foreign countries to enter Indonesia. In 2006, Indonesia exported 347,919 kg of nutmeg with a total export value of 1,861,232 USD. Then in 2017, Indonesia exported 11,505,972 kg of nutmeg with a total export value of 50,138,286 USD (Directorate General of Plantations, 2017).

Nutmeg consists of three parts namely flesh, mace, and nutmeg seeds. Nutmeg meat can be processed as a pickled fruit product. Whereas mace and nutmeg seeds are known to have high oleoresin content, but oleoresin has easily oxidized properties. Oleoresin extract from nutmeg has a high amount of demand for domestic and export needs so it requires more handling to maintain its quality. Therefore it need a method to produce optimal amount of oleoresin. The most commonly used method to obtain oleoresin is the extraction method. Ultrasound extraction has high mass transfer efficiency, less use of organic solvents, and higher extraction rates than the Soxhlet method.[1] At present ultrasound, extraction has begun to be applied in industries such as edible oil extraction. [2] Ultrasound extraction is more effectively applied for extraction at low temperatures which would be advantageous for extracting heat-sensitive compounds. [3]

This study aims to determine the effect of the ratio between the solvent and the ingredients, temperature, and extraction time on the effectiveness of the nutmeg oleoresin extraction process. In addition, this study also aims to determine the total phenolic content and antioxidant activity in nutmeg oleoresin extract.

2. Methods

2.1. Materials

The materials used in this study were nutmeg, n-hexane, *Folin-Ciocalteau* compound, sodium carbonate solution (7.5%), DPPH solution in ethanol 0.1 mM, gallic acid, KOH 0.1 N solution, *phenolphthalein* indicator, a solution of 0.5 N HCl, and distilled water.

2.2. Response Surface Methodology (RSM)

This method is used to analyze problems where several independent variables influence the response variable and create a model or experimental design to optimize the response. The relationship between the response of Y and the independent variable X is as follows:

$$Y = b0 + \sum_{i}^{k} bi.xi + \sum_{i}^{k} bii.xi^2 + \sum_{i}^{k} bi.xi.xii$$

Explanation:

Y= response variable

Xi= independent variable / factor (i = 1,2, ... k)

b= coefficient

The first step of the *Response Surface Method* (RSM) is to determine the relationship between response Y and response X through a first order polynomial equation and use a linear regression model. [4]

$$Y = b0 + \sum_{i}^{k} bi.xi + \sum_{i}^{k} bii.xii.$$

Y= response variable

Xi= independent variable / factor (i = 1,2, ... k)

b= coefficient

The full experimental design in this study can be seen as follows:

a) Independent Variable
- Temperature, connoted as X_1 with points: 390C, 450C, and 520C.
- Time, connoted as X_2 with points: 30 minutes, 45 minutes and 60 minutes.
- Ratio, connoted as X_3 with points: 5:100, 10:100, and 15:100.

b) Response Variable
- Percentage of yield
- Total phenolic
- Percentage of antioxidant activity

c) First order model: factorial design 2^3, each factor having a low level (code: -1), middle level (code: 0), high level (code: +1), and a level on the axial axis (code: -α and + α). The α value for the CCD design is:

$$\alpha = (2^k)^{1/4}$$

IOP Conf. Series: Materials Science and Engineering **854** (2020) 012029 doi:10.1088/1757-899X/854/1/012029

Explanation:
α= axial axis
k= number of factors or independent variables

In this study there are 3 (three) independent variables, then an α value of 1.682 is obtained. The magnitude of the variables with codes - α and + α can be calculated by the equation:

$- \alpha = (0) - 1,682\ [(0) - (-1)]$

$+ \alpha = (0) + 1,682\ [(0) - (-1)]$

With this equation, we obtain 3 (three) independent variables with 5 (five) level variables.

2.3. Oleoresin Extraction Process
The oleoresin extraction process begins by drying the nutmeg seeds using an oven at 60°C for 1 hour. Then nutmeg crushed using a blender. Furthermore, nutmeg seed powder is put into an Erlenmeyer tube and the n-hexane solvent is added, and then tightly closed with aluminum foil. After the Erlenmeyer tube is inserted in an *ultrasonic bath* that has been set in temperature and time. The extraction temperature range used is 39°C, 45°C, and 52°C. The extraction time range used is 30 minutes, 45 minutes and 60 minutes. The ratio of the sample to the solvent used is 5: 100, 10: 100, and 15: 100. Then the extract is filtered with a filter cloth. After that, the extract is evaporated with a *rotary vacuum evaporator* for 4-6 minutes at 40°C. With a speed of 50 rpm. Furthermore, the extract was put into a vial bottle and stored in a *freezer* so that it can be used for further analysis. [5]

2.4. Nutmeg Oleoresin Quality Testing
Tests carried out to measure the extraction efficiency of nutmeg oleoresin or rendemen[6], testing of antioxidant activity[7] and total phenolic testing[8].

3. Results and Discussion

3.1. Effectiveness of Oleoresin Extraction
The results of the optimization analysis of the yield of nutmeg oleoresin extraction can be seen in Figure 1, 2 and which is a *fitted surface* graph depicting the relationships between variables in three coordinate segments so as to form a 3D (three-dimensional) curve, namely (1) time and ratio variables (2) time and temperature (3) temperature and ratio, to the yield of nutmeg oleoresin yield. Based on the graph, the red area shows the optimal area. In this study the optimal point of extraction temperature was 41.27°C, extraction time was 55.30 minutes and the ratio of material was 11.79 grams to the yield of nutmeg oleoresin

IOP Conf. Series: Materials Science and Engineering **854** (2020) 012029 doi:10.1088/1757-899X/854/1/012029

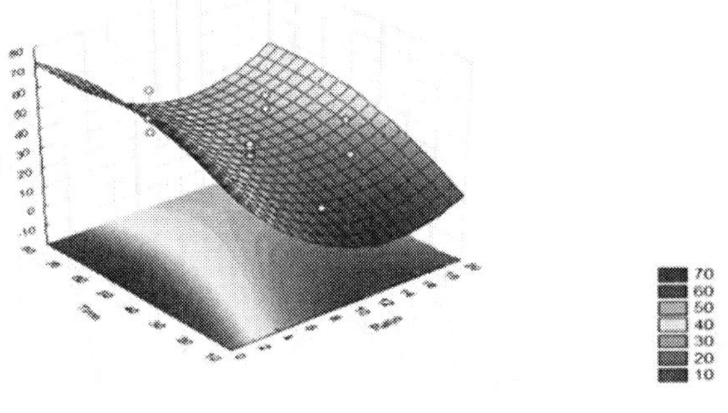

Figure 1. Optimization Analysis Results of Nutmeg Oleoresin Extraction Results Based on Time and Ratio

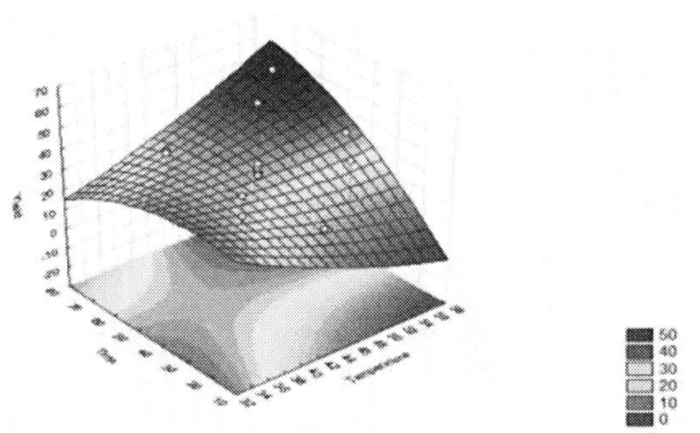

Figure 2. Optimization Analysis Results of Nutmeg Oleoresin Extraction Results Based on

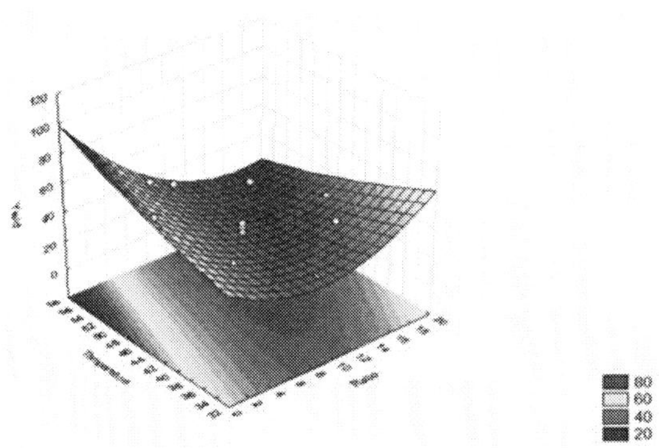

Figure 3. Analysis Results of Optimization Results of Nutmeg Oleoresin Yield Based on Temperature and Ratio

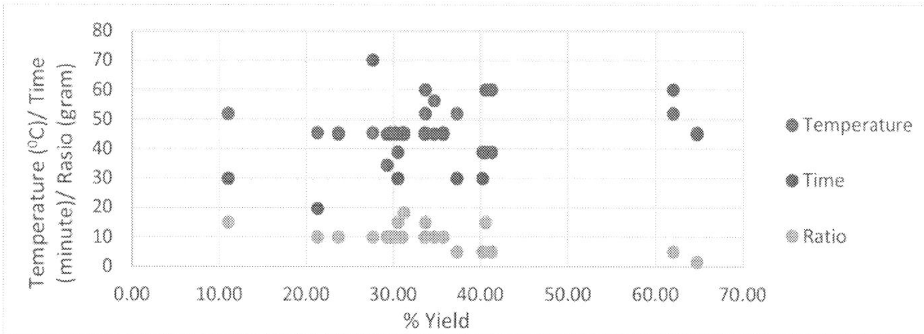

Figure 4. Results of Nutmeg Oleoresin Yield by Temperature, Time, and Ratio

Based on Figure 4, it can be concluded that the yield of nutmeg oleoresin obtained is fluctuating. Some points show that the higher the extraction temperature, the higher the yield produced and the other points show that the higher the extraction temperature, the lower the yield. This also occurs at the time of extraction, some points indicate that the longer the extraction time, the higher the yield produced and some other points indicate that the shorter the extraction time, the lower the yield produced. Whereas in the ratio of materials used, most points indicate that the less material used, the higher the yield produced.

Table 1. Results of Regression Analysis of Nutmeg Oleoresin Yield

Parameter	Regression Analysis Results	
	Coefficient	P Value
A constant	109,514	0,202
Time (X_2)	-1,185	0,263
Ratio (X_3)	0,214	0,942
$X_1.X_2$	-0,170	0,027
$X_1.X_3$	0,046	0,010
R^2	91,27%	

It can be seen in Table 1 that the interaction between temperature and time variables has a significant effect on the yield of nutmeg oleoresin because the p value is 0.027 or smaller than α (0.05). In addition, the interaction between temperature and ratio variables also had a significant effect on the yield of nutmeg oleoresin because the p value was 0.010 or smaller than α (0.05). Based on the results of the regression analysis, an equation is obtained:

$$Y= 109,514 - 1,185X_2 + 0,214X_1 - 0,170X_1.X_3 + 0,046\ X_1.X_2$$

Explanation:
Y= yield
X_1= temperature
X_2= time
X_3 = ratio

IOP Conf. Series: Materials Science and Engineering **854** (2020) 012029 doi:10.1088/1757-899X/854/1/012029

3.2. Antioxidant Activity

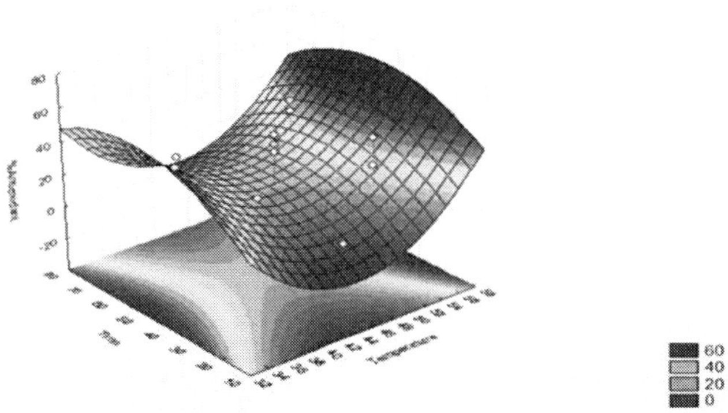

Figure 5. Optimization Analysis Results of the Nutrition Oleoresin Activity Test of Nutmeg Seeds Based on Temperature and Time

The graph above is a *fitted surface* graph that illustrates the relationship between variables in three coordinate segments so as to form a 3D (three-dimensional) curve that is the temperature and time variable of the antioxidant activity of nutmeg oleoresin. In this study, the optimal extraction temperature was obtained at 43.82°C, the extraction time was 51.92 minutes and the ratio of 1.59: 100 to the antioxidant activity of nutmeg oleoresin.

The longer the extraction time, the more extracts are produced. The longer time needed for the extraction process can make the contact between the sample with the solvent becomes longer so that the diffusion process can run optimally. The higher the temperature used in the extraction process will help the process of diffusion of solvents into the tissue material so that it can increase the activity of the solvent in hydrolyzing substances in the cell material and the extraction time can run faster. However, using too high a temperature and too long a time can damage the structure of the substance of the desired extract. The longer the extraction time and the more volume of the solvent used can produce a more concentrated extract. [9] Also the longer the extraction time can increase the specific gravity of the extract because the extract also evaporates along with the solvent. The high yield in extraction results can also be influenced by the presence of residues that carry over into the extract. Another factor that influences yield yields is the ratio of the ingredients to the solvent. In this study, the highest yield was found in the ratio of ingredients to solvents of 1.59: 100, followed by a comparison of materials with solvents of 10: 100. The more the amount of solvent added to the sample, the greater the ability of the solvent to extract the oil contained in nutmeg seeds. Howe ver, solvents have a maximum limit on the ability to extract oil in nutmeg seeds [10].

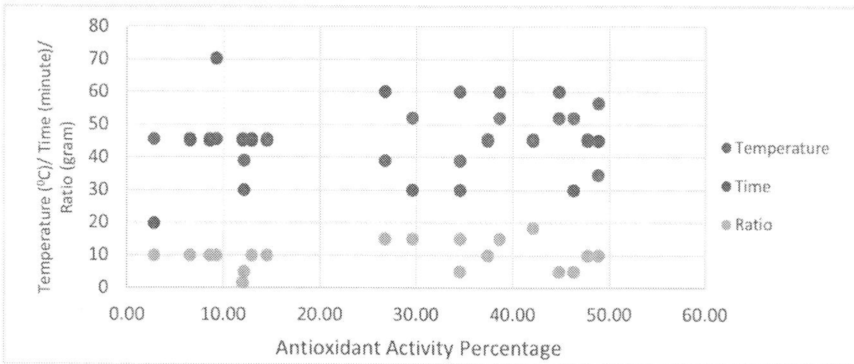

Figure 6. Test Result of Nutmeg Oleoresin Antioxidant Activity

Based on Figure 6 it can be concluded that the results of testing the antioxidant activity of nutmeg oleoresin extract obtained are fluctuating. Some points indicate that the higher the extraction temperature, the higher the yield produced and some other points indicate that the higher the extraction temperature, the higher the antioxidant activity. This also occurs at the time of extraction and the ratio of material used, some points indicate that the longer the extraction time and the more material used, the higher the antioxidant activity and some other points indicate that the shorter the extraction time and the less material used, the lower its antioxidant activity.

3.3. Total Phenolic

The graph below is a *fitted surface* graph that illustrates the relationship between variables in three coordinate segments so as to form a 3D (three-dimensional) curve that is the time variable and the ratio of total phenolic nutmeg oleoresin. Based on the graph, the red area shows the optimal area.

Overall, the highest total phenolic content was obtained at a temperature of 36.28°C for 35 minutes at a ratio of a material to a solvent of 9.76: 100. The longer the extraction time, the higher the extract produced. In addition, the use of temperatures that are not too high can prevent extraction damage due to heat, but the extraction time required is getting longer. Therefore, the higher the yield produced, the more antioxidants (phenols) contained in it so as to increase the total phenolic number.

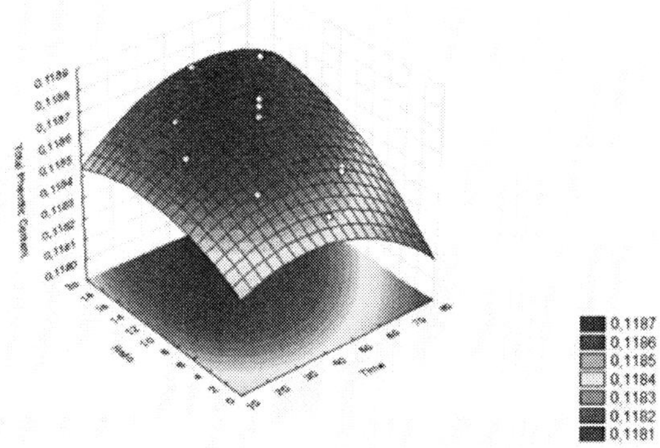

Figure 7. Analysis Results of Optimization of Total Test of Phenolic Oleoresin in Nutmeg Seeds Based on Time and Ratio

Based on Figure 8 it can be concluded that the total phenolic test results of nutmeg oleoresin extract obtained are fluctuating. Some points indicate that the higher the extraction temperature, the higher the polyphenols produced and the other points indicate that the higher the extraction temperature will reduce the polyphenol content in oleoresin extract This also occurs in the extraction time variable and the ratio of the material used, some points indicate that the longer the extraction time and the many ingredients used, the higher the polyphenol content in oleoresin extracts. While some other points show that the longer the extraction time and the many ingredients used will reduce the levels of polyphenols in nutmeg oleoresin extract.

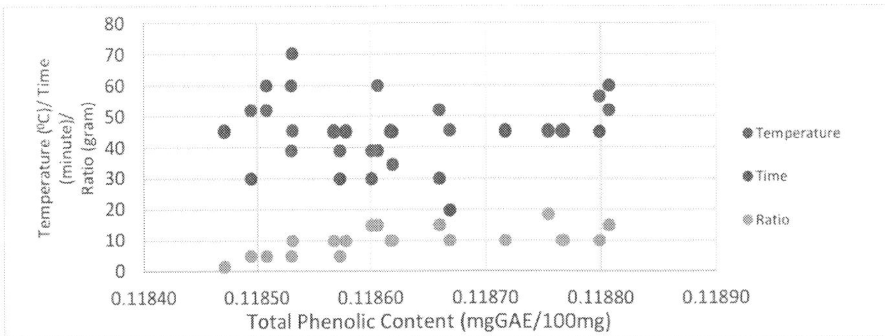

Figure 8. Analysis Results of Optimization of Total Test of Phenolic Oleoresin in Nutmeg Seeds

4. Conclusions

The combination of treatments to obtain the optimum percentage yield is at extraction temperature 41.27°C, extraction time for 55.30 minutes, and the ratio of 11.79:100. The combination of treatments to obtain antioxidant activity with optimum percentage is at extraction temperature of 43.82°C, extraction time for 51.92 minutes, and the ratio of 7.55: 100. The combination of treatments to obtain total phenolic with optimum amount is at extraction temperature of 43.82°C, extraction time for 51.92 minutes, and the ratio of 7.55: 100.

Based from the parameters tested, the yield is the response to the independent variables temperature, time, and ratio show that the yield has a mathematical model: $Y = 109{,}514 - 1{,}185X_2 + 0{,}214X_1 - 0{,}170X_1.X_3 + 0{,}046\ X_1.X_2$, with an R^2 value of 0, 9127. Based on the mathematical model determined, the optimum value of the independent variables from the results of the yield analysis will produce an antioxidant percentage of 94.71%. Based on the mathematical model determined, the optimum value of the independent variables from the total phenolic analysis yield results the percentage of 30.11%.

Acknowledgment

The authors are grateful to the financial support from *Ministry of Research, Technology and Higher Education of the Republic of Indonesia*, Nr: 010/L6/AK/SP2H.1/PENELITIAN/2019.

References

[1] Fang, Zhen, Richard L. Smith, Jr., dan Xinhua Qi. (2015). *Production of Biofuels and Chemicals with Ultrasound.* Springer. London.
[2] Chen, Dong, Sanjay K. Sharma, dan Ackmez Mudhoo. (2012). *Handbook on Applications of Ultrasound: Sonochemistry for Sustainability.* CRC Press. Boca Raton.
[3] Aguirre, Daniela Bermudez. (2017). *Ultrasound: Advances in Food Processing and Preservation.* Academic Press. UK.
[4] Nuryanti and Salimy, Djati. H. (2008). Response Surface Methodology and Its Application on Chemical Experiments. Risalah Lokakarya Komputasi Dalam Sains dan Teknologi Nuklir.

[5] Ni Made,R. O. A. (2015). Effect of Particle Size and Extraction Time on the Characteristics of Natural Colorant of Pandan (*Pandanus tectorius*). Universitas Udayana.
[6] Anggraini, Rini, Afghani Jayuska, dan Andi Hairil Alimuddin. (2018). Isolation and Characterization of Essential Oil from Black Pepper *(Piper nigrum* L.*) originated from Sajingan, West Borneo.* Jurnal Kimia Khatulistiwa. Volume 7 Nomor 4:124-133.
[7] Mardiana, F.M. (2015). Effect of Maltodextrin Addition on the Characteristics of Teak Leaves Powder Processed by Freeze Drying. Universitas Katolik Soegijapranata.
[8] Santos, Philipe, Ana C. Aguiar, Gerardo F. Barbero, Camila A. Rezende, dan Julian Martínez. (2015). *Supercritical Carbon Dioxide Extraction Of Capsaicinoids From Malagueta Pepper (Capsicum frutescens L.) Assisted by Ultrasound.* Ultrasonics Sonochemistry. Volume 22 nomor 1:78–88.
[9] Bustan, M. D., Febriyani, R. and Pakpahan, H. (2008). Effect of Time and Particle Size on the Yield of Ginger Oleoresin Produced by Different Concentration of Organic Solvents. Universitas Sriwijaya. Sumatera Selatan.
[10] Masud, Fajriyati dan Pabbenteng. (2016). Ratio of Rice Brand and Solvent on the Extraction of Paddy's Bran. Politeknik Negeri Ujung Pandang. Makassar.

IOP Conf. Series: Materials Science and Engineering **854** (2020) 012030 doi:10.1088/1757-899X/854/1/012030

Bioethanol Production from Alkali Steam Explosion of Oil Palm of Empty Fruit Bunch Fiber

Yanni Sudiyani[1], Eka Tri Wahyuni[1], Muryanto muryanto[1], Sephtian Marno[2], Nelizza Putri[2]

[1] Research Center for Chemistry, Indonesian Institute of Sciences (LIPI), Kawasan PUSPIPTEK Serepong, Tangerang Selatan 15314, Indonesia
[2] New and Renewable Energy, Research & Terchnology Center, PT Pertamina, Jakarta, Indonesia

Email: sudiyani@gmail.com

Abstract. Bioethanol is the most promising biofuels and can be generated from lignocellulosic biomasses. The study of bioethanol production from oil palm empty fruit bunch (OPEFB) pretreated by alkali steam explosion was carried out employing separate enzymatic hydrolysis (SHF) and Simultaneous Saccharification and Fermentation (SSF) approaches. Alkaline extraction after steam explosion pretreatment of OPEFB led to the partial removal of lignin, hemicellulose and other degradation products from fiber surface. The pretreatment process is of great importance to ethanol yield. In the present study, the pretreatment was processed using a steam explosion reactor at 150°C, 4 bars for 30 minute with alkaline concentrations of NaOH 10% (kg/L). After pretreatment, the OPEFB substrates 10% and 15% (w/v) was saccharification using 30 FPU of Ctec2 and Htec2 enzyme and fermentation using yeast of *Saccharomyces cerevisiae* for 72 hours. The glucose obtained after 72 hours was 137.80 g/L (subtrate 15%) and 97.90 g/L (subtrate 10%). Bioethanol production in SSF mode faster and slightly more efficient process than SHF.

1. Introduction

Sustainability and environmental safety are major necessities in the current world that requires the utilization of renewable and eco-friendly alternative resources. Lignocellulosic biomass is one of the most promising alternative resources to be exploited since its utilization will not disturb food supply [1]. Lignocellulose can be obtained from many sources such as crops, forest materials, agricultural wastes, and aquatic plants [2] [3]. They are primarily composed of three primary bio-polymers: cellulose, hemicellulose, and lignin. These bio-polymers can be converted into high-value products such as biofuels, energy sources, and chemicals [4] [5]. One of the most widely practised lignocellulose utilization is bioethanol production. Renewable and eco-friendly; bioethanol can be used to replace fossil fuels as primary energy source for transportation, industrial, and domestic uses. However, lignocellulose is relatively more recalcitrant to chemical process than sugar and starch due to many factors such as low available surface area, crystalline structure, protection of cellulose and hemicellulose by lignin, and heterogeneous nature of the biomass [6]. Therefore, bioethanol production from lignocellulosic materials requires three distinct processes: pretreatment of the biomass, hydrolysis/saccharification, and fermentation of monomeric sugars [2]. Proper understanding of these complex processes is necessary in order to break the lignocellulosic matrix and effectively convert the bio-polymers into fermentable monomeric sugars without producing excessive amount of side products or inhibitors that could disrupt the fermentation.

Pretreatment of lignocellulosic biomass can be performed through physical, chemical, physico-chemical, and biological methods. One of the most widely applied pretreatment methods is steam explosion due to its low environmental impact and modest cost [7].

Hydrolysis and fermentation is the next step conducted after pretreatment process for producing bioethanol. Separated Hydrolysis and Fermentation (SHF) was the conventional method that hydrolysis was carried out in the period and fermentation process after then. This process allowed hydrolysis process worked first to produce monosaccharide sugar, so sugar is ready when the fermentation begins. Another method of hydrolysis and fermentation was Simultaneous Saccharification and Fermentation (SSF). In SSF, hydrolysis and fermentation located in a single reactor, enzyme and yeast put together, so glucose is rapidly converted into ethanol [8]. Wyman *et al.* reported that SSF process gives higher yield of ethanol than SHF because low residual sugar relieves inhibition on the cellulase enzyme [9] [10].

The objective of this study was to investigate the effect of alkaline (sodium hydroxide, NaOH) pretreatment of oil palm of empty fruit bunch (OPEFB) with provide highest cellulose to glucose conversion during enzymatic hydrolysis for ethanol production using two methods of SHF and SSF.

2. Materials and Methods

2.1. Raw Materials

OPEFB fiber used in this study was obtained from palm oil plantation in Palembang, South Sumatera Indonesia Indonesia. It was dried (final moisture content 8.60%, measured by Moisture Analyzer OHAUS MB 45) and chopped (average particle size 6.5 mm) . Sodium hydroxide (NaOH) pellets used in the NaOH pretreatment with steam explosion was industrial grade. Cellic® CTec2 and Cellic® HTec2 enzymes were procured from Novozymes, Denmark. The activity of both enzymes were 144 FPU/mL.

2.2. Pretreatment (Base-Catalyzed Steam Explosion)

Pretreatment process of oil palm OPEFB was carried out in a steam explosion reactor CHEMEX method using dilute sodium hydroxide solution. Concentrations of sodium hydroxide (solution which used in this study was 10% (kg/L), a solid-liquid ratio of 1:5 then steam exploded at 150°C and 0.4 MPa for 30 minutes. The pretreated biomass was filtered with hydraulic press filter, then washed with water to neutral pH. The washed biomass was then dried in the oven at 50°C until the moisture content is below 10%, then used for as substrate in the enzymatic hydrolysis.

2.3. Separate Hydrolysis and Fermentation (SHF)

This procedure was performed on 10, and 15% (w/v) of pretreated EFB fiber in 0.05 M citrate buffer with a pH of 4.8. The flasks containing the substrate in citrate buffer were autoclaved at 121°C for 20 minutes to ensure sterile condition. After cooling down, 30 FPU/g of Cellic® CTec2 and 6 FPU/g of Cellic® HTec2 was added in each of the flasks. The hydrolysis was conducted in a shaking incubator at 150 rpm and 50°C for 72 hours. Sampling every 24 hour was employed to monitor producing sugar. Glucose and xylose were measured as a product in this hydrolysis. Data of glucose form HPLC was required to get the percentage of yield. The next process, fermentation, was conducted 72 hours after hydrolysis process. The temperature of shaking incubator was changed into 32°C. In the constant temperature, one percent (g/ml) of dry yeast, *Saccharomyces cereviceae*, was put in the each flask. Ethanol content, glucose, and xylose were monitored every 24 hour. The calculation percentage of yield in fermentation is as same as those in saccharification process i.e by comparing measured ethanol weight with the theoretical weight of ethanol. The anhydro correction is 0.51 (for glucose to ethanol)[11]. All process was conducted triplo.

2.4. Simultaneous Saccharification & Fermentation (SSF)
The procedure of SSF was similar with SHF, however 1% (w/v) of baker's yeast *S. cerevisiae* yeast was added along with the enzymes. The substrate loading of the process was 10 and 15 % (w/v); and the Cellic® CTec2 enzyme loading was 30 FPU/g while Cellic® HTec2 was added 20% (v/v) of Cellic® CTec2. The process was conducted in a shaking incubator under temperature condition 32°C with velocity agitation 150 rpm for 72 hours. Samples were taken every 24 hours for analysis of ethanol by HPLC.

2.5. Products Analysis
Analytical procedure for the measurement of carbohydrate (glucose and xylose) and lignin component on the raw material and pretreated biomass, was provided by National Renewable Energy Labotatory (NREL) [26]. Glucose, xylose, and ethanol concentrations was measured by High Performance Liquid Chromatography (HPLC) Waters, USA. Bio-Rad Aminex HPX-87H column was used as the stationary phase while the mobile phase is 5 mM H2SO4 at 0.6 mL/min. The detector is Waters 2414 refractive index (RI) detector. The oven temperature was maintained at 40°C in the column input and 65°C in the output.

2.6. Calculations
Glucose yields from enzymatic hydrolysis were calculated according to Eqs. (1) based on the glucose content (as cellulose) of the steam-exploded OPEFB fiber, while xylose yields were calculated according to Eqs. (2) based on the xylose content (as hemicellulose) of the raw OPEFB fiber.

$$\text{Glucose yield } (\%\,w/w) = \frac{\text{Glucose in hydrolyzate (g)}}{\text{Cellulose in substrate (g)} \times 1.11^{*}} \times 100\% \tag{1}$$

$$\text{Xylose yield } (\%\,w/w) = \frac{\text{Xylose in hydrolyzate (g)}}{\text{Hemicellulose in substrate (g)} \times 1.14^{*}} \times 100\% \tag{2}$$

*1.11 and 1.14 are correction factors to compensate the addition of water molecule after the breakage of cellulose into glucose monomers and hemicellulose into xylose monomers, respectively [7].

Ethanol yields from SSF were calculated according to Eqs. (3) based on the theoretical mass of produced ethanol assuming that all of the cellulose in steam-exploded OPEFB fiber was converted into glucose, while all of the produced glucose was converted into ethanol with maximum efficiency.

$$\text{Ethanol yield } (\%\,w/w) = \frac{\text{Ethanol in hydrolyzate (g)}}{\text{Cellulose in substrate (g)} \times 1.11 \times 0.51^{**}} \times 100\% \tag{3}$$

**0.51 is a correction factor to represent the maximum possible ethanol yield from glucose based on the stoichiometric ratio between ethanol and glucose in glucose fermentation [13].

3. Result and Discussion

3.1. Composition of EFB after Steam Explosion
The composition change of the OPEFB reflect the efficiency of pretreatment. Table 1 depicts the changes in the composition of untreated and pretreated OPEFB. The composition of the untreated was 36.59% cellulose, 24.97% hemicellulose, 26.53% lignin and 1.79% ash. When the OPEFB was pretreated by alkali steam explosion, the composition of carbohydrate (cellulose and hemicellulose) increased to 70%. The OPEFB is more carbohydrate enriched because the percent composition of lignin decreased. The percentage of lignin decreased due to an effect of alkaline pretreatment. Over than 70% lignin removed and as a consequent, the cellulose content increased. Lignin was considered as an inhibitor because they could restrict the hydrolysis enzyme to attack cellulose. Furthermore, using sodium hydroxide would induce swelling leading on the increase of surface area [12]. In the present study, alkaline pretreatment OPEFB was aimed to alter the structure of cellulosic biomass by removing

lignin and hemicelluloses, so that the cellulose became more accessible to the enzymes that convert carbohydrate polymers into fermentable sugars.

Table 1. Composition of EFB before and after steam explosion

No	Component	Before pretreatment (%)	After pretreatment (%)
1	Lignin	26.53	7.45
2	Cellulose	36.59	78.02
3	Hemicellulose	24.97	12.64
4	Ash	1.79	1.86

3.2. Enzymatic saccharification

The result of glucose and xylose during saccharification process obtained from HPLC analysis was shown in Figure 1. This analysis predicted the sugar released by enzyme performance. Cellulose and hemicellulose were converted into glucose and xylose respectively using combined enzyme. As shown in Figure 1 glucose concentration reached the highest concentration after 72 h were 137.80 g/L and 97.90 g/L for pretreated OPEFB substrate 15% and 10% concentration respectively. As expected, higher glucose concentration results in higher yield of hydrolysis. As like as glucose concentration, xylose also increased during saccharification. The combination of two enzymes not only converted cellulose to glucose, but it also changed hemicellulose to xylose. However, the xylose concentration was lower compared with glucose concentration. The discussion will be focused on the cellulose hydrolysis since *S. cerevisiae* only metabolizes hexose sugars to bioethanol and cannot metabolize xylose or any other pentose sugar. The glucose concentrations and yields from enzymatic hydrolysis of the steam-exploded OPEFB are shown on Figure 1.

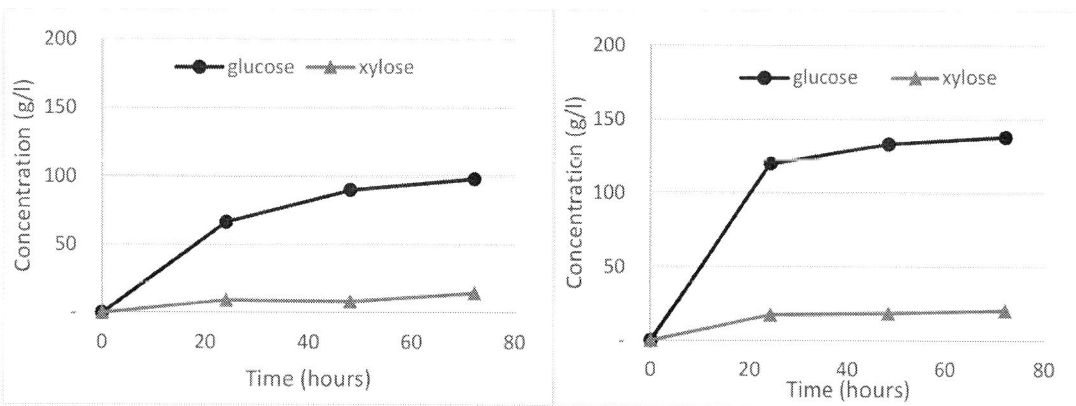

Figure 1. Sugar yield obtained after 24, 48 and 72 h of enzymatic hydrolysis of pretreated OPEFB (a) substrate 10%; (b) substrate 15%

3.3. Fermentation ; SHF versus SSF

Steam exploded alkali of OPEB was subjected to bioethanol fermentation by using *S. cerevisiae* yeast. Two different method were studied: a separate hydrolysis followed by fermentation (SHF) and a simultaneous saccharification and fermentation (SSF) process, in order to compare their efficiency in term, bioethanol concentration fermentation productivity, yield, and the overall time of the fermentation.

3.3.1. Separate hydrolysis and fermentation process (SHF)

SHF is one method of saccharification and fermentation process that involves two sequential steps: Enzymatic hydrolysis, often called saccharification and followed by fermentation process. First the cellulose in substrate was enzymatically hydrolyzed to glucose. Second, the glucose in the hydrolyzed was subsequently fermented to ethanol. For SHF study the previous saccharification was carried out for 72 h at 50^{0}C, according to the condition described at section 2.3. After 72 h of enzymatic hydrolysis, process was continued by fermentation, thus the total time processes were 144 h. The fermentation step was performed using baker's yeast *S. cerevisiae* at 32^{0}C giving satisfactory results. The fermentation profiles of sugar consumption and ethanol production obtained presented in the Figure 2. For SHF study, the concentration of glucose at the beginning of fermentation step was 138 g/l after 72 h fermentation ethanol produced reach in the concentration 69.84 g/l and there is no glucose after 72 fermentation. In theory, maximum ethanol yield is 0.51 g ethanol/g glucose. In this study ethanol results from the process is 69.84 g/l, by calculation the yield ethanol is 94.29% (Table 2). The ethanol concentration obtained in this study 69.84 g/l was higher than 47.4 g/l [8]. OPEFB pretreated by steam explosion under similar conditions. Xylose concentration remained practically constant along the fermentation as expected, since native *S.cerevisiae* cannot metabolize xylose.

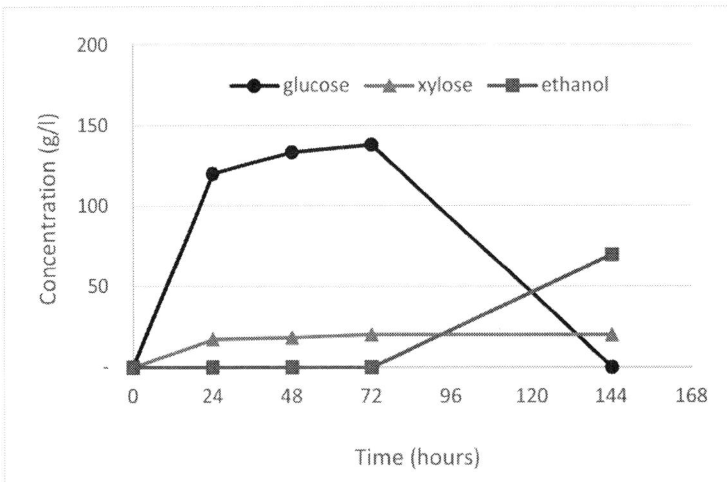

Figure 2. Ethanol concentration during fermentation of OPEB in the SHF process, substrate 15%.

3.4. Ethanol Concentration & Yield from SSF

SSF process allowed enzymes and yeast put together while working process was conducted at lower temperature. In this study, the used temperature was same with fermentation i.e., 32°C. More saving energy is one of the advantages of the SSF process. Time producing ethanol becomes shorter because the hydrolysis and fermentation were carried out at the same time. The enzymatic hydrolysis was accompanied by fermentation, almost from the beginning of this simultaneous process. At 24 h fermentation, glucose was completely exhausted and ethanol reached the concentration of 62.68 g/l for 15% substrate, however for 10% substrate 36.75% only. When enzyme produced glucose, the yeast, *S. cereviceae* changed to ethanol directly. It is why the concentration of glucose always zero in the 24, 48, 72 hour fermentation. The ethanol concentrations and yields from SSF are shown on Figure 3.

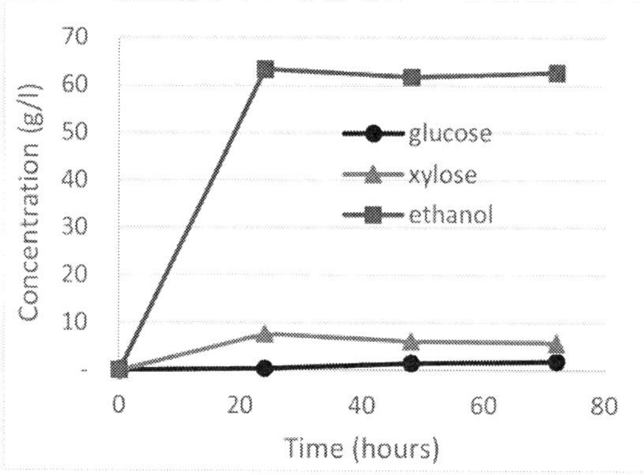

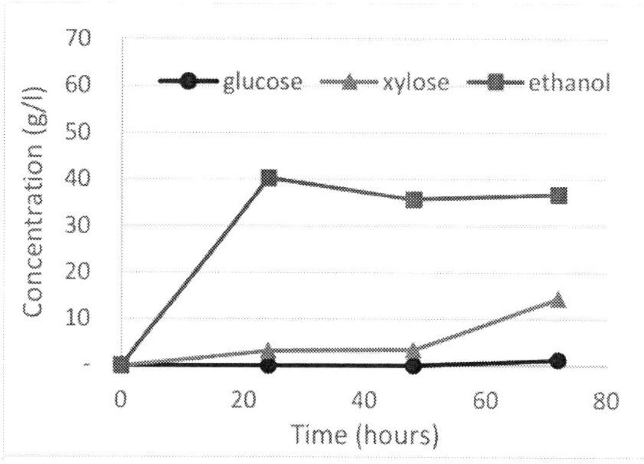

Figure 1. Producing Ethanol from OPEFB by SSF Process. Substrate (a) 15% and (b) 10%

Looking to the results obtained after 72 h of SSF of OPEB, the maximum ethanol yield was 84.29% (Table 2). The results from this study show that in the end of the assay the ethanol concentration obtained by SHF using 15% substrate concentration. The process of SHF for 144 h gave the highest yield of ethanol. However, it was not suitable time according to economic aspect. SSF method was considered as a better process than SHF due to rapidly ethanol production and the high concentration of produced ethanol.

Table 2. Yield of glucose and ethanol

	SHF 10%	SHF 15%	SSF 10%	SSF 15%
Glucose yield	97.89	94.88	-	-
Xylose yield		Not calculated		
Ethanol yield	40.74	94.29	66.84	84.62

4. Conclusion

Bioethanol production using pretreated of alkali steam explosion of Oil Palm Empty Fruit Bunch (OPEFB) was conducted in the area of hydrolysis and fermentation. Enzymatic hydrolysis and fermentation were studied in the two variation methods, SSF and SHF. The glucose obtained after 72 hours was 137.80 g/L (subtrate 15%) and 97.90 g/L (subtrate 10%). Using 15% of concentration substrate, it could be produce 69.84 g/l of ethanol in 72 hour fermentation by SHF process and 62.68 g/l of ethanol in 24 hour by SSF process. From this study, the SSF method was considered as a better process than SHF due to rapidly ethanol production and the high concentration of produced ethanol

Acknowledgements

This work was supported by PT. Pertamina, Indonesia.

References

[1] S. Nanda, R. Azargohar, A. K. Dalai, and J. A. Kozinski, "An assessment on the sustainability of lignocellulosic biomass for biorefining," *Renewable and Sustainable Energy Reviews*, vol. 50, no. 1, pp. 925-941, 2015.

[2] H. Zabed, J. N. Sahu, A. Suely, A. N. Boyce, and G. Faruq, "Bioethanol production from renewable sources: Current perspectives and technological progress," *Renewable and Sustainable Energy Reviews*, vol. 71, no. 1, pp. 475-501, May 2017.

[3] H. M. Kim, S. G. Wi, S. Jung, Y. Song, and H. -J. Bae, "Efficient approach for bioethanol production from red seaweed Gelidium amansii," *Bioresource Technology*, vol. 175, no. 1, pp. 128-134, January 2015.

[4] Y. -L. Loow, T. Y. Wu, J. M. Jahim, A. W. Mohammad, and W. H. Teoh, "Typical conversion

IOP Conf. Series: Materials Science and Engineering **854** (2020) 012030 doi:10.1088/1757-899X/854/1/012030

of lignocellulosic biomass into reducing sugars using dilute acid hydrolysis and alkaline pretreatment," *Cellulose*, vol. 23, no. 3, pp. 1491-1520, June 2016.

[5] C. Li, X. Zhao, A. Wang, and G. W. Huber, "Catalytic Transformation of Lignin for the Production of Chemicals and Fuels," *Chemical Reviews*, vol. 115, no. 21, pp. 11559–11624, October 2015.

[6] H. Zabed, J. N. Sahu, A. N. Boyce, and G. Faruq, "Fuel ethanol production from lignocellulosic biomass: An overview on feedstocks and technological approaches," *Renewable and Sustainable Energy Reviews*, vol. 66, no. 1, pp. 751-774, December 2016.

[7] Y. Y. Tye, K. T. Lee, W. N.W. Abdullah, and C. P. Leh, "Optimization of various pretreatments condition of kenaf core (Hibiscus cannabinus) fibre for sugar production: Effect of chemical compositions of pretreated fibre on enzymatic hydrolysability," *Renewable Energy*, vol. 99, no. 1, pp. 205-215, 2016.

[8] D. Dahnum, S. O. Tasum, E. Triwahyuni, M. Nurdin, and H. Abimanyu, "Comparison of SHF and SSF processes using enzyme and dry yeast for optimization of bioethanol production from empty fruit bunch," *Energy Procedia*, vol. 68, no. 1, pp. 107-116, 2015.

[9] Wyman, C.E., Spindler, D.D., Grohmann, K., 1992. Simultaneous saccharification and fermentation for several lignocellulosicfeedstocks to fuel ethanol. Biomass and Bioenergy Vol. 3, No. 5, Pp. 301-307

[10] J V. Rana, A. D. Eckard, and B. K. Ahring, "Comparison of SHF and SSF of wet exploded corn stover and loblolly pine using in-house enzymes produced from T. reesei RUT C30 and A. saccharolyticus," *SpringerPlus*, vol. 3, no. 516, pp. 1-13, December 2014.

[11] Jeon, H., Kang, K., Jeong, J., Gong, G., Choi, J., Abimanyu, H., Ahn, B., Suh, D., Choi, G., 2014. Production of anhydrous ethanol using oil palm empty fruit bunch in a pilot plant. Biomass and bioenergy 67:99-107

[12] S. Sun, S. Sun, X. Cao, and R. Sun, "The role of pretreatment in improving the enzymatic hydrolysis of lignocellulosic materials," *Bioresource Technology*, vol. 199, no. 1, pp. 49-58, January 2016.

[13] Sluiter et al., "Determination of Structural Carbohydrates and Lignin in Biomass," National Renewable Energy Laboratory (NREL), Colorado, Technical Report 2012.

IOP Conf. Series: Materials Science and Engineering **854** (2020) 012031 doi:10.1088/1757-899X/854/1/012031

Durability and Taste of Tofu with "Ie Asam Sunti" Coagulant

Salmyah[1], Fachraniah[1], Fajri[2], Maulizar[3]

[1] Chemical Engineering Department, Lhokseumawe State Polytechnic, Aceh, Indonesia
[2] Civil Engineering Department, Lhokseumawe State Polytechnic, Aceh, Indonesia
[3] Management Department, Lhokseumawe State Polytechnic, Aceh, Indonesia

Email: salmyah@pnl.ac.id

Abstract. One of the special spices in the Province of Aceh that is not available except "Pliek U" is "Asam Sunti". "Asam Sunti" is made from star fruit wuluh by means of drying in the sun and fermentation at night until the 4-5 days. Waste "Ie Asam Sunti" which can be used as a coagulating of soy protein and tofu preservatives because it contains some that are not contained in starfruit, among others: acids, citric, calcium citrate, flavonoids and anti-bacterial. The results showed that the "Ie Asam Sunti" can coagulating of soy protein into tofu, based on statistical analysis, Minitab software 14 optimal caking process at 5 hours immersion, 25 minutes coagulating time and volume of 250 ml "Ie Asam Sunti" get 120 - 140 % yield, content protein with the Kjedall tool obtained 6.89 – 10.54 %, the results of the microbial test showed the number of microbes <3APM / Gr and the test of the taste was very delicious.

1. Preliminary

Tofu is one of the foods made from soybeans and is widely known to the public. Like other traditional foods, tofu is generally processed on a small industrial scale, although at present large-scale tofu factories are also found in big cities, but the opportunity to know on a small scale is still there is.

Tofu is the result of the extraction of soybeans that contain high protein. The protein contained in soy has a quantity that has similarities with meat.

In the process of making tofu, usually added tofu (Ca2So4), acetic acid or glokkno delta laktono (GDI) into soybean extracts that serves to precipitate and agglomerate the protein contained in soybean extracts. Ca2So4 coagulating solids and acetic acid are widely used in the tofu industry (Hariyadi, Purnomo, Umaryadi, & Adawiyah, 1999).

2. Literature Review

The people of Aceh are known for their unique flavoring in Aceh which is the result of fermentation from starfruit, Sunti Acid. Tamarind is a spice in cooking with a distinctive taste and aroma. Making Sunti Acid is done in a very simple way that is wuluh starfruit which is ready to harvest soaked overnight, then salting in the afternoon, fermentation at night and drying in the afternoon, done repeatedly for 4-5 days. The next day star fruit extracted a fermentation solution for 4 days (600-700) ml in 5 kg starfruit for the addition of an average of 200 grams of salt (Salmyah, 2016). This solution is very limited in its use by the people of Aceh, namely in a small portion of the Pidie community as a companion to grilled fish and lime substitutes in Aceh's unique Keueng Asam.

According to (Mursito, 2002) and (Thomas, 2007), the content of starfruit is formic acid, citric acid, ascorbic acid (vitamin C), saponins, tannins, glucosides, flavonoids, and some minerals, especially calcium and potassium in the form of potassium citrate and calcium oxalate. Starfruit contains quite high levels of vitamin C. The content of vitamin C contained therein is around 25.8 mg / 100 grams. In addition, the presence of citric acid in starfruit has great benefits, citric acid can be used in food preservation, as an antioxidant and which prevents rancidity and retains color and aroma. Citric acid can also function as a sequestrant, a metal-binding chemical compound in the form of complex bonds (Tranggono, n.d.)

In vitro testing of the bacteria E, Coli, S, Aureua, M, Luteus and P, fluorescence shows the active potential as an anti-bacterial. The antibacterial potential contained in starfruit makes the opportunity to be developed as a natural preservative substitute for formaldehyde (Hayati, 2010).

Tofu is a processed soybean product that is very popular in Indonesia and is the most widely produced. As much as 40% of Indonesian soybean consumption is processed into tofu. Tofu has an original white color, compact texture but still soft and soft. The principle of making tofu is generally the extraction of soy protein with water and then agglomerated with coagulating material in the form of tofu stone (CaSO4) or acid (vinegar) and certain salts. Tofu is often called boneless meat because of its high nutritional content, especially the protein quality which is equivalent to animal meat. Even tofu protein is higher than soy protein and tofu has the best quality vegetable protein because it has the most complete amino acid composition and has a high digestibility or by 85% -98% (Widaningrum, 2015).

The addition of tofu stone as a clot has less effect on the preservation of the product so it is necessary to add other ingredients such as salt or even the addition of formalin which is very dangerous for health. Likewise with the addition of vinegar acid which sometimes produces products with a sour taste.

The use of Ie Kuloh Sira coagulating produces solid tofu and good taste because there is still residual salt (NaCl) because Ie Kuloh Sira is a residual water from the salt industry that has not completely crystallized (Salmyah, 2004).

The combination of nutritional content of starfruit fruit such as citric acid and the addition of salt in the process of making syringe opens the opportunity to make Ie Sunti Acid as a healthy clot and has a longer shelf life. In addition, Ie Asam Sunti coagulating is easily obtained because it is a waste from a very simple process.

The use of Ie Sunti Acid as a lump of soy milk and can be a tofu preservative because of the nutritional content contained in starfruit consisting of calcium, oxalic acid, citric acid and some antibacterial which is carried from the residual fermentation of injectable acids. The use of Ie Sunti Acid as a clot is very good because it can reduce the use of formalin as a preservative.

Reducing the effects of environmental pollution, especially in the Province of Aceh, which almost all regions make Sunti Asam sometimes on a large scale, sometimes because some people have starfruit plantations which are then sold as an Acehnese special spice. Increasing community income because the waste that has been thrown away so far, only by continuing the simple sterilization process can be sold to tofu entrepreneurs.

Waste of Sunti Acid can be used as a lump of soy protein as well as tofu preservatives because it contains several elements contained in starfruit, among others: acids, citric, calcium citrate, flavonoids and anti-bacteria. The results showed that Ie Sunti Acid can coagulate soy protein into tofu. Characteristic ie syringe acid based on the results of GCMS containing 2.3-Dihydro 3.5 dihydroxi-6methyl-4H Pyran-4 one is an anti-oxidant compound, Ethylic acic SS, Vinegar acid, Ethanoic acid and Ethanediamine, N-ethhyl-N-methyl SS; Ethylenediamine, N-ethyl-N-methyl which is useful for dealing with iodine disorders, bacterial infections, hourur and protozoa. Based on statistical analysis, Minitab software 14 optimum conditions of coagulating process not at Soaking Time 5 hours, clotting time 25 minutes and volume of 150 ml Sunti Acid obtained yield 121-135%, protein content 20-22.5 mg / l and texture 300-350 mm / det. (Salmyah, 2017)

Ie sunti acid which is a waste of the processing of sunti acid (a typical Aceh spice) that comes from the fermentation and drying process of starfruit with the addition of salt (NaCl). The process of tofu coagulating occurs by heating accompanied by the addition of acidic elements such as acetic acid and salt elements such as CuSO4. The acidic and salt elements contained in the syringe acid can agglomerate tofu, while some elements found in the results of research based on GCMS analysis results containing

IOP Conf. Series: Materials Science and Engineering **854** (2020) 012031 doi:10.1088/1757-899X/854/1/012031

2.3-Dihydro 3.5 dihydroxi-6methyl-4H Pyran-4 one is an anti-oxidant compound, Ethylic acic SS, Vinegar acid, Ethanoic acid and Ethanediamine, N-ethhyl-N-methyl SS; Ethylenediamine, N-ethyl-N-methyl which is useful for dealing with iodine disorders, bacterial infections, hourur and protozoa

Tofu is Indonesia's favorite food, especially even the world in general. Tofu which has an excellent vegetable protein content for health is produced by the community both large, medium industries and even household scale. Likewise, the province of Aceh has a number of tofu factories in the medium scale and households that have been using acetic acid and tofu stone as a clot while ie sunti acid which reaches 5 to 10 liters / KK from the Acehnese people has not been touched as an alternative lump which is much tastier he knows. Utilization of waste disposal process of making syringe acid can prevent the disposal of waste into the environment.

The compound as a preservative contained in ie sunti acid

1. 2.3-Dihydro 3.5 dihydroxi-6methyl-4H Pyran-4 one

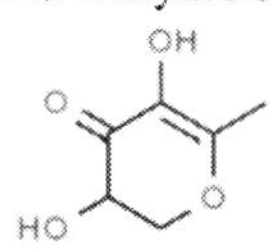

Formula: $C_6H_8O_4$

2. Ethylic acic SS
3. Vinegar acid

Vinegar or astenoic acid is an organic acid known as a sour taste in food or drinks. Vinegar has the chemical formula C2H4O2, this formula is often written CH3CO2H or CH3COOH or CH3-COOH. Glacial vinegar is a hygroscopic liquid that has no color and has a freezing point of 16.7 ° C. In the vinegar acid contained 3-9% by volume vinegar acid. Vinegar is a monoprotic weak acid with a pKa value of 4.76, but can attack the mill and is corrosive if dissolved in water. Liquid acetic acid is a hydrophilic polar or protic solvent, it is similar or the same as water and ethanol. Vinegar is produced as a precursor for polyvinyl acetate and cellulose acetate, produced for household consumption vinegar and in a variety of fibers and fabrics. (Berita et al., 2019)

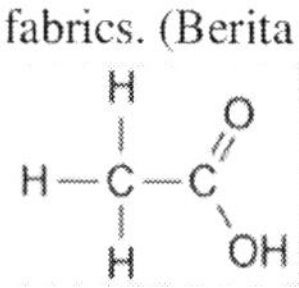

Formula : CH3COOH

4. Ethanoic acid (acetic acid)

Much of the ethanoic acid produced is converted into ethenyl ethanoate (vinyl acetate). The acid itself is the solvent for the liquid phase oxidation of 1,4-dimethylbenzene (p-xylene), leading to the production of polyesters. (ECI, 2016)

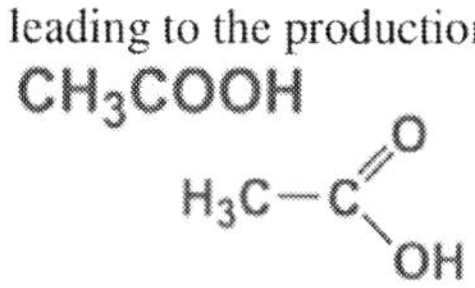

5. Ethanediamine

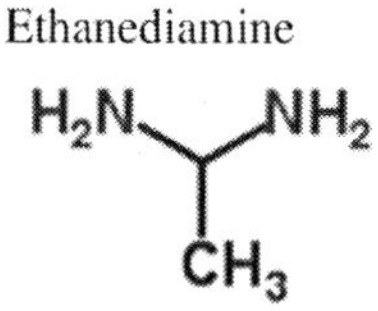

Formula : $C_2H_8N_2$

6. N-ethyl-N-methyl SS
7. Ethylenediamine
   Ethylenediamine is an alkane-alpha,omega-diamine in which the alkane is ethane. It has a role as a GABA agonist. It derives from a hydride of an ethane. (PubChem, 2019)

8. N-ethyl-N-methyl

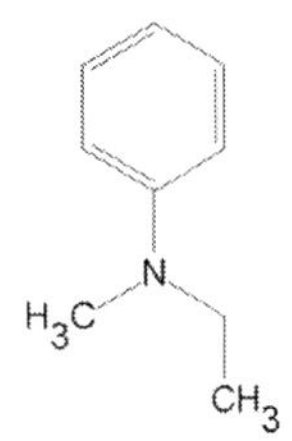

Formula : C9H13N

3. Research Method

This research method will be located in the Biotechnology and food laboratory majoring in Chemical Engineering at the Lhokseumawe State Polytechnic. Sample analysis will be conducted at the Analytical Chemistry Laboratory, Lhokseumawe State Polytechnic Chemical Engineering Department. The ingredients used are soybeans, Ie Kuloh Sira, water, and 25% acetic acid. While the equipment used is a stove, blender, glass beaker, measuring cup, stirrer, thermometer, erlenmeyer, wood mold, UV-VIS spectrophotometer Shimadzu type UV-1800 j and texture analyzer. The detailed process of making Tofu with "Ie Asam Sunti" coagulant is shown in Figure1.

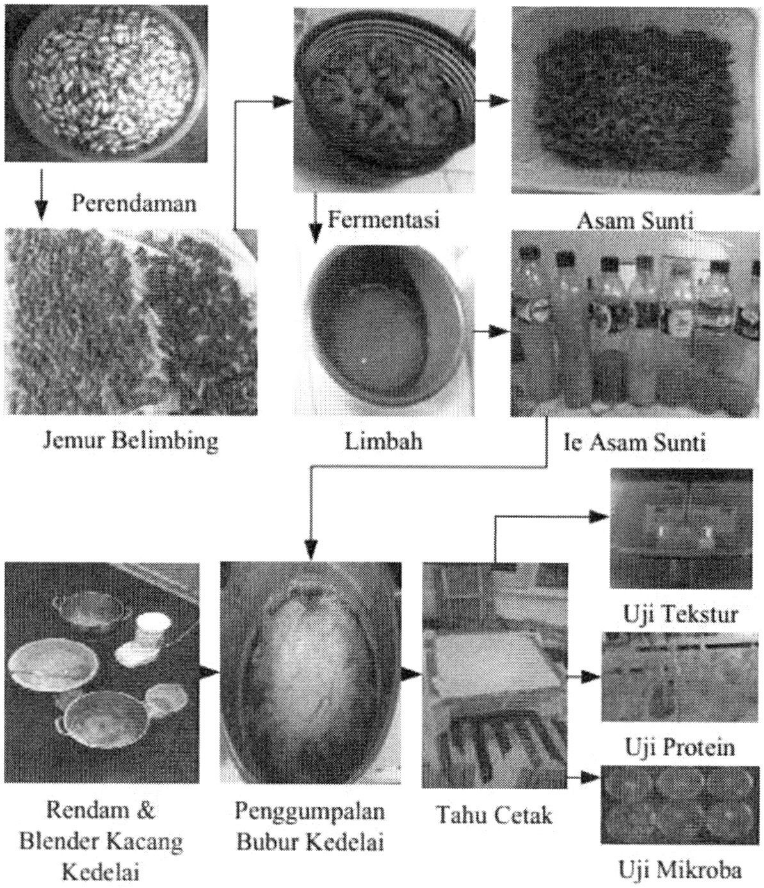

Figure 1. Process of Making Tofu with "Ie Asam Sunti" Coagulant

4. Discussion

Calculation Result Table (Minitab 14)

No	Run Order	Volume of "ie asam sunti" (ml)	Soaking Time (hour)	Coagulating time (minute)	Yield (%)	Protein Levels (%)	Texture (gr/mm)	Microbial Test
1	1	200,000	4,00000	20,0000	63,6	8,13	300,8	1
2	2	300,000	4,00000	20,0000	70,0	8,27	310,5	0
3	3	200,000	6,00000	20,0000	61,6	8,24	292,8	2
4	4	300,000	6,00000	20,0000	73,6	10,12	350,1	0
5	5	200,000	4,00000	30,0000	64,0	6,89	320,7	1
6	6	300,000	4,00000	30,0000	75,6	10,54	340,5	1
7	7	200,000	6,00000	30,0000	72,4	8,40	310,7	2
8	8	300,000	6,00000	30,0000	72,6	7,65	315,6	1
9	9	165,910	5,00000	25,0000	50,6	6,87	225,6	3

10	10	334,090	5,00000	25,0000	73,7	10,45	380,4	0
11	11	250,000	3,31821	25,0000	60,8	7,74	250,5	0
12	12	250,000	6,68179	25,0000	62,4	7,60	310,5	1
13	13	250,000	5,00000	16,5910	65,6	7,86	295,0	1
14	14	250,000	5,00000	33,4090	66,4	9,47	270,0	2
15	15	250,000	5,00000	25,0000	67,6	9,24	353,8	1
16	16	250,000	5,00000	25,0000	68,8	11,34	354,6	1
17	17	250,000	5,00000	25,0000	64,8	11,89	357,8	0
18	18	250,000	5,00000	25,0000	67,7	11,90	360,5	0
19	19	250,000	5,00000	25,0000	64,5	11,78	354,7	2
20	20	250,000	5,00000	25,0000	67,2	12,21	355,3	0

4.1. Yield

Regression Equation in Uncoded Units

$$\text{Yield (\%)} = -0.000135X_1 - 0.53X_2^2 + 0.0410X_3^2$$
$$-0.0145X_1X_2 - 0.0033X_1X_3 + 0.095X_2X_3$$
$$+0.324X_1 + 7.3X_2 - 1.45X_3 + 8.5$$

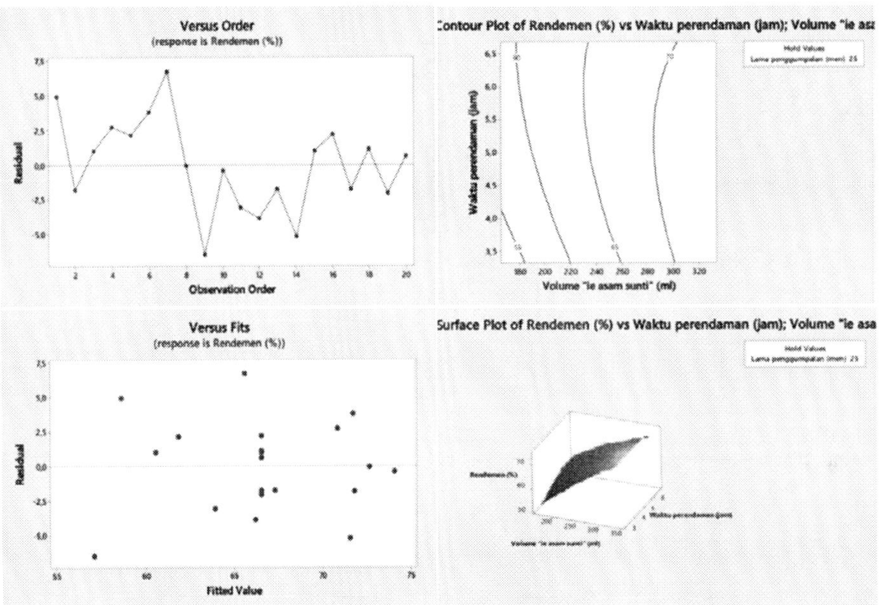

Based on the regression model obtained, it is seen that the chosen variable is significant to the model with R2 of 65.86% so that only 34.14% is due to other factors. The optimization process is at 250 ml clump volume, 5 hours soaking time and 25 minutes clumping duration. The yield obtained is at the level of 57.11 -65.66%.

IOP Conf. Series: Materials Science and Engineering **854** (2020) 012031 doi:10.1088/1757-899X/854/1/012031

4.2. Protein Levels

Regression Equation in Uncoded Units

Protein Levels (%) = $-0.00354X_1^2-1.235X_2^2-0.0353X_3^2$
$-0.00665X_1X_2+0.000X_1X_3-0.0835X_2X_3$
$+0.2153X_1+16.13X_2+2.096X_3-84.4$

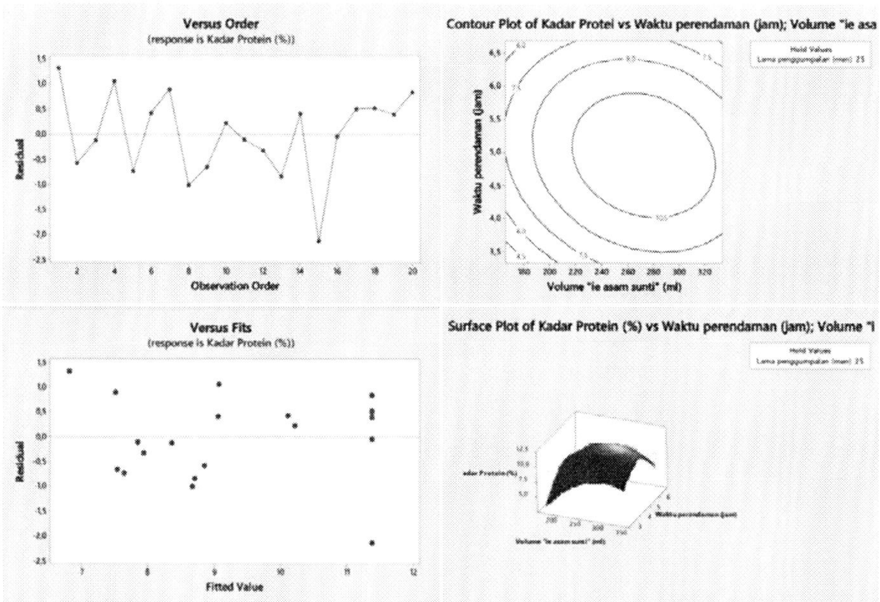

For the protein content significantly, the model contributing to the chosen variable was 78.70% with the results at the level of 6.81-11.38%

4.3. Texture

Regression Equation in Uncoded Units

Texture (gr/mm) = $-0.00471X_1^2-19.74X_2^2-0.761X_3^2$
$+0.082X_1X_2-0.0211X_1X_3-1.66X_2X_3$
$+299X_1+226X_2+51.5X_3$

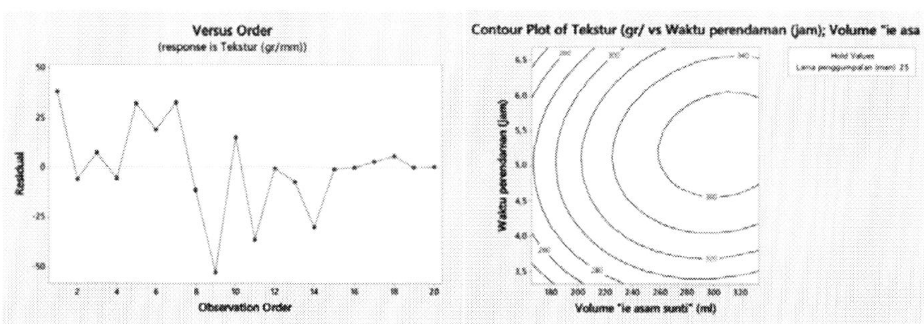

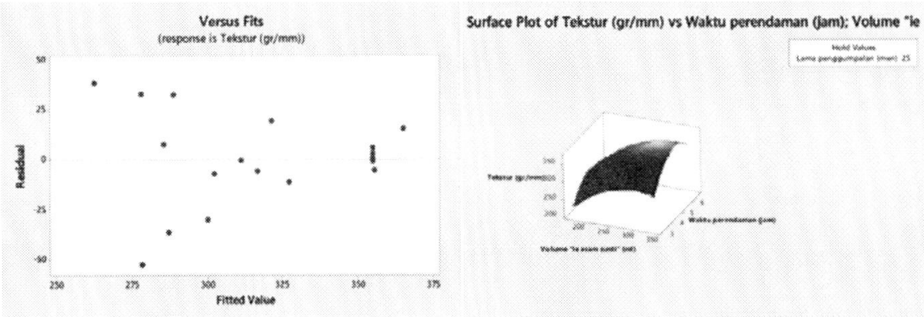

The results of the texture test with a limit of 262.7-278.4 gr / mm were also significant for the model with a contribution of 69.33%.

4.4. Response Optimization

4.4.1. Response Optimization (Maximum)

Response	Goal	Lower	Target	Upper	Weight	Importance
Texture (gr/mm)	Maximum	225,60	380,40		1	1
Protein Levels (%)	Maximum	6,87	12,21		1	1
Yield (%)	Maximum	50,60	75,60		1	1

4.4.2. Response Optimization (Minimum)

Response	Goal	Lower	Target	Upper	Weight	Importance
Texture (gr/mm)	Minimum		225,60	380,40	1	1
Protein Levels (%)	Minimum		6,87	12,21	1	1
Yield (%)	Minimum		50,60	75,60	1	1

Conclusion

1. Ie sunti acid can coagulate tofu protein
2. Optimization of the tofu coagulation process is in the volume of 250 ml ie acid injections, soaking time 5 hours and 25 minutes' coagulation
3. The research results obtained are yield of 57.11-65.66% wet basis, 6.81-11.38% protein content using the Kjedall tool and texture 262.7-278.4 gr / mm using a texture analyzer.
4. Based on the organeptic test, the taste is very good
5. Microbial test results for 24-hour storage time at room temperature are still in the standard that is <3APM / hour.

References

[1] Berita, S., Online, P., Smp, S. D., Smk, S. M. A., Tinggi, P., Terlengkap, T., … Cuka, R. A. (2019). Situs Berita Pendidikan Pengertian Asam Cuka dan Rumus Kimia Asam Cuka Sifat – Sifat Kimia, 1–5.
[2] ECI. (2016). The Essential Chemical Industry.
[3] Hariyadi, P., Purnomo, E. H., Umaryadi, M. E. W., & Adawiyah, D. R. (1999). Latihan Soal Prinsip Teknik Pangan. *Jurusan Teknoiogi Pangan Dan Gizi, IPB.*
[4] Hayati, E. K. (2010). Dibalik Mukzizat Tanaman Belimbing Wuluh (Averrhoa Bilimbi Linn) Sebagai Pengawet Alami.
[5] Mursito, B. (2002). *Ramuan tradisional untuk pengobatan jantung*. Penebar Swadaya.
[6] PubChem. (2019). *PubChem.* UK Food Standards Agency.
[7] Thomas, A. N. S. (2007). *Tanaman Obat Tradisional (Jilid 2). Kanisinus. Jakarta.*

[8] Tranggono, H. (n.d.). LAPORAN PRAKTIKUM PRINSIP TEKNIK PANGAN.
[9] Widaningrum, I. (2015). Teknologi Pembuatan Tahu yang Ramah Lingkungan (Bebas Limbah). *Jurnal Dedikasi, 12*.

Conclusion

1. Ie sunti acid can coagulate tofu protein
2. Optimization of the tofu coagulation process is in the volume of 250 ml ie acid injections, soaking time 5 hours and 25 minutes coagulation
3. The research results obtained are yield of 57.11-65.66% wet basis, 6.81-11.38% protein content using the Kjedall tool and texture 262.7-278.4 gr / mm using a texture analyzer.
4. Based on the organeptic test, the taste is very good
5. Microbial test results for 24 hour storage time at room temperature are still in the standard that is <3APM / hour.

References

[1] Berita, S., Online, P., Smp, S. D., Smk, S. M. A., Tinggi, P., Terlengkap, T., … Cuka, R. A. (2019). Situs Berita Pendidikan Pengertian Asam Cuka dan Rumus Kimia Asam Cuka Sifat – Sifat Kimia, 1–5.
ECI. (2016). The Essential Chemical Industry.
Hariyadi, P., Purnomo, E. H., Umaryadi, M. E. W., & Adawiyah, D. R. (1999). Latihan Soal Prinsip Teknik Pangan. *Jurusan Teknoiogi Pangan Dan Gizi, IPB.*
Hayati, E. K. (2010). Dibalik Mukzizat Tanaman Belimbing Wuluh (Averrhoa Bilimbi Linn) Sebagai Pengawet Alami.
Mursito, B. (2002). *Ramuan tradisional untuk pengobatan jantung*. Penebar Swadaya.
PubChem. (2019). *PubChem*. UK Food Standards Agency.
Thomas, A. N. S. (2007). *Tanaman Obat Tradisional (Jilid 2). Kanisinus. Jakarta.*
Tranggono, H. (n.d.). LAPORAN PRAKTIKUM PRINSIP TEKNIK PANGAN.
Widaningrum, I. (2015). Teknologi Pembuatan Tahu yang Ramah Lingkungan (Bebas Limbah). *Jurnal Dedikasi, 12.*

IOP Conf. Series: Materials Science and Engineering **854** (2020) 012032 doi:10.1088/1757-899X/854/1/012032

Characteristics of Wafer Originated from Coffee Waste as Ruminant Animal Feed

Milawarni[1], Arskadius[2], E.Elfiana[3], Yassir[1]

[1] School of Electrical Engineering, Politeknik Negeri Lhokseumawe
[2] School of Mechanical Engineering, Politeknik Negeri Lhokseumawe
[3] School of Chemical Engineering, Politeknik Negeri Lhokseumawe

Email: milawarni@pnl.ac.id

Abstract. Coffee waste is a by-product from processing coffee beans which has the potential to be used as a reinforcing feed material (concentrate) for ruminants. The waste of coffee fruit is physically quite large in composition, that is 48% of the pulp and the rest is waste. The disadvantages of this coffee waste to be used as feed are its high fiber content (21%), caffeine and tannin and its voluminous physical form. But all this can be overcome by fermentation using Aspergius Niger so as to increase its nutritional value. This research was carried out by making a feed composition including: P0 = 0% coffee husk waste (control) and 98.75% rice bran, P1 = 30% coffee husk waste and 68.75% rice bran, P2 = 60% coffee husk waste and 38.75 rice bran and P3 = 90% coffee husk waste and 8.75% rice bran. The variables observed were the physical characteristics of the feed wafer including: aroma, color and density and palability. Data was collected and analyzed using ANOVA and Duncan Range Test. The results showed the use of coffee husk waste should be used at a level of 30%, because it has a high palability value and low density and crumbs so it is easier for ruminant animals to consume.

Keywords: animal feed wafers, coffee husk waste, ruminant livestock, fermentation, composition (ratio)

1. Introduction

Feed has a very important role in livestock life. The cost of feed is the largest cost of the total production cost, reaching 70-80%. Feed limitations cause the capacity of livestock in an area to decrease or can cause disruption to normal production and reproduction. This can, among other things, be overcome if the potential of agriculture / industry and its waste are taken into account in the livestock business. Feed is one of the most important components for livestock growth. Because livestock require good food qualitatively and quantitatively to increase metabolic yield which can affect the development and growth of livestock. Utilization of agricultural and plantation waste is one way to find alternative sources of material. Efforts are being made to optimize these benefits as animal feed. One example of plantation waste that can be utilized is coffee husk waste.

Coffee is class of plants that produce large waste from processing. The waste is in the form of coffee husks, the amount of which ranges from 50-60% of the harvest. The nutritional content of good coffee husk can be used as ruminant animal feed, but it cannot be utilized optimally because there is a high crude fiber content of 21% [1]. Besides the obstacles faced in the use of coffee waste as animal feed is its voluminous nature, so it is still not widely used as animal feed.

One of the technologies applied to improve the benefits of coffee husk is to make it a source of fiber in the form of wafers. Wafer is a source of natural fiber feed which in its manufacturing process experiences compaction with pressure and heating so that it has the same shape and length and width [2]. In addition, it has nutritional content and uses relatively simple technology so that it is easily applied as an alternative to ruminant feed. Ruminant livestock is one of the cattle that has a complex digestive system compared to other livestock. This can be seen from the ability of ruminants to produce microbial protein in the rumen. Where microbial protein plays a major role in the availability of total protein available to livestock that will be used for the needs of livestock itself [3]. According to (Khalil, 2016), Giving amoniated coffee husk waste by 15% for 42 days can increase the body weight of boiler chicks [4], as well as giving 30% fermented coffee husks and giving leguminosa (gamal and calliandra) 70% provide growth which is good in PE goats [5]. Through the fermentation process, several high-fiber wastes including coffee husks can be increased in nutritional value. With Aspergillus niger inoculants, coffee husk protein levels can be increased from 9.8% to 12.43%. So that it can substitute for the needs of bran which has been widely used as reinforcing feed, but must be imported from outside the location and the price is relatively expensive. Plantation wastes such as coffee, cocoa and fermented cashew waste can be used as a source of concentrated feed ingredients for livestock. Coffee plant waste has the potential to be used as a reinforcing feed material (concentrate) for livestock. Through the process of processing the nutritional content of coffee waste can be increased, especially the protein content [6].

Diversification of the use of by-products which are often considered as waste from agricultural and plantation activities has been carried out. (Milawarni, 2017) has stated that coffee husk can be used as particle board material [7] [8] and block compost making technology [9]. At present coffee husk can be utilized as feed that can encourage the development of ruminant livestock agribusiness in an integrated production system with agricultural and plantation patterns through environmentally friendly biomass recycling or known as "zero waste production system". The development of the upstream agribusiness subsector, such as the feed industry, is one of the supporters in the development of ruminants which will directly help solve the problems of farmers in terms of feed availability. The reality on the ground shows that there are still many farmers who provide raw food without regard to quality requirements, quantity and efficiency of their supply. As a result, livestock productivity is not optimal, even many of the farmers who suffer losses due to inadequate feed.

In addition to have a large influence on livestock productivity, feed is also a significant production cost in livestock business. Thus producing feed is not only required for the feasibility of aspects of the quality and adequacy of nutrition, but also how to produce feed that is economical, cheap and affordable by the ability of farmers. For this reason, it is necessary to do an effective way to feed ruminant animals, namely food preservation and feeding in the form of wafers. The physical properties of the wafer affect the quality of the feed and the device design process. [10] [11]. Testing the physical properties of the wafer is used to design a processing, storage and transformation device for the processing of feed ingredients. This study aims to determine the physical characteristics of the wafer including: color and aroma, density and palability.

2. Material and Methods

Material used for making wafer feed consists of: coffee husk waste, rice bran, mollases, urea, NPK, salt (NaCl) and water. Treatment in wafer form was as follows: P0 = wafer with 0% coffee waste (control), P1 = wafer with 2.5% coffee waste, P2 = wafer with 5% coffee waste, P3 = wafer with 7.5% coffee waste.

The equipment used to make feed in the form of wafers is: wafer press machine, pressurized 200-300 kg / cm3, chopper machine, scales, media (where to mix the wafer feed ingredients) and tarpaulin. The composition of the treatment ingredients are: P0 = 98.75 Rice bran and 0% coffee husk, P1 = 30% coffee husk and 68.75% rice bran, P2 = 60% coffee waste and and 38.75% rice bran, P3 = 90 % coffee husk and 8.75% rice bran.

2.1. Production of Wafer based Feed

The procedure for making feed wafers is as follows: (a) Coffee husk is fermented using the fungus Aspergillus niger. This fungus is dissolved with water, added mollasis, urea and NPK, then incubated for 24 hours. Fermentation is carried out on floors covered with beons, with tile roofs. Waste that is ready for fermentation is sprinkled on the surface of the media as thick as 5-10 cm. Furthermore, watered with Aspergillus solution evenly. Watering is done with a shower (fat). Piles of material that has been watered again sprinkled with coffee waste 5-10 cm thick and watered back in the solution. And so on, so that the material has been piled up and splashed with Aspergius niger liquid. Piles of coffee husks are covered with a clean tarpaulin that is tightly closed and incubated for 5 days. After 5 days disassembled and dried. (b) Next to the reduction stage, the finished feed is reduced in size using a chopper, so that the size is homogeneous and easy to print, (a) Feeds that are already small in size are inserted into a rectangular wafer mold measuring 20 cm x 2 cm x 5 cm. After that hot pressing is carried out at a temperature of 1500C with a pressure of 200-300 kg / cm2 for 5-10 minutes, then (d) the cooling process of the wafer by placing the open air for at least 24 hours until the water content and weight are constant. (e) Wafers are stored or packaged.

2.2. Design Experiment

The experimental design applied was a Completely Randomized Design (CRD), consisting of 4 treatments and 4 replications. The treatment is P0 = Wafer with coffee waste 0% (control), P1 = wafer with 30% coffee waste, P2 = wafer with 60% coffee waste, P3 = wafer with 90% coffee waste. The data obtained were analyzed by analysis of variance (ANOVA) and if it gave tangible results it was followed by the Duncan Test. Variables observed were the physical properties of the wafer consisting of color, aroma and density indigo.

3. Results and Discussion

3.1. Physical Properties of Feed Wafers:

The physical properties of the wafer are the basic properties of a feed ingredient. Testing the physical properties of the wafer is used to design a processing and storage device as well as the transportation of the wafer material processing industry. The physical properties of materials are much influenced by the water content and particle size of the material. Besides that, it is also influenced by the particle size distribution, shape and surface characteristics of a material. Material properties and changes that occur in feed can be used to assess and determine feed quality. In addition, knowledge about physical properties is also used to determine the coefficients of a process of handling, processing and storage. The shape, size and color of the wafers made from coffee waste are shown in Figure 1.

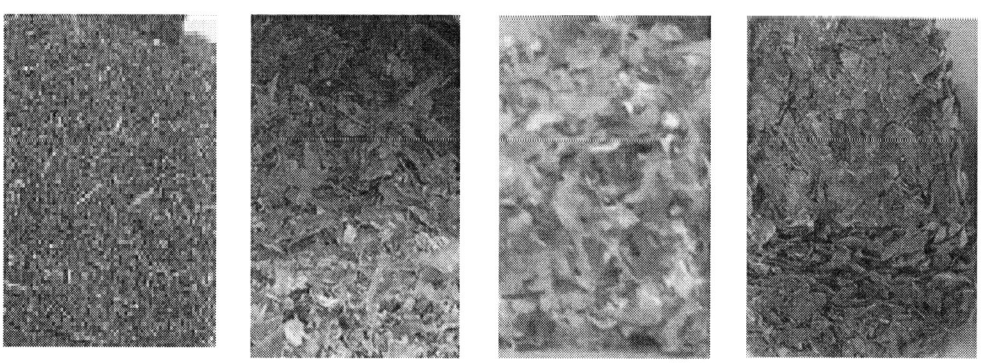

P0 (Control) P1 (30% coffee husk) P2 (60% coffee husk) P3 (90% coffee husk)

Figure 1. The physical form of feed wafers made from waste coffee

In this study, wafers made from raw coffee waste produced are light brown to brown. This is due to the large amount of coffee waste composition used as wafer preparation. The more use of coffee waste as a wafer preparation material, the color produced will also be more brown. Especially for wafers that

IOP Conf. Series: Materials Science and Engineering **854** (2020) 012032 doi:10.1088/1757-899X/854/1/012032

contain 90% coffee waste composition. The brown color of the wafer that results from the non-enzymatic browning reaction is the reactions between organic acids with reducing sugars and between amino acids and reducing sugars. This wafer is a special fragrant-scented ingredient of coffee husk. Pressure and heating on the raw material can cause a Maillard reaction, so that the resulting wafer has a distinctive aroma of the wafer base material.

3.2. Feed Wafer Density Value

Wafer density is a measure of compactness of the particles in the sheet and is very dependent on the density of the raw material used and the amount of pressure applied during the wafer width. Wafers that have a high density will provide a dense and hard texture making it easy to handle, both storage and shaking during transportation and is expected to be more durable in storage. On the other hand, low-density feed will show a form of wafer that is not too dense and texture that is softer and hollow (porous), so it is estimated that it can only survive in storage for some time. The average value of wafer density made from coffee waste from the four treatments is shown in table 1.

Results of variance showed that the density of wafers made from raw coffee husk waste there was a significant difference ($P < 0.05$) between treatments. Based on Duncan's test the difference in the value of the wafer density is in the treatment of P0 with P1, P2 and P3, but P1 is equivalent to P2 and P2 is equivalent to P3. The higher the use of coffee waste as a constituent of wafers, the more influence the value of the wafer density. The highest average wafer density value (0.73 ± 0.05) contained P3 treatment (90% coffee husk waste) and the lowest (0.54 ± 0.03) in P0 treatment (0% coffee husk waste).

Density greatly influences the physical form of the wafers produced and shows the density of wafers in the manufacturing technique. Raw material density is very dependent on the amount of compression given during the manufacturing process. Pressing pressure is carried out to create a bond between the surface of the adhesive and the adhesive material with the aid of a pressing device.

Table 1. Value of feed wafer density

Repetition	Treatment			
	P0	P1	P2	P3
1	0,57	0,58	0,62	0,73
2	0,51	0,57	0,64	0,72
3	0,58	0,55	0,62	0,78
4	0,52	0,51	0,57	0,70
Average	$0{,}54 \pm 0{,}03^{a}$	$0{,}55 \pm 0{,}04^{b}$	$0{,}65 \pm 0{,}03^{bc}$	0,73±0,05c

Note: Average values with different superscripts on the same line show significantly different effects ($P < 0.05$). P0 = wafer feed 0% coffee husk waste, P1 = wafer feed 30% coffee husk waste, P2 = wafer feed with 60% coffee husk waste and P3 = wafer feed 90% coffee husk waste.

Wafers with low density will have more free space or contact area between particles resulting in a large water absorption ability. Wafer density determines dimensional stability and physical appearance of wafer feed. The wafer density plays an important role in calculating the volume of space required for a material with a certain weight, for example filling silos, elevators and automatic dosing accuracy.

The wafer density can also be affected by the particle size value of the wafer compilation material. The density will increase with the increasing number of fine particles in a wafer feed. The high P3 and P4 density values are also due to the small particle size, while the large particle size in the P1 treatment causes the wafer density value to decrease.

3.2.1. Palatability of Feed Wafers
Palability value of wafer made from raw coffee waste in this study is used as a support or indicator to find out how much ruminants (cattle) like the feed in the form of the wafer. This palability test is carried out for 1 (one) month with ad lithium wafers. Palability test results for wafers made from raw coffee waste can be seen in table 2.

The results of variance showed that there were significant differences (P <0.05) on the value of the feed wafer palability. Table 2 shows the P0 treatment has a higher palability of 4475.0 ± 66.1b g / head) when compared to other treatments. Except with the P1 treatment there is no significant difference. The high palability in the P0 treatment was due to the value of the density of the feed wafer that it had in the P0 treatment was lower than the other treatments, making it easier for consumption by ruminants.

Table 2. Value of feed wafer palability

Repetition	TReatment			
	P0	P1	P2	P3
1	4300	4275	4240	4210
2	4270	4270	4300	4250
3	4340	4340	4235	4235
Average	4303,3±66,1^{b}	4295,0±76,4^{ab}	4258,3±38,2^{a}	4231,67±25,0^{a}

Note: Average values with different superscripts on the same line show significantly different effects (P <0.05). P0 = wafer feed 0% coffee husk waste, P1 = wafer feed 30% coffee husk waste, P2 = wafer feed with 60% coffee husk waste and P3 = wafer feed 90% coffee husk waste.

The density of wafer feed can affect the palability of livestock. Feed wafers that are too hard with high densities will cause the difficulty of livestock in consuming wafers directly so that it needs to be added with water when they are given and livestock in general prefer feed with low density. The results of this study indicate that the palatability of the wafer produced for all treatments is significantly different. This is likely to be greatly influenced by the type and composition of different coffee husks as a constituent of ruminant animal feed wafers.

4. Conclusion
Based on the physical test results of coffee wafers made from raw coffee waste and its palability in ruminants (cattle), it was concluded that feed wafers containing 60% and 90% coffee husk waste had higher density values compared to the control treatment, so as to reduce the palability value. The use of coffee husk waste should be used at a level of 30%, because it has a high palability value and low density and crumbs making it easier for ruminant animals to consume.

Acknowledgement
The authors recognize Kemenristekdikti for funding the 2019 PPTTG program, Director of Politeknik Negeri Lhokseumawe and P3M Politeknik Negeri Lhokseumawe for facilitating this activity. The same remarks by the authors convey to the Head of BPPS Bies District, Central Aceh Regency, the Village Head, Partners and the community of Bies District, Central Aceh Regency along with all the supporters of the implementation of this activity.

References
[1] M. Daud, M. Mulyadi, and Z. Fuadi, "Persentase Karkas Itik Peking yang Diberi Pakan dalam

Bentuk Wafer Ransum Komplit Mengandung Limbah Kopi," *J. Agripet*, vol. 16, no. 1, p. 62, 2016.

[2] Y. Retnani, S. Basymeleh, and L. Herawati, "Pengaruh Jenis Hijauan Pakan dan Lama Penyimpanan Terhadap Sifat Fisik Wafer," 2009.

[3] F. Peternakan and U. Hasanuddin, "Ekskresi Derivat Purin dan Estimasi Suplai Protein Mikroba pada Ternak Domba yang Mendapat Suplemen Protein Berbeda," vol. 12, no. 3, pp. 183–188, 2007.

[4] M. Khalil, "Pengaruh Pemberian Limbah Kulit Kopi (Coffea sp.) Amoniasi Sebagai Pakan Alternatif Terhadap Pertambahan Bobot Ayam Broiler," *J. Ilm. Mhs. Pendidik. Biol.*, vol. 1, no. 1, Aug. 2016.

[5] I. M. Londra, "UNTUK PERTUMBUHAN KAMBING PERANAKAN ETAWAH Effect of Fermented Coffee Skin and Leguminous Leaves to Increase the Growth of Etawah Grade Goats," vol. 3, pp. 45–51, 2013.

[6] H. L. Didanna, "A critical review on feed value of coffee waste for livestock feeding," *World J. Biol. Biol. Sci.*, vol. 2, no. 5, pp. 72–86, 2014.

[7] Milawarni, Nurlaili, and Sariyadi, "Characterization a binderless particleboard of coffee husk using Hydrogen Peroxide (H2O2) and Ferrous Sulfate (FeSO4)," in *IOP Conference Series: Materials Science and Engineering*, 2018, vol. 352, no. 1, pp. 1–6.

[8] Milawarni, N. Nurlaili, and E. Ernayusnianti, "Investigation on the feasibility of coffee husk (endocarp) as efficient filler material for enhancing physical and mechanical properties of styrofoam based particleboard," in *IOP Conference Series: Materials Science and Engineering*, 2018, vol. 334, no. 1, pp. 1–6.

[9] Milawarni;Moh.Arkardius;Elfiana, "TEKNOLOGI PEMBUATAN KOMPOS BLOK DARI LIMBAHPENGOLAHAN KOPI," in *SEMNAS 1 POLITEKNIK NEGERI LHOKSEUMAWE*, 2017, pp. 396–399.

[10] R. M. Salam, K. Kunci, : Bahan, and B. Lokal, "SIFAT FISIK WAFER DARI BAHAN BAKU LOKAL SEBAGAI BAHAN PAKAN TERNAK RUMINANSIA Physical wafer from local raw materials as animal Feed ruminant," *Ilmu Peternak.*, vol. 5, pp. 108–114, 2017.

[11] Y. Retnani, A. Saenab, and B. V Latulung, "WAFER PAKAN KOMPLIT LIMBAH SAYURAN PASAR UNTUK DI PETERNAKAN RAKYAT (Wafer Complete Feed of Vegetable Waste for Productivity Increasing of Sheep on Farm) PENDAHULUAN Pengembangan penyediaan protein hewani yang berasal dari ternak pedaging perlu mendapat," vol. 4, pp. 556–557, 2012.

IOP Conf. Series: Materials Science and Engineering **854** (2020) 012033 doi:10.1088/1757-899X/854/1/012033

Window Cleaning Robot by using Arduino as Microcontroller

Arni Munira Markom[1], Muhammad Asyraffuddeen Bin Arriffinjee[1], Muhammad Fikri Bin Haironi[1] and Zakiah Mohd Yusoff[1]

[1]Faculty of Electrical Engineering, Universiti Teknologi MARA, Cawangan Johor, Kampus Pasir Gudang, 81750 Masai, Johor Malaysia.

Email: arnimunira@uitm.edu.my

Abstract. Stunning and modern houses nowadays have been constructed with increasing number of curtain window glass walls and corresponds to the requirement of its maintenance, repair and care from dust and pollution. This window produces a lot of difficulties such as the window height and the exposure to the risk of hurt or injury during the cleaning process. A window cleaning robot by using Arduino Mega as microcontroller is developed with neodymium magnet, ultrasonic sensors, DC motor, servo motor, motor driver and buzzer. Neodymium magnet is used to attach the window robot vertically on the surface of glass wall whereas servomotor will move the robot accordingly. To detect an obstacle or the robot reached at the edge of panel at about 3 cm, an ultrasonic sensors are used to notify and enable the DC motor to change its current path. When all four sides of the edge of panel window is detected, the robot will turn on a buzzer continuously to notify user that glass window cleaning process is completed successfully.

1. Introduction

To date, most of buildings including houses have been constructed with modern and dazzling curtain glass window compare to Naco glass window as a traditional approach. This contemporary skyscrapers buildings and houses with the glass panel need periodic maintenance, repair and care especially window cleaning, due to excessive dust and pollution in the city. The conservative method for external maintenance such as the use of rope, gondola and a winch require experienced labours for safety precaution. The more dependent on human labour, the more risk be able to occur where they are inherently danger, and cause emergency conditions when earthquake or wind gust occurred. Figure 1 illustrates the existing human labour intensive cleaning methods that involve the risk of the worker falling using (a) scaffolding (b) a gondola and (c) a mobile elevated work platform [1]. Meanwhile at home, the human still exposes to the risk of being cut, hurt and injury due too high or any sharp objects on the surface of glass walls.

IOP Conf. Series: Materials Science and Engineering **854** (2020) 012033 doi:10.1088/1757-899X/854/1/012033

(a) (b) (c)

Figure 1. The risk of cleaning window human labour (a) scaffolding, (b) a gondola, and (c) a mobile elevated work platform (Kim, K., & Jun, Y. 2016)

Consequently, the demand for an external window cleaning robot had arisen. Several studies are being conducted on ways of automating the process of providing curtain wall care, such as the building maintenance robot (BMR). This robot is an automated window cleaning system from outer window and using infrared (IR) sensor to the contaminant level [2-3]. In Japan, they are using a commercial carrier systems that support a gondola that moves vertically and horizontally. The carrier is put at the top of roof of the building named CAFÉ robotic system [4-5]. Several research institutes and companies, including Nihon Bisoh in Japan and Fraunhofer IPA in Germany, are experimenting with the potential application of wall climbing building maintenance robots for cleaning external wall glasses [6-7]. There are also even experimenting with the use of renewable energy and solar panel to cut down the cost of power usage [8-10]. However, all of this designs are practical only to a large and skyscrapers building and impractical to a modern house. Therefore, the robot are bulky, heavy, complex, high energy and power consumption, and need a professional labour to handle appropriately. Obviously, it is expensive for both of designer and consumer.

Lee, Y. S. et. al., has designed a simple and automated robot using programmable logic controller (PLC) with combination of relays and switches as a brain of the window cleaning robot system [11]. A light-weight robot has been constructed with a prototype size of 700 mm x 500 mm x 100 mm and its weight is about 6 kg [12]. Then, a smaller size of cleaner robot is developed and approximately 300 mm x 300 mm x 100 mm and its weight is about 3 kg [13]. Both of this light-weight window cleaner robot and move the robot smoothly using suction cup. However, both suffer to clean at the edge of window panel due to errors on the angle and translation motion calculation.

Thus, this project propose is aim to develop an automated, simple and light-weight cleaning robot that practical to a modern house using a cost-effective microcontroller of Arduino as the brain of the robot. A project using Arduino gain huge interest not only for researchers, but also for university and school project due to flexibility, cheap and free Arduino software that available on their website [14]. Besides, we also propose a magnetic to attach the robot on the surface window with ultrasonic sensors to remotely control the robot movement and direction. To attach the robot vertically on the surface of window panel, a neodymium magnet is also known as NdFeB, NIB or Neo magnet are used. This rare-earth magnet is a permanent magnet made from an alloy of neodymium, iron and that widely used for motors in cordless tools, hard disk drives and magnetic fasteners [15-16].

2. Project Methodology

This project is divided by two sections; software and hardware. The two software used are Proteus Professional 8.5 and Arduino software. Proteus software is used to draw the full circuit of hardware and check the connectivity for simulation process. Then, printed circuit board (PCB) layout will be designed using the same software and followed by printing into a real PCB for hardware section. Meanwhile Arduino software is used to write a program the Arduino Mega 2560. This program is describes how the robot will be exucuted based on command given, also acts as a brain of microcontroller. After both of simulation are succeed, the project is continued to the hardware section.

Figure 2 shows a block diagram of the window cleaning robot project for hardware. There are three parts which are divided by input, microcontroller and outputs. The input for this project is ultrasonic sensor (HC-SR04) which is connected with the microcontroller. This ultrasonic sensor is used to detect an obstacle and avoid collision using ultrasonic sound wave that vibrates at a frequency above the range of human hearing. The microcontroller used is Arduino Mega 2560 as the brain of this robot. Finally, the microcontroller is interfaced to the outputs which is light emitting diode (LED), buzzer and motor driver shield L293D that are integrated to direct current (DC) and servo motor.

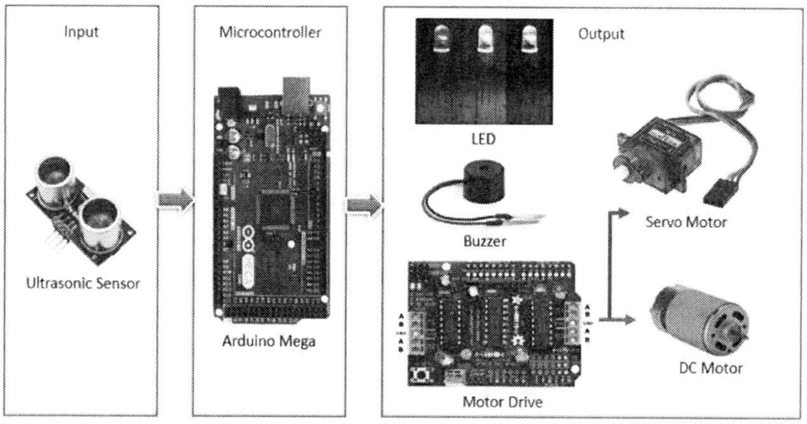

Figure 2. Window Cleaner Robot Block Diagram

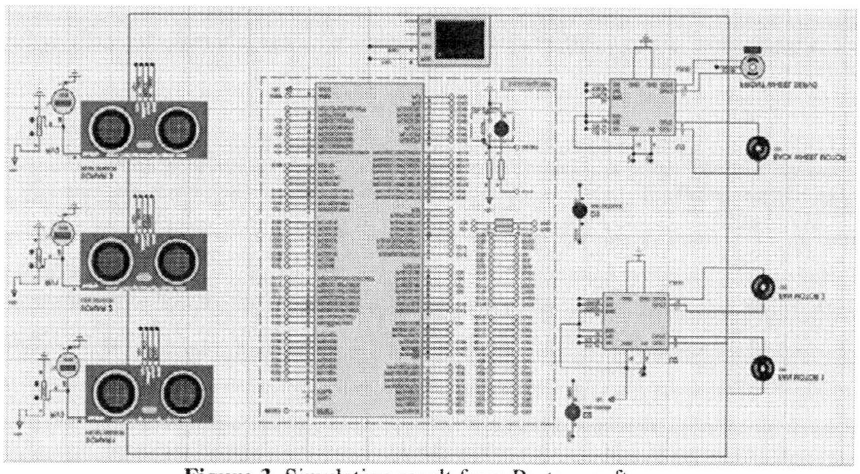

Figure 3. Simulation result from Proteus software

Figure 3 illustrates the simulation result from Proteus software to check and make sure the connectivity is good before implement to the hardware part. The main component for this connection is ultrasonic sensor, LED, DC motor and servo motor. This system shows an input from ultrasonic sensor whereas the outputs are LED, DC motor and servo motor. The RS, E, D4, D5, D6 and D7 pins of LCD are connected to the pin at IO7, IO6, IO5, IO4, IO3 and IO2 of Arduino. The VEE of LCD are connected to variable resistor to control the brightness of LCD. While the ECHO and TRIG pins of ultrasonic sensor are connected to pins IO8 and IO9 of Arduino. After the connectivity is good without error, then PCB layout is designed and implemented to the hardware.

IOP Conf. Series: Materials Science and Engineering **854** (2020) 012033 doi:10.1088/1757-899X/854/1/012033

Before hardware, one critical software must be done by Arduino software which is to initiate a pin connection code for motor driver, buzzer and ultrasonic sensor. At first, the Arduino will read the switch. When the switch is identified as on, the LED will light up and the DC motor starts running forward servo motor in 0 degree condition. When sensor detect an obstacle, this sensor will send an ultrasonic signal to Arduino. Then, Arduino will read this ultrasonic sensor notifications, and Trig pin is set high at 10 us. The signal will come back to Echo pin and its travel duration is kept in the duration variable. By applying the declaration as stated on Figure 4 (a), the distance of donated items can be calculated and stored in the variable named distance. In this project, the important part is ultrasonic sensor move program because it is acted as a remote control for the robot. Thus, there are 3 sensors have been fixed and named as FrontSensor, BackSensor and LeftSensor as shown in Figure 4 (b). The sensors are fixed to the minimum distance about 3 cm where if the minimum distance was detected, DC motor and servo motors direction will be changed.

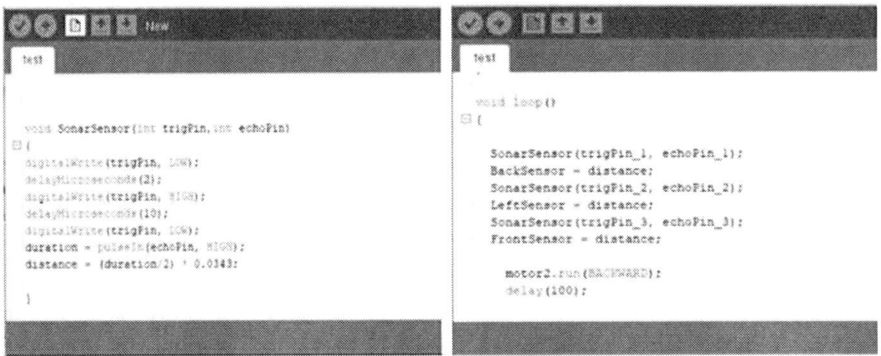

(a) (b)

Figure 4. (a) Ultrasonic sensor declaration (b) Ultrasonic sensor for robot movement instruction

3. Result and discussion

Figure 5 and 6 show the expected result from above and in front of robot without magnetic and cloth to clean a window. The robot is separated into two main sections which is the window cleaning robot itself and the neodymium magnetic part. All these two part was important to make sure the robot runs completely. Figure 7 on the right side of picture shows the magnetic section, only neodymium magnets have been used without any electronic component. Meanwhile the other section is cleaning cloth as shown in Figure 7 on the left side. To activate this robot, a switch should be turned on after connection of power source from battery. Then, a LED will turn on as indicator that all important component of motor fan, DC motor and ultrasonic sensors are activated. Meanwhile, the servo motor will turn 90 degrees. If the front sensors detect a minimum distance below 3 cm, the servo motor will turn to 45 degrees, rotate to backward and change the direction of the robot. This function to avoid a wall or obstacles ahead. Next, if the back sensors detect below 3 cm, the servo motor will turn to 90 degrees back, and the DC motor will rotate forward. Thus, the robot will move up and down on the window until it come to the end of the window's border. The side sensor will detect 3 cm and below and the DC motor and fan motor will stop while the buzzer will beep as a sign that the robot have finish the cleaning process.

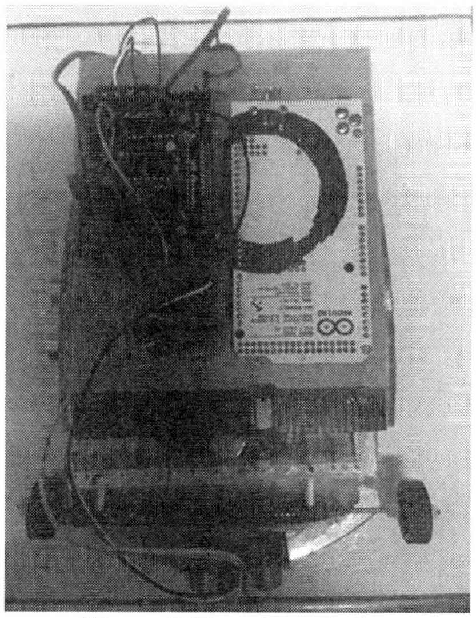

Figure 5. Top view of robot

Figure 6. At in front view of the robot

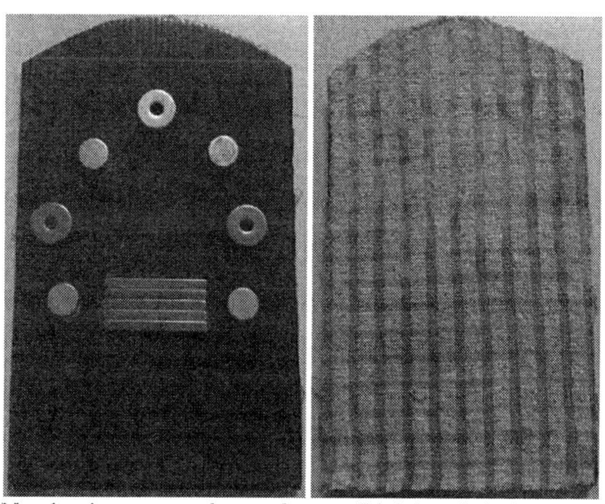

Figure 7. Neodymium magnetic attachment (Right) and cloth attachment (Left)

4. Conclusion

As a conclusion, a window cleaner robot is successfully designed for both of software and hardware sections. The Proteus 8.5 Professional is used for schematic, connectivity until the design of PCB part of hardware. Meanwhile Arduino software is used to program and instruct the robot that using Arduino Mega as their main controller. Other important components are ultrasonic sensors to control the movement of robot, to change direction when a wall or obstacles are met. However, this robot has a problem due to neodymium magnetic to attach the robot on the surface of window panel. This is because the neodymium magnetic is not strong enough to carry 7 kg prototype window cleaner robot. Besides, the battery power of 9V also insufficient to move the robot vertically as desired.

Thus, to improve this robot, a high DC motor should be used like 24V and above to power up the movement of robot vertically. The used of suction cup might be suitable compare to magnetic to move the small and light-weight robot smoothly. Moreover, window cleaner robot is also can be significantly improve using mobile application as user can control the robot via a smartphone for cleaning process. Further development for light-weight window robot that has capability to clean window panel vertically is still need to explore as almost all modern houses are using window glass curtain wall.

Acknowledgment

The authors gratefully acknowledge the supports given by Geran Penyelidikan BESTARI, Universiti Teknologi MARA Cawangan Johor.

References

[1] Kim, K., & Jun, Y. (2016). Development of a window-cleaning robot powered by new renewable energy capable of continuous operation during obstacle avoidance. International Journal of Sustainable Building Technology and Urban Development, 7(3-4), 214-218.

[2] Lee, Y. S., Kim, S. H., Lee, J. H., Kang, M. S., Sun, D. I., & Han, C. S. (2016). Development of a Sensor System for Detecting Window Contamination for the Building Maintenance Robot System. In ISARC. Proceedings of the International Symposium on Automation and Robotics in Construction (Vol. 33, p. 1). Vilnius Gediminas Technical University, Department of Construction Economics & Property.

[3] Moon, S. M., Shin, C. Y., Huh, J., Oh, K. W., & Hong, D. (2015). Window cleaning system with water circulation for building façade maintenance robot and its efficiency analysis. International Journal of Precision Engineering and Manufacturing-Green Technology, 2(1), 65-72.

[4] Kim, T., Yoo, S., Kim, H. S., & Kim, J. (2018, May). Design and Force-Tracking Impedance Control of a 2-DOF Wall-Cleaning Manipulator Using Disturbance Observer and Sliding Mode Control. In 2018 IEEE International Conference on Robotics and Automation (ICRA) (pp. 1-9). IEEE.

[5] Seo, T., Jeon, Y., Park, C., & Kim, J. (2019). Survey on Glass And Façade-Cleaning Robots: Climbing Mechanisms, Cleaning Methods, and Applications. International Journal of Precision Engineering and Manufacturing-Green Technology, 1-10.

[6] Van Gassel, F., Schrijver, P., & Lichtenberg, J. (2006). Assembling wall panels with robotic technologies. In The 23rd International Symposium on Automation and Robotics in Construction Japan: ISARC (pp. 728-733).

[7] Okada, N., Yamanaka, K., & Kondo, E. (2009, August). A wall climbing robot with simple suckers. In ICCAS-SICE, 2009 (pp. 5691-5694). IEEE.

[8] Kim, K. T., Jun, Y. H., & Shin, E. Y. (2018). Estimate of Economic Feasibility of a Window Cleaning Device. In ISARC. Proceedings of the International Symposium on Automation and Robotics in Construction (Vol. 35, pp. 1-4). IAARC Publications.

[9] Hassan, A. A., El-Habrouk, M., & Deghedie, S. (2018). Robotic and Mechatronic Applications Related to Renewable Energy–A Survey. International Journal of Robotics and Mechatronics, 5(1), 44-65.

IOP Conf. Series: Materials Science and Engineering **854** (2020) 012033 doi:10.1088/1757-899X/854/1/012033

[10] Ma, Z., Li, L., Fu, Y., Tao, S., Cui, Z., & Bi, L. (2015). Design of Wireless Remote Control Solar Cleaner. In MATEC Web of Conferences (Vol. 31, p. 08002). EDP Sciences.
[11] Lee, Y. S., Kim, S. H., Lee, J. H., Kang, M. S., Sun, D. I., & Han, C. S. (2016). Development of a Sensor System for Detecting Window Contamination for the Building Maintenance Robot System. In ISARC. Proceedings of the International Symposium on Automation and Robotics in Construction (Vol. 33, p. 1). Vilnius Gediminas Technical University, Department of Construction Economics & Property.
[12] Zhang, H., Zhang, J., Zong, G., Wang, W., & Liu, R. (2006). Sky cleaner 3: A real pneumatic climbing robot for glass-wall cleaning. IEEE Robotics & Automation Magazine, 13(1), 32-41.
[13] Miyake, T., Ishihara, H., Shoji, R., & Yoshida, S. (2006, June). Development of small-size window cleaning robot a traveling direction control on vertical surface using accelerometer. In 2006 International Conference on Mechatronics and Automation (pp. 1302-1307). IEEE.
[14] Asraf, H. M., Dalila, K. N., Hakim, A. M., & Hon, R. M. F. (2017). Development of experimental simulator via Arduino-based PID temperature control system using LabVIEW. Journal of Telecommunication, Electronic and Computer Engineering (JTEC), 9(1-5), 53-57.
[15] Croat, J. J. (2017). Rapidly solidified neodymium-iron-boron permanent magnets. Woodhead Publishing.
[16] Scheifers, J. P., Zhang, Y., & Fokwa, B. P. (2017). Boron: Enabling exciting metal-rich structures and magnetic properties. Accounts of chemical research, 50(9), 2317-2325.

IOP Conf. Series: Materials Science and Engineering **854** (2020) 012034 doi:10.1088/1757-899X/854/1/012034

Lateral Control of Functionally Graded Composite Beam under Static Load by using Piezoelectric Layer

Abdolreza Toudehdehghan

Department of Mechanical Engineering, INTI International University, Jalan BBN 12/1, Bandar Baru Nilai, 71800 Nilai, Negeri Sembilan, Malaysia.

Email: Abdolreza.toudehdehghan@newinti.edu.my

Abstract. Controlling the deflection of a composite cantilever beam under concentrate static force at the free end is investigated. The composite beam is made of ceramic and metal, which gradually change from ceramic to the metal in thickness coordinate. This kind of composite is called Functionally Graded Material (FGM). In FGM, material properties are presumed to differ as an exponential function. The governing equation of the rectangular composite beam is based on Euler-Bernoulli deformation theory. Two control models are considered in this article. First, two piezoelectric actuators are attached to the two sides of the composite beam which each of these actuators is applied the voltage separately. To control the deflection of each point of the composite beam, the difference between the two voltages is considered constant. Second, a piezoelectric actuator is attached to the free end of the composite beam, which also provides a relation to control any desired point of the composite beam. By comparing the two models, the superiority of the second model is proven.

Nomenclature

l = Length of beam
b = Width of beam
D_f = Flexural rigidity of FGM beam
D_{eq} = Flexural rigidity of FGM beam with piezoelectric layer
h_a = Thickness of actuator layer
h = Thickness of FGM layer
E_{fe} = Modulus elasticity of FGM
E_c = Modulus elasticity of ceramic
E_m = Modulus elasticity of metal
E_a = Modulus elasticity of actuator
x, z = axes coordinates
u = Displacement in x-axis
w = Displacement in z-axis
Q = Elastic coefficients matrix
e = Piezoelectric coefficients matrix
V_s = Shear force
d_a = Distance between actuator center to center of whole area
V_t = Operating voltage of top actuator
V_b = Operating voltage of bottom actuator
q = Uniformly distributed force
V = Operating voltage
M = Bending moment
M_y = Bending moment about y-axis
M^a = Bending moment of actuator
M_T^a = Bending moment of actuator on top
M_b^a = Bending moment of actuator on bottom

w_0 = Displacement in mid-plane
E = Electric field vector
ν =Poisson ratio
I_f = Second moment of FGM area
σ_{xx}, σ_{yy} = Normal stress in x, y axes
$\varepsilon_{xx}, \varepsilon_{yy}$ =Normal strain in x, y axes
x_s = Length of the beam without actuator
E_x, E_y, E_z =Electric field components
I_a =Second moment of actuator area
F =Concentrated force
σ_{xy} = Shear stress in x-y plane
ε_{xy} =Shear strain in x-y plane

1. Introduction

In the past two decades, the use of a distributed piezoelectric layer as an actuator and sensor has been extensively studied by researchers to control or damping the elastic deformation of structures and also energy harvesting [1-4]. Piezoelectric materials under a mechanical deformation generate the electric charge. This phenomenon is called the direct piezoelectric effect. Conversely, the electric field causes stress and strain, which is called the converse effect of piezoelectric [5]. This property of piezoelectric materials with other attractive advantages such as inexpensive, easily shaped, lightweight and very good attached to structures makes an excellent choice for static and dynamic control of structures as actuators and sensors. Piezoelectric layers together with non-piezoelectric layers of structures can be considered as composite structures [6]. Tylikowski and Hetnarski pioneered the study of the effect of the piezoelectric layer on transverse vibration control of Euler-Bernoulli two sides join beam under bending moment. They reduced the deflection of the beam by placing a piezoelectric layer in the middle of the beam and by applying appropriate voltage [7]. Tauchert et al. [8] examined a beam with piezoelectric layers on top and bottom which subjected to distributed loading and thermal loading. They analyzed this beam under dynamic loading by using the Fourier transform method. In order to find the appropriate voltage to reduce the deflection of the beam, they used the trial and error method. Donthireddy and Chandrashekhara developed a mathematical model of laminated composite beams with piezoelectric actuators by using a finite element method [9]. Then several researchers studied piezoelectric laminated beam and plate under mechanical and thermo-mechanical loads based on the finite element method [10-12]. Thomas Bailey et al. [13] studied design and analysis of the damping effect of distributed piezoelectric polymer layer on a thin cantilever beam by using an active damping model. Fumihiro Ashida et al. [14] examined the damping effect of piezothermoelastic layers on the vibration response of a composite beam under thermal loading by using a speed feedback control method. Osama J. Aldraihem et al. [15] compared control dynamic response of two different mathematical models (Euler-Bernoulli vs. Timoshenko) of cantilever composite beam based on Lyapunov control law. Although numerous researchers have studied different aspects such as mathematical model, kind of structure (beam, plate, shell, and so on), solving method, controlling model, loading and etc. of piezoelectric materials [16, 17].

Like piezoelectric materials, Functionally Graded Materials (FGMs) have received much attention from researchers over the past two decades [18, 19]. In FGMs, the volume fraction of composite in a specific direction is changing continuously [20]. To show the mechanical properties of FGM in the direction that volume fraction changes, several models and functions are provided [20-22]. However, three functions such as exponential, power-law, and sigmoid are the most notable. Due to characteristics of FGMs (material properties gradually change), disadvantages such as delamination, residual stress, and stress concentration under thermo-mechanical stresses are dramatically reduced [23]. B. L. Wang and N. Noda [24] are analyzed the stress discontinuity and thermal deformation of the FGM beam with two piezoelectric layers as actuator and sensor by using a finite element method solution. They revealed that the stress singularity and thermal deflection at the interface of the FGM layer and Piezoelectric layers can be controlled by using an exponential function for FGM material properties. Y. Kiani et al. [25] studied buckling analysis of FGM beam with piezoelectric layers under thermo-mechanical loading.

IOP Conf. Series: Materials Science and Engineering **854** (2020) 012034 doi:10.1088/1757-899X/854/1/012034

They considered the power-law function to show material properties of the FGM layer. The results of their work showed the effect of actuator voltage on buckling behavior of the Euler-Bernoulli beam.

The present work examines two models of static control of an FGM cantilever beam under concentrated load at the free end of the beam. In the first model, an FGM beam containing distributed piezoelectric actuators on top and bottom is developed. The second model containing a cantilever FGM beam with one-layer piezoelectric actuator at the free end of the beam is considered. The both models are based on the classical beam theory and using integration method to solve. The material properties of the FGM layer is considered as an exponential function. Poisson's ratio is assumed to be constant. In order to control the deflection of the composite beam, applying a voltage to piezoelectric actuators to control the cantilever beam deflection is provided a suitable relationship for the two model beams which are introduced.

2. Extracting Governing Equations

The displacement field of components of Euler-Bernoulli beam (EBT) are considered in the form below [26]:

$$u(x,z) = -z\frac{dw_0(x)}{dx} \qquad 1$$

$$w(x,z) = w_0(x) \qquad 2$$

Therefore, the axial strain component on the neutral axis of the EBT beam is

$$\varepsilon_{xx} = \frac{du}{dx} = -z\frac{d^2w_0}{dx^2} \qquad 3$$

The stress resultant on a cross-section of the beam by using the linear elastic constitutive relation for an FGM as

$$\sigma_{xx} = E_{fe}\varepsilon_{xx} = E_{fe}\left(-z\frac{d^2w_0}{dx^2}\right) \qquad 4$$

2.1. *Material properties of FGM*

The material properties of the FGM layer varying continuously based on exponential function across the thickness. Thus, the exponential modulus of elasticity of the FGM layer can be written as [27, 28]:

$$E_{fe}(z) = E_c e^{\left(\frac{1}{h}\ln\left(\frac{E_m}{E_c}\right)\left(z+\frac{h}{2}\right)\right)} \qquad 5$$

2.2. *Piezoelectric Equations*

The converse equation of piezoelectric [29] is

$$\begin{Bmatrix}\sigma_{xx}\\ \sigma_{yy}\\ \sigma_{xy}\end{Bmatrix} = \begin{bmatrix}Q_{11} & Q_{12} & Q_{16}\\ Q_{12} & Q_{22} & Q_{26}\\ Q_{16} & Q_{26} & Q_{66}\end{bmatrix}\begin{Bmatrix}\varepsilon_{xx}\\ \varepsilon_{yy}\\ \gamma_{xy}\end{Bmatrix} - \begin{bmatrix}0 & 0 & e_{31}\\ 0 & 0 & e_{32}\\ 0 & 0 & e_{36}\end{bmatrix}\begin{Bmatrix}E_x\\ E_y\\ E_z\end{Bmatrix} \qquad 6$$

Where σ_i, Q_{ij}, ε_i, e_{ij}, and E_i are the stress vector, the elastic coefficients, the strain vector, the piezoelectric coefficients, and the electric field vector respectively. The applied voltages (V) on the actuator layers are considered along their thicknesses (h_a). Therefore, the electric field vectors are defined as follows:

$$\{E\} = \begin{bmatrix} 0 & 0 & \frac{1}{h_a} \end{bmatrix}^T V \quad 7$$

Substituting Equations (3) and (7) into Equation (6), the converse piezoelectric equation of Euler-Bernoulli beam is as below:

$$\sigma_{xx} = Q_{11}\varepsilon_{xx} + e_{31}E_z \quad 8$$

The elastic coefficient is given as

$$Q_{11} = \frac{E_{fe}}{1 - \nu_{12}\nu_{21}} \quad 9$$

2.3. Governing Equation of beam

Equilibrium equations of an element of Euler-Bernoulli beam under transverse distributed load is [30]

$$\sum F_z = 0 \quad \Rightarrow \quad -\frac{dV_s}{dx} = q(x) \quad 10$$

$$\sum M_y = 0 \quad \Rightarrow \quad V_s - \frac{dM(x)}{dx} = 0 \quad 11$$

$M(x)$ is the bending moment and it can be defined in term of the stress (Equation (4)) on a cross section of the beam.

$$M(x) = \int_A \sigma_{xx} z dA = -D_{eq}\frac{d^2 w_0}{dx^2} \quad 12$$

Substituting Equation (12) into Equation (11) and then substituting Equation (11) into Equation (10), the fourth order governing equation of FGM beam based on Euler-Bernoulli assumption is

$$D_{eq}\frac{d^4 w_0}{dx^4} = q(x) \quad 13$$

In order to analyze the classical FGM beam with the embedded piezoelectric actuator, two models are considered:

The Model I: FGM cantilever beam as shown in Figure 1 with distributed piezoelectric actuator on top and bottom along the length of the beam subjected to concentrated force on the free end. Therefore, the governing equation of this beam is

$$D_{eq}\frac{d^4 w_0}{dx^4} = 0 \quad 14$$

$$D_{eq} = E_{fe}I_f + 2E_a I_a$$

By integrating four times from Equation (14), the governing equation is as follows:

$$D_{eq}w(x) = \frac{c_1}{6}x^3 + \frac{c_2}{2}x^2 + c_3 x + c_4 \quad 15$$

Boundary conditions of case I are as follow:

$$w(0)=0, \qquad \frac{dw}{dx}(0)=0$$
$$\frac{d^2w}{dx^2}(l)=\frac{M_T^a+M_b^a}{D_{eq}}, \qquad \frac{d^3w}{dx^3}(l)=\frac{F}{D_{eq}} \tag{16}$$

The values of M_T^a and M_b^a can be obtained by using Equation (8) are

$$M_T^a=-be_{31}d_aV_t$$
$$M_b^a=-be_{31}d_aV_b \tag{17}$$

Using the boundary conditions in the Equation (15), the following equation is obtained.

$$\Delta V=V_t-V_b=\frac{2}{be_{31}d_ax^2}\left(-D_{eq}w(x)-\frac{F}{6}x^3+\frac{Fl}{2}x^2\right) \tag{18}$$

To find the voltage difference needed to set zero deflection of FGM beam at any desired point, in the Equation (18) have to substitute zero deflection and desired length of the beam respectively.

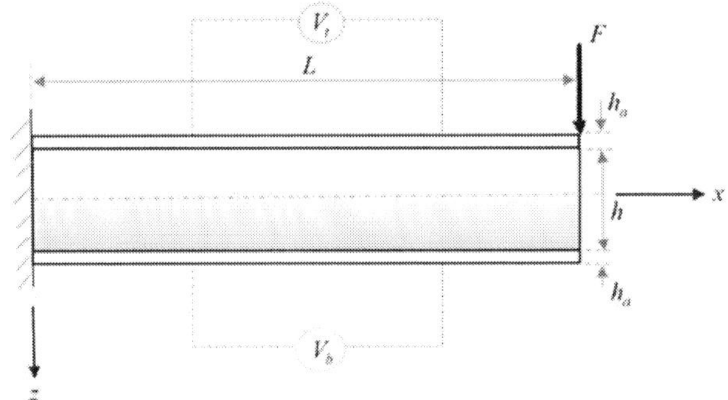

Figure 1. FGM cantilever beam with piezoelectric actuators under concentrated force.

The Model II: FGM cantilever beam as shown in Figure 2 with piezoelectric actuator on the bottom and the free end of the beam subjected to concentrated force on free end. Therefore, the governing equation of this beam is

$$D\,w(x)=\frac{c_1}{6}x^3+\frac{c_2}{2}x^2+c_3x+c_4 \qquad 0\le x\le x_s \tag{19}$$
$$D_f=E_{fe}I_f$$

$$D_{eq}w(x)=\frac{c_5}{6}x^3+\frac{c_6}{2}x^2+c_7x+c_8 \qquad x_s\le x\le l \tag{20}$$
$$D_{eq}=E_{fe}I_f+E_aI_a$$

Boundary conditions of Equations (19) and (20) are as follow:

$$w(0)=0, \quad \frac{dw}{dx}(0)=0$$
$$\frac{d^2w}{dx^2}(l)=\frac{M^a}{D_{eq}}, \quad \frac{d^3w}{dx^3}(l)=\frac{F}{D_{eq}} \tag{21}$$

The conditions of continuity of two Equations (19) and (20) are

$$w(x_s^-)-w(x_s^+)=0, \quad \frac{dw}{dx}(x_s^-)-\frac{dw}{dx}(x_s^+)=0$$
$$\frac{d^2w}{dx^2}(x_s^-)-\frac{d^2w}{dx^2}(x_s^+)=\frac{M^a}{D_{eq}} \tag{22}$$

By substituting the boundary conditions (Equation (21)) and the conditions of continuity (Equation (22)) into Equations (19) and (20), the voltage applied to the actuator to control the deflection of any arbitrary point of the FGM beam are as follow:

$$V=\frac{-1}{be_{31}d_a x^2}\left(D_f w(x)-\frac{F}{6}x^3+\frac{F}{2}x^2\right) \qquad 0\le x\le x_s \tag{23}$$

$$V=\frac{-1}{be_{31}d_a\left(\frac{x^2}{2}+x_s x-\frac{x_s^2}{2}\right)}\left(D_{eq} w(x)-\frac{F}{6}x^3+\frac{F}{2}x^2\right) \qquad x_s\le x\le l \tag{24}$$

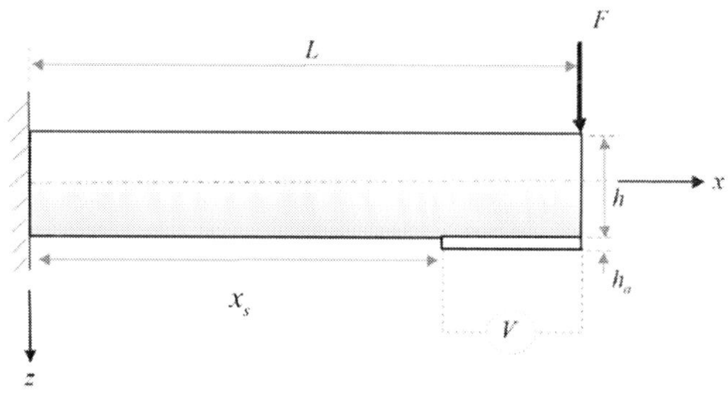

Figure 2. FGM cantilever beam with piezoelectric actuator at free end under concentrated force.

3. Results and Discussions

Two control models are presented in section 2 for controlling the deflection of composite cantilever beam (Equations (18) and (23, 24)). Based on these two models, it is possible to control the deflection of the FGM beam at any specific point. The geometrical dimensions and material properties of the piezoelectric and FGM layers used in static analysis of the composite beam are presented in Table 1. Figures 3 and 4 show the deflections of FGM beam with and without control model I. Since this model consist of two piezoelectric layers as actuators on top and bottom of the FGM beam, therefore, the voltage difference between these two layers is used to control the deflection of a specific point of the beam. Figures 3 and 4show the deflections of free end and point $x = 0.5$ of the FGM beam which are zero for the voltage difference $\Delta V = -530.4266$ and $\Delta V = -663.0332$ respectively. Figures 3 and 4 also demonstrate the control deflection at the free end and point $x = 0.5$ of the FGM beam for different voltages respectively.

Table 1. Dimensions and properties of FGM and piezoelectric materials

Parameter	Quantity	Parameter	Quantity	Parameter	Quantity
h_a (m)	0.0002	$\nu_{12} = \nu_{21}$	0.3	x_s (m)	0.7
h (m)	0.004	E_a (GPa)	63	e_{31} (m/V)	-190×10^{-12}
l (m)	1	E_m (GPa)	10	F (N)	1
b (m)	0.05	E_c (GPa)	100		

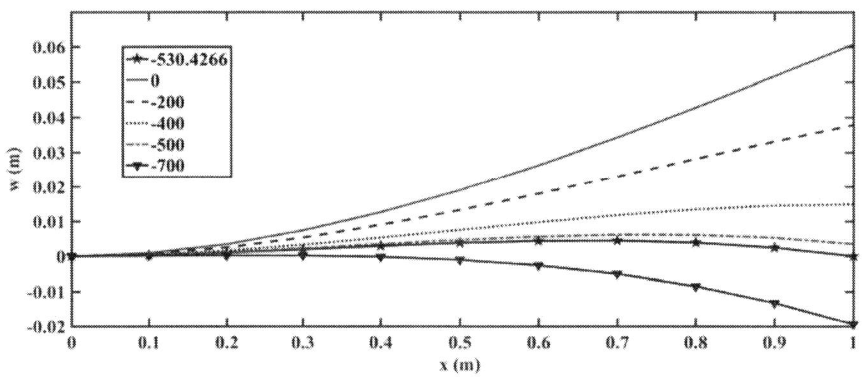

Figure 3. Control the free end of the beam in the model I for different voltages.

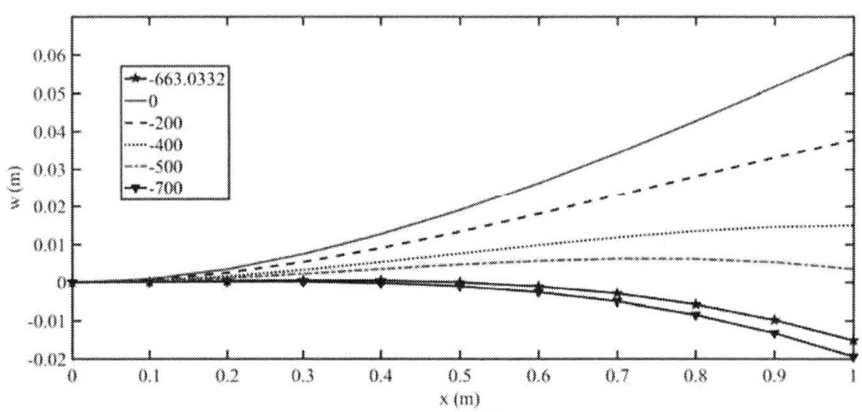

Figure 4. Control the deflection of point ($x = 0.5$) in model I for different voltages.

Figures 5 and 6 show the deflections of the FGM beam with and without control model II. Since this model consists of one piezoelectric layer as an actuator at the free end of the FGM beam, therefore, the FGM beam has two segments. The first segment $(x < x_s)$ of the FGM beam that has no piezoelectric layer and the second segment $(x \geq x_s)$ has a piezoelectric layer. In order to control the deflection of the FGM beam, if the desired point is in segment one the Equation (23) is used. If the specific point is in the second segment the Equation (24) is used. Figures 5 and 6 show the deflections of free end and point $x = 0.5$ of the FGM beam which are zero for the voltage $V = -277.7103$ and $V = -331.5166$ respectively. Figures 5 and 6 also demonstrate the control deflection at the free end and point $x = 0.5$ of the FGM beam for different voltages respectively.

By comparing the figures of the model I with those of model II, it is clear that the model II is much more efficient. As the second model requires less applied voltage and the much smaller piezoelectric layer. In this study, only mainly about the static control effect of two piezoelectric models on the FGM beam has been investigated and the advantages of using FGM in control of structural behavior with piezoelectric materials have not been addressed. Therefore, the author intends to research this area in the future.

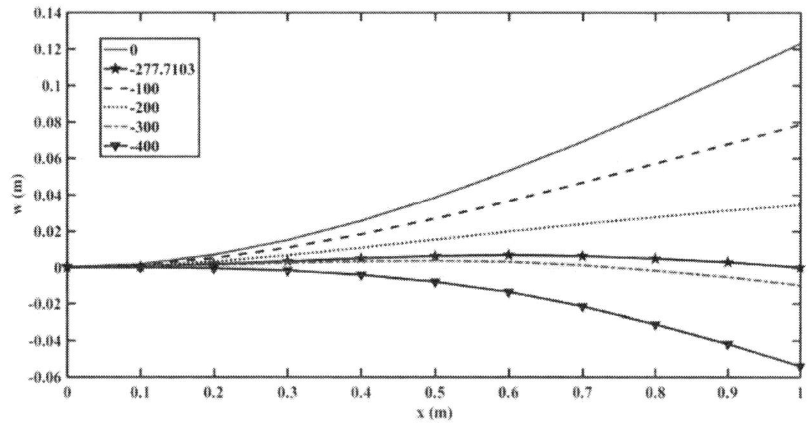

Figure 5. Control the free end of the beam in the model II for different voltages.

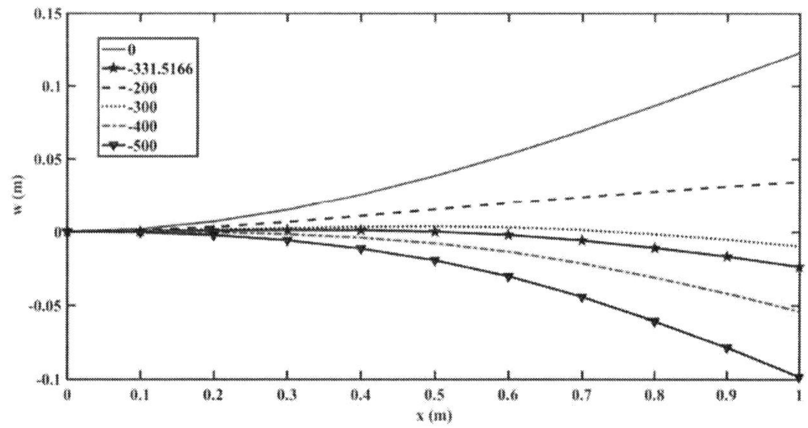

Figure 6. Control the deflection of point ($x = 0.5$) in model II for different voltages.

4. Conclusion

The piezoelectric effect on a functionally graded composite beam with cantilever condition subjected to a concentrated force at the free end is studied. The material properties of the FGM beam were considered based on exponential function. This study proposes two models to control the static deflections of thin composite beams with piezoelectric layers. In the first model, the composite beam has two distributed piezoelectric layers along the length on top and bottom of it. The second model, the composite beam has one layer piezoelectric on the free end. By considering the inverse piezoelectric effect on the thin composite beam, the following items are observed that,

- ➢ The converse equation of piezoelectric can be used to control any point of the beam in static conditions.
- ➢ In the first control model, the deflection of each point of the beam is directly related to the difference in applied voltage to the actuators on top and bottom.
- ➢ The applied voltage to the actuator is directly related to the deflection at each point of the beam and inversely related to squared distance the deflection point to the fixed edge of the beam.
- ➢ The second control model is much more efficient than the first control model.

References

[1] A. Erturk and D. J. Inman, *Piezoelectric energy harvesting*: John Wiley & Sons, 2011.

[2] S. Priya and D. J. Inman, *Energy harvesting technologies* vol. 21: Springer, 2009.

[3] T. R. Tauchert, F. Ashida, S.-i. Sakata, and Y. Takahashi, "Control of temperature-induced plate vibrations based on speed feedback," *Journal of Thermal Stresses,* vol. 29, pp. 585-606, 2006.

[4] H. A. Sodano, D. J. Inman, and G. Park, "A review of power harvesting from vibration using piezoelectric materials," *Shock and Vibration Digest,* vol. 36, pp. 197-206, 2004.

[5] Z. Wang, S.-h. Chen, and W. Han, "The static shape control for intelligent structures," *Finite elements in analysis and design,* vol. 26, pp. 303-314, 1997.

[6] J. N. Reddy, *Mechanics of laminated composite plates and shells: theory and analysis*: CRC press, 2003.

[7] A. Tylikowski and R. Hetnarski, "Passive and active damping of thermally induced vibrations," in *Proceedings of the Third International Congress on Thermal Stresses, Cracow Institute of Technology, Kraków,* 1999, pp. 573-576.

[8] T. Tauchert, S. Adali, and V. Verijenko, "Piezo-control of forced vibrations of a thermoelastic beam," in *Fourth International Congress on Thermal Stresses, Osaka, Japan,* 2001, pp. 477-480.

[9] P. Donthireddy and K. Chandrashekhara, "Modeling and shape control of composite beams with embedded piezoelectric actuators," *Composite structures,* vol. 35, pp. 237-244, 1996.

[10] H. Nguyen-Xuan, G. Liu, T. Nguyen-Thoi, and C. Nguyen-Tran, "An edge-based smoothed finite element method for analysis of two-dimensional piezoelectric structures," *Smart Materials and Structures,* vol. 18, p. 065015, 2009.

[11] E. Carrera, E. Zappino, and G. Li, "Analysis of beams with piezo-patches by node-dependent kinematic finite element method models," *Journal of Intelligent Material Systems and Structures,* vol. 29, pp. 1379-1393, 2018.

[12] P. Fatehi and M. Farid, "Piezoelectric Energy Harvesting from Nonlinear Vibrations of Functionally Graded Beams: Finite-Element Approach," *Journal of Engineering Mechanics,* vol. 145, p. 04018116, 2018.

[13] T. Bailey and J. E. Hubbard, "Distributed piezoelectric-polymer active vibration control of a cantilever beam," *Journal of Guidance, Control, and Dynamics,* vol. 8, pp. 605-611, 1985.

[14] F. Ashida, S.-i. Sakata, T. R. Tauchert, and Y. Takahashi, "Control of thermally induced vibration in a composite beam with damping effect," *Journal of Thermal Stresses,* vol. 29, pp. 139-152, 2006.

[15] O. J. Aldraihem, R. C. Wetherhold, and T. Singh, "Distributed control of laminated beams: Timoshenko theory vs. Euler-Bernoulli theory," *Journal of intelligent material systems and structures,* vol. 8, pp. 149-157, 1997.
[16] I. Izadgoshasb, Y. Y. Lim, N. Lake, L. Tang, R. V. Padilla, and T. Kashiwao, "Optimizing orientation of piezoelectric cantilever beam for harvesting energy from human walking," *Energy conversion and management,* vol. 161, pp. 66-73, 2018.
[17] C. Liu, L.-L. Ke, J. Yang, S. Kitipornchai, and Y.-S. Wang, "Nonlinear vibration of piezoelectric nanoplates using nonlocal Mindlin plate theory," *Mechanics of Advanced Materials and Structures,* vol. 25, pp. 1252-1264, 2018.
[18] D. Jha, T. Kant, and R. Singh, "A critical review of recent research on functionally graded plates," *Composite Structures,* vol. 96, pp. 833-849, 2013.
[19] V. Birman and L. W. Byrd, "Modeling and analysis of functionally graded materials and structures," *Applied mechanics reviews,* vol. 60, pp. 195-216, 2007.
[20] [20] H.-S. Shen, *Functionally graded materials: nonlinear analysis of plates and shells*: CRC press, 2016.
[21] T. Reiter, G. J. Dvorak, and V. Tvergaard, "Micromechanical models for graded composite materials," *Journal of the Mechanics and Physics of Solids,* vol. 45, pp. 1281-1302, 1997.
[22] A. Toudehdehghan, J. Lim, K. E. Foo, M. Ma'arof, and J. Mathews, "A brief review of functionally graded materials," in *MATEC Web Conferences*, 2017.
[23] A. Chakraborty, S. Gopalakrishnan, and J. Reddy, "A new beam finite element for the analysis of functionally graded materials," *International Journal of Mechanical Sciences,* vol. 45, pp. 519-539, 2003.
[24] B. Wang and N. Noda, "Design of a smart functionally graded thermopiezoelectric composite structure," *Smart Materials and Structures,* vol. 10, p. 189, 2001.
[25] Y. Kiani, S. Taheri, and M. Eslami, "Thermal buckling of piezoelectric functionally graded material beams," *Journal of Thermal Stresses,* vol. 34, pp. 835-850, 2011.
[26] C. Wang, J. N. Reddy, and K. Lee, *Shear deformable beams and plates: Relationships with classical solutions*: Elsevier, 2000.
[27] [27] S.-H. Chi and Y.-L. Chung, "Mechanical behavior of functionally graded material plates under transverse load—Part I: Analysis," *International Journal of Solids and Structures,* vol. 43, pp. 3657-3674, 2006.
[28] A. Toudehdehghan, M. M. Rahman, and F. Tarlochan, "Analyzing The Behavior of Classical Functionally Graded Coated Beam," in *MATEC Web of Conferences*, 2017, p. 03009.
[29] F. Heidary and M. Reza Eslami, "Dynamic analysis of distributed piezothermoelastic composite plate using first-order shear deformation theory," *Journal of Thermal Stresses,* vol. 27, pp. 587-605, 2004.
[30] J. N. Reddy, *Energy principles and variational methods in applied mechanics*: John Wiley & Sons, 2017.

IOP Conf. Series: Materials Science and Engineering **854** (2020) 012035 doi:10.1088/1757-899X/854/1/012035

The Shortest Path Finder for Tsunami Evacuation Strategy using Dijkstra Algorithm

A F Pratiwi[1], S D Riyanto[1], R Listyaningrum[2], and G M Aji[1]

[1]Department of Electrical Engineering, State Polytechnic of Cilacap, Indonesia
[2]Department of Informatics Engineering, State Polytechnic of Cilacap, Indonesia

E-mail: art_dhita@yahoo.com

Abstract. Having a tsunami early warning system during the tsunami disaster is important. The tsunami early warning system that is widely applied in Indonesia using loudspeakers and information board as a medium to convey information directly to the community during the disaster. The information provided is only about the status of the disaster and what needs to be done. People who panic, tend to choose the evacuation route through the main roads without thinking that the path will become very crowded resulting in total congestion. This paper discusses a tsunami evacuation strategy by finding the shortest path of evacuation route using dijkstra algorithm. This system contains information about the shortest evacuation routes that can be taken by the community towards a safe location, either vertically or horizontally. So, people can evacuate vertically or horizontally based on the geographical location of their residence. This system will be implemented in Cilacap districts. Shortest evacuation route is determined using the djikstra algorithm. Disaster-prone locations is the starting vertex and safe locations is finish vertex.

1. Introduction

An early warning is part of disaster risk reduction that is not only about giving accurate warning information, but also building a good understanding of risk from a warning and increasing the ability of authorities and communities to react appropriately to the early warning. The main priority for evacuation is people who live in risk areas [1]. Therefore, community preparedness in risk areas needs to be improved in order to improve understanding of evacuation routes, evacuation sites and evacuation buildings.

The tsunami early warning system that is widely applied in Indonesia based on straightforward approach using loudspeakers as a medium to convey information directly to the community during the disaster [2]. This system still has many disadvantages, which are very significant, the information delivered cannot be heard properly because of interference from the system or other external factors. To overcome this problem an electronic information board was developed which could display information on tsunami disasters that could be seen directly by the community [3,4].

Information that is conveyed on information boards or through loudspeakers, only conveys information about the status of the disaster and what must be done. This causes the people who obtain information on the potential tsunami disaster will immediately flee away from the coast to the highlands. People who panic, tend to choose the evacuation route through the main roads without thinking that the path will become very crowded resulting in total congestion. Even though there are actually other evacuation routes that have a closer distance to the safe location. Based on this condition, a system is needed to support the early warning system so that it does not only provide tsunami information or warnings, but also provide information on the shortest evacuation routes to safe locations.

In this paper, determining the shortest path for a tsunami evacuation strategy using dijkstra algorithm is discussed. This system contains information about the shortest evacuation routes that can be taken by the community towards a safe location, either vertically or horizontally. So, people can evacuate vertically or horizontally based on the geographical location of their residence. This system will be implemented in Cilacap districts which is one of the top 3 districts with the highest disaster threat of tsunami [5].

2. Tsunami Evacuation Strategy in Cilacap District

Tsunami is difficult to predict because of the earthquake that triggered it also did not predictable. The impact of the tsunami was very severe large so that earthquake-prone areas and suspected to be very prone to tsunami disaster must be conducted structural mitigation and non-structural. Evacuation of people living in risky areas is a top priority when the system Tsunami early warning starts operating [6].

Some mitigation steps should be done for the risky area, including identification of tsunami-prone area, development of tsunami early warning system, and emergency evacuation. Identification of tsunami-prone area is needed to provide the value of risk-index from the specified location. For the early warning system, Indonesia has an INA-TEWS (Indonesia Tsunami Early Warning System) which is controlled by National Meteorology, Climatology, and Geophysics Department (Badan Meteologi, Klimatologi, dan Geofisika, BMKG). BMKG provides an earthquake information or tsunami early warning information, and sends it to related institution, such as National Disaster Relief Department (Badan Nasional Penanggulangan Bencana, BNPB), Local Government, media, etc. But the most important steps in mitigation is the proper emergency evacuation procedure. In order to reduce the material losses and casualties, a comprehensive emergency management concept must be implemented [7]. Preparedness and responses are two significant elements in emergency evacuation concept. In addition, the transmission system for information dissemination, evacuation information, evacuation route, and tsunami evacuation areas are also needed.

In Cilacap District, tsunami handling procedures follow the standard procedures performed by the Local Disaster Relief Agency (Badan Penanggulangan Bencana Daerah, BPBD) in Cilacap District. In accordance with the procedure, if a tsunami strikes, the people who are close to the coast must immediately evacuate themselves to a higher place. The high places that can be used as evacuation sites are high buildings, high land locations, or places that far from the coast. However, in disaster conditions, the simultaneous movement of refugees slows the evacuation of people to safe evacuation sites.

3. Methods

This research consists of several stages to achieve the goal of obtaining the shortest path of tsunami disaster evacuation in Cilacap Regency as shown through the flow diagram in Figure 1. Currently, there are already many algorithms that look for the shortest path. The shortest path of the most famous Algortima is the dijkstra algorithm. The dijkstra algorithm is applied to find the shortest path in the directed graph. But this algorithm also applies to graphs that are not directed [8]. Dijkstra algorithm is an algorithm used in this research to obtain the shortest evacuation route.

IOP Conf. Series: Materials Science and Engineering **854** (2020) 012035 doi:10.1088/1757-899X/854/1/012035

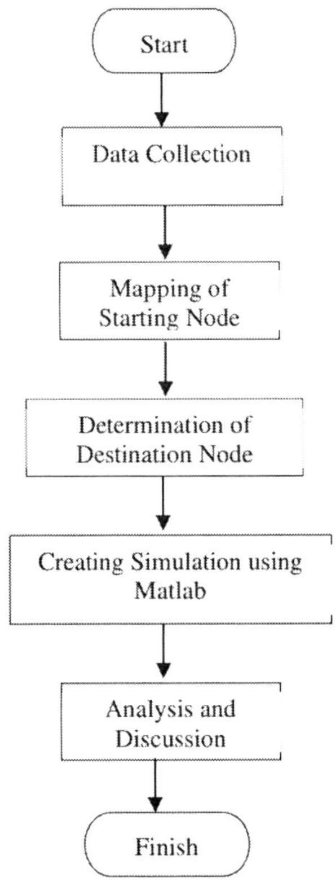

Figure 1. Flowchart of The Shortest Evacuation Path using Dijkstra Algorithm

The first step is data collection. In the calculation of the dijkstra algorithm the points and weights from the lines connecting these points is needed. The vertexes are from tsunami prone areas, crossroads, and safe locations, while the distance between intersections is assumed to be a weighted line. The data is obtained from several ways, namely field observations, interviews with related parties, literature studies, and data acquisition. If the data has been collected, the next step is creating a simulation using Matlab. In this simulation, the calculation of the dijkstra algorithm is applied to determine the shortest evacuation route for a tsunami disaster.

4. Results and Discussion

Based on the data that has been collected, a mapping has been obtained for the starting vertexes and the final vertexes are determined as a safe place for evacuation. The vertexes that are mapped as starting vertexes are divided into several points along the coastline as shown in Table 1, while in Table 2 are the determined end vertexes.

Table 1. List of Starting Vertexes

Code	Location name	Vertex Coordinate
A	Area 70 Pertamina Cilacap	-7.742328, 109.017578
B	Pantai Teluk Penyu Cilacap	-7.733707, 109.019516
C	Perum Shappire Regency	-7.728005, 109.017072
D	Rusunawa PPC	-7.726354, 109.022743
E	Puri Tegal Kamulyan	-7.716056, 109.030111

F	Perum Taman Gading	-7.707928, 109.033117
G	RSPC	-7.702403, 109.041512
H	Perum Griya Tegal Asri	-7.694164, 109.056547

Table 2. List of End Vertexes

Code	Location name	Vertex Coordinate
I	Lapangan Terbang Tunggul Wulung	-7.640680, 109.037433
J	Lapangan Jangrana Lebeng/ lapangan desa dondong	-7.619108, 109.066165

In addition to collecting data for start and end vertexes, it also collects a list that is traversed on the path used from the starting vertex to the end vertex. Table 3 shows the list of intersections along the path that was passed.

Table 3. List of Intersections along The Path

No	Name of Intersection
1	Jl. Dayung dan Jl. Kelapa Lima
2	Jl. Kelapa Lima dan Jl. Veteran
3	Area 70 Pertamina Cilacap
4	Jl. Kelapa Lima dan Jl. Karang
5	Jl. Karang dan Jl. Kol. Sugiono
6	Jl. Kol. Sugiono dan Jl. Brug Sentul
7	Jl. Brug Sentul dan Jl. Dr. Wahidin
8	Jl. Kolonel Sugiono dan Jl. Laut
9	Pantai Teluk Penyu Cilacap
10	Kantor Basarnas Cilacap
11	Jl. Lingkar Selatan dan Jl. Budi Utomo
12	Perum Shappire Regency
13	Jl. Dr. Wahidin dan Jl. Dr. Sutomo
14	Dafam Hotel
15	Jl. Jend. Sudirman dan Jl. S. Parman
16	Jl. Jend Sudirman dan Jl. Katamso
17	Alun-alun Cilacap
18	Jl. A. Yani dan Jl. Jend Sudirman
19	Pasar Gede
20	RS Santa Maria
21	Puskesmas Cilacap Selatan
22	Stasiun Cilacap
23	Jl. Veteran dan Jl. A. Yani
24	Pelabuhan Batu bara Wijayapura
25	Jl. Laut Jawa dan Jl. Syah Bandar
26	Pangkalan TNI AL Cilacap
27	Jl. Yos Sudarso dan Jl. Kluwih
28	Jl. Laut Jawa dan Jl. Selat Bali
29	Pelabuhan Tanjung Intan
30	PT. Pelindo III Persero
31	Kantor BEA Cukai

32	Pelabuhan Penyebrangan Seleko
33	Jl. Di. Panjaitan dan Jl. Jend. Sudirman
34	Jl. Di. Panjaitan dan Jl. LetJend. Suprapto
35	Jl. Katamso dan Jl. Tidar
36	Jl. S. Parman dan Jl. Gatot Subroto
37	Fave Hotel
38	Rusunawa PPC
39	Kantor kelurahan Tegal Kamulyan
40	TPI Pelabuhan Perikanan Samudra
41	Puri Tegal Kamulyan
42	Jl. Dr Sutomo dan Jl. Lingkar Selatan
43	PNC
44	Kantor kelurahan Sidakaya
45	Jl. Cerme dan Jl. Dr. sutomo
46	Yayasan Al Irsyad
47	Jl. Cerme dan Jl. Gatot Subroto
48	RSUD Cilacap
49	Jl. Kawi dan Jl. Gatot Subroto
50	AMN
51	Jl. Kendeng dan Jl. Flores
52	Jl. Flores dan Jl. Rinjani
53	Jl. Tidar dan Jl. Rinjani
54	Jl. Rinjani dan Jl. Ir. Juanda
55	Jl. Ir. Juanda dan Jl. Nusantara
56	RS Fatimah
57	Jl. Ir. Juanda dan Jl. Laban
58	Jl. Gatot Subroto dan Jl. Perintis Kemerdekaan
59	Terminal cilacap
60	Jl. Kalimantan dan Jl. Gatot Subroto
61	Pasar Tanjung
62	Jl. Dr. Sutomo dan Jl. Dr. Rajiman
63	Kantor kelurahan Gunung Simping
64	Kantor kecamatan Cilacap Tengah
65	Perum GSP
66	SMA N 3
67	Perum Taman Gading
68	Jl. Kalimantan dan Jl. Lingkar Selatan
69	TPI Kemiren
70	RSPC
71	Perum setia budi
72	Jl. Dr. Rajiman dan Jl. Setia Budi
73	Kantor kelurahan Kebon Manis
74	Jl. Setia Budi dan Jl. Perintis Kemerdekaan
75	Puskesmas Cilacap Utara 1
76	Jl. Perintis Kemerdekaan dan Jl. Tentara Pelajar

77	Kantor kelurahan Gumilir
78	Jl. Laban dan Jl. Tentara Pelajar
79	Jl. Damar dan Jl. Tentara Pelajar
80	Jl. Nusantara dan Jl. Tentara Pelajar
81	Jl. Bisma dan Jl. Urip Sumoharjo
82	Perum Rinenggo Asri
83	Perum Griya Tegal Asri
84	Jl. Lingkar Timur dan Jl. Kuda Laut
85	Pelabuhan perikanan Cilacap
86	TPI Lengkong
87	PT. Adhimix Precast Indonesia Plant
88	Jl. Lingkar Timur dan Jl. Laut
89	Jl. Sinta dan Jl. Urip Sumoharjo
90	Pasar Limbangan
91	Jl. Urip Sumoharjo dan Jl. Rawa Bendungan
92	Jl. Rawa Bendungan dan Jl. Cilacap-Wangon
93	**Lapangan Terbang Tunggul Wulung**
94	RS Afdila
95	Jl. Laut dan Jl. Raya Soekarno Hatta
96	Pasar wage karang kandri
97	Jl. Lingkar Timur dan Jl. Raya Soekarno Hatta
98	PLTU
99	Jl. Diponegoro dan Jl. Gerilya
100	**Lapangan Jangrana Lebeng/ lapangan desa dondong**

Then, the data is processed using the dijkstra algorithm in the Matlab software. The simulation produced in Matlab can be seen in Figure 2. In this simulation, there are two input variables, namely the starting vertex and the desired destination / end vertex. Furthermore, the simulation will produce the shortest path displayed in graph and distance values.

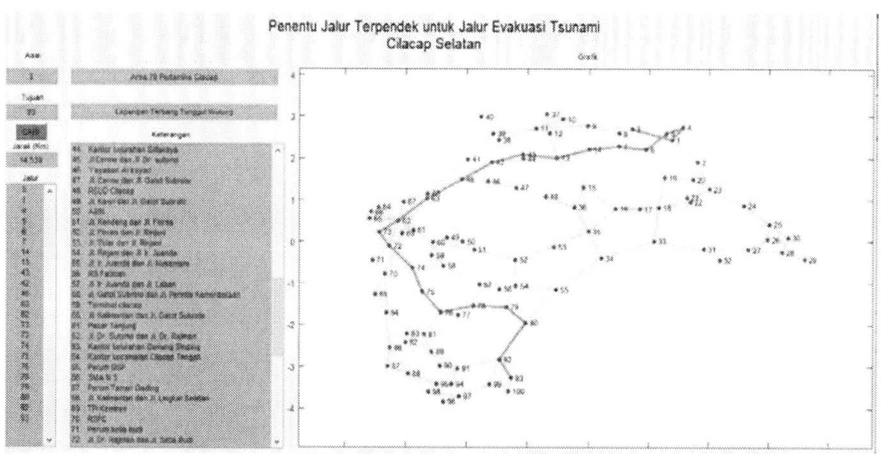

Figure 2. Simulation Results of The Shortest Path Using Dijkstra Algorithm

Based on the simulation results, the shortest distance from each starting vertex to the end vertex is obtained. The simulation has succeeded in determining the shortest path from the specified starting vertexes and end vertexes. Overall, the shortest distance from each point can be seen in Table 4.

Table 4. Simulation Results of the Shortest Path

Code	Starting Vertex	Distance (Km)	
		Tunggul Wulung	Jangrana
A	Area 70 Pertamina Cilacap	14,539	19,359
B	Pantai Teluk Penyu Cilacap	13,479	18,299
C	Perum Shappire Regency	12,439	17,259
D	Rusunawa PPC	12,9	17,6
E	Puri Tegal Kamulyan	11	15,7
F	Perum Taman Gading	9,898	14,718
G	RSPC	8,95	13,65
H	Perum Griya Tegal Asri	6,9	11,6

5. Conclusion

The shortest path finder using dijkstra algorithm has been successfully implemented in Matlab simulation. Based on the results of the test are obtained the shortest path from the starting vertex to destination/end vertex of 6,9 km to Tunggul Wulung and 11,6 km to Jangrana. This result is obtained because the location of these vertexes can directly access the vertical path for evacuation.

Acknowledgement
This work was partially supported by the Grant for Beginner Lecturer Research (Hibah Peneltian Dosen Pemula) FY2019 from the Ministry of Research, Technology, and Higher Education, the Republic of Indonesia

References

[1] Badan Meterologi Klimatologi dan Geofisika, InaTEWS Indonesia Tsunami Early Warning System, Konsep dan Impelementasi, Jakarta, 2010.
[2] Badan Meterologi Klimatologi dan Geofisika, Pedoman Pelayanan Peringatan Dini Tsunami, Jakarta, 2012.
[3] A.F. Pratiwi, G.M. Aji, Purwiyanto, Chairunnisa, and A. Munir, "Wireless Electronic Information Board for Tsunami Early Warning System Based on FM Radio," in Proceedings of 7th International Annual Engineering Seminar, Yogyakarta, 2017, pp. 139-142
[4] G. M. Aji, A. F. Pratiwi, Purwiyanto, Chairunnisa, and A. Munir, "Electronic Information Board Based on AVR Atmega128 for TEWS in Cilacap," Bulletin of Electrical Engineering and Informatics, Vo. 7, No. 2, June 2018, pp. 216-221.
[5] Badan Nasional Penanggulangan Bencana. Indeks Rawan Bencana Indonesia. Jakarta: BNPB. 2011.
[6] Mück, M., 2008. Tsunami Evacuation Modelling. Development and application of a spatial information system supporting tsunami evacuation planning in South-West Bali. Universität Regensburg.
[7] Federal Emergency Management Agency (FEMA). 2005. Tsunami Hazards FEMA Coastal Flood Hazard Analysis and Mapping Guidelines Focused Study Report. California, Amerika Serikat : FEMA
[8] Munir, R., 2016. Matematika Diskrit. Penerbit Informatika. Bandung.

IOP Conf. Series: Materials Science and Engineering **854** (2020) 012036 doi:10.1088/1757-899X/854/1/012036

Static Analysis of Functionally Graded Coated Plate on Elastic Foundation Based on Levy Method

Abdolreza Toudehdehghan

Department of Mechanical Engineering, INTI International University, Jalan BBN 12/1, Bandar Baru Nilai, 71800 Nilai, Negeri Sembilan, Malaysia.

Email: Abdolreza.toudehdehghan@newinti.edu.my

Abstract. Levy method for bending analysis of composite plate resting on a Winkler elastic foundation hypothesis is presented based on classical plate theory. Two models were considered for a composite coated plate. These two models were named conventional composite model and functionally graded coated model. The conventional composite model was consisting of a ceramic Homogeneous Coated (HC) layer and metal layer. The Functionally Graded Coated (FGC) model was consisting of Functionally Graded Material (FGM) as a composition of ceramic and metal on a metal layer. The material properties of FGM composite are assumed varying exponentially along the thickness direction. Also, the material properties of the nan-FGM layers are assumed homogeneous. Finally, the effect of the FGC layer, as well as, the elastic stiffness of the foundation is presented and compared with the results of the HC model.

Nomenclature

a = Length of plate

b = Width of plate

D = Flexural rigidity of coated plate

x, y, z = axes coordinates

w = Total displacement in z-axis

w_h = Homogeneous solution

w_p = Particular solution

ν = Poisson's ratio

B_n = Constant coefficient in homogeneous solution

β_n = Constant number

$g_n(y)$ = Fourier coefficient in particular solution

$p_n(x)$ = Fourier coefficient in applied load

h = Thickness of FGM plate

E = Modulus of elasticity

E_{fe} = Modulus elasticity of FGM

E_c = Modulus elasticity of ceramic

E_m = Modulus elasticity of metal

q = Uniformly distributed load

k = Elastic stiffness of foundation

n = number of the series member

A_n = Constant coefficient in homogeneous solution

α_n = Constant number

$f_n(y)$ = Fourier coefficient in homogeneous solution

IOP Conf. Series: Materials Science and Engineering **854** (2020) 012036 doi:10.1088/1757-899X/854/1/012036

1. Introduction

Nowadays, structures that perform multiple tasks have become increasingly accessible with the advancement of human knowledge in the construction of new materials. One of the advanced materials developed by Japanese scientists in the early 1980s is Functionally Graded Materials (FGMs)[1]. In FGMs, the composition of composite varies from one side to the other side based on requirements. For instance, an FGM coating layer has pure ceramic at the surface and pure metal at the other surface, in which the composition of the layer changes gradually from ceramic to metal [2]. In order to describe the effective material properties of the FGM layer, several functions were proposed [3]. However, the well-known functions are such as exponential function, power-law function, and sigmoid function.

Due to the material properties of FGM, the separation between the FGM coating layer and the metal substrate is significantly reduced compared to the homogeneous coating layer as conventional composites [4, 5]. Also, FGMs have several advantages over the conventional composites such as thermal barrier, resistant to corrosion, pitting, etc. with reducing thermal stress and stress concentration in the interface between composite layers [6-9]. Therefore, FGMs have a wide range of applications such as mechanical structures, aerospace, optics, electronics, nuclear, civil structures, and chemicals [10].

Analysis structures on foundations can be studied by geotechnical engineering. There are several foundation models introduced by geotechnical engineers such as Winkler, Umansky, Wieghardt, Shechter, Gorbunov-Posadov, Hetenyi, Shechter, Pasternak, Harr, Giround, Poulos, and Selvadurai which are used in the study of structures on elastic bases [11]. The Winkler model is the first base model and also the popular model used in the analysis of structures on elastic foundations. In the Winkler model, it is assumed the surface of the soil under the loading produces a settlement. The Winkler model is not accurate [11]. Because of the simplicity of this model, many researchers prefer to use this model than other precision models [12-15]. Ghannadiasl and Golmogany [16] considered an Euler-Bernoulli beam with several boundary conditions on the Winkler foundation. They used the B-spline collocation method to analyze the static behavior of the beam under a uniformly distributed load. M. Mofid and M. Noroozi [17] studied the static behavior of simply supported thin plate on a modified Winkler foundation.

In the present study, the Levy method is used to solve the Kirchhoff coated plate on Winkler elastic foundation problem. Two coated models as shown in Figure 1 under uniformly distributed load are considered. The first model is consisting of a ceramic homogeneous layer on top of the homogeneous metal layer. The second model is consisting of the FGM layer on top of the homogeneous metal layer. The geometric constraints of a coated plate are consisting of two opposite sides clamped and another two opposite sides pin. An exponential function represents the material properties of the FGM layer. The elastic coefficient of foundation and Poisson's ration are assumed constant. By increasing the elastic stiffness of the foundation underneath the coated plate, the foundation becomes a rigid layer, causing the deflection of the plate to decrease dramatically. Also, with increasing elastic base stiffness, the behavior of the HC model and the FGC model is very similar. Another case that is of interest in this study is the phenomenon of singularity between the coated layer and the substrate due to the different material properties of layers in the conventional composite (HC model). Using FGM as the coating layer, the de-bonding phenomenon at the interface of layers is completely eliminated. This research can show the priority of FGM as a coated layer of the plate on a rigid foundation.

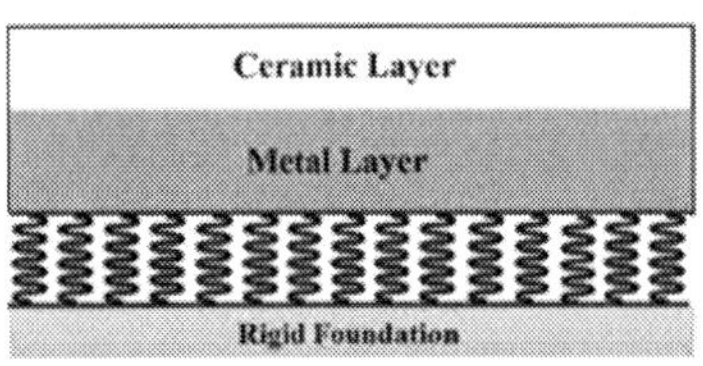

(a) Homogeneous Coated plate (HC)

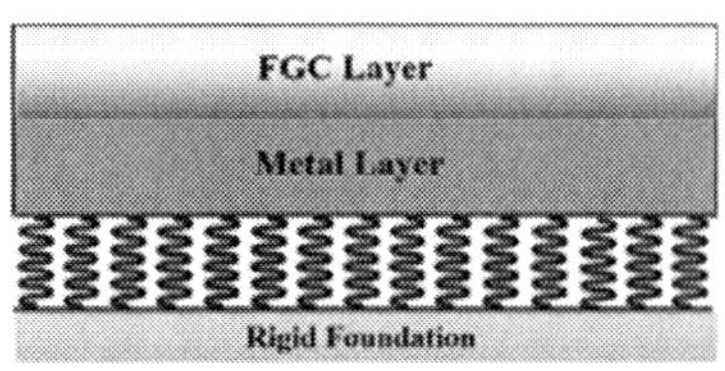

(b) Functionally Graded Coated plate (FGC)

Figure 1. Schematic composite plate on the Winkler base.

IOP Conf. Series: Materials Science and Engineering **854** (2020) 012036 doi:10.1088/1757-899X/854/1/012036

2. Formulations

The governing equation for the lateral deflection of a thin rectangular Plate on Winkler foundation model subjected to uniformly lateral loads is [18]

$$\frac{\partial^4 w(x,y)}{\partial x^4}+2\frac{\partial^4 w(x,y)}{\partial x^2\partial y^2}+\frac{\partial^4 w(x,y)}{\partial y^4}+\frac{k}{D}w(x,y)=\frac{q(x,y)}{D}$$

$$D=\frac{Eh^3}{12\left(1-\nu^2\right)} \tag{1}$$

The boundary conditions of coated plate on elastic foundation as shown in Figure 2 are

$$x=0,a \quad \Rightarrow \quad w=0, \quad \frac{\partial^2 w}{\partial x^2}=0 \tag{2}$$

$$y=\pm\frac{b}{2} \quad \Rightarrow \quad w=0, \quad \frac{\partial w}{\partial y}=0 \tag{3}$$

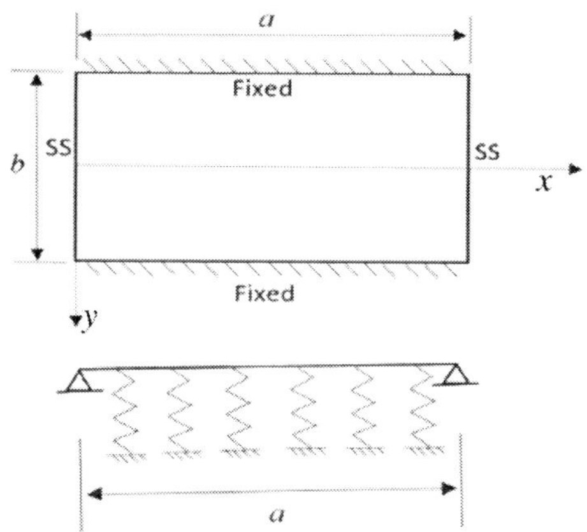

Figure 2. Plate on Winkler foundation with opposite edges fixed and other two simply supported.

2.1. Material properties of FGP

The FGP include ceramic and metal. The mechanical properties of the FGC layer vary as an exponential function from the ceramic surface to the other surface which is metal along the FGC thickness. Therefore, the Young modulus of the FGM can be determined as [19, 20]:

$$E_{fe}=E_c e^{\left(\frac{1}{h}\ln\left(\frac{E_m}{E_c}\right)\left(z+\frac{h}{2}\right)\right)} \tag{4}$$

Where $-h/2\le z\le h/2$. It should be noted that the Poisson's ratio is assumed constant [21, 22]. Figure 3 shows the exponential changes of the Young's modulus in FGP thickness. Assuming that the Young's modulus of ceramic and metal are 100 GPa and 10 GPa respectively.

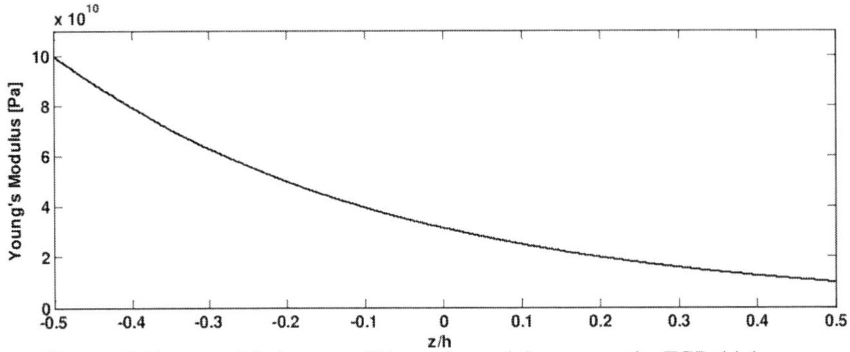

Figure 3. Exponential changes of Young's modulus across the FGP thickness.

2.2. *Levy solution of coated plate governing equation*

Equation (1) is a nonhomogeneous equation. The general solution of this equation is the summation of homogeneous solution and particular solution. Therefore, the general solution can be written as

$$w(x, y) = w_h(x, y) + w_p(x, y) \tag{5}$$

The homogeneous solution based on simply supported boundary conditions (Equation (3)) is written as

$$w_h(x, y) = \sum_{n=1}^{\infty} f_n(y) \sin \frac{n\pi x}{a} \tag{6}$$

Substituting Equation (6) into the homogeneous part of Equation (1) then after mathematically simplifying and considering that the deflections of FGP are symmetric to the x-axis. The homogeneous deflection of FGP is

$$w_h(x, y) = \sum_{n=1}^{\infty} \left[A_n (\sinh \alpha_n y)(\sin \beta_n y) + B_n (\cosh \alpha_n y)(\cos \beta_n y) \right] \sin \frac{n\pi x}{a} \tag{7}$$

Where

$$\alpha_n = \sqrt{\frac{1}{2}\left(\frac{n^2\pi^2}{a^2} + \sqrt{\frac{n^4\pi^4}{a^4} + \frac{k}{D_{fe}}} \right)} \tag{8}$$

$$\beta_n = \sqrt{\frac{1}{2}\left(\frac{n^2\pi^2}{a^2} - \sqrt{\frac{n^4\pi^4}{a^4} + \frac{k}{D_{fe}}} \right)} \tag{9}$$

A single Fourier series is expressed as the particular solution of the general solution (Equation (5)) as follows:

$$w_p(x, y) = \sum_{n=1}^{\infty} g_n(y) \sin \frac{n\pi x}{a} \tag{10}$$

Also, a single Fourier series is expressed as the applied load in Equation (1) as follows:

$$q(x, y) = \sum_{n=1}^{\infty} p_n(y) \sin \frac{n\pi x}{a} \tag{11}$$

With considering uniformly constant load in Equation (11), the Fourier coefficient $p_n(y)$ can be determined as:

$$p_n(y) = \frac{4q}{n\pi}, \qquad n = 1,3,5,... \tag{12}$$

IOP Conf. Series: Materials Science and Engineering **854** (2020) 012036 doi:10.1088/1757-899X/854/1/012036

Substituting Equation (12), (11), and (10) into Equation (1), the Fourier coefficient $g_n(y)$ can be determined as:

$$g_n(y)=\frac{4q}{D_{fe}n\pi}\times\frac{1}{\left[\frac{n^4\pi^4}{a^4}+\frac{k}{D_{fe}}\right]} \tag{13}$$

Substituting Equations (7) and (10) into Equation (5), the total deflection of the FGP becomes

$$w(x,y)=\sum_{n=1}^{\infty}\left[A_n\sinh\alpha_n y\times\sin\alpha_n y+B_n\cosh\alpha_n y\times\cos\alpha_n y+g_n(y)\right]\sin\frac{n\pi x}{a} \tag{14}$$

By using boundary conditions (Equation (3)), the constant coefficients A_n and B_n become

$$A_n=\frac{g_n}{\zeta_1\left(\cosh\alpha_n\frac{b}{2}\right)\left(\cosh\beta_n\frac{b}{2}\right)}$$
$$B_n=-A_n\left(\tanh\alpha_n\frac{b}{2}\right)\left(\tanh\beta_n\frac{b}{2}\right)-\frac{g_n}{\left(\cosh\alpha_n\frac{b}{2}\right)\left(\cosh\beta_n\frac{b}{2}\right)} \tag{15}$$

Where

$$\zeta_1=\frac{\zeta_2+\zeta_3}{\zeta_4-\zeta_5}-\left(\tanh\alpha_n\frac{b}{2}\right)\left(\tan\beta_n\frac{b}{2}\right)$$
$$\zeta_2=\alpha_n\cosh\left(\alpha_n\frac{b}{2}\right)\sin\left(\beta_n\frac{b}{2}\right)$$
$$\zeta_3=\beta_n\sinh\left(\alpha_n\frac{b}{2}\right)\cos\left(\beta_n\frac{b}{2}\right) \tag{16}$$
$$\zeta_4=\alpha_n\sinh\left(\alpha_n\frac{b}{2}\right)\cos\left(\beta_n\frac{b}{2}\right)$$
$$\zeta_5=\beta_n\cosh\left(\alpha_n\frac{b}{2}\right)\sin\left(\beta_n\frac{b}{2}\right)$$

3. Results and Discussions

As described in the previous section, two coated plate models are considered in this study. In order to solve the equations of coated plates on elastic foundation, a single series equation with two terms are used. The first coated plate model is consisting of the homogeneous ceramic as a coated layer on top of the metal layer (HC). The second coated plate model is consisting of the FGM as a coated layer on top of the metal layer (FGC). The geometrical dimensions, material properties of the ceramic layer, metal layer, FGM layer and elastic coefficient of foundation which used in static analysis of the coated rectangular plates on an elastic foundation are presented in Table 1. Figure 4 demonstrates the deflections of two coated models along the width of the coated plates with clamped-clamped boundary conditions under considering the various range of the elastic stiffness. With increasing the elastic stiffness (close to reality [18]) at the base, the difference deflection in two coated models are disappeared. Also, Figure 4 shows that the deflections along the width of plate and displacements at the constraints are dramatically reduced. Figure 5 shows the deflections of the two coated plate models along the half-length of the plates which the constraints are pin – pin. No displacement happens at pin

IOP Conf. Series: Materials Science and Engineering **854** (2020) 012036 doi:10.1088/1757-899X/854/1/012036

constraints. The behavior of two coated models plate with increasing the elastic stiffness of the foundation is very close. Also, the deflections dramatically decrease with increasing the elastic stiffness. It can be deduced from Figures 4 and 5 that the mechanical behavior of the two coated plate models on the elastic foundation is quite similar to considering the real elastic stiffness $\left(k > 10MPa\right)$. Figures 6 and 7 show the 3D deflections of the coated plate based on the data in Table 1 for the conventional coated plate model (HC) and the FGM coated plate model (FGC), respectively.

Table 1. Dimensions, material properties and elastic constant of foundation

Parameter	Quantity	Parameter	Quantity
h_m (m)	0.006	$\nu_{12} = \nu_{21}$	0.3
h_c (m)	0.004	k (Pa)	50000
h_{fe} (m)	0.004	E_m (GPa)	150
a (m)	1	E_c (GPa)	250
b (m)	1	$F\left(N/m\right)$	22000

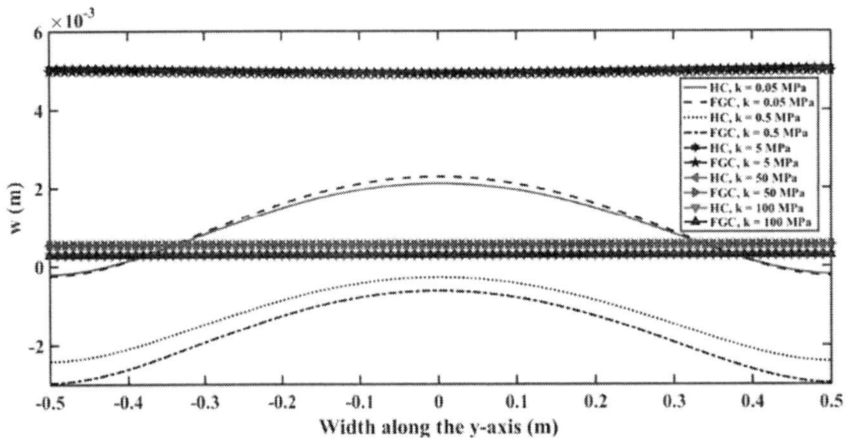

Figure 4. Deflection at the center of coated plates along the width based on various elastic coefficients.

IOP Conf. Series: Materials Science and Engineering **854** (2020) 012036 doi:10.1088/1757-899X/854/1/012036

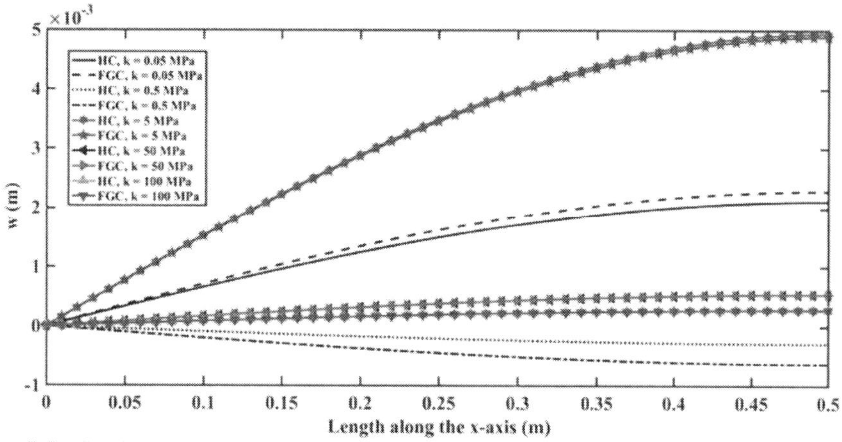

Figure 5. Deflection of half plates along the length at the center with considering various elastic coefficients.

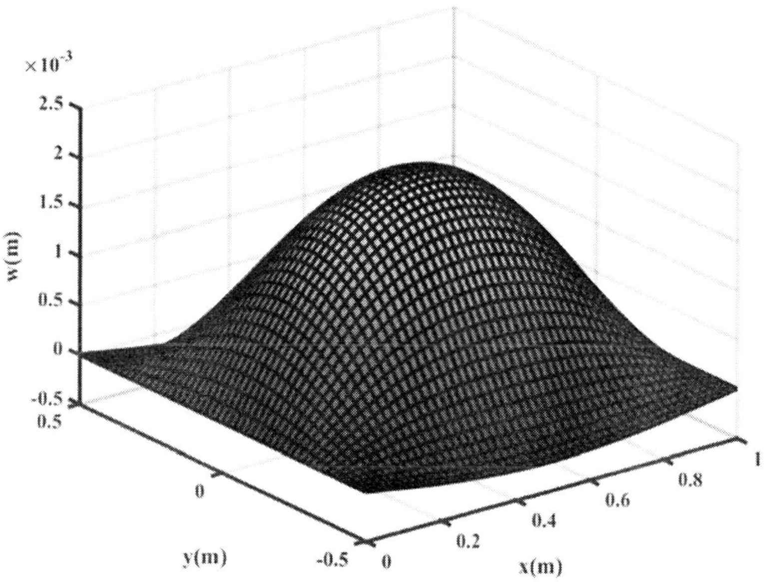

Figure 6. Deflection of homogeneous coated plate subjected to uniformly distributed load on elastic foundation.

IOP Conf. Series: Materials Science and Engineering **854** (2020) 012036 doi:10.1088/1757-899X/854/1/012036

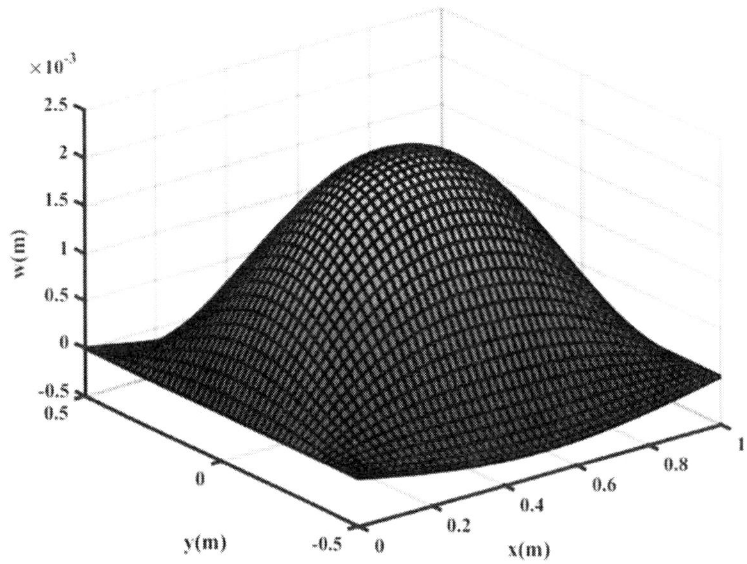

Figure 7. Deflection of Functionally Graded coated plate subjected to uniformly distributed load on elastic foundation.

Figures 8 and 9 show the stress at the center of the plate along the thickness of the coated plate, taking into account the minimum and maximum elastic stiffness values, respectively. Both figures show that using the FGM as a coating layer, the singularity phenomenon is completely eliminated between the coated layer and the substrate layer. Therefore, in the FGC model de-bonding phenomenon at the interface between layers will not occur. Figure 5 clearly shows that when the elastic stiffness is about $k = 0.05MPa$. The stress increased at the outer surfaces of the coated layer and substrate layer. This problem is eliminated by increasing the elastic stiffness to $k = 100MPa$ as shown in Figure 6.

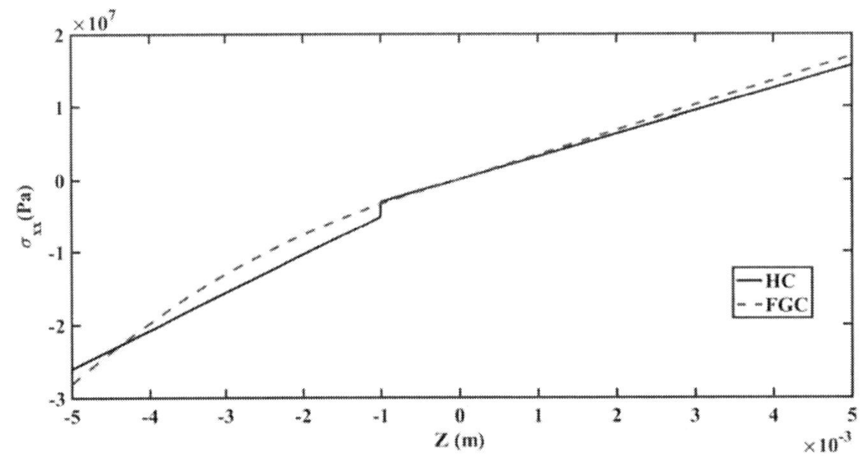

Figure 8. Stress across the thickness of the coated plates in two coated models on elastic foundation $(k = 0.05MPa)$.

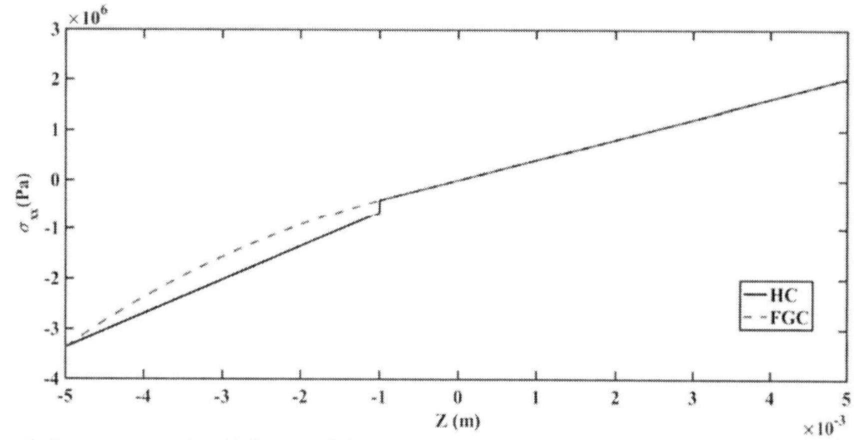

Figure 9. Stress across the thickness of the coated plates in two coated models on elastic foundation $(k = 100MPa)$.

4. Conclusion

The thin rectangular coated plate on Winkler foundation was solved by using single series equation for homogeneous solution, particular solution, and distributed loading on plate. Two terms of series equations are used to find the solution. Two models of coated plates are defined. In the first model, the coated layer is homogeneous ceramic (HC). The second one has FGM as a coated layer (FGC). The material properties of the FGM layer are defined based on exponential function. The results show that the priority of the FGC plate on a rigid enough foundation $(k > 10MPa)$.

References

[1] S. Chakraverty and K. K. Pradhan, *Vibration of functionally graded beams and plates*: Academic Press, 2016.

[2] H.-S. Shen, *Functionally graded materials: nonlinear analysis of plates and shells*: CRC press, 216.

[3] A. Toudehdehghan, J. Lim, K. E. Foo, M. Ma'arof, and J. Mathews, "A brief review of functionally graded materials," in *MATEC Web Conferences*, 2017.

[4] S.-H. Chi and Y.-L. Chung, "Mechanical behavior of functionally graded material plates under transverse load—Part I: Analysis," *International Journal of Solids and Structures*, vol. 43, pp. 3657-3674, 2006.

[5] H. Shodja, H. Haftbaradaran, and M. Asghari, "A thermoelasticity solution of sandwich structures with functionally graded coating," *Composites science and technology*, vol. 67, pp. 1073-1080, 2007.

[6] F. Ramirez, P. R. Heyliger, and E. Pan, "Static analysis of functionally graded elastic anisotropic plates using a discrete layer approach," *Composites Part B: Engineering*, vol. 37, pp. 10-20, 2006.

[7] A. Toudehdehghan, M. M. Rahman, and F. Tarlochan, "Mechanical and Thermal Analysis of Classical Functionally Graded Coated Beam," in *E3S Web of Conferences*, 2018, p. 01033.

[8] A. Toudehdehghan, M. M. Rahman, and F. Tarlochan, "Analyzing The Behavior of Classical Functionally Graded Coated Beam," in *MATEC Web of Conferences*, 2017, p. 03009.

[9] M. Zaki, F. Tarlochan, and S. Ramesh, "Two dimensional elastic deformations of functionally graded coated plates with clamped edges," *Composites Part B: Engineering*, vol. 45, pp. 1010-1022, 2013.

[10] D. Jha, T. Kant, and R. Singh, "A critical review of recent research on functionally graded plates," *Composite Structures,* vol. 96, pp. 833-849, 2013.
[11] E. Tsudik, *Analysis of structures on elastic foundations*: J. Ross Publishing, 2012.
[12] L. FRÝBA, "History of Winkler foundation," *Vehicle system dynamics,* vol. 24, pp. 7-12, 1995.
[13] A. Ghannadiasl and M. Mofid, "An analytical solution for free vibration of elastically restrained Timoshenko beam on an arbitrary variable Winkler foundation and under axial load," *Latin American Journal of Solids and Structures,* vol. 12, pp. 2417-2438, 2015.
[14] S. Binesh, "Analysis of beam on elastic foundation using the radial point interpolation method," *Scientia Iranica,* vol. 19, pp. 403-409, 2012.
[15] D. A. Dillard, B. Mukherjee, and R. C. Batra, "Reflections on the 150th Anniversary of Winkler's Foundation and its Profound Influence on the Field of Adhesion," in *40th Annual Meeting of the Adhesion Society*, 2017.
[16] A. GHANNADIASL and M. Z. GOLMOGANY, "Analysis of Euler-Bernoulli Beams with arbitrary boundary conditions on Winkler foundation using a B-spline collocation method," *Engineering Transactions,* vol. 65, pp. 423–445, 2017.
[17] M. Mofid and M. NOUROUZI, "A plate on Winkler foundation with variable coefficient," 2009.
[18] S. P. Timoshenko and S. Woinowsky-Krieger, *Theory of plates and shells*: McGraw-hill, 1959.
[19] A. Chakraborty, S. Gopalakrishnan, and J. Reddy, "A new beam finite element for the analysis of functionally graded materials," *International Journal of Mechanical Sciences,* vol. 45, pp. 519-539, 2003.
[20] M. Aydogdu and V. Taskin, "Free vibration analysis of functionally graded beams with simply supported edges," *Materials & design,* vol. 28, pp. 1651-1656, 2007.
[21] A. Toudehdehghan and M. M. Rahman, "Comparison conventional coated beam with functionally graded coated beam," *International Journal of Engineering and Technology (UAE),* vol. 7, pp. 713-721, 2018.
[22] Y. Hao, W. Zhang, J. Yang, and S. Li, "Nonlinear dynamic response of a simply supported rectangular functionally graded material plate under the time-dependent thermalmechanical loads," *Journal of Mechanical Science and Technology,* vol. 25, p. 1637, 2011.

IOP Conf. Series: Materials Science and Engineering **854** (2020) 012037 doi:10.1088/1757-899X/854/1/012037

Microsecond Pulse Generation using Bismuth Salenide as Saturable Absorber in 1.5 μm Region

R A M Yusoff[1], A A A Jafry[1], N F Zulkipli[2], N Kasim[1,*], S N M Rifin[1], Y Munajat[1], S W Harun[2]

[1] Department of Physics, Faculty of Science, Universiti Teknologi Malaysia, 81310 Skudai, Johor, Malaysia.
[2] Department of Electrical Engineering, University of Malaya, 50603 Kuala Lumpur, Malaysia

Email: k.nabilah@utm.my*

Abstract. A Q-switched Erbium-doped fiber laser (EDFL) was demonstrated utilizing Bismuth Salenide (Bi_2Se_3) as a saturable absorber. The saturable absorber (SA) is fabricated by mixing directly the Bismuth Salenide (Bi_2Se_3) into polyvinyl alcohol (PVA) aqueous solution to form a thin film. The SA is integrated into EDFL cavity to achieve a stable Q-switched at center wavelength of 1559.8 nm. The peak power is a steady increasing pattern from 2.52 mW to 3.21 mW as the pump power is increased from 86.1 mW to 116.6 mW. The highest pulse energy obtained was 36.9 nJ at the pump power of 116.6 mW.

1. Introduction

Pulsed laser has received tremendous focus among researchers within the field of laser because it plays crucial roles in wide selection of applications including optical communication, optical sensing, materials processing technology, life sciences, precision mechanics, national defence science and etc [1-3]. There are two techniques to realize pulsed laser which are active and passive. The latter is more favourable due to few advantages including ease of preparation, high flexibility and high efficiency. The generation of pulsed laser via passive means is highly dependent on the implementation of material as a saturable absorber inside a laser cavity. Recent years, many materials were implemented inside laser cavity to initiate pulsed laser such as semiconductor saturable absorber mirrors (SESAMs) [4] and carbon nanotubes (CNTs) [5].

For SESAMs, pulsing method require additional devices such as mirror, lenses and U-bench units which introduce extra losses to the cavity. Therefore, researcher starts to use carbon-based materials to generate pulsed laser. CNTs, which seems very promising in terms of electron mobility exhibit ultra-fast recovery time (< 1 ps) and wide absorption spectrum (1-2 μm) [6]. However, tenability of wavelength for CNTs are dependable on it shapes and sizes, thus provide complexity for pulse generation. For instance, structural, morphological and alignment control needed to be establish for the purpose of band gap alteration for CNTs [7]. Thus, 2-dimensional carbon allotrope (graphene) is introduced due to excellent electronic and optical properties. In terms of saturable absorption ability, graphene provide significant improvement compare to CNTs. Hexagonal honeycomb lattice of graphene exhibit wide-band operation due to its ultrafast recovery time (~200 fs) and point band gap structure [8]. However, graphene exhibits low modulation depth, typically lower than 2 % per layer. Since then, few Dirac-like materials are introduced as SA in laser cavity (Bi_2Se_3, Bi_2Te_3). Previously, Q-switched fiber lasers were demonstrated using Bismuth (III) Telluride (Bi_2Te_3) as a saturable absorber [9]. In this paper, we demonstrate a Q-switched fiber laser based on Bismuth (III) Salenide (Bi_2Se_3) as SA to generate laser in 1.5 micro region.

2. Preparation and Optical Characterisation of Bi_2Se_3 SA.

The commercially available few-layer Bi_2Se_3 powder with molecular weight of 654.84 g/mol was used to prepare the SA in this experiment. To prepare the host polymer, 1 g of polyvinyl Alcohol (PVA) (Sigma Aldrich) is dissolved in 120 ml de-ionized (DI) water with the aid of a magnetic stirrer at room temperature. Next, 14 mg of Bi_2Se_3 powder was mixed with 3 ml of the PVA solution before they are thoroughly mixed using a magnetic stirrer for three hours. Then the Bi_2Se_3-PVA solution was placed in ultrasonic bath for 10 minutes to make sure the Bi_2Se_3 material fully binds with the PVA. After that, the Bi_2Se_3 suspension was carefully poured onto petri dishes to avoid trapping any air bubble and is left to dry at room temperature for 48 hours to form Bi_2Se_3-PVA composite film.

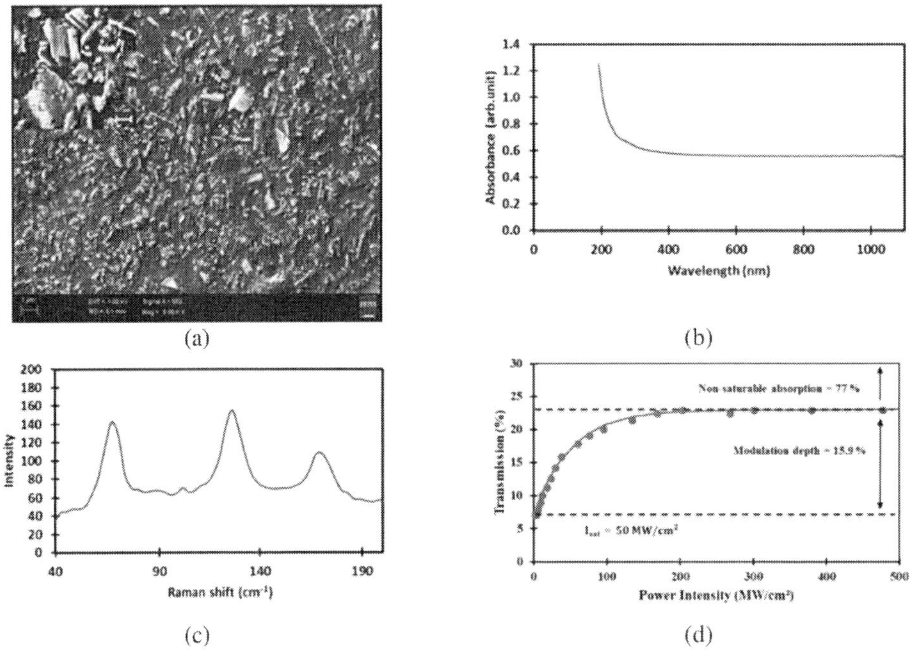

Figure 1. Optical characterisation (a) FESEM image, insert shows the image with higher magnification, b) Linear transmission, c) Raman Spectrum, d) non-linear absorption curve

Figure 1(a) shows the field emission scanning electron microscopy (FESEM) image of the fabricated Bi_2Se_3 film. As shown in the figure, the Bi_2Se_3 film has a high density of micro-rods and micro-grains, which can be clearly viewed on the substrate surface and distributed randomly on the substrate surface. These micro-rods and micro-grains are in irregular shapes with an average size of 0.3 to 1.66 μm. Figure 1(b) illustrates the absorbance spectrum of the prepared Bi_2Se_3-PVA film in the range of 200 nm to 1100nm. The figure attests a constant absorbance for the Bi_2Se_3-PVA film which indicates that it possesses a broadband resonance wavelength like graphene. Figure 1(c) shows the Raman spectrum of the fabricated Bi_2Se_3-PVA film. The spectrum indicates three distinct peaks at ~67 cm^{-1}, ~126 cm^{-1}, and ~170 cm^{-1}, which can be assigned to the A^1_{1g}, E^2_g and A^2_{2g} vibrational modes, respectively. The nonlinear optical response property for the Bi_2Se_3 film was also investigated to confirm its saturable absorption characteristic by applying dual optical power meter techniques. The pulse input source used a mode-locked fiber laser, which has femtosecond output pulse with a 17 MHz repetition rate and a 900 fs pulse duration, which the output power is approximately 5 mW. Figure 1(d) shows the transmission characteristic of the Bi_2Se_3 SA, which was fitted using the following simple saturable absorption model equation;

IOP Conf. Series: Materials Science and Engineering **854** (2020) 012037 doi:10.1088/1757-899X/854/1/012037

$$T(I) = 1 - a_0 \exp\left(\frac{I}{I_{sat}}\right) - a_{ns} \quad \text{Eq.1}$$

where $T(I)$, α_0, I, I_{sat} and α_{ns} stand for the transmission, modulation depth, input intensity, saturation intensity, and non-saturable absorption, respectively. The modulation depth, non-saturable absorption and saturation intensity were measured to be approximately 15.9 %, 77 % and 50 MW/cm^2, respectively as shown in Figure 1(d).

3. Experimental setup

Figure 2 shows the experimental setup of the erbium doped fiber laser with Bi_2Te_3 as a saturable absorber. The ring cavity consists of a 2.4 m long erbium doped fiber (EDF) as the active medium, a 980/1550 nm wavelength-division multiplexer (WDM), an optical isolator, a Bi_2Se_3 based SA, and a 90:10 output coupler. The Bi_2Se_3 as a SA is sandwiched between two FC/PC fiber connectors and inserted into the fiber cavity. The insertion loss of the SA is measured to be around 2 dB at 1550 nm. It is pumped by the 980 nm laser diode via a 980/1550 nm WDM. Isolator is located between coupler and gain medium to allow light travel in a one direction. The output of the laser is tapped out of the cavity through a 90:10 output coupler, which allows 10% of the output to be extracted for analysis.

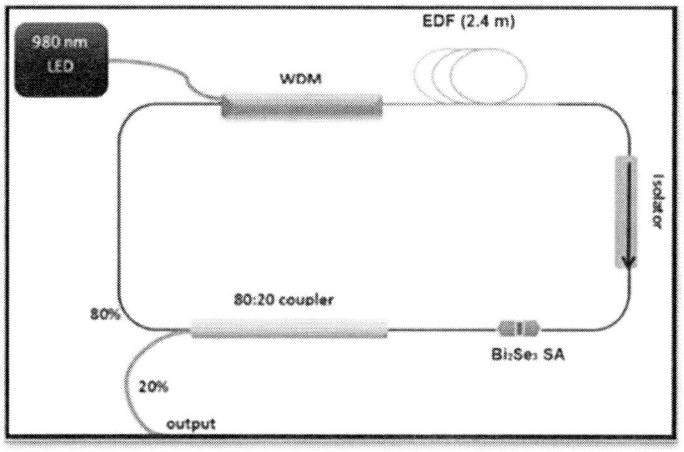

Figure 2. The experimental setup of the erbium fiber laser with Bi_2Se_3 as a passive saturable absorber.

4. Results and Discussion

In the experiment, the EDFL started to generate pulse at threshold pump power of 86.1 mW. As the pump power is further increased, the Q-switching operation was maintained up to 116.6 mW. Figure 3 shows the comparison of the output spectrum of the proposed EDFL with and without using Bi_2Se_3 as a SA at the pump power of 116.6 mW. As seen, the operating wavelength of the laser shifted to the left by 1.2 nm from 1561.0 nm to 1559.8 nm with the incorporation of the SA. The wavelength shift and higher intensity obtained by the proposed Q-switched EDFL is due to the integration of Bi_2Se_3 based SA device inside the cavity. The output spectrum with SA (solid line) was measured at 1559.8 nm. Without the SA, the ring cavity operates in a continuous wave (dotted line) method and output spectrum at 1561.0 nm.

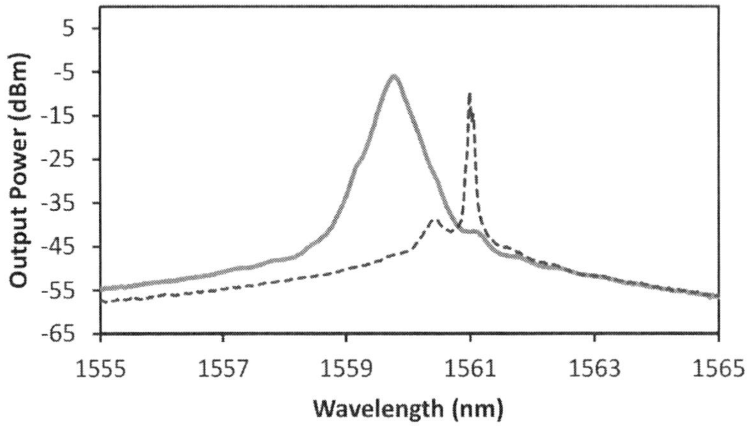

Figure 3. Output spectrum of the EDFL with and without Bi_2se_3 as a SA at pump power of 175 mW

The stable pulse trains with different repetition rate were observed when the pump sources were progressively increased from 86.1 mW to 116.6 mW. The EDFL started to produce a Q-switched pulse at threshold pump power of 86.1 mW. Figures 4 (a) shows typical oscilloscope trace of the Q-switched pulse train at the pump power of 116.6 mW. It shows the peak-to-peak duration of 11.49 μs, which is equal to the repetition rate of 87.03 kHz. Figure 4 (b) shows a single envelope of the Q-switching pulse at pump power of 116.6 mW. The pulse is observed to have an almost symmetric shape with a pulse width of approximately 5.11 μs.

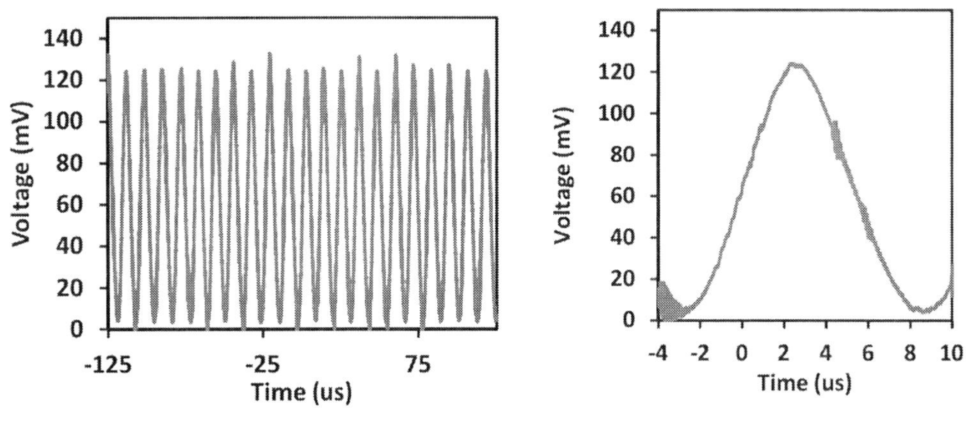

(a) (b)

Figure 4. (a) Oscilloscope pulse train at 116.6 mW pump power (b) a single envelop of the Q-switched at power pump of 116.6 mW

Figure 5 shows the relationship between the repetition rate and the pulse width against the pump power. The repetition rate of the proposed Q-switched EDFL can be tuned from 84.08 to 87.03 kHz as the pump power is raised from 86.1 to 116.6 mW. In the meantime, the pulse width reduces from 5.41 μs to 5.11 μs as the pump power is increased within the same range.

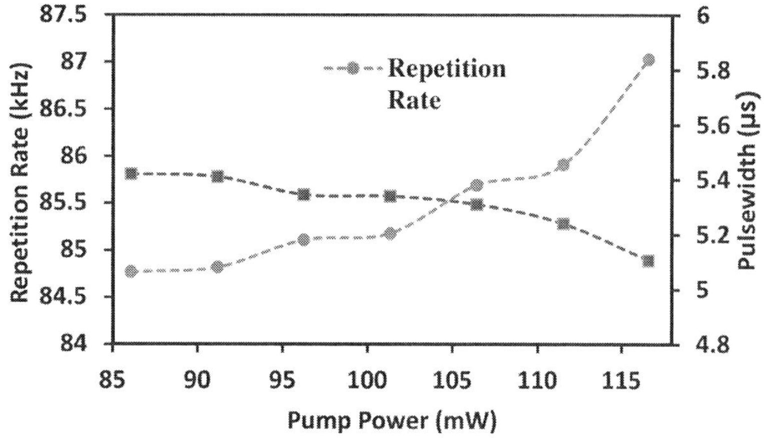

Figure 5. Repetition rate and pulse width of Q-switched EDFL within 86.1 mW to 116.6 mW

Figure 6 illustrates the relationship between the pulse energy and the peak power as the function of the pump power. The peak power is showing a steadily increasing pattern from 2.52 mW to 3.21 mW as the pump power is increased from 86.1 mW to 116.6 mW. The highest pulse energy obtained was 36.9 nJ at the pump power of 116.6 mW.

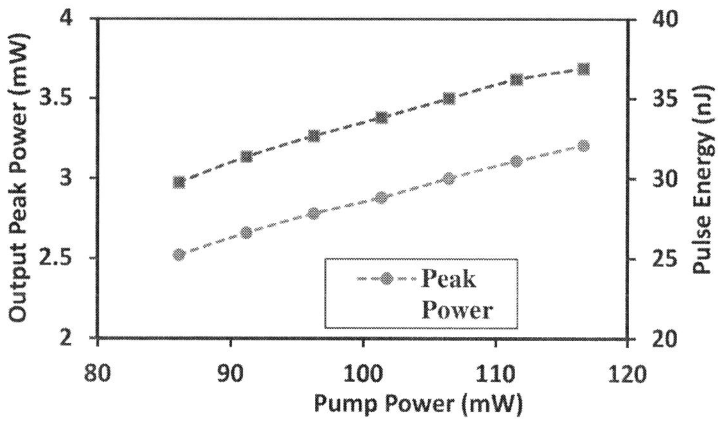

Figure 6. Repetition rate and output power as a function of 980 nm of pump power.

The radio frequency (RF) spectrum analyser at pump power of 116.6 mW was measured. Figure 7 shows the RF spectrum of the Q-switched laser output of the pumping sources. It shows a stable repetition rate of 87.03 kHz and peak to background ratio of about 60 dB. This indicates stable pulses are produced by our laser cavity.

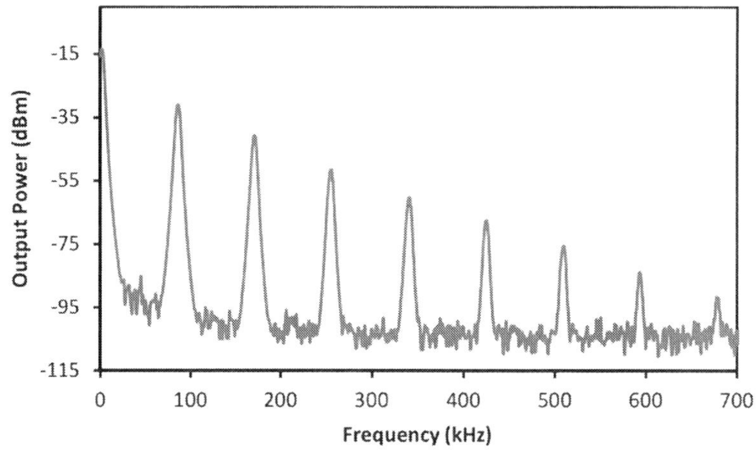

Figure 7. RF spectrum at pump power of 116.6 mW (SNR: 60 dB)

5. Conclusion

An EDFL is proposed and demonstrated using Bi_2Se_3 as a passive saturable absorber (SA). Bi_2Se_3 film is sandwiched between two FC/PC fiber connector to analyze the performance of erbium doped fibre laser based on Bi_2Se_3 as a saturable absorber. Without SA, the ring cavity operates in a continuous wave regime with output spectrum at 1561.0 nm. By inserting SA, the EDFL operates at 1559.8 nm. The lasers was self-started and generate stable pulse train by changing the pump power from 86.1 mW to 116.6 mW with repetition rate that can be tuned from 84.08 to 87.03 kHz. It operated at 1559.8 nm wavelength, which was slightly red-shifted from the CW laser without the SA due to insertion loss of the SA device. The spectral broadening was also observed due to self-phase modulation effect inside the laser cavity. The RF spectrum showed the signal noise to ratio of about 60 dB, which indicates Q-switched operates with an excellent stability. The output power increases from 2.52 to 3.21 mW as the pump power increased from 86.1 mW to 116.6 mW. The maximum pulse energy of 36.9 nJ was obtained at pump power of 116.6 mW. The lowest pulse width of 5.11 μs was also obtained at pump power of 116.6 mW.

Acknowledgements

This research is fully supported by FRGS grant, R.J130000.7826.4F928. The authors fully acknowledged Ministry of Education (MOE) Malaysia and Universiti Teknologi Malaysia for the approved fund which makes this important research viable and effective.

References

[1] S. N. Aziz and N. Arsad, ASE level enhancement of multiwavelength fiber laser at L-band assisted by un-pumped EDF Journal of Nonlinear Optical Physics and Materials. 23(2): 1–6.
[2] Snitzer, E. Proposed fiber cavities for optical lasers. J. Appl. Phys 32: 36-39
[3] Walter Koechner. Solid State Laser Engineering. 3(6): 35
[4] Keller, U., et al. Semiconductor saturable absorber mirrors (SESAM's) for femtosecond to nanosecond pulse generation in solid-state lasers. Selected Topics in Quantum Electronics, IEEE Journal, 2(3): 435-53.
[5] Harun, S. W., et al. A Q-switched erbium-doped fiber laser with a carbon
[6] nanotube based saturable absorber. Chinese Physics Letters, 29(11): 114202.
[7] S. Chen, Y. Chen, M. Wu, Y. Li, C. Zhao, S. Wen, Stable-Switched Erbium Doped Fiber Laser Based on Topological Insulator Covered Microfiber, IEEE Photonics Technology Letters, 26(10): 987-90.
[8] A. H. H. Al-Masoodi, F. Ahmad, M. H. M. Ahmed, H. Arof, S. W. Harun, Optical Engineering

56(5): 056103.

[9] H. Ahmad, M. R. K. Soltanian, L. Narimani, I. S. Amiri, A. Khodaei, and S. W. Harun, Tunable s-band q-switched fiber laser using Bi2Se3 as the saturable absorber, IEEE Photonics Journal. 7(3): 1-8.

[10] H. Haris, S. W. Harun, A. R. Muhammad, C. L. Anyi, S. J. Tan, F. Ahmad, R. M. Nor, N. R. Zulkepely, H.Arof, Passively Q-switched Erbium-doped and Ytterbium-doped fibre lasers with topological insulator bismuth selenide (Bi2Se3) as saturable absorber, Optics & Laser Technology 88: 121-27.

IOP Conf. Series: Materials Science and Engineering **854** (2020) 012038 doi:10.1088/1757-899X/854/1/012038

Nanosecond Pulse Erbium-Doped Fiber Laser based on Evanescent Field Interaction with Lutetium Oxide

A A A Jafry[1], R A M Yusoff[1], N Kasim[1,*], S N M Rifin[1], A R Muhammad[2], and S W Harun[2]

[1] Department of Physics, Faculty of Science, Universiti Teknologi Malaysia, 81310 Skudai, Johor, Malaysia
[2] Department of Electrical Engineering, University of Malaya, 50603 Kuala Lumpur, Malaysia

Email: k.nabilah@utm.my *

Abstract. The mechanism of evanescent field interaction is established between lutetium oxide (Lu_2O_3) and light on the surface of D-shaped optical fiber. The D-shaped optical fiber was prepared using rotating wheel technique with the improved two times polishing method. The side-polished fiber sample owns a remaining fiber diameter of 70 µm and polishing length of 1400 µm. Lutetium oxide was deposited onto the D-shaped optical fiber as a pulse initiator inside an erbium-doped fiber laser cavity. A Q-switched with operating wavelength of 1565 nm was generated. Stable pulses were recorded as the pump power raised from 146 to 162 mW with pulse width as short as 450 ns and repetition rate of 0.967 MHz. As the pump power varied, output power of 120 to 160 µW which corresponds to pulse energy of 124 to 166 pJ were obtained.

1. Introduction

Q-switched laser is one of the attractive techniques to compress and initiate the pulse in nanosecond to microsecond time domain. The avalanche applications of Q-switched laser including corrective eye surgery, tattoo removal technique and laser cutting technology had attracted enormous research effort in this area [1, 2]. Since the introduction of semiconductors saturable absorber mirrors (SESAMs) in 1992, the performance of Q-switched laser had been improved to several order of magnitude. SESAMs is able to generate pulses as short as 10 femtosecond due to its ability to produce regular pulse train in the laser cavity [3]. However, SESAMs possess narrow operating bandwidth which is in the range of 800 to 1600 nm, making it incompatible for longer near-infrared wavelength. The mechanism of pulses generation in SESAMs is also complex as it requires precise alignment of laser setup which will induce large signal loss inside the laser configuration. Therefore, an all-fiberized laser cavity is introduced as it manipulates the mechanism of saturable absorption inside the laser configuration. To date, few materials are utilized as a saturable absorber (SA) in the laser cavity.

Among materials, lutetium oxide (Lu_2O_3) exhibit nonlinear absorption of approximately 4%, non-saturable absorption of 13% and saturable intensity of 32.03 MW/cm^2 in 1.55 µm region. Therefore, lutetium oxide had been utilized as SA in erbium-doped fiber laser (EDFL) cavity for the generation of Q-switched and mode-locked [4, 5]. Lu_2O_3 which is used as a host material in laser, scintillator for X-ray imaging and production of ceramic possessed a wide absorption in near-infrared bandwidth [6-8]. Lu_2O_3 is proved to be an efficient additive to host materials (thulium, ytterbium and erbium) for rare-earth doped fiber laser as they are able to emit pulsed lasers in the 1- , 1.55- and 2-µm region [9-11]. In addition, Lu_2O_3 has a high thermal expansion and high melting temperature which is important to withstand high intensity laser illumination from laser pump.

In recent years, Q-switched laser is generated via the incorporation of a thin film inside an all-fiberized laser cavity. Researchers are able to initiate ultrafast pulse laser in the range of ps to sub-fs pulse duration in EDFL cavity [12]. However, thin film possesses a low damage threshold as most of them are produced using a polymer matrix as a host such as polyvinyl alcohol (PVA) and poly-methyl methacrylate (PMMA). PVA and PMMA has a melting temperature of approximately 200 °C and 160 °C, respectively, making them vulnerable to high intensity laser pulses. Here, we demonstrate a D-shaped optical fiber coated lutetium oxide as a SA in EDFL cavity. Our SA device has a long nonlinear interaction length and high optical damage threshold which makes it superior for the generation of Q-switched [13]. Nanosecond pulse erbium-doped fiber laser is induced via evanescent field interaction between Lu_2O_3 and light on the surface of the D-shaped optical fiber.

2. Preparation and characterization of D-shaped fiber-Lu_2O_3

The D-shaped fiber was prepared using polishing wheel technique, similar manner to Ahmad et al. [14]. An amplified spontaneous emission (ASE) source was launched onto the single-mode optical fiber (SMF-28) with output power meter (Thorlabs) connected to another end. Two fiber holders were used to clamp both ends of the single-mode optical fiber, ensure less vibration during polishing process. The waist region of the single-mode optical fiber was then placed onto the rotating wheel of the polisher. Approximately 2 mm long buffer was removed before polishing process. After 15 minutes of polishing process, a D-shaped fiber with 1400 μm long and 70 μm thick was produced.

The Lu_2O_3 solution was prepared by mixing isopropyl alcohol (IPA) with 99.99% pure Lu_2O_3 nanopowder purchased from Shanghai Xinglu Chemical Technology Co., Ltd. The process starts by stirring 50 mg of Lu_2O_3 nanopowder with 50 mL of IPA at room temperature for 24 hr. The mixture was then ultrasonicated for 6 hr at room temperature. A 3 μL of the prepared solution was drop onto D-shaped fiber for SA optimization. The linear absorption spectrum of the SA was shown in Figure 1 (a). The graph indicates ~1 dB absorption in 1565 nm wavelength, an optimum value for Q-switched generation. The modulation depth of prepared D-shaped fiber-Lu_2O_3 was measured using twin-balanced detector technique. Figure 1 (b) displayed a nonlinear absorption spectrum of SA device with 4% saturable absorption, 13% non-saturable absorption, and 32.03 MW/cm^2 saturable intensity.

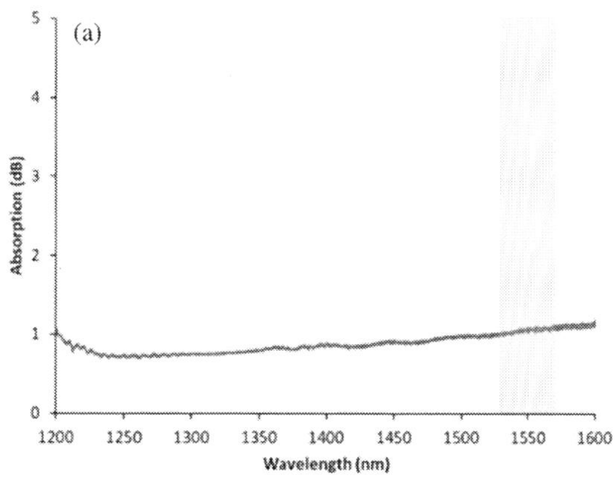

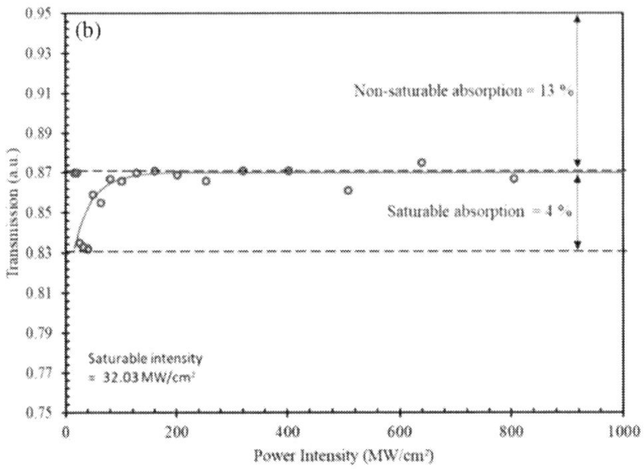

Figure 1. Optical properties of D-shaped fiber-Lu_2O_3; (a) linear absorption spectrum within the span of 1200 to 1600 nm, and (b) nonlinear absorption profile measured using twin-balanced detector technique.

3. Erbium Doped Fiber Laser Ring Cavity

The laser cavity was setup as drawn in Figure 2. A laser diode pump with 980 nm wavelength was launched onto the 980/1550 nm wavelength division multiplexer (WDM) as a wavelength-dependent beam splitter. The output port of the WDM was further connected to a 2.8 m erbium-doped fiber as a gain medium. The gain medium has an absorption coefficient of 23 dB/m, a core diameter of 4 μm, a cladding diameter of 125 μm and a numerical aperture (NA) of 0.16. Next, the light was converged to an optical isolator thus ensure unilateral light generated. The light was then allowed to propagate towards the D-shaped fiber-Lu_2O_3 as a SA device. The SA was connected to the 90/10 optical coupler with 90% of light cycled back to a 1550 nm port of WDM. A 10% of light was used for analysis purposes. An optical spectrum analyzer (Anritsu, MS9710C) with resolution of 0.05 nm and an optical power meter (Thorlabs) was used to characterize the output signal. A 350 MHz digital oscilloscope (Gwinstek, GDS-3352) and 9 kHz-7.8 GHz radio frequency spectrum analyser (Anritsu, MS2683A) connected via a 1.2-GHz InGaAs photodetector were used for temporal characteristics analysis.

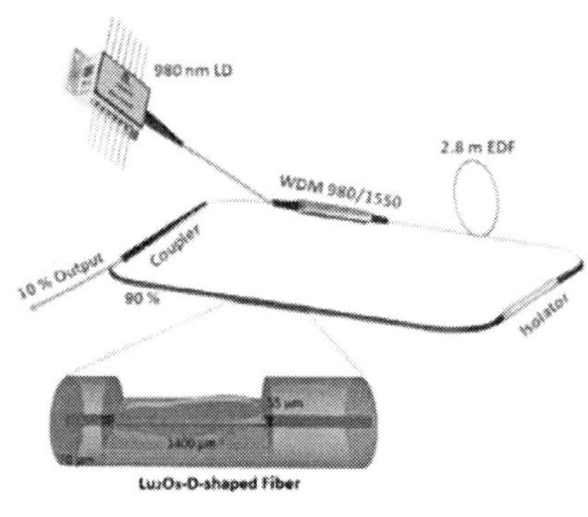

Figure 2. Experimental setup for nanosecond pulse generation in EDFL using Lu_2O_3-D-shaped fiber as saturable absorber.

4. Result and Discussion

A nanosecond pulse laser was generated at a threshold pump power of 146 mW. The optical spectrum analyzer measured a nanosecond pulse with a center wavelength of 1565 nm and a 3dB spectral bandwidth of 0.594 nm as depicted in Figure 3 (a). As the pump power raised from 146 to 162 mW, stable pulses observed. However, further increasing the input pump will causes the pulses to mitigate and eventually disappeared. Figure 3 (b) shows an oscilloscope trace of the nanosecond pulse generated at threshold pump power of 146 mW. The captured spectrum obtained a pulse width of 450 ns and a pulse period of 1034 ns. The signal was further analyzed using a radio frequency spectrum analyzer (RFSA) as shown in Figure 3 (c). The RFSA depicted a fundamental frequency with many harmonics within 6 MHz frequency span. The signal-to-noise ratio (SNR) of 46.24 dB was observed at a repetition rate of 0.967 MHz.

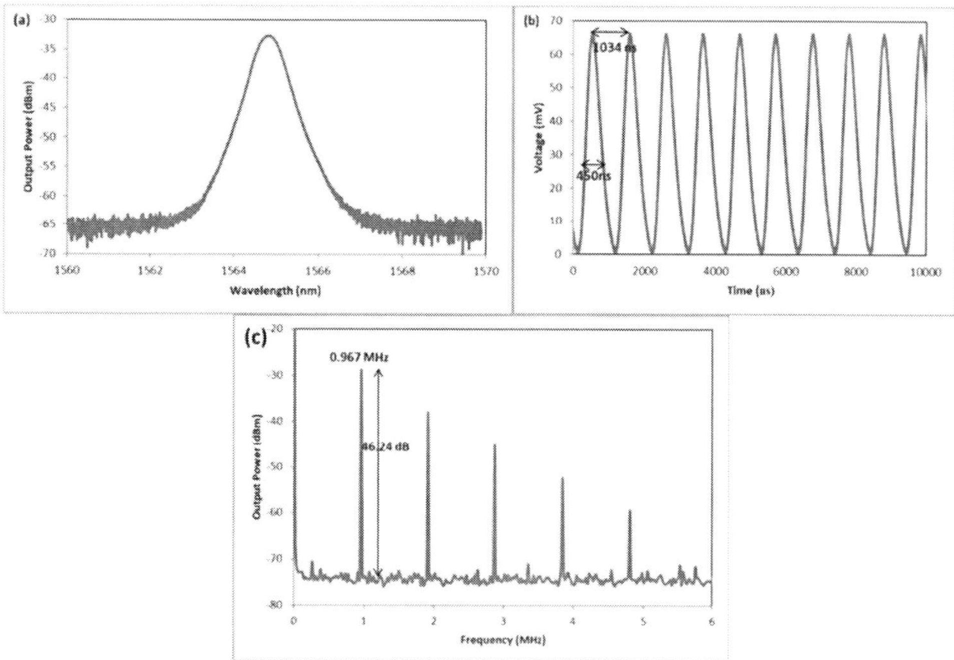

Figure 3. Spectral and temporal performances of the Q-switched laser; (a) output spectrum, (b) typical oscilloscope trace, (c) and RF spectrum at 146 mW pump power.

The signal was further investigated within the variation of multiple input pump power. The graph of repetition rate and pulse width against pump power was shown in Figure 4 (a). A nearly uniform pulse width and repetition rate was captured within 146 to 162 mW pump power. At a minimum attainable pump power, Q-switched generated obtained a pulse width of 450 ns and a repetition rate of 0.967 MHz. The graph of output power and pulse energy against pump power was also plotted in Figure 4 (b). A typical Q-switched behaviour was observed as the output power and pulse energy increased linearly with the increment of pump power. The Q-switched possesses an output power of 120 to 160 μW as the pump power raised within 146 to 162 mW. The laser generated exhibit pulse energy of 124 to 166 pJ within the same range of pump power.

(a) (b)

IOP Conf. Series: Materials Science and Engineering **854** (2020) 012038 doi:10.1088/1757-899X/854/1/012038

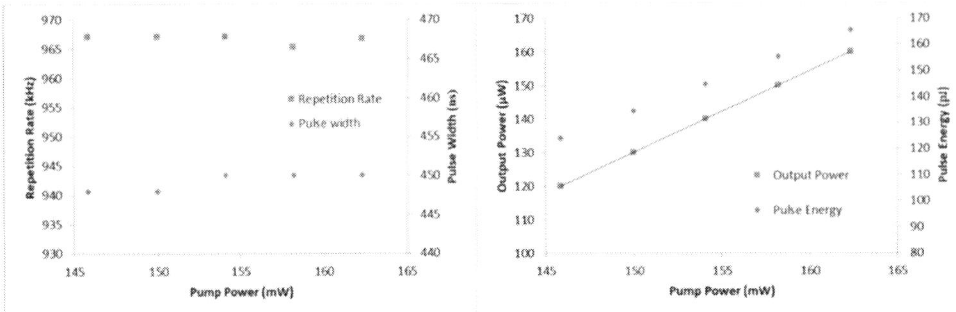

Figure 4. Q-switching performance; (a) repetition rate and pulse width against pump power (b) output power and pulse energy against pump power.

5. Conclusion

We present nanosecond pulse generation with Lu_2O_3 deposited onto D-shaped fiber as a pulse generator for EDFL cavity. The D-shaped optical fiber prepared using a rotating wheel technique owns a polishing length of 1400 nm and a diameter of 70 µm. The mechanism of the evanescent field between Lu_2O_3 and light was exploited to generate Q-switched in the EDFL cavity. With the appropriate amount of Lu_2O_3 deposited onto D-shaped optical fiber, stable Q-switched initiated within 146 to 162 mW. The generated pulses exhibit the highest repetition rate of 0.967 MHz corresponds to the shortest pulse width of 450 ns. SNR measured was 46.24 dB indicating the stability of the laser. The maximum output power and pulse energy generated was 160 µW and 166 pJ, respectively, at a maximum attainable pump power of 162 mW.

6. Acknowledgements

This research is fully supported by Fundamental Research Grant Scheme (FRGS), R.J130000.7826.4F928 and Research University Grant (RUG), Q.J130000.2654.16J01. The authors fully acknowledged Ministry of Education (MOE), Malaysia and Universiti Teknologi Malaysia for the approved fund which makes this important research viable and effective.

References

[1] Okamoto Y, Kitada R, Uno Y, Doi H. Cutting of Solid Type Molded Composite Materials by Q-switched Fiber Laser with High-Performance Nozzle. *J Adv Mech Des Syst Manuf.* **2(4)**:651-60.

[2] Shah S, Alster TS. Laser Treatment of Dark Skin An Updated Review. *Am J Clin Dermatol.* **11(6)**:389-97.

[3] Keller U, Weingarten KJ, Kartner FX, Kopf D, Braun B, Jung ID, Fluck R, Honninger C, Matuschek N, DerAu JA. Semiconductor saturable absorber mirrors (SESAM's) for femtosecond to nanosecond pulse generation in solid-state lasers. *IEEE J Sel Top Quantum Electron.* **2(3)**:435-53.

[4] Jafry AAA, Kasim N, Munajat Y, Yusoff RAM, Rusdi MFM, Mahyuddin MBH, Harun SW, Arof H, Apsari R. Passively Q-switched erbium-doped fiber laser utilizing lutetium oxide deposited onto D-shaped fiber as saturable absorber. *Optik.* **193**:162972.

[5] Baharom MF, Rahman MFA, Latiff AA, Wang P, Harun SW. Lutetium (III) oxide film as passive mode locker device for erbium-doped fibre laser cavity. *Opt Commun.* **446**:51-5.

[6] Martin T, Douissard PA, Seeley Z, Cherepy N, Payne S, Mathieu E, Schuladen J. New High Stopping Power Thin Scintillators Based on Lu2O3 and Lu3Ga5-xInxO12 for High Resolution X-ray Imaging. IEEE *Trans Nucl Sci.* **59(5)**:2269-74.

[7] Martinet C, Pillonnet A, Lancok J, Garapon C. Optical, structural and fluorescence properties of nanocrystalline cubic or monoclinic Eu : Lu2O3 films prepared by pulsed laser deposition. *J*

Lumines. **126(2)**:807-16.

[8] Vetrone F, Boyer JC, Capobianco JA, Speghini A, Bettinelli M. NIR to visible upconversion in nanocrystalline and bulk Lu2O3 : Er3+. *Journal of Physical Chemistry B*.**106(22)**:5622-8.

[9] Chen XT, Zhao SZ, Zhao J, Yang KJ, Li GQ, Li DC, Qiao WC, Li T, Zhang HJ, Feng TL, Xu XD, Zheng LH, Xu J, Wang YG, Wang YS. Sub-100 ns passively Q-switched Nd:LuAG laser with multi-walled carbon nanotube. *Opt Laser Technol.* **64**:7-10.

[10] Lagatsky AA, Antipov OL, Sibbett W. Broadly tunable femtosecond Tm:Lu2O3 ceramic laser operating around 2070 nm. *Optics Express.* **20(17)**:19349-54.

[11] Xu ZH, Lin JY, Sun YG, Ding F, Fan HT, Shi S, Fang QH, Bi YF, Gao Y. Facile Synthesis and Down-Conversion Emission of RE3+-Doped Lutetium Oxide Nanoparticles. *J Nanosci Nanotechnol.* **18(4)**:2850-5.

[12] Hasan T, Sun ZP, Wang FQ, Bonaccorso F, Tan PH, Rozhin AG, Ferrari AC. Nanotube-Polymer Composites for Ultrafast Photonics. *Adv Mater.* **21(38-39)**:3874-99.

[13] Park NH, Jeong H, Choi SY, Kim MH, Rotermund F, Yeom DI. Monolayer graphene saturable absorbers with strongly enhanced evanescent-field interaction for ultrafast fiber laser mode-locking. *Optics Express.* **23(15)**:19806-12.

[14] Ahmad MT, Rusdi MFM, Jafry AAA, Latiff AA, Zakaria R, Zainuddin NAM, Arof H, Ahmad H, Harun SW. Q-switched erbium-doped fiber laser using silver nanoparticles deposited onto side-polished D-shaped fiber by electron beam deposition method. *Opt Fiber Technol.* **53**:101997.

IOP Conf. Series: Materials Science and Engineering **854** (2020) 012039 doi:10.1088/1757-899X/854/1/012039

Pollutant Absorber Design to Minimize Air Pollution Emissions from Industry

Yayat Ruhiat*, Yudi Guntara

Dept. Of Physics Education, Faculty of Teacher Training and Education, Universitas Sultan Ageng Tirtayasa

Email: yruhiat@untirta.ac.id

Abstract. Pollutants emitted from industries vary depending on the type of fuel and the height of the chimney used. The higher the chimney used, the further the pollutant distance emitted. In this research, Pollutant Absorber was made with a condensation system using activated coconut shell charcoal. The research method uses Research & Development with the Design for Production (DFP) approach. In developing the model, pollutant emission analysis is carried out: before entering the chimney, during the chimney, and after exiting the chimney. Then, in designing tools to design absorbent devices, chimney designs with coconut shells are made as absorbent. In creating and making tools, it is done in two ways, namely simulation and testing. The results of the research are models of the distribution of air pollutants from industry, air pollutant absorbent designs emitted from industry, and prototype devices capable of absorbing SO_2 of pollutants emitted through the chimney.

1. Introduction

Industrial development in Indonesia continues to increase every year. Industry spread throughout Indonesia by using the capacity and type of fuel is different. The difference in size will have implications for the amount of pollutant concentration, while the difference in fuel will have impacts on the types of pollutants emitted [1]. Based on observations in industry Cilegon, capacity and type of fuel used was indeed different. The greater the size of the fuel used, the higher the chimney used. The type of fuel used include High-Speed Diesel (HSD), Marine Fuel Oil (MFO), Pyrolysis Fuel Oil (PFO), solar, residue, coal, and Fuel Gas (CNG). The various types of fuel used have implications for the types of pollutants emitted. Based on the results of ambient air measurements at 24 sample points in the Cilegon industrial area in the period July 2019, the types of pollutants emitted with these fuels, namely: nitrogen dioxide (NO_2), particles/ dust, sulfur dioxide (SO_2), carbon monoxide (CO) hydrocarbons (HC) and lead (Pb). The pollutant concentration varies between 0,03 -5600 $\mu g/m^3$, while the dust particles ranged from 102-686 $\mu g/m^3$. The type of pollutant with high concentration is CO and the smallest Pb, while the type of pollutant whose concentration shows an increase in SO_2. This increase occurred because, in the region, there are industries that use coal fuel with a capacity of 170-255 tons/hour. The large fuel capacity is assumed to have implications for the amount of pollutant concentration and the distribution distance emitted from the industry. SO_2 pollutants emitted from the Cilegon industrial area reach the Pulomerak region with concentrations ranging from 3,387-51,058 $\mu g/m^3$ [2].

The concentration of pollutants emitted, not only depends on the capacity of the fuel, will however depend also on the stability of the atmosphere, whereas the spreading distance depends on the height of chimney used. The higher the stack used, the further the pollutant distance emitted. In atmospheric stability A (very unstable), the concentration of pollutants emitted is very large, conversely, if the atmospheric stability is E (rather stable), the level of pollutants emitted is small [3]. The amount of

IOP Conf. Series: Materials Science and Engineering **854** (2020) 012039 doi:10.1088/1757-899X/854/1/012039

pollutant concentration and the distance emitted by the industry needs to be done to prevent the prevention of air quality. In terms of maintaining air quality, it is necessary to minimize pollution without disturbing the speed of industrial development. This can be done by making tools or modifying the chimney so that polluted emitters can minimized.

At this time, chimney modifications to reduce pollutants emitted had limited to industries that use coal fuel. Chimney modification had done by installing Electricity Catalytic Oxidation (ECO) SO_2 and electric coagulation. ECO-SO_2 functions to reduce sulfur (SO_2) waste and can reduce levels of NO_x, Hg (mercury), and particles that pollute water and fish and soil by up to 90% [4].The electrical coagulation function can coagulate smoke. before coming out of the chimney by using metal [5]. The metal edges in this tool will ionize the molecules in air. The smoke particles will be adsorbed into charged ions. The disadvantage of an electric coagulation device that requires a voltage of between 20,000-75,000 volts Minimizing the concentration of pollutants through chimney modification has inspired researchers to correct various weaknesses, including by making pollutant absorbent devices. Research related to the absorption of contaminants, including; gas treating with chemical solvents [6]. Pollutant detection by absorption using scattering noodles and topographic targets as retroreflectors [7], Effects on air pollutant removal by plant absorption and adsorption [8], Environmental Samples Analysis by atomic absorption spectrometry and inductively coupled plasma-optical emission spectroscopy [9], and Light absorption by pollution, dust, and biomass burning aerosols: a global model of study and evaluation with AERONET measurements [10]. Most of the previous studies focused on the reaction research, absorption rate, and CO_2 absorption efficiency system, which was carried out using absorber columns. Furthermore, anu researcher made a simulation of Air Pollutants Minimalization of Pollutant Absorber with Condensation. The simulation was carried out to determine the absorption of pollutants and absorbent endurance in absorbing air pollutants [11].

A device to absorb pollutants from the chimney continues to be strived to be able to minimize the concentration of polluted pollutants. In this research, a prototype of the Pollutant Absorber will be made with a condensation system, and the material used as an absorber is activated charcoal from a coconut shell. The charcoal is activated carbon, which is assumed to be able to absorb air pollutants before leaving the chimney. To keep the charcoal from burning, it is carried out by a condensation system.

2. Method

The research method uses Research & Development techniques with a Design For Production (DFP) approach. The DFP approach uses a multilevel system perspective from manufacturing activities to highlight the product development process. An important aspect of the DFP approach is the estimation of manufacturing system performance for products. Product design and manufacture of activated charcoal is carried out at the Sultan Ageng Tirtayasa University Laboratory of Engineering, while for testing, the equipment is carried out at UPB Suralaya. The research began by making activated charcoal derived from coconut shells. After the activated charcoal is finished, the charcoal is taken to the trial site, the UPB Suralaya chimney. The design of the prototype application of pollutant absorber in the UPB Suralaya chimney is presented in Figure 1.

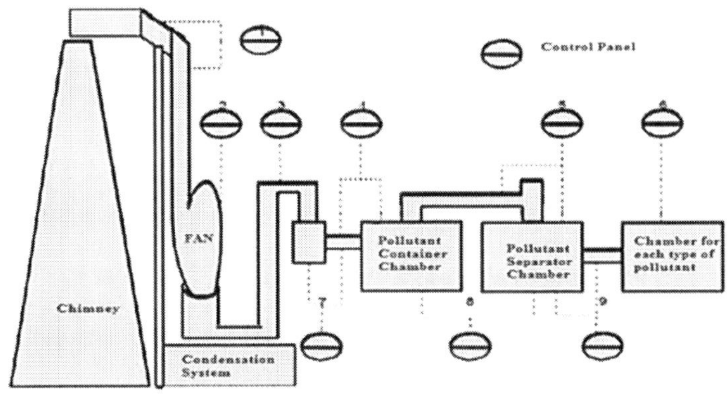

Figure 1. Pollutant absorber design

Based on the pictures 1, pollutants coming out of the chimney is absorbed by the filter and then accommodated in chamber 1. And then contaminants are separated in chamber 2, then every kind of pollutants are stored in the chamber 3. Activated charcoal is placed before the chamber 3. The flow of contaminants at any sub tools detected using a control instrument. To keep the flow of particles in each pipe made with a condensation system by maintaining the humidity tool. The working method of the device is to absorb all types of pollutants that come out of the chimney. Contaminants captured by the absorber filter are collected in chambers through capillary tubes. To maintain the change of separating and changing smoke particles into charged ions created by a condensation system by supporting energy through the enthalpy process. SO_2 concentration measurements were carried out for ten weeks.

3. Results and Discussion

In this research, the manufacture of pollutant absorber using local raw materials of coconut shell charcoal. This natural material is used as absorber because the cost is relatively inexpensive and easy to obtain in the Banten area. To maintain the resilience of the device to absorb pollutants is done by condensation. This is done to cool and dissolve pollutants. Coconut shell, as an absorber base, is first made into charcoal or called carbonization. The furnace used to make charcoal is a traditional ground furnace with a curing time of 3 days. Massa shell made charcoal is 100 kg. The charcoal produced is around 30 kg. Then the charcoal is broken into granular 2-3 cm diameter using a bat. The second step is to process dehydration using 2% NaOH salt. Charcoal that has been in granular form is put into a wall pond containing NaOH salt solution. Charcoal is soaked for 24 hours. After going through the process of immersion, then the charcoal is drained. Charcoal already drained then insert it into the activation space to be converted into activated charcoal. The principle in this third step is to open the pores of the charcoal to become more extensive. This activation is carried out for 48 hours. The dried charcoal is then put in a plastic bag to be transported to UPB Suralaya .

Most of the charcoal that is so, in the test beforehand whether the sample is in compliance with SNI 06 - 3730-1995 Deng's parameter is missing part in 950°C, levels of Water, levels of ash and absorption of I_2. The sample used is first converted into granules smooth ya ng sifted by size sieve 30 mesh. Moisture content and ash content are determined using the gravimetric method. Water content influences the absorption quality of activated carbon because high water content will reduce the absorption of activated carbon against gas or liquid. Determination activated carbon ash is done for me Knowing the content of the metal oxide d natural activated carbon. Ash in which obtained from the activated carbon results of research that is 5 % meet the standards SNI standard is a maximum of 10%. The adsorption capacity of activated carbon against I_2 correlates with the surface area of activated carbon. The pore surface area is an essential parameter in determining the quality of an activated carbon as an adsorbent. This is because the pore surface area is one factor that affects the adsorption capacity of an adsorbent. The activated carbon of research results has an absorption capacity of I_2 of 802 mg/g and meets the SNI set standards, which is a minimum of 750 mg/g.

Pollutant absorber of coconut shell charcoal with condensation system serves to absorb all kinds of pollutants emitted from industry, especially in this study is SO_2. By looking at picture 1, the product stored in chamber 1, then flows through the tube in chamber 2. In chamber 2, the absorbed pollutants are then separated with a separator through the enthalpies process by lowering the temperature and keeping energy from being lost. In this process, various types of particles derived from smoke will be obtained. After each particle is formed, the process is continued with adsorbs by the activated charcoal, which is carried out in chamber 3. In chamber 3, the smoke particles will be changed into charged ions. Because the production process takes place continuously, it will be obtained ions with large enough ions that can be used as a source of electrical energy.

The results of testing the activated charcoal on the prototype can be seen in Figure 2.

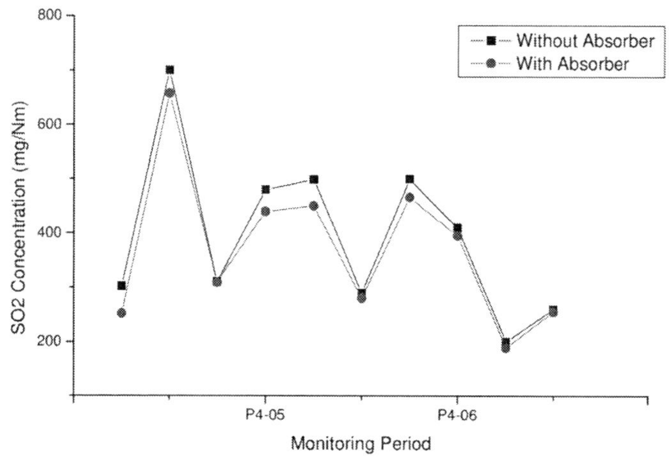

Figure 2. SO_2 concentration with and without absorber

The trial results show that activated charcoal can absorb SO_2. At the beginning of use, activated charcoal functions optimally in absorbing pollutants, especially SO_2, but after passing seven weeks, the charcoal starts to decrease its function. This is caused by a pretty drastic drop in temperature. Charcoal will be optimal at 900°C heating [12]. Because at that temperature the pores in the charcoal will be opened by the presence of high heat penetration and if there are pores that have opened it will cause an increase in pore diameter which causes complex compounds especially those that close the charcoal pore out so that the pollutant absorption will be higher. But it can be concluded that this prototype can be used as a step to reduce pollution emissions.

Pollutant absorber from coconut shell charcoal with condensation system is a control technique to reduce sea pollution. The difference between this tool and the tool previously seen in the position and function of the instrument. The feature of the previous device was only to absorb or separate coal-fueled pollutants, while this tool was able to absorb and separate pollutants from different fuels. From the function side of this tool can absorb SO_2 emitted. The absorbed products are then separated by a separator; after being separated, the pollutants are then absorbed so that they can be used as an energy source. The factors that influence adsorption include physical and chemical properties of adsorbent and adsorbate, liquid phase (pH, temperature), gas properties (temperature, pressure), adsorbate concentration, and contact time of adsorbent and adsorbate.

4. Conclusion

Coconut shell can be used and used as activated charcoal because it is relatively cheap and easy to obtain. Almost all parameters analyzed on the activated charcoal as a result of the research meet the requirements of the Indonesian National Standard now active. The use of activated charcoal as an ingredient absorbs SO2 can be applied to the industry, especially the use of coal fuel.

References

[1] D. Pope, V. Mishra, L. Thompson, A. R. Siddiqui, E. Rehfuess, and M. Weber, "Risk of Low Birth Weight and Stillbirth Associated With Indoor Air Pollution From Solid Fuel Use in Developing Countries," *Epidemiol. Rev.*, vol. 32, pp. 70–81, 2010.
[2] Y. Ruhiat, S. Sutisna, A. Hasani, M. Kuswantoro, S. Widiastuti, and R. Fahrizal, "A Formation Model in Distribution of Sulfur Dioxide (SO2) Concentrations in Cilegon City with the Use of Grid Technique," *Int. J. Interdiciplinary Environ. Stud.*, vol. 11, no. 3, 2016.
[3] Y. Ruhiat, "Model Prediski Distribusi Laju Penyebaran Pencemar SO_2 dan Debu dari Kawasan Industri Cilegon," Institut Pertanian Bogor, 2009.
[4] J. Wang, Y. Qiu, Y. Ma, S. He, and N. Liu, "Quantifying the geographical distribution effect on decreasing aggregated nitrogen oxides intensity in the Chinese electrical generation system," *J. Clean. Prod.*, vol. 222, no. x, pp. 856–864, 2019.
[5] M. T. Lim, A. Phan, D. Roddy, and A. Harvey, "Technologies for measurement and mitigation of particulate emissions from domestic combustion of biomass: A review," *Renew. Sustain. Energy Rev.*, vol. 49, pp. 574–584, 2015.
[6] G. Astarita, D. Savage, and A. Bisio, *Gas Treating with Chemical Solvents*. New York: Wiley, 1983.
[7] R. Byer and M. Garbuny, "Pollutant Detection by Absorption Using Mie Scattering and Topographic Targets as Retroreflectors," *Appl. Opt.*, vol. 12, no. 7, pp. 1–10, 1973.
[8] S. Fujii, H. Cha, N. Kagi, H. Miyamura, and Y. Kim, "Effects on Air Pollutant Removal by Plant Absorption and Adsorption," *Build. Environ.*, vol. 40, pp. 105–112, 2004.
[9] I. Popescu, C. Stihi, G. Cimpoca, G. Dima, G. Vlaicu, and A. Gheboianu, "Environmental Samples Analysis by Atomic Absorption Spectrometry (AAS) and Inductively Coupled Plasma-Optical Emission Spectroscopy (ICP-AES)," *Romammnia J. Physic*, vol. 54, pp. 1–6, 2009.
[10] M. Chin, T. Diehl, O. Dubovic, T. Eck, B. Holben, and A. Sinyuk, "Light Absorption by Pollution, Dust, and Biomass Burning Aerosols: A Global Model Study and Evaluation with Aeronet Measurements," *Ann. Geophys.*, 2009.
[11] Y. Ruhiat, F. C. Wibowo, and Y. Oktarisa, "Air Pollutants Minimalization of Pollutant Absorber with Condensation System," *J. Appl. Phy*, vol. 846, pp. 1–7, 2017.
[12] F. F. Polii, "Effects of Activation Temperature and Duration Time on the Quality of the Active Charcoal of Coconut Wood," *J. Ind. Has. Perkeb.*, vol. 12, no. 2, pp. 21–28, 2017.

IOP Conf. Series: Materials Science and Engineering **854** (2020) 012040 doi:10.1088/1757-899X/854/1/012040

Geographic Information System for Waste Management for the Development of Smart City Governance

N Tri Suswanto Saptadi, Phie Chyan, Andrew Christopher Pratama

Informatics Study Program - Universitas Atma Jaya Makassar, Indonesia

Email: ntsaptadi@yahoo.com, phiechyan@gmail.com, andrewchristhoper71@gmail.com

Abstract. The Era of the Industrial Revolution 4.0 has been able to change the paradigm of government towards the contribution of Information Technology in providing fast, precise and accurate services to the public for better information service needs. Trends in Smart City Governance will also increase public and government participation in utilizing application data so that they can provide suggestions and criticisms quickly, easily and accurately. Makassar has a population of 1,469,601 people and is the highest compared to other regions in South Sulawesi. Makassar has produced around 1,200 tons of waste per day that goes to Tamanagappa landfills. Waste generated by each district is different from other districts, so it is necessary to do integrated waste management. To overcome this problem, Makassar City Government applies the Smart City concept that focuses on Smart Environment. This concept aims to reduce and utilize waste and create better energy sources. The research objective is to develop a system that can provide information to the public so that the performance of the Makassar City Sanitation Office becomes effective and integrated. Geographic Information System (GIS) is a system designed to capture, store, manipulate, analyze, organize and display all types of geographic data. GIS components consist of data, software, hardware, users, and applications. Research using the Spiral method with tools or applications used by implementing ArcView, ArcGis, and Map Info. GIS tools used are based on Open Source such as Google Maps by providing various types of libraries that can be directly used. The results showed 90% of respondents agreed that this application could be used.

Keywords: governance, smart city, GIS

1. Background

The Industrial Revolution Era 4.0 has been able to change the paradigm of government towards the contribution of Information Technology to provide fast, appropriate and accurate services to the public. Nowadays people need better and more useful information services. Smart City governance can increase public and government participation in utilizing application data so that it will provide suggestions and criticisms quickly, easily and accurately.

Makassar has a population of 1,469,601 people. This number is relatively large compared to other cities in South Sulawesi such as Bone which only has a population of 746,973 inhabitants followed by Gowa 735,493 inhabitants.[1] With the large number of residents in Makassar, it is directly proportional to the amount of waste generated per day and has the potential for flooding. Total waste entering Tamanagappa landfill is around 1,200 tons per day.[2]

Of all the garbage that enters the Tamanagappa landfill each day, it must originate from a different sub-district as well as the volume of waste produced. The rubbish produced by each sub-district is different from one sub-district. The presence of waste is influenced by the population, geographical factors, community habits, and the collection system that is running. Waste management systems that are running must be able to work optimally and efficiently. To overcome this problem, Makassar City government began in 2014 started to implement the Smart City concept. Therefore in 2017 the Makassar City has been designated as the place to start the Movement Towards 100 Smart City programs in Indonesia by the Ministry of Communication and Information.

Smart City itself is defined as a city that is able to use HR, social capital, and modern telecommunications infrastructure (Information and Communication Technology) to achieve sustainable economic growth and high quality of life, with wise resource management through community participation-based governance.[3]

The concept of smart city can also increase public and government participation in utilizing application data, providing input and criticism easily.[4] In line with the development of smart cities in Indonesia delivered by the Director of Urban and Rural Ministry of PPN / Bappenas in the e-Indonesia Initiative (eII) and Smart Indonesia Initiatives (SII) Forum 1 in Bandung, October 15, 2015, urbanization is happening from village to city from year to year which shows quite significant figures. Based on data from the Central Statistics Agency (BPS) in 2014, the population in 2005: urban 48.39% and rural 51.61%, 2010: urban 54.19% and rural 45.81%, 2015: urban 59.35 % and rural areas 40.65%, and predicted in 2020: urban areas 63.84% and rural areas 36.16%.

One of the missions in the 2015-2045 sustainable city development goals is to build productive, creative and innovative, and efficient competitive economic and community activities in the city. Through the internet, information can be accessed and used for various purposes by anyone, anywhere and anytime. Various types of internet technology can be used. One of which is able to provide quality information in the form of text, images, sounds, and moving images. With this ability, the website becomes very famous and is growing very rapidly. Cities have a growing problem in spatial planning over time. Not infrequently the problems that arise due to the development of the city itself. To be able to prevent this, city management (governance) is needed through a sustainable planning concept approach, which is based on green technology.[5]

Part of the concept of Smart City is Smart Environment which aims to reduce and utilize waste and create better energy sources through the distribution and management of waste[6] so it is necessary to develop smart city governance based on Geographic Information Systems (GIS) of waste management. The lack of information about the area of waste distribution in the city of Makassar causes this waste to be disposed of in various places and in inappropriate times. To be able to maximize the performance of waste transportation in the city of Makassar, it is necessary to use GIS which is an information system with the ability to manage or process information that is bound or has a reference to time, space, or place.[7] GIS works by storing information about the earth as a collection of thematic map layers, all of which can be linked together. This method is expected to make data more flexible and dynamic so that it can be combined according to needs and usage. GIS automatically connects attribute data with maps, so there is a link between the two. The research is useful in providing waste information in every region or sub-district in Makassar so that it can improve the performance of the Makassar municipal government cleanliness department. The research is expected to be able to result in the development of GIS-based smart city governance in waste management so that it can know the area of waste distribution and can maximize waste collection in the city of Makassar.

2. Methods

2.1. Types of Research

Applied research is carried out with regard to practical realities, applications and the development of knowledge produced by basic research in real life. Applied research serves to find solutions about certain problems. The main objective of applied research is problem solving so that research results can be utilized for the benefit of humans either individually or in groups or for industrial or political purposes and not for scientific insight alone.[8] This research does not focus on developing an idea, theory, or ideas, but rather focuses on the application with a low level of abstraction so that the benefits or impacts can be felt directly by the government, especially the cleaning service and the city community that uses the application. To be able to improve performance and maximize better waste management, it is necessary to support information systems based on GIS in mapping and reporting recycled waste by type in each district and city of Makassar.

2.2. Design of Experiments and How It Works
The design method used in this study is spiral model, which is an evolutionary software model that compiles the iterative nature of the prototype by means of control and systematic aspects of the linear sequential model.[9] The spiral model has the potential to develop additional versions of the software quickly. Based on the spiral model, there are six phase that will be carried out in this study.

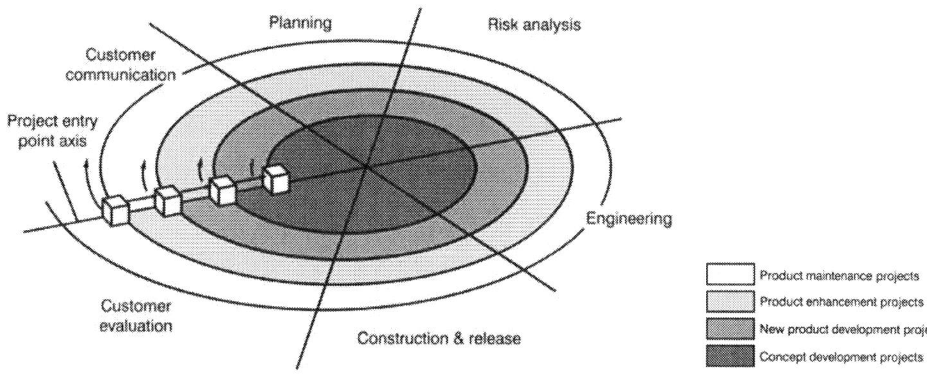

Figure 1. Spiral Model

a. Planning Phase
Planning phase to define resources, accuracy and performance in the form of information about the Makassar municipal waste management system.

b. Requirement and Analysis Phase
The requirement phase is carried out to determine the needs of the government and the community. Analysis will be carried out to trace and search for information on the Makassar City Sanitation Office and collect the necessary data such as the number of janitors in the field, the number of garbage transport vehicles both tricycle and trucks, Makassar city sub-district data and Makassar city daily waste volume data. The data was obtained from interviews from Makassar city Sanitation Office Staff and field surveys.

c. Design Phase
Design or depiction phase, is the phase of developing a system model that will be made based on GIS in an effort to manage waste in the Makassar city district in coordination with the Makassar city Sanitation Office. The design uses context diagrams, tiered, flowchart, ER-Diagrams, and system interfaces.

d. Construction Phase
Construction phase or the construction phase, to build a development of governance through information service distribution and waste processing. The application is created using the PHP programming language, and the MySQL database. Research using the spiral method with tools or applications used by implementing ArcView, ArcGis, and Map Info.

e. Testing Phase
Testing phase or the testing phase is carried out to test the system using the Black Box testing method and look for loopholes whether in the form of coding, design or database. As well as testing the system and is connected with the main system of the City Sanitation Department.

f. Deployment and Acceptance Phase
This phase will try the results of the development of GIS-based governance that is carried out at the construction phase, whether the mapping and waste reporting program that has been made in accordance with what is needed such as features and program user interfaces. This phase uses a web browser.

2.3. Data Collection Methods
The method used for data collection in developing smart city governance consists of:

a. Study of literature
Literature studies are used to collect data and information about the PHP programming language and MySQL database. Both in the form of written documents, electronic documents, scientific journals, video tutorials and websites related to PHP.

b. Interview
Interview is a process in obtaining information about the waste management process that runs in each district of Makassar city involving staff from the Sanitation Office as a resource in research.

c. Documentation
Documentation is a process of recording data obtained from the results of information ranging from writing, description, and important archives. In the documentation it is necessary to record information in the cleaning service needed by the data in the form of garbage transport vehicles, and data originating from districts in the city of Makassar.

2.4. Data Analysis
The data analysis phase uses qualitative techniques by conducting interview methods to find out whether the development of GIS-based smart city governance for waste mapping and management is made so that it can help manage waste in the city of Makassar.

3. Results and Discussion
Development of smart city governance is carried out through the stages of planning, requirements and analysis, design, construction, testing, as well as deployment and acceptance.

3.1. Planning
The Planning Phase uses GIS in Makassar through observation and interviews by looking at the work process from the sub-district to the Environmental Department. To find out the needs of two different agencies, an interview is needed. The results of this system planning are expected to provide information and images that can help the Smart City Makassar program and can provide direct information to the public regarding information about waste management in the city of Makassar.

IOP Conf. Series: Materials Science and Engineering **854** (2020) 012040 doi:10.1088/1757-899X/854/1/012040

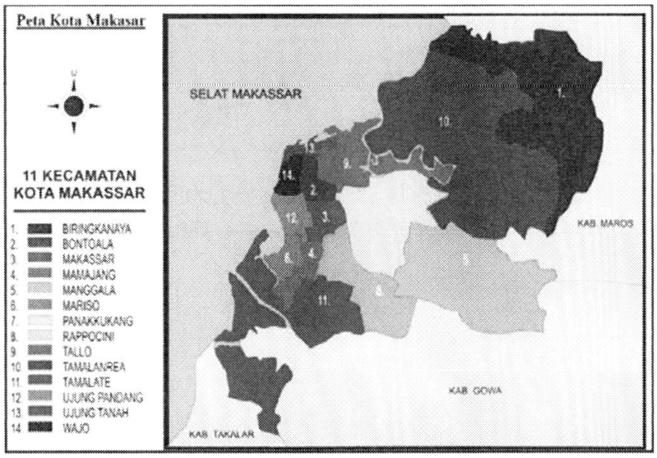

Figure 2. Makassar City Map

3.2. Requirement and Analysis
The research obtained data and analyzed some data that had been used to determine the needs in developing GIS-based governance. Where this system displays garbage data that has been previously entered by service officers into the system and then the data is processed for later display. The public can see directly the results of waste processing that takes place in the field in addition to being useful to the community this data can also be reused by the sanitation office to further optimize the performance of officers.

3.3. Design
Based on the results of the interview, the next stage is to design the development of the system created. The design starts by making a context diagram.

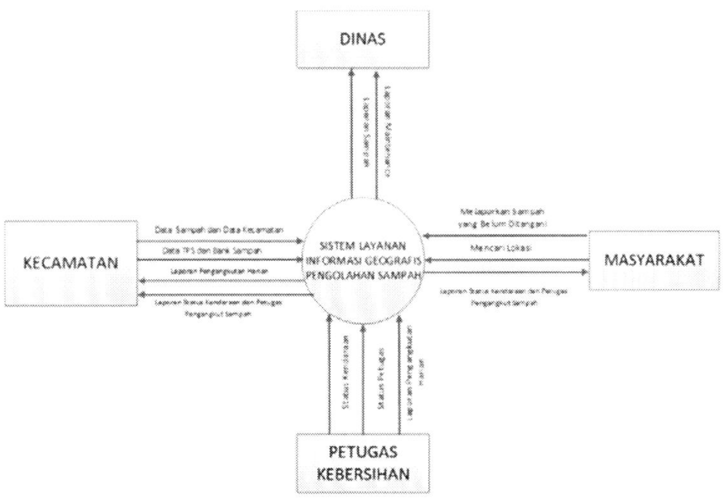

Figure 3. Context Diagram

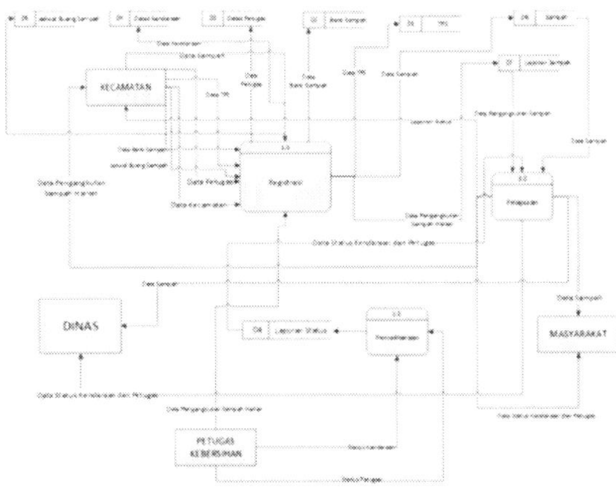

Figure 4. Level Diagram

Based on the context diagram, it can be seen that GIS for waste processing consists of 4 entities namely the Office of the Environment, Districts in Makassar city, Sanitation Officers, and Makassar City Communities. Admin comes from the district where inputting data such as subdistrict data and garbage data are stored in the database. Janitors input data such as the number of daily garbage transport reports, the status of the vehicle used to the status of workers in the field.

The Department of Environment and the people of Makassar City can see the results of reports from the two previous entities. The Office of the Environment can directly monitor the performance of officers from each sub-district to the status of the vehicle. The public can see waste information on the city of Makassar. The level diagram in Figure 4 shows the relationships that occur between entities, processes, and data storage needed.

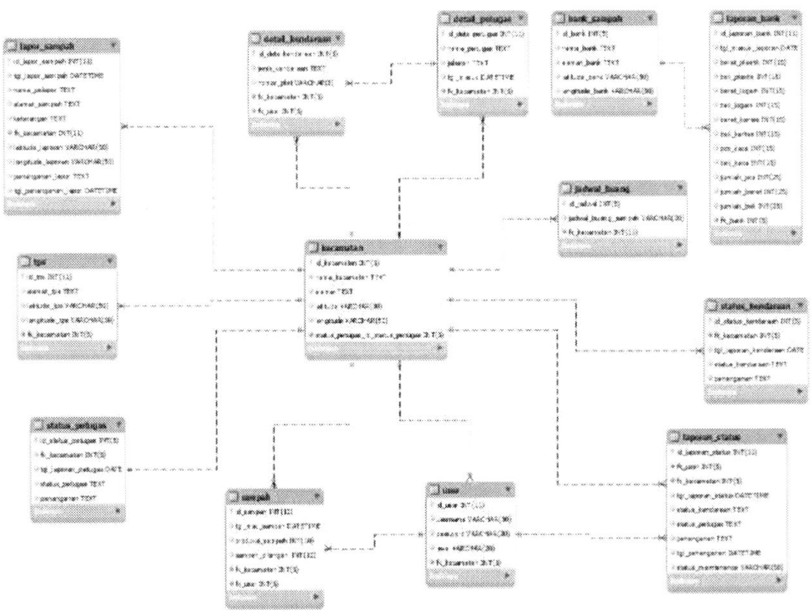

Figure 5. ER-Diagram

IOP Conf. Series: Materials Science and Engineering **854** (2020) 012040 doi:10.1088/1757-899X/854/1/012040

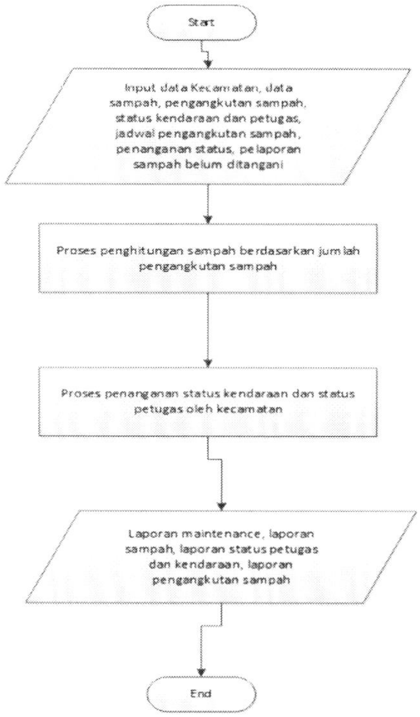

Figure 6. Flowchart

Figure 5 shows the ER-Diagram with 13 tables. The subdistrict table is a source of transactions for other tables and thus requires sufficient data support in order to provide accurate information. Figure 6 shows a system flowchart which explains that officials enter District data in the form of landfills, garbage banks, officer details, vehicle details. After that it continues the process of calculating the results of transportation of garbage that has been previously entered by the janitor to count. The process of handling the status by sub-district officials is in the form of vehicle status and personnel pre-filled by cleaning staff. Then the final filling is the process of making a system report in the form of maintenance, garbage, officer status and vehicle status, as well as waste transportation reports.

3.4. Construction

Based on the results of the interview, the next stage carried out by the author is to design the development of the system created.

IOP Conf. Series: Materials Science and Engineering **854** (2020) 012040 doi:10.1088/1757-899X/854/1/012040

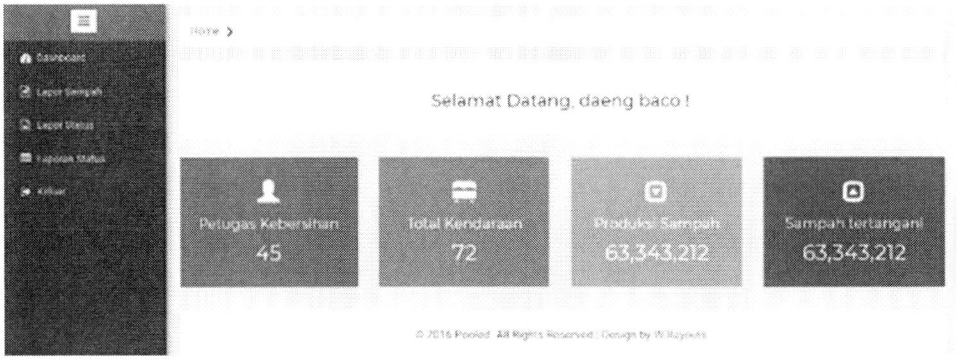

Figure 7. The page for the Janitor Admin

Figure 8. Officer Status Report

Figure 7 shows the appearance of the janitor admin page that contains the number of janitors, total vehicles, waste production, and garbage handled. Figure 8 shows the form for filling vehicle status, personnel, and maintenance status as a report.

IOP Conf. Series: Materials Science and Engineering **854** (2020) 012040 doi:10.1088/1757-899X/854/1/012040

```
{
 date_default_timezone_set("Asia/Makassar");
 $title = "Admin";
include "koneksi.php";
if (isset($_POST['btnkirim']))
{
$tanggal = date("Y-m-d");
$status_petugas = $_POST['status_petugas'];
$penanganan = $_POST['penanganan'];
$petugas = $_POST['petugas'];
$sql = "INSERT INTO status_petugas VALUES
('','$kecamatan','$petugas','$tanggal','$status_petugas','$penanganan')";
if(mysqli_query($koneksi, $sql)){
        header("Location: lapor_status_petugas.php");
    }else{
        echo mysqli_error($koneksi);
    }
}
```

Figure 9. Script Officer Status Report

Figure 10. Status Report List

Figure 9 shows the program script about the officer status report created with the PHP program. Figure 10 shows a list of vehicle and field officer status reports that are useful for knowing the date, time, and name of the officer.

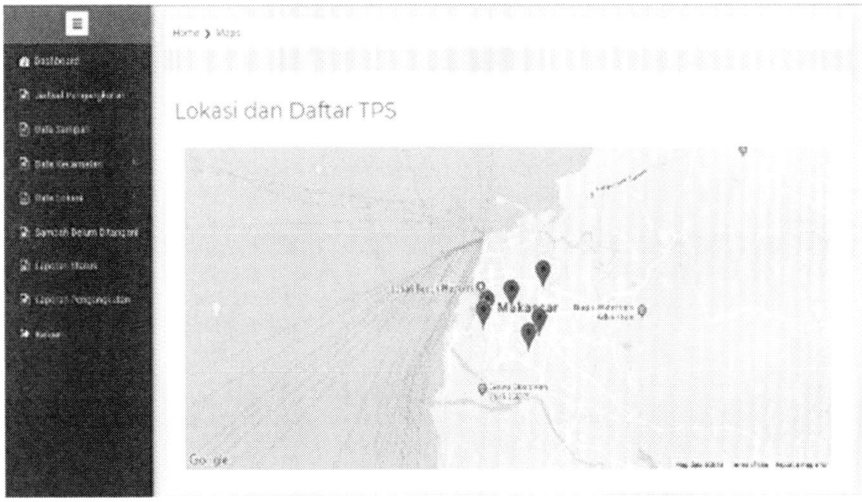

Figure 11. Location and List of Sub-District Landfills

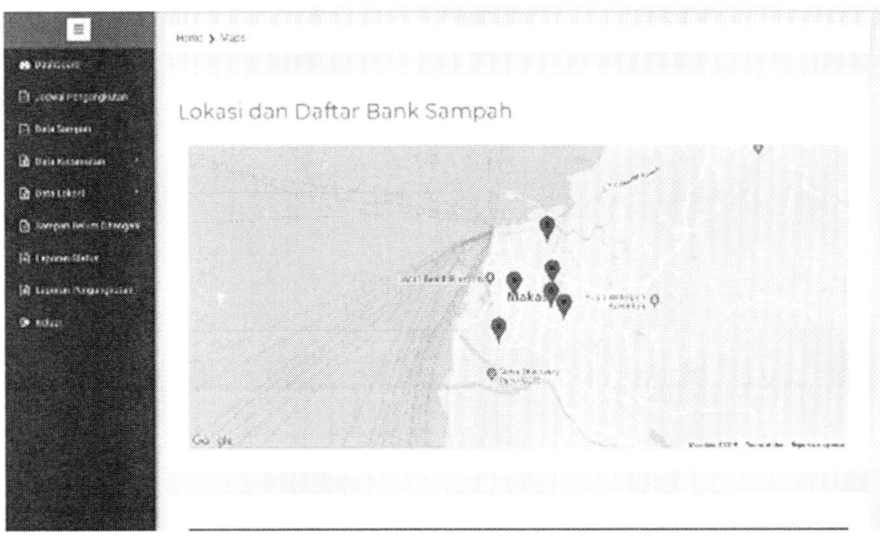

Figure 12. Location and District Garbage Bank

Figure 11 shows the location and list of the District Garbage Disposal Sites in the city of Makassar. The number of sub-districts is 14 and spread in various regions. Figure 12 shows the location and garbage bank located in each district of Makassar city. The location of the waste bank is needed so that the community knows the exact location of the location to facilitate management.

Figure 13. District Details

Figure 13 shows the community who wanted to obtain complete information on names, addresses, transportation schedules for the sub-district, waste production, transport vehicles, and the number of officers in the sub-district.

3.5. Testing
Testing is done by the Black Box method to find out the truth of the GIS by finding fault loopholes whether in the form of coding, design or database. The test focuses on the functional requirements of the system with the aim of trying to find errors of function errors, interface errors, data structure errors, or other performance errors.

Testing input data includes: admin login, district admin system, environmental office admin system, community system. The results showed 90% of respondents agreed that this application could be used.

3.6. Development and Acceptance
This stage will try the results of developing a GIS which will be carried out during the construction phase. Smart city governance development program for mapping and reporting waste that has been made in accordance with what is needed such as features and user interface programs. This stage uses a web browser. A system demonstration was conducted in front of several respondents to get a response back from the respondents. The results of system demonstrations and simulations of respondents indicate that:
1) The system has produced a service system that helps officers and the community.
2) System users are more enthusiastic about the report features, such as reporting waste that has not been handled up to vehicle and officer status reporting.
3) Existing features have been used by respondents and in accordance with the needs of respondents.

4. Conclusion
The development of GIS-based smart city governance can improve the performance of GIS-based services through attention and cooperation in the use of information technology. GIS has been able to assist the community in finding information about the process of managing and distributing waste in the city of Makassar to make people more aware of the ongoing waste management. Through this system the Makassar City Government can monitor the performance of waste management that occurs in each

district of Makassar city so as to maximize the performance of officers and support smart city programs in Makassar based on the respondents' assessment of 90%.

Acknowledgment
Thank you to The Ministry of Research, Technology and Higher Education of Republic of Indonesia for supporting the National Strategic Research (NSR) competition and Makassar City Government in completing the development of smart city governance. The author would like to acknowledge University of Malaya through the research and community service unit as the party that supports financially for this paper to be submitted to i-cosine 2019 in Mallaca, Malaysia.

References
[1] Central Statistics Agency of South Sulawesi. (2016). Projection of Population of South Sulawesi Province in 2001-2020. Makassar.
[2] Head of Service. (2017). Department of the Environment. Makassar.
[3] Caragliu, A; Del Bo, C. and Nijkamp, P. (2011). Smart Cities in Europe, Journal of Urban Technology.
[4] Awaluddin, M. (2016). Smart City Solution catalog. Telkom Solution.
[5] Hidayatulloh, Syarif. (2016). Internet of Things Bandung Smart City. Journal entry. Bandung. BSI University.
[6] Alawiah, Enok Tuti. (2017). Design of Mobile-Based Smart City Application to Improve the Quality of Public Services Case Study of Bogor City Government. Journal entry. Bogor. BSI University.
[7] Fair, Ahmad. (2017). Geographic Information System. Yogyakarta: Andi Publisher.
[8] Sukardi. (2003). Educational Research Methodology (Competencies and Practices). Jakarta: Earth Literacy.
[9] Munawar. (2005). Visual Modeling with UML. Yogyakarta: Graha Science.

Pore Pressure Prediction using Anfis Method on Well and Seismic Data Field "Ayah"

D Andrian, M S Rosid* and M R Septyandy

Geophysics, FMIPA Universitas Indonesia, Depok 16424, Indonesia

*Email: syamsu.rosid@ui.ac.id

Abstract. Nowadays oil and gas industries are focusing in production and development where well drilling is common and often done. One among many aspects need to be considered for drilling safety is pore pressure prediction. There are so many methods used in prediction the pressure including machine learning ANN but no one has ever done this with ANFIS which is combination of ANN and FIS machine learning. This study wants to use ANFIS for making a pore pressure distribution in 2D seismic data with 70% accuracy. Both pre-stack and post-stack seismic data are used here with well and RFT measurement. The modified Eaton-Azadpour and Eaton are two methods that applied to predict pore pressure. The methods are considered to be good in prediction pore pressure as they accommodate the rocks condition during drilling. The well parameters models are then distributed with ANFIS to find its correlation with P-impedance, S-impedance and density so we may find its distribution in 2D seismic data. The result shows that pore pressures are distributed very well but still need another study to give information regarding drilling safety.

Keywords: ANFIS; Azadpour; Eaton; Pore Pressure

1. Introduction

Indonesia is one of the big oil producer in the world and nowadays Indonesia is more focus on production and development steps where drilling will be more intensive. In drilling, the pore pressure prediction is usually done in advance to prevent any borehole breakout, kick, mud loss and any drilling hazard due to high pressure information. Pore pressure prediction usually done by many methods like Eaton [1] and Bowers [2] to predict pore pressure by using well and prestack velocity of seismic as its data but those methods and other methods also only use well data for predicting pore pressure but there is no method to combine both well and seismic data to predict pore pressure.

Nowadays, since the industry 4.0 era, machine learning is applied here to combine both seismic and well data to predict pore pressure in 2D seismic data. Recently, it is done by using PNN or Probabilistic Neural Network [3,4] and ANN or Artificial Neural Network [3] but there is almost no publication that apply ANFIS machine learning for predicting pore pressure. ANFIS has a greater accuracy than PNN and ANN because ANFIS is a combination of two machine learnings and they are ANN and FIS so it is hoped that ANFIS can give a better accuracy in pore pressure prediction [6]. In this study we want to discuss the accuracy of ANFIS, generating mechanism of overpressure in our study field are(ayah field) and the pore pressure distribution in 2D seismic section in study area as the result of ANFIS pore pressure distribution.

2. Geological Data and Methodology

The geology of study area can be explained by stratigraphy in the area as shown in Figure 1. There is Sembakung Formation that consist of sandstone and shale or claystone as this formation is made up from the subsidence (due to rifting process) during the basin formation and that is why it seem shale as the bottom of formation and shale again as the top of formation [4]. The Sujau Formation is made up of the cyclic sedimentation as the sea-level is very low so that the sand of Sujau Formation is deposited. As the sea-level rising and no sediment supply from land the Seilor Formation is formed where this formation is consisted of open marine shale and carbonate formation. No information regarding of what is the depositional environment of Sujau Formation but it could be suggested that this is shallow marine most probably continental shelf or lagoon. Next is Tempilan Formation that consist of shale and sandstone and the formation is formed due to sea-level fall so the coarsening upward is happened. Again no information regarding of how does there is an unconformity between Seilor and Tempilan but there might be an uplifted process combined with sea-level fall. Sea level again slowly rising and with a very calm, passive margin area where the Tabalar carbonate build-up is formed and during the formation of this build-up carbonate it could be possible the lagoon depositional environment. There is shale of Mesalai Formation and Naintupo Formation that is formed behind the carbonate platform that might be deposited from any sea life-form in that lagoon. The cessation of Tabalar carbonate build-up is shale on the Tabalar Formation which is formed due to slowly rising of the sea-level. Meliat Formation consist of sandstone and shale with an unconformity where at this time there is an uplifted happened from early of middle-miocene until late pliocene and pleistocene and all of the formation formed from this range of age is deposited as deltaic deposits due to uplift during the range of ages.

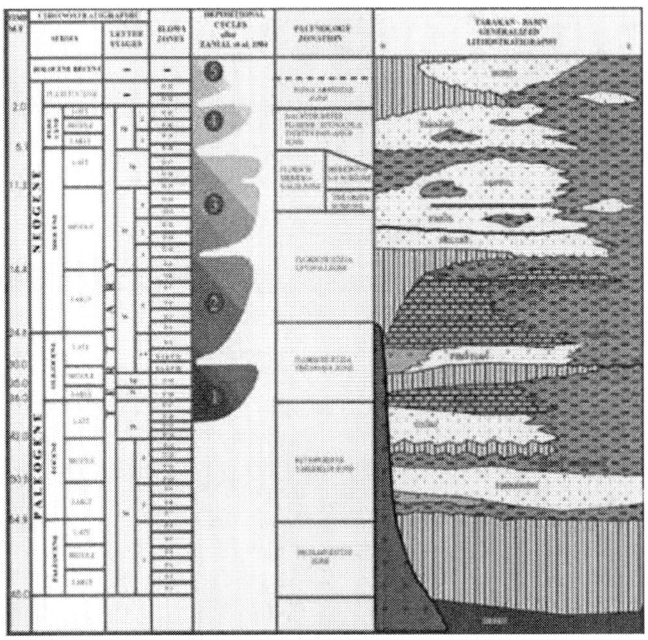

Figure 1: Stratigraphy of study area [7].

There are so many prediction methods in hands where each method has own limitation of its usage. Methods which is commonly used in industry are Bowers [1] and Eaton [2] methods (to be more precise it is called as modified Eaton method). Both methods are common due to its ability in distribute the value of pore pressure in 2D seismic section by using the prestack velocity data that is fitted with well velocity. The methods can be effectively used in pore pressure prediction of well section. However, the methods cannot combine two data of well and seismic data to predict pore pressure so machine learning is implemented here to combine two data to get accurate pore pressure prediction. First of all, the well tie seismic in order to do the seismic interpretation. Seismic picking is then done to pick the top of formation. Physically the overpressure could be occurred at shale formation because the water cannot be easily dewatered due to its low permeability and this might occur at shale formation of Santul, Meliat, Mesalah or Naintupo. According to well report, the overpressure occurs at Santul Formation so the picking is focused to the Santul top and Meliat top at seismic section.

Seismic interpretation clearly shows a deltaic progradation which the thickening horizon seaward and it is usually happened in deltaic formation. The formation is prone to disequilibrium compaction mechanism or clay diagenesis mechanism of overpressure so that why the Santul and Meliat is picked as top formations. The line chosen for pore pressure prediction is line A because the line has a well data as the validator of our ANFIS pore pressure model. The seismic picking is then verified by seismic QC and pre-inversion analysis. After the seismic simultaneous inversion is done, the ANFIS is then processed in Matlab. Before doing the ANFIS processing, the data pre-processing like combining training and validation data as one data and randomly choose the data for training is done. The result of testing modelled is then distributed the pore pressure using seismic data inversion. Modelling of ANFIS is done by taking the trace of inverted seismic data (Zp, Zs, and density) as a training data with pore pressure derived from well as output data. The data training is done by using Gaussian membership function and the result of training is then proven to be matched the field data. The model is then tested to real data and the result is also matched so that the model then may use to distribute pore pressure through the seismic data.

Before doing the simultaneous inversion, the pre-analysis must be done in order to see whether the inversion derived from software is matched with well data.This is necessary because if it does not match then the result of inversion is questionable. Software will compare the Zp, Zs and density generated from building model with inversion calculated from seismic directly. This inversion calculation is done by deconvolution of seismic seismogram to separate wavelet from acoustic impedance. The Zs then calculated via equation that relates density and Zp. This calculation should be matched with the actually measured (well log data).

3. Results and Discussion

From well data section of Figure 2 it can be seen the mechanism of overpressure tends to unloading mechanism. The sonic data is in reversal condition and the cause of this unloading mechanism is reducing clay content. The reducing of clay looks likely due to some processes like diagenesis, erosional, or hydrocarbon generation as the RHOB and Sonic crossplot data are confirmed that clay diagenesis occurs at this overpressure zone from depth 1360-1500 meter. Below this zone the lithology is dominated by carbonate rock and sandstone making it impossible to predict the pore pressure because the pore pressure prediction can only be done in shale or mud rock.

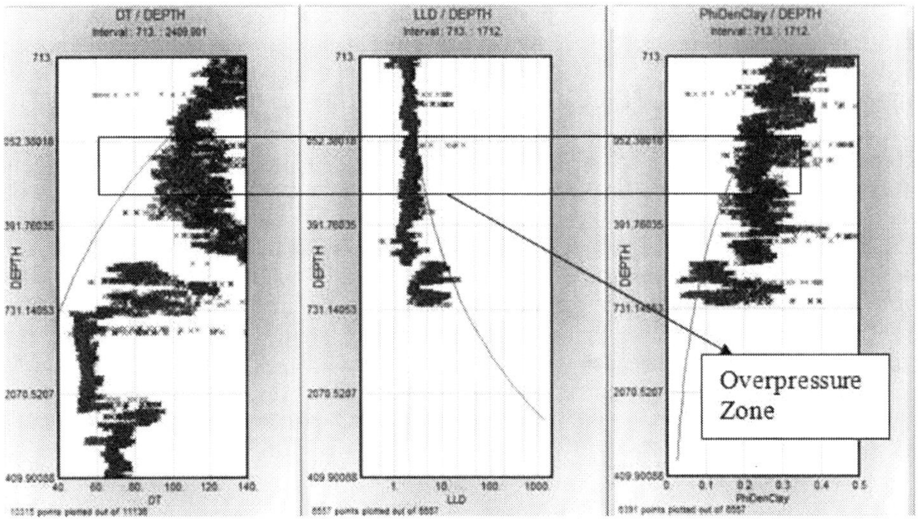

Figure 2: Well logs section at well A.

Figure 3 clearly shows that the diagenesis of clay may occur due to many points shows a higher content of illite near overpressure zone. This process might generate a certain amount of water and due to reaction of diagenesis, the smectite content in the shale might be reduced and will decrease the value of effective stress. However, according to the Figure 3 the process of clay diagenesis is still on progress because the data is slowly rising from smectite zone (under the red line in Figure 3) to illite zone (upper the red line in Figure 3) but comparing with Figure 3 at depth around 1400-1500 meter encounter very slow sonic velocity accompanied with low density but slowly decreasing resistivity. We suggest that the slow sonic velocity and low density is not only caused by water produced by clay diagenesis but also a gas is also reacting with the water produced by this diagenesis and the gas may come from the carbonaceous shale which might contain lignite. The reaction between gas and water combined with decreasing effective stress might create overpressure.

IOP Conf. Series: Materials Science and Engineering **854** (2020) 012041 doi:10.1088/1757-899X/854/1/012041

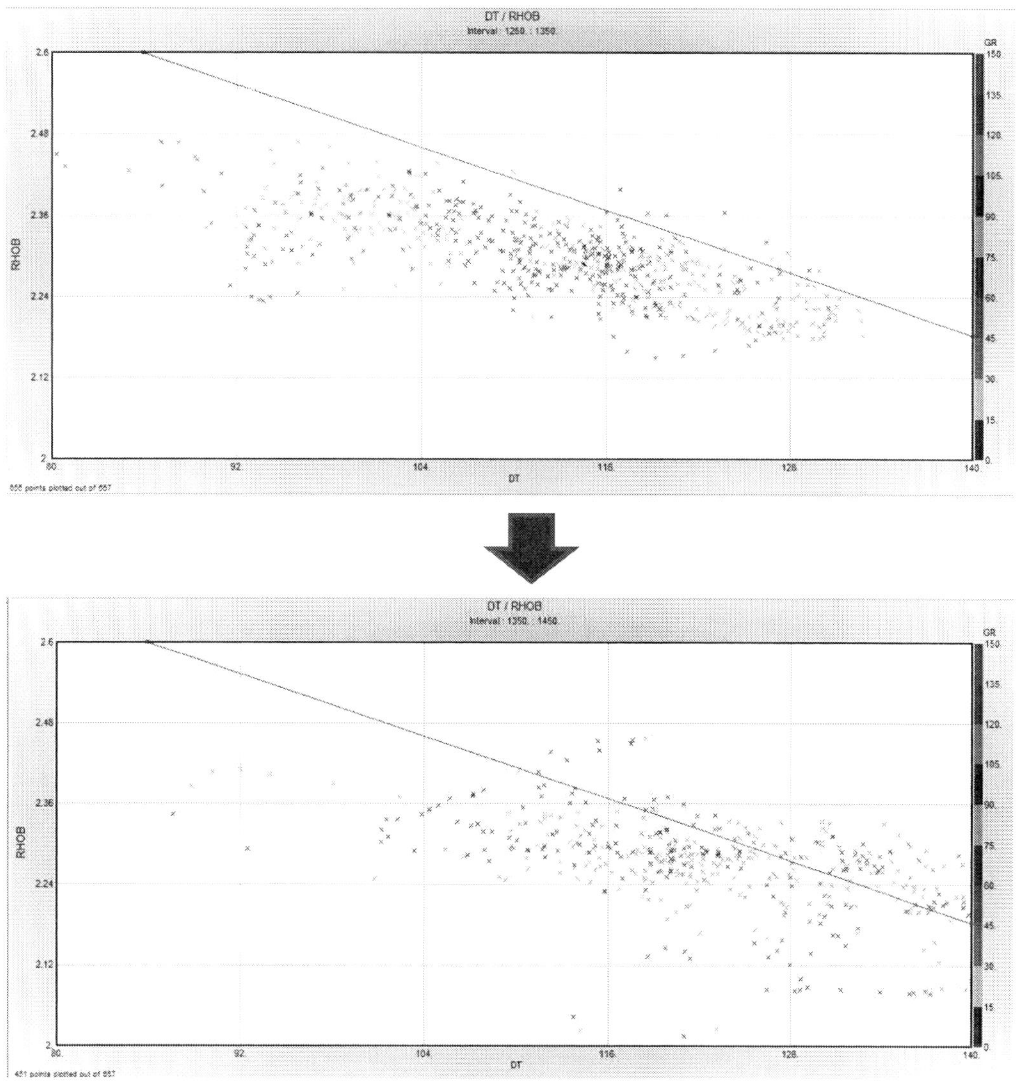

Figure 3: Clay diagenesis from well logs

In this study, Eaton pore pressure prediction method is modified to predict pore pressure in deltaic formation bounded by red squares and Azadpour is applied for carbonate rock bounded by blue square (see Figure 4). The Azadpour is considered a great method for predicting pore pressure in carbonate formation [5]. From the result in Figure 4 it can be seen the pore pressure prediction is strongly associated with RFT reading so the pore pressure prediction is, more or less, accurate. Overpressure happened at depth around 1360-1500 meter shown by pore pressure graph that is higher than hydrostatic pressure and caused by clay diagenesis.

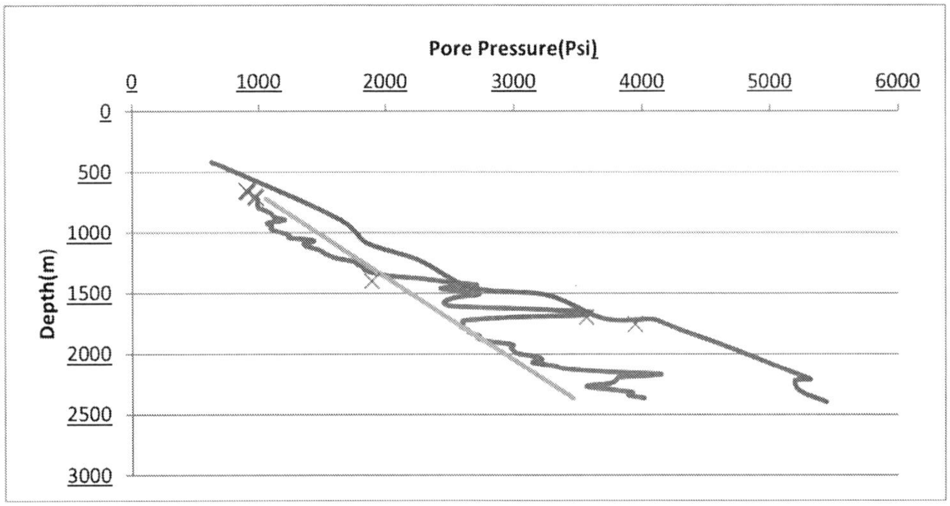

Figure 4: Modified Eaton prediction at well A.

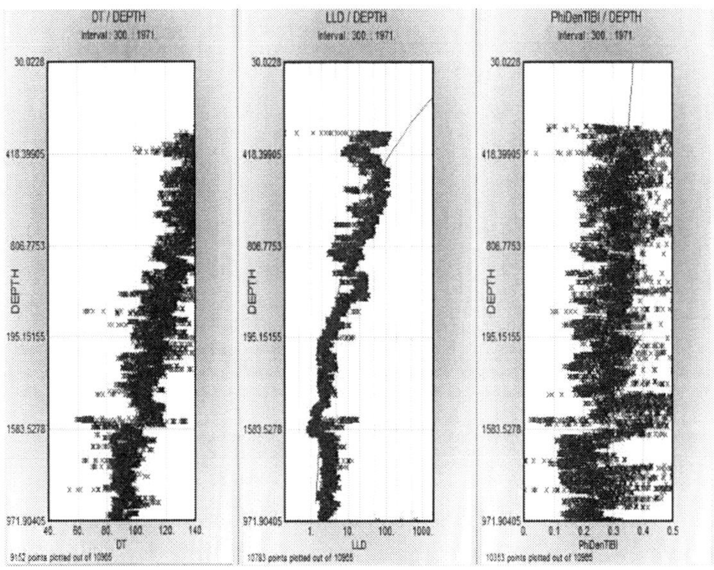

Figure 5: Well data section at well B.

In well B of Figure 5 it can be seen that the overpressure happens at depth near to 2000 meter (around 1900 meter). It is also confirmed by well report that suggest the overpressure occur at around 1900 meter. This is too thin to be predicted by seismic data so it may not see this overpressure phenomenon at seismic section. In this section the overpressure zone is indicated from constant sonic, density and resistivity value with increasing depth and the cause of this overpressure is might be fast sedimentation in deltaic environment because according to stratigraphy in regional geology the overpressure formation is deltaic and fast sedimentation in deltaic environment may inhibit the dewatering process. The overpressure zone might be thick but due to decision to stop drilling process the overpressure zone is thin.

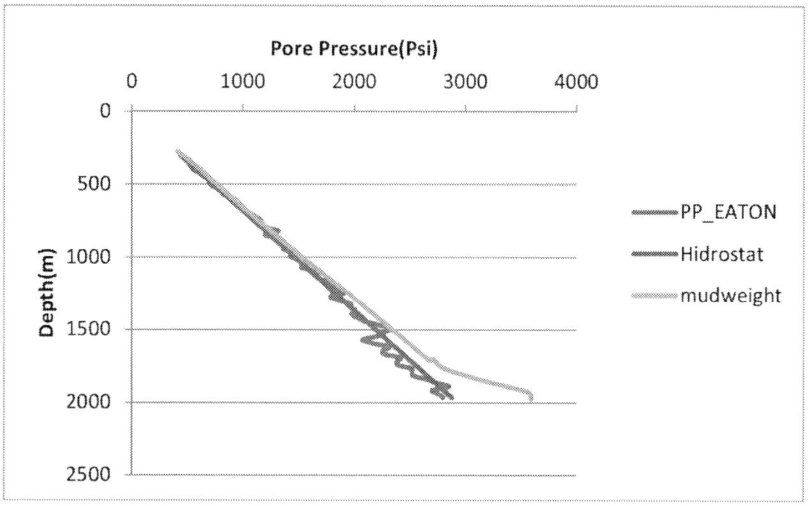

Figure 6: Pore pressure prediction using modified Eaton at well B.

Figure 6 of density and *Zp* picture shows the pore pressure is low around overpressure zone highlighted in the picture. The *Zp* value of overpressure zone is around 5000-8000 kg.s/m^2 and density around 2.1-2.2 g/cm^3 but it cannot be seen clearly because the overpressure zone is thin.

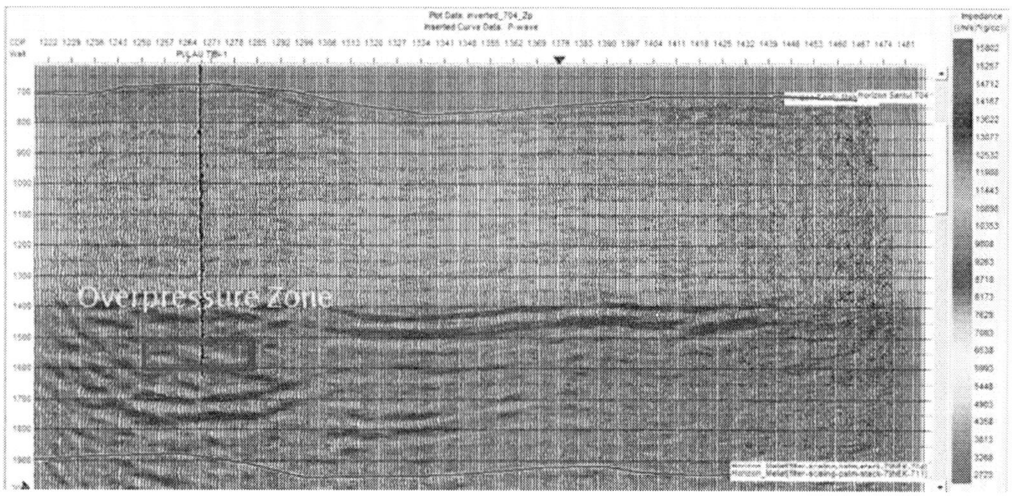

Figure 7: *Zp* of Line 704 shown a little value around 5000-8000 kg.s/m^3.

IOP Conf. Series: Materials Science and Engineering **854** (2020) 012041 doi:10.1088/1757-899X/854/1/012041

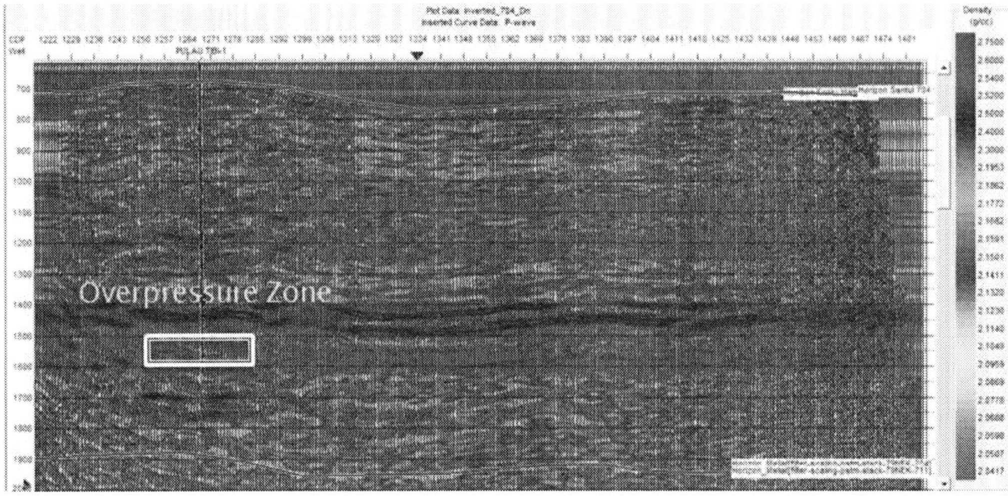

Figure 8: Density section of Line 704 shown low density around 2.11-2.16 g/cm^3.

Figure 7 and Figure 8 show the density and *Zp* values is low at depth below overpressure zone and it might be the reason why the overpressure zone might be thick. The low density and *Zp* suggest that rocks are not compacted normally and this might be caused by highly pressured water that hold the rock pore opened. The high water saturation might cause the density is low because water might dominate rock pores, thus, dominating the rock density. High water saturation might also cause the *Vp* decreasing due to fluid domination in the rock and with decreasing *Vp* and density the *Zp* will also decrease. The zone of overpressure below overpressure zone indicated by well is shown by green color in Figure 8 below the well zone and yellow zone in Figure 7.

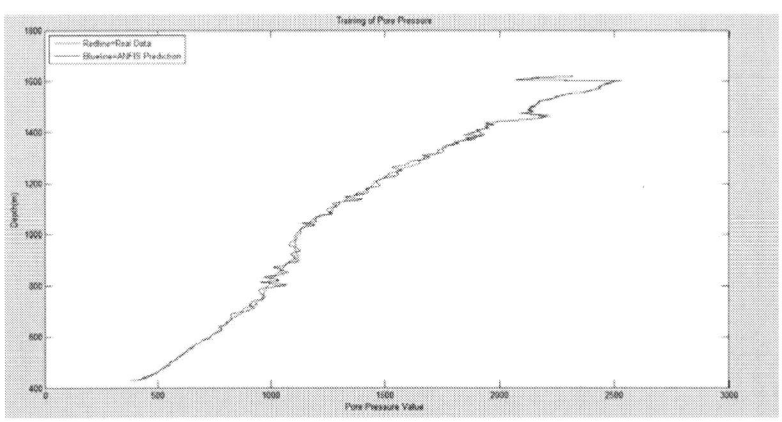

Figure 9: ANFIS training result.

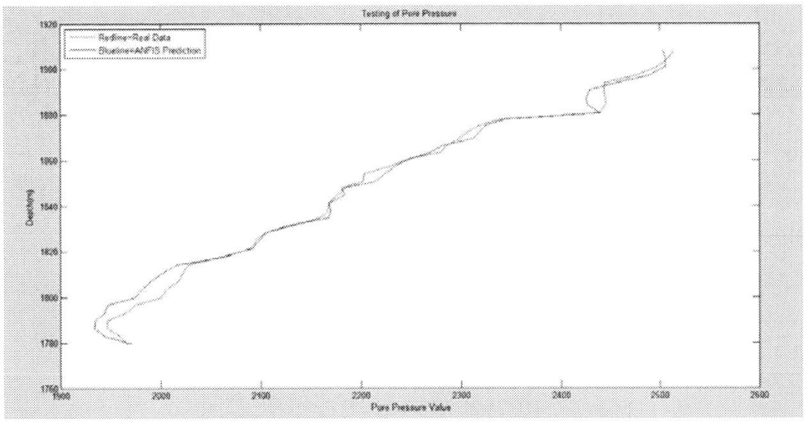

Figure 10: ANFIS testing result.

Figure 9 and 10 show us the modelling result of ANFIS using Gaussian membership function. It can be seen that the result is matched both in training and testing so the model is valid to be used in any kind of data.

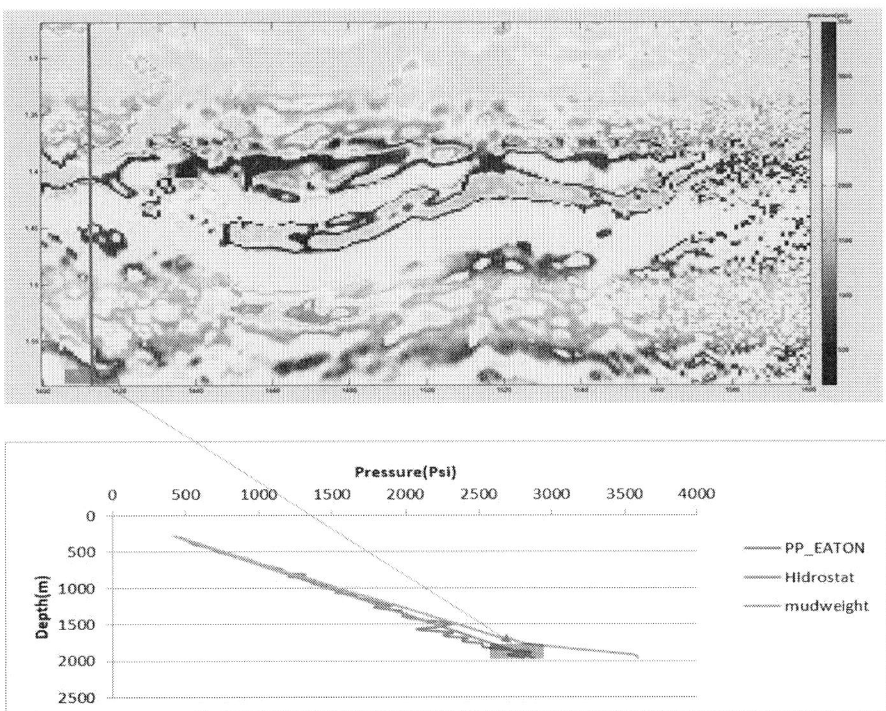

Figure 11: Two dimensional pore pressure model.

ANFIS pore pressure model shows a high value of pore pressure at the bottom of the well with value around 3000 Psi. This is matched to what is found at well report and this also supported by the fact that the Zp and density data is very low below the well bottom. According to ANFIS model, the pore pressure near the well is not high but at bit further there is the high pore pressure. At the middle depth and top, the pressure is not too high so it is not prone to overpressure.

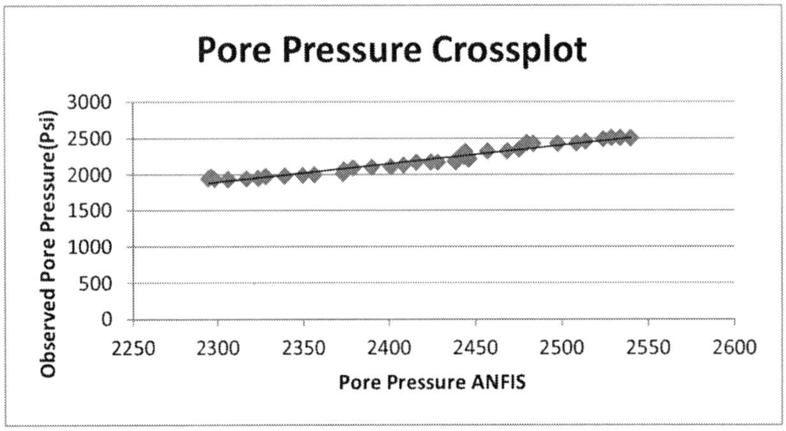

Figure 12: Comparison between real pore pressure value and ANFIS model.

Figure 12 shows that the accuracy of model is very high (correlation coefficient of $r^2 = 0.9536$) so we may conclude that the pore pressure model is accurate.

4. Conclusion

From explanation above we can see that ANFIS may be used to predict pore pressure as well as to make a pore pressure model with a high accuracy. So, it is very recommended to use this machine learning for future study. However, there are still many things to be corrected like anisotropy analysis, and comparison to other machine learning. The research also ignoring the mechanism of overpressure and still need to be considered for next study.

Acknowledgements

We would like to my geology lecturer for permission to use the data for publishing. Many thanks as well to DRPM Universitas Indonesia for supporting the study through the grant of PITTA 2019 No. NKB-0640/UN2.R3.1/HKP.05.00/2019.

References

[1] Eaton, B. A. 1975. The equation for geopressure prediction from well logs. *SPE*, Paper No. 5544, 11p.
[2] Bowers, G. L. 2001. Determining an appropriate pore-pressure estimation strategy. *Offshore Technology Conference*, paper OTC 13042.
[3] Singha, Dip & Chatterjee, Rima & Sen, Mrinal & Sain, Kalachand. (2014). Pore pressure prediction in gas-hydrate bearing sediments of Krishna–Godavari basin, India. *Marine Geology*. **357**. 1-11. 10.1016/j.margeo.2014.07.003.
[4] Haris, A., Sitorus, R. J., & Riyanto, A. (2017). Pore pressure prediction using probabilistic neural network: Case study of South Sumatra Basin. *IOP Conference Series: Earth and Environmental Science*, **62**(1), [012021]. https://doi.org/10.1088/1755-1315/62/1/012021.

[5] Darman, Herman. (2000). An outline of the geology of Indonesia. IAGI.
[6] Rahmanifard, H. and Plaksina, T. *Artif. Intell. Rev.* (2018). https. //doi. org/10. 1007/s10462-018-9612-8.
[7] Noon, S., Harrington, J., Darman, H., 2003, The Tarakan Basin, East Kalimantan: Proven Neogene Fluvio-Deltaic, prospective Deep-Water and Paleogene Plays in a Regional Stratigraphic Context, *Proceeding of the Indonesia Petroleum Association.*

IOP Conf. Series: Materials Science and Engineering **854** (2020) 012042 doi:10.1088/1757-899X/854/1/012042

Characterization of Lombok Earthquakes on July-August 2018 using Focal Mechanism Analysis

T Karima[1], S K Wijaya[2,*], M S Rosid[1,*], S Rohadi[3], Y H Perdana[3]

[1]Geophysics, FMIPA Universitas Indonesia, Depok 16424, Indonesia
[2]Physics Department, FMIPA Universitas Indonesia, Depok 16424, Indonesia
[3]Badan Meteorologi, Klimatologi, dan Geofisika, Jakarta 10610, Indonesia

[*]E-mail: skwijaya@sci.ui.ac.id, syamsu.rosid@ui.ac.id

Abstract. Focal mechanism study was carried out on 5 major earthquakes that occurred in Lombok on July-August 2018. The KIWI program was applied here to compare the parameters of the 5 earthquake sources. The principle of the KIWI program is inversion of wave components at various stations to obtain the moment tensor. Green's Function is also implemented to create synthetic waves representing the observation waves. Source parameters are taken from the synthetic waves from the five recorded events, and it was found that the 5 earthquakes on July 29th, 2018, August 5th, 2018, August 9th, 2018, and two earthquakes on August 19th, 2018 have strike values varying between 61°-74°, dip values between 18°-36°, and rake values between 75°-93°. These five earthquakes have similar source parameters, and have focal balls which state that the type of fault in these earthquakes are thrust faults. The similar source parameters, close range of time of occurrence, and hypocenter distances indicate that the earthquake events were most likely caused by the same fault.

1. Introduction

On Sunday, July 29th, 2018, at 05.47 WIB, Lombok experienced an earthquake with a magnitude of Mw 6.5 at a depth of 13 km. Exactly one week later, on Sunday, August 5th, 2018, the island was again rocked by an earthquake with a magnitude of Mw 6.9 at a depth of 15 km. Then, on August 9, 2018, Lombok was again struck by an earthquake by the magnitude of Mw 5.9. Exactly 10 days later, on August 19, 2018, two large earthquakes rocked Lombok with magnitudes of Mw 6.3 at 11.10 WIB and Mw 6.9 at 22.56 WIB [1]. These earthquakes have a close distance between the sources, and according to a study about earthquakes in Aceh, there is a possibility that close-distanced earthquakes are caused by the same fault [2].

Further analysis is needed to determine whether these earthquakes are generated from the same source. Focal mechanism analysis may be able to describe the source mechanism for the fault planes geometrically and mathematically when the earthquakes occurred [3]. The purpose of this study is to characterize those earthquakes as well as the fault planes that cause the earthquakes. Focal mechanism is used to evaluate an earthquake data to obtain the orientation of the fault movement that causes the earthquake. The analysis includes the determination of fault parameters, that are strike, dip, and slip angles [1]. Then, these parameters are compared to each other to find out whether the earthquakes have similar source mechanism characteristics.

2. Methodology

In order to analyze the focal mechanism of this earthquake sequence, the earthquake waveform data taken from BMKG catalog (*webdc*) is converted and the earthquake wave components, that show wave displacement in 3 directions, namely the Vertical (Z) direction, North-South direction (N), and East – West direction (E) are rotated into component of Vertical (Z), Radial, and Tangential directions. The reason why these components are rotated is because the N and E direction very rarely corresponds to the polarization of the S wave in the horizontal direction (SH) and in the vertical direction (SV) [4].

Band-pass frequency filters are then applied to eliminate the attenuation effect on high frequency waves. Aside of putting on filters, waves from seismographs are selected to eliminate stations or components that fail to record properly due to tool malfunctions when the earthquake occurs. The wave inversion process is then performed to obtain the focal beachballs and fault parameters for the Lombok Earthquake 2018. The parameters are then analyzed to see the similarity. The inversion was also carried out to differentiate compression and dilation quadrants by calculating synthetic waves for two possible solutions and cross-linking between synthetic waves and observation waves and applying the smallest misfit [5]. In addition, the distribution of aftershock from Lombok Earthquake 2018 sequence was also mapped to see the trends showing that could possibly indicate the presence of the faults.

3. Results and Discussion

The results obtained from processing the earthquake waveforms are in the form of source parameter values of faults, namely strike, dip, rake, and focal beachballs for each earthquake. There are two strikes, dips, and rakes values for each event, where both values represent 1 fault plane and 1 auxiliary plane. These two planes are perpendicular to each other.

The source parameters resulted from the Lombok Earthquake 2018 are shown in Table 1 as follows:

Table 1. The source parameters of the Lombok Earthquake 2018.

	29/07/18 (Mw = 6,5)		5/8/2018 (Mw = 6,9)		9/8/2018 (Mw = 5,9)		19/8/18 #1 (Mw = 6,3)		19/8/18 #2 (Mw = 6,9)	
Strike	64°	260°	61°	238°	62°	253°	74°	251°	67°	250°
Dip	21°	70°	28°	62°	36°	55°	18°	73°	29°	61°
Rake	75°	96°	93°	88°	81°	96°	93°	89°	87°	92°

The comparison of these parameters showed relatively similar values. To strengthen the analysis, the focal beachballs of the five strong earthquakes are also compared to see the similarities. The comparison is shown in Figure 1.

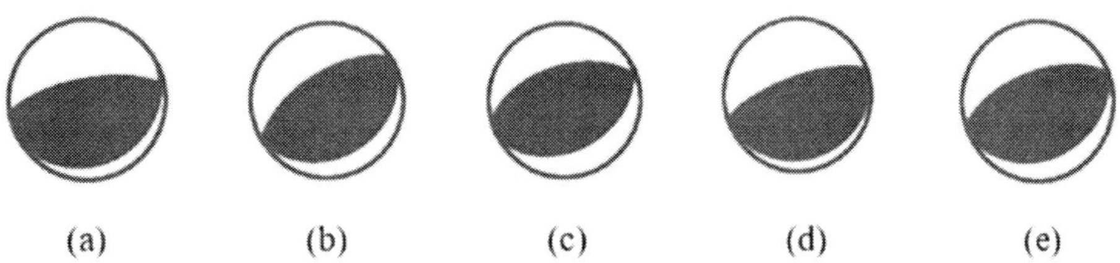

Figure 1. Focal beachball solutions of Lombok Earthquake 2018 on (a) July 29th 2018, (b) August 5th 2018, (c) August 9th 2018, and (d & e) August 19th 2018.

As shown in Figure 1, the 5 focal beachball solutions obtained from the Lombok Earthquake 2018 have a similar shape. These five beachballs indicate that the earthquakes occur due to the upward fault movement, widely known as thrust fault. Focal beachballs illustrate the direction of force, where the white part shows the maximum pressure zone and the red zone shows the minimum pressure zone [6]. The plotted beachballs for this earthquake sequence shows that the maximum force direction compresses the fault relatively from the North and the South, so that the fracture issues a minimum pressure to the top as shown in Figure 2.

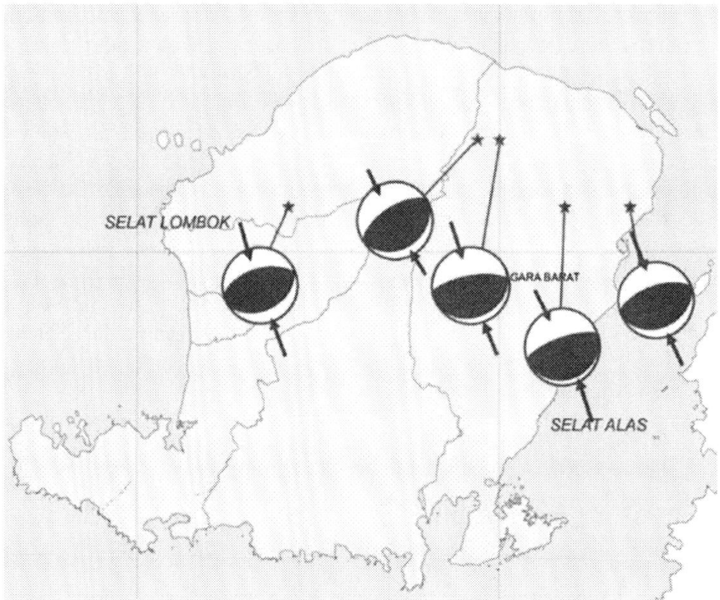

Figure 2. Beachball distributions of Lombok Earthquake 2018 and the relative direction of forces acting on the earthquake sources.

Figure 3 shows the distribution of earthquake epicenters in Lombok between July-September 2018 with magnitude above Mw 3.5.

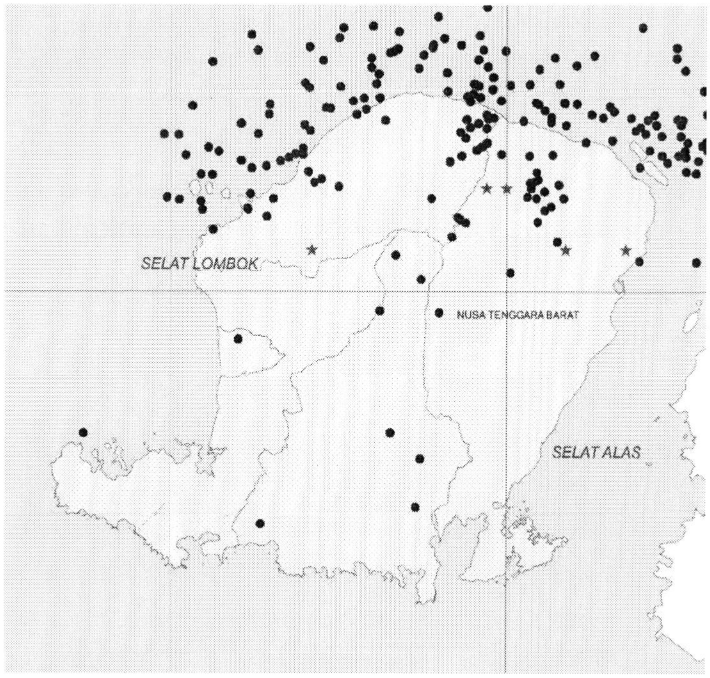

Figure 3. Epicenter distribution of Lombok Earthquake 2018 aftershocks on July-September 2018 ($Mw \geq 3.5$)

The epicenter distribution shown in Figure 3 tends to form a line in East-West orientation. This line shows a trend that could be interpreted as a fault, but it is difficult to determine whether this fault is the back-arc Flores Thrust or a local fault based on the available data. For this reason, a validation using secondary data in the form of tomographic study is necessary. The result of tomography study from Widyarta et al. [7] is shown in Figure 4.

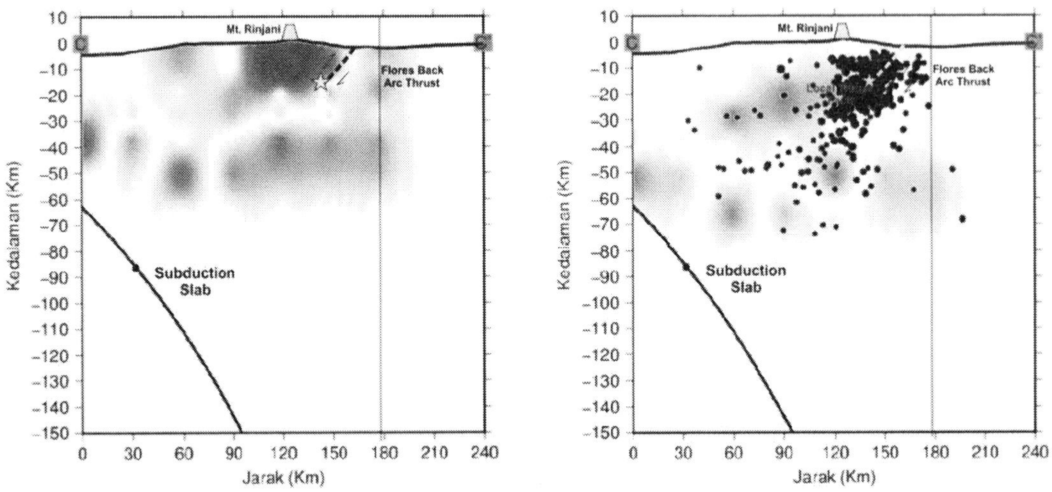

Figure 4. Tomography and hypocenter distribution of Lombok Earthquake 2018 on July-September 2018 [7].

From the tomography and hypocenter distribution shown in Figure 4, it can be seen that the hypocenter trend of aftershock distribution is not exactly in Flores Thrust, but is closer to the north coast of Lombok Island. Based on that data analysis, it can be said that this earthquake sequence was not caused by the activity of Flores Thrust, but rather by another local fault.

4. Conclusion

From this study, it can be concluded that the focal spherical patterns of the five Lombok Earthquake 2018 are able to characterize the faults. The earthquakes are caused by upward fault movements with similar focal spheres, similar parameter values, as well as the time of occurrence and adjacent distances. Therefore, the earthquake sequence is most likely caused by the same fault system. The sequence was most likely to be caused by the activity of a local fault along the north coast of Lombok Island, rather than the Flores Thrust.

Acknowledgment

We thank the Research and Development Center of Badan Meteorologi, Klimatologi, dan Geofisika (BMKG) for providing us the data and assistance for this study. We also thank DRPM Universitas Indonesia for supporting this study financially through PITTA's grant no. NKB-0666/UN2/R3.1/HKP.05.00/2019.

References

[1] Pusat Gempa dan Tsunami BMKG (2018) 5 Gempabumi Signifikan yang Mengguncang Lombok (Jakarta: BMKG).
[2] Qadariyah, Simanjuntak, A. V. H., dan Umar, M (2018) Analisis Mekanisme Fokal Menggunakan Inversi Waveform; Studi Kasus Gempa Pidie Jaya 7 Desember 2016, J. Aceh Phys. Soc. 7(3): 127–132.
[3] Gupta, H. K (2011) Encyclopedia of Solid Earth Geophysics (Dordrecht: Springer Science & Business Media).
[4] Stein, S., and Wysession, M (2003) An Introduction To Seismology, Earthquakes, and Earth Structures (Malden: Blackwell Pub).
[5] Domingues, A. L (2010) Kinematic Waveform Inversion – Study of Regional Earthquakes in Southwest Iberia (Thesis: Universidade Tecnica de Lisboa).

[6] Lay, T., and Wallace, T. C (1995) Modern Global Seismology (London: Academic Press).
[7] Widyarta, R., Wijaya, S.K., Rosid, M.S., and Rohadi, S. (2019) Identification of Fault Structure in Lombok Region, West Nusa Tenggara Using Tomography Lombok Earthquake Data of July-August 2018, The 2nd i-COSINE, Malaysia, in progress.

IOP Conf. Series: Materials Science and Engineering **854** (2020) 012043 doi:10.1088/1757-899X/854/1/012043

The Implementation of Augmented Reality in Increasing Student Motivation: Systematic Literature Review

Nur Farah Syuhada Mohamad Saadon, Ibrahim Ahmad, Ahmad Naim Chee Pee @ Che Hanapi

Fakulti Teknologi Maklumat dan Komunikasi, Universiti Teknikal Malaysia Melaka,

Email: nursyuhada0907@gmail.com, ibrahim@utem.edu.my, naim@utem.edu.my

Abstract. The purpose of this article is to analyze the implementation of augmented reality in increasing student motivation. A systematic literature review was conducted in March 2018 and updated in January 2019. This paper includes a literature review of relevant research of motivation in augmented reality since in year 2013. The object of research is the augmented reality, and the subject is the student motivation. As a result of the study, it has been found that augmented reality gives a positive result in increasing student motivation. The study show that augmented reality has a potential to have a significant demand to increase student motivation in learning and therefore the need to develop an appropriate tool to keep up student motivation in learning at schools and universities. Results have shown that students that used augmented reality is more confidence and show higher learning motivation in learning.

1. Introduction

Technology plays an important role in fostering a love for learning through design and production [1]. Nowadays, technology has become part of our life. With changing times, technology has also developed, and the way humans work, and thinking are different from the past. One of the developing technologies that has been achieved is Augmented Reality (AR). AR is a combination between real and virtual objects in a real environment; it is run interactively in three dimensions (3D) and in real-time [2]. AR provides an opportunity for educators to diversify their teaching techniques. Student can apply learning that involves the outside world without having to leave the classroom. AR is not limited to a specific technology, it can be applied to many different existing technology such as head-mounted display (HMD) [2], smartphones, computer & tablet [3].

Motivation is defined as the psychological process which activates, directs and maintains goal-oriented behaviour [4]. Motivation is divided into three category which is intrinsic motivation, achievement motivation and extrinsic motivation. Intrinsic motivation is where someone does something that they like or want for their own sake. Achievement motivation is a competition feeling where something is done to show the better performance. Lastly is extrinsic motivation which this type of motivation is a condition when a person do something in order to get some future value [5].

Learning can be done in many forms and one of them is by using AR technology. The technology has brought the learning style to be more interactive and attractive. Bringing students to learn and interacting with this world mostly is not easy. But with the capabilities of AR, teachers now have an option of bringing students to learn about the world without any hesitation. AR justifies the use of media with two-dimensional technology in education and this technology is flexible, easy to use, cheap and affordable [6]. There is so many type of AR. Table 1 shows the types and categories of augmented reality [7]:

Table 1. Types and categories of Augmented Reality.

Category	Type	Characteristics
Triggered	1a. Marker-based: Paper	Paper marker activates stimuli.
	1b. Marker-based: Object	Most objects can be made into markers.
	2. Location-based	Overlay of digital information on a map or live camera view. GPS may activate stimuli.
	3. Dynamic Augmentation	Meaningful, interactive augmentation with possible object recognition and/or motion tracking.
	4. Complex Augmentation	Augmented dynamic view and pull internet information based on location, markers and object recognition.
View-Based	5. Indirect Augmentation	Image of the real world augmented intelligently.
	6. Non-Specific Digital Augmentation	Augmentation of any camera view regardless of location.

Most of the reason in academic nowadays is student lack of motivation in the meantime they need more academic activities [8]. This statement is supported [9] where some of the academic strategies do not achieve student requirements. To achieve the target in learning process, a few factors should be taken care of and one of it is motivation. Motivation is one of the important factors that help the student performing their academic achievement and work productivity [10].

This study aims to explore the contribution of Augmented Reality application in increasing student motivation by investigating the effects of using Augmented Reality application in increasing student motivation.

2. Method

This study presents a systematic review of research examining the motivation in augmented reality. This systematic review included the research question, the selection of journals, the inclusion and exclusion criteria, the search strategy and databases, the data collection instruments and methods and the criteria for the analysis of the results which followed the recommendations of the PRISMA (Preferred Reporting Items for Systematic Reviews and Meta-Analyses) statement [11]. The declaration from PRISMA is the global and approved revised version of the declaration from QUORUM (Quality of Meta-analysis reporting). The review question was: "How student motivation can be increased by using augmented reality application?"

The search and the analysis of the paper identified in the first search conducted by two researchers. Data were organised in tables containing year, author, sample characteristics, method and main outcomes. Literature search was conducted using the following online database ScienceDirect, Google Scholar and Scopus Full Text. Publication year was restricted: it is covered the period from 2013 to 2018 and limited to studies written in English. Table 2 shows the Inclusion and Extraction Criteria:

Table 2. Inclusion and Exclusion Criteria.

Criteria	Eligibility	Exclusion
Literature Type	Indexed Journal (research articles)	Non-indexed journals, Systematic review journals, chapter in book.
Language	English	Non-English
Timeline	Between 2013 – 2018	Less than 2013

3. Results

To identify some published article describing representative development in augmented reality and motivation area, a systematic review was conducted. The search has focused on motivation in augmented reality.

Two studies had been conducted in Taiwan [11, 14], two in Spain [13,12] and two in Malaysia [15,8].In relation to sample and outcomes, three samples is from college students [11,13,15] , 1 from specific programme students [13] and 2 from primary schools [8,14]. Table 3 shows the paper reviewed from 2013-2018 in the field of motivation in augmented reality.

Table 3. Paper Reviewed from 2013 - 2018 in the field of Motivation in Augmented Reality.

Paper	Sample	Method	Outcome
Paper 1 [12]	College students in their second year (aged 18-20) (13 males & 20 females) :63	• Pre-Test • Post-Test • Experimental-Control Group (CG) required to attend classes using traditional outdoor instruction methods, where they were provided with a textbook. Experimental Group (EG) instructed to use mobile smartphones containing our proposed system to learn the same course materials in a real-world setting outside of the classroom. • Questionnaire	• Students who learned with the proposed system showed higher learning motivation and had better learning performance • Confidence was the most highly rated motivational factor among student. • Because they felt surer of their knowledge and performance.
Paper 2 [15]	Second year engineering students from La Laguna University (41 male, 22 female, mean age 21.30 years with a standard deviation of 2.69): 63	• Experimental • Questionnaire	• Intrinsic Motivation Inventory results (all values above 4.0 over 7.0) indicate that AR with tablets is a motivating activity for students. • Motivation and usability research with AR conducted in this experiment shows that this is a technology that can be implemented in formal teaching to facilitate and motivate the students' 3D understanding of the terrain represented with traditional

			2D cartographic techniques in a 3D scenario.
Paper 3 [13]	VET program of car's maintenance student. 100% male students. 23% aged 14-16 years old, 63% aged 17-19 years old, 9% aged 20-22 years old, 3% more than 25 years old: 35	• Experimental – Students used the application for 20 days as a learning process. • Questionnaire	• Mobile AR applications that are able to capture the interest of the students and increase their time on the learning activities, are the applications that better support student motivation.
Paper 4 [16]	Fifth grader from four classes: 111	• Pre-Test • Post-Test • Experimental - EG learned with the AR operating guidance system and discussions in class. CG guided by the teacher to finish the science project, worksheet and have discussions in the class. • Questionnaire - 20 minutes interview, 5 students from experimental group, 5 students from control group.	• With the help of the AR guiding mechanism, the students learning achievements, learning motivation, critical thinking tendency, and group self-efficacy were significantly improved.
Paper 5 [14]	First-semester social science undergraduates, with a mean age of 19.5 years: 120	• Pre-Test • Post-Test • Experimental- EG were allowed to use the learning application on their mobile phones. CG used a similar application on the desktop computer.	• Mobile augmented learning tool can be used as a learning application to help students learn anywhere, anytime, by capitalizing the mobility of mobile devices. • The effective use of such a novel tool may be mediated by several factors, such as gender.
Paper 6 [9]	Students in primary school (7 years old), 1 Indian male, 2 Malay male and 3 Malay females: 5	• Observation-To observe the motivation level that the participants shown during the lesson. • Semi-structured Interview- To confirm the results derived from the observation.	• Year one students perceive augmented reality (AR) pop-up book as being motivational through the attention, relevance, confidence, and finally the feeling of satisfaction as a result from the learning session

4. Discussion

Motivation is the most important concept in human behaviors [17]. Motivation is the factor that keep student achieve the learning target and one of the main factors that keep students want to learn. [18]. Motivation is important in create the engagement between learner and the things they learn. Motivation in learning is divided into two type which is Intrinsic (internal) motivation and Extrinsic (external) motivation [10]. Extrinsic motivations concern the drive of the utilitarian purpose [19], such as receiving benefits and rational function analysis while intrinsic motivation refers to an individual's motivation and the pleasure derived from the task itself or the sense of satisfaction in completing or working on the task itself [10][20].

In 1984, Keller has created a model to improve motivational attractiveness of teaching materials which is ARCS model [10]. According to Keller [21], there are four conceptual categories of human motivation for the ARCS model which are: (i) Attention, (ii) Relevance, (iii) Confidence and (iv) Satisfaction. All this four categories of motivation are needed and require deep understanding to bring an effective, efficient and attractive AR [22].

Students have been exposed to various technologies and some of them are attached to technologies. As an educator, they always need to think the best way to give the best learning experience to the students because the main objective for every educator is to make sure students understand what they are learning. Teacher must use appropriate methods to stimulate the interest of students and encourage them to develop a positive attitude towards an effective learning outcome [23].

AR gives students the opportunity to explore the world in a more interactive way [3]. Besides with imagination, now with the presence of AR students can be imaginative. Although technology is developing, not everyone can adapt to the advancement of technology that is fast growing. There are some people who still love to read books physically though there is a plenty of online book such as watt pad, eBook etc. According to [24], research show that some people still prefer physically existing because they love the physicality, transportability, robustness, flexibility etc. Following that, the interesting thing that arises in this technology is AR textbooks [6]. AR textbook is a combination of AR technology and textbook. A webcam is pointed to the normal printed book to bring the visualizations and interactions [6]. AR is not just a focus on one use of technology, but it involves a touch on the real thing.

In the course of AR activities, the interaction of learners with learning materials is increased without any environmental constraints or hazards, and this has a positive effect on their extrinsic motivation [9][25].

5. Conclusion

In this paper, a systematic literature review was reported. In conclusion, the result of existing study has showed that augmented reality application is a suitable application in increasing student motivation. Augmented reality success to increase student motivation through ARCS model which is attention, relevance, confidence and satisfaction. Results have shown that students that used augmented reality is more confidence and show higher learning motivation in learning.

Acknowledgement

We would like to express our gratitude to Universiti Teknikal Malaysia Melaka (UTeM) and Ministry of Higher Education (MOHE), for the support of research grant FRGS/2018/FTMK-CACT/F00389 in this research.

References

[1] C. Allan and H. Richard, *Rethinking Education in the Age of Technology: The Digital Revolution and Schooling in America*, Second Edi. New York and London: Teachers College Press, 2018.

[2] D. W. F. van Krevelen and R. Poelman, "A Survey of Augmented Reality Technologies, Applications and Limittions," *Int. J. Virtual Real.*, vol. 9, no. 2, pp. 1–20, 2010.

[3] M. Antonioli, C. Blake, and K. Sparks, "Augmented reality applications in education," *J. Technol. Stud.*, vol. 40, no. 2, pp. 96–107, 2014.
[4] T. A. Roberts and C. D. Sadler, "Letter sound characters and imaginary narratives: Can they enhance motivation and letter sound learning?," *Early Child. Res. Q.*, vol. 46, pp. 97–111, 2019.
[5] E. A. Locke and K. Schattke, "Intrinsic and extrinsic motivation: Time for expansion and clarification.," *Motiv. Sci.*, pp. 1–33, 2018.
[6] M. Kesim and Y. Ozarslan, "Augmented Reality in Education: Current Technologies and the Potential for Education," *Procedia - Soc. Behav. Sci.*, vol. 47, no. 222, pp. 297–302, 2012.
[7] A. Edwards-Stewart, T. Hoyt, and G. M. Reger, "Classifying different types of augmented reality technology," *Annu. Rev. CyberTherapy Telemed.*, vol. 14, no. December 2017, pp. 199–202, 2016.
[8] L. Legault, I. Green-Demers, and L. Pelletier, "Why do high school students lack motivation in the classroom? Toward an understanding of academic amotivation and the role of social support," *J. Educ. Psychol.*, vol. 98, no. 3, pp. 567–582, 2006.
[9] N. N. Mahadzir, "The Use of Augmented Reality Pop-Up Book to Increase Motivation in English Language Learning For National Primary School," *IOSR J. Res. Method Educ.*, vol. 1, no. 1, pp. 26–38, 2013.
[10] N.-C. Chang and H.-H. Chen, "A Motivational Analysis of the ARCS Model for Information Literacy Courses in a Blended Learning Environment," *Libri*, vol. 65, no. 2, pp. 129–142, 2015.
[11] D. Moher, A. Liberati, J. Tetzlaff, D. G. Altman, and P. Grp, "Preferred Reporting Items for Systematic Reviews and Meta-Analyses: The PRISMA Statement (Reprinted from Annals of Internal Medicine)," *Phys. Ther.*, vol. 89, no. 9, pp. 873–880, 2009.
[12] K. Y. Chin, C. S. Wang, and Y. L. Chen, "Effects of an augmented reality-based mobile system on students' learning achievements and motivation for a liberal arts course," *Interact. Learn. Environ.*, vol. 0, no. 0, pp. 1–15, 2018.
[13] J. Bacca, S. Baldiris, R. Fabregat, and Kinshuk, "Insights Into the Factors Influencing Student Motivation in Augmented Reality Learning Experiences in Vocational Education and Training," *Front. Psychol.*, vol. 9, p. 1486, Aug. 2018.
[14] H. F. Hanafi, C. S. Said, M. H. Wahab, and K. Samsuddin, "Improving Students' Motivation in Learning ICT Course with the Use of A Mobile Augmented Reality Learning Environment," *IOP Conf. Ser. Mater. Sci. Eng.*, vol. 226, no. 1, 2017.
[15] C. C. Carrera, J. L. S. Perez, and J. de la T. Cantero, "Teaching with AR as a tool for relief visualization: usability and motivation study," *Int. Res. Geogr. Environ. Educ.*, vol. 27, no. 1, pp. 69–84, 2018.
[16] S. C. Chang and G. J. Hwang, "Impacts of an augmented reality-based flipped learning guiding approach on students' scientific project performance and perceptions," *Comput. Educ.*, vol. 125, pp. 226–239, 2018.
[17] K. Li and J. M. Keller, "Use of the ARCS model in education: A literature review," *Comput. Educ.*, vol. 122, no. March, pp. 54–62, 2018.
[18] F. Paas, J. E. Tuovinen, J. J. G. Van Merriënboer, and A. A. Darabi, "A motivational perspective on the relation between mental effort and performance: Optimizing learner involvement in instruction," *Educ. Technol. Res. Dev.*, vol. 53, no. 3, pp. 25–34, 2005.
[19] R. M. Ryan and E. L. Deci, "Intrinsic and Extrinsic Motivations: Classic Definitions and New Directions," *Contemp. Educ. Psychol.*, vol. 25, no. 1, pp. 54–67, 2000.
[20] A.-N. Lay and K. Osman, "Developing 21 st century chemistry learning through designing digital games," *J. Educ. Sci.*, vol. 4, no. 1, pp. 81–92, 2018.
[21] J. M. Keller, *Motivational Design for Learning and Performance: The ARCS Model Approach.* 2010.
[22] C. S. Eskisehir, "Designing Motivational Learning Systems," *Online*, no. July, pp. 149–161,

2008.
[23] M. Mat Salleh, M. Ismail, and N. Abdullah, "Using PROFSTAR Game in Teaching the Periodic Table of Elements: Towards Enhancing Students' Engagement in Learning Chemistry," 2014, pp. 276–290.
[24] H. H. Otília Pasaréti and M. T.-S. Tamás Matuszka, András Jámbori, István Molnár, "Augmented Reality in Education," pp. 1–15, 2011.
[25] C. Erbas and V. Demirer, "The effects of augmented reality on students' academic achievement and motivation in a biology course," *J. Comput. Assist. Learn.*, vol. 35, no. 3, pp. 450–458, Jun. 2019.

IOP Conf. Series: Materials Science and Engineering **854** (2020) 012044 doi:10.1088/1757-899X/854/1/012044

Identification the Jakarta Fault using MS-SVD (Multi Scale - Second Vertical Derivatives) Method Gravity Data

K N Yudyawati and M S Rosid*

Geophysics, FMIPA, Universitas Indonesia, Depok 16424, Indonesia

*Email: syamsu.rosid@ui.ac.id

Abstract. Jakarta is a metropolitan city in Indonesia with high population. Geologically, its location has the potential to be affected by earthquakes from subduction zones as well as from local faults such as the Cimandiri, Lembang, and Baribis faults. Earthquake history showed that in 1780 and 1834 Jakarta was affected by an earthquake with magnitude Mw 8, destroying existing infrastructure. Several geoscientists believed the source of the fault is located in the south of Jakarta. Study with primary gravity data using the MS-SVD supported by the MS-HDVD method is a good method for identifying faults. From looking at the zero value shift on the MS-SVD graph, a fault zone is visible in the south of Jakarta. It is also confirmed by the CBA map that has contrast gravity between south and north of Jakarta. The faults have parallel-normal and parallel reverse type in East-West direction with dip angle for more than 79°. According to the fault that has been identified, south of Jakarta is evidently crossed by the fault, so the city safety needs to be considered especially for the potential of upcoming natural disasters. However, 2D forward modeling shows that subsurface Jakarta does not have density contrast between rock layers significantly.

1. Introduction

Jakarta is the capital city of Indonesia. From BPS (Badan Pusat Statistika) data online, Jakarta is densely populated reaches up 10.37 million people. The population always increases every year which also leads to infrastructure growth. For that reason, the city safety needs to be considered, such as the safety of potential natural disasters.

Geologically, Jakarta is not just above the fault zone. The conflux of the subduction plate between the Indo-Australian plate and Eurasian plate occurs in the southern part of Java Island [1]. Therefore, Jakarta has the potential to be shaken by the earthquake, especially the events that are generated by local faults around Jakarta.

History said, on January 22, 1780, the southeast of Sumatra and West Java were affected by an earthquake with an estimated strength up Mw 8.5 or larger [2]. Another history said, on October 10, 1834, the earthquake felt in Jakarta, Banten, Lampung, West of Sumatera, Karawang and also Tegal with an estimated strength up Mw 7.0 or larger [3]. Both of them were estimated to be caused by either Baribis Fault, or crustal or intraslab event [3]. Besides, on January 5, 1699, earthquake happened too, but originated from the intraslab or megathrust plates destroyed some of the existing infrastructures in DKI Jakarta with an estimated strength up Mw 7.5 or larger [3]. Talk about the presence of Baribis fault was estimated crosses south of Jakarta through GPS data analysis [4], the traces was estimated extension part of Kendeng fault in East Java [5]. However, the existence of the fault in Jakarta is still debated, so further scientific analysis is still needed. But the potential for its existence can be a concern and threat for Jakarta.

Modeling from Koulali *et al.* [4] in Figure 1 shows there are fault segments caused by the earthquakes in 1780 and 1834 with magnitude Mw 7.7. The two different segments are namely strike-slip and normal faults. This means, the fault is estimated to be a normal fault but the main segment is strike-slip. The blue and red-colored number in the red box are for the normal fault segment and the strike-slip segment, respectively. The strike-slip segment has a greater speed than the normal segment.

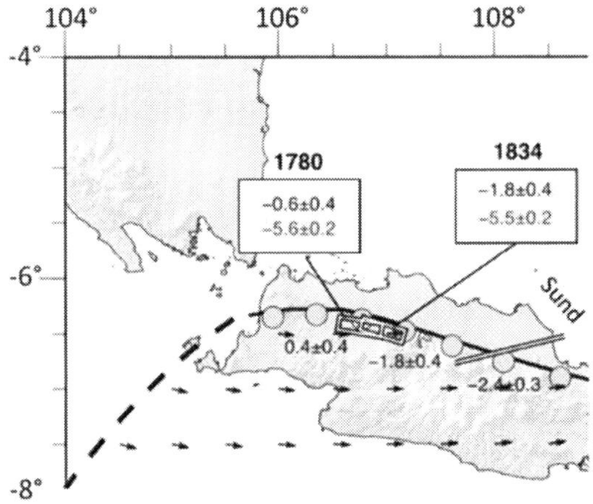

Figure 1. Koulali *et al.* (2016) modeling of deformation rate of each segment. The strike-slip fault in Jakarta (red number) has greater speed than the normal ones (blue number) [4].

When an earthquake occurs and its waves reach the surface, the vibration has the potential to damage anything on the surface, such as buildings and infrastructure. Potential hazards can be further investigated by looking at an amplification values in the geology area. Amplification is a magnification of soil waves that occurs due to significant rock density contrast, the ground wave will be enlarged when an event propagates from hard (high density) medium to a softer (low density) medium [6]. If the value of rock density contrast is high, then the level of damage is potentially high. If the density contrast between two layers is low, then the level of damage is potentially low as well. The amplification number can vary on the same rock depending on the level of deformation and weathering of the rock. The example, it can cause the phenomenon of the destructive earthquake with a small magnitude, has happened on Banjarnegara earthquakes of 19 April 2013 and 18 April 2018 [7].

One method that can examine the existence of the fault is gravity data. The gravity method can be used to identify and map geological structures by looking at variations in the value of gravity. This value is due to the difference in density values for each rock. The gravity method of processing MS-SVD (Multi Scale-Second Vertical Derivative) can be used to identify the presence fault in Jakarta and its characteristics such as direction and magnitude of dip as well as its strike.

This study is expected to provide information about the existence of a fault that crosses the south of Jakarta as well as the presence of density contrast. The density contrast may have detected by gravity 2D forward modeling. Therefore, it hopefully can help the society to be more aware of hazard potential and may help the local government in composing a city plan, also can help a practitioner to determine the type of building foundation that must be built (earthquake-resistant building).

2. Methodology

The method of this study is carried out by the MS-SVD method to identify the presence of fault structures in Jakarta. Then 2D forward modeling gravity data is made to illustrate the estimation of subsurface appearance and identify the possibility of a subsurface density contrast layer.

The CBA map gravity data were corrected beforehand, specifically the data is corrected by latitude, free air, Bouguer, and terrain correction. The CBA map is then done upward continuation to several high levels. On each level of upward continuation is carried out FHD and SVD filters to get FHD maps and depth-*n* SVD maps. The slicing data was done to get a graph of FHD and SVD-*n* depth.

The slicing data is north-south direction since the predicted fault probably in the relatively east-west orientation. When all maps of FHD and SVD depth-*n* have been sliced, then we get some combined graph of MS-HDVD and MS-SVD graph to determine the location and characteristics of the fault. The 2D forward modeling is also carried out after interpreting the values of MS-HDVD and MS-SVD.

3. Result and discussion

3.1. Gravity data

The gravity data as primary data has been acquired in 2018. Then it made as the corrected data. The average density of the study area that was obtained from processing is 2.3 g/cc. This number means that rocks below Jakarta are not too dense if compared to other sedimentary rocks [8]. For this map, we made three lines to extract the data from each area. Then, the data from each line are made into a graph. From the graph, we can interpret the fault.

It can be seen that the CBA map in Figure 2 has a gravity anomaly in the range of 39.0 mGal to 60.0 mGal. Low anomalies are dominating in the south and east area of Jakarta. High anomalies are dominating in the North of Jakarta. There may be a rock or anomalous structure that extends in that area [9] which causes a high gravity anomaly in the North of Jakarta. But, another hypothesis said it associated with subsidence and seawater intrusion [8, 10]. Meanwhile, the reduction in the subsurface mass in the southern part of Jakarta results in a low gravity anomaly.

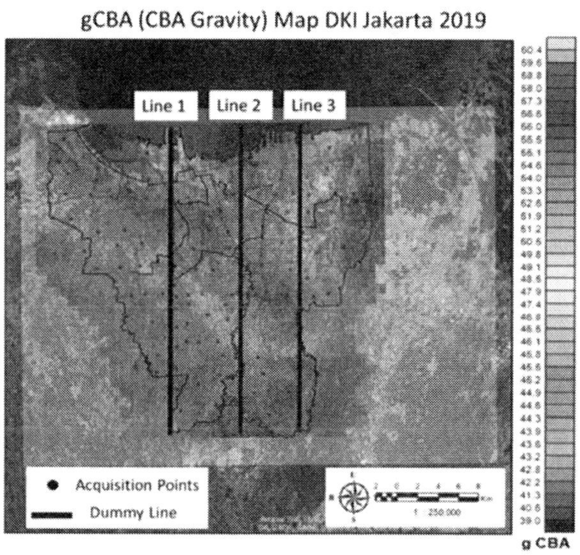

Figure 2. CBA map with distribution of gravity station is overlaid with satellite images of the Jakarta area.

3.2. Derivative analysis

The correlation graph between MS-SVD and MS-HDVD on line 1 has correlated. Three estimated zones (in axis-X) have a maximum FHD gravity with an SVD gravity has zero values. The result are first fault has 81.78° dip towards the south is a type of normal fault. The second fault has 59.74° dip towards the North with a normal fault type. The third fault has 67.46° dip towards the south with a type of normal fault.

The correlation graph between MS-SVD and MS-HDVD on line 2 (Figure 3), there will be estimated has two faults. Resulting in a first fault with a value of 81.97° dip towards the south with a type of normal fault. The second fault has 65.71 ° dip towards the North with a normal fault type. Figure 4 shows the plot between depth and location of fault.

The correlation graph between MS-SVD and MS-HDVD on line 3 is estimated to have four possibilities faults. The result are, the first fault has dip 79.47° towards the southern and has estimated reverse fault type. The second fault has 80.81° dip towards the North with a normal fault type. The third fault dip 77.91° towards the North with a normal fault type. The fourth fault has dip 76.65 ° towards the south with a normal fault type.

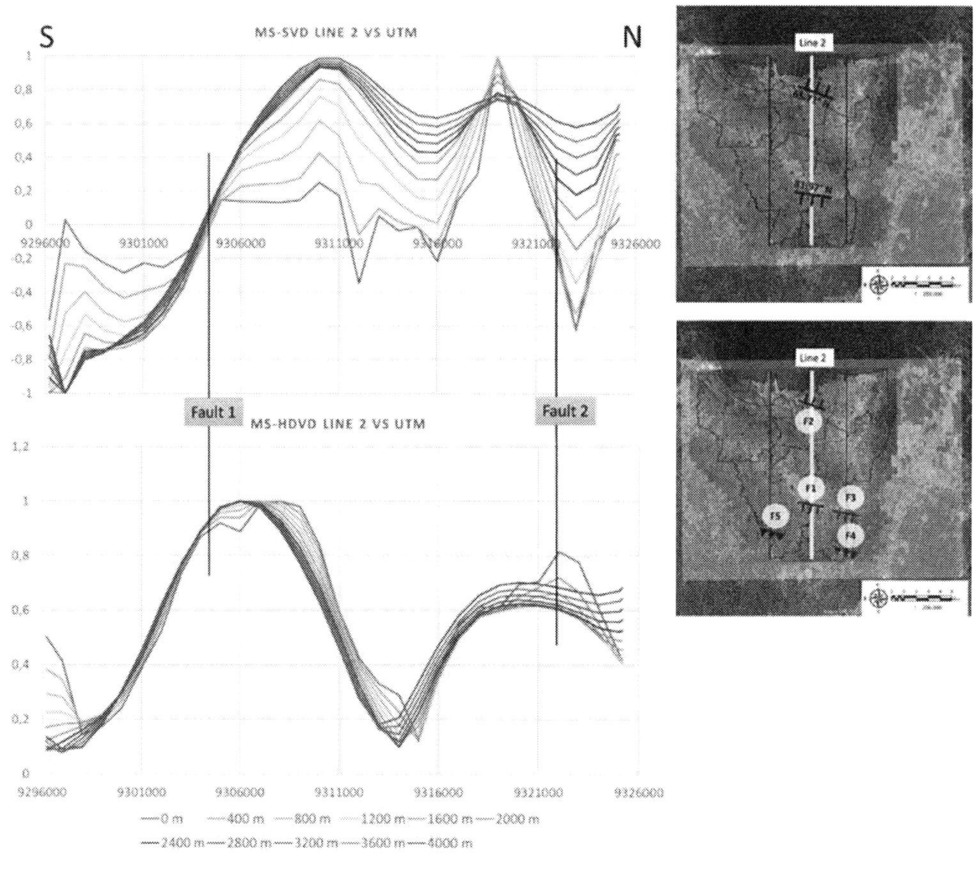

Figure 3. MS-SVD graph correlation with MS-HDVD in line 2.

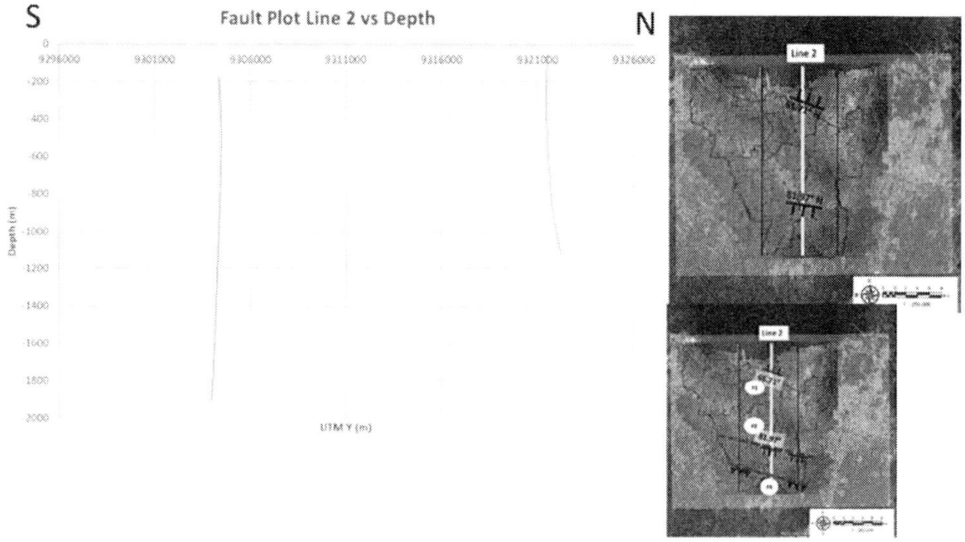

Figure 4. An estimation of two predicted faults conditions in line 2 (depth h vs UTM).

Table 1 shows a plot of the number taken from the zero point of the SVD graph on line 2. Then, the characteristics can be further analyzed and get the dip value and direction of the dip. Table 2 shows the characteristics of estimated fault in south of Jakarta.

Table 1. The parameter values to characterize fault in line 2.

Upward continuation depth-n		Fault 1		Fault 2	
	X	Y	h	Y	h
0	705240.7	9304368.1	-179.1	9321518.1	-49.3
400	705240.7	9304478.1	-458.5	9321478.1	-354.1
800	705240.7	9304448.1	-714.1	9321598.1	-635.1
1200	705240.7	9304378.1	-945.9	9321808.1	-892.3
1600	705240.7	9304298.1	-1154.0	9322238.1	-1125.8
2000	705240.7	9304238.1	-1338.4	9321518.1	-49.3
2400	705240.7	9304178.1	-1498.9		
2800	705240.7	9304118.1	-1635.7		
3200	705240.7	9304068.1	-1748.8		
3600	705240.7	9304018.1	-1838.0		
4000	705240.7	9303968.1	-1903.5		
Dip		81.97°		65.71°	
Dip direction		South		North	
$\|[\partial^2 g/\partial z^2]\|$		Max > Min		Max > Min	
Fault type		Normal		Normal	

IOP Conf. Series: Materials Science and Engineering **854** (2020) 012044 doi:10.1088/1757-899X/854/1/012044

Table 2. The parameter values to characterize estimated fault in South of Jakarta.

Upward continuation depth-n		**Fault 3**		**Fault 4**		**Fault 5**	
	X	Y	h	Y	h	Y	h
0	710726.8	9302375	-152.6	9296905	-130.3	9300955	-174.3
400	710726.8	9302455	-436.8	9297825	-418.6	9300845	-454.5
800	710726.8	9302615	-697.1	9297745	-683.1	9300785	-710.9
1200	710726.8	9302735	-933.7	9297685	-923.8	9300755	-943.7
1600	710726.8	9302815	-1146.5	9297655	-1140.8	9300735	-1152.6
2000	710726.8	9302855	-1335.6	9297625	-1334.1	9300695	-1337.8
2400	710726.8	9302865	-1500.9	9297615	-1503.5	9300645	-1499.2
2800	710726.8	9302845	-1642.5	9297595	-1649.2	9300585	-1636.9
3200	710726.8	9302805	-1760.3	9297575	-1771.2	9300505	-1750.7
3600	710726.8	9302745	-1854.3	9297555	-1869.3	9300415	-1840.9
4000	710726.8	9302655	-1924.5	9297535	-1943.7	9300295	-1907.2
Dip		80.81°		79.47°		81.78°	
Dip direction		South		South		South	
$\|[\partial^2 g/ \partial z^2]\|$		Max > Min		Max < Min		Max < Min	
Fault type		Normal		Reverse		Reverse	

3.3. Fault analysis

There are faults identified on three lines. The faults are estimated to extend relatively to East-West direction in the south of Jakarta. The faults can be seen clearly from CBA, SVD and residual maps of Figure 5 that the gravitational value is seen to be the contrast between the high and low gravity values, where this is in accordance with the principle of the gravity method that if there is a contrast of anomaly gravity then there is probably an anomaly body in the zone.

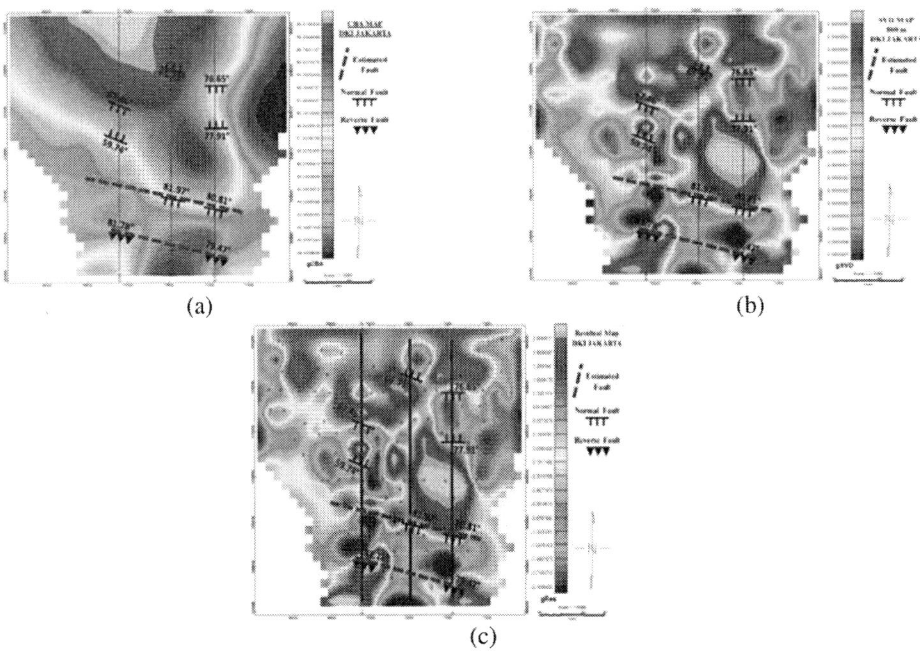

(a) (b) (c)

Figure 5. Plotting faults on 2D gravity maps; (a) CBA map, (b) SVD map, (c) residual map.

The fault was estimated (based model) crossed on south of Jakarta is oblique or normal movement with main segment strike-slip because the dip value is above 70°. The estimated the existence of a fault in Jakarta seems clearly from SVD Map of Jakarta (Figure 6) and residual map for supported data.

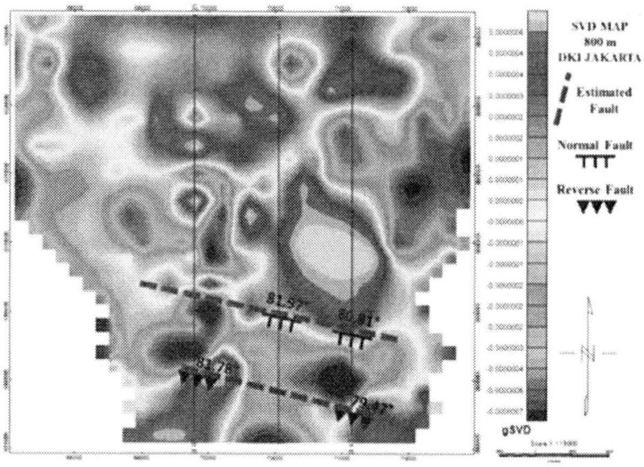

Figure 6. Estimated of the presence fault in south of Jakarta

The gravity anomaly of Jakarta can be associated to western Java sedimentary basins map shown in Figure 7 [11]. It is shown that Jakarta is the hill area, while the left and right areas are basins. The geology of this area (hill and basin) can potentially cause faults to form.

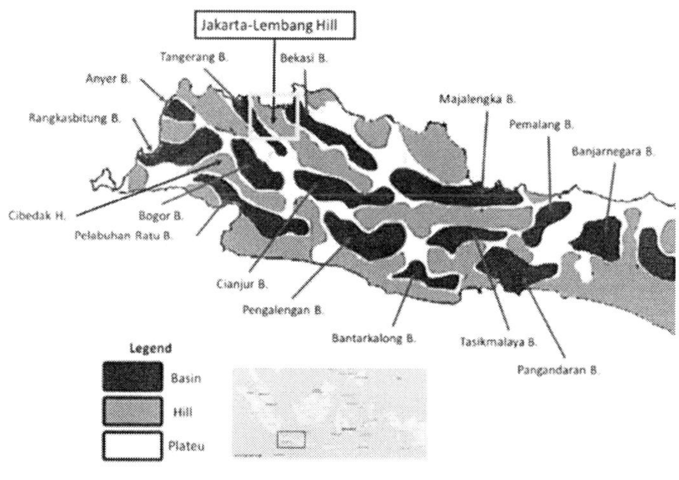

Figure 7. Map of West Java sedimentary basins [11]

3.4. *2D forward modeling*

The subsurface modeling in Figure 8, to estimate the section of subsurface on the 2nd line of the slicing CBA map. It can also be a potential determination of whether Jakarta is dangerous or not if there is an active fault below the surface.

The modeling refers to geological data and gravity data of Jakarta. There are several formations from bottom to top, there are Bojongmanik, Subang & Parigi, Serpong & Genteng, quaternary volcanic deposits [12]. This formation potential to be a rock basement of Jakarta because it forms in mid miosen age and it has a higher density because it consists of limestone (dominant) and also sandstone. The second formation are Subang and Parigi Formations consist of sandstone and silt. The third are Serpong & Genteng formation consists of sandstone, silt, conglomerate, and breccia. And the last is the top layer consists of alluvial, tuff and breccia [12].

The estimated density of each layer are estimated by seeing the kind of rock and other possibility. The estimated density of Bojongmanik Formation is 2.75 gr/cc because this formation had already compacted. The estimated density for the second formation is 2.6 gr/cc and the third formation is 2.5 gr/cc because this formation was estimated got the impact of seawater intrusion [8, 10]. The last layer (upper formation) is estimated for 2.0 gr/cc. We can see, high density contrast seems clearly between the third formation (2.5 gr/cc) and the layer above it (2.0 gr/cc). It also showed by Cipta, A. paper (2018) with HVSR and ANT method by looking at shear wave on its figure 11 [13].

In this line, there are two fault was estimated in south and north of Jakarta, and one additional fault in south of Jakarta. The characteristic of first fault is normal, heading to north and has dip 65.71°, the second fault is normal, heading to south and has dip 81.97°, and the additional fault was inferred from estimated fault on line 1 & 3.

We can see that Jakarta has faults and high density contrast. Even though Jakarta never has epicenter and the seismicity in Jakarta is classified as rare or almost nothing [14], but Jakarta has the potential to be affected by earthquake and has the potential hazard that will affect infrastructure in Jakarta. This 2D forward modeling is also needed to be considered and should to be always updated in the future, to make the data more comprehensive and accurate.

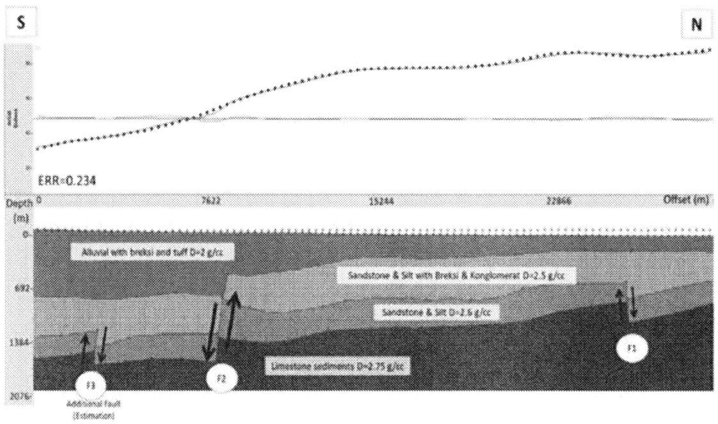

Figure 8. The section of 2D forward gravity modeling along line 2.

4. Conclusion

Based on the results, there is a suspected main fault in the south of Jakarta with East-West strike orientation. The dip of the fault is around >79° heading south. The faults are a normal fault and a reverse fault. Moreover, from the results of 2D forward modeling, Jakarta was estimated has significant density contrast between the third formation (2.5 gr/cc) and the layer above it (2.0 gr/cc). If the fault and density contrast in Jakarta is exist, this will make Jakarta has hazard potential from the earthquake. So, people have still must wary for their own personal and other protection. Besides, it would be nice if the building also began to be considered and built with earthquake-resistant foundations with further geotechnical study, while the existing building foundations are re-evaluated, to prevent if someday the buildings are affected by an earthquake. Especially, the building foundations in the south of Jakarta area. I realize that this research needs to be explore further using another supporting data to make the result more comprehensive and accurate.

Acknowledgement
We wish to thank DRPM Universitas Indonesia for funding the study through PITTA grant 2019 no. NKB-0641/UN2.R3.1/HKP.05.00/2019. Thank also to Naufal, Fadhil, Adit, and Syamil for interesting discussion during the analysis of MS-SVD technic.

References
[1] Bachri, Syaiful. (2014). Pengaruh Tektonik Regional Terhadap Pola Struktur dan Tektonik Pulau Jawa. JGSM 15 (4), 215-221
[2] Albini, P., Musson, R.M.W., Gomez Capera, A.A., Locati, M., Rovida, A., Stucchi, M., Viganò, D. (2013). Global Historical Earthquake Archive and Catalogue (1000-1903). In GEM Technical Report. GEM Foundation: Pavia, Italy
[3] Nguyen, N., Cipta, A., Cummins, P., & Griffin, J. (2015). Indonesia's Historical Earthquakes: Modelled Examples for Improving the National Hazard Map. Geoscience Australia.
[4] Koulali, A., McClusky, S., Susilo, S., Leonard, Y., Cummins, P., Tregoning, P., Meilano, I., Efendi, J., Wijanarto, A.B. (2016). The kinematics of crustal deformation in Java from GPS observations: implications for fault slip partitioning. Earth Planet. Earth and Planetary Science Letters, 458, 69–79
[5] Simandjuntak, T.O., and Barber, A.J. (1996). Contrasting tectonic styles in the Neogene orogenic belts of Indonesia. Geol. Soc. London Spec. Publ. 106 (1), 185–201
[6] Putri, A., Purwanto, M.S., dan Widodo, A. (2017). Identifikasi Percepatan Tanah Maksimum (PGA) dan Kerentanan Tanah Menggunakan Metode Mikrotremori Jalur Sesar Kendeng, Jurnal Geosaintek 3 (2)
[7] BMKG. (2019). Katalog Gempa Bumi Signifikan dan Merusak Tahun 1821-2018. Jakarta: Pusat Gempa Bumi dan Tsunami
[8] Rosid, M.S., Prastama, R.A., Daud, Y., and Riyanto, A. (2018). Identification of Subsidence in Jakarta Using 4D Microgravity Survey. The 43rd Annual Scientific Meeting Himpunan Ahli Geofisika Indonesia
[9] Apriani, M., Julius, A. M., Yusuf, M., Heryanto, D. T., & Marsono, A. (2018). Estimasi Ketebalan Sedimen dengan Analisis Power Spectral pada Data Anomali Gayaberat. Geomatika, 23(2), 65-74.
[10] Octonovrilna, L., & Pudja, I. P. (2009). Analisa Perbandingan Anomali Gravitasi Dengan Persebaran Intrusi Air Asin (Studi Kasus Jakarta 2006-2007). Jurnal Meteorologi dan Geofisika , 10 (1)
[11] Widianto, E. (2018, January). Petroleum system of Northwest Java basin based on gravity data analysis. In IOP Conference Series: Earth and Environmental Science. 106 (1), p. 012106.
[12] Fachri M., Djuhaeni, Hutasoit L.M., dan Agus M.R. (2002). Stratigrafi dan Hidrostratigrafi Cekungan Airtanah Jakarta. Buletin Geologi (Departemen Teknik Geologi ITB), 34(3). ISSN 0126-3498.
[13] Cipta, A., Cummins, P., Dettmer, J., Saygin, E., Irsyam, M., Rudyanto, A., & Murjaya, J. (2018). Seismic velocity structure of the Jakarta Basin, Indonesia, using trans-dimensional Bayesian inversion of horizontal-to-vertical spectral ratios. Geophysical Journal International, 215(1), 431-449.
[14] Sehah, S., Raharjo, S. A., & Dewi, R. Pemanfaatan Data Seismisitas untuk Memetakan Tingkat Resiko Bencana Gempabumi di Kawasan Eks-karesidenan Banyumas Jawa Tengah. In Seminar Nasional" Pengembangan Sumber Daya Pedesaan dan Kearifan Lokal Berkelanjutan II". Jenderal Soedirman University. 3(1).

IOP Conf. Series: Materials Science and Engineering **854** (2020) 012045 doi:10.1088/1757-899X/854/1/012045

Smart Bed Using Voice Recognition for Paralyzed Patient

Nurul Fadillah , Ahmad Ihsan

Faculty of Engineering, Samudra University

nurulfadillah@unsam.ac.id

Abstract. Smart beds are a way to simplify homework, especially for the elderly (elderly) and those who have physical disabilities, disabilities or paralysis. They can move the height of the bed according to their needs and comfort. The proposed system consists of a speech recognition module, an arduino, a relay circuit, and an adjustable bed. Voice recognition modules need to be trained before they are used to recognize commands. After successfully recognizing the voice command, Arduino directs the charge according to the help of the relay circuit. Adjustable bed height can be adjusted in three different modes according to the user's comfort and needs. The accuracy of the speech recognition module is measured in three bed height modes with different conditions Istirahat, Berbaring dan Bangun, Istirahat in a 145 degree position, Bangun 110 degrees and Berbaring 180 degrees. ON, OFF, Rest, Wake and Lie down. A total of ten experiments were carried out for each order listed in the table. Out of ten experiments, the speech recognition module correctly recognizes voice commands. The percentage accuracy of the voice recognition module in silence is 75%

1. Introduction

Smart bed is increasingly popular, because of its ease of use and extensive operating capabilities. Smart bed is a way to human homework using technology so that it can provide a sense of comfort, an easier life and more quiet time, especially for the parents (elderly) and those have disabilities or paralysis. Smart bed can also provide improved quality of life for people who might otherwise need caregivers or nurses. Integrating voice recognition technology into Smart bed makes the system more user-friendly and easy to operate. Smart bed is very necessary and provide will great assistance to meet the needs and comfort of people have physical disabilities, disabilities or paralysis.

2. Study of Literature

Proposed a Voice Recognition Based Home Automation System for Paralyzed People which system consists of a voice recognition module, Arduino UNO microcontroller, relay circuit to and an adjustable bed. The voice recognition module needs to be trained first before it can be used to recognize commands. Upon successful recognition of voice command, the Arduino drives the corresponding load with the help of the relay circuit. The adjustable bed elevation can be set to the three different modes as per the user's comfort and need. [1]

There have been several researches and developments on the home automation systems. The voice recognition-based home automation system uses the Microsoft speech API running on PC to recognize the voice commands. The RF transreceiver is used to send these commands to the controller to control the various electrical devices. The use of computer makes this system more expensive and difficult to handle [2].

Intelligent home navigation system for disabled and elderly person proposed a system which uses voice recognition module SR-07 for the speech recognition process, an Arduino controller, a wheelchair and a navigation module. The Arduino receives the command from the voice recognition module and move the wheelchair accordingly thus eliminating the need of any third person's assistance [3].

The voice recognition-based home automation system uses Lab VIEW to perform speech recognition and Zigbee module with a controller is used to control the devices wirelessly [4].

Proposed a home automation system which comprises a DSP processor for the voice recognition function, a microcontroller and relay module for the appliances control function like switching lights on-off etc. Zigbee wireless module is used which eliminates the need of additional wiring required for the signal transmission [5].

The home automation system proposed two ways to control home appliances that are by using timer or by using the voice commands. The software environment is developed on Virtual Basics 6.0 on PC and devices are controlled using PC parallel ports [6].

In this research, Smart Bed used voice recognition for patients suffering from quadriplegia or paraplegia (who cannot move their limbs but can speak and listen) in controlling and moving the height of the bed. With voice commands they can move the height of the bed according to their needs and comfort. The system consists of voice recognition modules, microcontrollers, relay circuits and adjustable beds. Voice recognition modules are trained in advance so they can be used to recognize commands. After successfully recognizing the voice command, the microcontroller directs the charge in accordance with the help of relay circuits.

The problem in this research considers the breadth of the problem, so this research is limited to the development of Smart Bed using speech recognition to move the bed height of patients suffering from quadriplegia or paraplegia (who cannot move their limbs but can speak and listen).

3. Research Method

The system design of the Smart Bed system uses voice recognition for paralyzed patients as shown in figure 1:

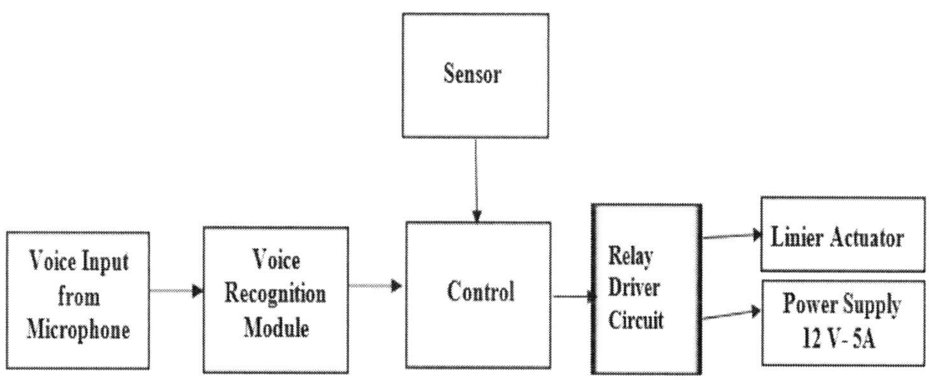

Figure 1. System planning of the Smart Bed uses voice recognition for paralyzed patients

Figure 1 explains the system process ie sound input from a microphone is given to the speech recognition module where the speech signal is compared to a sample of trained voice stored previously. After successfully obtaining a voice command, Arduino moves an electrical device that adjusts the height of the bed using the relay module. Data from the illumination sensor is processed in an Arduino controller and based on setpoint values, automatic control steps are taken to save energy. The hardware of the proposed system is as follows: Microphone and Voice Recognition Module, the microphone used to get voice commands to the speech recognition module is a type microphone in a simple direction with a 3.5 mm jack. Elechouse v3 voice recognition module is used for the speech recognition process as shown in Figure 1. Voice recognition modules need to be trained before known for recognized to recognize voice commands. Voice input from the microphone is given to the speech recognition module and compared to previously trained voice commands and if there is a match then the control action through the control circuit is taken. The v3 speech recognition module can store up to 80 commands, out of 80 only 7 commands can be entered into the recognizer for the recognition process. Thus only 7 commands are effective at one time and to add 7 other recognizer commands need to be cleaned first. This module has two ways to control Serial Port, General Input Pins, Shown in Figs. 2.

Figure 2. Elechouse v3 voice recognition module

3.1.1. Arduino Uno

The controller used for the proposed system is Arduino Uno. The Arduino platform provides a cheap and easy way for students and professionals to create devices that interact with their environment using sensors and actuators. Arduino is based on ATmega 328. It has 14 digital input/output pins (of 14 pins 6 its can be used as PWM output) and 6 analog inputs. Arduino works at 5V D.C and has a 16 MHz clock speed, shown in Figs. 3.

Figure 3. Arduino Uno

3.1.2. Relly

Control the patient's bed relay using Arduino. The relay used in the system is a 5V-5 pin relay as shown in Figs. 4. The relay remains in normally closed condition. When the relay coil is energized on a relay switch from a normal state which is normally closed to normal due to electromagnetic induction, shown in Figs. 4.

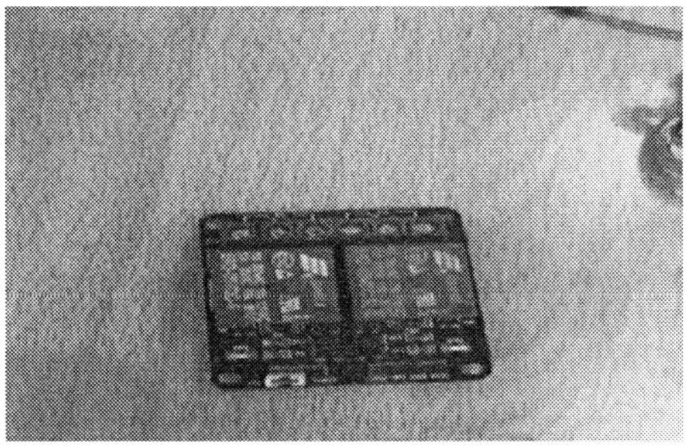

Figure 4. Relly

3.1.3. Linear Actuator

Linear Actuator circuit is to place the Linear Actuator as a driving force in detecting sound. Linear Actuator will automatically detects the directions up and down the bed with the sound. The linear actuator is connected with a sound sensor to detect the up and down direction of the bed. shown in Figs. 5

Figure 5. Linear Actuator

3.2. Smart Bed Design

Mechanical design is the first step to making a smart bed play a very important role in voice recognition for the position of the bed. As for some of the mechanical manufacturing stages of the bed is the selection of materials, the material used in the design of the bed is stainless with the aim of avoiding corrosion. Stainless material is cut and shaped as a component holder and bed frame, the frame is designed to be sturdy to support the patient and has a flexible motion function as needed. The number of motion functions is referred to as the degree of freedom (DOF). Overall mechanics can be seen in Figure 6 as follows:

Figure 6. Mechanical design

3.3. *Series Linear Actuator*

Linear Actuator Series is to place the Linear Actuator as a mover in detecting voice. Linear Actuator series will automatically detect the direction in accordance with voice commands. The linear actuator is connected to the Elechouse v3 voice recognition module to detect sound, Schematic Linear Actuator can be seen in Figure 7 below:

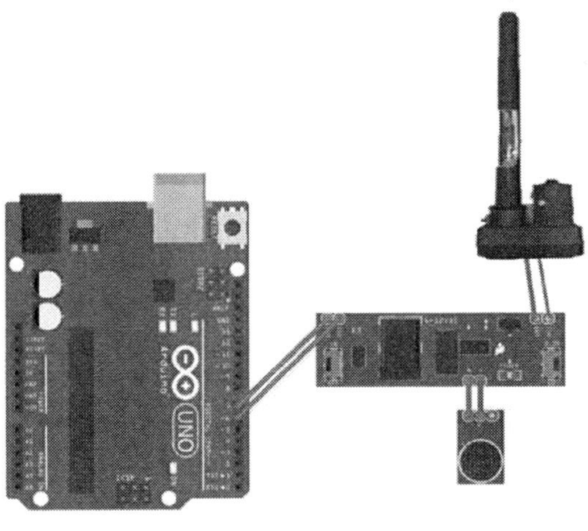

Figure 7. Visual Schematic Linear Driver Actuator

4. Results and Discussions

The main objective of this research is to design a smart bed system using speech recognition to control and model a bed with height in accordance with the voice commands given by the user. Recognized voice commands make Arduino change the relay and change the direction of the motor because which jack lifts the bed or brings the bed back to a lower altitude angle. The bed can be adjusted to fit three positions, *tidur* position, *istirahat* position, and *bangun* position. *Tidur* position is a position where the bed is at a height of 180 degrees to the ground. The person can *tidur* comfortably in this position. In a resting position, the bed is *istirahat* at 145 degrees to the ground and one can comfortably relax in this position. In the *bangun* position, the height of the bed is almost equal to 110 degrees and almost the person is in a *bangun* condition. the current drawn by the motor when lifting a bed or lowering at different weights.

5. Commands Voice and Their Functions

Table 1. Voice Commad And Their Functions

No.	Voice Commands	Position bed	Experimental Trial										Total Responses
			1	2	3	4	5	6	7	8	9	10	
1	On	-	1	1	1	1	1	1	1	1	1	1	10
2	Off	-	1	1	1	1	1	1	1	1	1	1	10
3	*Istirahat*	145 degrees	0	1	0	0	1	1	1	1	0	1	6
4	*Bangun*	110 degrees	1	0	0	0	1	1	1	0	0	1	5
5	*Tidur*	180 degrees	1	1	1	1	0	0	1	1	1	1	8

Table 1 shows the results of the path tests carried out in the speech recognition module on the smart bed system, to determine accuracy in determining each condition namely, ON, OFF, Rest, Wake and Lie down. A total of ten experiments were carried out for each order listed in the table. Out of ten experiments, the voice recognition module correctly recognizes voice commands. The percentage accuracy of the speech recognition module in silence is 75%.

In the noise conditions of the Ten trials, only three were recognized correctly so that the accuracy of the voice recognition module in the noise conditions was 65%.

References

[1] Mukesh Kumar and Shimi S .L, "Voice Recognition Based Home Automation System for Paralyzed People," vol. 4, no. 10, pp. 2508–2515, 2015.
[2] T. Kirankumar and B. Bhavani, "A Sustainable Automated System for Elderly People Using Voice Recognition and Touch Screen Technology," vol. 2, no. 8, pp. 265–267, 2013.
[3] R. K. Megalingam, R. N. Nair, and S. M. Prakhya, "Automated Voice based Home Navigation System for the Elderly and the Physically Challenged," pp. 603–608, 2011.
[4] A. . E and M. Jagadeeswari, "Control of Electrical Appliances through Voice Commands," *IOSR J. Electr. Electron. Eng.*, vol. 9, no. 1, pp. 13–18, 2014.
[5] S. K. Gopy, G. Hurry, and T. T. Gopaul, "A Study on Smart Home Control System through Speech," vol. 69, no. 19, pp. 30–39, 2013.
[6] S. M. A. Haque, S. M. Kamruzzaman, and A. Islam, "A System for Smart-Home Control of Appliances Based on Timer and Speech Interaction," pp. 4–7, 2006.
[7] S. M. Metev and V. P. Veiko, *Laser Assisted Microtechnology*, 2nd ed., R. M. Osgood, Jr., Ed. Berlin, Germany: Springer-Verlag, 1998.
[8] J. Breckling, Ed., *The Analysis of Directional Time Series: Applications to Wind Speed and Direction*, ser. Lecture Notes in Statistics. Berlin, Germany: Springer, 1989, vol. 61.
[9] M. Wegmuller, J. P. von der Weid, P. Oberson, and N. Gisin, "High resolution fiber distributed measurements with coherent OFDR," in *Proc. ECOC'00*, 2000, paper 11.3.4, p. 109.
[10] R. E. Sorace, V. S. Reinhardt, and S. A. Vaughn, "High-speed digital-to-RF converter," U.S. Patent 5 668 842, Sept. 16, 1997.
[11] (2002) The IEEE website. [Online]. Available: http://www.ieee.org/
[12] M. Shell. (2002) IEEEtran homepage on CTAN. [Online]. Available: http://www.ctan.org/tex-archive/macros/latex/contrib/supported/IEEEtran/*FLEXChip Signal Processor (MC68175/D)* Motorola, 1996.
[13] "PDCA12-70 data sheet," Opto Speed SA, Mezzovico, Switzerland.
[14] A. Karnik, "Performance of TCP congestion control with rate feedback: TCP/ABR and rate adaptive TCP/IP," M. Eng. thesis, Indian Institute of Science, Bangalore, India, Jan. 1999.
[15] J. Padhye, V. Firoiu, and D. Towsley, "A stochastic model of TCP Reno congestion avoidance and control," Univ. of Massachusetts, Amherst, MA, CMPSCI Tech. Rep. 99-02, 1999.

IOP Conf. Series: Materials Science and Engineering **854** (2020) 012046 doi:10.1088/1757-899X/854/1/012046

Intelligent System of Automatic Shrimp Feeding

N Novianda, F Liza, I Ahmad

Informatics Engineering Study Program, Faculty of Engineering, Samudra University, Meurandeh Langsa Lama Village, Langsa City, Aceh 24354

Email: novianda_tif@unsam.ac.id, lizafitria@unsam.ac.id, ahmadihsan@unsam.ac.id

Abstract. Feeding on shrimp culture is important to the shrimp growth process so that the expected yields are achieved. Feeding must be balanced and timely, there are no advantages or disadvantages in feeding because all the feed provided is not necessarily completely consumed by shrimp, therefore it is necessary to have technology that can control shrimp feeding. Automatic shrimp feed technology can be a solution to the problem of feeding to support the productivity of shrimp culture. Automatic shrimp feeder is built using an arduino microcontroller module, RTC module (Real Time Clock), and feed throwing system. The shrimp feeding method with the working principle of shrimp feeder is shrimp feed thrown horizontally due to centrifugal force. With the influence of the centrifugal force from the propeller spreader, the feed will be thrown throughout the surface of the pond. Feeding is controlled using Arduino by setting the time. Shrimp feeding is done every 4 hours and the appliance is on for 7 minutes to spread 10 kg of feed for the number of shrimps 1,500 and the width of kolan 400 m2 with a minimum throwing distance of feed 1 meter and a maximum of 7 meters

1. Introduction

Shrimp is one of the fisheries commodities that provide a significant contribution in increasing local revenue. Aceh is one of the provinces that is very rich in marine and fisheries potential resources. The maritime and fisheries sector will be a leading sector that can be relied on in terms of economic and political strength in lifting the dignity and dignity of the people of Aceh to achieve the welfare they aspire to when utilized optimally. One of the fisheries sectors that can be utilized is shrimp ponds. The large number of shrimp ponds in Aceh makes people think about how to increase the productivity of shrimp harvests and minimize failure in harvesting shrimp.

Langsa City is one of the cities in Aceh, Indonesia. Langsa City is a former part of East Aceh Regency. Located approximately 400 km from the city of Banda Aceh. Langsa City Area 262, 41 Km2. Langsa City is a coastal city which has a coastline of 16 km. A very heterogeneous population - Aceh, Java, Malay, Gayo Batak, and Karo - is only 246 km from Medan City. In the field of marine and fisheries, Langsa City has the potential to be developed. Based on data from the Aceh Marine and Fisheries Service, in 2016 the productive pond area in Langsa City reached around 4,647 Ha, producing 112.5 tons of Windu shrimp. Therefore, development is needed to increase shrimp production in the town of Langsa which is also one of the raw materials in making shrimp paste which is already very well known.

One effort to increase shrimp production in intensive cultivation (intensification) systems is feeding. Feeding system by manually spreading without controlling and paying attention to the level of feed that has been given will be fatal for the survival of shrimp. Feed can be an 'enemy' if it is not managed properly because the feed given is not entirely consumed, the remnants of the food that is not consumed will experience decomposition and decay. Feed spoilage can pose a risk of disease so that it can reduce

IOP Conf. Series: Materials Science and Engineering **854** (2020) 012046 doi:10.1088/1757-899X/854/1/012046

shrimp production. In addition to disease risk, economically the feed consumes the highest budget in shrimp farming if it is not properly and efficiently controlled. Therefore, it is very necessary to control the shrimp feeding system automatically which can determine the amount of feed in and out

2. Water Quality Parameters

Water quality parameters of the media must be at optimal conditions. The parameters that influence the cultivation are pH, dissolved oxygen, ammonia, temperature, salinity, and nitrite. Optimum level and range of water quality that can be accepted by vannamei shrimp (Table 1)

Table 1. Range of optimum values for water quality parameters in shrimp rearing

No	Water parameters	Optimum value
1.	Temperature	28.5 – 31.5 ºC
2.	Salinity	15 – 25 ppt
3.	Brightness	30 – 45 cm
4.	Dissolved oxygen	>3.5 mg/l
5.	pH	7.5 – 8.5
6.	CO_2	< 25 mg/l
7.	Ammonia	<0.01 mg/l
8.	Nitrite (NO_2)	0.01 mg/l

3. Feed thrower design

This feed thrower design considers the efficiency of shrimp feeding. The aim is to produce technology that can help farm workers in providing shrimp feed consistently, on time and in a measurable manner. To answer this question, the design of an automatic shrimp feeder was produced, which consisted of a feed container, a timer using RTC (Real Time Clock), an accumulator (battery), a food thrower and a microcontroller as the control center and control of feeding automatically. The main block of the feed thrower can be seen in Figure 1.

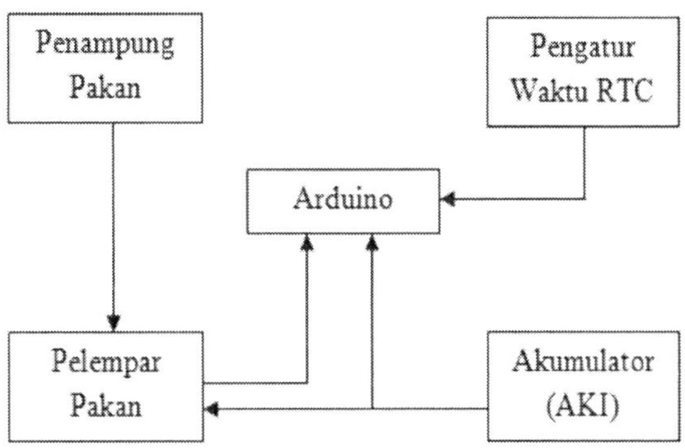

Figure 1. Block Design of an automatic feed thrower

4. Feed Algorithm

Motor drive, and feeding schedule stored on the eprom. The algorithm as shown in Figure 5 starts the RTC time setting and the feeding schedule. Every time the microcontroller will read the time from the RTC then compare it to the schedule stored in the eprom. If the RTC time and schedule are the same, the feed throwing process starts with a duration of 70 seconds for 1 kg of feed. The process will be recorded continuously every 4 hours.

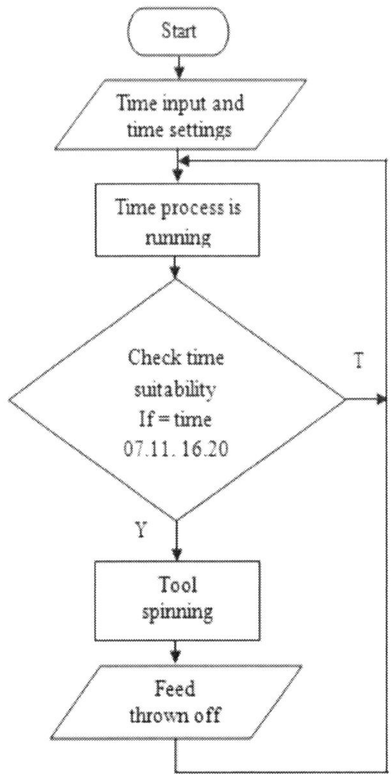

Figure 2. Block diagram of automatic feeding algorithm

5. Component Assembly

This research, automatic shrimp feeder device is designed from start to finish, supporting components are needed to design the tool so that it can work according to their respective functions. The following are the components of an automatic feed tool.

Figure 3. Components of automatic feed tools

6. Results and Discussion

Based on the results of research conducted, it is obtained the results of an analysis that the manufacture of automatic shrimp feed equipment has been successfully made and tested manually to ensure the tool can rotate properly or not. After the first test is done on the instrument traced error in the rotary propeller which is stuck by a pile of feed pellets so that the motor movement becomes slow. For this error, the rotary propeller is reopened and then repaired so that it functions and can rotate properly and feed can come out through the holes in the rotary propeller. The finished tool is then installed a system that is able to work in real time to control the feeding time of the shrimp automatically so that the work will be more efficient because the feeding works automatically at the specified time. In Figures 4 and 5 are shown tools that have been assembled shrimp feed.

.

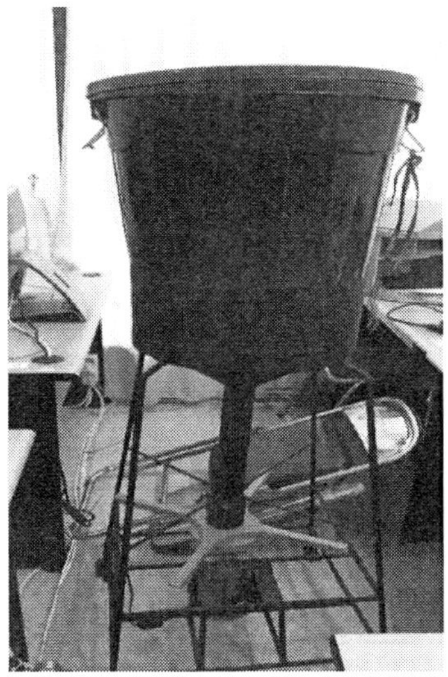

Figure 4. Shrimp feed equipment

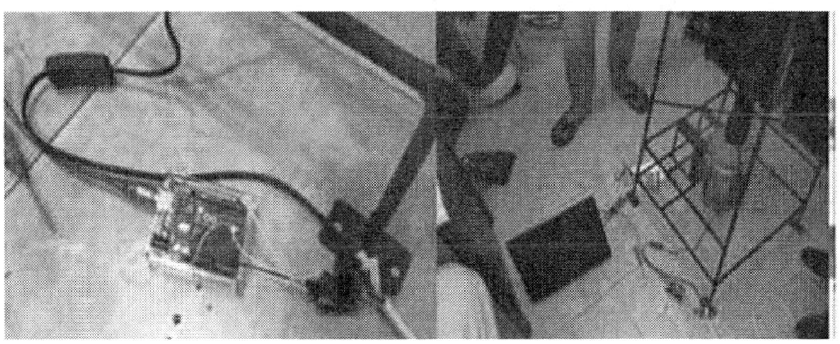

Figure 5. Test tool

In the picture it is explained that the devices that have been assembled are always in the program to be able to work automatically using Arduino controls. All components that have been put together can work according to their functions, so that the automatic feed tool has been completed and the results can be applied to shrimp ponds. In table 2, it is explained how the tool works in removing feed

Table 2. Working sizes of automatic feeders

IOP Conf. Series: Materials Science and Engineering **854** (2020) 012046 doi:10.1088/1757-899X/854/1/012046

Automatic Timing		Pool area (M^2)	amount Feed In (Kg)	amount Feed Out (Kg)	amount of Shrimp (tail)	Time (seconds)	Throwing distance (meters)		
Time on	**Time off**								
07.00.00	07.00.07	400	40	10	1500	70	1	3	7
11.00.00	11.00.07	400	30	10	1500	70	1	2	5
16.00.00	16.00.07	400	20	10	1500	70	1	2.5	6.5
20.00.00	20.00.07	400	10	10	1500	70	1	1.5	6

In table 2 is a working size chart of the automatic feeder tool, the tool works regularly at a predetermined time. The appliance starts at 07.00 and dies at 07.00.07 seconds. At 11.00 the appliance is on and at 11.00.07 the appliance is dead, and so on. In table 2 the size of the pond, the amount of feed coming in and going out, the number of shrimps in the pond, the number of seconds and the throwing distance of the feed varied, but the minimum distance of the feed was thrown at 1 meter and a maximum of 7 meters. This may change depending on the tools and propellers used, the area of the pond is proportional to the number of shrimps stocked.

7. Conclusion

The position of the tool is placed in the middle of the pond so that when the tool rotates the feed does not come out of the pool area, the tool rotates for 70 seconds, every 4 hours the tool rotates to release as much as 10 kg of food, automatic shrimp feeder equipment makes it easy for the farm owner to feed on time and automatic shrimp feeding device is completed designed and working Real Time Clock (RTC)

References

[1] S.Subaashri*, M.Sowndarya , D.K.S. Sowmiyalaxmi , S.V.Sivassan, C. Rajasekaran, 2017. Automatic Pet Monitoring and Feeding System Using IoT, International Journal of ChemTech Research. Vol.10 No.14, pp 253-258. ISSN: 0974-4290

[2] Dileep Kumar Appana1*, Mohammad Wajih Alam2 and Bigyan Basnet3, 2016, A Novel Design of Feeder System for Aqua Culture Suitable for Shrimp Farming. International Journal of Hybrid Information Technology. Vol.9, No.4

[3] Dimas Angga Hedianto, Astri Suryandari, dan Didik Wahju Hendro Tjahjo, 2016. Population Dynamic And Exploitation Rate Of Of The Indian Tiger Prawn Penaeus Monodon (Fabricus, 1789) In The East Aceh Waters, Aceh Province. Jurnal penelitian perikanan indonesia Volume 22 Nomor 2 e-ISSN: 2502-6542

[4] Siti Aisyah Binti Saahri. 2015 Design And Fabrication Of An Automatic Fish Feeding System For Home Aquarium. Tesis of Faculty Of Manufacturing Engineering Universiti Malaysia Pahang

IOP Conf. Series: Materials Science and Engineering **854** (2020) 012047 doi:10.1088/1757-899X/854/1/012047

Effect of Addition of Chitosan-ZnO as an Antibacterial on The Thermal Stability of PLA-Based Nanocomposites for Biofilm Packaging

T Rihayat [1], S Riskina[1], Nurhanifa[1], R Syahyadi[2]

[1] Department of Chemical Engineering, Politeknik Negeri Lhokseumawe, 24301, Aceh Indonesia

[2] Department of Civil Engineering, Politeknik Negeri Lhokseumawe, 24301, Aceh Indonesia

Email: teukurihayat@yahoo.com

Abstract. This research is about Composite Polylactic Acid with chitosan and zinc oxide reinforcement (ZnO) in the manufacture of packaging biofilms. The purpose of this study is to produce biofilms with good characteristics that can be applied to food packaging. The method used is heating by varying the concentration of chitosan (1, 3 and 6%) and the concentration of ZnO at (1,2,3 and 4%). The resulting biofilm was analyzed by tensile test, thermal gravimetric, and FTIR. The results showed that the optimum tensile strength value obtained at the addition of Chitosan 6% - ZnO 3%, amounting to 1.22 MPa. The functional groups contained with the addition of ZnO and chitosan are functional groups C-H, C = O, O-H, and N-H. Biofilm samples underwent single decomposition with biofilm degradation temperatures ranging from 220-300 °C, the highest On Set temperature was 238.24°C, while the highest End Set was 295.16 °C. ZnO reinforcement added to the resulting composite can increase the tensile strength.

1. Introduction

The development of technology that accelerated this century pushed the industry to include food. Currently available food packaging that is practical and durable in a long time. Unfortunately, this food packaging is made from ingredients that cannot be degraded in nature, thus encouraging environmental pollution. These non-degradable materials are generally material from petroleum products that are harmful to the environment. The solution to this is to use biopolymer materials that can be degraded naturally by nature. Many researchers have improved the material from the biopolymer, one of which is to use it the concept of nanocomposites [1]. Nanocomposites are promising solutions for biopolymer problems.

Bio-composite is an environmentally friendly and biodegradable material, this material consists of two or more materials, the matrix of which is a Polylactic Acid (PLA) biopolymer and nanofiller used as chitosan and zinc oxide (ZnO). Because several materials used can increase the durability of the packaging.

The PLA matrix is synthesized by the fermentation of renewable agricultural sources, including corn, cellulose, and other polysaccharides. Although some of its characteristics are unfavorable (for example poor melting properties, mechanical fragility, low heat resistance, and slow crystallization), there are routes potential to overcome this deficiency. Then the PLA matrix can be included in the copolymerization, mixing, modification of the plasticization, or the addition of the reinforcing phase for example, chitosan (Cs), cellulose, and starch) [2] and is well applied for food security.

In a study conducted by Lin et al [3], which mixed PLA and Chitosan with variations in the concentration of 0.2; 0.4; 0.8% by weight after 2 hours of incubation in 10 times simulated body fluid (10 × SBF). X-ray diffraction results show that it can increase more nucleation and growth calcium phosphate in the condition of chitosan concentration of 0.4%. These results indicate that this composite can mimic the structural, composition and biological functions of the original bone and can function as a good candidate for bone tissue engineering (BTE).

Silva et al [4], conducted research between PLA and ZnO having great potential for antimicrobial activity, but poor interface interactions between PLA and ZnO caused poor mechanical and thermal properties. The ZnO / PLA film selects the mechanical properties of the inferior by proving that 5% of the mass of ZnO-PLA increases the tensile strength and modulus of elasticity of each 30% and 65%, respectively. PLA nanocomposite films with ZnO exhibit exceptional antimicrobial properties against E. coli and S. aureus, where bacterial counts are greater than 98% in a day and there are no dead bacterial colonies in high fillers.

Based on the reference of the composite merge research above, the focus of this research is on mixing PLA, chitosan and ZnO to be able to improve the mechanical and thermal properties. Because in this case, chitosan used in this research as an amplifier of the film produced and the addition of ZnO here is very influential, because ZnO has properties such as an antibacterial effect or intensive ultraviolet absorption. ZnO is also a multifunctional and active ingredient for packaging (to increase the shelf life of food products) as well as textile applications with exceptional antimicrobial properties.

2. Material and Method

2.1. Manufacture of PLA-Chitosan-ZnO nanocomposites

PLA pellets are dissolved in 30 ml of chloroform solution at 60 °C for 2 hours. After the PLA dissolved, chitosan and zinc oxide (ZnO) were added with variations in the concentration of zinc oxide (ZnO) 1%, 2%, 3%, and 4% and variations in the concentration of chitosan 1%, 3%, and 6%. Then it is heated at 80 °C and stirred constantly so that the mixture is homogeneous for 15 minutes. After all homogeneous, added 2 ml of glycerol. The composite solution is put into a mold, then dried in the oven at 40 °C for 24 hours to dry. Then the composite formed is tested by tensile strength and FTIR tests.

2.2. Characterization of PLA-Chitosan-ZnO nanocomposites

2.2.1. Tensile Strength Test (ASTM. D 638-02 type 4)

Testing of mechanical properties is carried out by tensile strength testing of composite materials using ASTM D 638 carried out by cutting dumbbell-shaped samples using typical cuts following specified dimensions: 1. Samples that have been printed according to size are placed at both ends of the stapler that are positioned on the pulling tool, 2. The pull engine switch and the chart recording switch are turned on together, according to the engine pull speed, 3. From the results of testing the test machine will be obtained by the relationship between the tensile force to the length increase. And here is a picture of a tensile strength test specimen [5] as in figure 1.

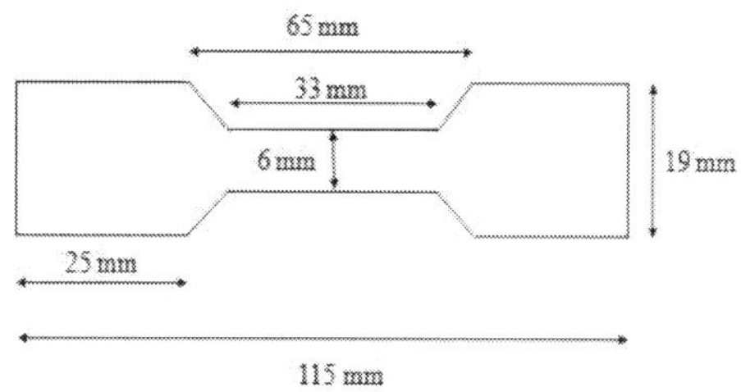

Figure 1. Tensile Strength Test Sample Forms With ASTM D 638 Type IV standard

2.2.2. FTIR analysis

Infrared Fourier Transform (FTIR) analysis is the spectra determined in the spectral range from 600 cm^{-1} to 3600 cm^{-1} with a resolution of 4 cm^{-1}. Data were analyzed with FTIR Spectrum (Perkin Elmer) software.

2.2.3. XRD analysis (Thermografimetric Analyzer)

The thermal stability of the sample is determined using a thermogravimetric analyzer, using the Perkin Elmer Thermogravimetric analysis at a heating rate of 10 °C/min under atmospheric nitrogen, from 50 °C to 700 °C. About the weight of each sample is analyzed and consequently, the weight loss of the sample is determined.

3. Result and Discussion

3.1. Effect of Chitosan and Zinc Oxide Concentration Variations on Mechanical Characteristics of PLA-Chitosan-ZnO Composites

The mechanical characteristics of PLA-Chitosan-ZnO composites were analyzed through testing the tensile strength of the sample using a UTM (Universal Testing Machine), The specimen dimensions used follow the specimen dimensions in ASTM D 638-99 type IV where each composite to be tested is given a tensile load of 200 kgf, with the test specimen being pulled parallel until the composite material is broken. According to Farah [6], the value of the tensile strength of a material is shown based on the value of UTS (Ultimate Tensile Strength) which is the maximum stress value that can be received by the material before the test material is damaged or broken.

The result of the high tensile strength (mechanical properties) is related to research conducted by Wafiroh [7], on biofilms that have high mechanical properties will increase the ability and strength of biofilms to improve the quality of the products it provides. This tensile strength test was made by the biofilm constituent components published by PLA, chitosan and Zinc Oxide (ZnO), the homogeneous level of the dissolution of this biofilm and the relationship of tensile strength with the concentration of Chitosan and ZnO are presented in Figure 2.

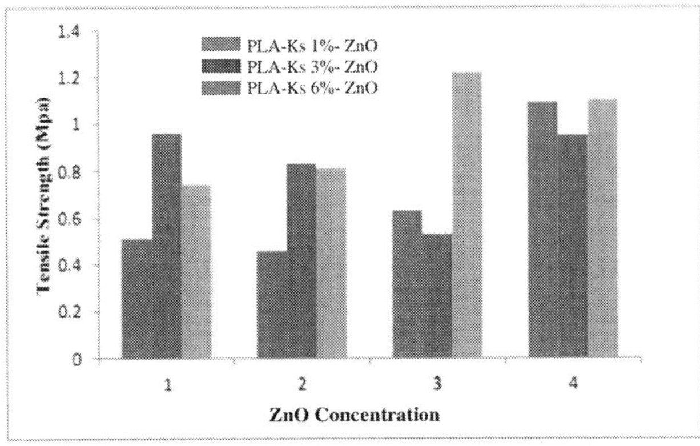

Figure 2. Tensile strength of various composites

The variation in the tensile strength of PLA-Chitosan-ZnO composites is a function of optimal dispersion and good inter-component interactions between PLA, Ks and ZnO. This shows that uniform dispersion in the presence of fillers causes good interaction with the matrix. The polymer then causes an increase in mechanical integrity and thermal stability of the composite. It is recognized that the tensile strength of the PLA matrix shows improvement with the addition of ZnO and chitosan in composites with a certain charge.

From the picture, it can be seen that the tensile strength value produced during testing is not constant, this result is not in accordance with research conducted by Amni [8], where the results produced are increasingly added to ZnO reinforcement, then the tensile strength value produced by biofilms will be the higher, because more ZnO amplifiers will affect the structure of the biofilm. This is caused by the effect of the addition of plasticizer Glycerol as plasticity, this is reinforced by research conducted by Yuniarti [9], which obtained an unstable tensile strength test value of starch-based bioplastics with the addition of glycerol resulting from. Glycerol causes a tendency to unite between atoms and molecules in stretchable bioplastics, where this is one of the important factors that affect the mechanical properties of bioplastic materials is activity between its constituent components which is a phenomenon where atoms or molecules tend to unite and bind [10].

ZnO and Chitosan were put into the PLA matrix, the composite tensile strength was obtained well, so that the optimal tensile strength value was obtained from the highest value obtained in the composition with a ratio of 6% chitosan and 3% ZnO.

3.2. FTIR analysis of PLA-Chitosan-ZnO nanocomposites

FT-IR analysis was carried out to identify the functional groups contained in the PLA-Chitosan-ZnO Composite sample by using a Shimadzu IR Prestige - 21 Fourier Transform Infrared Spectrophotometer (Serial No. A210048 02519). This analysis also aims to determine whether the addition of chitosan and ZnO concentrations affect existing functional groups and to observe the possibility of reactions occurring at the processing stage marked by the emergence of new functional groups that were not previously detected or vice versa.

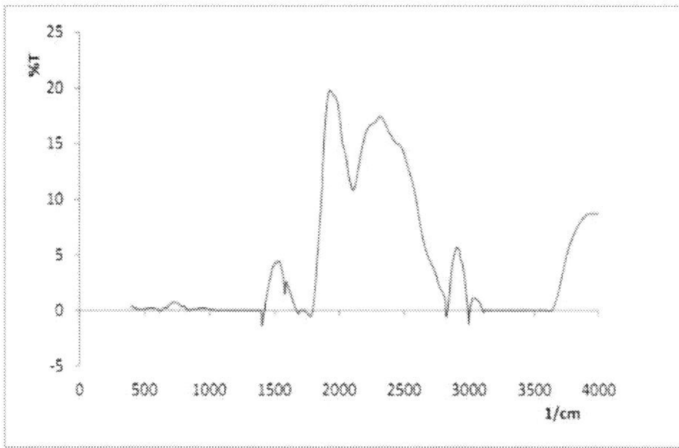

Figure 3. Graph analysis of PLA-Chitosan-ZnO Composite functional groups

Table 1. Wave Number with Alleged Compounds

Functional Groups	Wave Number cm^{-1}
N-H	3336,85
O-H	3630,03
C-H	3294,42
C-O	1793,80

In figure 3, a graph of the PLA-Chitosan-ZnO composite functional group analysis graph is displayed. The graph shows the FTIR test results that from the optimum sample the tensile strength at PLA-Ks 6% -ZnO 3% contained N-H, C-H, O-H, and C = O. Where the NH group is in the wavenumber 3336.85 cm^{-1} with a wavelength range 3300-3400 cm-1, the OH group is in the number 3630.03 cm^{-1} in the wave range 3584 - 3700 cm^{-1}, the CH group is in the number 3294.42 cm^{-1} with a wave range of 3267-3333 cm^{-1}, and the C = O group is found at number 1793.80 cm^{-1} with a wave range 1540-1870 cm^{-1}. This means that no new functional groups have been found so that this PLA biofilm has properties such as its constituent components.

3.3. TGA analysis (Gravimetric Thermal) of PLA-Chitosan-ZnO nanocomposites

Thermogravimetric analysis is a method used to measure the overall temperature and weight loss due to the influence of temperature on the composite material when the material is heated there will be a structural change resulting in a change in heat capacityor the thermal energy of the material. TGA testing aims to determine qualitatively the thermal stability of biocomposite PLA-Chitosan with ZnO amplifier. Effect of addition of chitosan and zinc oxide on resistance properties thermal composite materials can be seen in the following image.

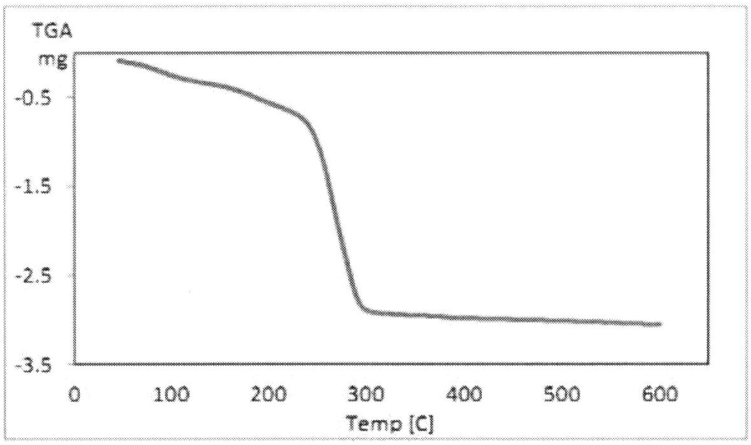

Figure 4. TGA composite PLA-Ks-ZnO test

From the figure 4. above it can be seen that the decrease in mass on the y-axis and increase in temperature on the x-axis. The graph shows that biofilm samples undergo a single decomposition because on set and end set only occur once, where on set is the temperature at which the sample begins to be degraded thermally and the end set is the temperature at which the sample temperature holds its mass during the thermal combustion reaction process. Biofilm degradation temperature at this research ranges in the range 220 - 300 °C. Based on the picture of 4 biofilms that experienced on set the highest thermal degradation is a biofilm with the addition of ZnO of 238.24 °C, this shows that the thermal resistance of biofilms has higher thermal stability to temperature. As for figure 4, the oldest degradation process (End Set) is obtained biofilm namely the degradation process ends at 295.16 °C. End Set is the end of the degradation process which is the stage where the process of evaporation of a substance or termination of group chain fish will cause the release of these compounds in biofilms. And at the stage after the End Set sample will undergo a carbonization process (ashing) because there are no more compounds and water in the evaporated biofilm.

4. Conclusion

From the result of this study some conclusion can be drawn : (1) The optimum tensile strength value obtained on the addition of Chitosan 6% - ZnO 3%, which is equal to 1.22 MPa. (2) The FT-IR test results contained C-H, O-H, N-H and C = O groups and showed no new functional groups. (3) Biofilm samples underwent single decomposition with biofilm degradation temperatures ranging from 220 - 300 °C, the highest On Set temperature was 238.24 °C, while the highest End Set was 295.16 °C. (4) ZnO reinforcement added to the resulting composite can increase the tensile strength.

References

[1] Marra, A., Silvestre, C., Duraccio, D., & Cimmino, S. (2016). Polylactic acid/zinc oxide biocomposite films for food packaging application. International Journal of Biological Macromolecules, 88, 54–262

[2] Elsawy, M. A., Kim, K. H., Park, J. W., & Deep, A. (2017). Hydrolytic degradation of polylactic acid (PLA) and its composites. Renewable and Sustainable Energy Reviews, 79, 1346-1352.

[3] Lin, C. C., Fu, S. J., Lin, C. Y., Yang, K., & Gu, Y. (2014). Chitosan-coated electrospun

PLA fibers for rapid mineralization of calcium phosphate. International Journal of Biological Macromolecules, 68, 39-47.

[4] Silva, R. T. D., Pasbakhsh, P., Mae, L. S., & Kit, A. W. (2015). ZnO deposited/encapsulated halloysite–poly (lactic acid) (PLA) nanocomposites for high performance packaging films with improved mechanical and antimicrobial properties. Applied Clay Science, 111, 10-20.

[5] Tumpal, O. R., Hansen. A. M. S. (2016). Analisis Fraksi Volume Serat Pelepah Batang Pisang Bermatriks Unsaturated Resin Polyester (UPR) Terhadap Kekuatan Tarik dan SEM. J urnal Mechanical, Vol-6, 01.

[6] Farah, S., Anderson, D. G., & Langer, R. (2016). Physical and mechanical properties of PLA, and their functions in widespread applications - A comprehensive review. Advanced Drug Delivery Reviews, 107, 367-392.

[7] Wafiroh, S., Ardiato, T., Agustin, E, T. (2017). Pembuatan Dan Karakterisasi Edible Film Darikomposit Kitosan-Pati Garut (Maranta Arundinaceae L) Dengan Pemlastis Asam Laurat. Media Matematika dan Ilmu Pengetahuan Alam, Vol-1, 04.

[8] Amni, C., Marwan., Mariana. (2015). Pembuatan Bioplastik Dari Pati Ubi Kayu Berpenguat Nano Serat Jerami Dan Zno. Jurnal Litbang Industri, Vol-5, No.2.

[9] Yuniarti, L, I., Gatot, S, H., Abdul, R. (2014). Sintesis Dan Karakterisasi Bioplastik Berbasis Pati Sagu (Metroxylon Sp). Agrotekbis, Vol-2(1), 38-46.

[10] Teuku, R., Suryani., Teuku, F., Harry, A., B. Wirjosentono., Syafruddin., Helmi., Zulkifli., Pocut, N, A., Muhammad, S. 2017. Mechanical properties evaluation of single and hybrid composites polyester reinforced bamboo, PALF and coir fiber. IOP Conference Series: Materials Science and Engineering. Vo.334.

[11] Teuku, R., Suryani. (2010). Synthesis and Properties of Biobased Polyurethane/Montmorillonite Nanocomposites. International Scholarly and Scientific Research & Innovation, Vol. 4 (5).

[12] Teuku, R., Saari, M., Mahmood, H,M., Yunus, W,Z,M,W., Suraya, R,A., Dahlan, M,H,Z,K., Sapuan, M,S. (2007). Mechanical Characterisation of Polyurethane/Clay Nanocomposites. Polymers & Polymer Composites. Vol. 15 (8).

IOP Conf. Series: Materials Science and Engineering **854** (2020) 012048 doi:10.1088/1757-899X/854/1/012048

Preparation, Characterization of Palm Oil Based Polyols Reinforced with Bentonite-Chitosan as A Sustainable Raw Material for Anti-Bacterial Polyurethane Coatings in Medical Devices

T Rihayat [1], Nurhanifa[1], S Riskina[1], Y Yunus[2]

[1] Department of Chemical Engineering, Politeknik Negeri Lhokseumawe, 24301, Aceh Indonesia
[2] Department of Civil Engineering, Politeknik Negeri Lhokseumawe, 24301, Aceh Indonesia

Email: teukurihayat@yahoo.com

Abstract. In this study, polyurethane synthesis was carried out using the prepolymer method using the reaction of toluene diisocyanate (TDI) with palm oil-based polyols. To provide heat and anti-bacterial properties in polyurethane coatings, the addition of bentonite (B) and chitosan (K) into polyurethane samples was carried out. The concentration of bentonite and chitosan used were 1.5%, 3.5%, and 5.5%, respectively. Polyurethane-bentonite-chitosan (PU-B-K) samples were analyzed using FTIR to analyze the structure of chemical compounds. Preparation of bentonite purification was carried out by adding cetyltrimethylammonium bromide solution to the bentonite solution. The expected results in this study are the nature of the bacterial resistance produced, while to see the characteristics FTIR tests were performed. Meanwhile, to see the heat resistance TGA test will be performed.

1. Introduction

Polyurethanes are thermoplastic polymers produced from the reaction between polyols and isocyanates [1]. Currently, polyols in polyurethanes are synthesized from petroleum. The availability of petroleum which is increasingly narrow now encourages other parties to look for new alternatives as a substitute for petroleum. One alternative that can be used is vegetable oil. Synthesis that has been carried out with various vegetable oils such as canola oil, camelina oil, lanola oil, sunflower oil, nulin oil [2], neem oil [3]. Then the synthesis of polyurethanes from vegetable oils obtained from various plants such as jatropha, palm oil, soybeans, and cotton [4]. In other words, vegetable oil can replace petrochemicals as the raw material for making polyols, besides that the use of polyols from vegetable oils is superior because of their abundant availability in nature and easily decomposed.

In this study, oil palm is used for the synthesis of polyols. At this time information on polyols derived from palm oil is still limited. The main components of palm oil are triglycerides or triglycerols which are triesters of glycerol and unsaturated fatty acids. Because the content of fatty acids is not saturated, palm oil can carry out epoxidation and hydroxylation reactions to form polyols which will be used for the manufacture of polyurethanes.

The advantage of polyurethane is that it is safe to use in the medical world. Polyurethanes are usually applied to paints and coatings, the automotive industry, fibers, construction materials, and synthetic

IOP Conf. Series: Materials Science and Engineering **854** (2020) 012048 doi:10.1088/1757-899X/854/1/012048

leather. With the addition of chitosan which is an alloy in modifying polyurethane, it can become anti-bacterial in polyurethane. According to the results of research showing that the intercalation of chitosan through the cation exchange process can increase thermal stability and antimicrobial activity of nanocomposites. Chitosan properties such as biodegradability, biocompatibility, non-toxic and anti-bacterial have become an attraction for the industry.

In this research, palm oil-based polyurethane will be made as a raw material for polyols modified with the addition of bentonite to improve the physical and mechanical properties of polyurethane. The resulting polyurethane coating was further modified with the addition of chitosan to provide anti-bacterial properties, so that the coatings obtained would not only have strength, stiffness, heat resistance, corrosion resistance, and chemicals, but also have resistance to bacteria.

1.1. Polyurethane
Polyurethanes are copolymer blocks that contain low molecular weight polyester or polyester segments associated with urethane groups (-NHCO-O-). This polymer is formed from the reaction between the polyisocyanate with a polymer that contains a hydroxyl group (polyester or polyether polyols) and chain elongation such as glycol and diamin [4].

Polyurethane is one of the classes with high deformation presentation, where polyurethane has a good degree of flexibility, requires a low degree of interaction between molecules and almost perfect crosslinking. Good crosslinking is caused by hydrogen double bonds. Chemically, crosslinking bonds are actually caused by three or more compounds involved in it. Once formed, this crosslinking bond is very unlikely to be moved by the act of heating even, except in some instances where the chemicals used are very unstable so they can act irreversibly. Therefore, polymers that have the ability to be separated from crrosslinking by heat treatment become a polymer that is widely used and polyurethane is one of them [5].

1.2. Isocyanate
In the formation of polyurethanes it is very important to choose the right isocyanate to react with polyols because it will be able to determine the final results such as biuret, urea and urethane. Isocyanates can react with alcohol to form carbamate, with water to form urea and CO2 gas, with amines to form urea, with urea to form urethane and isocyanate. Isocyanate is the main material for making polyurethanes. Commonly used as diisocyanate are 2,4-toluene diisocyanate (TDI), 4,4-methylene-bis phenylisocyanate (MDI), 1,6-hexamethyl diisocyanate (HDI), 2,2,4-trimethyl-1,6- hexametyl diisocyanate (TMDI), 1,5-naphtacene diisocyanate (NDI), as shown in Figure 1 [5].

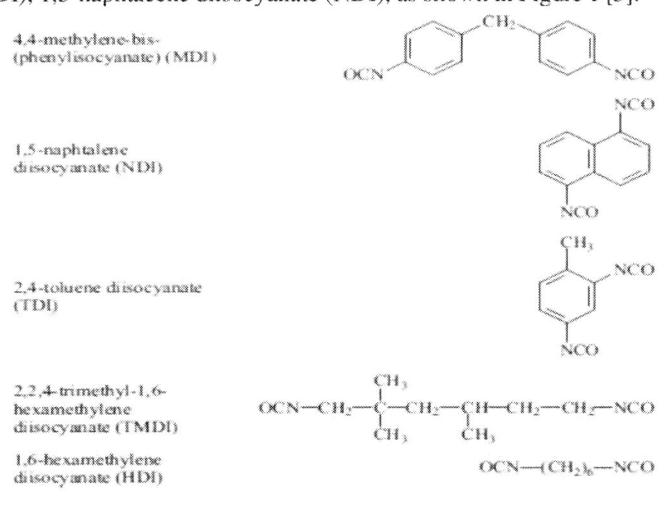

Figure 1. Examples of Some Isocyanate Compounds [3].

2. Research Methods

2.1. Materials
A set of polyurethane synthesis tools, Magnetic stirrers, Penangas, Centrifuges, Petri dishes, Filter paper, Analytical scales, Fourier Transform Infra-Red spectrophotometers (FTIR), Thermal Gravimetry Analysis (TGA). Oleic acid based on palm oil, Galetic acetic acid, Aquadest, Bentonite, Commercial chitosan, Diisocyanate toluene (TDI), Cetyl trimethyl ammonium bromide (CTAB), 30% H_2O_2, concentrated H_2SO_4, Methanol, Nutrient agar, Steel plate.

2.2. Synthesis of polyols

2.2.1. Epoxidation
Polyol synthesis is carried out in a 350 mL size 3 neck flask equipped with a mechanical stirrer and cooling system. then put into the reactor as much as 60 mL glacial CH3COOH and 30 mL 30% H2O2 slowly while stirring. Through the dropper funnel, add 2 mL of concentrated H2SO4 and stir slowly at 30°C for 1 hour. Furthermore, through the dropper funnel slowly added 100 ml of palm oil oleic acid. The temperature is maintained at 30°C and continues to stir for 3 hours. The reaction product is an oleic acid epoxidation compound, which is cooled to room temperature and the separation of the oil phase as an oxidized oil which will then be used in the hydroxylation process.

2.2.2. Hydroxylation
100 mL of methanol were added 50 mL of glycerin, a concentrated H2SO4 catalyst of 2 mL and 5 mL of water into a 350 mL three neck flask, heated to 40°C. The mixture was added to an oxidized oil solution into a three neck flask, stirring at 50°C for 2 hours. Then it is cooled to room temperature and transferred to a separate flask to separate the formed polyols and then stored in a glass bottle. Furthermore, it was analyzed by FTIR to find out the OH group in polyols.

2.3. Processing of Bentonite
18.2 grams of cetyl trimetyl ammonium bromide (CTAB) dissolved with 250 mL aquades in 500 mL beaker glass, this solution is heated at 80°C for 1 hour. Separately 20 grams of bentonite were dissolved with 250 mL of distilled water in a 1000 mL beaker glass. Subsequently the dispersion of bentonite solution was put into CTAB solution and stirred for 1 hour. Filtered bentonite then washed with distilled water several times until there are no more bromides. The filtrate was tested by dropping 1 M AgNO3 until a white precipitate formed. Bentonite is put into the oven at 60°C, then filtered using a sieve tray with a size of 100 μm.

2.4. Making Polyurethane / Bentonite / Chitosan Coatings
The polyol, bentonite, chitosan and TDI mixed into the beaker glass were mixed with a magnetic stirrer at 200 rpm for 1 hour. In this procedure, a number of bentonite and chitosan are used, each 1.5, 3.5 and 5.5 weight percent (wt%). The resulting polyurethane is then cooled to room temperature. Then the chemical structure of polyurethane, bentonite and chitosan were analyzed using FTIR.

3. Results

Table 1. Experimental design of polyurethane-bentonite-chitosan synthesis on inhibition of the bacteria Staphylococcus Aureus (a) and Escherichia Coli (b).

	Inhibition zone area (mm)

Bentonite: Chitosan (% b/b)	Bacteria *Escherichia Coli*	Bacteria *Staphylococcus Aureus*
-	Identified	Identified
1.5 : 1.5	0	1.6
1.5 : 3.5	1.3	2.8
1.5 : 5.5	1.8	3.6
3.5 : 1.5	0	2
3.5 : 3.5	2.4	3.2
3.5 : 5.5	3	3.8
5.5 : 1.5	1.3	2
5.5 : 3.5	2.6	3
5.5 : 5.5	3.2	4

Polyol formation process will occur if the alcohol reacts with the epoxide group. Therefore, a mixture of methanol and glycerin is used, where the methanol used also aims to avoid cross linkage in polyol products. In this hydroxylation process, the results obtained visually observed are products that consist of two layers, namely the upper layer which is turbid yellow, and the lower layer is turbid white. The obtained polyol product is expected to meet the standard characteristics of the polyol as the raw material for making polyurethane where the analysis shows that the number of hydroxyl numbers is determined because the polyol product of its hydroxyl group will react with isocyanate to form polyurethane as in table 1.

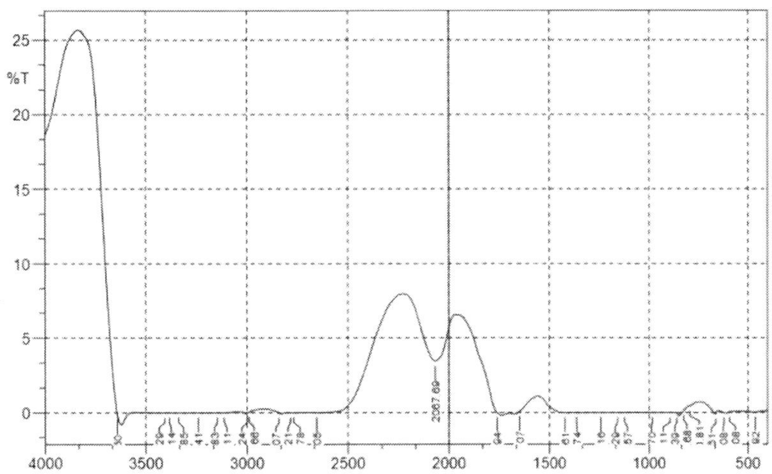

Figure 2. FTIR spectrum of palm oil based polyols

Table 2. Results of the characterization of palm oil polyols by FTIR

	Peak	Intensity	Corr. Intensity	Base (H)	Base (L)	Area	Corr. Area
1	1651.07	0.053	0.043	1654.92	1562.34	209.487	0
2	1762.94	0	-0.006	1766.08	1759.08	1.543.053	379.485
3	3336.85	0.006	0.001	3348.42	3302.13	195.01.00	1.769
4	3383.14	0.006	0	3387	3367.71	81.39.00	0.058
5	3406.29	0.005	0.001	3417.86	3390.86	114.971	0.3951

The presence of a hydroxyl group is demonstrated by transmission in the wave number 3300-3600 cm^{-1}. FTIR analysis results which have shown at figure 2 and table 2 that hydroxyl groups have formed on palm oil epoxide compounds as evidenced by the absorption of OH wave numbers that widen at 3336.85 cm^{-1}, 3383.14 cm^{-1}, and 3406.29 cm^{-1}, it has been stated that the reaction of the compound the compound has formed the desired polyol product. The results of the wavelength measurement of hydroxyl groups in previous studies were 3384.90 cm^{-1} [2], 3412.38 cm^{-1} [6], 3445 cm^{-1} [7].

3.1. *Results of Polyurethane Characterization with Bentonite and Chitosan Fillers*

Polyurethanes are generally polymer compounds whose main chain constituents are urethane groups (-NHCOO-). This polymer has been widely used in the paint industry, printers, foams, elastomers and others. In the polyurethane coating industry is widely used because it has good scratch resistance, stiffness, hardness, good flexibility, resistant to abrasion, resistant to chemicals and corrosion, and has mechanical characteristics that can be applied in various fields.

Polyurethanes are made by the addition reaction between polyols and isocyanates. Initially the main raw material for the manufacture of paint came from petroleum-based materials. However, because the solvents used pose risks to human health, environmental pollution and cause global warming if they evaporate into the environment, other alternatives are used to prevent this by replacing petroleum into vegetable oils. Vegetable oil can be used because it is renewable and environmentally friendly.

Vegetable oil used in this study is palm oil. Which is used as a material for making polyols is oleic acid contained in the palm oil. To form polyurethane one more important component is isocyanate. Where the reaction of polyols and isocyanates will produce urethane groups. So that the polyurethane produced has better physical properties, it is added with additional fillers in the form of bentonite and chitosan.

Bentonite provides high thermal resistance properties, because bentonite has its constituent material called montmorillonite. Cethyl Trimethyl Ammonium Bromide (CTAB) is used as a surfactant to modify bentonite to form organobentite by changing its nature from hydrophilic to more hydrophobic. With the insertion of the suffractant into the bentonite, d-spacing in the bentonite will become larger (intercalated). The use of chitosan as a filler in polyurethanes aims to provide anti-bacterial properties. Chitosan contains a functional group of positively charged amines which can inhibit bacterial growth and are non-toxic, because microbial membranes have a negative charge.

Palm oil based polyurethane coatings are made by reacting two main components namely polyols and isocyanates. The isocyanate used is toluene diisocyanate (TDI). Polyols will first be mixed with fillers in the form of bentonite and chitosan which will be stirred until all the homogeneous mixture is stirred for 60 minutes and then added touluene diisocyanate (TDI) in a ratio of 1:9 between the addition of TDI and polyols. Stirring is carried out to obtain a thick and homogeneous result. The resulting polyurethane product is then applied to the surface of the stainless steel plate provided, dried at room temperature for ± 2 days until it dries. Then the product is tested for characteristics with TGA and FTIR instruments. Meanwhile, to see the anti-bacterial activity test in quantity on the product is tested by weight loss techniques.

4. Conclusion

The addition of bentonite to polyurethanes to improve the resulting resistance properties depends on the weight percent of the bentonite added, the more bentonite added to the polyurethane, the higher the thermal resistance properties produced. The more weight percent chitosan added to polyurethane, the greater the bacterial resistance properties produced by the polyurethane.

References

[1] Gonzales, M., Levi, M., Turri, S., (2017). Development of polyester binders for the production of sustainable polyurethane coating. Politenico di Milano, 164: 171-178

[2] Ourique, P., Krindges, I., Aquzzoli, C., Figueroa, C., Amalvy, J., Wanke, C., Bianchi, O., (2017).

IOP Conf. Series: Materials Science and Engineering **854** (2020) 012048 doi:10.1088/1757-899X/854/1/012048

Synthesis properties and application of hybrid polyurethane urea obtained. Journal of Organic Coatings, 108: 15-24.

[3] Petrovic, Z. S., & Ferguson, J. (1991). Polyurethane elastomers. Progress in Polymer Science, 16(5), 695–836.

[4] Badri, K.H., Othman, Z. and Ahmad, S.H., (2012). Rigid Polyurtehane foams from oil palm resources. J. Matter. Sci, 39, 5541-5542.

[5] Gala, S. (2011). Sintesa Poliol dari Minyak Sawit dengan Reaksi Epoksidasi dan Hidroksilasi. Jurnal Chemical,12, 36-43.

[6] Locke, T., Sally, K., Andrew, W., Rory, M. (2013). Microbiology Infections Diseases. PT Indeks, Jakarta. 110.

[7] Manawwer., Akram, D., Sharmin, E., Zafar, F., & Ahmad, S. (2014). Vegetable oil based eco-friendly coating materials: A review article. Arabian Journal of Chemistry, King Saud University, 7, 469-479.

[8] Teuku, R., Suryani. (2013). Morphology Properties of Polyurethane/clay Nanocomposites Base on Palm Oil Polyol Paint. Biomaterial and Bioengineering, Vol. 647. Pp. 701-704

[9] Teuku, R., Suryani., Teuku, F., Harry, A., B. Wirjosentono., Syafruddin., Helmi., Zulkifli., Pocut, N, A., Muhammad, S. 2017. Mechanical properties evaluation of single and hybrid composites polyester reinforced bamboo, PALF and coir fiber. IOP Conference Series: Materials Science and Engineering. Vo.334

[10] Teuku, R., Suryani. (2010). Synthesis and Properties of Biobased Polyurethane/Montmorillonite Nanocomposites. International Scholarly and Scientific Research & Innovation, Vol. 4 (5).

[11] Teuku, R., Saari, M., Mahmood, H,M., Yunus, W,Z,M,W., Suraya, R,A., Dahlan, M,H,Z,K., Sapuan, M,S. (2007). Mechanical Characterisation of Polyurethane/Clay Nanocomposites. Polymers & Polymer Composites. Vol. 15 (8).

IOP Conf. Series: Materials Science and Engineering **854** (2020) 012049 doi:10.1088/1757-899X/854/1/012049

Synthesis of Castor Oil-Based Polyols Reinforced with Organoclay-Chitosan as Nanocomposite Polyurethane Coatings in The Application of Environmentally Friendly Medical Equipment

T Rihayat[1], K Y Putri[2], Y Yunus[2], S Riskina[1]

[1] Department of Chemical Engineering, Politeknik Negeri Lhokseumawe, 24301, Aceh Indonesia

[2] Department of Civil Engineering, Politeknik Negeri Lhokseumawe, 24301, Aceh Indonesia

teukurihayat@yahoo.com

Abstract. In this research, castor oil is used as raw material for making polyols. This review provides an overview of polyurethane coatings for the application of anti-bacterial medical devices. The anti-bacterial properties used in this review are chitosan. Chitosan is an amino polysaccharide that has a great structural tendency to produce bioactive materials with innovative properties and a variety of applications, especially in the biomedical field. To improve the physical and mechanical properties of polyurethane such as heat resistance, modifications were made with the addition of bentonite. This study with a variation of polyurethane/bentonite/ chitosan ratio of 2, 3 and 4 (% w / v), as well as tests conducted in the form of FTIR, TGA, and anti-bacterial tests. The results obtained for the best level of thermal stability in polyurethane / bentonite / chitosan 4%: 4% with a temperature of 600.94 °C and remaining weight 7.5 mg, while the anti-bacterial test obtained the largest inhibitory zone in polyurethane/bentonite/ chitosan 4%: 4% for Staphylococcus Aureus bacteria at 7.66 mm and Escherichia Coli bacteria at 5.8 mm.

1. Introduction

The consumption of polyurethane polymer materials has increased every year. Specifically in the medical field, polyurethanes are used as face shields, blood bags, infusion tubes. The main medical instrument used in medicine is a stethoscope. However, in this case, several problems arise that become obstacles to health, namely the use of a stethoscope not only for one patient but many patients who use it so that the resulting transmission of bacteria is increasingly easy.

Therefore, it is necessary to modify polyurethanes on antibacterial medical devices. Polyurethanes are produced from the reaction between polyols and isocyanates. Generally, polyurethanes are made from petroleum-based polyols. However, the current condition of petroleum is running low. Environmental problems and risks are a major concern as a reduction in conventional petroleum reserves. Therefore, it is necessary to divert raw material for making polyols, such as vegetable oils such as soybean oil, sunflower oil, canola oil, flaxseed oil, and castor oil. Utilization of vegetable oils is currently in the spotlight of the chemical industry because they are one of the most important renewable chemical platforms because of their universal availability, biodegradability, low prices, and exceptional environmental credentials (ie, low levels of environmental toxins and low toxicity to humans). Now natural ingredients are taking advantage of in research and development, with vegetable oils derived from polymers/composites used in various applications including coatings, adhesives, and biomedicine [1].

R Sudrajat said in his research using Jatropha curcas oil has a hydroxyl number of polyols obtained at 3461cm -1. The length of time the reaction has an effect on increasing the hydroxyl number, i.e. the longer the reaction time, the hydroxyl number obtained will be even greater. In this study using castor oil-based polyols with ricinoleic acid content which represents 90% of the total fatty acid content [2] with a methylation method and a longer reaction time.

To improve the mechanical properties of polyurethane products as a coating on medical devices used fillers such as bentonite and chitosan. Chitosan has various applications in the biomedical field such as wound healing, antibacterial coatings, tissue engineering scaffolding, membrane separation, stent coatings, sensors, and drug delivery systems [3]. Chitosan has the potential for a wide variety of applications in biocompatibility (non-toxic), biodegradability, antimicrobial properties, and friendly natural environment so as to provide a good opportunity for future progress. One of the main disadvantages of coating vegetable oil-based polymers is not being able to withstand high temperatures [4]. Therefore, the current development of science about nanotechnology not only produces materials that are anti-bacterial but also able to withstand heat. So, in this study utilizing the use of nanocomposite bentonite as a filler material in a polyurethane matrix as a heat-resistant reinforcement and an increase in the mechanical properties of the coating material.

This research will make castor oil-based polyurethane as a paint and coating material. To improve the physical and mechanical properties of polyurethane such as having strength, heat resistance, corrosion resistance, and chemicals as well as having resistance to bacteria, modifications were made by adding bentonite and chitosan with various concentrations.

1.1. Vegetable Oil-Based Castor Oil Polyols

Vegetable oil is an ingredient available in nature, easily available at affordable costs. Vegetable oil sourced from nature is biodegradable and has potential as a renewable natural resource for environmentally friendly materials. Vegetable oil has been applied as an ingredient for making biodiesel, lubricants, cutting fluid (liquid cutting) (liquid for cutting and forming metal materials), paints and coatings [5]. The use of vegetable oil in this study is due to seeing that oil is running low and is expected to be a new alternative to produce polyurethane. One of the natural ingredients that have the potential to be used as polyols is castor oil or so-called castor oil which in its molecular structure contains three hydroxyl groups. Jatropha oil is widely used in the chemical industry because of the hydroxyl group in the middle of the carbon chain at C12 and the existence of double bonds which allows it to produce chemical derivative products from the oil. Castor oil can be considered a polyol because there are several hydroxyl groups.

Polyols are organic compounds which have more than one hydroxyl group and, in the material industry is widely used both as reagents and additives. Polyol compounds can be obtained directly in nature such as starch, cellulose, sucrose, and lignin or even industrial chemicals. Polyols from vegetable oils have been widely developed to be able to replace petroleum-based polyols in the manufacture of polyurethanes and polyesters, and have also been widely used as plasticizers in the polymer matrix to produce a material, as well as softeners and stabilizers with the aim of obtaining certain hardness and softness so that the material it is easily formed into various types of goods as needed [6].

There are two methods of making polyols from vegetable oils such as castor oil through the method of epoxidation and hydroxylation. Epoxidation reaction is the reaction of converting unsaturated bonds of a compound into a saturated bond in the form of an oxidant group by oxidizing the compound. The hydroxylation reaction is the reaction of opening the oxidant group to the hydroxyl group which is carried out on an epoxy compound. Hydroxylation is a reaction that adds a hydroxyl group to an organic compound. In this reaction, it can also be called the opening reaction of the oxygen ring on the epoxide compound. The process of hydroxylation of epoxy oil is carried out by reacting epoxy and methanol compounds. The results of the hydroxylation process are organic compounds in the form of hydroxyl compounds or also called polyol compounds.

1.2. Polyurethane Paint and Coating

Polyurethane is a polymeric material that has the characteristic presence of urethane (-NHCOO-) functional groups in the main polymer chain. The urethane function group is produced from the reaction between compounds containing hydroxyl groups (OH) commonly called polyols and compounds containing isocyanate groups (-NCO-). Polyurethanes are generally synthesized from vegetable oils that are converted into derivatives, for example, alkyd resins and alkyd based polyols. The polyols are then reacted with different diisocyanates to obtain polyurethane coatings [7]. Polyurethanes (PU) are block copolymers containing hard and soft segments in which hard semicrystalline segments are formed by the reaction of polyisocyanates and extender chains (diol/diisocyanate) and amorphous soft segments consisting of long polyol chains [8].

Polyurethanes are block copolymers formed from the reaction of polyamide between (or poly-) isocyanate (toluene 2,4-diisocyanate [TDI], diphenylmethane diisocyanate, naphthalene 1,5-diisocyanate, hexamethylene diisocyanate, isophorone diisocyanate) and (di-poly or alcoholic acid) or other compounds containing active hydrogen atoms, as raw material for polyurethanes. Isocyanates can react with hydroxyl groups such as alcohol to form urethane, the reaction mechanism of isocyanates with hydroxyl groups of alcohol compounds is determined by the reactivity of various types of hydroxyl groups. Diol amides and polyols from vegetable oils have been used as the main ingredients to synthesize polyurethanes with aliphatic and aromatic isocyanates [9]. The longer the reaction time for making polyurethane, the more bonds are formed, so the higher the temperature needed to break the bonds, in other words, the mechanical properties are also increasing [10].

Polyurethane (PU) coatings have excellent resistance in terms of toughness, low temperature, flexibility, corrosion resistance, and chemical resistance, so it is used in various applications such as the maintenance industry for chemical resistance [7] and biomedical devices.

2. Research Methods

2.1. Materials

Polyols based on castor oil, acetic acid 2% and glacial 100%, Aquadest, Chitosan, Bentonite 100 mesh, Diisocyanate toluene (TDI),Cetyl trimethyl ammonium bromtontide (CTAB), Formic acid, H_2SO_4 96%, NaOH 0.1 N, Methanol 98%, Molecular sieve, Nutrient agar, Steel plate. A set of polyurethane synthesis tools, Magnetic stirrer, Penangas, Centrifuge, 50 ml Beaker glass, 100 ml, 500 ml, and 1000 ml, 5 ml and 25 ml volume pipettes, Petri dishes, Filter paper, Analytical scales, Analytical Fourier Transform Spectrophotometer Equipment FTIR), Thermal Gravimetry Analysis (TGA) Equipment.

2.2. Synthesis of polyols

2.2.1. Epoxidation

30 ml of 30% H_2O% is inserted into the reactor then added CH_3COOH 100% as much as 50 ml and H_2SO_4 96% as much as 2 ml. Stirred at a speed of 200 rpm and a temperature of 40 °C for 1 hour. Then 100 ml of castor oil were added and stirred at a speed of 200 rpm at 40 °C for 3 hours. and cooled to room temperature and separation of the oil phase as the oxidized oil for use in the hydroxylation process.

2.2.2. Hydroxylation

Entered 100 ml of methanol, 50 ml of glycerin, H_2SO_4 96% 2 ml catalyst and 5 ml of water into a 350 ml three-neck flask heated to 40 °C, then added the solution of the oxidized oil and stirred for 2.5 hours at 50 °C, and cooled to temperature room, then transferred to a separate flask and separated polyols, analyze with FTIR to find out the OH group in the polyol.

2.3. Processing of Bentonite

Dissolved 18.2 grams of cetyl trimethyl ammonium bromide (CTAB) with 250 ml of distilled water into 500 ml of beaker glass, then heated at 80 °C for 1 hour, Dissolved 20 grams of bentonite with 250 ml of distilled water into 1000 ml glass beaker, Dispersion of bentonite solution is put into a temperature of 80°C for 1 hour in CTAB solution and stirred for 1 hour, filtered, and bentonite washed with distilled water and dried at 60 °C.

2.4. Making Polyurethane / Bentonite / Chitosan Coatings

Montmorillonite and chitosan were mixed 2, 3, 4 (% w / v) into 16 ml of castor oil polyols on beaker glass using 600 rpm magnetic stirred for 1 hour, then added 4 ml of toluene and stirred for 5 minutes, the resulting polyurea applied to metal specimens, chemical structure analysis of polyurethane coatings, bentonite, chitosan by using FTIR, heat resistance analysis of coatings by using TGA, Bacterial susceptibility testing by disk diffusion method.

3. Results

Table 1. Design data for analysis of polyurethane by weight percent of bentonite and chitosan on the inhibition zone area of Escherichia Coli and Staphylococcus Aureus bacteria.

Bentonite: Chitosan (% b/v)	**Inhibition zone area (mm)**	
	Bacteria *Escherichia Coli*	**Bacteria *Staphylococcus Aureus***
-	Identified	Identified
2 : 2	0	1.3
2 : 3	2	2
2 : 4	3.2	5.3
3 : 2	0	1.6
3 : 3	2	2
3 : 4	5.3	6.73
4 : 2	1.3	1.6
4 : 3	4.5	5.3
4 : 4	5.8	7.66

Based on table 1 shows the sample that has the most powerful anti-bacterial activity, namely in S. aureus with inhibition zone area of 7.66 mm while the strongest antibacterial activity on E. coli with inhibition zone area of 5.8 mm. The data in table 1 shows that the inhibition produced by E. coli bacteria is smaller than that of S. aureus, this is because E. coli is a gram-negative bacteria that have thinner and more complex cell walls with high lipid content. so it is difficult to penetrate. While S. aureus is a gram-positive bacterium that has a simple and thick cell wall, single layer, low lipid content, layers of a cytoplasmic membrane composed of peptidoglycan and teichoic acid in the form of water-soluble polymers so that gram-positive bacteria are more easily penetrated by substances that come from preparations. Overall, the results of antibacterial polyurethane research that provided the strongest activity were polyurethane research results with S.aureus bacteria, with an area of inhibition zone of 7.66 mm.

Epoxide oil from castor oil is obtained from the reaction between parasitic acid ($C_2H_4O_3$) and castor oil, while parasitic acid is produced from the reaction between hydrogen peroxide (H_2O_2) and glacial acetic acid (CH_3COOH) with the help of a concentrated sulfuric acid catalyst (H_2SO_4). Hydrogen peroxide is used to oxidize double bonds into epoxide groups during the epoxidation process. The castor oil epoxy results obtained have a brighter color compared to the color of castor oil ricinoleic acid [11]. The formed epoxide oil is an intermediate compound which can react further to form diol compounds because it has two reactive sites, the carbonyl group which can connect glycerides with fatty acids and also epoxide groups. So the process of forming polyols will occur if the epoxide group reacts with alcohol. The reaction of the formation of epoxide compounds in vegetable oils has been carried out by [4],[5],[10].

Table 2. FTIR spectrum results data

	Peak	Intensity	Corr. Intensity	Base (H)	Base (L)	Area	Corr. Area
1	686.66	0	548	690.52	682.8	771	774
2	709.8	0	2980	717.52	705.95	766	779
3	1458.18	0	2525	1477.47	1454.33	769	790
4	3421.72	41	80	3660.89	3101.54	124	172

Based on table 2 shows FTIR analysis results have shown the formation of N-H urethane groups in castor oil polyurethane compounds, the reaction lasts for 1 hour at room temperature as evidenced by the absorption of the NH wave number that widens at 3402.43 cm^{-1} and 3483.44 cm^{-1} measurement results NH group wavelengths in previous studies were 3300 cm^{-1} [12], 3311 cm^{-1}, 3315 cm^{-1} and 3316 cm^{-1} [9]. While uptake of wave number C=O urethane widened at 1681.93 cm^{-1} and 1728.22 cm^{-1}, in the previous study of 1720 cm^{-1} [12], while the CH group at absorption was 2854.65 cm^{-1}, 2908.65 cm^{-1} and 2947.23 cm^{-1} in the previous study of 2924.10 cm^{-1} [9]. The analysis shows that the formation of polyurethanes is seen from the absorption of wavelengths in the functional groups N-H, C=O urethane, and C-H. Detected functional groups represent polyurethane functional groups (-NHCOO-) in the form of wavelengths absorbing the bonds of separate functional groups.

Based on the analysis results it can be seen that the functional groups of pure PU and PU / Bent / Ch did not experience changes in functional groups. Mixing PU with bentonite and chitosan materials did not affect the wavelength absorption of the polyurethane functional group itself, the same as the results of previous studies on FTIR analysis of commercial polyurethanes (PUK), palm oil polyurethanes (PU-MKS) and palm oil + polyurethane (PU) polyurethanes -MKS- Clay) [8]. This is due to the process of mixing bentonite and chitosan with polyurethane which occurs only in physical bonds rather than chemical bonds. Thus, no chemical bond changes occur so that the wavelength absorption of the PU / Bent / Ch functional groups detected remains the same.

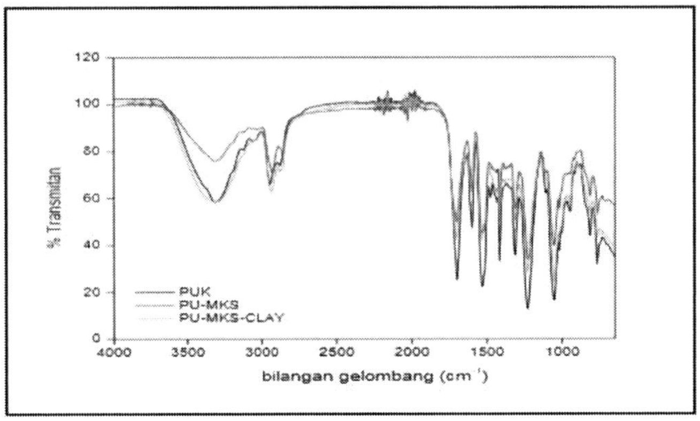

Figure 1. Results of FTIR Commercial Polyurethanes (PUK), Polyurethane Palm Oil (PU-MKS) and Polyurethane Palm Oil + Clay (PU-MKS-Clay)

Sources of bacterial contamination in the hospital area can come from patients, the environment (air, space, and facilities available for operations), surgery materials and equipment, and members of the surgery team. To protect and prevent bacteria from being contaminated, efforts required include coating medical devices with anti-bacterial polyurethane paint. Surgical wound infections can also be caused by bacteria such as Staphylococcus aureus, and Escherichia coli [11]. The types of bacteria used in this study are Staphylococcus aureus and Escherichia coli. These bacteria are pathogenic and can cause diarrheal diseases, meningitis, skin infections and cause sores such as ulcers presented in Figure 1.

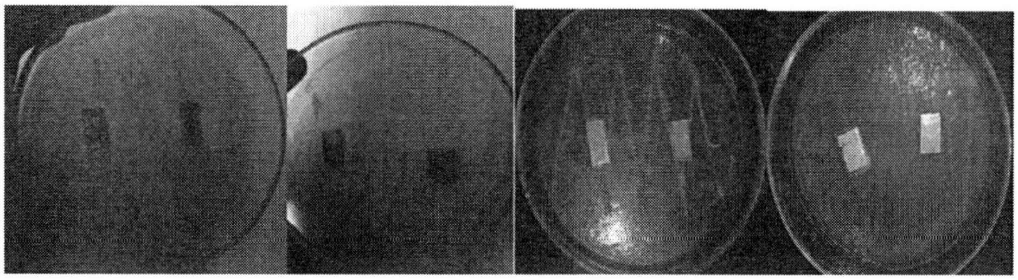

Figure 2. Antibacterial analysis results a) PU / B / K with S. Aureus bacteria. b) PU / B / K with E.Coli bacteria. c) Pure PU and Blank Plate with S. Aureus bacteria. d) Pure PU and empty plates with E.Coli bacteria.

Bacteria representing Gram negative categories or groups in this study were E. coli while those representing Gram positive bacteria were Staphylococcus aureus. Bacteria E. coli and Staphylococcus aureus are often found in medical devices, for which both bacteria were tested in this study. The method used in this antibacterial test is the halo zone. To find out whether this polyurethane has antibacterial properties, the inhibition zone observation was continued for 24 hours. After being observed at 24 hours the bacteria experienced inhibitory regions. In figure a planted PU / B / K in S. Aureus bacteria has a wider inhibition zone compared to the media in picture b which is planted PU / B / K in E. coli bacteria. Whereas in pictures c and d planted with pure PU and empty plate samples do not have inhibitory zones but bacteria grow in the surrounding area. That is because there is no chitosan content in the sample presented in Figure 2.

4. Conclusion

From the anti-bacterial test results, the largest inhibitory zone area was polyurethane / bentonite / chitosan 4%: 4% with area for Staphylococcus Aureus bacteria of 7.66 mm while for Escherichia Coli bacteria it had inhibition zone area of 5.8 mm. From the results of the analysis of thermal stability using TGA. The best thermal resistance in 4%: 4% polyurethane / bentonite / chitosan is at a temperature of 600.94 °C with a remaining weight of 7.5 mg. From the results of the analysis of functional group characteristics using FTIR, there is no change in chemical bonds that occur between pure polyurethanes and polyurethanes / bentonite / chitosan so that the absorption wavelengths of the detected functional groups remain the same.

References

[1] Adak, B., Butola, B. S., & Joshi, M. (2018). Effect of organoclay-type and clay- polyurethane interaction chemistry for tuning the morphology, gas barrier and mechanical properties of clay/ polyurethane nanocomposites. Applied Clay Science, 161, 343-353

[2] Alagi, P., Choi, Y. J., Seog, J., Hong, S. C. (2016). Efficient and quantitative chemical transformation of vegetable oils to polyols through a thiol-ene reaction for thermoplastic polyurethane. Industrial Crops and Products, 87,78-88.

[3] Gala, S. (2011). Sintesa Poliol dari Minyak Sawit dengan Reaksi Epoksidasi dan Hidroksilasi. Jurnal Chemical,12, 36-43.

[4] Locke, T., Sally, K., Andrew, W., Rory, M. (2013). Microbiology Infections Diseases. PT Indeks, Jakarta. 110.

[5] Manawwer., Akram, D., Sharmin, E., Zafar, F., & Ahmad, S. (2014). Vegetable oil based eco-friendly coating materials: A review article. Arabian Journal of Chemistry, King Saud University, 7, 469-479.

[6] Marlina. Et al. (2017). Sintesis membrane poliuretan berbasis bahan alam. Banda Aceh: Universitas Syiah Kuala.

[7] Sudradjat, R., Yulita, R. I., & Setiawan, D. (2010). Pembuatan poliol dari minyak jarak pagar sebagai bahan baku poliuretan. Jurnal Penelitian Hasil Hutan, 28, 231-240.

[8] Salwiczek, M., Yue, Q., James, G. 2014. Trends in Biotechnology. Emerging Rules for Effective Antimicrobial Coatings. Vol. 32. No. 2.

[9] Sing, Y,R., Mubarak, N,M., Faisal, A,T. 2017. Journal of Environmental Chemical Engineering.

IOP Conf. Series: Materials Science and Engineering **854** (2020) 012049 doi:10.1088/1757-899X/854/1/012049

Structure-Property Relationship of Cellulose Nanowhiskers Reinforced Chitosan Biocomposite Film. Vol. 5. Hal: 6132-6136.

[10] Gregory, C., Liudmila, N., Sabine, P. 2016. Applied Clay Science. How Efficiently Combine Sonochemistry and Clay Science?, Vol.119, Hal: 193-201.

[11] Chen, L., Jian, B, L. 2017. International Journal of Biological Macromolecules. Preparation of Chitosan-Ferulic Acid Conjugate: Structure Characterization and in the application of Pharmaceuticals. Vol. 105. Hal: 1539-1543.

[12] Gang, X., Pingya, L., Qiang, D. 2017. Applied Clay Science. Investigation of The Inhibition Mechanism of The Number of Primary Amine Groups of Alkylamines on The Swelling of Bentonite. Vol. 136. Hal: 43-50.

[13] Teuku, R., Suryani. (2013). Morphology Properties of Polyurethane/clay Nanocomposites Base on Palm Oil Polyol Paint. Biomaterial and Bioengineering, Vol. 647. Pp. 701-704.

[14] Teuku, R., Suryani., Teuku, F., Harry, A., B. Wirjosentono., Syafruddin., Helmi., Zulkifli., Pocut, N, A., Muhammad, S. 2017. Mechanical properties evaluation of single and hybrid composites polyester reinforced bamboo, PALF and coir fiber. IOP Conference Series: Materials Science and Engineering. Vo.334

[15] Teuku, R., Suryani. (2010). Synthesis and Properties of Biobased Polyurethane/Montmorillonite Nanocomposites. International Scholarly and Scientific Research & Innovation, Vol. 4 (5).

[16] Teuku, R., Saari, M., Mahmood, H,M., Yunus, W,Z,M,W., Suraya, R,A., Dahlan, M,H,Z,K., Sapuan, M,S. (2007). Mechanical Characterisation of Polyurethane/Clay Nanocomposites. Polymers & Polymer Composites. Vol. 15 (8).

IOP Conf. Series: Materials Science and Engineering **854** (2020) 012050 doi:10.1088/1757-899X/854/1/012050

Chitosan Modified Bio-Fibre Based Board as Antimicrobial and Anti-Crack Board

Suryani[1], Fitria[2], T Rihayat[1], Nurhanifa[1], T Y R Hasnah[1]

[1] Department of Chemical Engineering, Politeknik Negeri Lhokseumawe, 24301, Aceh Indonesia
[2] Departement of Dermato Venereology , Medical Faculty, Universitas Syiah Kuala, 23111, Aceh

Email: suryani@yahoo.com

Abstract. Bio-based products represent a key component of global bio-economic growth. In the process of forming a bamboo fibre-based bio board in this study begins with the preparation of bamboo fibres to make it compatible as a filling material in polyester polymers. The mixing of bamboo fibre is done by considering the composition of the fibre material with the polymer used, namely polyester at a ratio of 90%:10%, 80%:20% and 70%:30% of the total weight of the mixture. This mixture is added with a natural substance that has been investigated to have resistance to microbial contamination (anti-microbial), chitosan at a ratio of 1%, 3% and 5% by weight of the total mixture of all components. Bio board printing is carried out using a hot press (moulding) tool. The functionality of bio-based and conventional natural fibre-based materials was investigated for a series of bio boards. The tests include tensile, antimicrobial activity and morphological testing using Scanning Electron Microscopy (SEM). The overall test results put bio board with a ratio of 70%:30% with the addition of chitosan as much as 5% wt as the best sample with a tensile strength value of 147 Mpa, the flexural value of 0.09 kN. The sample shows a good interface appearance of the bond that occurs between the polymer and the filler under the electron microscope.

1. Introduction

In recent years, the joint is more environmentally friendly while synthetic fibers can be a source of environmental pollution. Composite is currently considered the most promising material in society because of its unique mechanical properties, biodegradable and abundant raw material availability. Many researchers have proven increasing interest in developing composite biodegradable polymer (BFRP) reinforced fibers as conventional fuels, especially in the automotive, marine, packaging, furniture and building construction industries. The production of natural fibers has become very attractive and beneficial, due to global problems and reduced petroleum. Natural fibers play an important role in developing biodegradable composites to solve environmental problems and current problems [1].

Natural fibers such as bamboo have been studied as reinforcement and fillers in composites. Bamboo is a plant that can grow quickly, high mechanical properties and specific gravityt he low one. so it has the potential to be an abundant source of fiber, including the Bambusae family, under the genus Gramineae [2]. There is increasing public demand for a self-binding bio-board that is produced to have antimicrobial functions such as antibacterial and without high formaldehyde emissions with the emergence of bamboo-based panels [3]. The antimicrobial bio-board can be improved by chitosan modified bamboo fibers before hot pressing. Chitosan (β-1,4-D-glucosamine) is a natural polymer obtained from chitin which originates mostly in skeletons of insects and crustaceans [4]. Chitosan has antimicrobial activity due to positive amino acid groups. Chitosan has antimicrobial activity due to positive amino acid groups the leakage of protein and intracellular components of the microorganism that causes these amine groups to react with the cell membrane of negative microorganisms [5].

The present study prepared a antimicrobial and anticrack board by hotpressing, and examined its properties by tensile, microbial activity and morphological testing using Scanning Electron Microscopy (SEM).

2. Material and Method

2.1. Preparation of chitosan modified bamboo fiber.
Cut bamboo and crusher using a crusher machine to obtain bamboo powder. Wash bamboo powder using water then immersed in sodium hydroxide (NaOH) at a concentration of 6% of the volume of water for 3 hours at room temperature. The fiber is then washed twice with water, then mixed with chitosan with a variation of concentration of 1%, 3%, and 6% and dried at room temperature for 8 hours then dried at 50 °C for 2 days. Dry fibers are stored in sealed plastic bags to avoid atmospheric moisture contamination before the bio-board is formed.

2.2. Stage of Making Bio-board Materials
The mixing of bamboo fibre is done by considering the composition of the fibre material with the polymer used, namely polyester at a ratio of 90%:10%, 80%:20% and 70%:30% of the total weight of the mixture. The appropriate amount of fiber and matrices are mixed into the glass and then stirred slowly with a spoon until evenly distributed. Add catalyst in the ratio according to the resin volume fraction. Stir gently until it is mixed for about 2 minutes. Pour the mixture into the mold that has been formed according to STM D 638 Type I standard for tensile test and ASTM D 695 for flexural test. Flatten the surface of the mixture on the mold, wait for it to dry for about 24 hours. The dried specimen is removed from the mold and then the surface parts are smoothed with a miser. Bio-board specimens that have been smoothed and measured in geometry were initially said to be specimens ready for testing.

2.3. Characterization of Bio-Board

2.3.1. Tensile Strength Test (ASTM. D 638-02 type 4)
This mechanical property test is carried out by tensile strength testing of composite specimens using ASTM D 638 Type I. Tensile testing equipment is conditioned at a load of 200 kgf with a with drawal speed of 20 mm / min, the specimen is observed to break, recorded maximum stress (F max) and its strain.

$$\sigma = \frac{Load}{A} \quad \text{...} (1)$$

2.3.2. Antimicrobial Activity
The antimicrobial effect of chitosan modified bio-fiber (bamboo) samples was measured according to the inhibition zone [6].

2.3.3. SEM analysis
To analyze the effect of treatment on the surface structure of the fiber is done with a SEM microscope. The surface structure of the fiber was observed using a JEOL-T20 microscope. Scanning electron analysis is carried out at 5-20 KV [7].

IOP Conf. Series: Materials Science and Engineering **854** (2020) 012050 doi:10.1088/1757-899X/854/1/012050

3. Result and Discussion

3.1. Tensile strength analysis on chitosan modified bio-board.

Tensile testing of bio-board material was strengthened by natural fibers and chitosan was analyzed using Universal Testing Machine (UTM) Model E43 with the sample specimen used, namely ASTM D-638. The influence of fiber volume fraction, chitosan, and bamboo polyester bio-board matrix on the tensile test characteristics as shown in table 1. where the mixture of filler: matrix and chitosan fibers at a ratio of 70%:30% with chitosan concentration 5% yields a maximum value of 147 Mpa, while at a ratio of 90%:10% with a chitosan concentration of 1% produces a minimum value of 73 MPa. Composite stiffness increases significantly with increasing matrix content.

Table 1. Wave Number with Alleged Compounds

Fraksi Volume Filler:Matrix	Tensile Strength (Mpa)		
	Chitosan 1%	Chitosan 3%	Chitosan 5%
90%:10%	73	83	107
80%:20%	89	90	108
70%:30%	96	114	147

Theoretically, bamboo fiber reinforced polyester (bio-board) composites have good tensile properties due to their high cellulose content in bamboo fibers. Help the alkali process by immersing the bamboo fiber in a 6% NaOH concentration as reported by Abdul Chalil [8], that the optimum value was found and produced the best mechanical properties for composite bamboo. This increase is also caused by the removal of hemicellulose, wax, lignin and other impurities resulting in a rough topographic surface on bamboo fibers and offers better interface adhesion properties to the fiber / matrix. Alkaline treatment also causes an increase in the surface area of the fibril fiber so that it is more effective when moistened with resin.

3.2. Antimicrobial activity of Bio-board

In applications, bio-board is easily destroyed by insects and bacteria, and greatly reduced . To enhance the antimicrobial effect of bio-board, bamboo was treated with chitosan and then hot pressed to make chitosan modified bio-fibre (bamboo). The characteristics of the inhibition zone of chitosan modified bio-fiber (bamboo) for microbial growth were observed and are listed in Figure 1.

In figure 1, that chitosan used for the modified bio-fiber treatment was shown to enhance its antimicrobial properties. Different concentrations of chitosan used on the fiber had different effects. The results demonstrated that the treatment with chitosan can significantly enhance the antimicrobial activity such as antibacterial of a bio-fiber against Escherichia coli and Staphylococcus aureus [9].

In the case of bacterial pathogens, bio-fiber modified with 5% chitosan showed the highest antibacterial activity against gram-negative pathogens of E. coli with a 17 mm inhibition zone. Meanwhile, antibacterial activity against S.aerus sp. the diameter of the inhibition zone is 15 mm. Gram negative bacteria have a relatively simpler structure with thick cell walls (15-80 nm) so that antibacterial compounds more easily enter the cell and find targets to work [10].

IOP Conf. Series: Materials Science and Engineering **854** (2020) 012050 doi:10.1088/1757-899X/854/1/012050

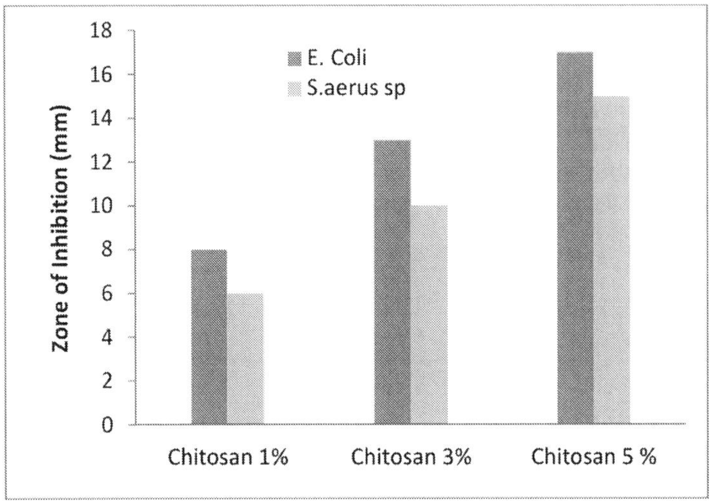

Figure 1. Antimicrobial activity of chitosan modified bio-fiber

3.3. Results of SEM Analysis

The scanning electron microscope (SEM) analysis is used to study the surface structure of a material. In this study SEM analysis was carried out on composites that had the best and lowest mechanical properties in order to see the failures that occurred in the composite.

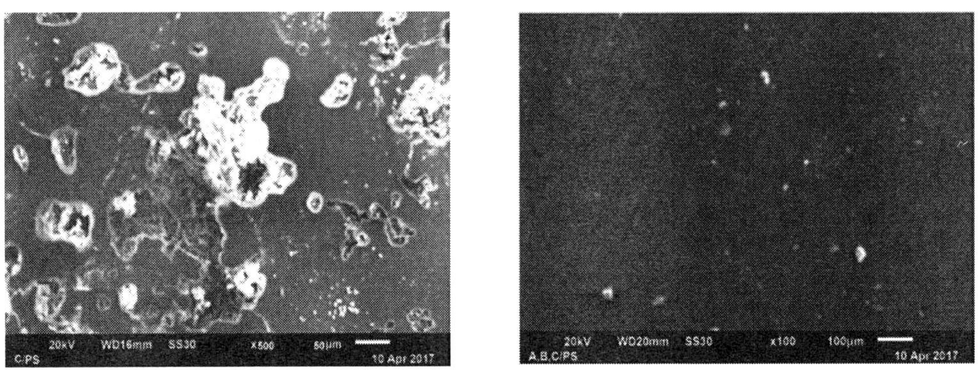

(a) (b)

Figure 2. (a) Filler:matrix at a ratio of 90%:10% with chitosan concentration 1% at 100x Zoom, (b) Filler:matrix at a ratio of 70%:30% with chitosan concentration 5% at 100x Zoom,

From the results of the analysis published in Figure 2. (a) it is clear what is in the polyester in the modified fibers with chitosan 1% which is calculated between the fibers and above to effectively compile the voltage transfer from matrix to fiber, where the maximum The use of fiber strength in the composition is approved and affects the mechanical properties of the composite which have been caused by the results of the tensile test in the previous discussion. The morphology of polyester composites in the modified fibers with 5% chitosan is shown in Figure 2. (b) observed at 100x magnification shows an increase in fiber and matrix adhesion that is distributed evenly throughout the composite so that it shows the optimal attachment of the interface between the fiber and the matrix. Interlocking is one of the bonding mechanisms that occurs between fibers and fiber-reinforced polymer matrices. The interlocking properties exhibited by polyester composites in the modified fibers with 5% chitosan involve diffusion of the polymer matrix into the porous fiber surface. The polymer matrix will flow to the surface of the porous fiber and the polymer is embedded and compacted so as to form a strong bond between the fiber and the matrix so that it shows the optimum value as has been reported in the flexural test results previously.

4. Conclusion

From the result of this study some conclusion can be drawn : (1) Optimal tensile strength values obtained on the bio board with a ratio of 70%:30% with the addition of Chitosan 5%, which is equal to 1.47 MPa. (2) Bioboard modified with Chitosan is able to control the activity of microbes such as bacteria (E.coli and S. Aerus sp). (3) The morphological structure of the bio-board shows the presence of voids and interlocking the filler interface with the matrix thereby affecting the mechanical properties of the bio board.

References

[1] Kasim, A. N., Selamat, M. Z, Daud, M. A. M., Yaakob, M. Y., Putra, A., Sivakumar, D., 2016. Mechanical Properties of Polypropylene Composites Reinfprced with Alkaline Treated Pineapple Leaf Fibre from Josapine Cultivar. International Journal of Automative and Mechanical Engineering. Vol.13, hal:3157-3167.

[2] Abdul Khalil, H. P. S., Bhat, I., Jawaid, U. H. M., Zaidon A., Hermawan, D., Hadi, Y. S. 2012. Bamboo Fibre Reinforced Biocomposites: A review. Materials and Design. Vol.42, hal:353–368.

[3] Sun, Y.C., Lin, Z., Peng, W.X., et al., 2014. Chemical changes of raw materials and manufactured binder less boards during hot pressing: lignin isolation and characterization. Bioresources 9 (1), 1055–1071.

[4] Casariego, A. B. W. S., Souza, B. W. S., Cerqueira, M. A., Teixeira, J. A., 627 Cruz, L., Díaz, R., & Vicente, A. A. 2009. Chitosan/clay films' properties as 628 affected by biopolymer and clay micro/nanoparticles' concentrations. Food 629 Hydrocolloids, 23(7), 1895-1902.

[5] Islam, S., Bhuiyan, M. R., & Islam, M. 2017. Chitin and chitosan: Structure, properties and applications in biomedical engineering. Journal of Polymers and the Environment, 25(3), 854 –866.

[6] N. F. Ghazali and N. A. B. N. Mahmood, "Composite material making from empty fruit bunches of palm oil (EFB) and Ijuk (Arengapinnata) using plastic bottle waste as adhesives Composite material making from empty fruit bunches of palm oil (EFB) and Ijuk (Arengapinnata) using plastic bottle waste as adhesives."

[7] P. Pradeep, M. Mittal, R. Chaudhary, M. Mittal, and R. Chaudhary, "Mechanical properties evaluation of single and hybrid composites polyester reinforced bamboo , PALF and coir fiber Mechanical properties evaluation of single and hybrid composites polyester reinforced bamboo , PALF and coir fiber."

[8] Teuku, R., Suryani., Teuku, F., Harry, A., B. Wirjosentono., Syafruddin., Helmi., Zulkifli., Pocut, N, A., Muhammad, S. 2017. Mechanical properties evaluation of single and hybrid composites polyester reinforced bamboo, PALF and coir fiber. IOP Conference Series: Materials Science and Engineering. Vo.334.

[9] Teuku, R., Suryani. (2010). Synthesis and Properties of Biobased Polyurethane/Montmorillonite Nanocomposites. International Scholarly and Scientific Research & Innovation, Vol. 4 (5).

[10] Teuku, R Saari, M., Mahmood, H,M., Yunus, W,Z,M,W., Suraya, R,A., Dahlan, M,H,Z,K., Sapuan, M,S. (2007). Mechanical Characterisation of Polyurethane/Clay Nanocomposites. Polymers & Polymer Composites. Vol. 15 (8).

IOP Conf. Series: Materials Science and Engineering **854** (2020) 012051 doi:10.1088/1757-899X/854/1/012051

Refining Citronella Oil (Cymbopogon Nardus L) by Utilizing Sunlight Using Solar Cells (Photovoltaics)

N Safitri[1], R Syahputra[1], K Y Putri[2], T Rihayat[3], Nurhanifa[3]

[1] Department of Electrical Engineering, Politeknik Negeri Lhokseumawe, 24301, Aceh Indonesia
[2] Department of Civil Engineering, Politeknik Negeri Lhokseumawe, 24301, Aceh Indonesia
[3] Department of Chemical Engineering, Politeknik Negeri Lhokseumawe, 24301, Aceh Indonesia

Email: nellysafitri@pnl.ac.id

Abstract. Sunlight is one of the very abundant energy. In its use, many things can be applied such as converting the sunlight into electrical energy with the help of solar cells. In this research, 4 solar cells are used with a capacity of 1000 wp, so in a day they are able to produce 28,000 watts of electricity. This is a very potential electrical energy because it is commonly used to facilitate the process of refining essential oils. The results and the quality of oil obtained in refining research by utilizing sunlight using solar cells shows that the refining time of 6 hours is the best time for the refining process, while the 1 day drying time is better than 2 and 3 days. This is consistent with the oil yield obtained. The yield that has been produced is tested based on Indonesian National Standard (SNI) 2385-2006 through density and Bias Index tests.

1. Introduction

Essential oils are one of Indonesia's export commodities usually used in the perfume, cosmetics, pharmaceutical and food industries. This oil is known as etheric oil, essential oil or volatility oil because it contains organic compounds which are volatile terpenes at room temperature without being decomposition. Essential oils have a bitter taste (pungent taste) and smell good in accordance with the smell of the original plant [1]. Citronella is a plant that has many benefits, including as a producer of essential oils. This plant is easily cultivated and does not require special treatment, so it is easy to develop. Citronella plants (Citronella), have several biological effects including antioxidant, antifungal, anti-bacterial and insect repellent activity [2].

The process of obtaining essential oils is known as distillation or distillation of oil-producing plants. In the commercial world, the method of distillation / distillation of essential oils can be in 3 ways, including:

1. Distillation with a Boiled System (Water Distillation)
2. Water and Steam Distillation
3. Direct Steam Distillation

The application of the method is based on several considerations such as the type of raw materials, oil characteristics, the process of oil diffusion with hot water, oil decomposition due to heat effects, production efficiency and reasons for economic value and production effectiveness [3].

IOP Conf. Series: Materials Science and Engineering **854** (2020) 012051 doi:10.1088/1757-899X/854/1/012051

Solar cells are a set of modules to convert solar energy into electrical energy. Photovoltaic is a technology that functions to convert or convert solar radiation into electrical energy directly. PV is usually packaged in a unit called a module. In a solar module consists of many solar cells that can be arranged in series or in parallel [4], [5]. Meanwhile, what is meant by solar is a semi-conductor element that can convert solar energy into electrical energy on the basis of the photovoltaic effect.

Solar cells are a set of modules to convert solar energy into electrical energy. Photovoltaic is a technology that functions to convert solar radiation into electrical energy directly. PV is usually packaged in a unit called a module. In a solar module consists of many solar cells that can be arranged in series or in parallel [6], [7]. Meanwhile, what is meant by solar is a semi-conductor element that can convert solar energy into electrical energy on the basis of the photovoltaic effect. Solar cells have become popular lately, in addition to the depletion of fossil energy reserves and the issue of global warming. The energy produced is also very cheap because energy sources (solar) can be obtained free of charge. Schematic image of a solar cell can be seen in figure 1.

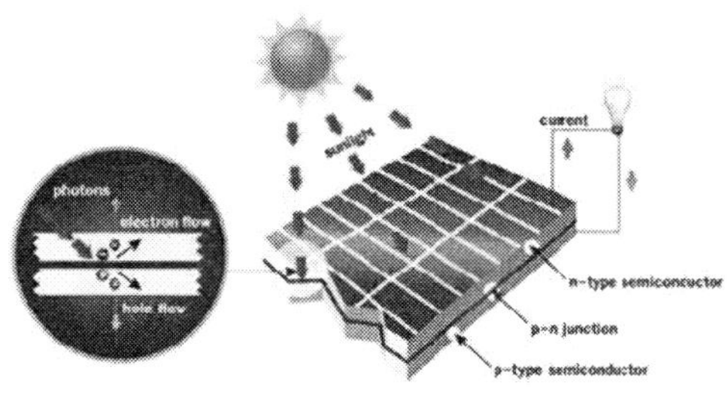

Figure 1. Schematic of a solar cell

Renewable energy has a very important role in fulfill energy needs. This is due to the use of fuel for conventional power plants over a long period of time which will deplete the source of petroleum, gas and coal which is running low and can also cause environmental pollution [8]. One of the efforts that has been developed is the Solar Power Plant (PLTS). PLTS or better known as solar cells (photovoltaic cells) will be more in demand because they can be used for a variety of relevant purposes and in various places such as offices [9], factories [10], housing [11], and others. In Indonesia, which is a tropical region has a very large solar energy potential with an average daily insulation of 4.5 - 4.8 $KWh/m^2/day$ [12]. However, the electrical energy produced by solar cells is strongly influenced by the intensity of sunlight received by the system.

The calculation amount of power consumption, the number of solar panels and the number of batteries used can be mention in thi following section [13]:

a. Calculation of power requirements (calculation of the device's electrical power can be seen on the label on the back of the device, or read from the manual).

b. calculation of the number of modules needed:

$$\frac{planned\ power}{power\ capacity\ module\ x\ intensity\ of\ sunlight\ in\ a\ day}$$

c. Minimum battery requirements (batteries are only used 50% to meet electricity needs), thus we multiply the power needs 2 x.

$$\frac{planned\ power}{battery\ capacity}\ \mathrm{x}\ 2$$

2. Material and Method

2.1. Refining of Citronella Oil
Photovoltaic equipment arranged according to the established scheme. The water is put into the pot as much as 25 liters based on principle 1:5. The chopped lemongrass leaves are put into a distilled pot. The pump is turned on to drain the cooling water then the heater is turned on until it reaches the refining temperature of 100°C and then the distillation process is stopped according to the specified time. After the distillation time is complete, the process is stopped [14].

2.2. Characterization of Citronella Oil

2.2.1. Density Test
Weigh the Empty Pycnometer. Fill the empty pycnometer with refined oil then weigh it again. Note the initial weight and finally weight then calculated using the formula:

$$\rho = \frac{m}{v}$$

2.2.2. Refractive Index Test
The autorefractometer is turned on. The calibration process is carried out using distilled water, then citronella essential oil is dropped in the column. Press the "measure" button on the tool to measure the refractive index of the oil..

3. Result and Discussion
The distillation process of essential oils usually generated by utilizes the artificial heat generated from the combustion process both firewood and stoves. Therefore this study utilizes sunlight as a substitute for heat sources for distillation or commonly called photovoltaic. Photovoltaic technology always has more advantages compared to other technologies because it is pollution-free and unlimited solar energy [14]. Another advantage with photovoltaic technology is that it does not emit greenhouse gases during operations and is environmentally friendly. Intermittency of solar radiation can be a limitation for technology because it cannot supply electricity continuously during periods of sunlight, but this problem can be overcome by using battery storage. However, there is a need to understand the application of this technology to be feasible for its users [15].

3.1. Effect of Drying Time and Refining Time for Citronella Oil Yield
From the figure 2 below, it can be seen that the amount of yield produced has fluctuated. The highest yield was 0.84% with 6 hours distillation treatment and 1 day drying time. The efficient drying time for citronella oil is 1 day because the longer the drying time the oil content in plants will decrease. The best distillation time is 6 hours, this happens because the longer a material receives heat, the more even the diffusion process causes the more efficient distillation process.

The effect of drying time on yield proves that the condition of the material that produces a large% yield is when the condition of the material starts to wilt compared to the fresh material. So the condition and treatment of these materials can improve the process of % essential oil yields in accordance with the literature which states that the process of withering aims to reduce the water content in the material glands, so that the extraction process is easier to do and enumeration is an effort to expand the area of evaporation and contact with water so that the lemongrass oil fragrance is easier to extract.

IOP Conf. Series: Materials Science and Engineering **854** (2020) 012051 doi:10.1088/1757-899X/854/1/012051

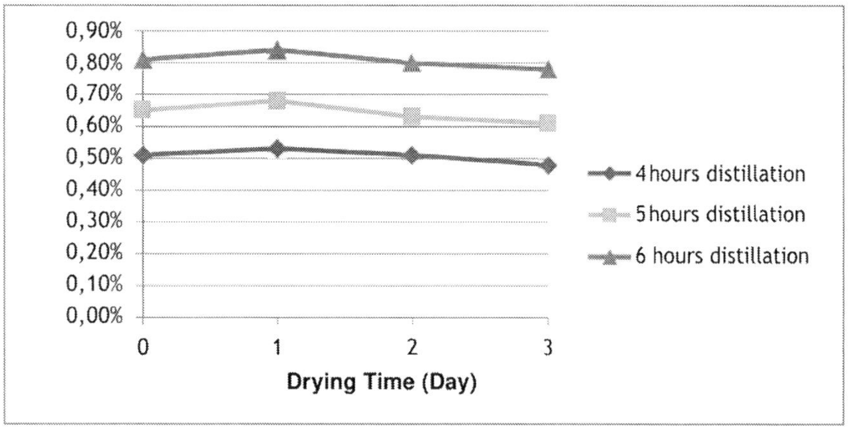

Figure 2. Yield Value Against Operating Time

Drying is the process of reducing the water content of a material until it reaches a certain water content. The basis of the drying process is the evaporation of material water into the air due to differences in the moisture content between the air and the dried material. For a material to become dry, air must have a moisture content or moisture lower than the material to be dried [16].

3.2. Density Test For Citronella Essensial Oil

Density defined as the ratio between mass and volume of material. In general, density is related to Viscosity, which is a denser liquid (high density) having a higher viscosity compared to a liquid whose its low. The value odensity of citronella oil ranged from 0.8757. The effect of refining time on the density of lemongrass oil shows that at the three heating time of material showed a significant effect on the density value.

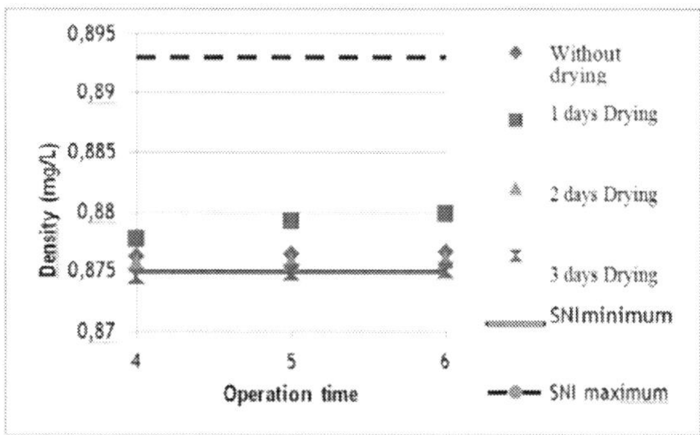

Figure 3. Density Value Against Operating Time

From the figure 3, can be seen that the density value does not exceed the value of the SNI range and the density value above greatly affects the operation time and the drying time. The highest value of the density of citronella oil obtained at the time of 1 day drying with 6 hours operating time. While the lowest density of citronella oil values obtained at 3 days of drying at 4 hours of operation.

3.3. Refractive Indexs Test for Citronella Essensial Oil

It can be seen in the graph that from 12 observed trial runs almost have the same refractive index value which is around 1.465, the refractive index obtained is in accordance with the Indonesian National Standard (SNI). From all the experiments, there were 2 trials whose refractive index value was higher than the others, with a refractive index value of 1.474, which was run at a distillation time of 5 and 6 hours at 3 days of drying. This shows that the drying time will affecting the quality of the essensial oils. At the highest citronellal volume, a refractive index of 1.465 was obtained in the experiment on one drying day and at 6 hours of refining time.

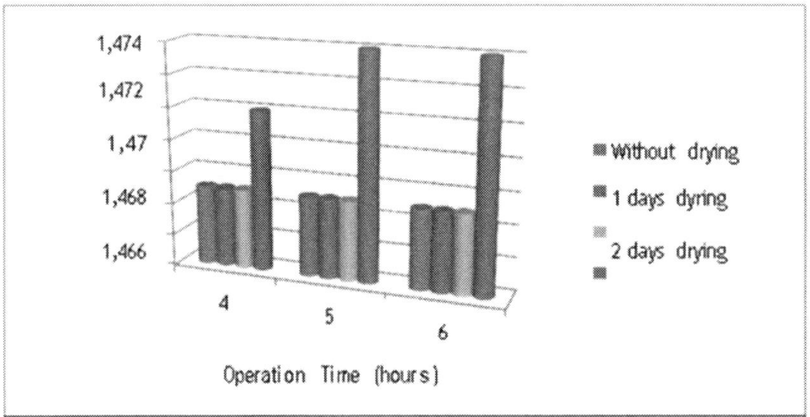

Figure 4. Refractive Indexs Value

The presence of high water content in oil produces a small refractive index because light is easily refracted in water compared to oil due to differences in liquid density. Therefore, essential oils with large refractive index have better purity and quality presented in figure 4.

4. Conclusion

From the result of this study some conclusion can be drawn : (1) The optimum time for refining essensial oil were 6 hours distillaton and 1 days drying time. (2) . The highest value of the density of citronella oil is 0,88 obtained at the time of 1 day drying with 6 hours operating time. (3) At the highest citronellal volume, a refractive index of 1.465 was obtained in the experiment on one drying day and at 6 hours of refining time.

References

[1] Sugiyama, "Solar Energy Materials and Solar Cells Degradation and fault diagnosis of photovoltaic cells using impedance spectroscopy," Sol. Energy Mater. Sol. Cells, vol. 194, no. January, pp. 130–136, 2019

[2] I. Mathews, S. N. Kantareddy, T. Buonassisi, and I. M. Peters, "Technology and Market Perspective for Indoor Photovoltaic Cells," Joule, pp. 1–12, 2019.

[3] L. Hernández-callejo, S. Gallardo-saavedra, and V. Alonso-gómez, "A review of photovoltaic systems : Design , operation and maintenance," Sol. Energy, vol. 188, no. June, pp. 426–440, 2019.

[4] H. Fu, G. Li, and F. Li, "Performance comparison of photovoltaic / thermal solar water heating systems with direct-coupled photovoltaic pump , traditional pump and natural circulation," Renew. Energy, vol. 136, pp. 463–472, 2019

[5] P. Gonçalves, V. Sampaio, M. Orestes, A. González, R. Monteiro, D. Vasconcelos, M. Aylla, J. Carlos, D. Toledo, J. Paulo, and P. Pereira, "Photovoltaic technologies : Mapping from patent

analysis," Renew. Sustain. Energy Rev., vol. 93, no. May 2017, pp. 215–224, 2018.

[6] E. Tatsi and G. Gri, "Solar Energy Materials and Solar Cells Polymeric materials for photon management in photovoltaics," vol. 196, no. January, pp. 43–56, 2019.

[7] A. Chauhan, V. V Tyagi, and S. Anand, "Futuristic approach for thermal management in solar PV / thermal systems with possible applications," Energy Convers. Manag., vol. 163, no. February, pp. 314–354, 2018.

[8] Y. Jia, G. Alva, and G. Fang, "Development and applications of photovoltaic – thermal systems : A review," Renew. Sustain. Energy Rev., vol. 102, no. December 2018, pp. 249–265, 2019.

[9] N. Safitri, "Different Trends of Hybrid Solar And Raindrops Energies to Generate Different Trends of Hybrid Solar And Raindrops Energies to Generate Photovoltaic," 2019.

[10] N. Safitri and F. Abdurrahman, "Integration of DC Households System Generated by Single-Phase Rooftop PVs into Unbalanced Three-Phase Residential Feeder," vol. 4, no. 2, 2017.

[11] S. Seme, K. Sreden, Š. Bojan, and M. Had, "Analysis of the performance of photovoltaic systems in Slovenia," vol. 180, no. January, pp. 550–558, 2019.

[12] Z. Zhang, M. Yao, X. Li, Q. Deng, Q. Peng, and J. Zhong, "Solar Energy Materials and Solar Cells Simultaneous functional and structural imaging for photovoltaic devices," Sol. Energy Mater. Sol. Cells, vol. 193, no. January, pp. 101–106, 2019.

[13] T. Rihayat, "Composition on Essential Oil Extraction from Lemongrass Fragrant by Microwave Air Hydro Distillation Method to Perfume Dermatitis Production Composition on Essential Oil Extraction from Lemongrass Fragrant by Microwave Air Hydro Distillation Method to Perfume Dermatitis Production," pp. 6–11, 2019.

[14] H. Agusnar, B. Wirjosentono, S. Salim, T. Rihayat, and T. Fauzi, "Synthesis and Characterization of Chitosan with Addition of Patchouli Oil to Improve Mechanical Properties Biofilm Synthesis and Characterization of Chitosan with Addition of Patchouli Oil to Improve Mechanical Properties Biofilm," 2018.

[15] Jaafar, Jamiluddin., Januar, P.S·, Mohd, B.M.P., Tezara, C., Sharmiza, A., Rihayat, T. "Influence of Selected Treatment on Tensile Properties of Short Pineapple Leaf Fiber Reinforced Tapioca Resin Biopolymer Composites". Journal of Polymers and the Environment. 2018.

[16] Rihayat, T., Suryani, T. Fauzi., H.Agusnar., B. Wirjosentono., Syafruddin., Helmi., Zulkifli., P..Alam., M.Sami. "Mechanical properties evaluation of single and hybrid composites polyester reinforced bamboo, PALF and coir fiber". IOP Conference Series: Materials Science and Engineering. Vol.334. Pp 1-8, 2018.

[17] T. Rihayat, M. Saari, M. H. Mahmood, W. M. Z. W. Yunus, A. R. Suraya, K. Z. H. M. Dahlan, and S. M. Sapuan, "Mechanical Characterisation of Polyurethane / Clay Nanocomposites," vol. 15, no. 8, pp. 647–652, 2007

[18] T. Rihayat, J. P. Siregar, and M. Yunus, "Synthesis and Characterization of North Aceh CEC Bentonite Determination with Methylene Blue Method and Increased D-Spacing after Addition of Surfactants CTAB-SDS Synthesis and Characterization of North Aceh CEC Bentonite Determination with Methylene Blue Method and Increased D- Spacing after Addition of Surfactants CTAB-SDS," 2019.

IOP Conf. Series: Materials Science and Engineering **854** (2020) 012052 doi:10.1088/1757-899X/854/1/012052

Detection of Empty Slots in Car Parking System Using Neural Network Method

Ahmad Ihsan, Nurul Fadillah, C R Gunawan, Mursyidah Mursyidah

Faculty of Engineering, Samudra University, Departement of Information Technology and Computers, Politeknik Negeri Lhokseumawe, Indonesia

ahmadihsan@unsam.ac.id

Abstract. There are many transportation equipment such as a car, a motorbike in Indonesia. Increasing the volume of vehicle users will increase parking activities in every public place. Parking is a condition of a vehicle that is not temporary. The current parking system still uses a conventional parking system that only uses parking lots and parking attendants who control each vehicle that enters, and often do not pay attention to the capacity of parking space that has been provided and often drivers go around first to find an empty parking lot. Therefore, detection of empty slots in the parking system is needed. To obtain this information, this study using a neural network method. The neural network is one of the information processing systems designed by imitating the workings of the human brain in problem-solving. The purpose of this research is to determine which car parking slots filled or empty parking slots. The results of detection of car parking slots using a neural network method were carried out experiments on 20 objects with as many as 12 slots and with three slots that are rectangles, squares, and traingles. In the square slot with an area of 784 cm^2 get accuracy results 100%, rectangular slots with an area of 3.608 cm^2 obtained accuracy 95% and a traingle slot with an area of 1.804 cm^2 obtained an accuracy 90%.

1. Introduction

Technological developments and increasing human needs are things that influence each other. The increased human need to trigger technology developments, while technology developments trigger negative effects of new technology. An example needs to use transportation equipment such as a car, motorbikes, and other public transportation equipment. The total volume of vehicle users such as cars and motorbikes in Indonesia is very much. The increasing volume of vehicle user will increase parking activities in every public place such as offices, campuses, shopping centers.

The current parking system still uses a conventional parking system that only utilizes parking lots and parking officers who control each vehicle that enters, and also often does not pay attention to the capacity of parking space owned by a building. This cause loss from the vehicle owner because the driver does not know where parking lot empty and forced to leave if not find an empty parking lot. Parking lots are important facilities that must be owned by a public place. Most of the parking lots have large land so drivers are difficult to reach parking lots.

The process of detecting the availability of parking slots will be obtained using a neural network method. A neural network is a method whose pattern follows the work human nervous system. Where the main processing of the human nervous system lies in the brain. In utilizing a neural network method, visitors will find it easier to get a parking space.

IOP Conf. Series: Materials Science and Engineering **854** (2020) 012052 doi:10.1088/1757-899X/854/1/012052

2. Study of Literature

2.1. Introduction to Parking
Parking is the condition of a vehicle that is not temporary, while parking facilities are locations that are determined as a temporary stop for vehicles to carry out activities in a period. [1]

2.2. Slot
The system can detect the condition of parking slot whether parking slot in a parking lot is still empty or has been filled. Parking space is used as input from a system. [11]

2.3. Artificial Intelligence
Artificial intelligence is one of the computer sciences that makes the machine (computer) can do the work as well as what humans do. Artificial Intelligence is a branch of computer science which in representing knowledge uses more forms of symbols than the number and processes information based on heuristic methods or based on several rules. [6]

2.4. Deep Learning
Deep learning is one of the fields of machine learning that utilizes many nonlinear information processing layers to perform feature extraction, pattern recognition, and classification. [3] Deep learning is an approach to solving problems in computer learning systems that use hierarchy concept. Hierarchy concept makes computers capable of learning complex concepts by combining simpler concepts. If a graph is illustrated how the concept is built on another concept, this graph will be deep with many layers, this is reason referred to as deep learning. [5]

2.5. Neural Network
Neural Network is a method whose pattern follows the work of the human nervous system. Where the main processing of the human nervous system lies in the brain. The smallest part the human brain is nerve cells which are the basic units for information processors. This unit is called a neuron. Around 10 billion neurons in each human brain and around 60 trillion connections between neurons in the human brain. [2]

2.6. The Architecture of Artificial Neural Networks
The arrangement neurons and the pattern of interconnections between layers are called net architecture. Artificial neural network architectures are classified as a single layer, multilayer, and competitive layer. To determine many layers used, input layer not included as the layer used. The number of layers included in artificial neural networks shows the number of weight values associated between these layers because weight values were important in artificial neural networks. [10]

2.6.1. Single-layer-net
The single-layer net has one layer to connect its weight value. Input neurons are directly related to output neurons. This network only receives information and directly processes it into output without going through the hidden layer. The characteristics of this single-layer-net have only one input layer and output layer.

2.6.2. Multilayer net
Multilayer net is a network that has one or more additional layers (hidden neurons) between the input and output layers. This multi-layer network to solve more complex problems than networks with one layer.

2.6.3. Competitive layer net
Competitive layer net consist of two or more artificial neural networks. This network architecture can connect one neuron with another neuron.

3. Research Method

3.1. General Architecture
Following is the general architecture of the application:

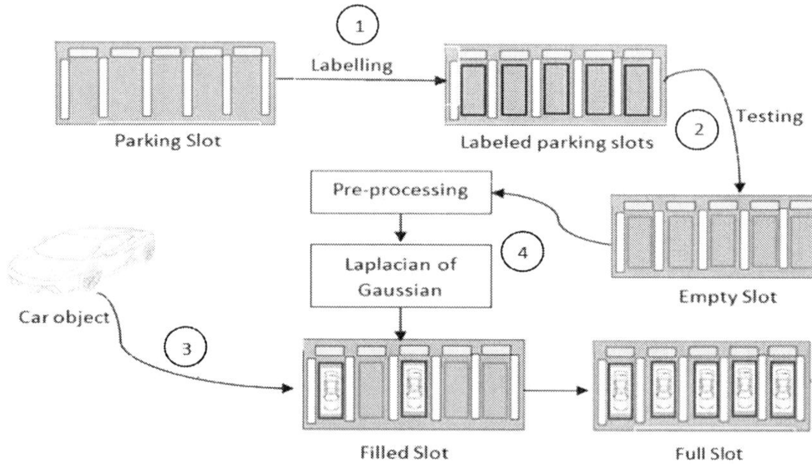

Figure 1. General Architecture

Based on figure 1, it can explain that: (1) The first process is labeling for each available parking slot, labeled parking slots marked with blue slots. The purpose of this labeling is to be easily recognized by each slot, (2) After the labeling process, it was followed by testing, the image shows an empty slot, where the empty slot is indicated by a green slot. In this testing phase, there is a pre-processing stage that is followed by laplacian of gaussian operators where pre-processing is a process used to improve the quality of image display to have a better format so that the image becomes easy to process to be good edge detection. The gray scaling process is used as pre-processing. The gray scaling process is the process of converting images that have colors to images that have a gray level. The edge detection process here uses a laplacian of gaussian operator developed from the second derivative, the gaussian followed by the Laplace operation. (3) Next, there is a car object entering the available parking slot. Then the car object goes to slot 1 and slot 3. if the car object enters an empty slot, the parking area will be full. (4) After the car object enters the parking lot. Then the previous green slot turns red, the process of changing the color of this slot because it has passed the pre-processing stage using the Laplacian of Gaussian operator.

3.2. Ilustration of a Car Parking System
Slot manipulation:

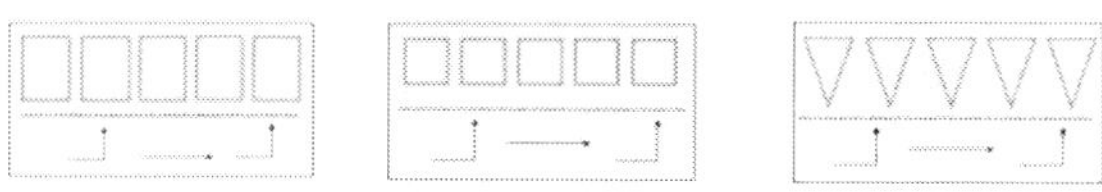

(a) (b) (c)

Figure 2. Empty Parking Slot (a) Rectangular Slots (b) Square Slots and (c) Triangle Slots

Parking lots will be to design with five slots, slots marks in green if the parking lot is empty. If the car is parked, it turns red.

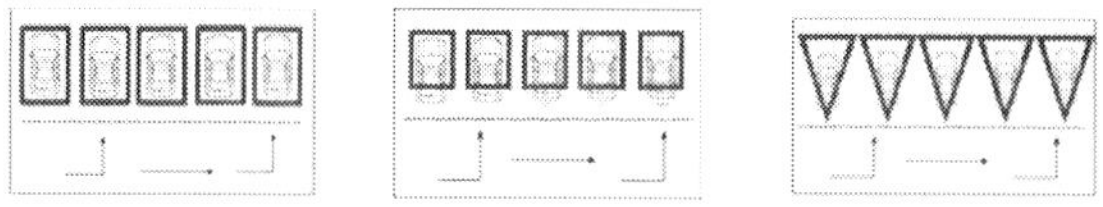

(a) (b) (c)

Figure 3. Full Parking Slots (a) Rectangular Slots (b) Square Slots and (c) Triangle Slots

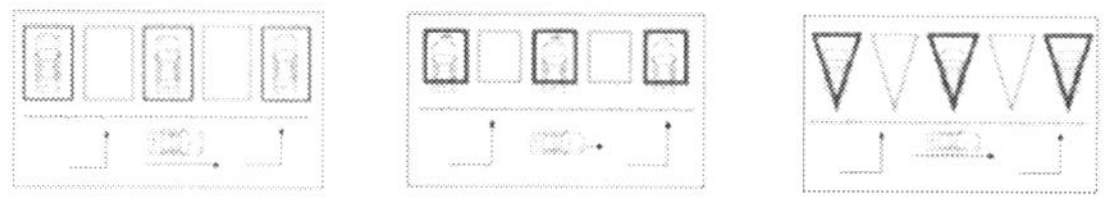

(a) (b) (c)

Figure 4. Filled Slots (a) Rectangular Slots (b) Square Slots and (c) Triangle Slots

Based on figure 4 above, some slots are filled which are marked with red slots and green slots indicating empty slots. The color will change if the parking area has been placed by the visitor.

3.3. Car Parking System Using a Neural Network

The mechanism of the car parking system using the neural network is as follows:

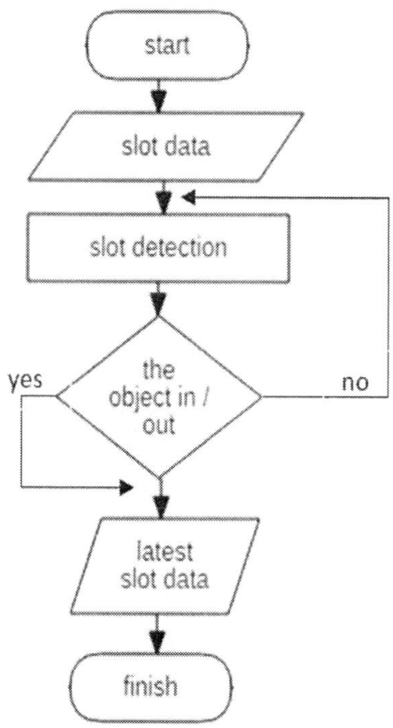

Figure 5. The Mechanism of the Car Parking System Using a Neural Network

The initial data slot that the system receives is in the form of coordinate parking slot data. Where the coordinates of the parking slot are obtained from the size of the pixel image used. The slot detection process is where the object will go to the available slot on the parking lot. The empty slot is marked with a green slot and the slot is filled with a red slot. In this process there is a pre-processing stage, where pre-processing is the initial stage in this research process which is used to apply several effects to be processed in order to continue the next stage and the pre-processing stage is also used to improve the quality of the image display to have a format better so that the image becomes easier to process further to detect objects.

3.3.1. Empty Parking Classification Process

At this stage input the image of the empty parking slot, then the user can specify the parking slot where the parking slot can be rectangular, square and triangle. At the pixel calculation stage, initially, it is seen from the pixel of the entire image that has been inputted before. Example: pixel of an image of 520 x 320 which is 520 as the width and is the coordinate point of the x-axis and 320 as the height which is the coordinate point of the y-axis. Then in the Laplacian of Gaussian process there is a pre-processing process, where pre-processing is the initial stage in the research process that is used to apply several effects to be processed to continue the next stage and the pre-processing stage is also used to improve the image display quality to have a better format so that the image becomes more easily processed further to detect objects. the process used in the pre-processing stage is the Laplacian of the Gaussian operator. Where laplacian of gaussian is an edge detection operator developed from derivative of two processes, namely, Gaussian followed by Laplace operation. The function of the gaussian is to reduce noise while the Laplace operation is used to minimize the possibility of error detection. The approach with the Laplacian of Gaussian operator starts with an image blurring process where the blurring process is used to reduce noise with blur, then followed by a grayscale process, after going through the grayscale process, then the final stage is the laplacian stage. The slot data obtained is in the form of coordinate data from each slot that has been describing in the previous process as in figure 6.

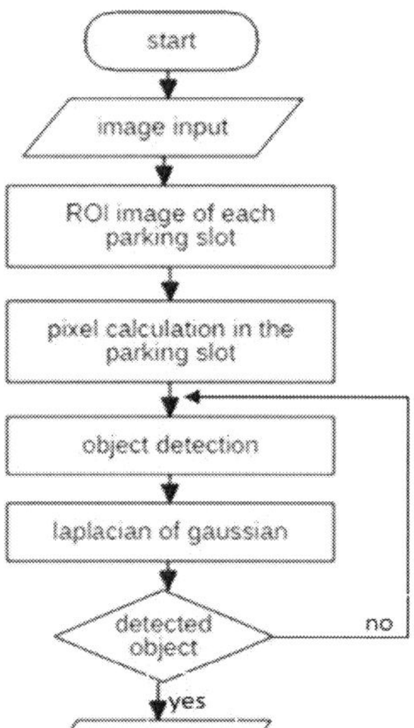

Figure 6. Empty Parking Classification

IOP Conf. Series: Materials Science and Engineering **854** (2020) 012052 doi:10.1088/1757-899X/854/1/012052

3.3.2. Neural Network Process
As follows:

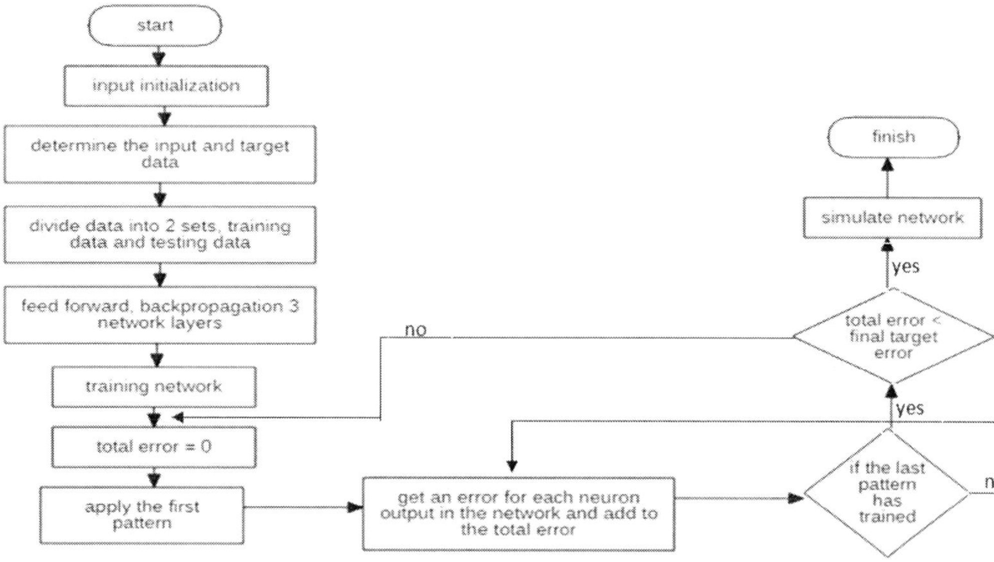

Figure 7. Flowchart Neural Network

In this process, the first step is a random input initialization. Random initialization is the method most often used in weight initialization, where weights are initialized randomly without using a scale factor. Then specify the input data and target data. Also, specify training data and test data. Then create a feed-forward (back feed), backpropagation 3 network layers. Backpropagation consists of many layers (multilayer), namely the input layer, at least one hidden layer, and the output layer. In the output layer and hidden layer, each one is given a value that can be 1. Next, to train the network, there are 4 processes in this process, namely calculating gradient errors in the output layer, updating the hidden layer weight to the output layer, calculating the gradient error in the hidden layer and update the input layer weight to the hidden layer.

4. Results and Discussion

4.1. Parking Slot
The parking slots contained in the system are rectangular, square slots and triangle slots.

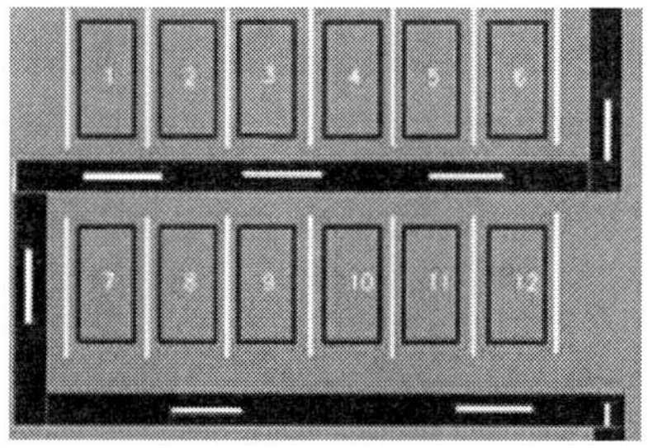

Figure 8. Rectangular Slots

IOP Conf. Series: Materials Science and Engineering **854** (2020) 012052 doi:10.1088/1757-899X/854/1/012052

The rectangular slot above has an area of 3,608 cm^2 with a length of 82 cm and a width of 44 cm. Each slot is given a number 1 to 12 marks that are used to get to know the slot. Square slots are as follows:

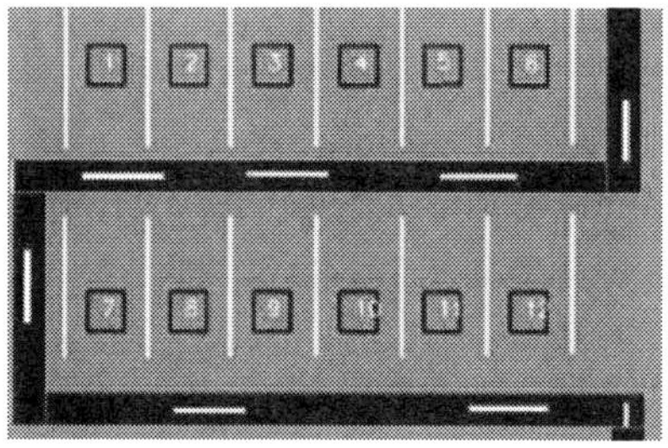

Figure 9. Square Slots

The square slots above have an area of 784 cm^2 with each side length of 28 cm. Triangle slots are as follows:

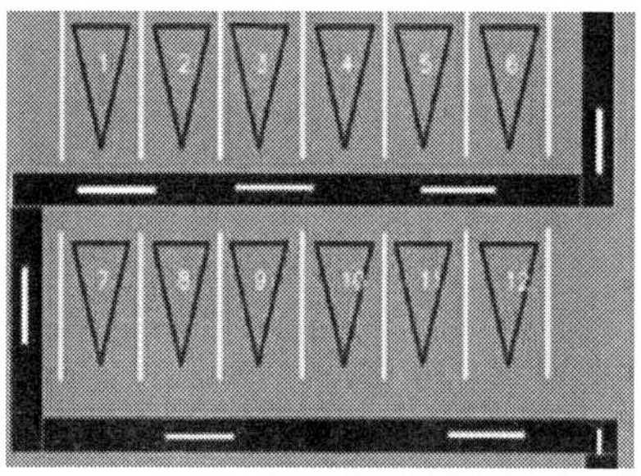

Figure 10. Triangle Slots

The triangle slot above has an area of 1.804 cm^2 where the base is 44 cm and height is 82 cm.

4.2. Object of Car

The data used is data determined by the researcher. Data consists of 20 car objects.

Table 1. Object of Car

Object	Figure
1st Object	
2nd Object	
3rd Object	
4th Object	
5th Object	

IOP Conf. Series: Materials Science and Engineering **854** (2020) 012052 doi:10.1088/1757-899X/854/1/012052

6th Object	
7th Object	
8th Object	
9th Object	
10th Object	
11th Object	
12th Object	
13th Object	
14th Object	
15th Object	
16th Object	
17th Object	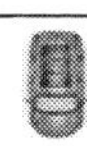
18th Object	
19th Object	
20th Object	

IOP Conf. Series: Materials Science and Engineering **854** (2020) 012052 doi:10.1088/1757-899X/854/1/012052

4.3. Test on the Parking Slot

4.3.1. Parking Slot Testing Based on Slot Area

This testing process uses 20 car objects with 12 slot slots. The following are the results of testing a rectangular slot with a slot area of 3.608 cm^2:

Table 2. Results of Testing Rectangular Slots with Slot Area 3.608 cm^2

No.	Slot	Total	Recognizable Object	Unrecognize d Object	Figure
1.	Slot 1	2	2	0	
2.	Slot 2	1	1	0	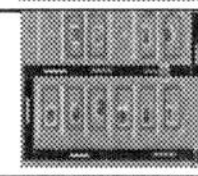
3.	Slot 3	2	2	0	
4.	Slot 4	2	2	0	
5.	Slot 5	2	2	0	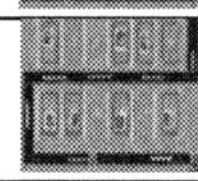
6.	Slot 6	1	0	1	
7.	Slot 7	1	1	0	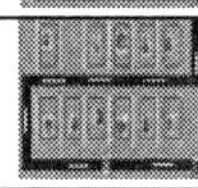
8.	Slot 8	2	2	0	
9.	Slot 9	2	2	0	
10.	Slot 10	2	2	0	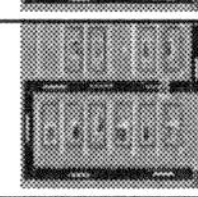
11.	Slot 11	1	1	0	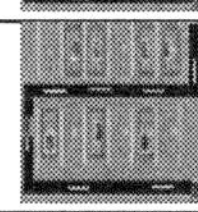
12.	Slot 12	2	2	0	
Total		20	19	1	

The following is the percentage of accuracy from the results of testing table 2:

$$\frac{20-1}{20} x\ 100\% = 95\%$$

Based on the calculation of the data, it was found that the average accuracy for the testing process of objects in slots was rectangular with an area of 3.608 cm^2 slots of 95%.The following are the results of testing a square slot with a slot area of 784 cm^2:

Table 3. Results of Testing Square Slots with Slot Area 784 cm^2

No	Slot	Total	Recognizable Object	Unrecognized Object	Figure
1.	Slot 1	2	2	0	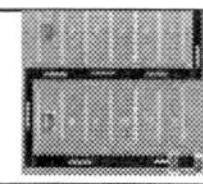
2.	Slot 2	1	1	0	
3.	Slot 3	2	2	0	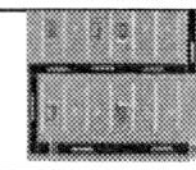
4.	Slot 4	2	2	0	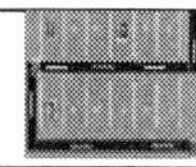
5.	Slot 5	2	2	0	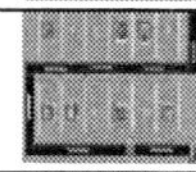
6.	Slot 6	1	1	0	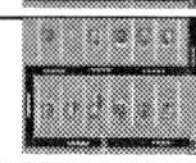
7.	Slot 7	1	1	0	
8.	Slot 8	2	2	0	
9.	Slot 9	2	2	0	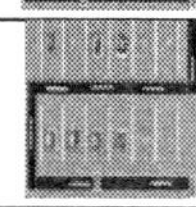
10.	Slot 10	2	2	0	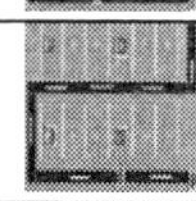
11.	Slot 11	1	1	0	
12.	Slot 12	2	2	0	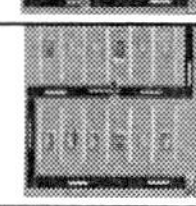
Total		20	20	0	

The following is the percentage of accuracy from the results of testing table 3:

$$\frac{20-0}{20} x\ 100\% = 100\%$$

Based on the calculation of the data, it was found that the average accuracy for the testing process of objects in slots was a square shape with an area 784 cm^2 slots of 100%. The following are the results of testing a triangular slot with a slot area 1.804 cm^2:

IOP Conf. Series: Materials Science and Engineering **854** (2020) 012052 doi:10.1088/1757-899X/854/1/012052

Table 4. Results of Testing Triangular Slots with Slot Area 1.804 cm^2

No.	Slot	Total	Recognizable Object	Unrecognized Object	Figure
1.	Slot 1	2	2	0	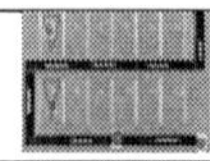
2.	Slot 2	1	1	0	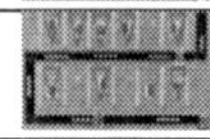
3.	Slot 3	2	2	0	
4.	Slot 4	2	2	0	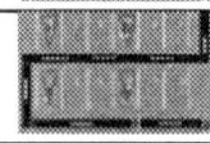
5.	Slot 5	2	2	0	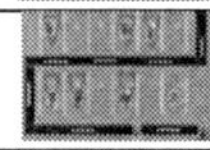
6.	Slot 6	1	1	0	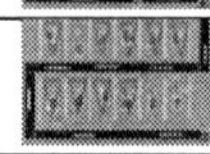
7.	Slot 7	1	1	0	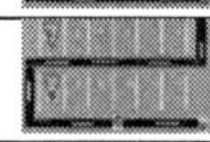
8.	Slot 8	2	2	0	
9.	Slot 9	2	2	0	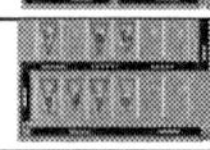
10.	Slot 10	2	2	0	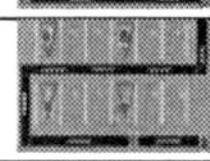
11.	Slot 11	1	0	1	
12.	Slot 12	2	1	1	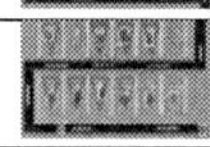
Total		20	18	2	

The following is the percentage of accuracy of the test results table above:

$$\frac{20-2}{20} x\ 100\% = 90\%$$

Based on the calculation of the data, the average accuracy for the object testing process in a triangle-shaped slot is obtained with a slot area of 1.804 cm^2 of 90%.

4.3.2. Test Results Based on Accuracy Value

The test results are based on the accuracy values obtained from the average accuracy for the testing process in a rectangular, square and triangular slot. Can be shown in the form of a diagram as follows:

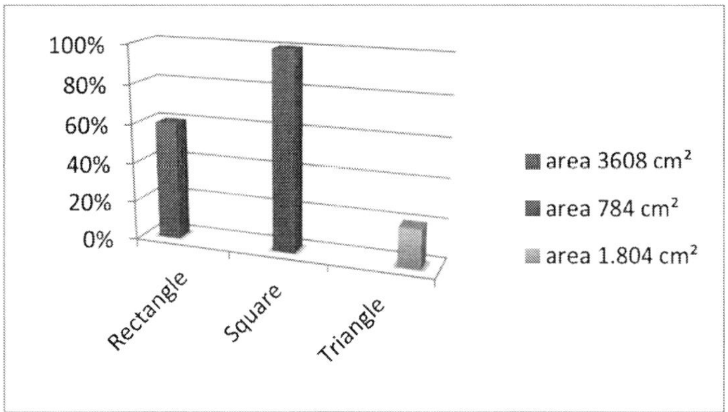

Figure 11. Test Result Diagram Based on Accuracy Value

Based on the graph (Figure 11) shows that if it has a small slot area (for example 784 cm^2), the accuracy of the object is detected very well. The test is done by the state of the object stop at the same place, but only the shape of the slot is different. In rectangular slots, the value of accuracy is 95% where there is 1 car object that is not recognized by a rectangular slot. In the triangle-shaped slot with an area of 1.804 cm^2, the value of accuracy obtained is 90% because there are 2 car objects from 20 unrecognized car objects. Square slots with a slot area of 784 cm^2 all car objects can be recognized by a square slot so that it reaches an accuracy value of 100%.

4.4. Car Object Testing Based on the Middle Point of the Car Object Coordinate

4.4.1. Start Position and End Position of the Car Object

The following is the test result based on the coordinate midpoint of the beginning and end of a car object:

Table 5. Test Results Based on the Midpoint of the Start and End Position of a Car Object

No.	Object to Slot	Start Position (x,y)	End Position (x,y)
1.	1st object to Slot 3	(505, 303)	(80, 215)
2.	2nd object to slot 1	(505, 303)	(74, 44)
3.	3rd object to slot 4	(505, 306)	(286, 48)
4.	4th object to slot 10	(506, 303)	(287, 214)
5.	5th object to slot 3	(507, 303)	(216, 52)
6.	6th object to slot 8	(507, 304)	(152, 210)
7.	7th object to slot 9	(506, 304)	(213, 215)
8.	8th object to slot 12	(506, 303)	(424, 211)
9.	9th object to slot 5	(506, 306)	(357, 52)

10.	10th object to slot 3	(505, 304)	(217, 52)
11.	11th object to slot 9	(506, 303)	(223, 195)
12.	12th object to slot 11	(506, 303)	(357, 214)
13.	13th object to slot 6	(506, 304)	(422, 50)
14.	14th object to slot 2	(506, 303)	(152, 52)
15.	15th object to slot 4	(506, 304)	(287, 48)
16.	16th object to slot 1	(506, 303)	(78, 39)
17.	17th object to slot 12	(505, 303)	(434, 206)
18.	18th object to slot 8	(507, 306)	(151, 210)
19.	19th object to slot 5	(505, 305)	(358, 49)
20.	20th object to slot 10	(506, 304)	(285, 206)

4.4.2. Test Results Based on Start Position and End Position of the Car Object

The following are the test results based on the starting and ending positions of a car object, which can be illustrated with a graph as follows:

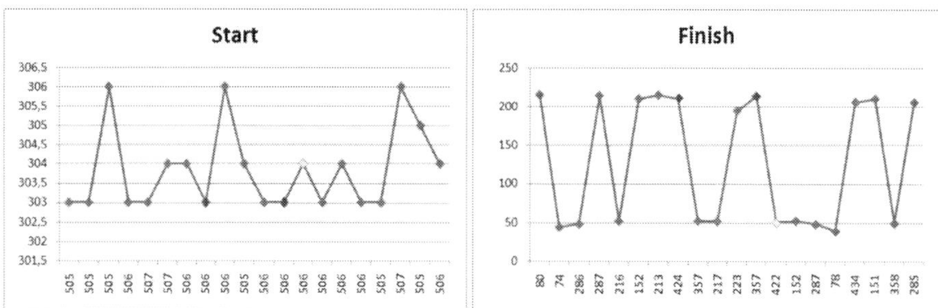

Figure 12. Coordinate Graph of the Start Position and End Position of a Car Object

Based on the graph above, it can be analyzed that there are 20 car objects with each position of the starting and ending points of the car object coordinates, where 20 car objects will be detected with different forms of car parking slots, namely rectangular slots, square slots, and triangular slots. Base on the coordinate points of the car objects detects with rectangular slots there is 1 undetectable object, namely the 13th object with the coordinates of the initial position is (506, 304) and the 13th object stops in slot 6 where the coordinates of the final position is (422, 50), where the chart marked with a yellow plot. On car objects detect with square slots, each object from 20 car objects can detect properly In the car object detected with a triangle-shaped slot, there are 2 undetectable objects, namely the 8th object and the 12th object which are marked with a red plot, where the 8th object stops in slot 12 at the start coordinate point (506, 303) and the position of the object stops at the coordinate point (424, 211) and the 12th object stops at slot 11 at the start coordinate point (506, 303) and stops at slot 6 with the coordinates of the final position (357, 214).

5. Conclusion

Based on the results of the research conduct, it can be concluded that if the slot has an area with small size it can detect the location of the car properly. Level of accuracy obtained by 100% in the square parking slot with a smaller area than the rectangular slot and triangle slot. whereas if the parking slot has a larger area, for example at a rectangular slot has an accuracy rate of 95% and at the triangle-shaped slot has an accuracy of 90%. The start and final position of the car object also affect the object detected or not if the initial position is at the coordinate point (506, 303) and the position of the car object stops (357, 214) based on test the object cannot recognize. Based on the size of the car object, the object of the car does not affect the slot detected or not detected, because the thing that affects the success of this research is the area of each parking slot itself and the position of the stop of a car object.

References

[1] A, S. R. I., & Triyanto, D. (2014). Model Sistem Parkir Informatif Berbasis Programmable Logic Controller (PLC). *Coding Jurnal Komputer Dan Aplikasi Untan*, *02*(1), 1–9.

[2] Abas, M. I., Syukur, A., & Soeleman, M. A. (2017). Prediksi Rentet Waktu Jumlah Penumpang Menggunakan Algoritma Neural Network Berbasis Genetic Algorithm. *Jurnal Teknologi Informasi*, *13*, 101–114.

[3] Ahmad, J., Farman, H., & Jan, Z. (2019). Deep Learning Methods and Applications. *SpringerBriefs in Computer Science*, 31–42. https://doi.org/10.1007/978-981-13-3459-7_3

[4] Alqurni, R. P., & Muljono. (2016). Pengenalan Tanda Tangan Menggunakan Metode. *Directory Open Access Journals*, *15*(4), 352–363.

[5] Bengio, Y., Goodfellow, I., & Courville, A. (2016). Deep Learning（中文） 深度学习, 785. Retrieved from http://www.deeplearningbook.org

[6] Budi Serasi & Ramadhan, F. G. (2018). Perancangan Game Become a King Berbasis Artificial Intelligence. *Methomika: Jurnal Manajemen Informatika Dan Komputerisasi Akuntansi*, *2*(1), 25–35.

[7] Informatika, J. T., Sains, F., & Teknologi, D. A. N. (2016). Sistem Parkir Otomatis Mengidentifikasi Identitas Pengendara Dengan Biometrik Dan Kendaraan Dengan RFID. https://doi.org/10.3157/021.121.0306

[8] Kusumodestoni, R. H., & Zyen, A. K. (2015). Prediksi Kecepatan Angin Menggunakan Neural Network Untuk Mengetahui Besar Daya Listrik Yang Dihasilkan, *6*(1).

[9] Lubis, M. R. (2018). Analisis jaringan saraf tiruan back propgation untuk peningkatan akurasi prediksi hasil pertandingan sepakbola 1. *TECHSI - Jurnal Penelitian Teknik Informatika*, *10*(1), 51–62. https://doi.org/10.1016/0169-5347(94)90266-6

[10] Majumder, M. (2015). Artificial Neural Network, 49–54. https://doi.org/10.1007/978-981-4560-73-3_3

[11] Maulana, R., Fitriyah, H., & Prakasa, E. (2018). Implementasi Sistem Deteksi Slot Parkir Mobil Menggunakan Metode Morfologi dan Background Subtraction. *Jurnal Pengembangan Teknologi Informasi Dan Ilmu Komputer (J-PTIIK) Universitas Brawijaya*, *2*(5), 1954–1959.

[12] R, A. D., Imamah, F., Andre, Y. M., & Ardiansyah. (2018). Aplikasi Chatbot (Milki Bot) Yang Terintegrasi Dengan Web CMS Untuk Customer Service Pada UKM MINSU. *Jurnal Cendikia*, *XVI*, 100–106.

[13] Samiudin. (2016). Peran metode untuk mencapai tujuan pembelajaran. *Jurnal Studi Islam*, *11*(2), 94–97.

IOP Conf. Series: Materials Science and Engineering **854** (2020) 012053 doi:10.1088/1757-899X/854/1/012053

Fault Plane Estimation Through Hypocentres Distribution of the July-August 2018 Lombok Earthquakes Relocated by using Double Difference Method

M S Rosid[1], R Widyarta[1], T Karima[1], S K Wijaya[2], and S Rohadi[3]

[1] Geophysics, FMIPA Universitas Indonesia, Depok 16424 Indonesia.
[2] Physics Departement, FMIPA Universitas Indonesia, Depok 16424 Indonesia.
[3] Research and Development Center, Meteorological, Climatological, and Geophysical Agency (BMKG), Jl. Angkasa 1 No. 2, Kemayoran, Jakarta, Indonesia.

Email: syamsu.rosid@ui.ac.id

Abstract. Indonesia is actually located in collisions between three large plates of Eurasia, Indo-Australia, and Pacific. It causes earthquakes to occur frequently in several regions, one of which occurs sequentially the 5 big shocks in Lombok in period of July-August 2018. There is a Flores thrust fault in the North of the Lombok Island region. The objective of study is determining whether the earthquakes sequence is related to the fault. The study is done through the analysis of relocated 321 of 489 hypocentres of the earthquakes acquired from 13 BMKG stations in July-October 2018. The hypocentres are relocated by using the double-difference method. The events are focused only with magnitude > 3.5 M. The relocation result is validated by the RMS residual histograms with the average value close to 0. The results show that the seismic activity mostly spread in the northern part of the Island of Lombok. The distribution of the epicenter points tends to form a pattern oriented from West to East in the northern region of the island which is thought to be the zone of the faults. The events most likely have not generated and not laid along the Flores Fault.

1. Introduction

Bali, Lombok and the Islands of Nusa Tenggara are one of the regions in Indonesia that have a high level of vulnerability to earthquakes. One source of the earthquake that has been identified is an earthquake originating from a subduction zone in the South of Indonesia [1]. This zone originates from collisions between the Indo-Australian plate (South) and Eurasia (North). Other causes of earthquakes can also come from the movement of fault structures in the earth. Such events that occur in this region are earthquakes in the Lombok region, Nusa Tenggara in July-August 2018. From the data of the National Disaster Management Agency (BNPB) there are ± 555 people died and material losses worth ± 5.04 trillion rupiahs. The earthquake is indicated to occur due to the movement of the thrust fault in the northern part of the island arc [2].

To analyze and prove the seismicity caused by tectonic activities as in the Lombok case, it is necessary to determine the exact location of the hypocenter point. One technique to relocate earthquake hypocenter points is the double-difference method. The data used in this method is wave travel time data between pairs of earthquake events with observer stations. The principle of this method assumes that if the distance between two paired earthquakes is relatively small compared to the distance between each earthquake and the paired observer station, then the raypath and the second waveform are considered almost the same [3]. Thus the velocity model error can be minimized without using station correction [4]. The purpose of this study is relocating the hypocenter point of the Lombok 2018 earthquake using the double difference method and to find out the estimated of the fault structure from the hypocenter distribution pattern. This study, therefore, can be used as a parameter for subsequent earthquake mitigation analysis.

2. Data and Methodology

The hypocenter relocation method used in this study is the double difference. The method runs to minimize the residual travel time of calculations and observations. The data used in the study are 1086 earthquakes obtained from BMKG catalog from July 1, 2018 till Sept 15, 2018 with magnitude 3.5 - 9 M and depth 1-150 km. The study area in Lombok, Nusa Tenggara located in -6.62 S – 9.83 S and 113.12 E - 122.6 E. The study results is validated by looking at the value of residual histogram where the difference between travel time calculation and observation is small.

Double difference method assumes if there are two adjacent earthquake events, with smaller distances between events compared to the distance between the two events to the recording station, it is considered raypath these two events tend to be the same [5]. Travel time difference between two events recorded at the station is a function of distance between earthquakes. The difference in travel time of observation and calculation of two events is called residual time, which is written in the form:

$$dr_k^{ij} = (t_k^i - t_k^j)^{obs} - (t_k^i - t_k^j)^{cal} \quad (1)$$

The equation above is the double difference equation, which t_k^j is the first term of observation travel time for two event (*ij*) in *k* station and the second term is for the calculation travel time. The processing of this method is run by HypoDD software.

3. Results and Discussion

The results of the earthquake hypocenter relocation indicate a change in the location of the coordinates of both the latitude, longitude and depth. The new hypocenters are different from the hypocenter coordinates before relocated. The changes of locations are clearly seen in the epicenter map. The results indicate that the distribution of relocated epicenter is probably closer to the geological interpretation of the study area.

Figure 1 shows the distribution of epicenter point of the earthquake which is divided based on depth parameters. Red points indicate shallow depth events (<60 km), yellow indicate moderate depth (60-300 km) and green indicate deep depth events (> 300 km). It can be seen that if the epicenter point is seen approaching the area that is suspected to be the fault line, the epicenter point forms an East-West oriented pattern (yellow circle in Figure 1b). Then if we look further at the vertical sections (see Appendix 1), it can be seen that before the data is relocated, the distribution of hypocenter spreads at a depth of 10 km, which is an effect of recording data that is less than perfect or commonly known as fixed depth. After relocation, the distribution of the hypocenter becomes more diffuse and can be more interpreted. It can be seen if the hypocenter distribution (of mainshock and aftershock) that had been clustered at a depth of 10 km began to spread and form a pattern or trend (yellow circle) which is assumed to be shaped like a structure that slopes towards the south (Figure 3).

Figure 2 shows a graph of comparison histogram between residual values before and after the relocation process. The residual value is the difference in travel time between the travel time of observation and the travel time of calculation. In the graph, the residual value of the relocated event is mostly displayed close to 0 (more toward the middle). It really different from the residual value of events before the relocation process where there is still a lot of data with high residual. The residual value close to 0 can be interpreted that the difference time of observation and calculation is not much different, which means that the relocation results are good [6].

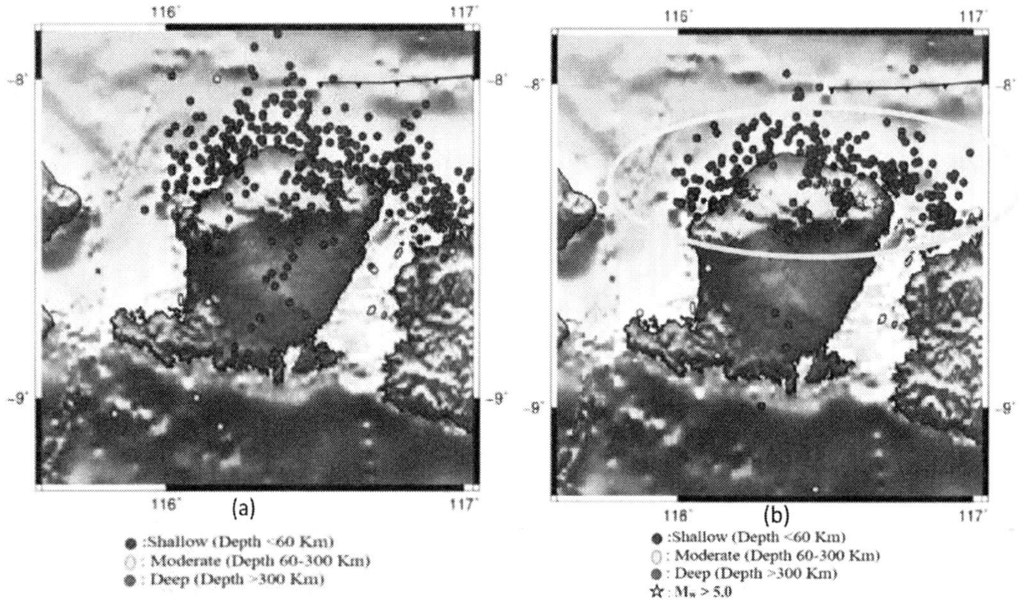

Figure 1. Epicenter distribution from the July-September 2018 Lombok earthquake data (a) before and (b) after relocation. The events are dominated by shallow earthquake and tends to spread in east-west orientation.

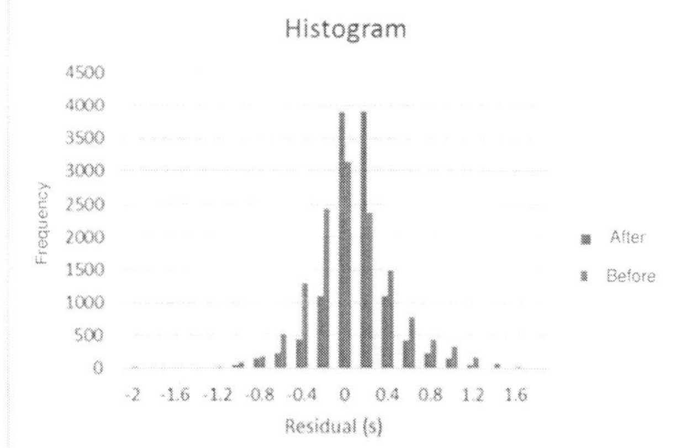

Figure 2. The histogram of residual value before and after the relocation process.

From the series of figure in the appendix and Figure 3, appear that most of the events occur around the line of BB', CC', and DD'. Along the line AA' and EE' only quiet seismicity. The hypocenter trend here indicates a thrust fault with the dip to southwards. This is clearly visible on the section of CC' and DD'. From a hypocenter spread, the alleged fault was at a depth of several km to 50s km. This fault plane seems next to the north Island of Lombok and in the south of Thrust Flores Fault and tends to align with the Flores Fault.

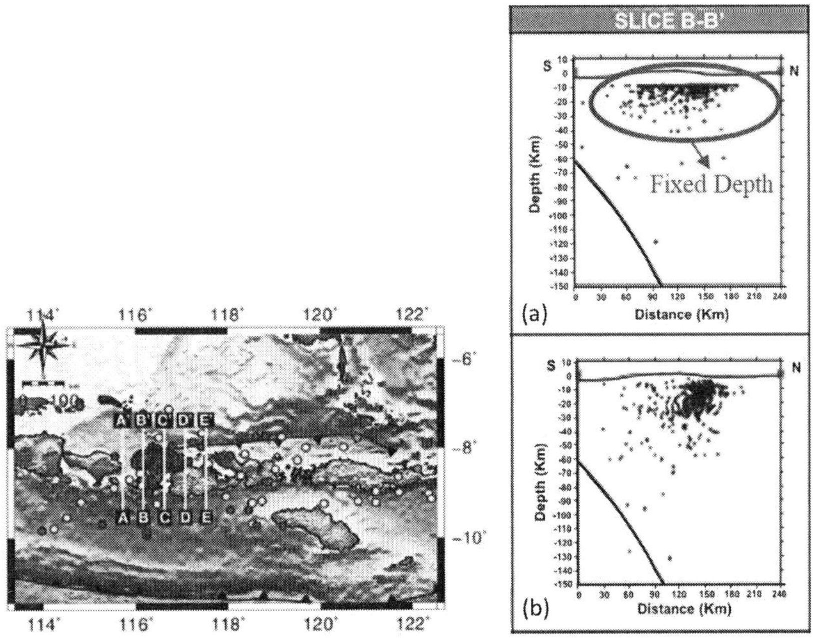

Figure 3. The vertical section at the BB' line shows the hypocenter distribution (a) before relocation and (b) after being relocated.

4. Conclusion

The results of the relocation Lombok earthquake data that are taken from July 1 to October 15, 2018, show the seismic activity occurring in the study area mostly spread in the northern part of the Island of Lombok. Mostly the shallow epicenter points are gather to form a pattern oriented from West to East which is thought to be the zone of the presence of fault plane. The active faults most likely in the northern region of the island as a local fault rather than a tectonic fault.

References

[1] PUSGEN. 2017. Peta Sumber Dan Bahaya Gempabumi Indonesia Tahun 2017. Pusat Penelitian dan Pengembangan Perumahan dan Pemukiman. Kementerian PUPR. Bandung.

[2] Hidayati, N., Trisnawati., Sativa, O., Wallansha, R., Sakti, A., Pramono, S., Permana, D. 2018. Ulasan Guncangan Tanah Akibat Gempabumi Lombok Timur. Bidang Seismologi Teknik – BMKG.

[3] Randi, C., Hakim, A. R., Azhar, T. and Setiadi, P. 2017. Estimasi Kedalaman Sesar di Wilayah Papua Menggunakan Metode Double Difference. Jurnal Riset Geofisika. Jakarta

[4] Waldhauser, F., & Ellsworth, W. L. 2000. A double-difference earthquake location algorithm: Method and application to the northern Hayward Fault, California. Bulletin of the Seismological Society of America, 90 (6), 1353-1368.

[5] Kayal, J. R. 2008. Microearthquake Seismology and Seismotectonics of South Asia. Capital Publishing Company and Springer. India.

[6] Hakim, A. R., Rohadi, S., and Jatnika, J. 2017. Relokasi Hiposenter Dan Tomografi Gelombang

IOP Conf. Series: Materials Science and Engineering **854** (2020) 012053 doi:10.1088/1757-899X/854/1/012053

P Menggunakan Metode Inversi Simultan Di Wilayah Papua. Jurnal Meteorologi Dan Geofisika, 18 (2), 65-71.

Appendix

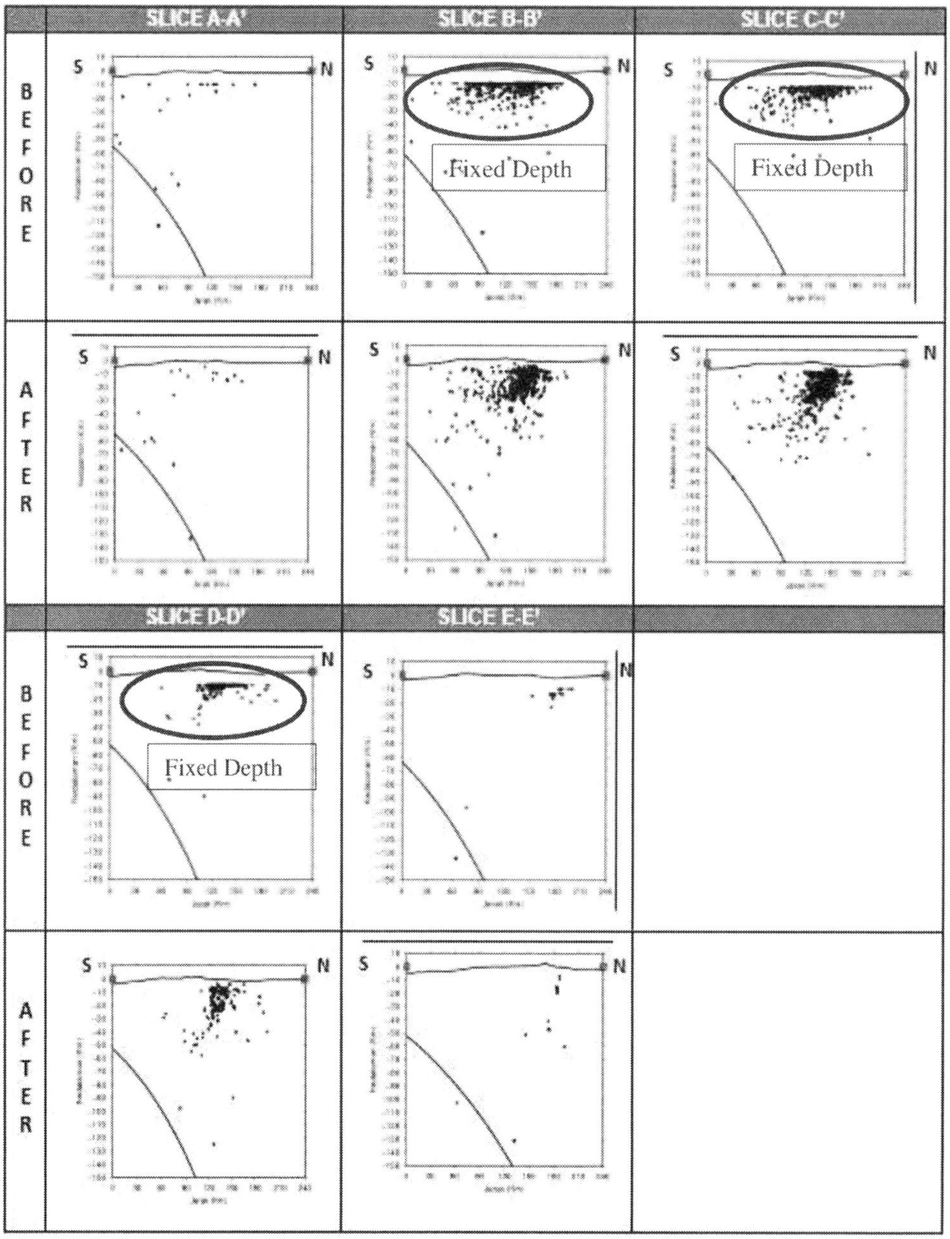

The section of hypocenter distribution map before-after relocation along line of AA', BB', CC', DD' and EE' of Figure 3.

IOP Conf. Series: Materials Science and Engineering **854** (2020) 012054 doi:10.1088/1757-899X/854/1/012054

Identification of Fault Structure in Lombok region, West Nusa Tenggara using Tomography Lombok Earthquake Data of July-August 2018

R Widyarta[1], S K Wijaya[2,*], M S Rosid[1,*], and S Rohadi[3]

[1] Geophysics, Universitas Indonesia, Depok 16424 Indonesia.
[2] Physics Department, Universitas Indonesia, Depok 16424 Indonesia.
[3] Research and Development Center, Meteorological, Climatological, and Geophysical Agency (BMKG), Jl. Angkasa 1 No. 2, Kemayoran, Jakarta, Indonesia.

*Email: skwijaya@sci.ui.ac.id and syamsu.rosid@ui.ac.id

Abstract. Lombok and Nusa Tenggara are one of the regions in Indonesia that have quite complex tectonic arrangements. With this order, not infrequently in the region there are often natural disaster phenomena. One of the most striking things is the occurrence of a series of earthquakes of magnitude Mw > 5.0 that rocked the northern region of the Island of Lombok on 29 July 2018 (M_w = 6.4), 5 August 2018 (M_w = 6.9), 9 August 2018 (M_w = 5.9), 19 August 2018 (M_w = 6.3 and 6.9) and 25 August 2018 (M_w = 5.5). This study aims to identify the condition of tectonic structures under the surface of the earthquake occurring region using the tomography method. This method utilizes earthquake travel time data recorded at 15 BMKG seismic stations. The tomogram results show the contrast of anomalous values in the *Vp* and *Vs* models. The contrast of the anomalous is associated with the presence of fault structures that have an allowance angle of about ± 20-30° and a thrust fault type, verified by focal mechanism ball data. This thrust fault is then indicated as the cause of the occurrence of earthquakes with magnitude M_w > 5.0, which shook the northern part of the island of Lombok in July-August 2018. According to the tomography result, the fault is located closer to the north side of Lombok Island than Flores Back arc Thrust.

1. Introduction

Lombok Island and the Islands of Nusa Tenggara are one of the archipelago regions in Indonesia that have a high degree of vulnerability to earthquakes. One source of earthquakes that has been clearly identified is the active subduction zone that plunges from the southern part of Indonesia [1]. This zone originates from collisions between the Indo-Australian plate (south) and Eurasia (north). Another cause of earthquakes is the presence of active faults below the surface of the earth. One of the earthquakes caused by the fault movement was the occurrence of a series of earthquakes with M_w > 5 in the Lombok region, West Nusa Tenggara which happened in the range of July - August 2018. The earthquake was quite horrendous, because it caused a lot of damage and claimed many lives. From the data of the National Disaster Management Agency (BNPB) there are about 555 people died and material losses worth for more than 5 trillion rupiahs.

Earthquakes in July-August 2018 are classified as unique because there are 6 earthquakes with M_w > 5.0 that occur in the adjacent period, namely on 29 July 2018 (M_w = 6.4), 5 August 2018 (M_w = 6.9), 9 August 2018 (M_w = 5.9), 19 August 2018 (M_w = 6.3 and 6.9) and 25 August 2018 (M_w = 5.5) [2]. To

IOP Conf. Series: Materials Science and Engineering **854** (2020) 012054 doi:10.1088/1757-899X/854/1/012054

find out more about the structure that caused the earthquake, the tomography method was applied. It is hoped that the study can provide further information about the tectonic structures that exist in the Lombok region, West Nusa Tenggara, especially for further application in terms of disaster mitigation.

2. Methodology

Seismic tomography is a method to reconstruct the structure of the earth's surface by using waveform data or travel time data from seismic waves. In tomography the time of propagation seismic waves are slowness integrals that are passed by a beam that connects the source and receiver. Thurber defines wave propagation (T_{ij}) from source (i) to receiver station (j) in form of [3]:

$$T_{ij} = \int_{source}^{receiver} u ds \quad (1)$$

where, ds is the length of the segment of the seismic wave that is passed and u is the slowness function which is a position function. This integration function depends on the location of the hypocenter as the source and the location of the station as a receiver as well as an unknown 3-dimensional velocity structure. The arrival time of a seismic wave in a receiver can be written with the equation:

$$t_{ij} = \tau_i + T_{ij} \quad (2)$$

where t_{ij} is the wave arrival time, τ_i is the time of the earthquake event. The tomography equation is a non-linear equation so it is necessary to do an inversion process where the process of this method is carried out using a Simulps12 software [4]. The steps of this study is drew in workflow diagram in Figure 1.

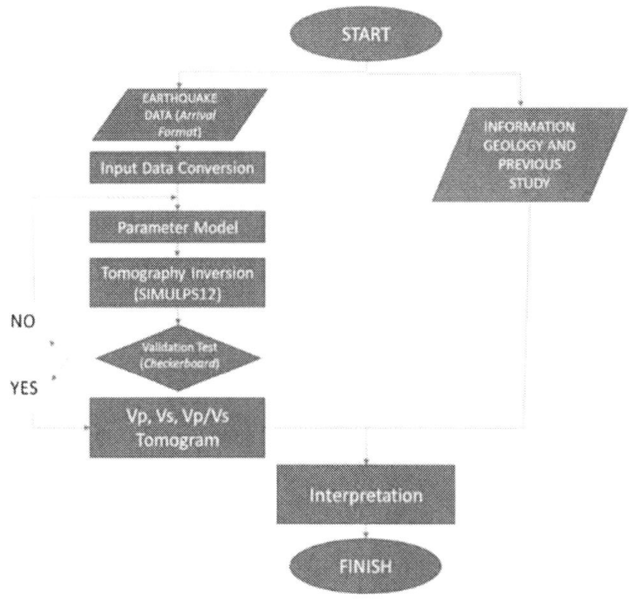

Figure 1. Workflow diagram

3. Results and Discussion

The tomography inversion process using Simulps12 resulting cross sections of velocity structures *Vp*, *Vs and Vp/Vs* both horizontally (Figure 2) and vertically (Figure 3). The process of interpretation, carried out by analyzing the height and value of perturbation *Vp* and *Vs* models which then made the model of *Vp/Vs* ratio. The process is carried out based on the review in a mathematical formula that is matched with the possibility of geology in the study area. Mathematically, the P and S wave equations can be expressed as:

$$Vp = \sqrt{\frac{K + \frac{4}{3}\mu}{\rho}} \tag{3}$$

$$Vs = \sqrt{\frac{\mu}{\rho}} \tag{4}$$

Then the Equation 3 is substituted into Equation 4 to be

$$Vp^2 = \frac{K}{\rho} + \frac{4}{3}Vs^2 \tag{5}$$

where ρ is a density value, with K is a bulk modulus which has a value of λ as a component related to incompressibility (fluid ability holds pressure where water has a higher value of incompressibility than steam/gas). So, it can be said physically that the value *Vp* is related to fluids content in rocks. Whereas in *Vs*, there is a rigidity component (μ) which physically talks about the rock matrix contained. So, from the equation (5) an analysis of the possible types of geological structures that cover the study zone can be done.

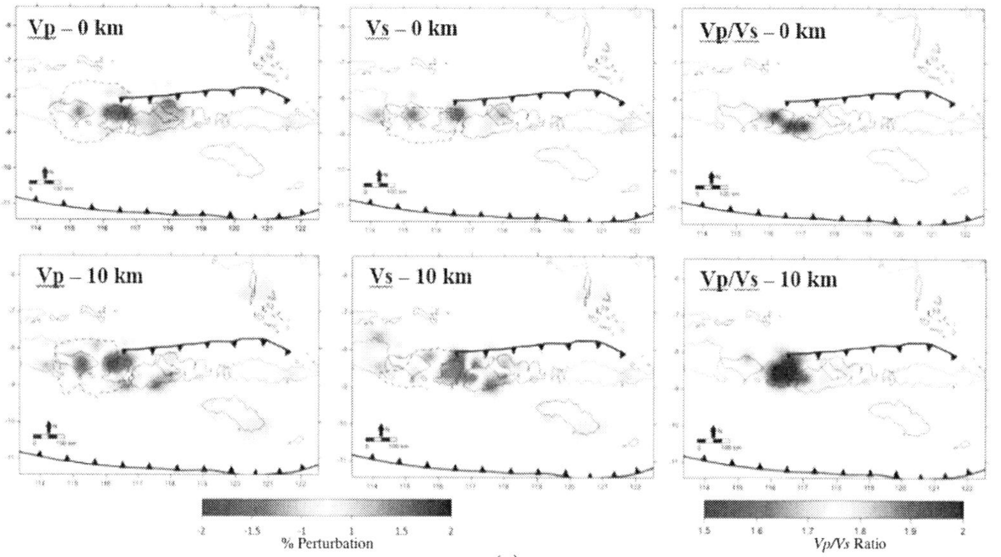

(a)

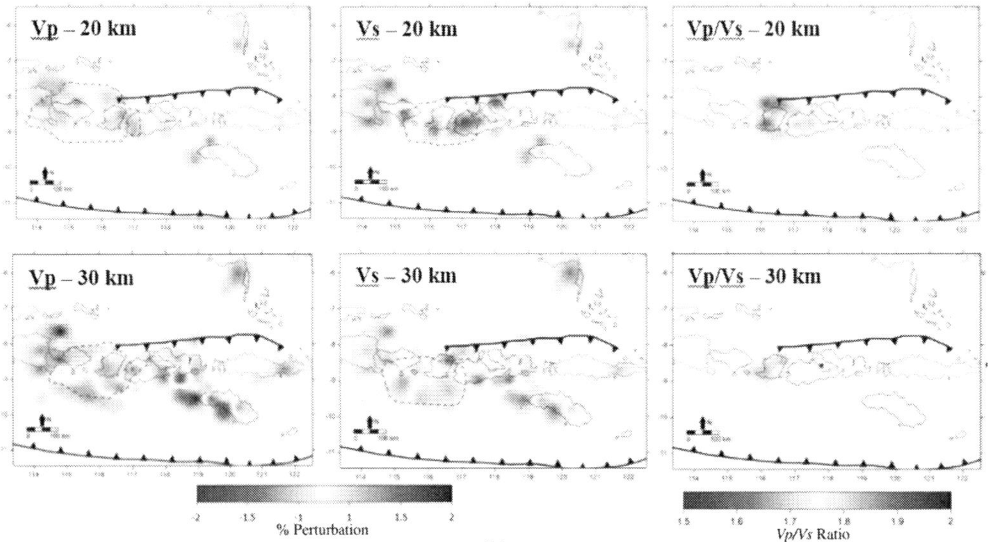

(b)

Figure 2. Horizontal section of *Vp, Vs* and *Vp/Vs* tomographic inversion model at the depths of a) 0 and 10 km; b) 20 and 30 km.

In the horizontal tomography model, both at the depths of 0, 10, 20 and 30 km, the tomographic inversion results in the *Vp* model showed a low perturbation anomaly in the Lombok Island region. The values of *Vs* model are also showed there are contrast area low results (red in the South) and high value (blue in the North). According to Kusnandar, the presence of high-low contrast values can be indicated as the presence of fault zones [5]. Then the results of high *Vp/Vs* ratio is obtained. It indicates that there is a considerable gap between the values of *Vp* and *Vs*. In this case, if referring to the mathematical analysis in equation (5), it can be assumed that rocks in the Lombok region have a high level of incompressibility and a low level of rigidity (attachment between rock matrices), which means rocks in the zone are weak rocks.

Structure interpretation model is also made based on the *Vp* and *Vs* tomography modeling on the CC' line (Figure 3a). There is low anomaly (in red color) in the region which is allegedly associated with the possibility of a weak zone. The zone could be the magma flow zone of Mt. Rinjani. The blue area on the other hand shows a high anomaly that looks contrasting to each other forms a zone that refers to the existence of fault-like zones. The fault zone is likely associated with the Flores back arc fault line which has a slope of up to ± 30 ° and the possibility of a new fault formed with similar allowance angles that are thought to be the cause of the Lombok earthquake that occurred in July - August 2018. This is confirmed by the results of the Widiyantoro et al. tomogram model [6] which shows the contrast which is thought to be a fault structure and the results of a study by Yadnya et al. [7] who had studied similar patterns in the Northern region of Bali.

From these indications, we tried to make a baseline (dashed black line) which is supposed to be a slope with a slope of approximately 20-30° in the anomalous contrast zone in the CC' cross sectional model. As supporting data, then the fault line data was verified by Karima et al. [8] which examined the focus mechanism on the Lombok earthquake. The earthquake (yellow star) which occurred on August 19, 2018 (M_w 6.9) happened not far from the CC' line. The results of the plot indicate that the location of the hypocenter of the earthquake is right in the line of the fault. The results of the focal spherical mechanism also indicate that the earthquake originated from a thrust fault mechanism that has a dip value of 29° with a depth of 16.1 km below the surface.

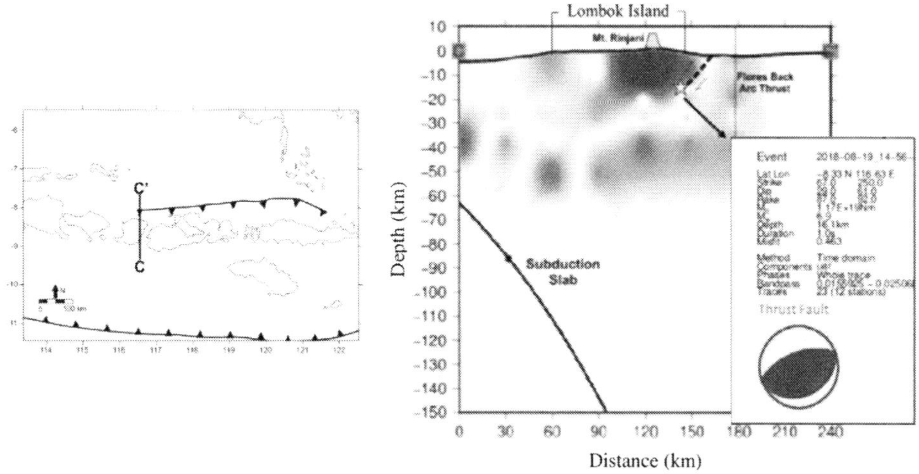

(a)

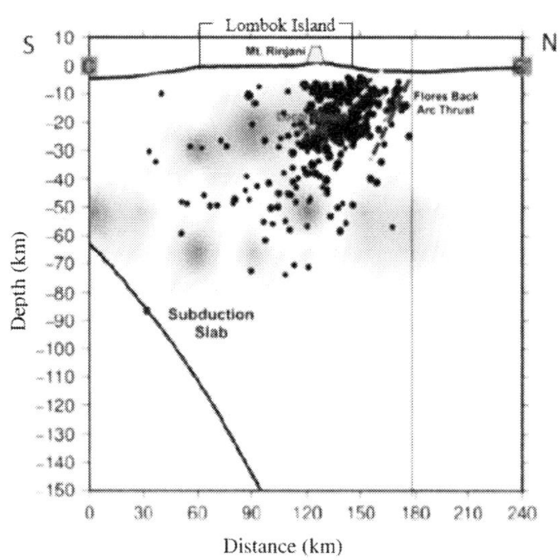

(b)

Figure 3. a) An interpretation model based on the *Vp* tomographic inversion model on the CC' line is justified by the focus mechanism model of Karima et al. [8] b) The interpretation model made based on the results of *Vs* tomographic inversion modeling on CC' line.

The same results were also shown when the same hypocenters data in the CC' cross section model are plotted (Figure 3b). Here the supporting data in the form of hypocenter distribution from Rosid et al. [9] are added. The result is the fault line that was drawn on the first model (*Vp* tomogram model) in accordance with the hypocenter distribution which is seen to be in the area around the pull of the fault line. In addition to the faults indicated as the cause of the August 19, 2018 earthquake, the dashed red line is also tried to draw. The line thought to be related to the existence of Flores's rear fault in the *Vs* tomogram model of CC' cross section. The drawing is also based on the distribution of hypocenter which forms a structural pattern near the location of the Flores reverse fault, where the results of the drawing form around an angle of approximately 30°. Refer to Daryono study [10], the Flores Fault is continued in the northern region of the Island of Bali. The predicted faults are appropriate because the faults in North Bali have fault angle allowances that vary from 20-35°. Figure 4 is a simple illustration that illustrates the results of the mechanism of the increase in the fault that allegedly caused the earthquake in the Lombok region.

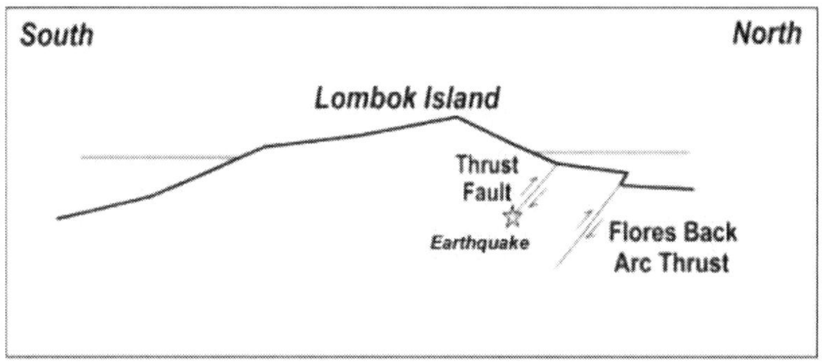

Figure 4. Illustration of the fault model for the Lombok 2018 earthquake that is made based on the results of tomographic inversion modelling.

Based on the results of the tomography model above, the researcher draws conclusions, if the cause of several earthquakes with a $M_w > 5.0$ that occurs in the northern region of the Lombok Island, presumably due to the existence of fault which presence of compressive forces between the Indo - Australia plate in the South and the Eurasian plate in the North, giving rise to a force of stress which then slowly breaks and forms a fault zone with a faulty mechanism with an angle of approximately 20-30° which plunges southward to at a depth of more than 15 km.

4. Conclusion

The tomogram results show the contrast of low anomaly values and high in the *Vp* and *Vs* models which after being matched with focal spherical data, it is indicated if the anomalous contrast values are associated with the existence of a fault structure to the south around ± 20-30° and a type of upward fault. This upward fracture is then indicated as the cause of $M_w > 5.0$ magnitude earthquake that shook the North Island of Lombok in July-August 2018 (29 July 2018 ($M_w = 6.4$), 5 August 2018 ($M_w = 6.9$), 9 August 2018 ($M_w = 5.9$), August 19, 2018 ($M_w = 6.3$ and 6.9) and August 25, 2018 ($M_w = 5.5$).

Acknowledgments

We thanks to the Research and Development Center, Meteorological, Climatological, and Geophysical Agency (BMKG) for providing and permission to use the data. We also many thanks to DRPM Universitas Indonesia for beneficial financial support in publishing the result of study through the PITTA's grant No. NKB-0666/UN2.R3.1/HKP.05.00/2019.

References

[1] PUSGEN. 2017. Peta Sumber Dan Bahaya Gempabumi Indonesia Tahun 2017. Pusat Penelitian dan Pengembangan Perumahan dan Pemukiman. Kementerian PUPR. Bandung.

[2] Hidayati, N., Trisnawati., Sativa, O., Wallansha, R., Sakti, A., Pramono, S., Permana, D. 2018. Ulasan Guncangan Tanah Akibat Gempabumi Lombok Timur. BMKG. Jakarta.

[3] Thurber, C. H. 1993. Local earthquake tomography: velocities and Vp/Vs theory, Seismic Tomogrphy: Theory and Practice. Chapman and Hall. London. pp. 563– 583.

[4] Evans, J. R., Eberhart-Philips, D., Thurber, C. H. 1994. User's Manual for SIMULPS12 for Imaging Vp and Vp/Vs : A Derivative of the "Thurber" Tomographic Inversion SIMUL3 for Local Earthquakes and Explosions.

IOP Conf. Series: Materials Science and Engineering **854** (2020) 012054 doi:10.1088/1757-899X/854/1/012054

U.S. Geological Survey. USGS-OFR-94-431.

[5] Kusnandar, R. 2013. Tomografi Seismik Gelombang P Wilayah Bali dan Sekitarnya. Skripsi. Fakultas Teknik Pertambangan dan Perminyakan, Institut Teknologi Bandung. Bandung.

[6] Widiyantoro, S., Pesicek, J. D., and Thurber, C. H. 2011. Subducting Slab Structure Below The Eastern Sunda Arc Inferred From Non-Linear Seismic Tomographic Imaging, Geological Society. London, Special Pub. vol. 355, pp.139-155.

[7] Yadnya, P. K., Nugraha, A. D., Rohadi, S. 2012. Pencitraan Struktur 3-D Vp, Vs, Rasio Vp/Vs Menggunakan Tomografi Double Difference di Wilayah Bali. Jurnal Geofisika. Vol. 13, no. 1.

[8] Karima, T., Wijaya, S. K., Rosid, M. S., Rohadi, S., and Perdana, Y, H. 2019. Characterization of Lombok Earthquake on July-August 2019 Using Focal Mechanism Analysis. The 2nd i-COSINE, Malaysia, in progress.

[9] Rosid, M. S., Widyarta, R., Karima, T., Wijaya, S. K., and Rohadi, S. 2019. Fault Plane Estimation through Hypocenters Distribution of the July-August 2019 Lombok Earthquake Relocated by Using Double Difference Method. The 2nd i-COSINE, Malaysia, in progress.

[10] Daryono. 2011. Identifikasi Sesar Naik Belakang Busur (Back Arc Thrust) Daerah Bali Berdasarkan Seismisitas dan Solusi Bidang Sesar. Artikel Kebumian, Badan Meteorologi Klimatologi dan Geofisika.

Structure identification of geothermal field "X" using ML-SVD method of gravity data

M A Naufal and M S Rosid*

Physics Magister Programme, FMIPA Universitas Indonesia, Depok 16424, Indonesia

*Email: syamsu.rosid@ui.ac.id

Abstract. Understanding the main geothermal components such as clay cap, reservoir, heat source and faults structure can reduce the risk at the drilling stage. The role of the faults is very important since it can be a recharge/discharge medium of geothermal fluid. One method that can be used to identify the faults is the gravity method. The combination of the upward continuation and second vertical derivative (ML-SVD) methods is carried out to characterize the geological structure both the slope direction, depth estimation, dip estimation and type of structure. Upward continuation is carried out from a height of 0 - 2000 meters with an interval of 250 meters. SVD filter is applied at each height level. ML-SVD result shows that the fault structures at the study area have a value of dip > 70° with various fault configurations. Some faults are estimated at a certain depth under a layer of rock. ML-SVD shows a circular pattern that indicated as a caldera structure. The fault orientation of the study area tends SW-NE and NW-SE with dip perpendicular to the strike.

1. Introduction

A fault is one aspect that is sought when conducting geophysical investigations. The faults in the case of geothermal acts as a discharge/recharge medium of geothermal fluid [1]. Aside from being a medium, faults sometimes act as barriers to the geothermal system. The barrier formed by the deposition of geothermal minerals that fills the fracture gap so that becomes an impermeable zone. Understanding fault structure in the geothermal field can improve the information of geothermal fluid flow, the structureted to the geothermal reservoir and another potential development area. Faults identification needs to be done to reduce the potential risk in geothermal well drilling. The aim of this study is to understand the characteristic of fault at X area that associated with geothermal. One geophysical method that can identify the fault existence is the the gravity method.

Gravity data processing methods have been developed by many researchers for subsurface analysis. Gravitational data analysis methods are derivative analysis [2,3], Euler deconvolution analysis [4], and anomaly separation analysis [5]. Euler deconvolution method aims to determine the type of subsurface objects and its depth but we must choose the type of structure input. The anomalous separation method aims to separate regional anomalies and residual anomalies. The processing method that can identify the presence of fault is horizontal derivative analysis. A horizontal derivative can detect any kind and shape of structure because of it was sensitive to lateral density contrast. The purpose of this study is to understand the characteristic of faults in the study area associated with the geothermal system using the ML-SVD method.

The study area consists of several structure that is *A* Caldera that has an important role in the study area. Inside the caldera structure, there is a permeability zone which can be a geothermal upflow zone.

Usually, hydrothermal or geothermal that associated with caldera limited to the ring structure [6]. Another structure that has an important role in this study area is a 6 km wide graben that possibly associated with *WAL* fault at the east part of graben. East structure of graben has a roll as a tunnel so that geothermal fluid can ascend to the surface. Based on geological information, *KG* fault is shown on the west part of the study area and bordering the northern part of the caldera.

2. ML-SVD Methods

The ML-SVD method is a a combination of upward continuation filter and SVD (Second Vertical Derivative) filter. This method developed by Rosid & Naufal [7]. ML-SVD is a band-pass filter because the upward continuation passes the gravity anomaly associated with the low frequency while the SVD passes the high-frequency anomaly. Low-frequency anomalies are associated with the deep source while high-frequency anomalies associated with shallow or faults objects. SVD can detect the fault because SVD is sensitive to the lateral density contrast. Also based on SVD we can calculate the depth of fault from the surface. This is because fault at certain depth gives some unique respond so that the depth of fault below the surface can be calculated [7]. The principle of this method is performing SVD operation on the results of upwards continuation. This is like looking for a fault at a certain depth.

In practice, ML-SVD is carried out by applying SVD operations at each level of height. Illustration of ML-SVD is shown in Figure 1.a. Figure 1.a seen that with increasing the height of the continuation the location of fault based on SVD will be shifted according to the dip direction. Upward continuation is carried out on gravity data that has been transformed into the frequency domain by applying equation (1). Where $F[U]$ is gravity data in the frequency domain, $e^{-\Delta z|k|}$ is the filter function where z and k is the height of continuation in meter and wave number respectively, $F[U_U]$ is the result of upward continuation in the frequency domain. SVD operations are performed on gravity data in the spatial domain because gravity satisfies Laplace's equation [9]. Equation (1) shows that the higher the continuity, the frequency passed is dominated by low frequencies. The dominance of low frequencies passed reflects deeper subsurface conditions. The combination of SVD and upward continuation shows that there is a relationship between the height of upward continuation and depth [7]. The relationship is shown in Figure 1.b which means that the higher the height of continuation, the deeper the detected faults.

$$F[U_u] = F[U]e^{-\Delta z|k|} \tag{1}$$

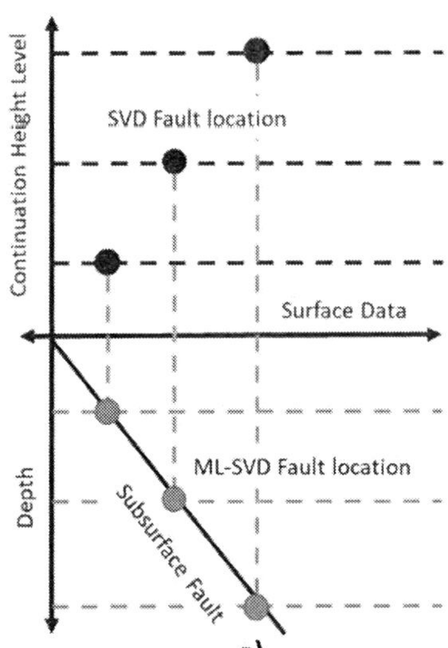

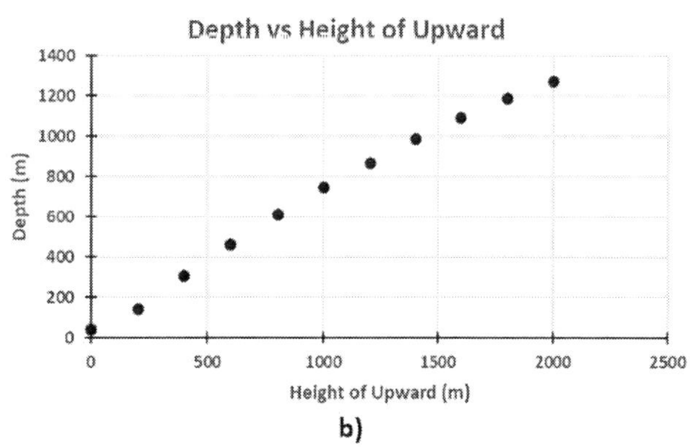

Figure 1. a) Illustration of ML-SVD concept. b) Relationship between the height of upward continuation and depth [7].

The study site is located in an area that has a geothermal system associated with volcanos. There are 200 gravity measurement stations with a distance of 250 meters covering an area of 31 km^2. A series of corrections to gravity data is performed to obtain the gravity caused by density variations in the subsurface. The ML-SVD concept that has been carried out by Rosid & Naufal [7] is used to determine the fault characteristics of the study area. ML-SVD analysis is performed by filtering SVD at each continuity level with a height interval of 250 meters.

3. Results and Discussion

Before implementing ML-SVD to the data, a series of corrections is applied to gravity data to obtain CBA (Complete Bouguer Anomaly) value. To obtain Bouguer correction, the average density of the study area is calculated. The calculation of the average density is done using the Parasnis and Nettleton methods, which is an average of 2.687 g/cc. The density calculation is coincident with the geological conditions of the study area which is a volcanic area with the dominance of igneous rocks like andesite and basaltic lava.

CBA values in the study area varied from 60.5 mGal to 70.5 mGal. High CBA anomaly (> 67.5 mGal) is found in the southwest of the study to the middle area. The high CBA anomaly is thought to be associated with relatively higher density rocks. There is a circular pattern showing a low CBA anomaly (< 63.5 mGal) in the northern part of the caldera and the eastern part of the C fault. A low anomaly is thought to be associated with a mountain conduit structure around the anomaly. The CBA distribution map is shown in Figure 2.

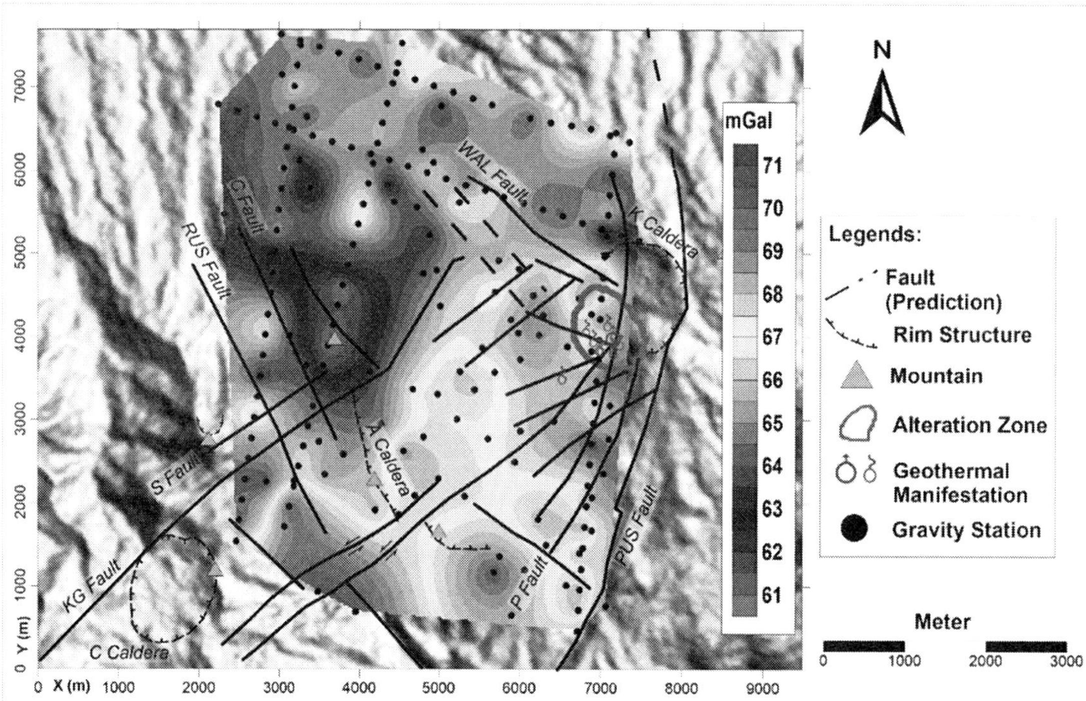

Figure 2. CBA map at the area of study.

3.1. *The fault of the Research Area based on ML-SVD*

Fault's characterization at X geothermal field was performed using the ML-SVD gravity data method. ML-SVD did by the continuation of upwards to a height variation of 0-2,000 meters with an interval 250 meters. The plot of the location of SVD0 (SVD = 0 mGal/m^2) values for each continuity height is shown in Figure 3. The value of SVD = 0 mGal/m^2 is indicating the location of the fault.

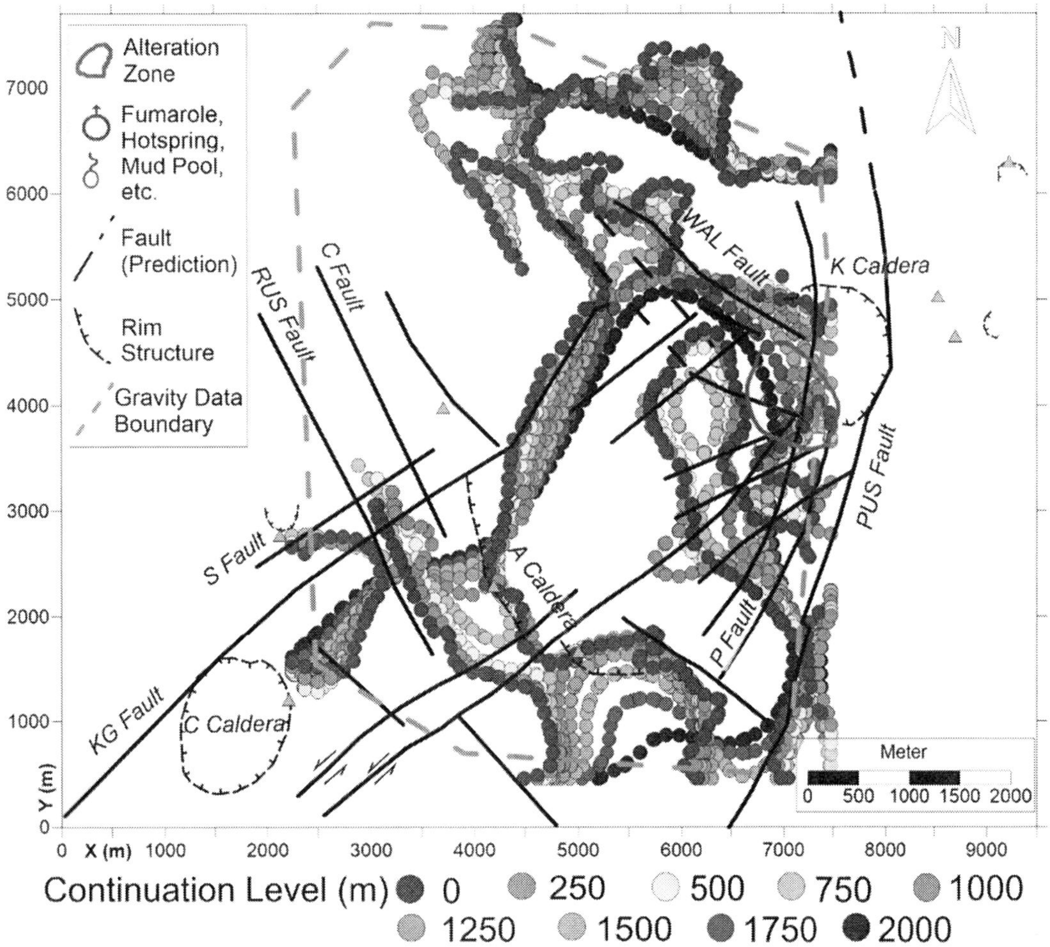

Figure 3. ML-SVD results to figure out faults at the study area.

Figure 3 shows the digitization of SVD0 which is indicated by the colored round symbol and the simplification of the ML-SVD results shown in Figure 4. The difference in color reflects the level of upward continuation that has been done. Round symbols of the same color appear to form a fault pattern. The red color represents the lowest level of continuity, the green color represents the medium level of continuity while the blue color represents the highest level of continuity. The picture shows that there is a shift of location as the height of the upward continuation increases. The higher the height of continuity reflects the deeper subsurface conditions [7]. The fault location shifting as the height of continuity increase that indicated the direction of faults slope (dip). Simplification of the display is done by drawing a fault pattern line based on the ML-SVD for easy interpretation of the fault structure (see Figure 4).

The ML-SVD pattern on the west of *A* caldera structure is thought to be an ML-SVD response due to *RUS* fault (number 1 in Figure 4). The ML-SVD pattern coincides with the *RUS* fault's pattern and has the same strike direction based on geological information (northwest-southeast). The ML-SVD *RUS* fault pattern appears to the level of continuity 750 meters. The dip direction of the *RUS* fault based on ML-SVD is northeast with a dip of 75.84°. The calculation of rock layer thickness covering the *RUS* fault based on the SVD pattern is 280 meters. The density of footwall at *RUS* fault greater than the density of hanging wall. The vertical location of the RUS fault estimated at 1,300 – 900 meters above sea level (masl).

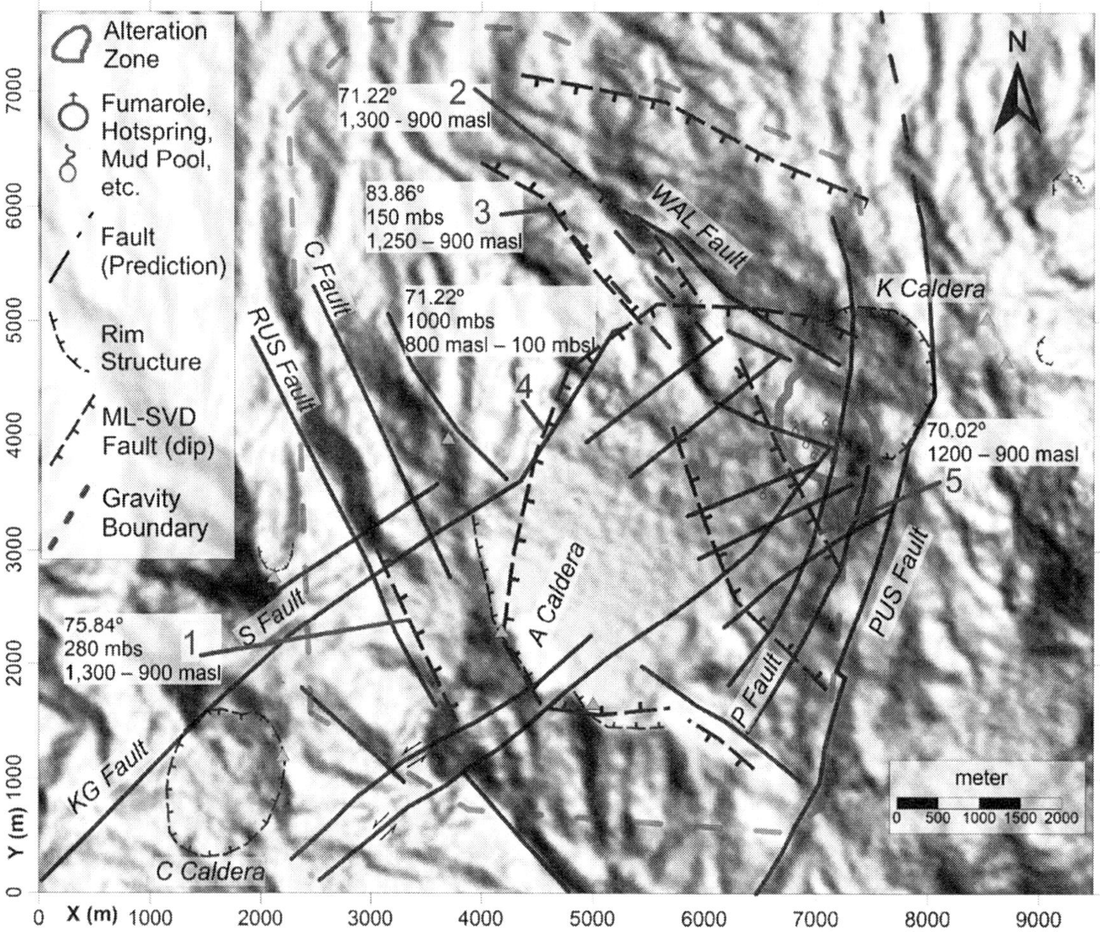

Figure 4. Fault interpretation based on ML-SVD overlaid with the SRTM map.

The fault information based on geology in the north is not much but the ML-SVD shows a fault pattern in the north. Fault in the north is dominated by the strike direction northwest-southeast with the direction of dip southwest and northeast. The ML-SVD fault pattern is suspected to have an association with *WAL* fault (pattern number 2). Pattern number 2 cannot provide the exact fault location because the ML-SVD fault pattern is not clear. The obscurity of ML-SVD fault pattern allegedly due to the dominance of noise associated with high frequency. The ML-SVD pattern at a continuation level > 500 meter appears more clear. The ML-SVD pattern at continuation level > 500 m indicates the alleged ML-SVD pattern correlates with the *WAL* fault. The ML-SVD which correlates with the *WAL* fault appears to 1,000 m continuation level. We tried to calculate the fault dip obtained at an angle of 71.22°.

The ML-SVD pattern number 3 is to the west of the ML-SVD pattern number 2 (or *WAL* fault). Geological data show the existence of a fault structure that correlates with pattern number 3. Based on the ML-SVD, it shows that the fault is under the rock layer with a thickness of 150 meters. Calculation of dip obtains estimated value 83.86°. The fault pattern numbers 2 and 3 are forming a graben structure with a width of about 500 meters. The fracture pattern number 2 and 3 is limited by a circular structure marked by pattern number 4. It is suspected that the structure of numbers 2 and 3 is the continuation of the ML-SVD pattern number 5 in the southeast. Fault number 2 & 3 estimated at 1300 – 900 masl.

The ML-SVD fracture pattern number 4 is an interesting one. The ML-SVD pattern number 4 forms a circular pattern at the center area with the eastern part open. The south and west sections of pattern number 4 coincides with the rim structure. The northern part of the number 4 pattern is thought to be

IOP Conf. Series: Materials Science and Engineering **854** (2020) 012055 doi:10.1088/1757-899X/854/1/012055

strongly related to the shifting of the rim structure north section. Shifting rim structure in the past is thought to have been caused by *KG* fault activity. *KG* fault in the past is a fault that has a lateral direction shift [7].

ML-SVD number 4 shows that the dip structure tends to point towards the inside of the circle. The rim structure appears to the level of continuity ML-SVD 2000 meters. The SVD curve in pattern number 4 shows the pattern |-SVD| > |+SVD| which indicates that the density outside the circular structure is lower than the density inside the circular structure. Characteristics of pattern |-SVD| > |+SVD| along the pattern number 4 indicate that the northern circular structure is part of the rim (*A* caldera) structure. Obscurity in pattern number 4 is seen in several parts. The obscurity is thought to be the complexity of the rim's structure. The complexity of the rim structure is caused by volcanic events that occur in the study area, namely the emergence of younger volcanoes along with the rim structure.

The fault dip calculation in pattern number 4 obtained with value 71.22°. Another calculation based on the MEQ method shows that *A* caldera having an estimated angle of 76.65° [8]. The calculation results obtained have a difference of 6% from ML-SVD calculation. The SVD signal provides information that the *A* caldera is under a layer of rock with a thickness of 1,000 meters at the western part and 200 meters at the northern part. This is assumed at a depth of 1,000 meters to the surface, contrast density in the fault area is very low. The alleged low-density contrast is proven by geological information showing the presence of a caldera structure on the surface and SRTM imagery showing a circular pattern. *A* caldera is located from -110 – 800 masl at the western part and 300 – 1400 masl at the northern part.

ML-SVD cannot provide *KG* fault information to the west of the rim structure. This is thought to be related to the initial history of the formation of the *KG* fault (the *KG* fault is shifted laterally). The density contrast of lateral fault is generally so low (in fact there is almost no density contrast) that the ML-SVD cannot detect lateral faults. This also applies to the two lateral faults that are in the southwest part of the *A* caldera. The southwest part of the *A* caldera has two faults in the northeast-southwest strike direction which shifts the caldera laterally. The two faults in the southwest are not identified by the ML-SVD method. Very low-density contrast produces relatively homogeneous gravitational anomalies so that indications of a fault are not visible when an SVD filter is performed.

There is an ML-SVD pattern that forms a fault pattern inside the caldera structure (fault pattern number 5). The two faults form a pattern in the northwest-southeast direction with the NE and SW dip. The western fault has a dip to the northeast while the east fault is directed southwest to form a graben structure. Both faults are thought to be associated with a fault in the east N 140° E strike direction found on the past study [8]. The fault in the western graben structure is thought to have a connection with the alleged fault that is located west of the *WAL* fault. If a line is drawn between pattern numbers 2 and 3 with pattern number 5 the two faults will form a continuity. This fault pattern confirms the faults in the direction of the N140°E strike in the study area forming a graben structure with a width of ± 6 km. The western part of the graben structure ± 6 km in this study was not found because the area of gravity data did not cover the western structure. The eastern graben structure is thought to act as a geothermal fluid discharge as indicated by the presence of geothermal manifestations on the surface.

Based on ML-SVD calculation, the subsurface fault model can be done. The subsurface fault model is shown in Figure 5. Figure 5.a shows the cross-section of the fault model in SW-NE direction while Figure 5.b shows the cross-section of the fault model in NW-SE direction. The depth of the fault location is obtained by converting the height of the continuity value into a depth value. Rosid & Naufal argues that there is a relationship between continuity height and depth so that the conversion can be done by considering the dip of the fault [7]. The top of the reservoir line provided from resistivity modeling result of the magnetotelluric method [8]. From ML-SVD calculation *A* Caldera is located about 1,000 m below surface and by combining magnetotelluric data *A* Caldera is shown at reservoir zone. ML-SVD cant sho the shallow part of the caldera is possible because of low-density contrast at shallow part. The resume of ML-SVD calculation shown in Table 1.

Table 1. ML-SVD calculation result

Fault no	Dip (°)	Depth from Surface (mbs)	Vertical Location	Association
1	75.84	280	1,300 – 900 masl	*RUS* Fault
2	71.22	100	1,400 – 900 masl	*WAL* Fault
3	83.86	150	1,250 – 900 masl	un
4	71.22	1000	800 masl – 100 mbsl	*A* Caldera
5	70.02	300	1,200 – 900 masl	East graben?

un = unidentified mbs = meter below surface masl / mbsl = meter above/below sea-level

ML-SVD method still needs further development. This method requires more simulation by considering the faults in the real case. The equation to estimate dip of fault, depth of fault beneath rock layer and upward height-depth conversion can be complete if more simulation and more real cases have been done. At this point, this method can tell the direction of fault dip well enough. However, the method is difficult to identify the shear fault due to the small density contrast.

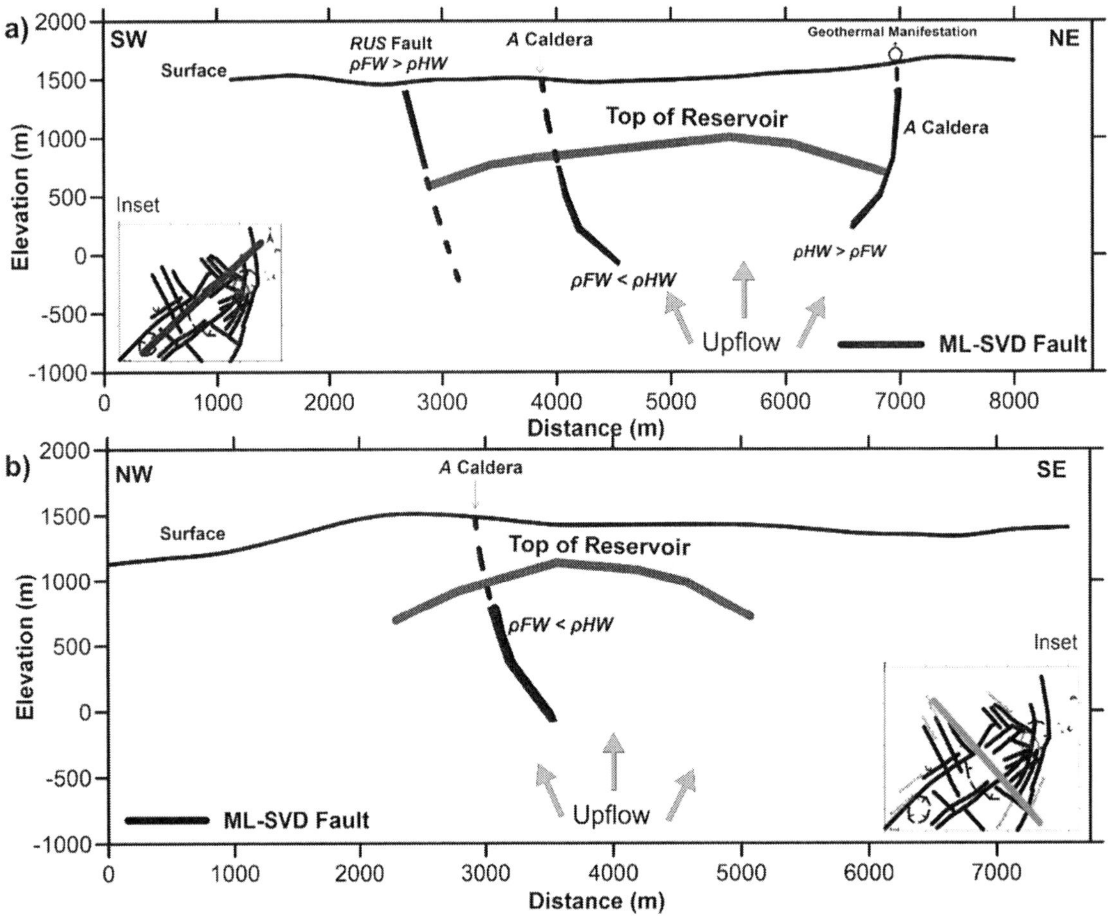

Figure 5. Cross-Section of ML-SVD Fault Model, a) Southwest-Northeast direction, b) Northwest-Southeast direction.

4. Conclusion

ML-SVD method can be implemented to figure out some fault structures in geothermal fields. The faults identified by ML-SVD are *RUS* faults, *A* Caldera, *WAL* faults that are associated with graben structure. The fault that can be identified with the ML-SVD method mostly has dip > 70°. The *A* caldera is a rim that has open structure at the east section with dip direction to the inside of the circle. The northern part of the caldera is thought to be buried 1000 meters below the surface. Faults in the study area are under a rock layer with a certain depth indicating low-density contrast on the surface. Some faults which cannot be detected by the ML-SVD method are caused by low-density contrast in the fault.

Acknowledgments

The publication is funded by DRPM Universitas Indonesia through a grant of PITTA 2019 No. NKB-0641/UN2.R3.1/HKP.05.00/2019. Therefore, we would like to thank the DRPM for the user support.

Reference

[1] Goff F., Janik C.J. 2000. Geothermal System, in: Sigurdson (Ed), Encyclopedia of Volcano, ch. 49 (817-834). Academic Press.

[2] Elkins T.A. 1951. "The Second Derivative Method of Gravity Interpretation." *Geophysics*, **16**(1), 29-50. https://doi.org/10.1190/1.1437648

[3] Rosenbach O. 1953. "A Contribution to the Computation of the "Second Derivative" from Gravity Data." *GEOPHYSICS*, **18**(4), 894-907. https://doi.org/10.1190/1.1437943

[4] Reid, A.B. Allsop, J.M. Granser, H. Millett, A.J. and Somerton, W. 1990. Magnetic Interpretation in Three Dimensions Using Euler Deconvolution, *Geophysics*, v. **55** no. 1, p. 80-91.

[5] Blakely, R.J. 1996. Potential Theory in Gravity and Magnetic Applications. Cambridge: Cambridge University Press. doi: 10.1017 / CBO9780511549816.

[6] Wohletz, K., Heiken, G. 1992. *Volcanology and Geothermal Energy*. Berkeley: University of California Press. http://ark.cdlib.org/ark:/13030/ft6v19p151/

[7] Rosid M.S. and Naufal M.A. 2019."Characterization of Geological Structure Based on Multi Level Second Vertical Derivative (ML-SVD) Gravity Data", Journal of Geophysical Research, in progress.

[8] Naufal, M.A, 2019. Identifikasi Struktur Geologi Sistem Geotermal Daerah X Berdasarkan Metode Multi Level Second Vertical Derivative (ML-SVD) Data Gravitasi. *Thesis*. Universitas Indonesia.

[9] Telford, W.M., Geldart, L.P., and Sheriff, R.E. 1990. *Applied Geophysics* Second Edition. Cambridge University Press. New York.

IOP Conf. Series: Materials Science and Engineering **854** (2020) 012056 doi:10.1088/1757-899X/854/1/012056

Identification of Potential Distribution of Gold Mineralization Zone of High Sulfidation Epithermal System using Time-Domain Induced Polarization (TDIP) and Magnetic Method at "N" Mountain Qrea

M F Khoir and M S Rosid

Geophysics, FMIPA, Universitas Indonesia, Depok 16424, Indonesia

*E-mail: syamsu.rosid@ui.ac.id

Abstract. The high gold commodity as well as the Pongkor production field that has existed since decades ago made it necessary to find a new potential zone for gold mining production. To find out the zone of gold mineralization and its continuity, geophysical studies then conducted by using the methods of Time-Domain Induced Polarization (TDIP) and magnetic. TDIP data consisting of resistivity and chargeability data are interpreted to determine alteration zones and subsurface mineralization that are correlated with magnetic data that has been inversed in 3D to obtain susceptibility parameters. Based on the value of resistivity in the study area, the argillic alteration zones are thought to has a value of < 50 Ohm.m. Silicification alteration zones supposedly have the value > 50 Ohm.m. The value of the chargeability at the mineralized zone indicated has a value of 100-810 msec. Also, study in the area of the mineralized zone have a susceptibility value in the range of -0.073 to -0.021 cgs. From that parameter of the mineralized zone, it is assumed that the mineralization zone is in the eastern part of the study area and oriented North-South direction with the orientation of the main fault.

1. Introduction

Pongkor Field has been operating to produce gold since 1994 [1], so development is needed to find new reserves. Mount "N" is located close to the mining site that produces in Antam Pongkor. It is thought to have new reserve potential. Geological mapping shows the texture of vuggy quartz rocks on Mount "N", which indicates high sulfide epithermal deposits [2].These deposits have unique characteristics that distinguish them from other deposits. High sulfide epithermal deposits have more acidic properties than low sulfide, this is because the constituent fluid is dominated by magmatic fluid and only a few meteoric fluids. The process of deposition of gold minerals occurs in the cavity of the rock structure that causes the presence of this mineral is not concentrated in one place but spread[3]. Geophysics methods are needed to solve this problem. This study is used to determine the presence of mineralized zones with characters spread and also the relation with geological conditions of Mount "N" Area.

IOP Conf. Series: Materials Science and Engineering **854** (2020) 012056 doi:10.1088/1757-899X/854/1/012056

2. Geological Setting

Based on Geological observation, Rocks in the study area are dominated by pink tuff crystal and red andesite igneous rocks and also dacitic rocks at the edges of the study area. The volcanic rocks indicate there used to be a magmatism process which is closely related to the hydrothermal process [4]. Based on geological data, there are 4 faults that cut all units of existing rock [5]. Three of the faults are shown in Figure 1. From the existing structural patterns in the study area, there are 2 different deformation stages. The first stage causes a north-south oriented right fault, then the second phase of the fault is left-aligned which intersects the main fault and is oriented southwest-northeast. These two structures are thought to be the pathways for the release of several andesitic intrusions which are located following the distribution of the fault pattern.

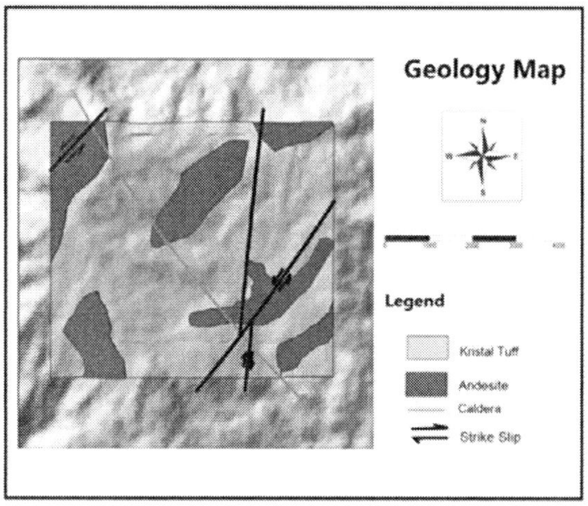

Figure 1. A geological map at Mount "N" shows the 3 faults in the study area.

3. Method

The data used in this study consists of resistivity, chargeability, and magnetic data. Resistivity and chargeability data are obtained from the measurement of the TDIP method using a dipole-dipole configuration because this configuration is very good at mapping the distribution of vertical and horizontal subsurface layers [6]. There are 3 measurement lines that have a distance of 100 meters for each line. The data was then processed by using software to create the resistivity and chargeability profile.

Meanwhile, the magnetic method is used by using the base-rover technique with a space between the measurement points is 25 meters and north-south oriented. The data is processed to get the 3D inversion and forward modeling display. In addition, the inversion will produce susceptibility parameters that will be classified as a gold mineralization zone.

4. Results and Discussion

In this study, the TDIP method is used to determine the presence of metal sulfide minerals carrying elements of gold (Au) and non-sulfide minerals in high epithermal systems. Gold deposits in the epithermal system can be identified by the presence of argillic and silicification alterations. The maps of Figure 1 shows that the study area is dominated by alteration of argillic silicification and alteration. Based on the value of resistivity and chargeability, the study area is divided into two alteration zones, namely alteration silicification, advanced argillic alteration and argillic alters which the authors combine into one argillic alteration unit. The division of alteration values can be seen in Table 1.

IOP Conf. Series: Materials Science and Engineering **854** (2020) 012056 doi:10.1088/1757-899X/854/1/012056

	Resistivity (Ohm.m)	Chargeability (m.sec)	Metal Factor (ms/ohm.meter)
Silisification Zone	> 50	> 100	> 2500
Argillic Zone	< 50	> 100	> 2500

Tabel 1. The physical parameter's value for TDIP interpretation.

The section of the 2D model shows the mineralized zone has a high value of chargeability and magnetic response, whereas the value of resistivity is low. The mineralization zone seems in a small area only (see Figure 2). The mineralization can occur in rocks that experience silicified and argillic alteration [7]. On-Line N-1 the mineralized zone is expected to occur in alteration silicification (which is marked in the red box of Figure 2) and argillic alteration (which is marked in black ones). The changeability value for path 1 has a value of 100 msec to 810 msec, the magnetic susceptibility has a low to moderate value in the range of -0.073 to -0.005 cgs.

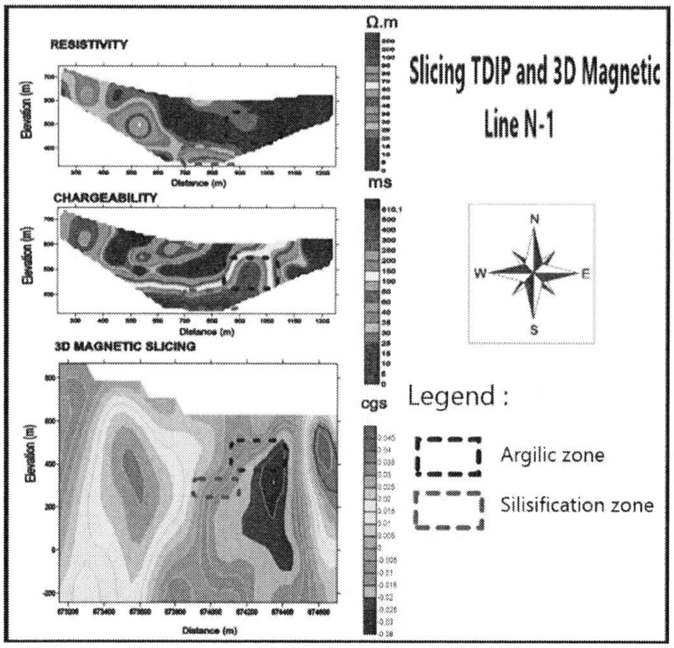

Figure 2. Expected of mineralization zone in Line N-1.

In-Line N-2 (Figure 3) there is a body with resistive, presumably the presence of Vuggy Quartz with alteration silicification which causes a very high resistivity value. The Vuggy Quartz texture contained iron oxide ore minerals such as hematite minerals and gold-bearing sulfide minerals (Au) namely alunite minerals which have a high chargeability value. This alteration occurs in a range of 600-900 m and with a height of 400-500 m. Whereas in the East the dominant argillic alteration occurs with a low resistivity value. The value of metal factors on this track ranges from 500 to 10000 mohs/meter. Also in Line N-2 shows, that argillic alteration patterns have susceptibility values which tend to be low and alteration silicification as well. Silicified alteration with low susceptibility at a distance of 600 to 900 m is thought to have quartz content and become an association mineral from the presence of gold minerals.

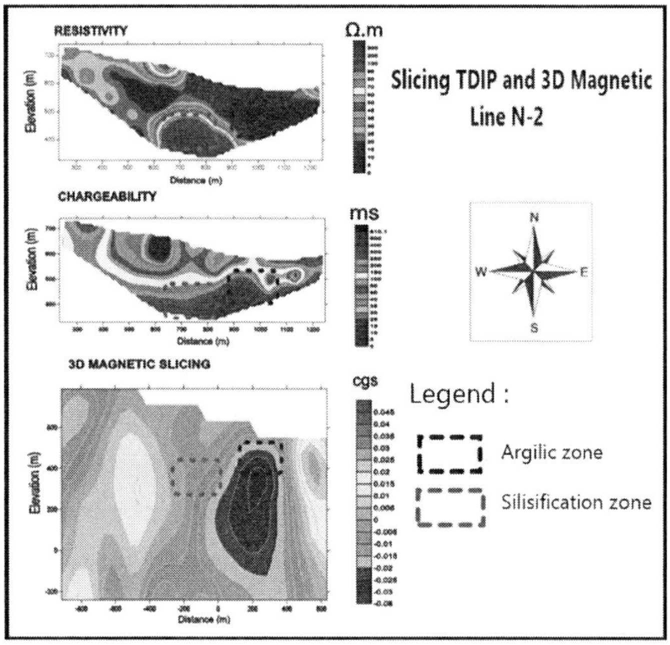

Figure 3. Mineralization zone from interpretation of TDIP and magnetic data in Line N-2.

The Line N-3 (Figure 4) has a high resistivity value of above 50 to 100 ohm.m, which are located at a distance of 600-900 meters with a height of 400-500 meters with chargeability worth 100 to 500 ms. Whereas argillic alteration is on the East surface with a resistivity value of less than 50 ohm.m and with chargeability around 100 ms. The value of the metal factor on this track is 2000 to 10000 mho/meter.

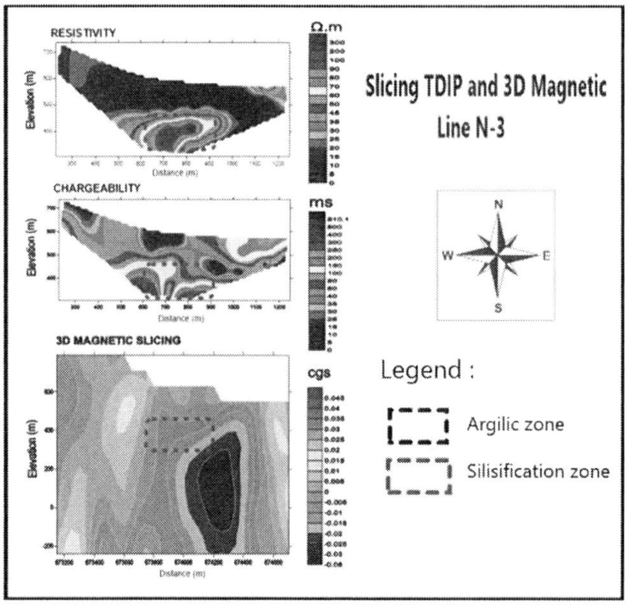

Figure 4. Interpretation of TDIP and magnetic methods Line N-3.

The values of these three parameters are thought to be mineralized zones. Mineralized zone if seen in Figure 4. In two alteration zones, namely argillic alteration and silicification zone, this zone is thought to contain sulfide ore minerals such as enargite, hematite, magnetite, and pyrite minerals and hydrothermal clay minerals such as alunite, kaolinite, dickite illite which is dominant and is thought to have gold content [2].

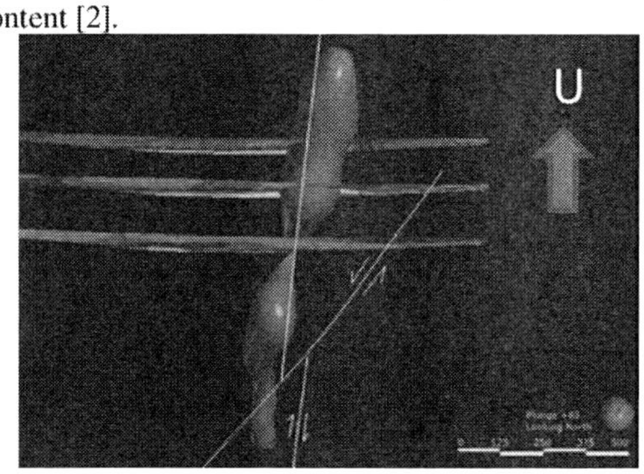

Figure 5. Correlation each line

The 3D model in figure 5 can provide geometry and relation between parameter values that indicate mineralized zones. In 3D models using isosurface models from the results of magnetic 3D inversion and a cross-section of changeability. Also, High susceptibility indicates levels of iron content[8]. This is done because these two parameters are considered dominant in determining the mineralized zone. High susceptibility also shows If plotted into the structure map in the study area, it will show continuity from North to South in the same direction as the right-hand fault. the pattern can be seen in figure 5.

5. Conclusion

There is indicated a mineralized zone in the study area. The mineralized zone with a high sulfide deposit is characterized to have a low susceptibility value (-0,073 to -0,005 cgs), high chargeability (more than 100 ms) and low to high resistivity (less than 50 Ohm.m is the argillic alteration and resistivity more than 50 Ohm.m for silicified alteration). The subsurface model shows the suspected mineralized zone in the study area which is thought to be in the east of the measurement and is thought to be controlled by a horizontal fault oriented North-South.

Acknowledgments
We would like to thank PT Antam Tbk for permission to publish the data. Many thanks as well to DRPM Universitas Indonesia for financial support through the grant of PITTA 2019 No. NKB-0640/UN2.R3.1/HKP.05.00/2019.

References

[1] Basuki, A., Sumanagara, D. A., and Sinambela, D. (1994). The Gunung Pongkor gold silver deposit, West Java, Indonesia. *Journal of Geochemical Exploration*, **50** (1-3), 371-391. https://doi.org/10.1016/0375-6742(94)90032-9.
[2] White N.C, and Hedenquist JW. (1995). Epithermal gold deposits: styles, characteristics and exploration. *SEG Newsletter*, **23**:8–13.
[3] Irvine, R., & Smith, M. (1990). Geophysical exploration for epithermal gold deposits. *Journal of Geochemical Exploration, 36(1-3), 375-412.*
[4] Milesi, J. P., Marcoux, E., Sitorus, T., Simandjuntak, M., Leroy,J., and Bailly, L. (1999) Pongkor (West Java, Indonesia): A Pliocene supergene-enriched epithermal Au-Ag-(Mn) deposit. Mineralium-deposits, **34**(2). 131-149. https://doi.org/10.1007/s001260050191.
[5] Kadarisman, D. S. (2013). Alterasi dan Mineralisasi Daerah Gunung Dahu dan Sekitarnya Kecamatan Nanggung, Kabupaten Bogor, Provinsi Jawa Barat. *Jurnal Teknologi*, **Vol II,** 22nd ser., 34-42.
[6] Loke, M. H. 1995. Least Squares Deconvolution of Apparent Resistivity Pseudosection. *Geophysics*. Malaysia.
[7] Rahmah, Siti. (2009). Pencitraan dua dimensi data resistivity dan induced polarization untuk mendelineasi deposit emas sistem epithermal di daerah "X". *Skripsi*. Universitas Indonesia: Depok
[8] Telford, W. M., Geldart, L. P., and Sheriff, R. E. (1990). *Applied geophysics* (2nd ed.). Cambridge: Cambridge University Press.

IOP Conf. Series: Materials Science and Engineering **854** (2020) 012057 doi:10.1088/1757-899X/854/1/012057

Performance Analysis of CUDA by Implementation of Hypergraph Partitioning for Parallelizing Sparse Matrix-Vector Multiplication Using Quadro K4200

Murni* **and T Handhika**

Computational Mathematics Study Center, Gunadarma University, 100 Margonda Raya Street, Pondok Cina, Depok, West Java, 16424, Indonesia

*E-mail: murnipskm@staff.gunadarma.ac.id

Abstract. This paper addresses the performance analysis of sparse matrix-vector multiplication through hypergraph partitioning techniques using CUDA GPU-based parallel computing. Quadro K4200 is the GPU that is used in this paper. On the implementation of matrix-vector multiplication, various sizes and types of matrices are attempted. Our results show that on the average scenarios with 2 partitions, 4 partitions, 8 partitions, 16 partitions and 32 partitions in 1024 threads, CUDA performs up to 700 × better than sequential programming.

1. Introduction

CUDA (Compute Unified Device Architecture) is one of the parallel computing platforms and programming models that have been developed by NVIDIA for general computing on the graphical processing unit (GPU). CUDA was first introduced NVIDIA in 2006. As one of the graphics processor companies, the company is motivated to develop a platform that can be used to maximize GPU usage. This platform is named GPGPU (General Purpose Computation on GPU). However, due to constrained API, then in 2007 NVIDIA issued a parallel programming standard model on a GPU called CUDA. CUDA is a minimal extension of the C / C ++ programming language to direct the process of computing from the CPU to GPU (GPU computing). With CUDA, developers are able to dramatically speed up computing applications by harnessing the power of GPU. Since then, CUDA GPU-based parallel programming has been one of the effective solutions to the constraints faced in MPI and OpenMP implementations, where the resulting speed-up reaches tens to hundreds of times [1].

GPU implements a multi-core architecture consisting of thousands of threads on hundreds of cores. This is certainly different from the CPU that implements multi-core architecture consisting of several cores of multiples of 2. With thousands of threads on hundreds of cores in the GPU, massive parallel programming can be applied to achieve greater performance improvements [4].

This paper will discuss the implementation of sparse matrix-vector multiplication. This implementation based on hypergraph partitioning models in shared memory parallel programming [2, 3, 12, 13, 14]. The benefits obtained by applying hypergraph partitioning using shared memory architecture can minimize communication and cache coherence overhead. The algorithm used in this paper in determining partition of hypergraph partitioning is the Fiducia-Matheyesses-Sanchis algorithm (or known as the FMS algorithm). After the number of partitions is formed, then the matrix problem will turn into a block of matrix. After that, parallel calculations performed on the GPU using CUDA.

IOP Conf. Series: Materials Science and Engineering **854** (2020) 012057 doi:10.1088/1757-899X/854/1/012057

2. Hypergraph Partitioning

Hypergraph partitioning is a difficult optimization problem. Then, for solving this problem we can use optimization algorithms. The algorithm used in this paper is KL and FM algorithms developed by Sanchis (or known as FMS algorithms) [5,11]. FMS algorithm is able to partition to *K*-way partition [15]. The concept of this FMS algorithm is to swap out one by one vertex on each iteration. First of all, define the initial state, i.e. the set of vertices are randomly partitioned into *K* parts. Furthermore, each vertex will change position from one partition to another partition with the concept of gain, which is a value indicating a reduction cutsize. The positive gain will reduce cutsize while the negative gain will increase cutsize. Vertex with the greatest gain will be moved to the other position. Then, vertex has been moved to other partitions and is not included in the calculation of the gain on the next iteration. In addition, the movement is causing changes in the value of the gain on the other vertices. This process is carried out continuously repeated such that there are no vertex switching positions again. Thus, the optimal solution of the FMS algorithm is the partition with the smallest cutsize.

Hypergraph partitioning will be implemented on the parallelization of sparse matrix-vector multiplication. The basic technique to parallelize sparse matrix-vector multiplication is a way to partition the rows and columns of the matrix into parts. This paper will discuss the process of calculating the sparse matrix-vector multiplication in parallel with adopting hypergraph partitioning techniques.

In hypergraph partitioning, there are two models for the decomposition of the sparse matrix. These models are column-net and row-net models. Suppose given $m \times n$ matrix and $n \times 1$ vector as follow:

$$A = \begin{bmatrix} a_{11} & a_{12} & \cdots & a_{1n} \\ a_{21} & a_{22} & \cdots & a_{2n} \\ \vdots & \vdots & \ddots & \vdots \\ a_{m1} & a_{m2} & \cdots & a_{mn} \end{bmatrix}, \bar{x} = \begin{bmatrix} x_1 \\ x_2 \\ \vdots \\ x_n \end{bmatrix}$$

In the *column-net* model, matrix *A* is represented as a *hypergraph* $\mathcal{H}_R = (V_R, N_C)$ where each row of the matrix A is represented by the vertices in V_R and each column of the matrix A is represented by nets in N_C. whereas, in a *row-net* model, matrix *A* is represented as a *hypergraph* $\mathcal{H}_C = (V_C, N_R)$ where each row of the matrix A is represented by the vertices in V_C and each column of the matrix A is represented by nets in N_R. Thus, there is one vertex v_i and one net n_j for each row *i* and column *j* in column-net or row-net model.

Matrix A subsequently converted into blocks as $K \times K$ matrix as much by adopting hypergraph partitioning techniques [6, 7, 8, 9]. Block matrix A can be represented as follows.

$$A = \begin{bmatrix} A_{11} & A_{12} & \cdots & A_{1K} \\ A_{21} & A_{22} & \cdots & A_{2K} \\ \vdots & \vdots & \ddots & \vdots \\ A_{K1} & A_{K2} & \cdots & A_{KK} \end{bmatrix},$$

where K is the number of processors with the size of each matrix A is $m_i \times n_j$, such that $\sum_{i=1}^{K} m_i = m$ dan $\sum_{j=1}^{K} n_j = n$.

Furthermore, vector $\bar{x}$ also divided into several sub-vectors corresponding as follows:

$$\bar{x} = \begin{bmatrix} \bar{x}_1 \\ \bar{x}_2 \\ \vdots \\ \bar{x}_K \end{bmatrix},$$

where $\bar{x}_i$ consists of n_i elements $\left(\sum_{i=1}^{k} n_i = n\right)$.

In hypergraph partitioning techniques, blocks of matrix A is not located on the main diagonal are arranged such that the majority of the elements of the matrix block a lot of zeroes. It means that communication between processors is reduced. Furthermore, submatrix A and subvector $\bar{x}$ are multiplied in parallel using CUDA on GPU.

Problems matrix *A* and the vector *x* above have been simplified using hypergraph partitioning into sub-matrices *A* and sub-vector $\bar{x}$ with the size $K \times K$ and $K \times 1$. Thus, the multiplication of sub-matrix *A* and sub-vector *x* will produce $K \times 1$ sub-vector $\bar{y}$, as given in the following equation:

IOP Conf. Series: Materials Science and Engineering **854** (2020) 012057 doi:10.1088/1757-899X/854/1/012057

$$\bar{y} = \begin{bmatrix} \bar{y}_{11} \\ \bar{y}_{21} \\ \vdots \\ \bar{y}_{K1} \end{bmatrix}.$$

The next procedure on the parallelization of sparse matrix-vector multiplication using hypergraph partitioning on the GPU is determining the number of blocks. Determination of the amount of this block in accordance with the size of the multiplication result sub-matrix A with sub-vector $\bar{x}$, that is $K \times 1$. This is what distinguishes the procedure on the GPU parallelization of matrix multiplication. The number of block-based parallelization procedures of sparse matrix multiplication on GPU determined from the maximum limit thread owned machine GPU, while the number of block-based procedures parallelization of sparse matrix multiplication using hypergraph partitioning on the GPU is determined from the size of the sub-matrix multiplication results. After that, the procedure followed was to determine the total of the block size for each block. The number of threads in one block should not exceed the number of threads that exist on the machine GPU. If the thread in one block exceeds the number of threads that exist on the machine GPU then the process will occur sequentially in one block. To overcome this, the number of partitions to be enlarged so that the number of blocks obtained will be more and more. This is done repeatedly to obtain the block number where the number of threads in each block meets the existing number of threads on the GPU. Once this is completed then the process parallelization of sparse matrix-vector multiplication using hypergraph partitioning can be processed using CUDA on GPU [10, 14].

3. Implementation

In the GPU there is a grid, block and thread hierarchy [16]. Therefore, before proceeding the sparse matrix-vector multiplication procedure, it is necessary to explain the specifications CPU and GPU to be used. The parallelization program of matrix-vector multiplication will be run on the CPU and GPU with the following specifications:

CPU:

Model Identifier	: ASUS MAXIMUS VII HERO
Processor Name	: Intel Core i7
Processor Speed	: 3.6 GHz
Number of Processor	: 1
Total Number of Cores	: 4
L2 Cache (per Core)	: 256 KB
L3 Cache	: 6 MB
System Bus	: Intel Direct Media Interface 5GT/s
Software	: OS X 10.9.1 (13B42)

GPU:

Mesin	: Quadro K4200
Architecture	: Kepler
CUDA Cores	: 1344

For Quadro K4200 is known that machines used have the Kepler architecture, a block may have at most 1024 threads. Given a various matrix A and vector $\bar{x}$. In this paper, five scenarios have been done, such as Hypergraph Partitioning with 2 partitions, 4 partitions, 8 partitions, 16 partitions and 32 partitions. All these scenarios use 1024 threads. The results can be seen in the following figure.

IOP Conf. Series: Materials Science and Engineering 854 (2020) 012057 doi:10.1088/1757-899X/854/1/012057

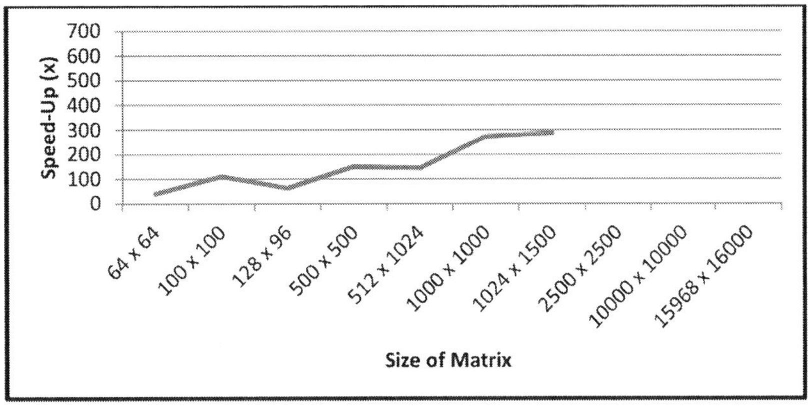

Figure 1: Hypergraph Partitioning with 2 Partitions

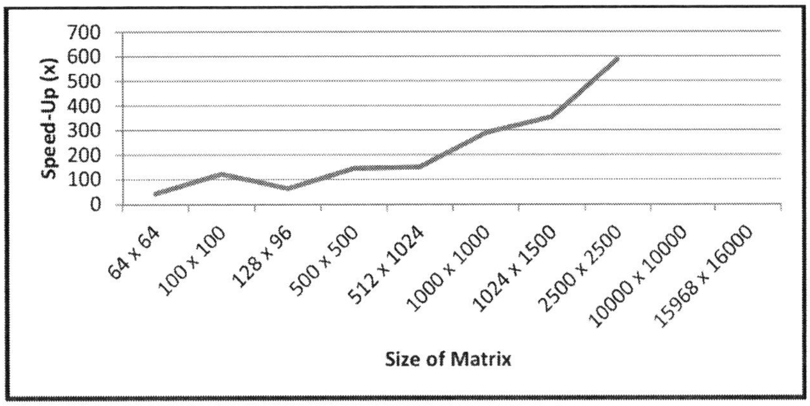

Figure 2: Hypergraph Partitioning with 4 Partitions

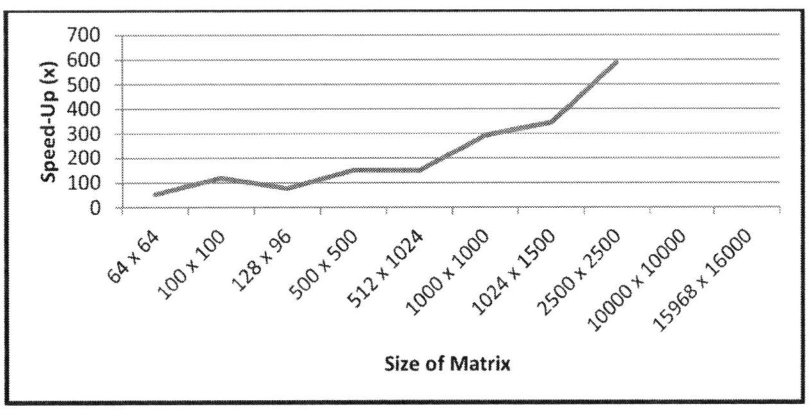

Figure 3: Hypergraph Partitioning with 8 Partitions

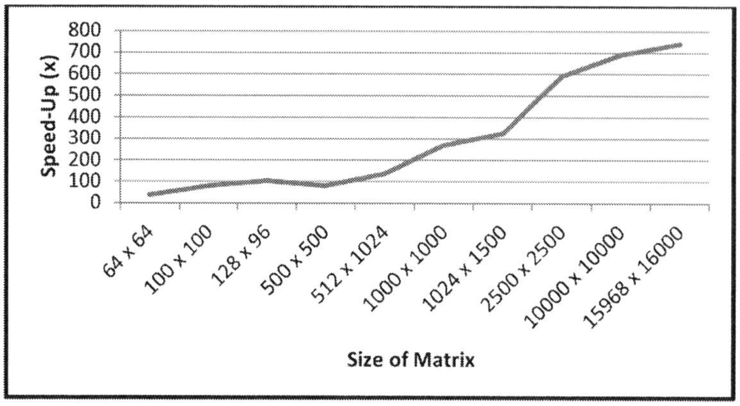

Figure 4: Hypergraph Partitioning with 16 Partitions

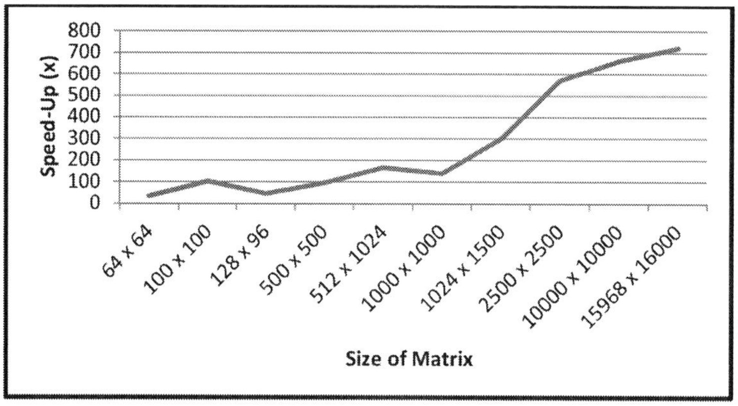

Figure 5: Hypergraph Partitioning with 32 Partitions

Based on the results in Figure 1, Figure 2, Figure 3, Figure 4 and Figure 5 we find that the fastest speed up for Sparse Matrix-Vector Multiplication in this study is obtained in Figure 3, which is Hypergraph Partitioning with 8 partitions. The speed-up movement increases with the increasing size of the matrix. Overall we achieved CUDA performs up to 700 × better than sequential programming.

4. Conclusion and future works

Based on the discussion of this paper, it can be concluded that the Hypergraph Partitioning in shared-memory parallel programming in sparse matrix-vector multiplication using CUDA with GPU can perform up to 700 × better than sequential programming. This result is obtained by using five scenarios on various matrixes. The speed-up movement increases with the increasing size of the matrix.

As future work, the use of different types of GPU devices is possible, more high specification to process the more large matrix that used. Furthermore, the parallel reduction can be performed on the parallelization of sparse matrix-vector multiplication to obtain better performance.

References

[1] A. Bustamam, G. Ardaneswari, H. Tasman and D. Lestari. 2014. Performance Evaluation of Fast Smith-Waterman Algorithm for Sequence Database Searches using CUDA GPU-based Parallel Computing, *Journal of Next Generation Information Technology*, **5**, pp 38-46

[2] U.V. Catalyurek and C. Aykanat. 1996. Decomposing Irregularly Sparse Matrices for Parallel Matrix-Vector Multiplication, *Parallel Algorithms for Irregularly Structured Problems, Irregular '96, In Lecture Notes in Computer Science*, **1117**, pp 75-86

[3] U.V. Catalyurek and C. Aykanat. 1999. Hypergraph-Partitioning-Based Decomposition for Parallel Sparse-Matrix Vector Multiplication, *IEEE Transactions on Parallel and Distributed Systems*, **10**, pp 673-693

[4] S. J. Eggers. 1991. Simplicity Versus Accuracy in a Model of Cache Coherency Overhead, *IEEE Transactions on Computer*, **40**, pp 893-906

[5] C.M. Fiduccia and R.M. Mattheyses. 1982. A Linear-Time Heuristic for Improving Network Partitions. *Proc. 19th ACM/IEEE Design Automation Conf.*, pp 175-181

[6] B. Hendrickson. 1998. Graph Partitioning and Parallel Solvers: Has the Emperor No Clothes?, *Solving Irregularly Structured Problems in Parallel, Irregular '98, Number 1457 In Lecture Notes in Computer Science*, pp 218-225

[7] B. Hendrickson and T.G. Kolda. 1998. Partitioning Sparse Rectangular Matrices for Parallel Computations of Ax and A^Tx, *Applied Parallel Computing in Large Scale Scientific and Industrial Problems, PARA '98, Number 1541 In Lecture Notes in Computer Science*, **1541**, pp 239-247

[8] B. Hendrickson and T.G. Kolda. 2000. Graph Partitioning Models for Parallel Computing, *Parallel Computing*, **26**, pp 1519-1534

[9] B. Hendrickson and T.G. Kolda. 2000. Partitioning Rectangular and Structurally Unsymmetric Sparse Matrices for Parallel Processing, *SIAM J.*, **21**, pp 2048- 2072

[10] R. Hochberg. 2012. Matrix Multiplication with CUDA – A basic introduction to the CUDA programming model, *Shodor*.

[11] B.W. Kernighan and S. Lin. 1969. An Efficient Heuristic Procedure for Partitioning Graphs, *The Bell System Technical Journal*, pp 291-307

[12] Murni, T. Handhika, I. Sari and D. Indarti. 2016. Implementasi Graph Partitioning pada Paralelisasi Perkalian Matriks-Vektor, *in Proceedings of National Seminar in Mathematics and Statistics (SEMASTAT) Universitas Negeri Padang*, pp 194-199

[13] Murni and T. Handhika. 2016. Implementasi Hypergraph Partitioning pada Paralelisasi Perkalian Matriks-Vektor, *in Journal of Computer & Information System*, pp 1-6

[14] Murni, A. Bustamam , Ernastuti, T. Handhika and Kerami D. 2016. Hypergraph partitioning implementation for parallelizing matrix-vector multiplication using CUDA GPU-based parallel computing, *in Proceedings of the international symposium on current progress in mathematics and sciences*, pp 1-6

[15] L.A. Sanchis. 1989. Multiple-Way Network Partitioning, *IEEE*, **38**, pp 62-81

[16] J. Sanders and E. Kandrot. 2008. *CUDA by Example: An Introduction to General Purpose GPU Programming*, New Jersey, Addison-Wesley

IOP Conf. Series: Materials Science and Engineering **854** (2020) 012058 doi:10.1088/1757-899X/854/1/012058

Ultrasound Assisted the Degradation of Methylene Blue Using WO_3-Deoiled Spent Bleaching Earth as a catalyst

Noor Hindryawati, R.R. Dirgarini Julia N.S, Aman S. Panggabean and Wilsoma

Chemistry Department of Mathematics and Science Faculty, Mulawarman University, Jl. Barong Tongkok No. 4 GunungKelua, Samarinda, Kalimantan Timur, Indonesia

Email: hindryawati@gmail.com

Abstract. Spent Bleaching Earth (SBE) that has been impregnated with WO_3 can degrade methylene blue through sonocatalytic. The stages of this research are SBE preparation, activation, WO_3-SBE modification through wet impregnation process and sonocatalytic test. WO_3-SBE is characterized using SEM and XRD. In the SEM results, there are cavities in the surface morphology of SBE after activation and after impregnation, the cavities are covered. WO_3-SBE XRD diffraction pattern shows the presence of WO_3 at an angle of $2\theta = 28.855°$; 34,813°. The percentage of degradation of methylene blue was 99.12%, using 0.15 grams catalyst, 100 ppm concentration and 15 minutes contact time.

1. Introduction

People often complain of various problems, one of them is water pollution caused by textile dye industrial waste. Synthetic dyes often used in batik cloth, paper, office and cosmetic industries. The development of the textile industry has increased the use of dyes that can pollute the environment. The dye that is often used in this industry is methylene blue [1]. Methylene blue (MB) is a very important and relatively inexpensive base dye compared to other dyes. [2] Some alternative technologies are used to treat wastes that contain dyes such as coagulation, flocculation and absorption techniques. Color removal by the process of coagulation, flocculation, and absorption only transforms dyes from the liquid into the solid phase. Those processes do not separate the complex compounds of color. The color particles coagulants need further processing so that it will cause further pollution or new waste. [3] Another alternative that can be developed is the sonocatalytic method to degrade dyes.

One of the more effective methods to degrade dyes is Advance Oxidation Processes (AOPs). [4] AOPs technology is a method of oxidation of a solution phase which is mainly based on the formation and use of hydroxyl radicals (OH) as a by-product that can cause the destruction of pollutant compounds. [5] The Sonocatalytic Method is one of the AOPs that is environmentally friendly because it produces CO_2 and H_2O. This method works to degrade organic dyes in water media by using ultrasonic vibrations with the help of a catalyst (semiconductor). This method comes from the cavitation effect bubbles produced by liquids under ultrasonic radiation. This can absorb sound energy and collapse to release energy in a very short time, high temperature and pressure. Catalysts that are often used are TiO_2, ZnO, CdS, Fe_2O_3 and WO_3. TiO_2 has been extensively studied to degrade harmful pollutants but it is only active under ultraviolet radiation. In contrast to Tungsten trioxide (WO_3) is a semiconductor material that can accelerate the degradation process under visible radiation. WO_3 has a narrow bandgap of 2.7-

2.8 eV. [6] The semiconductor catalyst is long-lasting and can be used repeatedly. One of the catalyst support used is Spent Bleaching Earth (SBE). Based on the background described above, research of WO_3 impregnation in Spent Bleaching Earth (SBE) as a sonocatalytic to degrade methylene blue is conducted.

2. Experimental Procedure

2.1. Spent Bleaching Earth Preparation
At first, the SBE is weighed as many as 300 grams using an analytical scale. Then the SBE is sieved using 140 sized mesh and then put in an oven at a temperature of 120°C for 2 hours.

2.2. SBE Activation
Spent Bleaching Earth (SBE) is a mixed with$H_2SO_4$1N solution with a ratio of 1:10 (b/v) and then homogenized. The mixed container is tightly closed. The mixture is then treated with ultrasonic wave irradiation at the optimum irradiation temperature of 50-60°C in 60 minutes of activation time. After that, the sample is filtered and the SBE obtained is washed using distilled water to neutralize the pH. Then the SBE is dried in an oven at 110°C for 3 hours and sifted in 140 mesh. [7]

2.3. WO_3 Impregnation on SBE
SBE weighed as many as 50 grams and mixed into 20 grams of WO_3, then added 400 mL of 7 M NH_4OH solution. The mixture is refluxed for 5 hours at 60-70°C. After that, the sample is filtered and the WO_3-SBE obtained is washed using distilled water to neutralize the pH. Then put it in the oven at 110°C in 3 hours. After drying, WO_3-SBE is crushed and then sieved using a 140 mesh sieve. Furthermore the solid is calcined at 500°C for 2 hours. [8]

2.4. Sonocatalytic Test on Methylene Blue Using WO_3-SBE
Sonocatalytic tests are performed using variations in contact time and concentration of methylene blue. 0.15 grams of WO_3-SBE is added to 50 mL of 100 ppm methylene blue. Then put into an ultrasonic cleaner at a temperature of 60°C with a time variation of 5, 15, 45 and 60 minutes. Then it is filtered. Furthermore, absorbance is measured using a UV-Vis spectrophotometer. Then the degradation percentage is calculated. This is also conducted at variations in the concentration of 50, 100, 150 and 200 ppm.

2.5. Material Type Variation Test
In order to find out the performance of WO_3-SBE, a comparison of material types is conducted using the previous optimum condition with variations of WO_3, SBE and WO_3.SBE material.

3. Results and Discussions

3.1. Impregnation and Characterization of WO_3-SBE
SBE is activated by using H_2SO_4 1N that can increase the surface area by exchanging Na^+, Mg^{2+}, Ca^{2+}, Mn^{2+} and others contained in the interlayer into H^+ion. H_2SO_4 1N is the optimum concentration; using concentrations above 1N can damage the *montmorillonite* structure. This is because Al^{3+} can dissolve H_2SO_4. Meanwhile, the concentration under 1 N makes the activation process takes longer. [9]

The base impregnation is conducted by using NH_4OH as a solvent that can dissolve WO_3. Because the material modification is shorter, the cost is relatively cheap and the success of modification is greater. At first, WO_3-SBE is refluxed and assisted by constant heating and stirring at the optimum temperature of 60-70°C. Then the WO_3-SBE obtained is washed using distilled water to reach pH 7. Furthermore, calcination is performed to tighten the distribution of the particles so that they are not easily separated, remove the remaining impurities and convert metals to their oxides.

Then the WO_3-SBE characterization is conducted using XRD to find out the peak of WO_3 diffraction on the impregnated SBE. The XRD analysis results for WO_3-SBE can be seen in Figure 1.

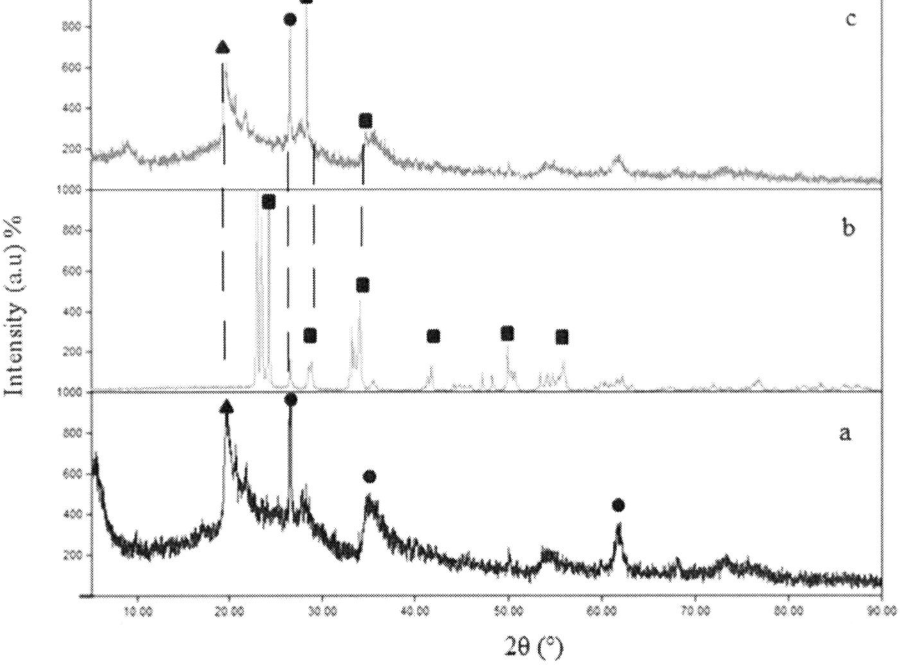

Figure 1. XRD results (a)SBE (b) WO_3dan (c) WO_3-SBE
(WO_3= ■ ; SiO_2 =▲ ; Al_2O_3 =●).

From the XRD data in figure 1(a) a diffraction pattern is obtained with SiO_2 peak at 2θ = 21.460° and Al_2O_3 peak at 2θ = 26.559°; 35,164°; 62,131°. Based on research by Fadillah [10] there are similarities in diffraction patterns that indicate the mineral is montmorillonite. SiO_2 peaks are found at 2θ = 21,8714° and 26,6552°. Al_2O_3 peaks are at 2θ = 26,6552° and 50,1340°. After conducting WO_3 impregnation, XRD data is obtained in Figure 1(b) showing WO_3 diffraction patterns peak at 2θ = 23.267°; 26,619°; 28,958°; 34,176°; 49,880°; 55,159°. Figure 1(c) shows diffraction patterns of WO_3 peaks are 28,855°; 34,813°. SiO_2 2θ = 21,460° and Al_2O_3 2θ = 26,540°. Based on Putri [11], the similarity of diffraction patterns peak at 2θ = 23,1571°; 24,4047° and 34,2186° which show the typical WO_3 diffraction. Then the SEM analysis is performed to determine the morphological form of SBE after activation and WO_3-SBE which can be seen in Figure 2.

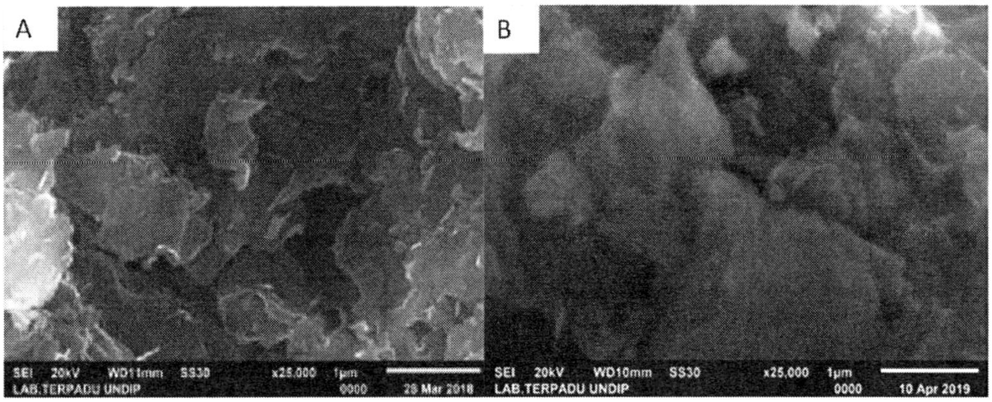

Figure 2. SEM results in 25.000 times magnification (a) SBE after activation, (b) WO_3-SBE.

In figure 2(a) SBE after activation of the morphological form shows the presence of pore and heterogeneous shape of the particle. While in figure 2(b) WO_3-SBE, shows some of the pores has been covered with the suspected entry of WO_3 in SBE.

3.2. Sonocatalytic Test on Methylene Blue using WO_3-SBE

3.2.1. Contact Time Variation towards Methylene Blue Degradation

In this research, the variation of contact time is conducted in order to find out the optimum contact time needed by WO_3-SBE to degrade Methylene Blue. The following graph shows the relationship between contact time and degradation percentage as follows (Figure 3).

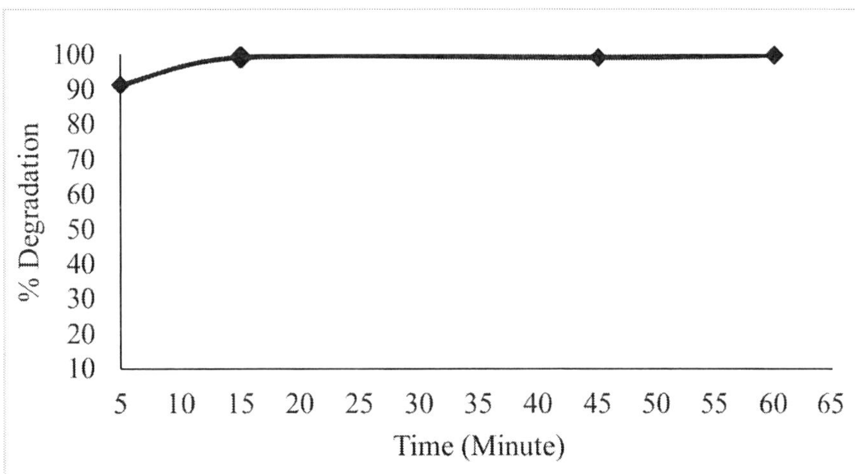

Figure 3. Contact time variations towards methylene blue degradation, 0.15 gram catalyst.

On the graph, the 5th minute obtains 91.14% degradation. Then the percent degradation is increased in the 15 min to 99.12%, indicating that it is the optimum contact time of WO_3-SBE in degrading methylene blue. Then the 45 min is 99.08% and 60 min is 99.81%, which indicates the percent degradation is running constantly. According to Zhou et al. [12] in general, the semiconductor catalyst (WO_3) will produce sonocatalytic electrons (e-) and holes (h^+). After that, it will cause the electrons in the valence band to be excited towards the conduction band. This has been explained in the mechanism of sonoluminescence. Sonoluminescence involves light entering through the recombination of free radicals formed from the bubbles of cavitation effects. The cavitation effect gives emission of light, so that energy can move electrons [13]. Then OH radicals are formed which can degrade methylene blue. The longer the time needed for WO_3-SBE, the more OH radicals are formed. It is causing a greater catalyst ability to degrade methylene blue. The more OH radicals produced, the more methylene blue is degraded.

3.2.2. Optimum Concentration Variation towards Methylene Blue Degradation
In this research, variations in concentration are conducted in order to determine the optimum concentration required by WO_3-SBE to degrade methylene blue. The following graph shows the relationship of concentration with degradation percentage of methylene blue (Figure 4).

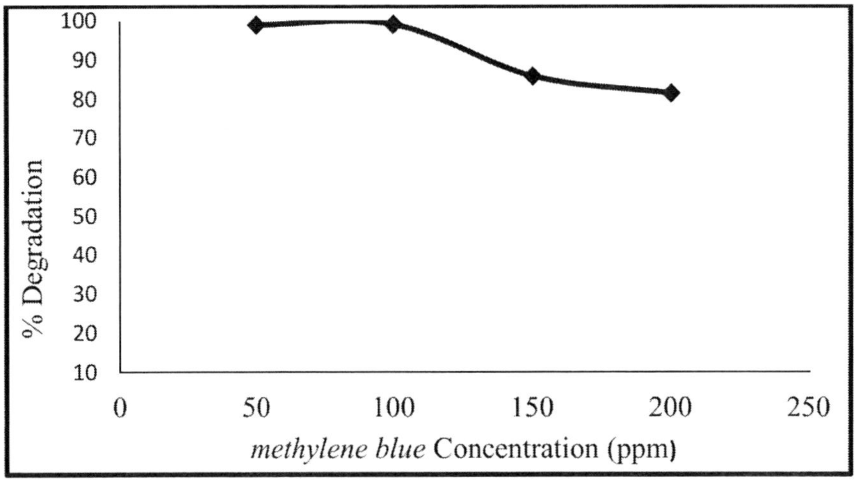

Figure 4. Concentration variations towards methylene blue degradation, 0.15 grams catalyst.

Based on the graph, the use of 50 ppm concentration obtains 99.02% degradation. Then the degradation percentage has increased at 100 ppm to 99.12% which indicates that it is the optimum concentration of WO_3-SBE in degrading methylene blue. When the concentration of methylene blue gets smaller the adsorption rate and the degradation rate will be directly proportional. This is because SBE and WO_3 have the same performance. At a concentration of 150 ppm decreases to 85.91% as well as a concentration of 200 ppm gets 81.51%. At high concentrations, it is possible that the adsorption rate is getting smaller because the methylene blue that is absorbed is not in accordance with the SBE capability. It causes the degradation process in WO_3 to be hampered. [14]

3.2.3. Variations on Type of Material
In this research, variations in material type aim to determine the process of adsorption or degradation of WO_3-SBE. The following is a graph of the relationship of variations in material types with degradation percentage of methylene blue (Figure 5).

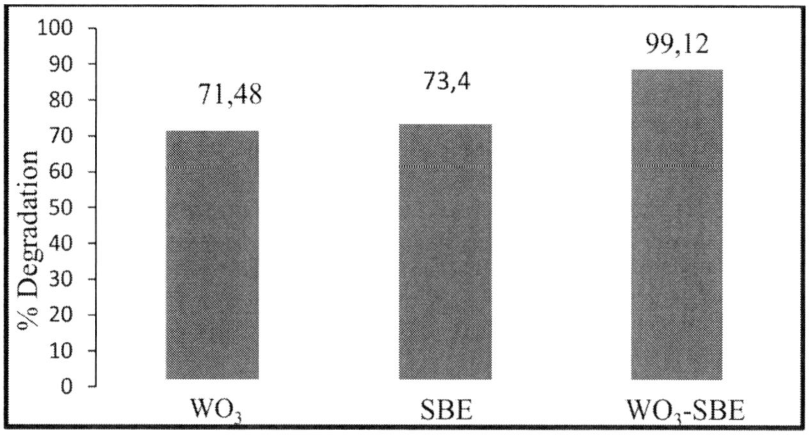

Figure 5. Material types variations towards methylene blue degradation.

The optimum weight is 0.15 gram, the contact time is 15 minutes and the concentration of methylene blue is 100 ppm. At $WO_3$the degradation percentage is 71.48%, this is considered as the process of degradation. The SBE degradation percentage is 73.45%, this is due to the adsorption process. Then the WO_3-SBE percentage is 99.12% this is due to the existence of two processes, adsorption and degradation. The first is the adsorption process on SBE then the second is the degradation process at WO_3. SBE which has a large surface area causes the active adsorbent to be greater so that methylene blue is absorbed on the surface simultaneously and is degraded with WO_3. Then the degraded methylene blue causes the active adsorbent to return vacant. Then the process of adsorption and degradation continues. This is a possible greater synergistic effect. If the adsorption rate is greater means the rate of degradation will be greater.

4. Conclusion

In the SEM results, the surface morphology of SBE after impregnation with WO_3 shown the cavities are covered. The XRD results show WO_3 in SBE. The methylene blue degradation percentage with an optimum 0.15 grams WO_3-SBE catalyst, 100 ppm concentration and 15 minutes contact time obtain 99.12%. The degradation percentage of $WO_3$is 71.48%; SBE is 73.45% and WO_3-SBE is 99.12%.

References

[1] Andari, N. D. and Wardhani, S. 2014. TiO2-Zeolite Photocatalysis in Degrading Methylene Blue. *Journal of Chem. Prog Vol. 7, No.1.*

[2] Fitriani, D., Oktiarni, D. and Lusiana. 2015. Utilization of Banana Peel as Methylene Blue Color Adsorbent. *Gradient journal. Vol. 11, No. 2, Page. 1091–95.*

[3] Widiyandari, H. and Burhanudin, S. 2012. Degradation of Methylene blue (MB) Dyes Using WO_3 / Fe_2O_3 Photocatalysis by Sunlight. *Journal of Science and Mathematics. Vol. 20, No. 2, Page. 26-29.*

[4] Kumar, M. S., Sonawane, S. H. and Aniruddha, B. 2017. Degradation of methylene blue dye in Aqueous Solution Using Hydrodynamic Cavitation Based Hybrid advanced Oxidation Processes. Journal of Chemical Engineering and Processing.

[5] Cahiskan, Y., Yatmaz, H. C. and Bektas, N. 2017. Photocatalytic Oxidation of High Concentrated Dye Solutions Enchanced by Hydrodynamic Cavitataion in a Pilot Reactor. Journal of Process Safety and Environment Protection.

[6] Zhang, X. Hao, C., Ma, C., Shea, Z., Gun, J. Sun, R. 2019. Study on Sonokatalytic Degradation of Rhodamin B in Aqueous Solution. *Ultrasonics Sonochemistry.*

[7] Helbianurramadan, M., Hindryawati, N. and Julia N. S, R. R. D. 2017. Activation of Deoiled Spent Bleaching Earth (SBE) by Using Ultrasonic Methods to Adsorb Pb^{2+} Metal Ion. *Atomic Journal. Vol. 02, No. 2, Page. 241-47.*

[8] Rayendra, A.F., Wardhani, S. and Tjahjanto, R.T. 2014. Effect of TiO_2-Bentonite Composition on Methylene Blue Degradation. *Journal of Chemistry. Vol. 2, No. 2, Page. 555–61.*

[9] Widyastuti, L. 2007. Methanolysis Reaction of Jatropha Seed Oil into Methyl Esters as Substitute Fuel for Diesel Oil Using KOH Catalyst. Essay. Faculty of Mathematics and Natural Sciences Semarang.

[10] Fadillah, N. D. 2017. Pilarisasi TiO_2 Pada Deoiled Spent Bleaching Clay dan Uji Fotokatalitiknya pada Methylene Blue. *Skripsi*. Jurusan Kimia, Fakultas Matematika dan Ilmu Pengetahuan Alam, Universitas Mulawarman.

[11] Putri, R. A. 2018. Pilarisasi WO_3 pada Deoiled Spent Bleaching Earth (DSBE) sebagai Fotokatalisis untuk Degradasi Zat Warna Methylene Blue Dari Limah Pewarna Kain Tenun. Skripsi. Jususan Kimia, Fakultas Matematika Dan Ilmu Pengetahuan Alam, Universitas Mulawarman

[12] Zhou, M., Yang, H., Xian, T., Li, R.S., Zhang, H.M., and Wang, X. 2015. Sonocatalytic Degradation of RhB over $LuFeO_3$ Particles under Ultrasonic Iradiation. Hazard Mater Journal, Page. 149-157.

[13] Hassani, A., Eghball, P., and Metin, O. 2018. Sonocatalytic Removal of Methylene Blue From Water Solutions by Cobalt Ferrite / Mesoporous Graphtic Carbon Nitride ($CoFe_2O_4$ / mpg-C_3N_4) Nanocomposites: Response Surface Methodology Approach. Journal of Environmental Science and Pollution Research.

[14] Sun, Z., He, X., Du, J. and Gong, W. 2016. Synergistic Effects of Phototocatalysis and Adsorption of Nano-TiO_2 Self-Assembled Onto Sulfanyl / Activated Carbon Composite. Journal of Environ Sci Pollut Research.

IOP Conf. Series: Materials Science and Engineering **854** (2020) 012059 doi:10.1088/1757-899X/854/1/012059

Q-switched Erbium-Doped Fiber Laser Incorporating Multi-Walled Carbon Nanotubes as a Saturable Absorber

C Huda[1,*], M Yasin[2], SW Harun[3], A A A Jafry[4], and A H A Rosol[3]

[1] Department of Physics, Faculty of Science and Technology, Airlangga University, Surabaya (60115) Indonesia; Department of Physics Education, Faculty of Science and Technology, Universitas Kanjuruhan Malang, Malang, Indonesia.
[2] Department of Physics, Faculty of Science and Technology, Airlangga University, Surabaya (60115) Indonesia.
[3] Department of Electrical Engineering, University of Malaya, 50603 Kuala Lumpur, Malaysia.
[4] Department of Physics, Faculty of Science, Universiti Teknologi Malaysia, 81310 Skudai, Johor, Malaysia.

Email: *choirul@unikama.ac.id

Abstract. We demonstrate the generation of Q-Switched erbium-doped fiber laser (EDFL) by employing a multi-walled carbon nanotubes-polyvinyl alcohol (MWCNTs-PVA) as a saturable absorber (SA). The SA was prepared by embedding MWCNTs into PVA via stirring and ultrasonication. A stable Q-Switched pulse train operating at 1559.4 nm was successfully generated by employing MWCNTs-PVA SA into a laser cavity. The repetition rate and pulse width of the laser were 52.85-151.9 kHz and 6.313-2.395 μs, respectively. The maximum output power and maximum peak power obtained are 52 μW and 142.94 μW, respectively at a pump power of 63 mW. This demonstration proves that the MWCNTs-PVA based SA is suitable for the generation of Q-Switched fiber laser at 1.55 μm region.

1. Introduction

Pulsed laser is a potential device for vast industrial applications including biomedical, micromachining, and laser cutting technology. It can be realized using passive or active technique. The latter suffer from few drawbacks including bulky, costly and frequent maintenance configuration. Therefore, the passive approach is preferable due to its simplicity, maintenance-free operation and portable. Passively Q-switched is initiated using the modulation of Q-factor in the laser cavity by a saturable absorber (SA). Various materials are incorporated inside an all-fiberized laser cavity as a SA such as nanoparticles [1], inorganic compound [2], and 2-dimensional materials [3, 4]. However, those materials possess few limitations such as low-operating bandwidth, complex fabrication procedure and immature SA device.

Carbon nanotubes (CNTs) is one of the most investigated material in recent years [5, 6]. CNTs is an atomically arranged carbon with sp^2 hybridized hexagonal honeycomb lattice structure. It owns an outstanding physical and chemical characteristics alike graphene. The superiority of CNTs including; a very thin structure, strong material, high electrical conductivity ($0.34–1.10^{-4}$ Ωcm), high melting temperature, lightweight material, tuneable electronic structure, flexible, and possesses a high thermal conductivity than graphene (3000 W/K) [7]. CNTs can be classified into two; single walled carbon nanotubes (SWCNTs) and multi-walled carbon nanotubes (MWCNTs). SWCNTs has a diameter of approximately 1-2 nm whereas MWCNTs owns a diameter of 2-50 nm. CNTs had been manipulated in various field such as sensor, membrane, energy storage for lithium ion-batteries, electrodes of supercapacitor, fuel cell components, candidates for molecular electronics, electron-field emitter, composite fibers, biomedical applications, saturable absorber (SA), etc.

SWCNTs, CNTs and graphene was the most investigated SA as they own unique optical and physical properties. Vast research works demonstrated Q-switched generation using MWCNTs as a passive SA. MWCNTs was proposed to be an efficient SA in ytterbium-doped fiber laser (YDFL) [8], erbium-doped fiber laser (EDFL) [9] with nanotube diameter less than 8 nm, thulium-doped fiber laser (TDFL) [10], multi-walled carbon nanotubes-polyethylene oxide (MWCNTs-PEO) for Q-switched EDFL [11], high power Q-switched TDFL [12], MWCNTs doped bismuth telluride as SA for multiwavelength generation [13], and dual-wavelength generation [14]. Here, we demonstrated Q-switched erbium-doped fiber laser incorporating multi-walled carbon nanotubes-polyvinyl alcohol (MWCNTs-PVA) as SA. Our SA generates stable Q-switched with an operating wavelength of 1559.4 nm, maximum repetition rate of 151.9 kHz, and minimum pulse width of 2.395 μs.

2. Preparation of MWCNTs-PVA

The MWCNTs used has a 99.99% purity, a 1-2 μm length, and a 10-20 nm diameter. Synthesis of SA starts with the preparation of functionalizer solution. The solution was produced by dissolving 4 g of natrium dodecyl sulphate in a 400 mL of deionized (DI) water. The functionalizer act as a solvent for the MWCNTs to mixed with water. Next, approximately 250 mg of MWCNTs was added to the functionalizer solution. The solution was then ultrasonicated at 50 W for 6 min, thus a homogeneous dispersion of MWCNTs was produced. The prepared solution was further centrifuged at 1000 rpm to get rid of the large particles from MWCNTs. The distributed suspension was stable for few weeks. The MWCNTs-PVA composite was fused with PVA to produce a free-standing thin film. PVA was chose as a host as it owns excellent film forming, good adhesive properties, ease to emulsify, high tensile strength, flexible, contain high oxygen and aroma barrier. First, the PVA solution was prepared by dissolving 1 g of PVA into 400 mL of DI water. The solution was further mixed with the prepared MWCNTs suspension. A 2:3 composition was used to produce MWCNTs-PVA. The mixture was then ultrasonicated for 1 hr. The ultrasonication allows the production of a precursor with enough viscosity for thin film preparation. Finally, the precursor was distributed over a glass substrate and left to dry at room temperature for few days. The dried MWCNTs-PVA was then peeled off from the glass and employed into the laser cavity for pulse generation.

3. Erbium Doped Fiber Laser Ring Cavity

Erbium-doped fiber laser cavity for the generation of Q-switched was drawn in Figure 1. The incorporation of SA inside an EDFL cavity was done by sandwiching a 1 mm x 1 mm thin film of MWCNTs in between two fiber-ferrule. The fiber-ferrule surface was applied with a sol-gel before the SA was incorporated to ensure less friction between the two surfaces of fiber-ferrule. A light was pumped via a 980 nm port of wavelength-division multiplexer (WDM) using a 980 nm laser diode pump. An EDF with an absorption coefficient of 23-27 dB/m, a numerical aperture (NA) of 0.16, a core diameter of 4 μm and a cladding diameter of 125 μm was used as a gain medium. The isolator was connected to the EDF thus allowing unidirectional light to propagate inside the laser setup. The light was then converged to the MWCNTs-PVA as SA for the purpose of Q-factor modulation inside the cavity. Then, the light was allowed to pass through an optical coupler with the signal division ratio of 90/10. A 90% of the signal cycles back into the cavity via 1550 nm port of WDM. Remaining 10% of the light was tapped out of the cavity for analysis purposes. The stability of the signal was evaluated using radio frequency spectrum analyzer (RFSA) (ANRITSU, MS2683A). Digital oscilloscope (GWINSTEK, GDS-3352) was used to measure the temporal performance of the pulses. Both oscilloscope and RFSA was connected to the 10% output from the cavity via a 1.2-GHz InGaAs photodetector. The spectral performance was recorded using optical spectrum analyzer (YOKOGAWA, AQ6370D) with the resolution of 0.03 nm. The output power was also recorded using an optical power meter.

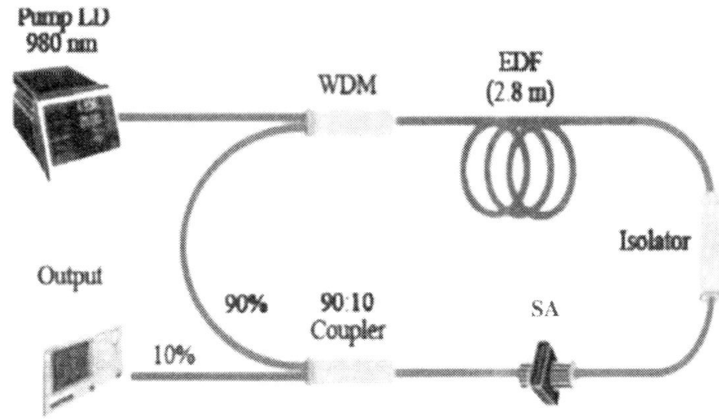

Figure 1. Experimental setup for Q-switched erbium-doped fiber laser using MWCNTs-PVA as saturable absorber.

4. Result and Discussion

The laser cavity starts to generate continuous wave (CW) at a pump power of 20 mW. Figure 2 (a) displayed a CW spectrum of laser cavity without the incorporation of SA at a pump power of 30 mW. The CW has a center wavelength of 1571.6 nm. As the SA was deployed into the laser cavity, the center wavelength shifted to 1559.4 nm. This is due to the increment of loss induced by the MWCNTs-PVA as SA in the EDFL cavity. The Q-switched EDFL possessed a 3dB spectral bandwidth of 1.4 nm as depicted in Figure 2 (b). The graph of output power against wavelength was obtained at a pump power of 59 mW. Oscilloscope trace of Q-switched at a pump power of 59 mW was shown in Figure 2 (c) with inset of two pulses envelope was also plotted. It was observed that the pulse generated has a full-width half maximum (FWHM) of 2.48 μs with a peak-to-peak distance of 7.3 μs. RF spectrum shown in Figure 2 (d) exhibit signal-to-noise ratio (SNR) of 34.12 dB which indicates stability of the pulses generated. The fundamental frequency measured was 137.1 kHz, which is in a good agreement with the oscilloscope trace.

IOP Conf. Series: Materials Science and Engineering **854** (2020) 012059 doi:10.1088/1757-899X/854/1/012059

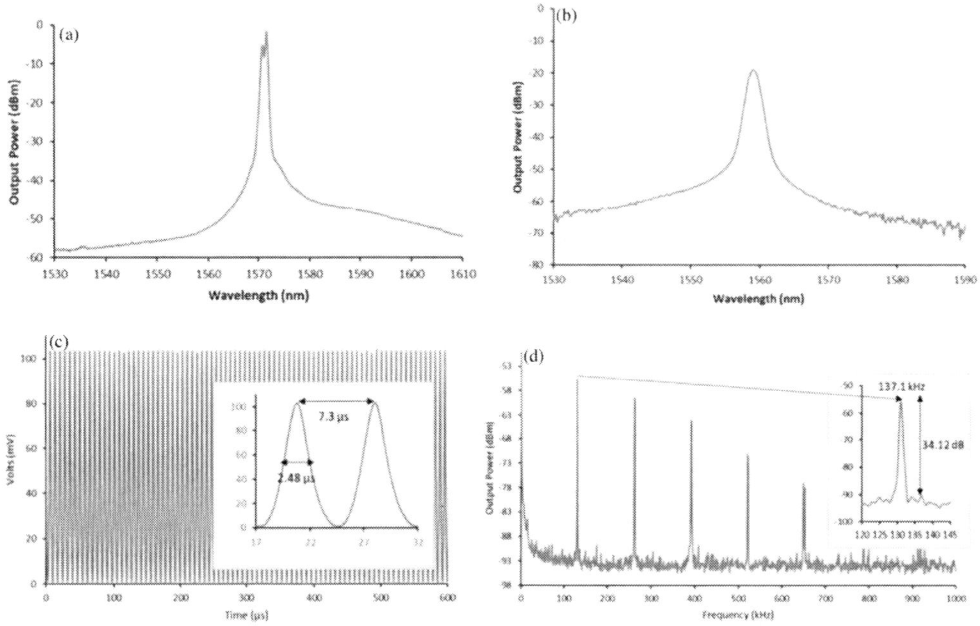

Figure 2. Spectral and temporal performances; (a) continuous-wave spectrum, (b) Q-switched spectrum, (c) typical oscilloscope trace with two pulses envelops, and (d) RF spectrum of the Q-switched laser at 59 mW pump power.

The Q-switched pulse was investigated over a variation of pump power. As the pump power raised from 30 to 63 mW, stable pulses observed on the digital oscilloscope. The pulses minimized and eventually disappeared as the pump power adjusted above 63 mW. As the pump power increased from 30 to 63 mW, repetition rate of the pulses increased from 52.85 to 151.9 kHz whereas pulse width decreased from 6.313 to 2.395 µs. The relationship between pulse width and repetition rate with pump power was shown in Figure 3 (a). Figure 3 (b) shows measured output power which was 17 to 52 µW plotted against the pump power. Whereas, calculated peak power was also plotted in the same figure as the pump power increased from 30 to 63 mW. The Q-switched generated exhibit a peak power of 50.95 to 142.94 µW.

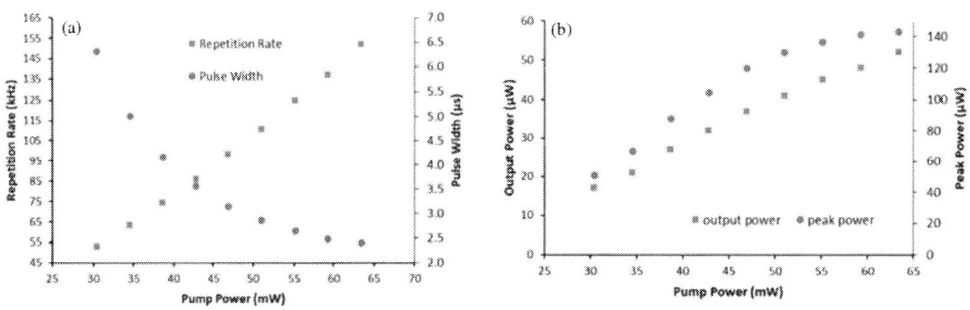

Figure 3. Q-switching performance; (a) repetition rate and pulse width against pump power and (b) output power and peak power against pump power.

5. Conclusion

We presented Q-switched erbium-doped fiber laser using MWCNTs-PVA incorporated into a laser cavity as a saturable absorber. The SA was prepared by embedding MWCNTs with PVA. A spectrum with an operating wavelength of 1559.4 nm captured on the optical spectrum analyzer. A stable Q-switched generated pulse width of 6.313 to 2.395 μs within the pump power of 30 to 63 mW. The repetition rate recorded within the same pump power was 52.85 to 151.9 kHz. SNR measured was 34.12 dB indicating stability of the laser. The maximum output power generated was 52 μW with the maximum peak power of 142.94 μW.

Acknowledgements

This work was financially supported by LPDP (Indonesia endowment fund for education). This work was also supported by Photonics Engineering Laboratory, Department of Electrical Engineering, University of Malaya and Universitas Kanjuruhan Malang of Indonesia.

References

[1] Lokman MQ, Yusoff S, Ahmad F, Zakaria R, Yahaya H, Shafie S, Rosnan RM, Harun SW. Deposition of silver nanoparticles on polyvinyl alcohol film using electron beam evaporation and its application as a passive saturable absorber. *Results Phys.* **11**:232-6.

[2] Xian T, Zhan L, Gao L, Zhang W, Zhang W. Passively Q-switched fiber lasers based on pure water as the saturable absorber. *Opt Lett.* **44**:863.

[3] Li J, Zhang ZL, Du L, Miao LL, Yi J, Huang B, Zou YH, Zhao CJ, Wen SC. Highly stable femtosecond pulse generation from a MXene Ti3C2Tx (T = F, O, or OH) mode-locked fiber laser. *Photonics Res.* **7(3)**:260-4.

[4] Li S, Yin Y, Lewis E, Garrell G, Wang P. A twelve-wavelength Thulium-doped fibre laser based on a microfibre coil resonator incorporating black phosphorus. *Opt Commun.* **437**:342-5.

[5] Liu HH, Chow KK, Yamashita S, Set SY. Carbon-nanotube-based passively Q-switched fiber laser for high energy pulse generation. *Opt Laser Technol.* **45**:713-6.

[6] Sun Z, Hasan T, Ferrari AC. Ultrafast lasers mode-locked by nanotubes and graphene. *Physica E.* **44(6)**:1082-91.

[7] Saeed K, Khan I. Carbon nanotubes–properties and applications: a review. *Carbon letters.* **14**:131-44.

[8] Kasim N, A-Masoodi AHH, Ahmad F, Munajat Y, Ahmad H, Harun SW. Q-switched ytterbium doped fiber laser using multi-walled carbon nanotubes saturable absorber. *Chin Opt Lett.* **12(3)**:4.

[9] Zuikafly SNF, Ahmad F, Ibrahim M, Harun SW. Passively Q-switched of EDFL employing multi-walled carbon nanotubes with diameter less than 8 nm as saturable absorber. *EPJ Web of Conferences.* **162**:01014.

[10] Ahmad H, Muhamad A, Sharbirin AS, Samion MZ, Ismail MF. Tunable Q-switched thulium-doped Fiber Laser using multiwall carbon nanotube and Fabry-Perot Etalon filter. *Opt Commun.* **383**:359-65.

[11] Ahmad F, Haris H, Nor RM, Zulkepely NR, Ahmad H, Harun SW. Passively Q-Switched EDFL Using a Multi-Walled Carbon Nanotube Polymer Composite Based on a Saturable Absorber. *Chin Phys Lett.* **31(3)**:034204.

[12] Chernysheva M, Mou C, Arif R, AlAraimi M, Rümmeli M, Turitsyn S, Rozhin A. High Power Q-Switched Thulium Doped Fibre Laser using Carbon Nanotube Polymer Composite Saturable Absorber. *Scientific Reports.* **6**:24220.

[13] Zuikafly SNF, Ahmad F, Shafie S, Harun SW. APPLICATION OF MULTI-WALLED CARBON NANOTUBES-BISMUTH (III) TELLURIDE (MWCNTS-Bi2Te3) BASED PASSIVE SATURABLE ABSORBER IN PASSIVELY Q-SWITCHED FIBER LASER. *SOLID STATE SCIENCE AND TECHNOLOGY.* **(2)**:1-8.

[14] Zuikafly SNF, Ahmad F, Ibrahim M, Abdul Latiff A, Harun SW. Dual-wavelength passively Q-

IOP Conf. Series: Materials Science and Engineering **854** (2020) 012059 doi:10.1088/1757-899X/854/1/012059

switched Erbium-doped fiber laser with MWCNTs slurry as saturable absorber. *Photonics Letters of Poland.* **8**:98.

IOP Conf. Series: Materials Science and Engineering **854** (2020) 012060 doi:10.1088/1757-899X/854/1/012060

The simulation of evacuation from multistorey building using NetLogo

D P Lestari[1,*], A Sabri[2], T Handhika[1], Murni[1], I Sari[1], and A Fahrurozi[1]

[1]Computational Mathematics Study Center, Gunadarma University, 100 Margonda Raya Street, Pondok Cina, Depok, West Java, 16424, Indonesia
[2]Department of Informatics, Gunadarma University, 100 Margonda Raya Street, Pondok Cina, Depok, West Java, 16424, Indonesia

E-mail: dewi_putrie@staff.gunadarma.ac.id

Abstract. The occupants of a multistorey building are very vulnerable when disaster happens (e.g. fire, earthquake, etc.). In the worst case, the only way to escape is through the emergency stairways, in which they scramble with the occupants coming from the higher floors. This situation often creates bottlenecks along the way, which in turn extends the evacuation time. In addition, the panic situation increases the risks of accidents, causing injuries or even death. In this research, we use NetLogo to simulate the evacuation process from a simplified multistorey building based on some parameters, which are the placement of the emergency stairways, the number of occupants in each floor, the number of stories, and the height of the floor. We set simulation scenarios to analyze the relation between the number of stories with evacuation time in each configuration. In our experiment context, we conclude that when the time-limit applies, the total evacuation through stairways is not effective for buildings having more than three storeys.

1. Introduction

The construction of high-rise buildings has increased exponentially in the cities due to scarcity of space for population growth [1]. Critical attention for high-rise buildings management and the government is to provide a safer building evacuation strategy and improve the safety aspects that affect the decision-making of people, staff and rescue operators [2]. In Jakarta for instance, there are about 160 buildings with height more than 100 meters. Several earthquakes there in recent years realize the importance of the very short-time total evacuation strategies and building designs.

In general, the occupants of high-rise buildings are always vulnerable to disasters. In case of fire or earthquake, they only have very limited time to escape in a very limited (and probably the only) way: the stairways. Several evacuation strategies were proposed, some are using mathematical analysis (for instance [3, 4]), and others using simulation (for instance [5, 6, 7]).

In this paper, we approach the problem from the building configurations. We run agent-based simulations to get initial insights about how the stairway placements affect the evacuation time, based on our pre-defined building configuration and initial conditions. We apply behaviors on the agents that cause three phenomenons: arching, clogging, and follow-the-crowd. Our scenarios are to make a total evacuation from the buildings and compare the results between those configurations. We choose NetLogo to simulate our scenarios, considering it is very adaptable to various configurations and more convenience in programming the agent behaviors.

2. The model

The parameters of our models are the number of storeys, the number of agents (occupants), and the floor height. We apply our model on two building classes, one is with a single stairway and the other with double stairways. Each of these classes consists of two different stairways placement: at the center of one side or at one corner of the building. We call these configurations after the building class followed by the placement of the stairway: single-center (S1), single-corner (S2), double-center (D1), and double-corner (D2).

Each floor consists of a single square room with no obstacles (no furniture, no partition, good visibility, etc). Each floor is divided into $(n+1) \times (n+1)$ grids, where n is even so that the stairway position can align with the middle grids. Each grid on the floor represents an area of 0.5×0.5 square meters. Initially, all agents are distributed uniformly on all floors, and every grid can only contain at most one agent. See Figure 1 for an instance of the floor.

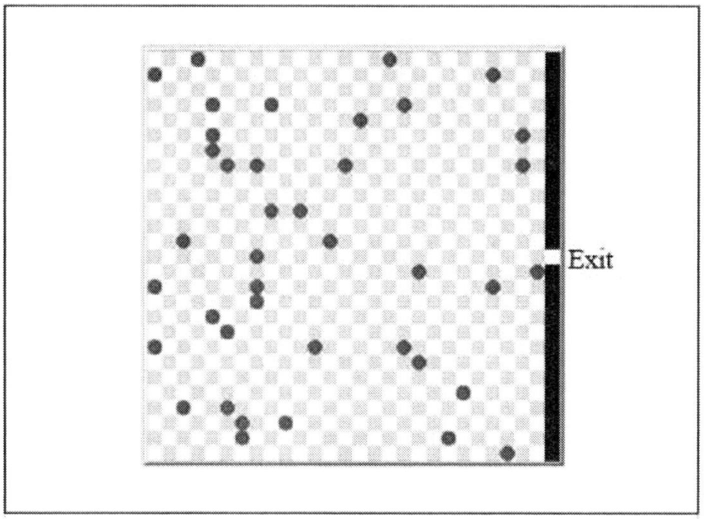

Figure 1. A floor of size 27 × 27 grids occupied by 40 agents.

The wide of the stairways is enough for only one agent (we consider wider stairways give faster evacuation time). We assume that it needs six steps to go down 1 meter (each step is about 16.6667 cm of height). A 1-meter-tall stairway is represented by vertical 2 rows × 3 columns grids. If the agent moves to the next grid, it means he goes down one step. Descending 1 meter is simulated by an agent entering the first grid on the first row, then move horizontally to the two next grids, then go down perpendicularly to the grids on the second row and move in the opposite direction until the last grid. See Figure 2 for illustration.

For experiments purpose, we set all the storeys in the building to 4 meters in height. Figure 3 shows all configurations of 3-storeys building containing agents distributed uniformly on the floors, where each floor consists of 19 × 19 grids.

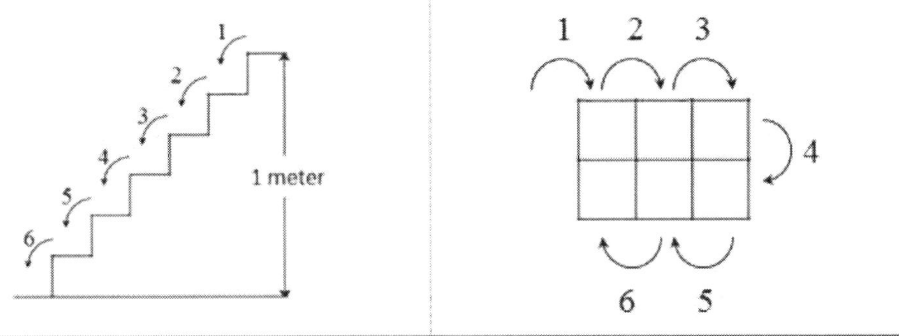

Figure 2. Descending 1 meter through the stairway steps (left) and its representation in the model (right).

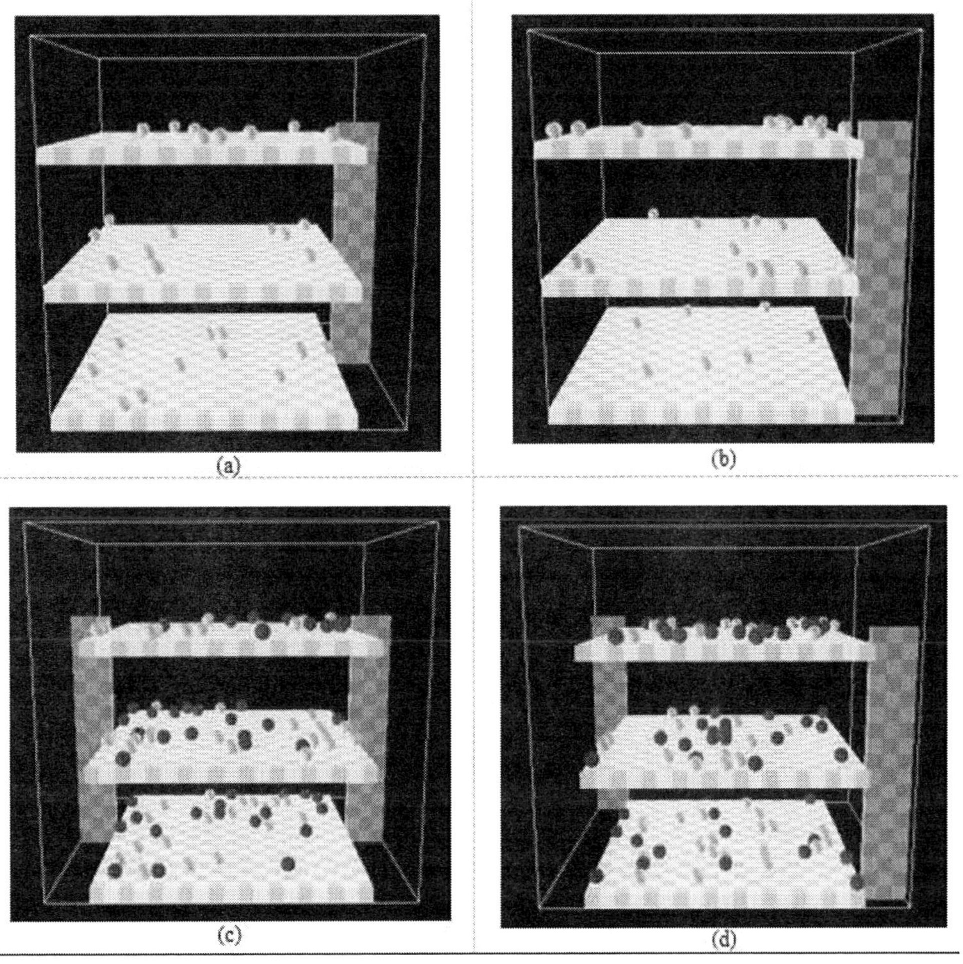

Figure 3. The configurations of simplified 3-storeys building: (a) single-center (S1); (b) single-corner (S2); (c) double-center (D1); (d) double-corner (D2).

Every agent can move to one of eight adjacent grids, as long as it is empty. Once the emergency situation is declared, all agents move towards their pre-chosen exit door on their respective floors. To reach the exit, they move to the adjacent grid that falls in their direction. However, if this grid is occupied, then they choose randomly an available grid in the range from 9 o'clock to 3 o'clock relative to their current direction (see Figure 4). In case there is no available grid to move to, the agent stays freeze. This behavior implies the arching phenomenon. To simulate a follow-the-crowd phenomenon, we set that if an agent is surrounded with 4 or more agents going to the other exit door, then this agent switches his direction.

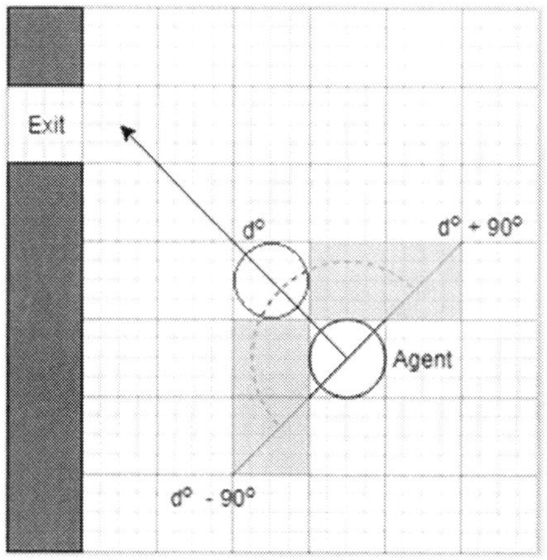

Figure 4. The alternative grids to move to if the agent cannot move forward (shown by the dark grids).

The agent in front of the exit door, let say agent a, has to struggle against those in his left and right (if any) to pass the door. These two agents push agents away from the exit door. On the other hand, the agent behind (if any) pushes the agent towards the exit door. If the cumulating power of the left and right agents exceeds that of a and of his behind, then a cannot move. The struggle is simulated by increasing the power of a randomly in the next timestamp until it can move forward. This behavior implies a clogging phenomenon in the simulation.

3. Implementation and analysis

We set the simulation interface as shown in Figure 5. There we can set the parameters of the model: the number of occupants, the number of storeys, the height of the floor, and the stairway placement (center or corner). If the "Setup" button is clicked, then the initial situation is set and displayed. The agents are distributed uniformly on all floors and their pre-chosen exit-door (uniform-randomly assigned by the program) is shown by the color of the agents (in this case, blue for agents preferring the left door and yellow for those preferring the right door). Clicking the "Go" button declares the emergency situation and all agents, controlled by their behavior, move towards their respective exit doors.

During the evacuation process, the simulation shows arching and clogging (Figure 6) and follow-the-crowd phenomenons (Figure 7). The clogging is indicated by the agent with black color. This agent cannot move forward since his forward power is not enough to overcome the pressure from his left and right.

An example of evacuation simulation is illustrated in Figure 8. The simulation is set for 3 storeys building holding 300 occupants with 42 × 42 grids floor of height 4 meters and double-corner building configuration. In the earlier ticks (the time stamp in NetLogo), chaotic moments happen when occupants block each other due to their movement to opposite directions. After 20 ticks the moment's end, the occupants begin to assemble in arch formation in front of exit doors. Agents in red indicate that they stay freeze. Finally, the building is clear after 293 ticks.

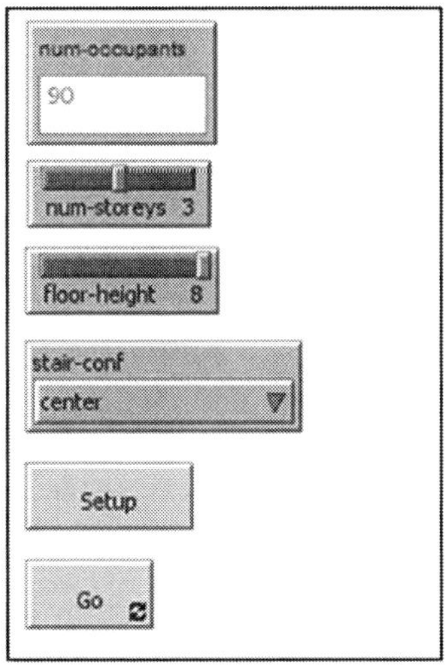

Figure 5. The interface.

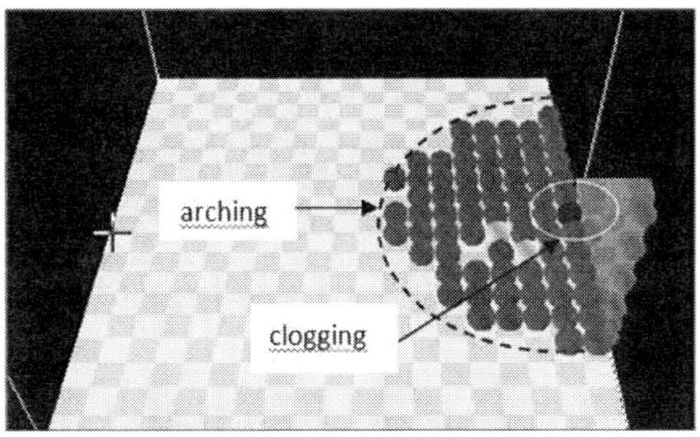

Figure 6. Arching and clogging phenomenon.

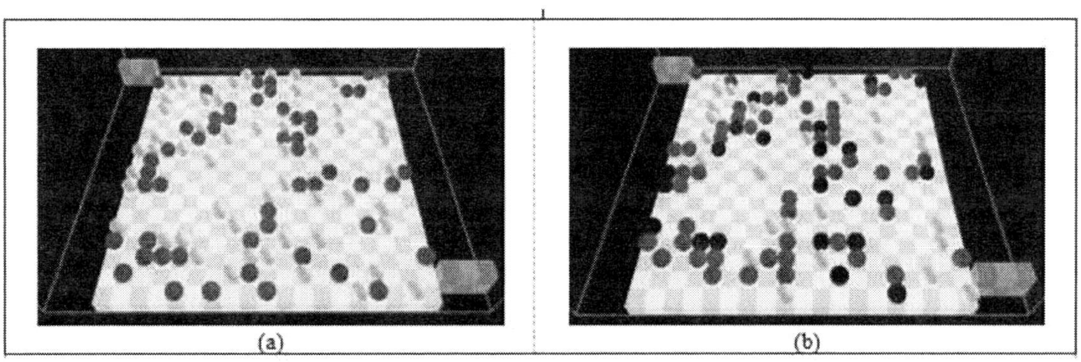

Figure 7. The follow-the-crowd phenomenon, the black agents on the (b) are those who change their courses compared to the (a): (a) previous state; (b) current state.

We apply some experiments on our model to get some insights about the relation between the parameters with the time required to clear the building. The experiment scenarios are to simulate evacuation from 2, 3, 4, and 5 storeys building holding 200, 300, 400, and 500 occupants, respectively, which uniformly distributed over 36 × 36 grids in all floors. Intuitively, double stairways configuration always gives faster evacuation time than the single one (and our experiments show this tendency). Therefore, we compare the evacuation time (the number of ticks needed) between stairways placement in the same building class. The results show that the relation between the number of storeys and evacuation time tends to be linear in each configuration. The double configurations (D1 and D2) give about two times faster than the single one (S1 and S2). On average the evacuation time is not significantly different for the higher number of storeys (≥ 3 for single and ≥ 4 for double stairways). See Table 1 and Figure 9 for the results obtained.

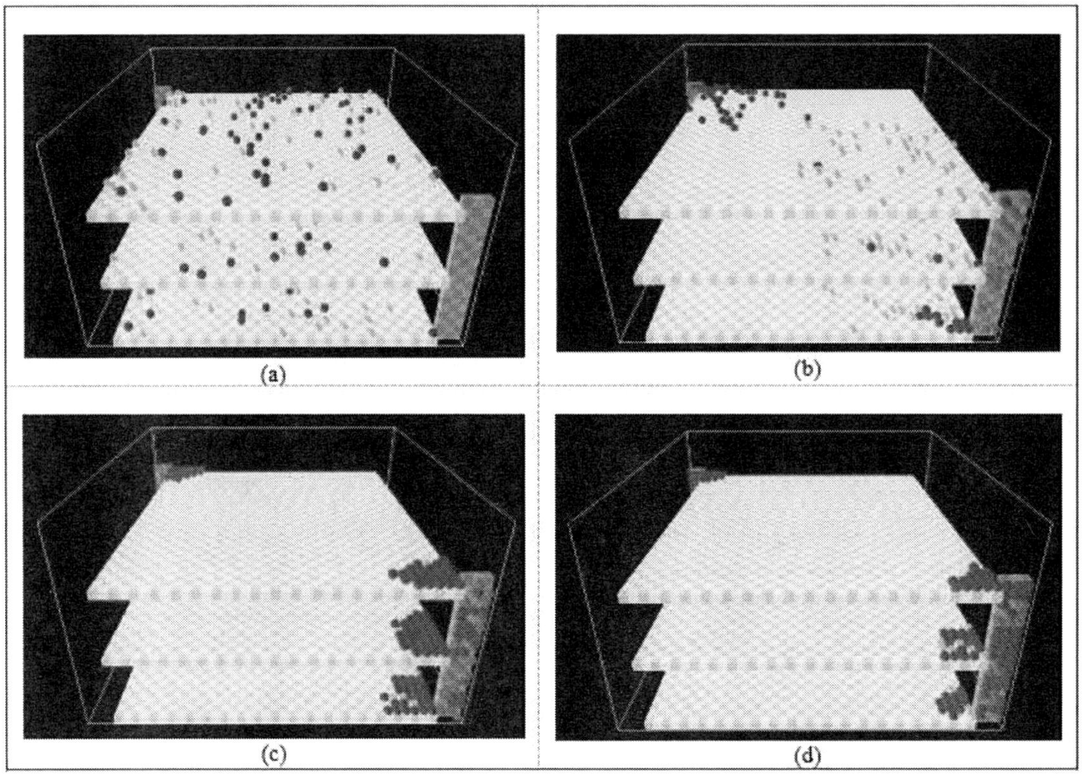

Figure 8. The simulation of evacuation from 3 storeys building containing 300 occupants: (a) initial; (b) after 20 ticks; (c) after 50 ticks; (d) after 100 ticks.

Table 1. The evacuation time (in ticks) according to the number of storeys in each configuration.

	Single-Center (S1)	Single-Corner (S2)	Double-Center (D1)	Double-Corner (D2)
2	493	393	252	204
3	622	586	328	302
4	823	788	408	403
5	1019	989	507	503

We observe as well that in the single-center and double-center experiments, the bottlenecks develop slowly on the stairway due to the flow resistance caused by the clogging phenomenon in each floor. However, the more the storeys, the more likely bottlenecks to happen. In the single-corner, the clogging phenomenon is less frequent, since the pressure received by the agent in front of the exit comes only from one side. This less-restricted flow develops bottlenecks more often on the stairway.

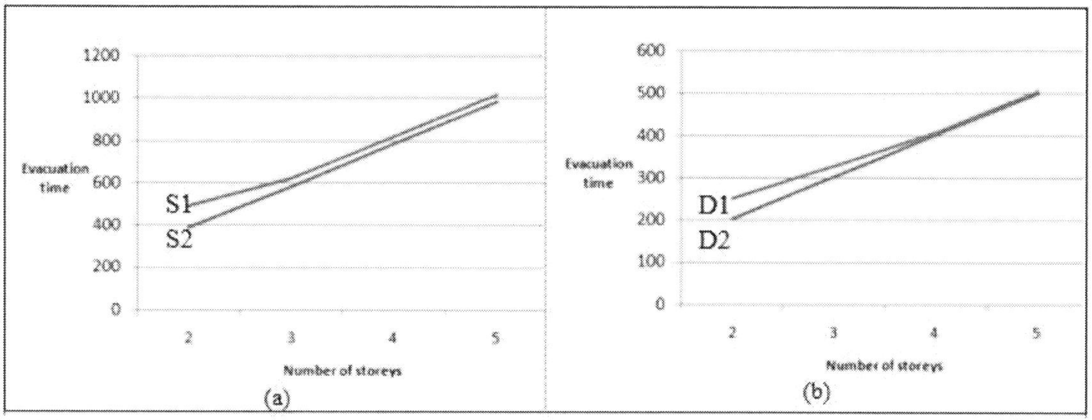

Figure 9. The graph of evacuation time to the number of storeys for each model: (a) single stairway; (b) double stairways.

4. Conclusion and future works

Based on the experiment results, we conclude that our simulation scenarios give significant difference evacuation time for the building having maximum of three storeys. In addition, if the time limit is enabled, then potentially many occupants do not survive when the time is up. The evacuation from a high-rise building is only effective up to the third storey, even when there are no obstacles. We suggest doing the real-world experiments to validate the model. For further developments, we suggest adding more behavior and various configurations to the model.

Acknowledgments

This research was funded by National Competitive Research Grant 2019 in Penelitian Terapan (PT) scheme, Direktorat Riset dan Pengabdian Masyarakat, Direktorat Jenderal Penguatan Riset dan Pengembangan Kementerian Riset, Teknologi, dan Pendidikan Tinggi (DRPM KEMENRISTEKDIKTI) Indonesia to the Gunadarma University No. 15/AKM/PNT/2019.

References

[1] Sheeba A A and Jayaparvathy R 2019 Performance modeling of an intelligent emergency evacuation system in buildings on accidental fire occurrence *Safety Science* **112** 196-205

[2] Aleksandrov M, Cheng C, Rajabifard A and Kalantari M 2019 Modeling and finding optimal evacuation strategy for tall buildings *Safety Science* **115** 247-255

[3] Liu C, Mao Z and Fu Z 2016 Emergency evacuation model and algorithm in the building with several exits *Procedia Engineering* **135** 12-18

[4] Tissera P C, Printista M and Errecalde M L 2007 Evacuation simulations using cellular automata *JCS&T* **7** 14-20

[5] Milla A G, Borges F, Suppi R and Luque E 2015 Crowd evacuations SaaS: an ABM approach *Procedia Computer Science* **51** 473-482

[6] Shi J, Ren A and Chen C 2009 Agent-based evacuation model of large public buildings under fire conditions *Automation in Construction* **18** 338-347

[7] Almeida J, Kokkinogenis Z and Rossetti R 2012 NetLogo implementation of an evacuation scenario *In Fourth Workshop on Intelligent Systems and Applications (WISA)* (Madrid Spain: IEEE Publisher)

IOP Conf. Series: Materials Science and Engineering **854** (2020) 012061 doi:10.1088/1757-899X/854/1/012061

Travel Time Estimation for Destination In Bali Using kNN-Regression Method with Tensorflow

Murni, R Kosasih, A Fahrurozi*, T Handhika, I Sari, D P Lestari

Computational Mathematics Study Center, Gunadarma University, 100 Margonda Raya Street, Pondok Cina, Depok, West Java, 16424, Indonesia

*E-mail: achmad_fahrurozi@staff.gunadarma.ac.id

Abstract. On a tour activity, travel time estimation is needed so that the travel itinerary goes according to the plan. Travel time estimation is very important so we can estimate the time needed to arrive at the destinations in the travel itinerary. Therefore we need a method that can estimate travel time from one place to another. In this study, we propose the k-Nearest Neighbors Regression (kNN-Regression) method with Tensorflow to construct an estimation model. The proposed number of features in our estimation model is 8 features, i.e. zone information, time information, day information, weather information, temperature information, wind speed information, humidity information, and precipitation information. The data obtained from travel information from Ngurah Rai airport to Kuta Beach using GPS and weather information using weather application in real-time. We divide our data into two groups: a historical group consisting of 177 data and a testing group consisting of 51 data. In the testing stage, kNN-Regression will find the historical data closest to the testing data, so that the estimation value of the travel time of some testing data is not much different from the value of the nearest historical data. As a result, our proposed model gives the Mean Absolute Error (MAE) of 2.196078, Root Mean Square Error (RMSE) of 2.977036294 and accuracy rate 88.1819%.

Keyword: KNN-Regression, Features, MAE, RMSE, Accuracy rate

1. Introduction

Travel time estimation is something that is suggested to be done before going on a trip. One of the problems in travel time estimation is traffic jams. If a traffic jam occurs it will cause a longer travel time. So that we need a method that can estimate travel time from one place to another place. Travel time estimation is very important so that we can estimate the time needed to arrive at the destination. This estimation can also be used by travel agents such as travel agents to be able to estimate when the departure time and which route will be chosen. So that the travel agent can estimate how many destinations you want to visit in one day.

In planning a trip, there are several features that can affect travel time. For example, are time information. Departure time will affect travel time, for example, is the number of people who go to work in the morning by using a vehicle will cause traffic jams, or the number of people who want to go on vacation on a holiday will also cause traffic jams. The other features are weather conditions such as rain can cause the vehicle rate to be slower because the road becomes more slippery so that it can cause a traffic jam. The next feature is the temperature, if the temperature in the area is very extreme, it will cause people to do not want to leave the house so that it can cause traffic to be smooth.

Several studies have been carried out such as D'Angelo et al. Predicting travel time using a nonlinear time series model and arguing that models that only use speed data are superior to multivariable models that use speed, occupancy, and volume of data obtained from loop detectors [1]. Next Kwon et al predicted travel time on the freeway by using flow and occupancy data from a single loop detector and historical travel time information [2]. They use linear regression and tree-based methods to show that the current traffic data collected from the loop detector is a good predictor for the near future (up to 20 minutes). Zheng Wang uses a regression problem to predict vehicle travel time based on floating car data by considering account factors such as spatial information, temporal information, traffic information and personal information [3]. However, this study only considers 4 features. Therefore in this research, we will find other features that can affect travel time. We proposed a model with 8 features such as zone information, time information, day information, weather information, temperature information, wind speed information, humidity information, and precipitation (rainfall) information by using KNN regression method with TensorFlow.

2. Literature Review

This section describes the theory of travel time estimation using KNN regression with Tensorflow.

2.1. K Nearest Neighbor (KNN)

K Nearest Neighbor (KNN) algorithm is a method of classifying objects based on their similarity to other objects. The KNN is the classification technique without having to know about the distribution of the data [4-7]. In this method, each sample should be classified similarly to the surrounding samples. Therefore, if the classification of a sample is unknown, then it could be estimated by considering the classification of its nearest neighbor samples [7]. The Nearest Neighbor rule (NN) is the simplest form of KNN when K = 1 [7].

For example, suppose there is one object that wants to be classified using training data. The first stage is to calculate all distances from the object to all training data. The distance with the smallest value corresponds to the sample in the training set closest to the unknown data. Therefore, the unknown sample may be classified based on the classification of this nearest neighbor [6-10].

The stages of KNN algorithm can be seen in Figure 1.

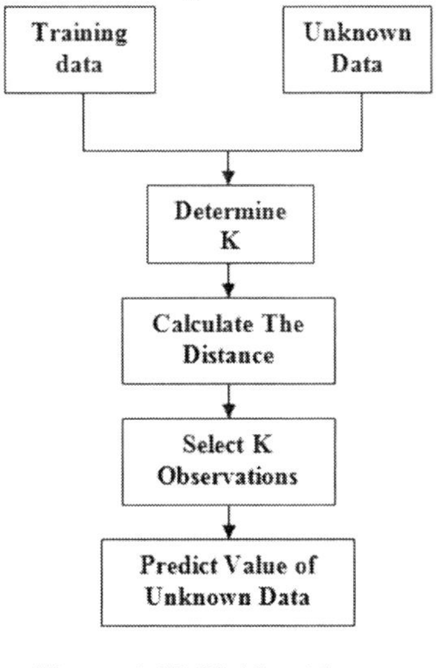

Figure 1. KNN Algorithm.

In Figure 1, we determine the value of K and calculate the distance from the unknown object to all training data. After that, select K observations in the training data closest to the unknown data. The last step is to predict the value of unknown data by using the value of the K-nearest neighbor object that was selected in the previous step.

2.2. Tensorflow

TensorFlow is a machine learning system that operates at large scale and in heterogeneous environments. Tensorflow is a library from a language of python that used for numerical computation that makes machine learning faster and easier. TensorFlow uses dataflow graphs to represent computation, shared state, and the operations that mutate that state. Each node in the graph represents a mathematical operation, and each connection or edge between nodes is a multidimensional data array or *tensor*.

TensorFlow can train and run deep neural networks for handwritten digit classification, image recognition, word embeddings, recurrent neural networks, sequence-to-sequence models for machine translation, natural language processing, and PDE (partial differential equation) based simulations. Best of all, TensorFlow supports production prediction at scale, with the same models used for training.

3. Research Methodology

This section describes the stages of the method of research that has been done. The stages of research can be seen in Figure 2.

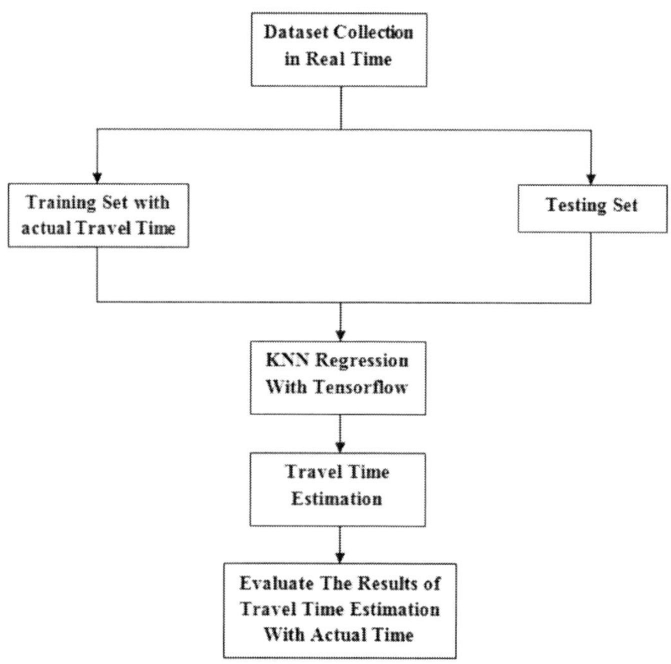

Figure 2. General frameworks for this research.

In Figure 2, the first step in estimating travel time is the collection of datasets in real-time. The data is divided into 2 namely training data (historical data) with actual travel time and testing data used to estimate travel time. To estimate travel time we use the KNN Regression method with tensorflow. Next, we evaluate the estimation results by using Root Mean Squared Error (RMSE), Mean Absolute Error (MAE) and Level of Accuracy (Acc).

3.1. Dataset Collection

In this study, travel information is taken through GPS and the weather information dataset is taken through weather application in real-time. The sample of travel information from GPS and weather information can be seen in Figure 3 and Figure 4.

Figure 3. Travel information from GPS in real-time

Figure 4. Weather information in real-time

The dataset is divided into two namely 177 training data (historical data) and 51 testing data. The training data was collected manually through the application from 25 July 2019 until 1 August 2019 in the Bali region especially from Ngurah Rai Airport to Kuta beach tourist attraction. After that testing data was collected manually through the application from 9 August 2019 until 12 August 2019. Data collected is

- Zone Information
  Zone information is one of the features that can be used to estimate travel time. In this study, there are 3 zones that can be passed from the Airport to Kuta Beach. For example, different zones will produce different travel times.
- Time of The Day Information
  Time information can affect travel time estimation. That is because at certain times there is a pattern of traffic. For example in the morning, many people who do activities so that traffic jams occur
- Day of The Week Information
  Day information can affect travel times estimation. For example, on weekends many people go on vacation to attractions, causing traffic jams
- Weather Conditions
  Weather conditions can also affect travel time estimation. For example, if extreme weather occurs, such as extreme heat, it will cause people to stay at home so there is no traffic jam
- Temperature Information
  Temperature conditions can affect travel time estimation. Such extreme temperatures tend to keep people inside the house so that traffic congestion decreases and causes travel times to be faster than usual
- Wind Speed Information
  Wind speed can affect vehicle speed. If the wind speed is very high, it will cause the vehicle speed to decrease, causing traffic jams
- Humidity Information
  The higher the humidity the air will cause the level of pollution is higher so as to make people stay in the house so that traffic density decreases and causes travel times to be faster than usual

- Precipitation Information
  if the precipitation (rainfall) is very high it will cause people to stay inside the house so that traffic density decreases and causes travel times to be faster than usual.

3.2. KNN Regression With Tensorflow
In addition to classification, K Nearest Neighbors (KNN) can also be applied for regression. The stages of KNN regression can be seen in Figure 5.

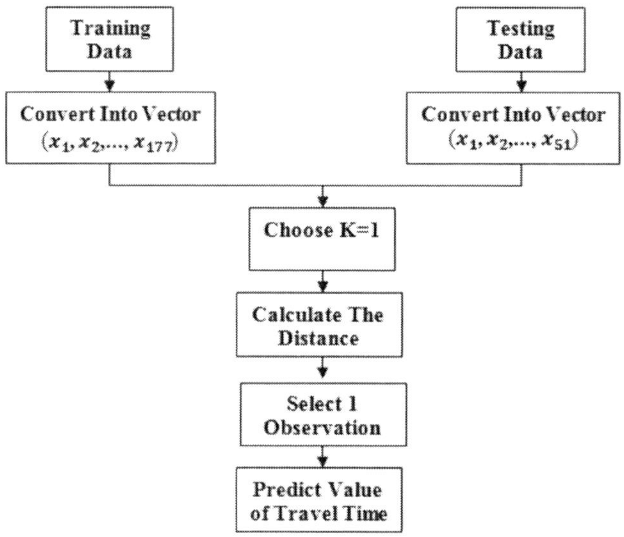

Figure 5. KNN Regression Algorithm.

In Figure 5, we use 8 features to predict travel time from Ngurah Rai Airport to Kuta Beach. Furthermore, each data that has 8 features is assumed to be a vector x_i with dimension 8. In this study, we use 177 training data (Historical Data). Therefore there are 177 vectors $(\boldsymbol{x_1}, \boldsymbol{x_2},..., \boldsymbol{x_{177}})$. Each vector represents a point in 8 dimension space.

The next stage is to predict travel time by using 51 testing data. After that, we use the KNN Regression to predict travel time. In this study, the K used was 1, in other words, 1 was chosen closest to the training data. Furthermore, calculating the distance from every testing data to all training data using Manhattan distance. The Formula of Manhattan Distance can be seen in equation (1).

$$MD = \sum_{i=1}^{k} |a_i - b_i| \quad (1)$$

Where a_i is testing data and b_i is training data.

KNN will find the training data (historical data) closest to the testing data so that the predicted travel time is not much different from the value in the closest historical data.

3.3. Evaluate
After estimating the travel time by using the KNN regression, the next step, we evaluate the results of travel time estimation using Root Mean Squared Error (RMSE), Mean Absolute Error (MAE) and Level of Accuracy (Acc).

The formula of Root Mean Squared Error (RMSE) [7] can be seen in equation (2).

$$RMSE = \sqrt{\frac{\sum_{i=1}^{N}(y_i - \hat{y}_i)^2}{N}} \quad (2)$$

and the formula of Mean Absolute Error (MAE) [11] can be seen in equation (3).

$$MAE = \frac{\sum_{i=1}^{N}|y_i-\hat{y}_i|}{N} \tag{3}$$

and the formula of Accuracy (Acc)[11] can be seen in equation (4).

$$Accuracy = \left(1 - \frac{\sum_{i=1}^{N}\left|\frac{y_i-\hat{y}_i}{y_i}\right|}{N}\right) \times 100\% \tag{4}$$

Where y_i is actual travel time, $\hat{y}_i$ is the result of travel time estimation and N is the number of dataset testing

4. Experimental Result and Discussion

In this study, we use two models to estimate travel time. The first model is a model with 5 features (model 1) and the second model is a proposed model with 8 features.
Model 1 can be seen in equation (5)

$$y = \sum_{i=1}^{5} x_i + \varepsilon \tag{5}$$

Where x_1 = Zone information, x_2 = Time of The Day, x_3 = Day of The Week, x_4 = Weather Conditions, and x_5 = Temperature Information
Futhermore, proposed model can be seen in equation (6)

$$y = \sum_{i=1}^{8} x_i + \varepsilon \tag{6}$$

Where x_1 = Zone information, x_2 = Time of The Day, x_3 = Day of The Week, x_4 = Weather Conditions, x_5 = Temperature Information, x_6 = wind speed information, x_7 = Humidity Information, and x_8 = precipitation information

The next stage is to find the predicted value of two models by using the KNN regression method. After that, we evaluate two models by calculating the value of MAE, RMSE, and Accuracy which can be seen in Table 1.

Table 1. Evaluate Model

	MAE	RMSE	Accuracy
Model 1	2,254902	3,0901488	86,2103 %
Proposed Model	2.196078	2,977036294	88,1819 %.

In Table 1, there are two models being compared, namely a Model 1 with 5 features and a Proposed Model with 8 features. From table 1 it can be seen that the value of Mean Absolute Error (MAE), Root Mean Squared Error (RMSE) of the proposed model is smaller than the value obtained using a model with 5 features and for the level of accuracy of the proposed model has a higher value compared to the accuracy of model 1 with 5 features. This shows that the Proposed Model is better than Model 1 with 5 features.

5. Conclusion

Travel time estimation is needed by someone or a travel agent who wants to travel. Therefore we need a method that can estimate travel time from one place to another. In this study to estimate travel time, we propose using the K Nearest Neighbors (KNN) regression method by considering features such as zone information, time information, day information, weather information, temperature information, wind speed information, Humidity information, and precipitation information. Data retrieval of these features is obtained by using GPS data and weather information in real-time through the weather application. The data is divided into two namely 177 training data (historical data) and 51 testing data. In testing, KNN will find the historical data closest to the testing data so that the predicted travel time is not much different from the value in the closest historical data. In this study, we use two models namely a model with 5 features and a proposed model with 8 features. From the evaluation results using Mean Absolute Error (MAE), Root Mean Squared Error (RMSE) and level of Accuracy, it can be seen that the proposed model is better than the model 1 with 5 features. For the next experimental research, we will add a sample of training to get higher accuracy.

Acknowledgment
This research was supported by National Competitive Research Grant 2019 in Penelitian Terapan (PT) scheme, DRPM KEMENRISTEKDIKTI Indonesia to the Gunadarma University.

Reference

[1] D'Angelo M P, Al-Deek H M, and Wang M C 1999 Travel-Time Predictiion for Freeway Corridors. *J. Transp. Res. Board* **1676** 184–191.
[2] Kwon J Coifman B, and Bickel P 2000 Day-to-Day Travel-Time Trends and Travel-Time Prediction from Loop-Detector Data *J. Transp. Res. Board* **1717** 120–9.
[3] Wang Z, Fu K and Ye J 2018 Learning to Estimate the Travel Time *Proc. Int. Conf. on Knowledge Discovery & Data Mining (London)* vol 24 (New York : USA ACM) p 858-66.
[4] Devroye L 1981 On the equality of Cover and Hart in nearest neighbor discrimination *IEEE Trans. Pattern Anal. Mach. Intell.* **3** 75-8.
[5] Devroye L, Gyorfi L, Krzyzak A and Lugosi G 1994 On the strong universal consistency of nearest neighbor regression function estimates *Ann. Statist* **22** 1371–85.
[6] Devroye L and Wagner T J 1982 Nearest neighbor methods in discrimination, In Classification Pattern Recognition and Reduction of Dimensionality *Handbook of Statistics* vol 2 (Amsterdam : Elsevier) p 193–7.
[7] Imandoust S B And Bolandraftar M 2013 Application of K-Nearest Neighbor (KNN) Approach for Predicting Economic Events: Theoretical Background *J. Eng. Res. and App.* 3 (**5**), 605-10
[8] Domeniconi C, Peng J and Gunopulos D 2002 Locally adaptive metric nearest-neighbor classification *IEEE Trans. Pattern Anal. Mach. Intell.* 24(**9**) 1281–5.
[9] Dudani S A 1976 The distance-weighted k-nearest neighbor rule *IEEE Trans. Sys. Man. Cyber.* **6** 325-7.
[10] Enas G G and Choi S C 1986 Choice the smoothing parameter and efficiency of K-Nearest Neighbor classification *Comp. Maths. Apps.* 12(**2**) 235-44.
[11] Kumar B A, Jairam R, Arkatkar S S and Vanajakshi L 2019 Real time bus travel time prediction using k-NN classifier *Int. J. Transp. Res.* 11(**7**) 362-72.

IOP Conf. Series: Materials Science and Engineering **854** (2020) 012062 doi:10.1088/1757-899X/854/1/012062

A Study on Nitriding Alloy Steel in INTI Plasma Focus Machine

A Singh[1], R Jeyagopi[1], T O Teh[1], S Manmeshpal[1], S H Saw[2,3] and S Lee[1,3,4]

[1] Faculty of Engineering and Quantity Surveying, INTI International University, Persiaran Perdana BBN, Putra Nilai,71800 Nilai, Negeri Sembilan, Malaysia
[2] First City University College, No.1, Persiaran Bukit Utama, Bandar Utama, 47800 Petaling Jaya, Malaysia
[3]Institute for Plasma Focus Studies, Chadstone, VIC3148, Australia
[4]University of Malaya, Kuala Lumpur, Malaysia

Email: arwinders.jigiris@newinti.edu.my

Abstract. The INTI Plasma Focus machine was used to irradiate AISI 304 stainless steel samples with nitrogen ion beams to improve the hardness of the sample surface. The increase in hardness was plotted as a function of pressure at a constant 12 kV operation. Optimum hardness was found at 1.5 Torr using the Vickers Hardness test. Computation with the Lee codes produces a plot of ion beam energy versus pressure. The plot shows that the optimum ion beam energy of 7.3 Joules occurs at 1.5 Torr. Thus we have correlated the optimum hardness of the treated sample surface to the optimum ion beam energy.

1. Introduction

Nitriding is a process in which nitrogen ions are diffused into the surface of a metal to create surface hardening. The surface hardening is important because it is directly proportional to the stress at the location of an imposed strain to prevent any deformation on the surface. Nitriding can be done via gas nitriding, salt bath nitriding and plasma nitriding [1]. In a research article exploring nitriding on low carbon steel material by using the dense plasma focus machine, the researcher [2] observed that the distance of the sample from the anode has an inverse relationship with the hardness and the further the distance of the sample from the anode, the less was the hardness obtained. Another researcher, Al-Hawat et.al obtained results that showed that the surface hardness increased with increasing shot number and decreased with increasing distance from the anode [3].

In this study, the surface hardening of AISI 304 alloy steel (T300 Series Alloy steel) also known as chromium nickel alloy steel was carried out. This study has two main objectives which are as follows:

- To study the effect of operating the plasma focus in different nitrogen gas pressure on surface hardness of AISI 304 alloy samples treated in the plasma focus
- To study the effect of ion beam and its energy using Lee codes [4] on the surface hardness of AISI 304 alloy

2. Methodology Used

In this research, the following methodology is used. First, the sample is prepared. It is then nitrided by exposing to the plasma focus radiation, and finally, the results are analyzed. The flow chart is as shown in Figure 1.

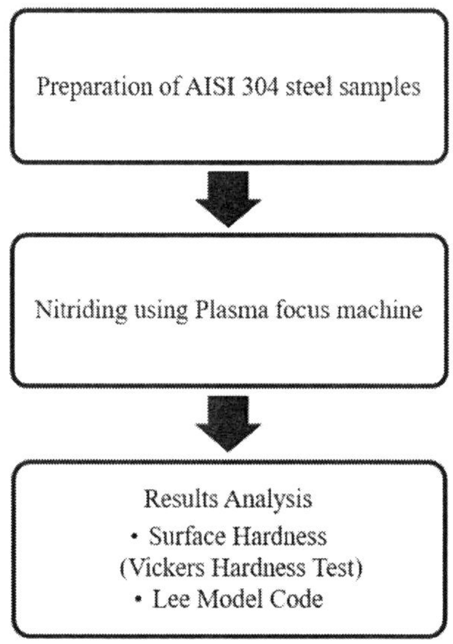

Figure 1: General description of the whole process

To prepare the sample, the steel bar (AISI 304 alloy) was cut into the small square sample (25mm × 25mm × T 8 mm) that fitted in the sample holder (placed 40 mm from the anode tip [2]). The samples were then milled to remove the outer layer (remove the oxide layer). The samples were then gradually heated from room temperature to the required temperature (over a period of 3 hours) and then kept constant at 1060^{0}C for 1 hour. The samples were then allowed to cool down for 24 hours [5]. The samples were then polished and grinded to remove any oxide on the surface layer of the steel [6]. The hardness of the steel was measured using a Vickers Hardness test. This process is shown in Figure 2.

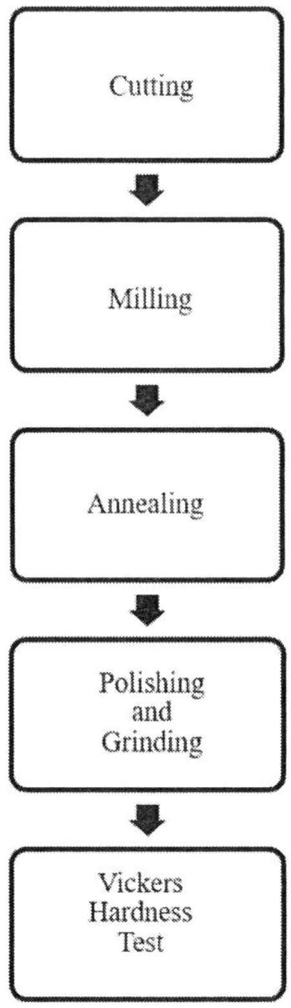

Figure 2: The process of preparing the samples

The samples were then placed into the Plasma focus machine. A Franklin vacuum pump was used to evacuate the 6 litre INTI plasma focus chamber to a base pressure of 0.01 Torr. It was then filled with the required nitrogen gas to a higher pressure than required before being evacuated until the desired pressure (0.5, 1, 1.5 and 2 Torr) was reached.

The capacitor bank was then charged up to 12 kV before its energy was discharged into the nitrogen gas. The nitrided steel sample after 4 shots is shown in Figure 3.

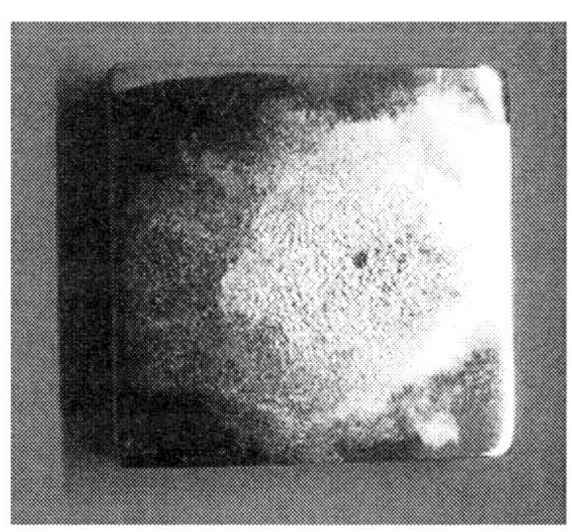

Figure 3: The alloy sample after nitriding.

Vickers Hardness test was carried out and the change in hardness was recorded.

At the same time, a Digital Phosphor Oscilloscope [Tektronix-TDS3034C] was used to observe the current derivative waveform obtained using a Rogowski coil. This waveform was numerically integrated to obtain the waveform of the discharge current. The voltage waveform was also monitored. The presence of an intense voltage spike indicates the strong focusing (plasma pinch) action. A sample of the oscilloscope screen is shown in Figure 4.

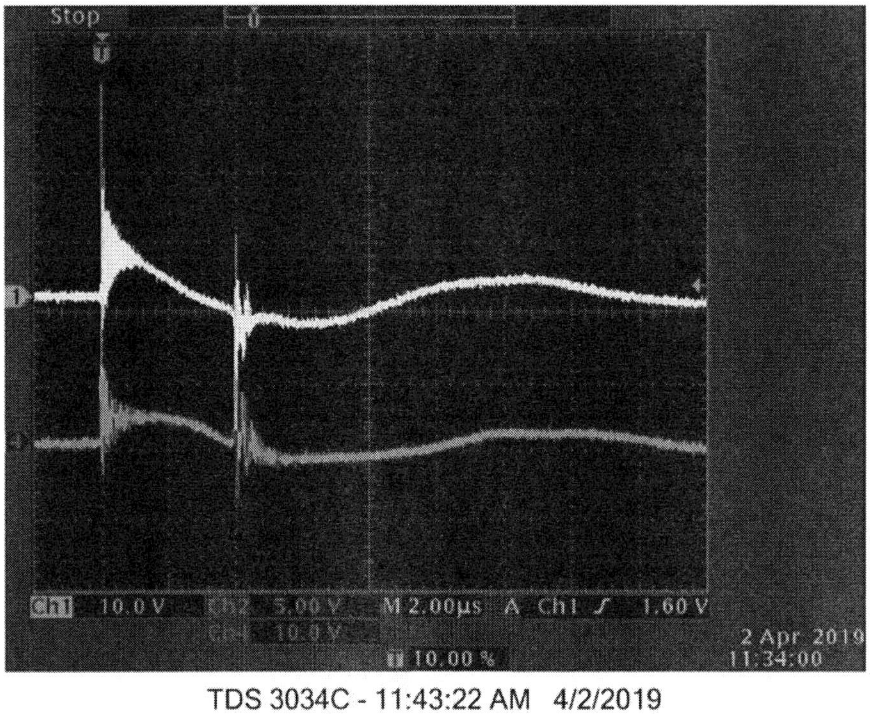

Figure 4: The current derivative (top yellow trace) and voltage waveform from INTI Plasma Focus machine at 1.5 Torr nitrogen gas.

The current derivative data file obtained from the Digital Storage Oscilloscope (DSO) was integrated and the current waveform was obtained. This was then inputting into Lee 6 phase code [7].

To use the Lee code, the bank, tube and operational parameters were input into the codes. The bank parameters consist of the values for the capacitance, inductance and unavoidable stray resistance. The length and radius of the anode, as well as the radius of the cathode, were also input into the code.

For the operational parameters, the voltage of 12 kV, nitrogen gas and pressure of operation (0.5, 1, 1.5 and 2 Torr) respectively were keyed in the input.

Finally, the model parameters (mass and current factor in both axial and radial phase) were obtained by fitting the computed current waveform to the measured current waveform. The flow chart for this is shown in Figure 5. The current fitting is shown in Figure 6.

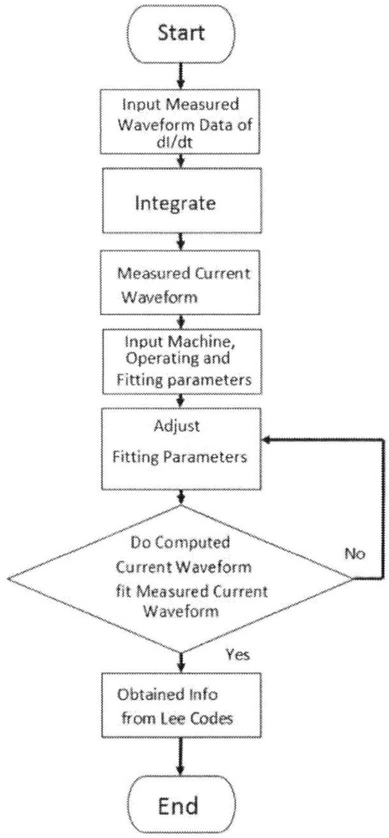

Figure 5: The steps in using the Lee Codes.

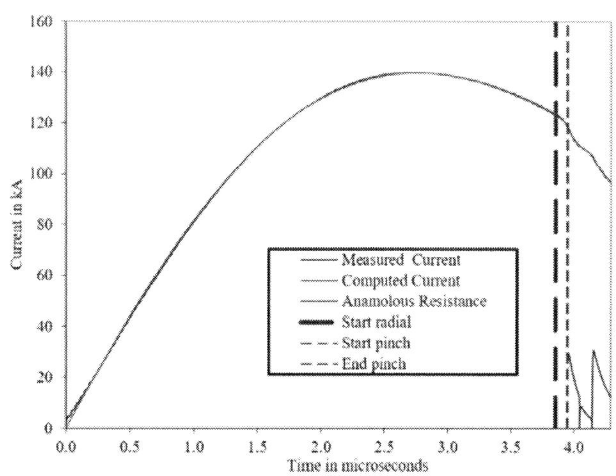

Figure 6: Computed current waveform of INTI machine at 12 kV, 1.5 Torr nitrogen gas fitted to the measured current waveform.

IOP Conf. Series: Materials Science and Engineering **854** (2020) 012062 doi:10.1088/1757-899X/854/1/012062

Once the current waveform had been fitted, the plasma dynamics and properties of the focus for that particular shot were computed within the code and also printed in the code output.

3. Results and Discussion

We measured the average of the hardness value of the sample. The pattern of plasma treatment is represented in Figure 7. The hardness reading at the centre of the pattern was taken. The average was calculated by obtaining the hardness value from the centre, right, left, above and below. The outer rings were not measured since these apparently were caused by shock wave interaction whereas the central region was primarily due to ion beam interaction.

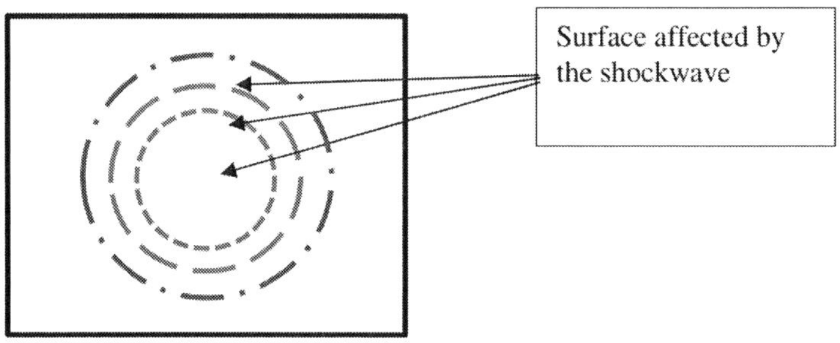

Figure 7: The formation of rings on the test sample

The results obtained from the Vickers Hardness Test is shown in Table 1.

Table 1: The Vickers Hardness Test at the centre of the pattern (see Figure 7).

Sample No. (pressure)	Surface Hardness at the center (HV 0.3 kgf)					Average value after nitriding (HV 0.3)	Average value before nitriding (HV 0.3)	Percentage improvement in hardness
	Center	Right	Left	Above	Below			
1 (0.5 Torr)	260.7	244.5	250.8	225.2	219.0	240.0	154.6	55.2
2 (1.0 Torr)	309.5	303.7	294.0	295.4	289.9	296.7	158.3	87.4
3 (1.5 Torr)	351.2	434.1	306.6	389.4	402.0	376.6	158.1	138.2
4 (2.0 Torr)	298.1	337.5	274.7	353.0	327	318.2	156.5	103.3

Table 1 is plotted in Figure 8, representing the percentage improvement in hardness at different pressures. The best pressure for improving the hardness was at 1.5 Torr which improved the hardness by 138%.

IOP Conf. Series: Materials Science and Engineering **854** (2020) 012062 doi:10.1088/1757-899X/854/1/012062

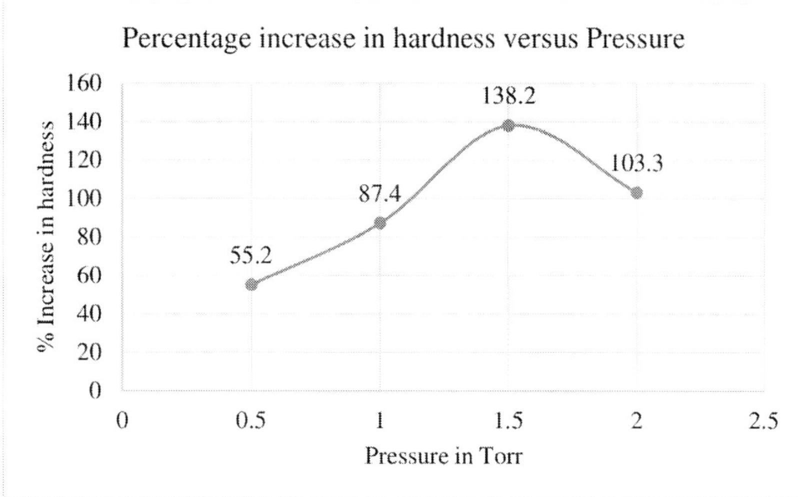

Figure 8: Improvement in hardness at different pressures

From Lee code, when the computed current waveform was fitted to the measured current waveform, the information on the ion beam energy [8,9] was extracted to find the pressure where the maximum value occurred. This data is shown in Table 2.

Table 2: The information of Ion Beam Energy extracted from the Lee code.

Sample No. (Pressure)	Ion Beam Energy (J)
1 (0.5 Torr)	2.55
2 (1.0 Torr)	4.44
3 (1.5 Torr)	7.25
4 (2.0 Torr)	6.23

Figure 9 shows a plot of the values of beam energy at different pressures.

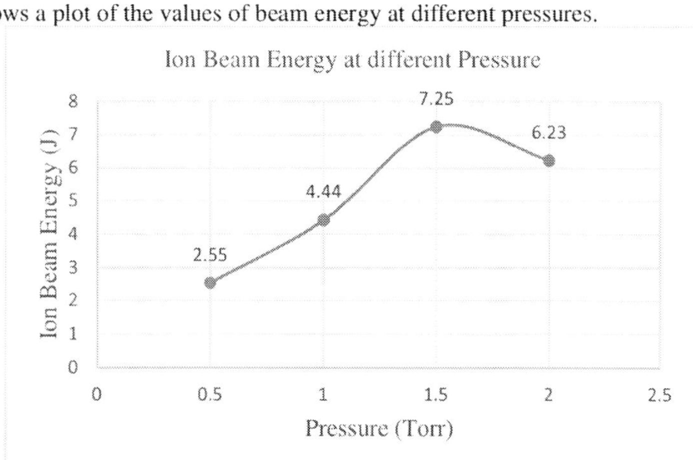

Figure 9: Beam energy at different pressure

When the beam energy and average hardness is plotted together as shown in Figure 10, it shows that the hardness and maximum beam energy occurs at the same optimum pressure of 1.5 Torr. There is good correlation between the hardness of the face of the treated sample and the ion beam energy produced by the plasma focus.

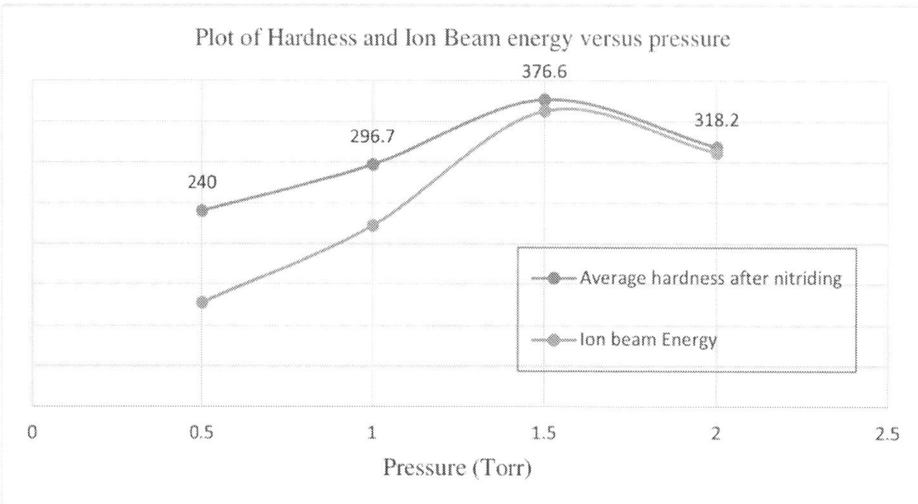

Figure 10: Plot of beam energy and hardness after nitriding versus pressure

(Note: Magnitude not using same scale)

Extracting information from Tables 1 and 2, the waveform of the percentage improvement in hardness versus Beam Energy was plotted as shown in Figure 11.

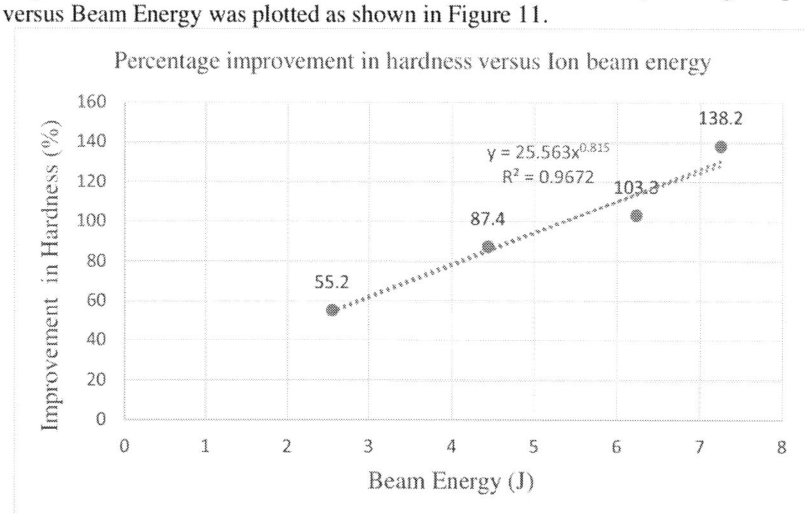

Figure 11: Plot of the percentage improvement in hardness vs beam energy

The figure shows that as the beam energy increases from 2.5 to 7.3 Joules, the hardness improves from 55 % to 138 %. The plot in Figure 11 reveals that for this sample, the hardness relates to the beam energy, by the equation:

$$\%\ \text{improvement in hardness} = 25.6\ E_{\text{beam}}^{0.82}$$

4. Conclusion

Nitriding improves the hardness of the steel due to the penetration of nitrogen ions into the surface of the steel. The ions are emitted in a narrow beam along the axis of the device. This narrow beam strikes the central region of the interaction pattern. The high density of energy deposited by the ion beam in this central region causes a high temperature and localized melting. The results of computation show that the beam energy correlates with the hardness of the irradiated sample. This is consistent with the concept of increase of hardness due to the diffusion of nitrogen ions into the surface of the metal. This paper is a preliminary study in investigating the relationship between beam energy and improvement in hardness of the AISI 304 stainless steel alloy. Further studies should be conducted to determine the upper limit of hardness and the beam energy required to achieve the upper limit. The effect of the number of shots required to diffuse the required number of nitrogen ions should also be considered.

References

[1] E Menthe, A Bulak, J Olfe, A Zimmermann, K T Rie, (2000), "Improvement of the mechanical properties of austenitic stainless steel after plasma nitriding". Surface and Coatings Technology. 133-134, 259-163.

[2] Thiam Oun TEH, Arwinder SINGH, Jalil ALI, Chee An NG, Xue Yinn NG, Jun Wen WONG, Sor Heoh SAW and Sing LEE (2019), "A Study on the Surface Hardness Obtained by Nitriding with a Plasma Focus Machine" Walailak Journal of Science and Technology 16 (6):379-384.

[3] S Al-Hawat, M Soukieh, MA Kharoub and W Al-Sadat (2010)," Using Mather-type plasma focus device for surface modification of AISI304 steel". Vacuum 84: 907-12.

[4] S Lee (2014), Radiative Dense Plasma Focus Computation Package: RADPF, 2014.Avaliable at www.plasmafocus.net

[5] S.A. Tukur, M.S Dambatta, A. Ahmed, N.M. Mu'az (2014), "Effect of Heat Treatment Temperature on Mechanical Properties of the AISI 304 Stainless Steel" International Journal of Innovative Research in Science, Engineering and Technology,3 (2):9516-9520.

[6] F Sonmez and A Demir (2007). "Analytical relations between hardness and strain for cold formed parts". Journal of Materials Processing Technology, (186), 163-173.

[7] S Lee, S H Saw, A E Abdou and H Torreblanca (2010), "Characterizing Plasma Focus Devices-Role of Static Inductance- Instability Phase Fitted by Anomalous Resistances", J Fusion Energy, 30:277-282.

[8] S Lee and S H Saw (2012), "Plasma focus ion beam fluence and flux—Scaling with stored energy', Phys. Plasmas 19, 112703.

[9] S Lee and S H Saw (2017), "The Plasma Focus-Numerical Experiments, Insights and Applications", Chapter 3 p113-232 in Plasma Science and Technology for Emerging Economies-An AAAPT Experience, ed Rajdeep Singh Rawat. ISBN 978-981-10-4216-4, ISBN 978-981-10-4217-1, DOI 10.1007/978-981-10-4217-1, © Springer Nature Singapore Pte Ltd. 2017

IOP Conf. Series: Materials Science and Engineering **854** (2020) 012063 doi:10.1088/1757-899X/854/1/012063

Reservoir Characterization with Acoustic Impedance Inversion and Multi attribute Method on "Essen" Field, Talang Akar Formation, Sub-Ciputat Basin, North West Java

E S Nugroho[1], A Riyanto[1], M Suardana[2]

[1] Department of Geosciences, FMIPA, Universitas Indonesia, Depok, Indonesia, 16424.

[2] Pertamina UTC, Gambir, Kota Jakarta Pusat, Daerah Khusus Ibukota Jakarta 10110

Email: ezranugroho@Gmail.com

Abstract. Acoustic impedance and multi-attribute seismic inversion seismic are a number of seismic methods that can be used to map sandstone reservoir distribution. Using this method, we can separate the sandstone and shale in the Talang Akar Formation found on the "Essen" Field, Ciputat Sub-Basin. Both of these methods will be compared with each other in order to get more valid results in sandstone reservoir mapping. The inversion method used in this study is a model-based method. While the multi-attribute method used is the neural network in mapping gamma-ray volume, shale volume, and porosity. The inversion results are not able to image the sandstone distribution good enough because the range is too large and there is overlapping on the acoustic impedance values of sandstones with ranges (8000-12000) (m/s)*(g/cc). The results of the gamma-ray multi-attribute, shale volume and porosity, have been shown to consistently show the distribution of sandstones that have a trend in the distribution of NW-SE reservoir zones. From the results of the analysis there are several potential areas that have the potential to become the next development area, namely the distribution of sandstones in the northern part with high effective porosity, and good seals; In the south with a large volume of sandstones; as well as the distribution formed on the channel surrounding the fault.

1. Introduction

In hydrocarbon exploration, seismic reflection is the main method used nowadays. The widely used seismic method can provide a geological structure and bottom surface coating with high detail and accuracy. In addition, penetration power is also very suitable to see the potential for adequate hydrocarbons [7]. In order to be able to increase the success ratio in drilling, reservoir characterization needs to be done, i.e. further analysis not only in the reflector field but also in its rock properties.

One effort that can be made to characterize the reservoir is the use of the inversion method. The inversion method enables us to transform seismic amplitude into a seismic attribute called Acoustic Impedance. Seismic inversion can provide a description of the fluid and lithology of the reservoir [8].

For a better reservoir characterization, a multi-attribute analysis might be necessary. Multi-attribute analysis, in this case, adopts a statistical approach that uses more than one attribute for the prediction of several physical properties of the earth [1]. It is expected that a good lateral resolution can be obtained while identifying the distribution of the existing rocks.

2. Regional Geology

According to Padmosukismo [2], the North West Java Basin regionally is a back-arc system located between the Sunda Microplates and the Indian-Australian plate supports. The North West Java Basin is affected by a block faulting system that runs north-south. This north-south trending fault system divides the North West Java Basin into a graben or several sub-basins from west to east, namely the Ciputat sub-basin, the Pasir Putih sub-basin and the Jatibarang sub-basin. Each sub-basin is separated by height (the block rises from the fault) as shown in **Figure 1**.

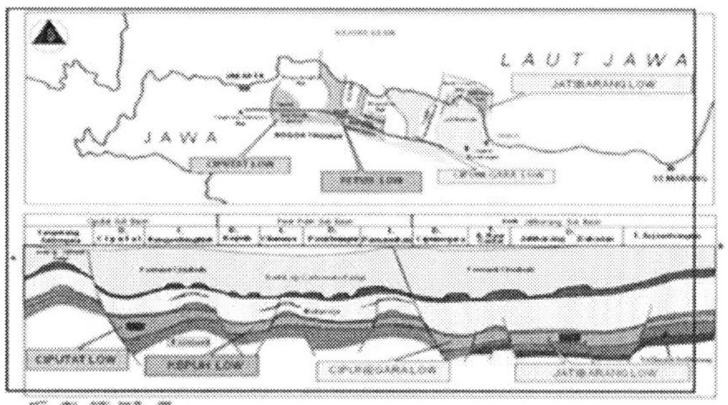

Figure 1: Regional geology of the North West Java Basin (Martodjojo, op. Cit. Nopyansyah, 2007)

Sedimentation in The North West Java Basin has a range of ages from the Middle Eocene to the Quaternary. The oldest deposit is in the Middle Eocene, namely in the Jatibarang Formation which was deposited unconformably on the Rock Base. The regional stratigraphy sequences from the oldest to the youngest are Batuan Base, Jatibarang Formation, Lower Cibulakan Formation (Talang Akar, Baturaja), Upper Cibulakan Formation (Massive, Main, Pre-Parigi), Parigi Formation and Cisubuh Formation, as illustrated in **Figure 2**.

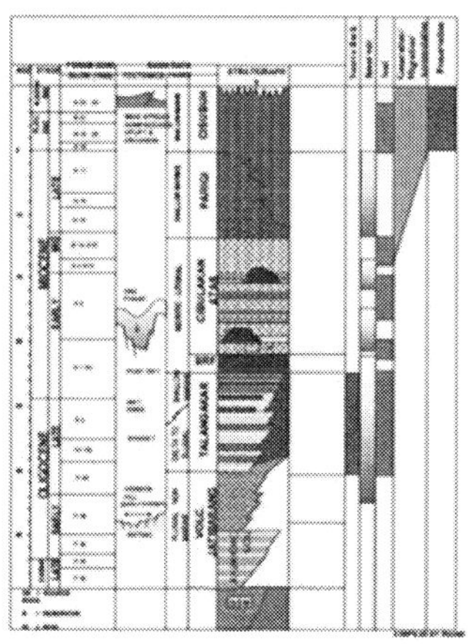

Figure 2: Stratigraphic Columns of the North West Java Basin (Arpandi and Padmosukismo, 1975

3. Data & Method

3.1. Data

The data used in this study are seismic and well data. Seismic data used in this study are post-stack time migration with 972 (861-1832) inline numbers, 904 (5001-5904) cross line numbers, with a sampling rate of 2ms with zero phases. This is determined based on the reflection coefficient on the boundary

IOP Conf. Series: Materials Science and Engineering **854** (2020) 012063 doi:10.1088/1757-899X/854/1/012063

between layers in the well data, where the increase in the acoustic impedance is shown as the peak in the seismic. The good data used in this study are PDM-1, PDM-2, and PDM-4, where each well has a variety of log data that will be used in this study, among them are gamma-ray, neutron porosity, density, sonic and check shots data.

3.2. Method

In this study, two methods will be used namely post-stack seismic inversion, and multi-attribute analysis. The seismic inversion method is basically a process of transforming seismic amplitude value to impedance value. This is done by deconvolution process which transforms seismic traces to earth reflectivity. The post-stack seismic inversion methods use stacked (zero-offset) seismic data to produce images of the AI in depth or time. AI is one of the rock-physics parameters, which is influenced by the type of lithology, porosity, fluid content, depth, pressure and temperature.

An updated approach to inversion is a model-based inversion where an initial low-frequency model is modified iteratively to give the best fit to seismic data. It avoids the problems of recursive inversion by iteratively changing a model to give a least-squares fit to the seismic data [5]. The main step in the inversion procedure includes the data preparation and data input into the software, calibration by tying well logs to the seismic data, estimation of the wavelet, generation of a low resolution initial model, inversion analysis, and inversion.

Multi-attribute seismic analysis is one of the statistical methods using more than one attribute to predict multiple physical properties of the earth. In this analysis, the relationship between logs and seismic data at the location of wells is sought and then uses these relationships to predict or estimate the volume of the log property at all locations in the seismic volume [4]. Statistics in reservoir characteristics are used to estimate and simulate the relationship of variable spacing to the desired value at locations that do not have measurable sample data. This is based on the fact that it often occurs in nature that the measurement of a variable in an adjacent area is the same.

4. Result and Discussion

4.1. Log Analysis & Cross-plot

The first step taken at this stage is to determine the formation boundaries and correlate them by making a top and bottom markers for each well. After that, the type of layer is identified.

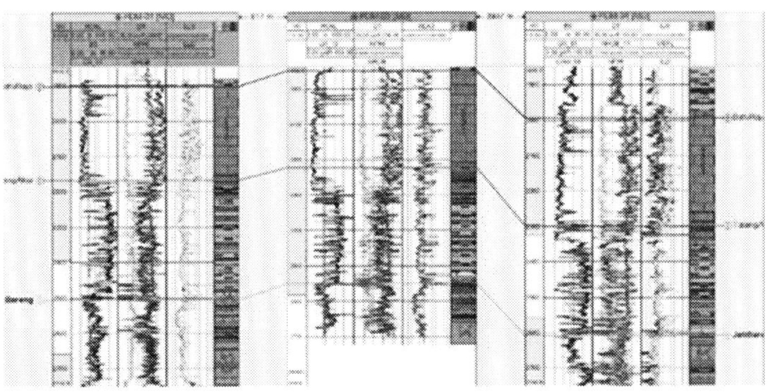

Figure 3: Target zone quick-look on PDM-01, PDM-03, and PDM-04 well

From **Figure 3** above the sandstone, a layer is shown in yellow, the carbonate layer is shown in blue and the shale layer is shown in green. The interesting zone is bounded by TAF (Talang Akar Formation) and JTB (Jatibarang Formation) markers. Cross plot that has been carried out including p-impedance -

IOP Conf. Series: Materials Science and Engineering **854** (2020) 012063 doi:10.1088/1757-899X/854/1/012063

gamma ray which is used to see whether the results of the acoustic impedance inversion, can later separate the layers of sandstone and shale.

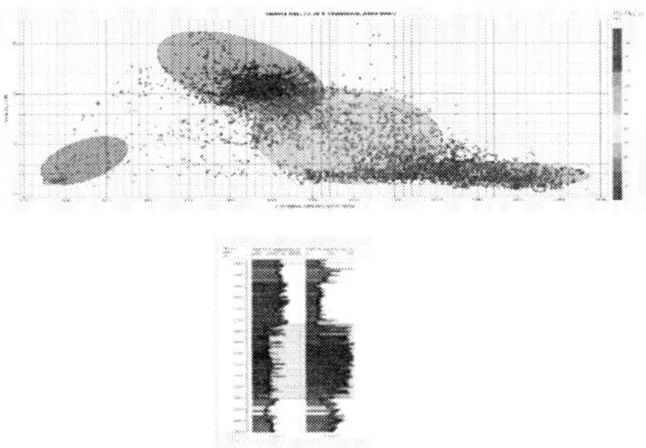

Figure 4: Cross plot Analysis: *Acoustic Impedance – Gamma Ray*

In **Figure 4**. It can be seen that sandstones with shale are quite separable even though there is some overlapping. From the gamma-ray cross plot analysis, there are 4 zones that can be separated. The area circled in red is coal, the blue area is the shale layer, while yellow is a sand layer characterized by gamma-ray values lower than shale (<75 API), and has a concentration of P-Impedance values between 8500 - 12000 (m / s) * (g / cc). The last zone which is purple is the carbonate layer.

4.2. Acoustic Impedance Inversion Result

In conducting inversion analysis, the first thing to do after the well-tie is to make an initial model. The initial model is a model made based on the results of the good tie & picking horizon. The frequency parameter used is high cut with frequencies ranging from 10/15 Hz. After that, the inversion result is conducted in the area of interest. The result is using 60% upper and lower hard constraint, with *average block size* 0ms, *pre-whitening* 1%, and 10 times iteration. Here is the impedance inversion result is shown in **Figure 5**.

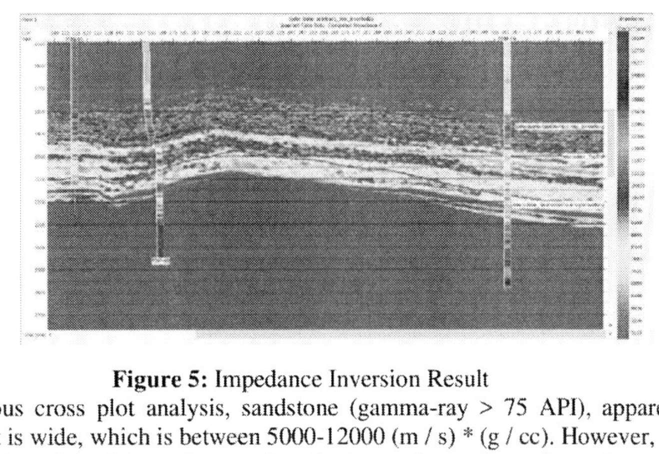

Figure 5: Impedance Inversion Result

Based on the previous cross plot analysis, sandstone (gamma-ray > 75 API), apparently has a P-impedance value that is wide, which is between 5000-12000 (m / s) * (g / cc). However, by referring to the higher concentration of the P-impedance values in the sandstone layer, it can be assumed that the value tends to fall in the number 8500-12000 (m / s) * (g / cc), which indicates that the sandstone layer in the "Essen" field is probably compacted.

IOP Conf. Series: Materials Science and Engineering **854** (2020) 012063 doi:10.1088/1757-899X/854/1/012063

After that, the slicing is performed on the volume of inversion based on the model. The slicing parameter used is in accordance with the results of the interpretation in the previous log analysis, which is to do the slicing by averaging the amplitude values that exist along with the TAF -35 ms marker.

While referring to the results of the cross plot, it can be seen that the sandstone layer will show p-impedance values ranging between 8500-12000) (m / s) * (g / cc), while the shale layer has a lower acoustic impedance value. As for the coal seams in this formation, they are considered negligible and will not damage the desired results. That is because the coal layer in this formation is a very thin layer.

By comparing the results of the slicing inversion with the seismic map slicing in **Figure 10a**, we can see that the sandstone reservoir is concentrated on the basin area shown on the black line circle. In the results of slicing inversion, it is seen that the spread of sandstone reservoirs is in several areas: in the north, south, and there is also a spread around the fault.

However, the slice resulting from this inversion has not been able to clearly and surely give the spread of sandstones, due to the wide range of acoustic impedance values in sandstones and overlapping with other rock types as we saw in the previous crossplot analysis. To help clarify the distribution of sandstone, multi-attribute analysis is necessary.

4.3. Multiattribute analysis result

In this analysis, an important parameter to be considered is the maximum number of attributes to use. In this study, the author uses the *Emerge* module of HRS software to look for groups of attributes that can be combined to predict targets. This is done by the process called step-wise regression. This means that the first search is determined by a single attribute from the list that predicts best by itself. The criterion for evaluating predictions is an RMS (root mean square) error. In other words, *Emerge* tries every attribute, calculates RMS errors, and determines the best single attribute as the attribute with the lowest error. After finding the single best attribute, *Emerge* looks for the best attribute pair, assuming that one of the pairs is the single attribute that has just been discovered. Once again this is done by trial and error, solving the system of equations as many attributes to be paired.

4.3.1. Vshale multiattribute result

	Final Attribute	Training Error	Validation Error
1	(inverted_Zp)**2	19.298887	21.063429
2	Filter 15/20-25/30	18.307864	20.437079
3	Filter 25/30-35/40	17.708730	20.157714
4	Derivative	16.769289	19.630282
5	Integrated Absolute Amplitude	16.226346	19.418142
6	Time	16.159779	19.554333
7	Quadrature Trace	15.605051	19.796360
8	Filter 35/40-45/50	15.471149	19.423365
9	Amplitude Weighted Phase	15.209735	19.148041

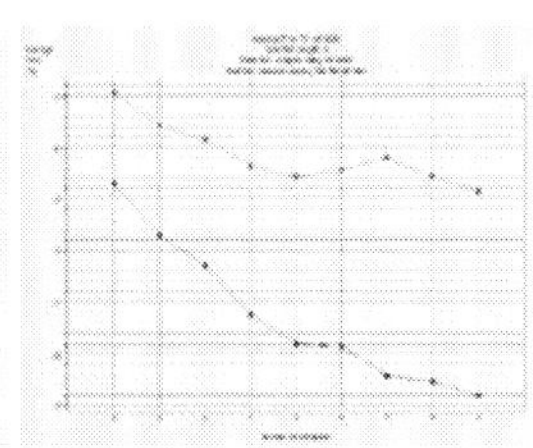

Figure 6: List of the attribute used in predicting shale volume values as well as error graphics and error validation

IOP Conf. Series: Materials Science and Engineering **854** (2020) 012063 doi:10.1088/1757-899X/854/1/012063

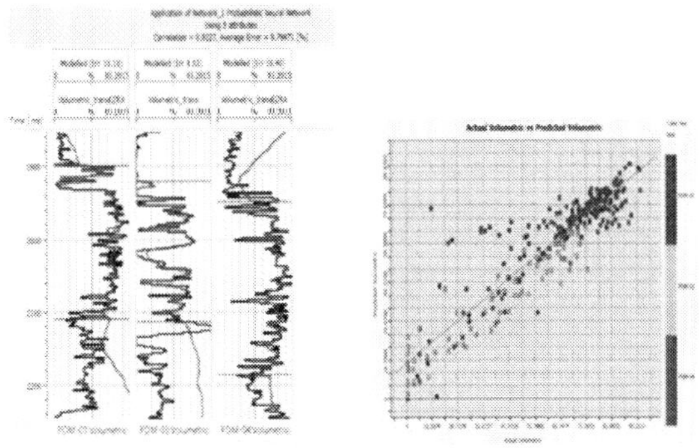

Figure 7: Similarities of trends and crossplot results between predicted Volume Shale and actual Volume Shale

From the results of the analysis in **Figure 6**, we can see that 5 attributes at once will give the best result because it provides the best validation and error. We can also see from **Figure 7** that the correlation result is 0.92 with an error of 9.8%. From these results, it can be assumed that the multi-attribute seismic method is good enough to be validated and justified in characterizing the reservoir zone.

In **Figure 10b,** we can see that the low shale volume in the area indicates that it contains little shale in the zone and is good for the reservoir zone. From the shale volume parameter, it can be assumed that the presence of sandstones is indicated by a vshale value below 54%. Because the shale volume parameter is directly proportional to how much the shale content if the shale volume value is higher, it also means that the area of lithology contains a lot of shales [2].

Shale volume below 54% are considered as an indication of sandstones shown in green-yellow-red color. Based on this map of shale volume attributes, the trend of sandstone distribution leads to the NW-SE, and is spread in the lowlands of the south, highlands in the north, and there are also spread around the fault.

4.3.2. Effective Porosity Multi-attribute Results

	Final Attribute	Training Error	Validation
1	Integrated Absolute Amplitude	5.691301	5.846390
2	Integrate	5.605270	5.799940
3	Filter 35/40-45/50	5.548992	5.790800
4	Filter 25/30-35/40	5.497370	5.896100
5	Filter 45/50-55/60	5.478043	5.979939
6	Amplitude Weighted Frequency	5.327437	6.100619
7	Instantaneous Frequency	5.234065	6.180234
8	Amplitude Weighted Phase	5.178748	6.319828
9	Sqrt(inverted_Zp)	5.082108	6.459013
10	Filter 55/60-65/70	4.987753	6.568771

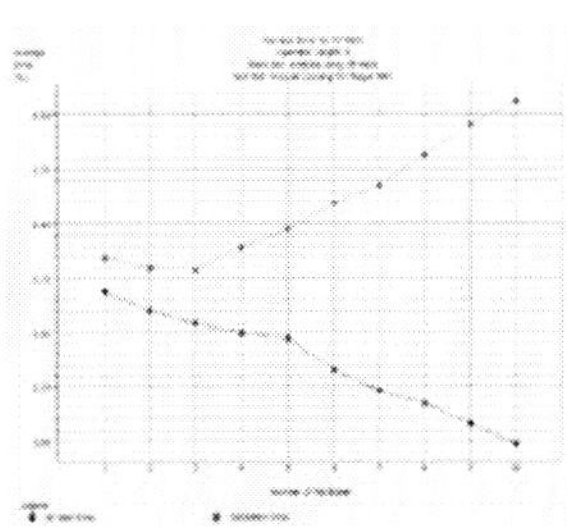

Figure 8: List of the attribute used in predicting porosity values as well as error graphics and error validation

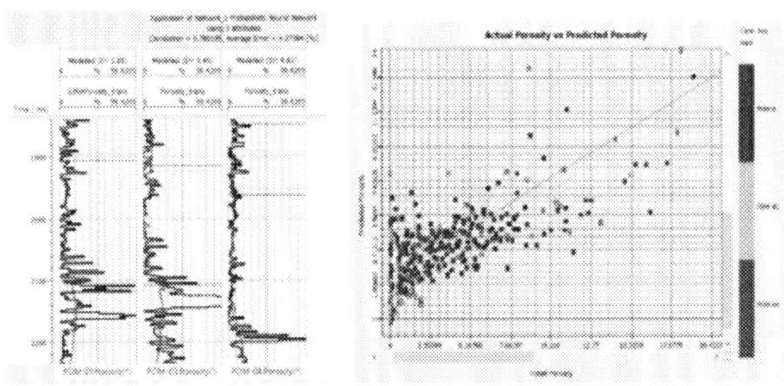

Figure 9: Similarities of trends and cross plot results between predicted Porosity and actual Porosity

From the result above in **Figure 8,** we can see that effective porosity results provide sufficient results to be able to provide log predictions. The author decides to use up to three attributes because if added 4 attributes the validation results will be reduced. From **Figure 9** we can see the correlation results obtained are 0.78 with an error of 4.07%. Cross plot results and similarity results of trends provide sufficient results.

The porosity map in **Figure 10c** was obtained after a multi-attribute analysis. It can be assumed that the sand porosity value ranged from 8.7-13% which is marked as color blue to purple (good enough) [5]. Based on the results, it can be concluded that the sandstone in this area is dominated by compacted sandstone. Areas that have high porosity values will be ideal as reservoirs because hydrocarbon fluids will enter the reservoir through rock pores and will be trapped in the reservoir if the reservoir is surrounded or limited by impermeable rocks such as shale [5]. It can also be seen that the area suspected to be a sandstone is flanked or surrounded by impermeable rocks so that reservoirs can form.

It can be seen that sandstone distribution based on the porosity map leads to NW-SE along with the normal fault in the SE direction, and also gather a lot in the low closure which is indicated by the red circle boundary.

4.3.3. Interpretations from inversion and multiattribute slicing maps

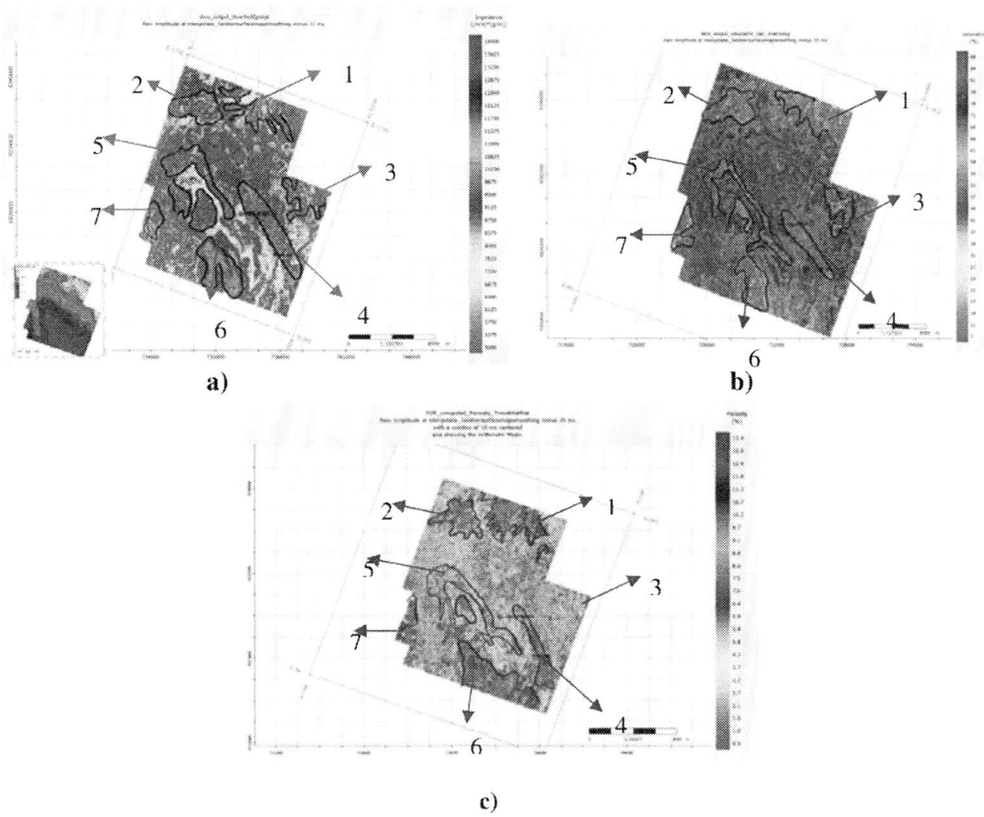

Figure 10: a) Impedance Inversion Slicing b) multi-attribute slicing of shale volume c) multi-attribute slicing of effective porosity all at an altitude of 35 ms above the Jatibarang horizon.

We can see from the inversion slicing map (**Figure 10a**) that the sandstone reservoir is consistently concentrated in several locations. We can see several concentration on the faults as shown in numbers 4 and 5. In the northern part shown by numbers 1,2 and 3, we can see the spread of deltaic fluvial sandstones that may originate from a source in the northern part of the map. This is appropriate when compared to the Jatibarang horizon map because fluvial deltaic is in the highlands. At numbers 4 & 6 there are several sandstone distributions that may originate from sources on the southeast, while at number 7 there are also sandstone distributions that may originate from sources on the western part, but this needs to be confirmed again due to limited seismic data that is owned. In areas marked on numbers 4 & 5, it is seen that it forms a channel or river which is possibly an ancient river that is controlled by a fault but has not been connected/disconnected with the source due to deposition of sediment.

However, the slice resulting from this inversion has not been able to clearly and surely give the spread of sandstones, due to the wide range of acoustic impedance values in sandstones and overlapping with other rock types as we saw in the previous crossplot analysis. To help clarify the distribution of sandstone, multi-attribute analysis is needed.

If you look at the vshale multi-attribute map (**Figure 10b**), there is consistency of sandstones distribution as obtained from the seismic inversion map. With numbers 1, 2, and 3 in the form of deltaic fluvial. In area number 6 on the vshale (**Figure 10b**), and porosity map (**Figure 10c**), it can be seen more clearly that there is a distribution of sandstones that may have resulted from sources in the southeast. In the area No. 5 it is possible that a shallow marine deposition that flows from the north may emerge, resulting in thinner and less continuous sand. Area Number 7 appeared to be controlled by faults which are shown on the jatibarang horizon map. Numbers 4 and 5 on the vshale map (**Figure 10b**) appear to have low values and forms a channel. It is assumed that the river may be controlled by a fault that leads north-west so it looks curved around the fault. But on the porosity map (**Figure 10c**), areas number 5 and 3 have a porosity result that is not too good, so that it may be taken into consideration for future decisions. The results of the multi-attribute analysis are proven to validate the results of seismic inversion. The main structure in this field is a high closure and 2 normal faults in the south. Reservoir traps contained in the "Essen" field are structural and stratigraphic traps where fault, fracture, fold, and stratigraphy affect the sandstone distribution.

From the results of the analysis, there are several potential areas that may be considered to be the next drilling area. First is on the north with the deltaic fluvial depositional environment, a fairly high effective porosity, and closed with shale, so there is good potential for reservoirs. The second area which is quite potential is in the southern part where there is a large sandstone distribution area. However, there is a lack of seal from shale, so the possibility of leakage/migration is quite high. The third area which may have the highest risk is the area around the fault where there is good distribution and seal, but it has a tighter sand content.

5. Conclusions

From this study, we can conclude that the Acoustic Impedance inversion method is less effective in distinguishing lithology because the range is too large and the existing overlapping on the value of the acoustic impedance, so that the lithology of the sandstone cannot be dissipated. The multi-attribute analysis provides the distribution of shale volume values with a range of values <0.54%, and porosity values ranging from 8.7 to 13% (porosity is good enough). The application of the multi-attribute method to the "Essen" field gives the trend of hydrocarbon reservoir zone distribution in the NW-SE (South West-North East) direction and converges on the low closure and fault area. There are several areas that have the potential to become the next drilling area, namely the distribution of sandstones in the north with good porosity and seals, In the south with a large volume of sandstones, as well as the distribution that forms in the channel surrounding the fault.

Acknowledgments

We would like to thank PT. Pertamina – UTC for permission to use the data in this study. We would like to also thank Made Suardana for his permission for using his seismic inversion data as a parameter in this study.

IOP Conf. Series: Materials Science and Engineering **854** (2020) 012063 doi:10.1088/1757-899X/854/1/012063

References

[1] Chopra, S., & Marfurt, K. J. (2005). *Seismic attributes —A historical perspective. GEOPHYSICS, 70(5), 3SO–28SO.*

[2] Harsono, A., 1997. Pengantar Evaluasi Log, Schlumberger Data Services. Jakarta:Schlumberger Oil Field Service.

[3] Narpodo, J., 1996. Studi Konversi Kedalaman dengan Metode Stacking Velocity dan Layer Cake di daerah Jawa Barat Utara , Skripsi-S1 Geofisika FMIPA UGM, Yogyakarta.

[4] Nawaz, U. S., "Acoustic and Elastic Impedance Models of Gullfaks Field by Post-Stack Seismic Inversion," M. Sc. Thesis, Petroleum Engineering and Applied Geophysics, Norwegian University of Science and Technology, 2013.

[5] Koesoemadinata, R.P., 1980, Geologi minyak dan gas bumi Jilid 1 Edisi ke II, Institut Teknologi Bandung, Bandung.

[6] Russel, B., 1997. Multiatribute Seismic Analysis. The Leading Edge, Vol.16, pp.1439-1443.

[7] Russell, B., Hampson, D., "A Comparison of Post-stack Seismic Inversion," 61st Annual International Meeting, SEG, (1991), pp. 876-878.

[8] Sukmono, S., "An Introduction to Seismic Reservoir Analysis, in Seismic Inversion and AVO Analysis for Reservoir Characterization," Geophysical Engineering, ITB, Bandung (2002) Another reference

[9] Telford, W. M. (1976). *Applied geophysics*. London: Cambridge University Press.

IOP Conf. Series: Materials Science and Engineering **854** (2020) 012064 doi:10.1088/1757-899X/854/1/012064

Thermal Conductivity of Some Marble Stones Available in South Aceh District

L Lindawati[1,2], N F Yuliza[1], I Irwansyah[1]

[1]Mechanical Engineering Department, Polytechnics of South Aceh
[2]Mechanical Engineering Department, Abulyatama University

Email: lindawati203@gmail.com

Abstract. Thermal conductivity is a fundamental property for insulating purposes when they are used as a building material. It controls the temperature distribution within the building. In this paper, the thermal conductivities of marble samples from the South Aceh District are reported. The samples were collected from three different areas in South Aceh District. The thermal conductivities of the samples were determined by using laboratory measurements. The results showed that samples of marble stones existing in South Aceh District have low thermal conductivity values. It revealed that marble stones existing in South Aceh District are appropriate to be applied as thermal insulating material.

1. Introduction

Thermal conductivity is a value defining how well heat is conducted through a material [1]. It is a fundamental property for a material when it is used as a building material for an insulating purpose [2]. It controls the temperature distribution within the building. The highest level of consumed energy within buildings generally comes from air conditioning. Thus, the selection of appropriate building materials for energy efficiency is needed in the early stage of the building design process.

Natural stones are one of those materials used as a building material with high preference because of their decorative appearance [3-6]. Generally speaking, using marble as a building material can reduce the thermal loads in buildings. Since the last decade, marble has been utilized for decorative and construction material. Marble floors are considered able to absorb heat from the environment. However, a scientific report on the thermal properties of marble floors is still limited. Marble is a natural rock produced by limestone metamorphosis due to high temperatures and pressures in the innards of the earth. The South Aceh District is one of the promising areas of abundant marble deposits. The appearance and mineral composition of South Aceh marble were stated in the previous study [7-8]. Still, the information about the thermal properties of South Aceh Marble has not been published yet.

There are many studies regarding the thermal conductivity properties of building material [9-15]. The findings show that all materials have thermal conductivity even if in a small number. Materials with low thermal conductivities and densities can be obtained by porous insertion. The thermal conductivity of material generally depends on some factors such as mineralogical composition, microstructure, porosity, saturation, pressure and temperature [16-17]. From a number of these studies, almost no one discussed the thermal conductivity of natural stones, especially marble. In the present study, the thermal conductivity of marble stones from the South Aceh District is reported.

IOP Conf. Series: Materials Science and Engineering **854** (2020) 012064 doi:10.1088/1757-899X/854/1/012064

2. Material and Method

In this study, the thermal conductivity of marble stone from three different areas in the South Aceh District is measured. **Figure 1** representing the marble collected from Gunung Kerambil (GK), Meukek (M), and Alur Kering (AK). Those marble samples were prepared in a square shape with a flat surface, in accordance with the apparatus test that is 15 cm in length, 15 cm in width and 1cm in thick.

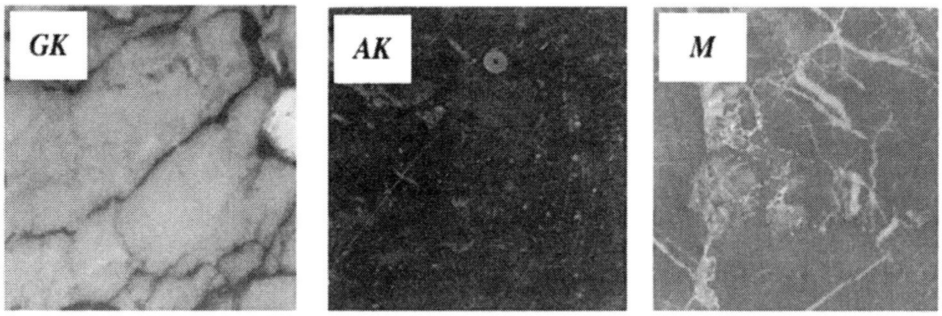

Figure 1. Test Material

The thermal conductivity of the sample was determined based on laboratory measurements using the heat flow method. The process and devices used to measure the thermal conductivity of materials are displayed in **Figure 2.**

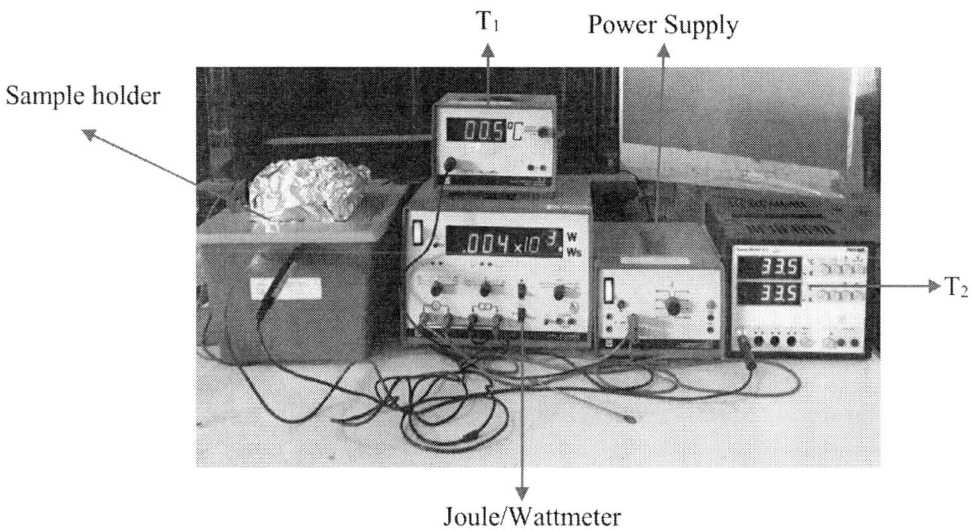

Figure 2. Thermal Conductivity Measurement Process

The measurements were conducted by turning on the heating element and measuring the temperature rise over time. Heat transfer occurs through the surface of test material with surface area (A) as the object is given heat with a temperature difference (T1-T2). The parameter measured is the amount of heat flowing (Q) through the material and temperatures different between the two surface sides of the material. Data on the heat flow rate (Q) and temperature difference (ΔT) were recorded every 1 minute for 30 minutes with two times of repetition. The value of thermal conductivity is calculated using the empirical formula of Fourier's Law, which is expressed in equation 1 as follows:

$$\frac{\Delta Q}{\Delta t} = -kA\frac{\Delta T}{\Delta x} \quad (1)$$

Where $\frac{\Delta Q}{\Delta t}$ is the amount of heat transferred per unit time (W), k is conductivity (W/m.K), A is the cross-sectional surface area (m^2), ΔT is the temperature difference (K) and Δx is the thickness of the marble sample(m) [5]. Based on the thermal conductivity data, the thermal resistance (R) of the sample test can be calculated using the following relation in equation 2 [8].

$$R = \frac{l}{k} \quad (2)$$

3. Results and Discussion

3.1. Heat transferred (Q)

The thermal conductivity *(k)* is defined as the rate of heat transferred (*Q*)in an object with a temperature gradient (ΔT) [18]. The heat transferred measured are tabulated in **Table 1.**

Table 1. Heat Transferred per Unit Time

Time (s)	Q(x10^{4}W)		
	GK	AK	M
0	0,04	0,04	0,04
5	1,38	1,32	1,29
10	2,71	2,63	2,62
15	4,05	3,96	4,26
20	5,38	5,28	5,63
25	6,72	6,61	6,94
30	8,05	7,92	8,15

Based on data in Table 1, it observed that the heat flow rate is significantly increased with time. The amount of heat transferred per time ($\frac{\Delta Q}{\Delta t}$) can be graphically demonstrated in **Figure 3**. As Expected, a high correlation (R^2=1) between the rate of heat flow and time was determined. It can be seen that the heat flow rate in the material test from Gunung Kerambil, Meukek, and Alur Kering are increasing with the time.

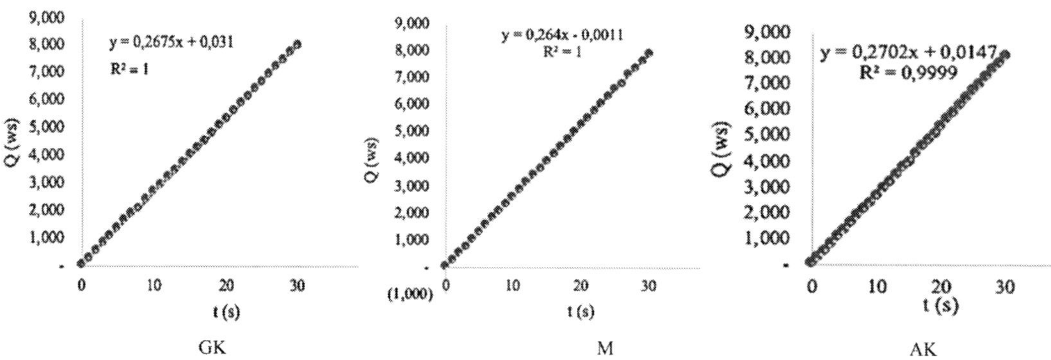

Figure 3. Relationship of Heat Flow Rate and Time

3.2. Temperature Differences (ΔT)

Heat transferred in building materials occurs when there is a temperature difference between the outer surface and the inner surface. **Table 2** shows the temperature difference of the sample test during the thermal conductivity measurement.

Table 2. Temperature Differences of the Sample

Time (s)	**ΔT (K)**		
	GK	**AK**	**M**
0	12,05	8,49	15,60
5	11,78	7,90	15,70
10	11,28	7,62	15,70
15	10,88	7,34	15,70
20	10,58	6,95	15,70
25	10,37	6,79	15,70
30	10,16	6,69	15,60

Data in Table 2 demonstrates that the temperature difference generally decreases with time. The higher the temperature difference, the greater the heat flow and respectively. Heat insulation itself is a way to reduce heat by using an insulating material.

3.3. Thermal Conductivity (k)

The thermal conductivity calculated for each marble sample from South Aceh District is displayed in Table 3.

Table 3. Thermal Conductivity of Samples

Sample Code	***k* (W/mK)**	***R (m²K/W)***
GK	2,02	0,22
AK	2,30	0,20
M	1,28	0,35

In this study, the thermal conductivity of each marble from South Aceh District is 2,09 W/mK, 2, 30 W/mK and 1,28 W/mK. The thermal resistance obtained are 0,22 m^2K/W, 0,2 m^2K/W, and 0,3 m^2K/W. In reference, the thermal conductivity of marble ranges between 1,59 and 4,0 W/mK [18]. Thus, the measured thermal conductivity value for marble from Gunung Kerambil and Alur Kering area is close to the thermal conductivity value of marble stated in reference. However, marble from the Meukek area shows a lower thermal conductivity value. A high thermal conductivity value indicates a large number of heat energy transferred within materials. The material that has a high thermal conductivity value is a good conductor while those that have a low thermal conductivity k-value are good insulators. Consequently, marble samples from South Aceh are bad conductors and good isolators. Those are appropriate to be used as a building material for insulating purposes.

4. Conclusion

In this study, the data of the measurement of thermal conductivity South Aceh marble were presented. The thermal conductivity of each marble from South Aceh District is 2,09 W/mK, 2, 31 W/mK and 1,28 W/mK. The thermal resistance obtained are 0,22 m^2K/W, 0,2 m^2K/W, and 0,3 m^2K/W. Based on the results, it can be concluded that the marble sample from South Aceh District has a low thermal conductivity which means they are bad heat conductors and a good insulator. It revealed that marble stones existing in South Aceh District are appropriate to be applied as the thermal insulating material

Acknowledgment
Authors thank the Indonesian Ministry of Research, Technology and Higher Education for funding this research through the 2019 Beginner Lecturer Research Grant.

References
[1] Andolfsson, T. 2013. Analyses of thermal conductivity from mineral composition and analyses by use of Thermal Conductivity Scanner: A study of thermal properties in Scanian rock types. Dissertations in Geology at Lund Uni-versity, No. 361, 27 pp.45
[2] H. T. Özkahraman & E.C.Işık. 2003. Determination of Thermal Conductivity of Building Stones from P-Wave Velocity, 18'" International Mining Congress and Exhibition ol Turkey-IMCET 2003, ISBN 975-395-605-3.
[3] B. Tufan and M. Kun. 2014. Thermal Insulation Performance and Thermal Conductivity Evaluation of Natural Stones by Infrared Thermography. Proceedings of the International Conference on Mining, Material and Metallurgical Engineering Prague, Czech Republic, August 11-12.
[4] M.B. Baker. 2017. *The Application of Marble and Granite as Building Materials in Jordan.* Jordan Journal of Civil Engineering Volume 11. No. 2
[5] Daku, S.S., Wazoh, H.N and Ojo, O.A. 2017. Engineering Properties of Jakura Marble and Its Suitability for Tiles Production. IOSR Journal of Applied Geology and Geophysics (IOSR-JAGG) e-ISSN: 2321–0990, p-ISSN: 2321–0982.Volume 5, Issue 1 Ver. I
[6] F Asdrubali, G Baldinelli, F Bianchi, A Presciutti, F Rossi and S Schiavoni. 2015. *Thermal and Optical Characterization of Natural and Artificial Marble for Roof and External Floor Installations.* Journal of Physics: Conference Series 655 012017
[7] Lindawati Lindawati, Nuzuli Fitriadi, Afdhal Afdhal. 2018. *Analysis of Noise Level Generated by Stone Cutter Machine*, A Case Study in Marble Production Unit, South Aceh, Jurnal Inotera. 3 (1).
[8] Lindawati Lindawati, Mursal Mursal, Irwansyah rwansyah, Nuzuli Fitriadi, Edi Saputra. Mineral Identification of South Aceh Marble by Using X-Ray Diffraction Method, IOP Conference Series: Materials Science and Engineering. 536. 012005
[9] Carmeane Efftinga*, Saulo Güthsb*, Orestes Estevam Alarcon. 2007. *Evaluation of the Thermal Comfort of Ceramic Floor Tile.* Materials Research, Vol. 10, No. 3, 301-307.
[10] E. Rambaldi, F. Prete, G. Timellini. 2014. *Thermal and Acoutics Performances of Porcelain Stoneware Tiles.* Qualicer'14. Castellon (Espana). 1-4.
[11] Samheri A. Almuradi . 2017. *Effect of Different Types of Thermal Insulation Used in Ground Building Tiles Material.* International Journal of Applied Engineering Research. ISSN 0973-4562. 12 (16). 6267-6279.
[12] Sarkawt.A.Saeed, Shler S. Qadir, Ryadh .H. Shaan. 2012. *Thermal Insulating Concrete Tiles.* International Journal of Engineering and Technology. 2 (11). 1877-1880.
[13] Dana Koňáková, Monika Čáchová, Eva Vejmelková, Martin Keppert, dan Robert Černý. 2014. *Thermal Properties of Selected Timbers.* Advanced Materials Research. ISSN: 1662- 8985, Vol. 982, 100-103.
[14] Boning T., Chuanging Z., Ming X., Tiange C., and Shengbiao H. 2019. *Thermal Conductivity of Sedimentary Rocks in the Sichuan Basin, Southwest China.* Energy Exploration & Exploitation. ol. 37(2) 691–720.
[15] Małgorzata L. and Krzysztof L. 2018. *Thermal Conductivity and Diffusivity of Fine-Grained Sedimentary Rock.* Journal of Thermal Analysis and Calorimetry (132:1669–1676
[16] Qiang S., Shen-en C., Qiao G., Weiqiang Z., Jishi G., Yuliang Z. 2017. *Analysis of Factor Influencing Sandstone Thermal Conductivity.* Acta Geodyn. Geomater. Vol. 14, No. 2 (186), 173–180
[17] Shusaku G. and Osamu M. 2009. *Relations between the Thermal Properties and Porosity of Sediments in Theeastern Flank of the Juan de Fuca Ridge.* Earth Planets Space,61, 863–870

[18] P. Mielke, K. Bär, and I. Sass. 2017. *Determining the Relationship of Thermal Conductivity and Compressional Wave Velocity of Common Rock Types as a Basis for Reservoir Characterization.* Journal of Applied Geophysics 140 (2017) 135–144.

IOP Conf. Series: Materials Science and Engineering **854** (2020) 012065 doi:10.1088/1757-899X/854/1/012065

Study of Dynamic Stability Analysis in the Sigura-gura Hydro Power Plant using the Power Stabilizer System as Compensation

Zulfikar

Electric Engineering Study Program, Politeknik Negeri Lhokseumawe

Email: zul_elka@pnl.ac.id

Abstract. To maintain the stability of the power plant system, effort are needed to keep the generator in sync. This study was conducted to determine the performance of each power plant in the Sigura-gura power system when interference with the transmission network. By reviewing the response curve through simulations with the EDSA software, it can be seen the changes in rotor angle, voltage oscillation and power for each generating system when a disturbance occurs. The application of PSS as a compensation to improve the character of the power system's stability against interference by observing changes in rotor angle, voltage and power. The observations show that the installation of PSS controllers in excitation as compensation to the Sigura-gura electric power system can contribute to better system performance by improving the rotor angle oscillation time, voltage and overshoot respectively 7.8%, 66.6% and 37.9%.

Keywords: stability, overshoot, oscillation, stabilizer system, compensation

1. Introduction

Power system stability is the state of the system to respond to disruptions during normal operation by returning to conditions where normal operation returned [1][2] Interference in the system can cause oscillation of system variables, such as voltage, frequency, and power. While system requirements require that the variable be fixed at a certain operating point. Therefore, the problem in system stability is how to make oscillations that occur due to interference can quickly return to normal conditions. For this reason, large systems that have been built must be reviewed and improved every time there is a development in the system itself. As is the case with the Medium Electricity System in North Sumatra and Nanggroe Aceh Darussalam, now this area is expanding because demand for higher load services causes resources or power plants to service loads or uses that are located far from each other and cover large areas, so the central load center is carried out through a transmission channel with limited capacity. Such operations often cause problems with power system dynamics that cause the steady-state to be disrupted. Real conditions that often occur is a short circuit on the transmission line.

IOP Conf. Series: Materials Science and Engineering **854** (2020) 012065 doi:10.1088/1757-899X/854/1/012065

Conditions like this can cause the power system to be disrupted. The worst thing that happens is that the power system becomes unstable either by shifting the angle of the rotor, oscillating the voltage and reducing the electrical power in the system. To anticipate these things, a gradual evaluation needs to be done to find out how the behavior of each interconnection installation in North Sumatra if there is interference in one of the transmission lines and the extent of system reliability and what actions can be taken to overcome the unbalanced system. For this reason, evaluating the stability of Sigura-gura plants is the priority analyzed.

2. Basic Systems in Electric Power Systems

In order to understand the nature of power plant systems and control designs for performance improvement, it is necessary to know the basic components of power generation. [3]

The basic components of a power plant are systematically shown in Figure 1. The picture shows the turbine and governor with feedback speed, SG generator, VR voltage regulator excitation with voltage feedback, transformer and transmission network.

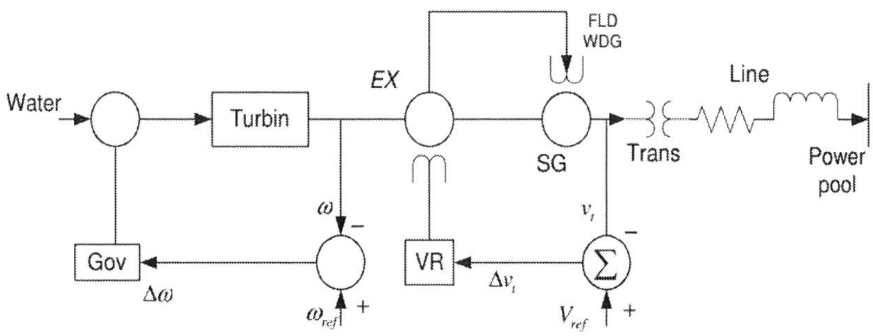

Figure 1. The basic components of a power plant

2.1. *Turbine*

Conversion of mechanical energy with a steam turbine is a thermodynamic process, where steam is normally expanded through low, medium and high pressure turbines all on one shaft. The high pressure steam energy and high temperature of the boiler are converted to mechanical energy through the turbine fins and diverted to the shaft connected to the generator.

2.2. *Governor*

The function of the governor is to maintain a constant speed, which is the synchronous speed of the turbine-generator set. When the speed decreases, in order to increase the output of electric power it will send a signal to the governor to increase the mechanical power input to the turbine and when the speed rises then the mechanical input power is reduced to maintain a constant speed. In large generators the governor provides power and frequency control functions, from areas that are in large interconnections.

2.3. *Swing Equations*

Under normal operating conditions the relative position of the rotor axis and the magnetic field axis are fixed. The angle between the two is called the power angle or torque angle. During disruption, the rotor will slow down or accelerate the synchronous rotation of the magnetomotive force (mmf) air gap, and start relative motion [4] The equation that describes the relative motion is the swing equation. If after the oscillation period, the rotor returns to synchronous speed, the generator will maintain stability. If interference is not followed by a change in power, the rotor returns to its original position. If the disturbance is followed by changes in generation, load or network conditions, the rotor will move to a new operating angle relative to the rotating field that is synchronized again. Two-cylinder cylindrical generator illustrative diagram in Figure 2.

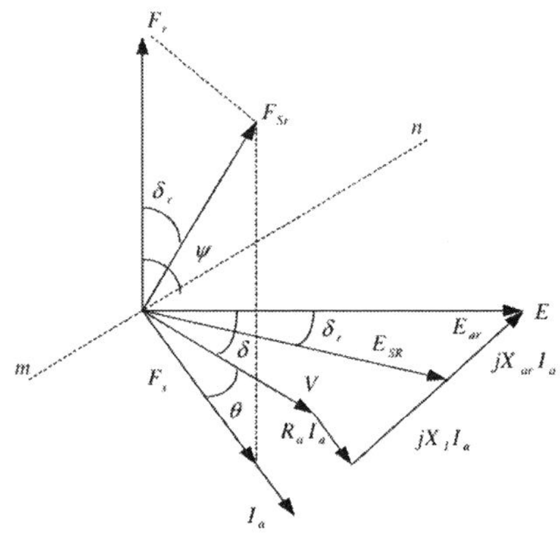

Figure 2. Diagram of single-phase phasor cylindrical rotor generators

The power angle is the angle between the mmf rotor and the resultant air gap mmf, both rotating at synchronous speed. Also the angle between the emf generated without load and the resultant stator voltage. If the leakage flux and generator anchor resistance is ignored, the angle between E and terminal voltage V is the power angle.

Synchronous generator that generates electromagnetic torque and rotates with speed and mechanical torque on the rotor, so the steady state operation by ignoring losses is

$$T_m = T_e \quad (1)$$

If is a combination of the moment of inertia of the prime mover and the generator, ignoring the friction and torque of the dampers, the rotation law is obtained as

$$J\frac{d^2\theta_m}{dt^2} = T_a = T_m - T_e \quad (2)$$

Swing equation with the inertia constant

$$M\frac{d^2\delta_m}{dt^2} = P_m - P_e \quad (3)$$

In the form of perunit

$$\frac{H}{180}\frac{d^2\delta_m}{dt^2} = P_m - P_e \quad (4)$$

$$P_e = \frac{|E'||V|}{X_{12}}\sin\delta \quad (5)$$

With the transient voltage behind the transient reactance (bus-1), the voltage at infinity (bus 2), the reactance between bus 1 and bus 2, this is a simple form of the power flow equation and a basis for understanding stability problems. The relationship of the power transferred depends on the transfer reactance and the angle between the two voltages [5]

Increased generator power output possible until maximum electrical power is transferred. This maximum power is expressed as a steady state stability limit, and occurs during the shrinkage of 90^0.

$$P_{\max} = \frac{|E'||V|}{X_{12}} \tag{6}$$

2.4. Equation of the excitation system

Excitation control systems generally consist of several components, namely rectifiers, voltage regulators and comparators, amplifiers and exciter.

The excitation system reviewed for generators from the electric power system in North Sumatra is as follows

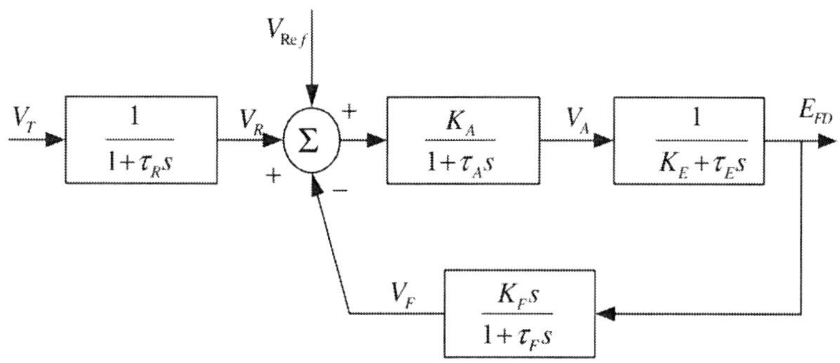

Figure 3. Block diagram of the excitation system

The description of the state space of the excitation system is:

$$\dot{V}_R = \frac{1}{\tau_R} V_T - \frac{1}{\tau_R} V_R$$

$$\dot{V}_F = \frac{K_F}{\tau_F} E_{fd} - \frac{1}{\tau_F} V_F$$

$$\dot{V}_A = \frac{K_A}{\tau_A}\left(V_{ref} - V_F\right) - \frac{1}{\tau_A} V_A$$

$$\dot{E}_{fd} = \frac{1}{\tau_E} V_A - \frac{K_E}{\tau_E} E_{fd} \tag{7}$$

for:

τ_R: regulator imput time constant

K_F : strengthening the regulator stabilizer circuit

τ_F : time constant of the regulator stabilizer circuit

K_A: regulator strengthening

E_{fd} : Field voltage

τ_A: regulator time constant

IOP Conf. Series: Materials Science and Engineering **854** (2020) 012065 doi:10.1088/1757-899X/854/1/012065

K_g : exciter self-excitation at full load field voltage

2.5. Power Sistem Stabilizer (PSS)

The basic function of the Power stabilizer system (PSS) is to provide damping oscillation of the rotor generator by controlling the excitation signal [1] To get a good damping the stabilizer converts the rotor speed to electric torque. An overview of the Power Stabilizer System (PSS) is shown in Figure 4.

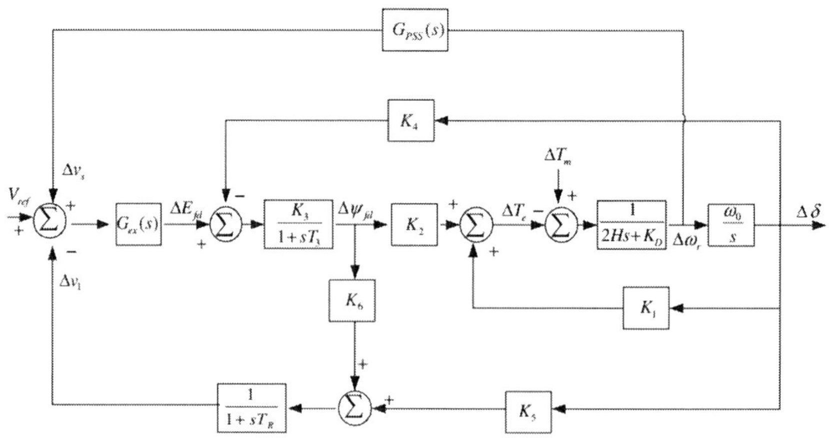

Figure 4. Block diagram with AVR and PSS

From above figure, if the transfer function and generator transfer function is the basic gain, the feedback will add up the damping torque as compensation for the Powers System Stabilizer (PSS).

By illustrating the principle of using the Power System Stabilizer (PSS) by following the parameters exemplified to determine the effect of the excitation system in Figure 3. From the block diagram 4 if ignored, the PSS can be written

$$\frac{\Delta\psi_{fd}}{\Delta v_s} = \frac{K_3 K_A}{sT_3 + 1 + K_3 K_6 K_A} \quad (8)$$

3. Sigura-gura parameter

Table 1. Sigura-gura parameter

Gen	X_q (PU)	X_q' (PU)	X_d (PU)	X_d' (PU)	T'$_d$	T'$_{q0}$
18	0.67	0.67	1	0.16	4	0.33
19	0.67	0.67	1	0.16	4	0.33
20	0.67	0.67	1	0.16	4	0.33
21	0.67	0.67	1	0.16	4	0.33

IOP Conf. Series: Materials Science and Engineering **854** (2020) 012065 doi:10.1088/1757-899X/854/1/012065

4. Results and Discussion

4.1. The response of the rotor angle

The expected rotor angle before and after *Power System Stabilizer (PSS)* installation is shown in Figure 5(a) and 5(b).

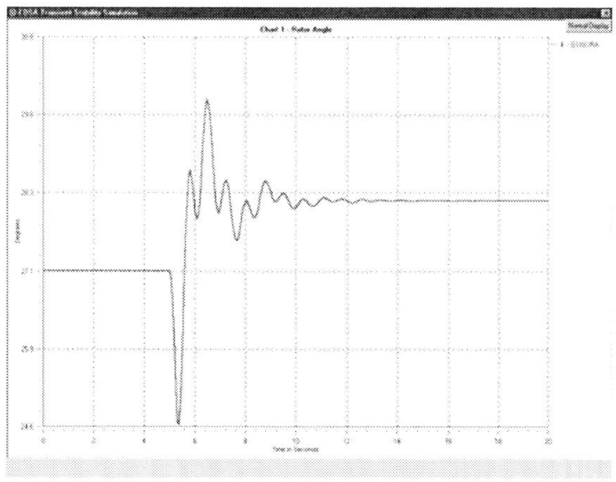

Figure 5(a). Rotor angle before using PSS

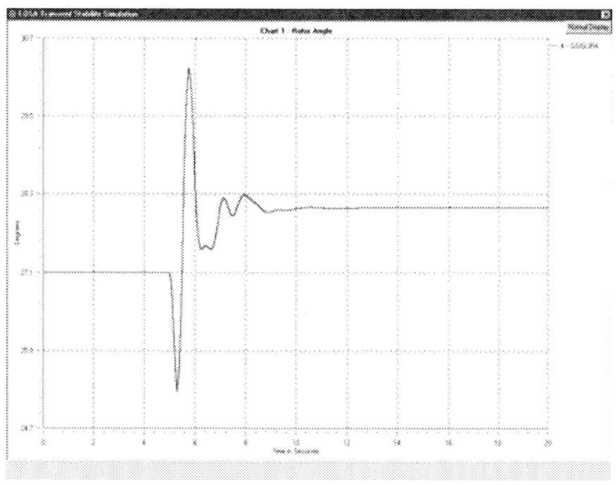

Figure 5(b). Rotor angle after using PSS

4.2. Voltage response

Voltage responses before and after the installation of the Power System Stabilizer (PSS) are shown in Figures 6(a) and 6(b).

Before installing the Power System Stabilizer (PSS) in Figure 6(a) shows that when the load was released on the Langsa bus. The voltage at the Paya Pasir generator was oscillated for 9.1 seconds with an overshoot of 0.003 pu. After the PSS installation in Figure 6(b) shows that the system experienced an oscillation time reduction from 9.1 seconds (Figure

6(a) to 4.65 seconds (Figure 6(b), resulting in an improvement of 6.05 seconds or 37.6%. 0.003 pu to 0.001 pu, or improved by 0.002 pu. This shows that installing PSS can make time for oscillations and damping oscillations in the Sigura-gura generator.

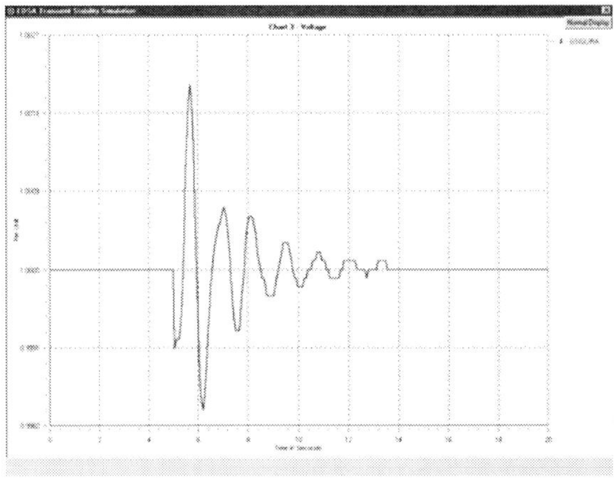

Figure 6(a). The voltage response of the Sigura-gura generator before using PSS

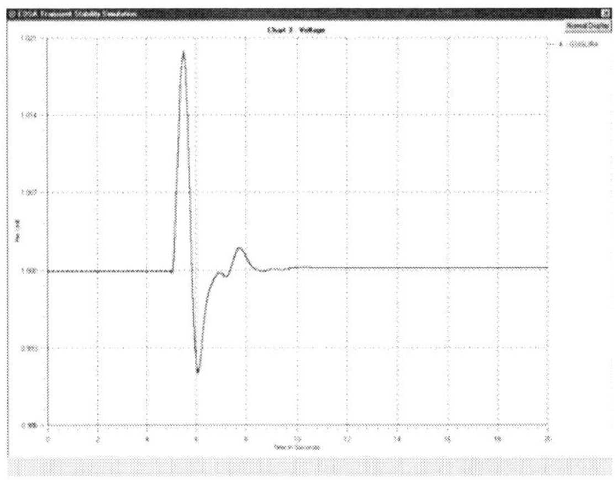

Figure 6(b). Responses to the Sigura-gura power plant after using PSS

Table 2. Response of generator systems before PSS installation

Generator	Response rotor angle		Voltage response	
	Rotor angle	Time oscillation (sec)	Overshoot (pu)	Time oscillation (sec)
Sigura-gura	1. 2^0	7.3	0.003	9,1

Table 3. Response of generator systems after PSS installation

Generator	Response rotor angle		Voltage response	
	Rotor angle	Time oscillation (sec)	Overshoot (pu)	Time oscillation (sec)
Sigura-gura	1.12^0	6	0.001	5.65

From Table 2 shows the rotor angle shift 1,120, oscillation time 7.3 / sec, voltage overshoot 0.0022 and voltage oscillation time 8.56 in the Sigura-gura hydropower plant system due to interference on the transmission network but after installing PSS as compensation in Table 3, they can reduce each: rotor angle oscillation 7.8%, Overshoot 66.6%. 37.9% oscillation time

5. Conclusion

The performance of each hydropower plant in Sigura-gura against one phase ground disturbance is currently quite good, in terms of the response of the rotor angle and the voltage of the plant when the disturbance occurs. Changes in the rotor angle for the whole plant range from 1,120 and the oscillation time to 7.3 sec and the length of the oscillation time of the voltage ranges between 8.36 sec. rotor angle, voltage and overshoot respectively 7.8%, 66.6% and 37.9%.

References

[1] Kundur ,P, 1993, " *Power System Stability and Control*" McGraw-Hill, inc, New York
[2] Rahmad Syaputra Lubis 2003 " *Perbaikan Dinamik Sistem Tenaga dengan Static Syncranous Series Compensator* (SSSC) " tesis Program Pasca Sarjana Program studi Teknik Elektro UGM Yogyakart.
[3] Yu Yao-nan ,1993 *Electrical Power System Dynamic*, Academic Press New York London
[4] Anderson, P.M. Fuad,A.A,1977 " *Power System Control and Stability,* " the Lowe State University Press, Ames, Lowa, U.S.A
[5] Allen J. Wood, Bruce F. Wollenberg 1996 " *Power Generation, Operation andControl* ," John Willi & Son New York.
[6] Powell, L., 2005, "*Power System Load Flow Analysis*", McGraw-Hill, USA

IOP Conf. Series: Materials Science and Engineering **854** (2020) 012066 doi:10.1088/1757-899X/854/1/012066

Analysis of the Effect of Frame Variations on Improving the Quality of Streaming Network Services on the Local Network

Atthariq, Mursydah, Azhar

Department of Information Computer Technology, Politeknik Negeri Lhokseumawe, Lhokseumawe, 24301, Indonesia

E-mail: atthariq.huzaifah@pnl.ac.id , mursyidah@pnl.ac.id , azhar.tik@pnl.ac.id

Abstract. The quality of network services on streaming services is influenced by several aspects, one of which is the quality of the video used, the variation of frames and the right video format in streaming communication is expected to improve the quality of streaming services. In this study using IEEE 802.11n Wireless LAN devices as transmission media and 2-layer switches. The switch will classify incoming data packets at the input port based on access-list here incoming packets will be defined and marked according to the access-list attachment that will give priority so that higher-class packages will be prioritized first. This research uses HTTP protocol in streaming communication with H264 / MPEG-2 video format and frames 320x576 pixels, 352x576 pixels, 426x284pixel, 544x576pixel and 704x576pixel. In this study using WireShark as a data analysis tool and service quality on streaming networks. From the test results obtained an average delay of 86.14484 ms for MPEG-2 video codec with a resolution of 426x284 pixels 25 fps. The jitter of 13,96518221 ms, packet loss of 15.22%.

1. Introduction

Wireless multimedia communication via Wireless Local Area Networks (WLAN) has received attention in recent years, this is due to the agreement of technologies such as Bluetooth, IEEE 802.11, 3G, and WiMAX which specifically emerged as a technology applicable to broadband wireless access because it supports real-time Multimedia conversation applications such as Voice and video conferencing. Local Area Networks Wireless Networks, providing high-level data services at a low cost because they work on the 2.4GHz frequency which is license-exempt to be used in the fields of education, industry, scientific, and medical.

The use of (WLAN) as a means of streaming communication has become a trend for now, the IEEE 802.11n-2009 network commonly abbreviated as IEEE 802.11n is a standard of wireless networks (wireless networking standard) which uses many antennas to improve performance. With IEEE 802.11n, bandwidth can be increased from 54 Mbit / s to 600 Mbit / s. Channel width also increased from the previous 20 MHz to 40 MHz. In IEEE 802.11n added support for MIMO (Multiple Input Multiple Output). and frame aggregation. Thus the use of the IEEE 802.11n network is very supportive of streaming services.

In building a video streaming service network over a network, we need to take into account the bandwidth requirements for the success of a video conference. Generally, there are two bandwidth requirements that need to be fulfilled, the first is the bandwidth requirement for sending picture/video signals and the second is the bandwidth requirement for sending audio/video signals. Between the two requirements above, the need for video sending bandwidth is very important because it will take up most of the existing communication bandwidth. It is not surprising that techniques for data compression have become very strategic which allows telecommunications bandwidth savings.

The problem faced in building streaming services is the process of capturing and live to decode on the server-side. In addition to problems with the server, the biggest problem faced by this technology is bandwidth limitations. Streaming media applications that require a high enough bitrate cause network load to increase, causing the services provided to not run properly. The urgency of this research is to find out how much influence the frame rate in the network, with the available bandwidth.

2. Method

Noriaki Kamiyama [1] has conducted research related to streaming HD video on the internet. When transmitting the bit-rate generated is quite large, so that traffic on the network is too high and takes data traffic. Yung-Sung Huang [2] conducted a study related to video streaming services using webcam media on hospitals that were applied to patients and all-important medical data was transmitted using a 3G-wireless network communication system to various client devices.

There are two standards bodies responsible for locating video coding standards, the International Organization of Standards (ISO) and the International Telecommunications Union (ITU), which have developed a set of standards that have shaped the development of the visual communication industry. ISO JPEG, MPEG-1, MPEG-2, and MPEG-4 standards have the biggest impact: JPEG has become one of the most widely used formats for image storage and MPEG-2 forms the basis of video system formats on digital television and DVDs [4]

The ITU H.261 standard was originally developed for video conferencing via ISDN, but H.261 and H.263 are now widely used for real-time video communication over various networks including the Internet. H.264 from ITU-T is known as the International Standard video encoding, which is the latest standard in the H.261 video coding standard sequence [5]. Each international standard also has a common approach to meet these goals. Video coding standards explain the syntax for representing compressed video data [6].

2.1. MPEG-2

MPEG-2 is a determination for a group of coding and compression for audio and video, which is approved by MPEG and published as an international standard ISO / IEC 13818. MPEG-2 is usually used to encode audio and video for broadcast signals, including direct broadcast satellites and cable television. MPEG-2 with some modifications is also the coding format used in commercial DVD films. Using MPEG2 it is necessary to pay license fees to patent holders through the MPEG Licensing Association. The MPEG-2 standard is designed to provide the ability to compress, encode, and transmit multimedia signals on high-quality multichannel, satellite distribution, and broadband networks [7]. MPEG-2 consists of three main parts namely: video, audio (based on MPEG-1 audio coding).

2.2. H264

H.264 / AVC is the latest motion-compensation codec standard. This codec standard was developed by ITU-T Video Coding Expert Group (VCEG) together with ISO / IEC Moving Picture Expert Group (MPEG) and is a partnership effort known as the Joint Video Team (JVT). The ITU-T H.264 standard and the ISO / IEC MPEG-4 AVC standard (official, ISO / IEC 14496-10 - MPEG-4 Part 10, Advanced Video Coding) are developed together so that they have identical technical content. H.264 is used in applications such as Blu-ray Discs, videos from YouTube and iTunes Store, DVB broadcasts, live broadcast satellite television services, cable television services, and real-time video conferencing.

The goal of the H.264 / AVC standard is to be able to provide good video quality at a price slightly lower than the previous standard (for example half or less the MPEG-2 H.263 bit rate, or MPEG-4 Part 2), without increasing the design complexity so much so it would be impractical or too expensive to implement.

Additional objectives are to provide flexibility to be applied to various applications on various networks and systems, including low and high bit rates, low and high-resolution video, broadcasting, DVD storage, RTP / IP packet networks, and multimedia ITU-T telephone systems.

The first H.264 / AVC standardization was completed in May 2003. The JVT then developed an extension to the original standard known as the Range Fidelity Extension (FRExt). This extension enables high-quality video by supporting precise coding with increased color sharpness and higher resolution, including the sampling structure known as Y'CbCr 04:02:02 (= YUV 04:02:02) and Y'CbCr 04:04: 4. Some other features are also included in the Range Fidelity Extensions project, such as adaptive changing between 4×4 and 8×8 integers.

The name H.264 follows the ITU-T naming convention, where the standard is a member of the H.26x VCEG video coding standard; The MPEG-4 AVC name relates to the naming conventions in MPEG ISO / IEC, where the standard is part 10 of ISO / IEC 14496, which is a standard known as MPEG-4. [8]

The H.264 video format has a very wide range of applications which include all forms of digital video compression from low-bit streaming internet applications, to broadcast HDTV and Digital Cinema applications. By using H.264, save a bit rate of 50% or more. The quality of digital satellite TV can be achieved at 1.5 Mbit / s, compared to the current operating point of MPEG-2 video at around 3.5 Mbit / s. To ensure compatibility and adoption that is trouble-free, many standard bodies have changed or added H.264 / AVC to their video standards so users can use H.264 / AVC.

3. Experimental scenario

3.1. Network Testing Scenarios

In this study applying two different video codecs H.264 and MPEG-2. Video streaming files are transmitted using two different protocols namely HTTP and UDP. Transmission performance will be measured using the WireShark tool. In streaming network testing, several scenarios are performed which aim to analyze the performance of services using Wireshark software. This software will capture all data passing through and then analyzed by calculating the value of delay, jitter, packet loss, and throughput.

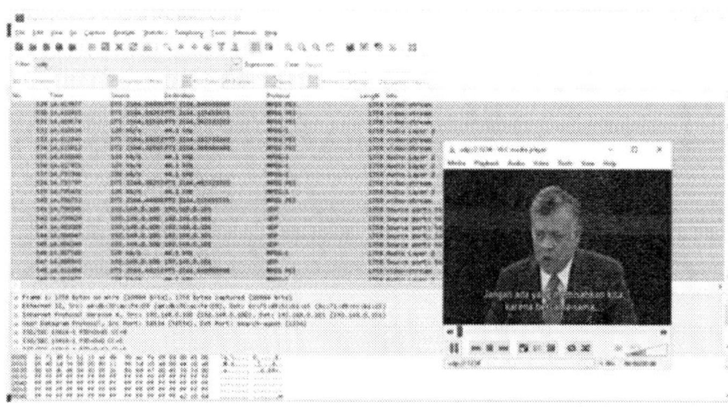

Figure 1: The results of capturing streaming video from the VideoLan application using the Wireshark tool that analyzes packets on the UDP protocol.

For the first scenario video streaming from server to client will be done with the VideoLan Client software with the H.264 video codec using the HTTP protocol. Testing in this scenario uses different video resolutions with frame number 25. Then proceed with the second scenario testing using the UDP protocol, the same video file format as the first scenario, the third scenario testing using the same method with the first scenario only format the file is replaced by the MPEG-2 codec and the fourth scenario tests using the UDP protocol and video format the same as the third scenario. Then Wireshark that has been installed on the client will be used to capture packets that pass through the network, namely UDP packages, Figure 1 shows the results of the Wireshark capturing tool.

3.2. Measurement and Analysis

In this section analysis will be made of the performance of the video streaming network, this analysis will be carried out when streaming video from the server to the client. The analysis process is carried out on the QoS parameters, namely delay, military, packet loss and throughput obtained from each scenario. For discussion of the scenario has been explained in the previous section. Data retrieval is done by using two files with different formats, namely H.264 and MPEG-2 with different file sizes.

3.2.1. Delay Measurement and Analysis

Delay is the time taken by the packet from the sender to the receiver. Delay is a parameter needed to determine the performance of a streaming video network quality. Based on the ITU-T G.1010 standard for good video streaming quality, the delay must be <150ms.

This test will stream video from the server to the client with the VideoLan Client software. Packets that pass on the network will be captured from the client-side using Wireshark software. The results of the measurement of the delay parameters can be shown in Tables 3.1 and 3.2

Table 3.1: Delay Measurement Results for the H.264 video codec

Video format and resolution	Protocols	
	HTTP	UDP
H.264 320x576 pixel 25 fps	86.3545	29.4403
H.264 352x576 pixel 25 fps	84.1125	29.4307
H.264 426x284 pixel 25 fps	86.8876	29.5205
H.264 544x576 pixel 25 fps	86.8741	30.9401
H.264 704x576 pixel 25 fps	86.4955	30.7741

Table 3.2: Delay Measurement Results for MPEG-2 codec

Video format and resolution	Protocols	
	HTTP	UDP
MPEG-2 320x576 pixel 25 fps	86.4686	29.5762
MPEG-2 352x576 pixel 25 fps	90.1779	30.3425
MPEG-2 426x284 pixel 25 fps	98.9255	39.4704
MPEG-2 544x576 pixel 25 fps	95.1976	61.0027
MPEG-2 704x576 pixel 25 fps	85.4732	65.6233

From the measurement results show the difference in the value of delay for each of the data in the test delay generated from the results of the H.264 video code data testing with testing on the HTTP and UDP protocols can be seen the results of an average delay difference of 64.96% between these protocols. The increase in the delay value occurs in the HTTP protocol. The biggest increase in delay value is owned by the test data with MPEG-2 video codec at a resolution of 426x284 pixels 25 fps. The magnitude of the measured delay is due to the small number of packets that can reach the client within the timeframe of sending video streaming packets. When there is communication there is a disruption so there is a delay in the arrival of packets. An increase in the value of the delay can also occur due to the increasing number of packets that come. The increase in the value of this delay results in a decrease in the value of throughput in accordance with the analysis of the average throughput can be seen in table 3.8.

IOP Conf. Series: Materials Science and Engineering **854** (2020) 012066 doi:10.1088/1757-899X/854/1/012066

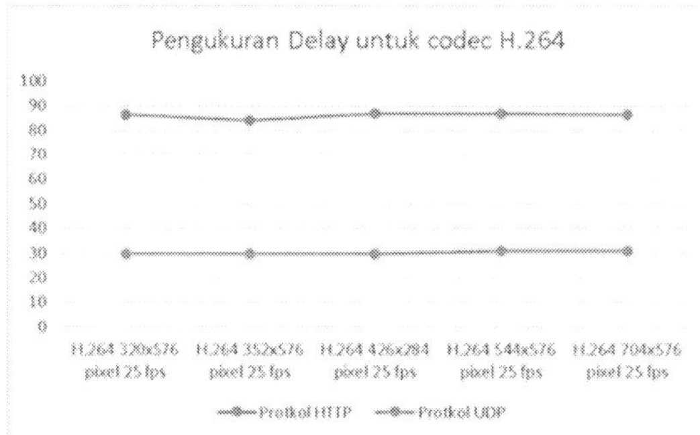

(a)

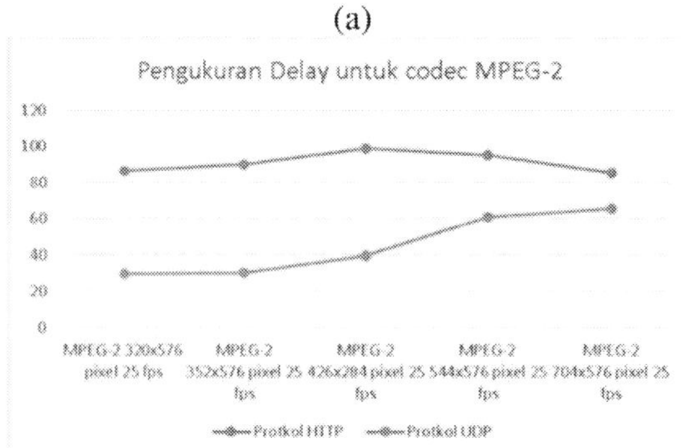

(b)

Figure 2: (a) Delay Measurement Results for H.264 codec with different video resolutions
(b) Delay Measurement Results for MPEG-2 codec with different video resolutions

Based on Figure 2 it can be seen that each delay has changed, the largest delay value occurs in the testing of the HTTP protocol with MPEG-2 test data resolution of 426x284pixel 25 fps, which is equal to 98.9255ms file size that is streamed at 15.7 MB. The delay value from the implementation results is still within the tolerance limits according to the International Telecommunication Union (ITUT G.1010) recommendation [3] for video streaming services. Therefore, the delay value from the implementation is still acceptable for video streaming services.

3.2.2. Jitter Measurement and Analysis

Jitter is a variation of the time of arrival of each package. Jitter can be measured from the time between the packet received now and the package received previously. Jitter is caused by different trajectories between packages, variations in queue length, and data processing times. Telecommunication Union (ITUT G.1010) [3] recommending a good jitter is <30 ms. At the time of the trial, the measured jitter is the average jitter (jitter) of the jitter of all video packages captured by the Wireshark tool. In the measurement of jitter also performed the same scenario with the measurement of delay. The following is an example of jitter calculation and the results in table Table 3.3 and Table 3.4.

IOP Conf. Series: Materials Science and Engineering **854** (2020) 012066 doi:10.1088/1757-899X/854/1/012066

$$Average\ jitter = \frac{total\ delay\ variation}{total\ package\ received - 1}$$
$$= \frac{176173.210155.82286}{12404 - 1}$$
$$= 13.96518221\ ms$$

Table 3.3 Jitter Measurement Results for the H.264 codec

Video format and resolution	Protocols	
	HTTP	UDP
H.264 320x576 pixel 25 fps	13.9652	12.3407
H.264 352x576 pixel 25 fps	14.0442	13.1459
H.264 426x284 pixel 25 fps	13.9803	12.3815
H.264 544x576 pixel 25 fps	13.9626	13.2417
H.264 704x576 pixel 25 fps	13.8932	13.3817

Table 3.4 Jitter Measurement Results for MPEG-4 codecs

Video format and resolution	Protocols	
	HTTP	UDP
MPEG-2 320x576 pixel 25 fps	9.8862	12.7646
MPEG-2 352x576 pixel 25 fps	9.1892	13.4986
MPEG-2 426x284 pixel 25 fps	14.8361	12.4412
MPEG-2 544x576 pixel 25 fps	6.816	5.8651
MPEG-2 704x576 pixel 25 fps	5.0938	4.7006

In table 3.3 and table 3.4 there is an increase in the value of jitter 14.8361 ms in the test data with MPEG-2 video codec resolution of 426x284 pixels 25 fps. The increase in the value of jitter for files measuring 15.7 MB Large value of jitter is influenced by the delay that occurs in the packet while in the router can be seen in Table 3.2. in testing with this test data the result is smaller than the value of 30 ms which is a good jitter standard. The results of this experiment stated the value of jitter still met the standard for Quality of Service. This happens because of the buffer used by the real-time video streaming application. Therefore, jitter does not greatly affect the video that is run by real-time video streaming applications.

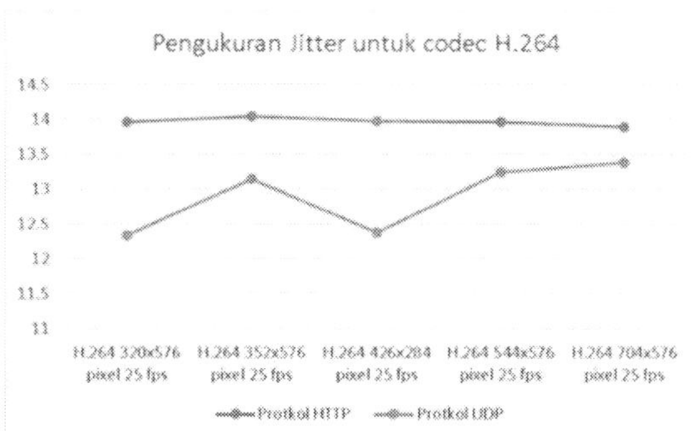

(a)

IOP Conf. Series: Materials Science and Engineering **854** (2020) 012066 doi:10.1088/1757-899X/854/1/012066

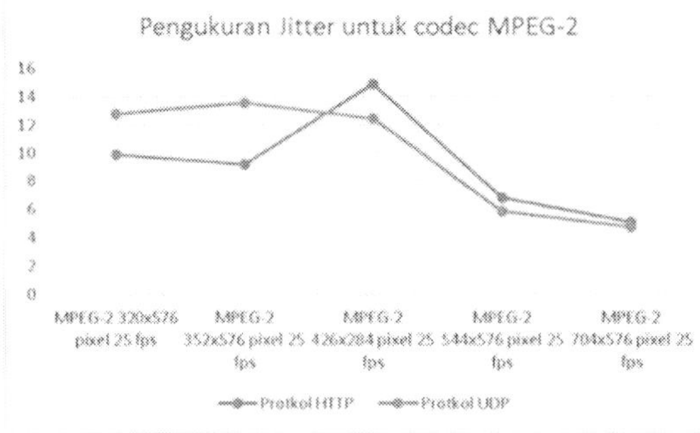

(b)

Figure 3: (a) Jitter Measurement Results for H.264 codec with different video resolutions (b) Jitter Measurement Results for MPEG-2 codec with different video resolutions

Based on Figure 3 it can be seen that the jitter value of each time changes. The size of jitter is in the range of 4,706 ms to 14,8361 ms. The value of jitter is caused by buffers used by real-time video streaming applications and the influence of unstable wireless-based network conditions. Wireless speed and signal strength from access points can change at any time.

3.2.3. Measurement and Analysis of Packet Loss

Packet loss aims to determine the amount of packet loss when streaming video from source address to destination address. The greater packet loss causes the quality of the video received. The results of the measurement of packet loss parameters are shown in table 3.5 and table 3.6. Give an example of the calculation of test results using the UDP protocol with H.264 video format 320x576 pixel 25 fps resolution.

$$Packet\ Loss = \frac{(Packets\ sent - packets\ received)}{packets\ received} x100\%$$

$$Packet\ Loss = \frac{(14556 - 14376)}{14556} x100\%$$

$$= 1.236603462489695\%$$

Table 3.5: Jitter Measurement Results for H.264 codec

Video format and resolution	Protocols	
	HTTP	UDP
H.264 320x576 pixel 25 fps	0	1.23
H.264 352x576 pixel 25 fps	0	1.35
H.264 426x284 pixel 25 fps	0	1.36
H.264 544x576 pixel 25 fps	0	1.46
H.264 704x576 pixel 25 fps	0	1.41

Table 3.6: Jitter Measurement Results for MPEG-4 codecs

Video format and resolution	Protocols	
	HTTP	UDP
MPEG-4 320x576 pixel 25 fps	0	1.38
MPEG-4 352x576 pixel 25 fps	0	1.45
MPEG-4 426x284 pixel 25 fps	0	1.3
MPEG-4 544x576 pixel 25 fps	0	0.61
MPEG-4 704x576 pixel 25 fps	0	0.5

Figure 4 can explain that the average packet loss when the system is streaming video ranges from 0.0% to 1.46%, where the packet loss is still within the tolerance limit. Packet loss below 10% is still permitted. Packet loss that occurs is caused at the time of data retrieval, as well as existing traffic conditions. From Figure 4 it can be seen that the large packet loss is owned by the H.264 video codec test data with a resolution of 544x576 pixels 25 fps with UDP protocol at table 3.5. The number of packets that are lost on network streaming results in a smaller measurable throughput of data seen in table 3.6. A large packet loss rate can reduce the value of throughput. A large packet loss rate in real-time video streaming applications results in certain parts of the video being interrupted. If there are a lot of broken parts, then the information will also be reduced.

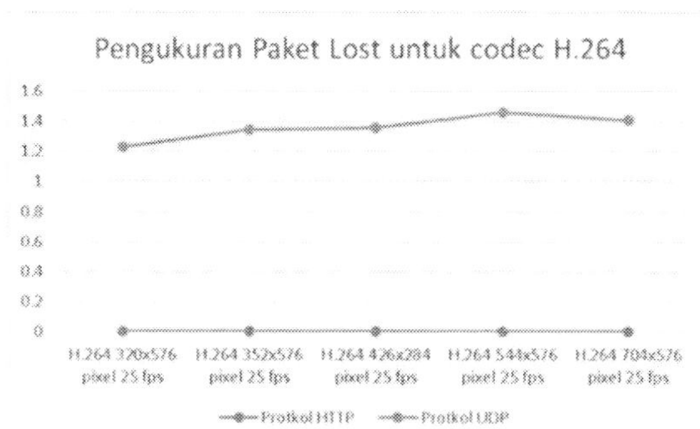

(a)

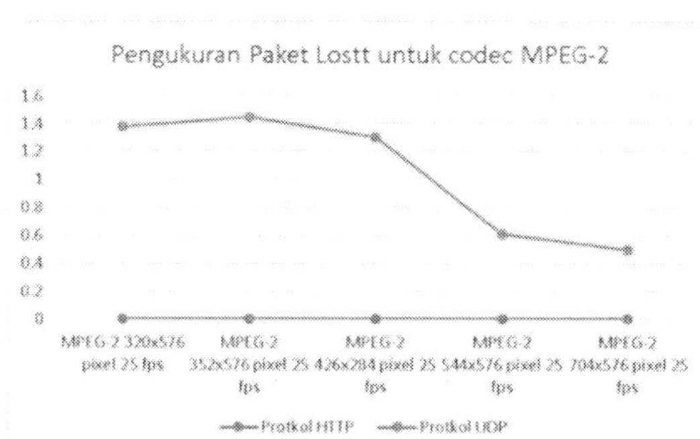

(b)

Figure 4: (a) Results of lost packet measurements for H.264 codecs with different video resolutions (b)Results of lost packet measurements for MPEG-4 codecs with different video resolutions

IOP Conf. Series: Materials Science and Engineering **854** (2020) 012066 doi:10.1088/1757-899X/854/1/012066

3.2.4. Measurement and Analysis of Throughput

Throughput is measured based on packet transmission speed. The experiment was conducted several times to find the average (average) throughput. In the trial of measuring throughput for video transmission from server to client on this network, the average value of throughput is obtained. The throughput values are shown in Table 3.7 and Table 3.8. Data throughput in this study is taken from the packet capture with the Wireshark sniffing tool, which can be seen in Figure 5.

Table 3.7: Throughput Measurement Results for the H.264 codec

Video format and resolution	Protocols	
	HTTP	UDP
H.264 320x576 pixel 25 fps	0.858	0.834
H.264 352x576 pixel 25 fps	0.844	0.791
H.264 426x284 pixel 25 fps	0.855	0.826
H.264 544x576 pixel 25 fps	0.858	0.771
H.264 704x576 pixel 25 fps	0.845	0.806

Table 3.8: Throughput Measurement Results for the MPEG-2 codec

Video format and resolution	Protocols	
	HTTP	UDP
MPEG-2 320x576 pixel 25 fps	1.159	0.82
MPEG-2 352x576 pixel 25 fps	1.281	0.756
MPEG-2 426x284 pixel 25 fps	0.817	0.834
MPEG-2 544x576 pixel 25 fps	1.763	1.798
MPEG-2 704x576 pixel 25 fps	2.251	2.271

In tables 3.7 and table 3.8 it can be seen that the comparison of throughput on the results of all tests. Based on Figure 6table

from the measurement results of two file formats namely H.264 and MPEG-2 that the smallest throughput occurs in the test file with MPEG-2 format resolution of 352x576 pixels 25fps in table 2.8. which is 0.756Mbps, while the largest throughput occurs in the MPEG-2 test file with a resolution of 704x576 pixel 25 fps which is 2,271 Mbps. The size of the data packet received is due to different file sizes. This is in accordance with the understanding of throughput, namely the greater the data sent, the greater the data received.

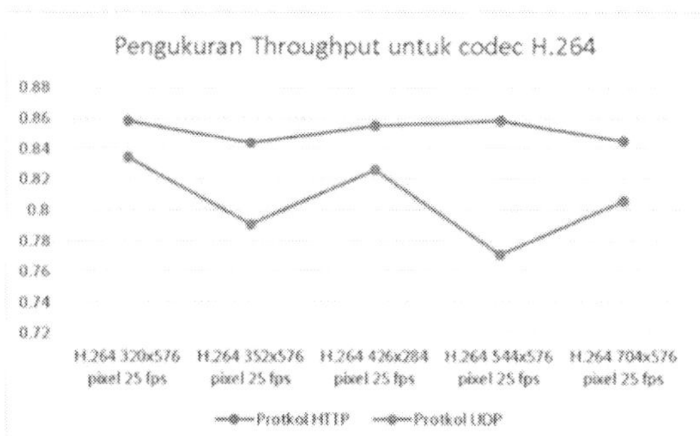

(a)

IOP Conf. Series: Materials Science and Engineering **854** (2020) 012066 doi:10.1088/1757-899X/854/1/012066

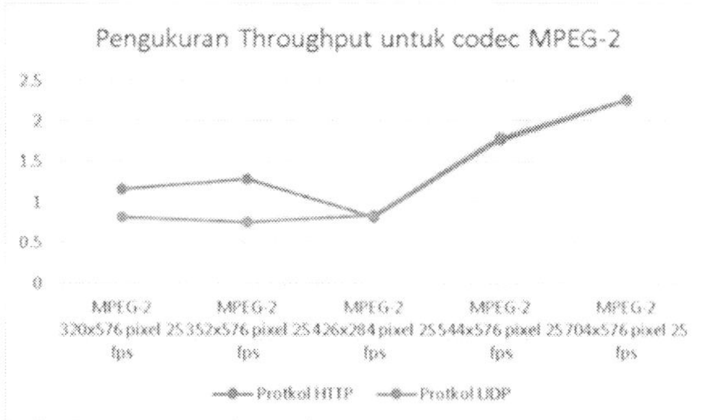

(b)

Figure 5: (a) Results of Measurement of Throughput for codec H.264 (b) Results of Measurement of Throughput for codec MPEG-2

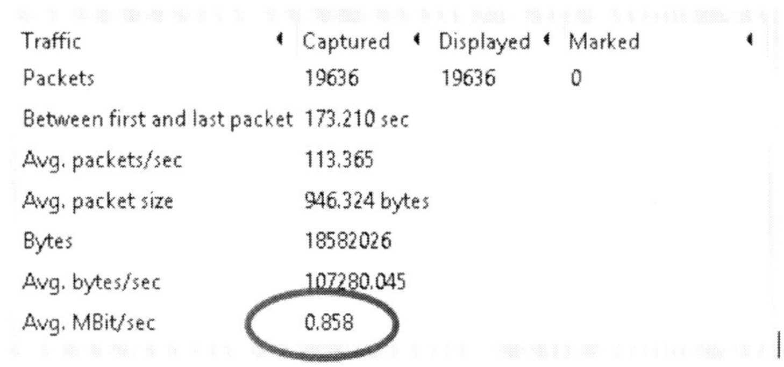

Traffic	Captured	Displayed	Marked
Packets	19636	19636	0
Between first and last packet	173.210 sec		
Avg. packets/sec	113.365		
Avg. packet size	946.324 bytes		
Bytes	18582026		
Avg. bytes/sec	107280.045		
Avg. MBit/sec	0.858		

Figure 6: Wireshark summary to see the average packet size streamed, average streaming speed and bandwidth used by H.264 544x576 pixel 25 fps data with HTTP protocol

4. Conclusion

From the test results obtained an average delay of 86.14484 ms for the MPEG-2 video codec with a resolution of 426x284 pixels 25 fps. The jitter of 13,96518221 ms, packet loss of 15.22%. The results of the experiment show that MPEG-2 jitter time is less than H.264. So the MPEG-2 protocol is better than H.264 through the UDP protocol. In contrast, H.264 jitter time is less than MPEG-2 via the HTTP protocol. So H.264 is better than MPEG-2 via the HTTP protocol. But in terms of appearance H.264 is better than MPEG-2. H.264 image quality is also better and smoother playback than with MPEG-2 basic compression. But the most interesting feature is the lower bit-rate needed for network transmission.

References

[1] Noriaki Kamiyama, Ryoichi Kawahara, Tatsuya Mori, Shigeaki Harada, Haruhisa Hasegawa, "Parallel video streaming optimizing network throughput", Elsevier, 2010.

[2] Yung-Sung Huang, "A portable medical system using real-time streaming transport over 3G wireless networks", Springer-Verlag, 2010.

[3] International Telecommunication Union SERIES G: Transmission Systems and Media, Digital Systems and Networks Quality of service and performance (ITUT G.1010) End-user multimedia QoS categories 2001.

[4] Iain E.G. Richardson, Video Codec Design- Developing Image and Video Compression Systems, John Wiley & Sons Ltd, 2002.

[5] Jörn Ostermann, Jan Bormans, Peter List, Detlev Marpe, Matthias Narroschke, Fernando Pereira, Thomas Stockhammer, and Thomas Wedi, Video coding with H.264/AVC: Tools, Performance, and Complexity, IEEE CIRCUITS AND SYSTEMS MAGAZINE, 2004.

[6] Jerry D. Gibson, Multimedia Communications Directions and Innovations, Academic Pres, 2001.

[7] K. R. Rao, Zoran S. Bojkovic, Dragorad A. Milovanovic, Introduction to multimedia communications Applications, Middleware, Networking, JOHN WILEY & SONS, INC., 2006.

[8] ITU-R RECOMMENDATION BT.500-11: Methodology for the subjective assessment of the quality of television pictures, Geneva, 2002.

IOP Conf. Series: Materials Science and Engineering **854** (2020) 012067 doi:10.1088/1757-899X/854/1/012067

Mapping of basement layer in field "X" petroleum system using ESA – MWT technique gravity data

M S Rosid and J Andrian

Department of Physics, FMIPA Universitas Indonesia, Depok 16424, Indonesia

E-mail: syamsu.rosid@ui.ac.id

Abstract. Land gravity data acquired over part of North Sumatera Basin, Indonesia, was used to estimate depth to multiple horizons of density interface. Recently, gravitational data has difficulty to map the depth of subsurface layer structure except by using modeling. Here, we apply the energy spectral analysis–multi window test (ESA–MWT) technique to map the subsurface horizons. The estimated depth was computed from energy spectral analysis as the transformation of the gridded gravity data to 2D Fourier transform (Fast Fourier Transform). The energy spectrum analysis is performed at a test point by windowing the gridded complete Bouguer anomaly (CBA) at that test point with the square window size constantly increasing 1 km width from the previous window size. The distance between the test points is about 2 km on each of the gravity measurement lines. Plateau depths are then obtained which represents the density interfaces of the plot result between the depths of the anomaly to the window size. The results of the mapping of density interfaces correspond to sedimentary interfaces, those are the economic basement as Top Tampur Formation (Horizon 1), the top of several other sedimentary interfaces (Horizon 2-6), one of them is Top Belumai Formation (Horizon 2).

1. Introduction

Petroleum system of field "X" is the part of one of the back-arc basin in Northern Sumatera, Indonesia, formed by the convergence subduction activity of Indo-Australian oceanic and Eurasian continental plates.

Gravity data integrated with the seismic section can be used in petroleum exploration. The horizon of the sedimentary interfaces as contrast density can be mapped by applying energy spectral analysis – multi window test (ESA-MWT) technique gravity data. The technique will be useful especially in an area where seismic has difficulties, such as the onshore petroleum system "X". The ESA–MWT technique has been successfully applied in some fields. Offshore Andaman Sea basement has been mapped clearly and was confirmed by regional geology and geophysical data from the nearest field (Kivior *et al.* 2012). Markham *et al.* (2011) also mapped onshore field of Delta Baram, Sarawak that has a good correlation with seismic data, and onshore field of Canning Basin (Kivior *et al.*, 2018) that has good correlation with seismic and well data. The objective of the study is to map the basement layer as well as other horizontal layers in the field "X" petroleum system by using ESA-MWT technique gravity data.

The location of the study in the field "X" has the fault structures that overlap (complex). This makes the depth of the sedimentary layer varied, which becomes a challenge in the process of structure mapping. Contrast density between the basement and the above sediments is the main reason for the success of mapping the top basement horizon. Gravity data were acquired over a 50x70 km^2 area with 11 measurement lines and the tie lines. The distance between the gravity measurement points is about 1 km on each of the gravity measurement lines.

2. Regional Geological Setting

The petroleum system "X" is part of a NW-SE oriented back-arc basin formed by convergence tectonic activity in the perpendicular direction (figure 1). This back-arc basin is a sedimentation depocenter as the Pre-Tertiary granitic and metamorphic basement. The basin platform is a boundary that borders the shallow basement area exposed as a Pre-Tertiary outcropped basement as a mountain in the Southwest and basin slope complex and basin deep in the Northeast. Tertiary sediments begin to be deposited as carbonate platforms (until Late Eocene) as the economic basement that makes up Tampur Formation. Then above the Tampur Formation, is overlaid by clastic that dominated by sandstones in the formation above it (Oligocene to recent) (Doust and Noble, 2008).

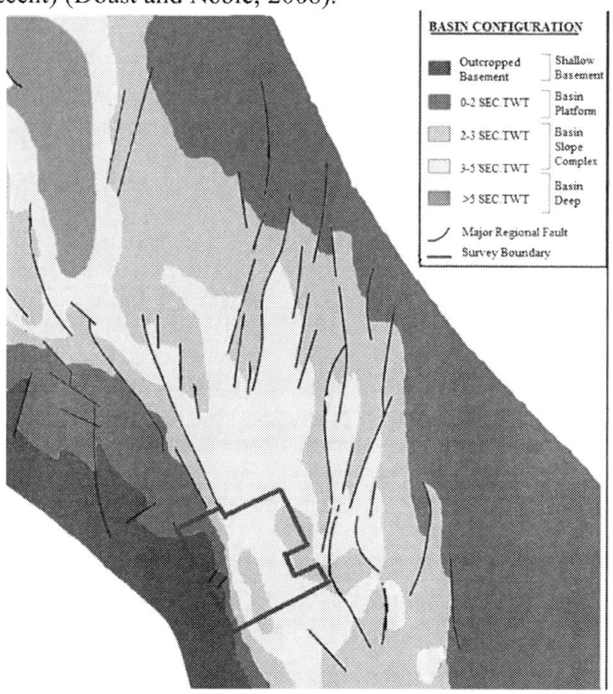

Figure 1. Basin configuration with a prospect area which bounded by blue lines (modified from Doust and Noble, 2008).

3. Gravity Data Reduction

The reduction process of gravity data is done to obtain a complete Bouguer anomaly (CBA) using equation (1). Some corrections are applied to the observed gravity data (G_{obs}). The corrections are normal gravity correction (G_N), free-air correction ($FAC = 0.3086\ h$), Bouguer correction ($BC = 0.04192\ \rho h$), and terrain correction ($TC = \rho T$). The parameter of h is an elevation of gravity station in meters, ρ is the average density of the area in gr/cc, and T is the geometrical factor of terrain correction in mgal.cc/gr.

$$CBA = G_{obs} - G_N + FAC - BC + TC \tag{1}$$

Estimation of the average density of the area can be calculated using Parasnis method by modifying CBA's equation to be a linear equation $y = mx + c$, where an independent variable x as $(0.04192\ h + T)$, a dependent variable y as $(G_{obs} - G_N + FAC)$, and the slope of the linear equation m as the density ρ in gr/cc. The separation of the density counting area was done as shown in figure 2 because the density calculation was not successful in total area. The density value of the total area of 3.66 gr/cc is not scientifically acceptable. The results of the calculation of the average density show a good correlation with the geological information in figure 1. The higher gravity anomaly area in the Southwest direction relates to a higher density area of 2.53 gr/cc compared to the lower gravity anomaly in the Northeast direction of 2.27 gr/cc. Shallow basement in the high gravity area resulting in a density of the area is dominated by basement rocks that have a higher density. In the final calculation, we obtain a value of the average density estimation of 2.37 gr/cc as shown in equation (2) which is then used for gravity data reduction.

$$\bar{\rho} = \frac{\sum \rho_i A_i}{\sum A_i} = \frac{(2.5274 \times 925715549) + (2.27 \times 1572297619)}{(925715549 + 1572297619)} = 2.365\ \text{gr/cc} \tag{2}$$

where ρ_i are the densities of each area in gr/cc, and A_i is the areas in m^2. The gridded CBA contour is processed with 100 m grid cell size, resulting in a CBA map.

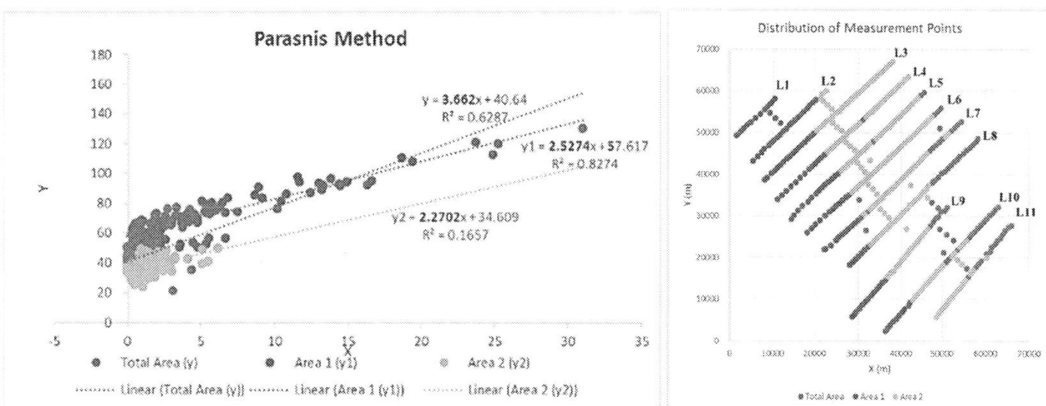

Figure 2. (a) Parasnis method for calculating the average density area, and (b) distribution of measurement points (separation points) to calculate the estimation of the average density of the area, L1-L11 is the measurement lines.

4. Methodology

ESA – MWT technique is applied to the mapping horizon in the area of study. The technique is based on spectral energy analysis (ESA) that utilizes 2D Fourier transforms based on the original work of Bhattacharyya (1966). Fourier transforms work mathematically to transform the function of the anomaly amplitude in the spatial domain into the function in the frequency domain (wavenumber). Lower wavenumber related to deeper anomaly components, this is identical to the regional gravity component at conventional interpretation. And higher wavenumber related to shallower anomaly component. By following Spector and Grant (1970), a causative body is modeled by a statistical multi-prism with

IOP Conf. Series: Materials Science and Engineering **854** (2020) 012067 doi:10.1088/1757-899X/854/1/012067

uniform density or susceptibility. The curve of the log spectrum energy to the radial frequency (the radially averaged power spectrum/RAPS curve) can be interpreted to determine the depth of the causative body as,

$$Depth = -\frac{s}{4\pi} \tag{3}$$

where s is the slope of RAPS curve.

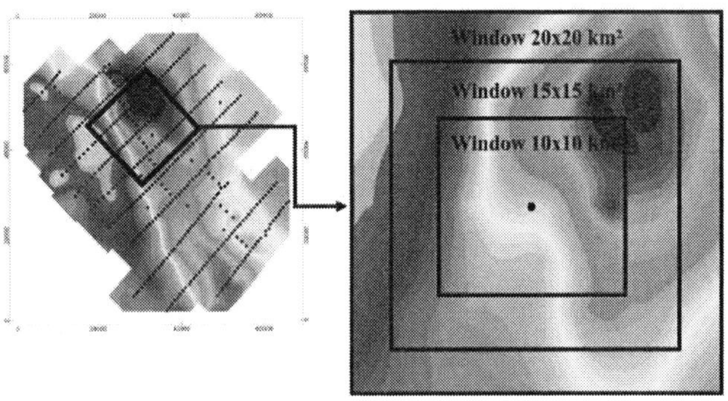

Figure 3. The windowing CBA map at a test point (30412, 45141).

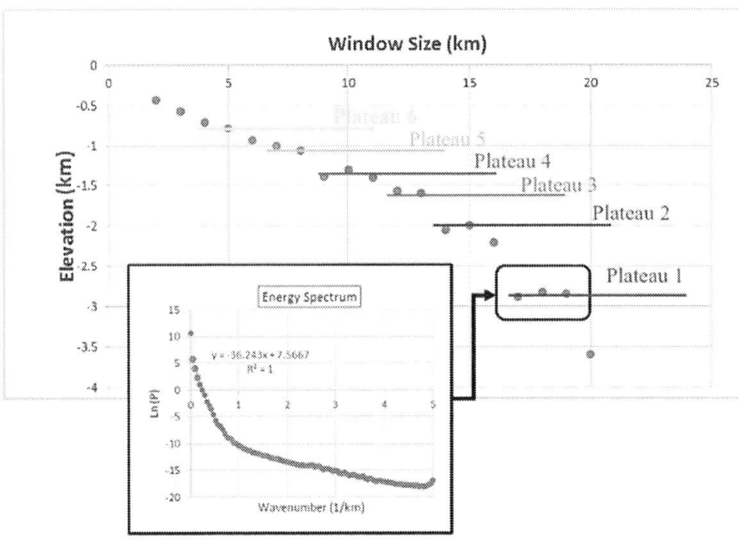

Figure 4. The multi-window test (MWT) graph at the same test point (30412, 45141) of figure 3.

Multi-window technique is applied to a test point as the centre of the window. Anomaly limiting values aim to localize anomalies where the size of the window corresponds to the depth of the anomalous body, the larger the window, the deeper the anomaly is contained in the window. By calculating the energy spectrum in that window series, where the window size varies constantly increases. Depth plateau is detected when the depth of the source ensemble does not change significantly between two or more window sizes (Kivior *et al.*, 2012). The depth plateau is used to approximate density interfaces. The processing procedures are shown in figure 3 and figure 4.

ESA – MWT technique is performed at a test point by windowing the CBA map at that test point with the square window size constantly increasing 1 km width from the previous window size. The

IOP Conf. Series: Materials Science and Engineering **854** (2020) 012067 doi:10.1088/1757-899X/854/1/012067

window size starts from 4x4 km^2 to 20x20 km^2. The distance between the test points is about 2 km on each of the gravity measurement lines.

5. Results and Discussion

The ESA-MWT mapping results in 6 sedimentation horizons. Integration with estimated horizons from a seismic section on the L6 measurement line (figure 5), Horizon 1 is suspected to correlate with top economic basement as top Tampur Formation, Horizon 2 is suspected to correlate with top Belumai Formation which is containing limestone and shale, and dominated by sandstone. Horizon 3-6 are suspected as layers of clastic sediments above Belumai Formation (intra-sedimentary and shallower sediment), Horizon 4 is suspected to correlate with top MBS Formation as the top Intra-Sediment 2. In general, the pattern of the horizon mapping results from ESA - MWT correlates with the seismic section. However, there is some deviation on Horizon 1 and 2 which is a deeper depth there is a shift to the estimated seismic horizon. In the middle (anticlines) shifts upward about 1 km and to the right, and deviation at some other location.

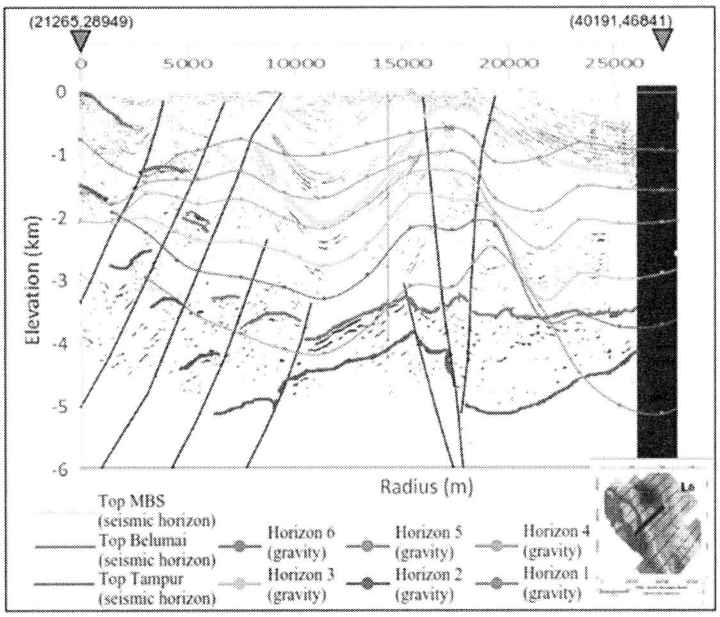

Figure 5. Correlation between horizons from ESA–MWT with estimated horizons from the seismic section at measurement line L6.

The existence of a highly complex fault structure associated with the sedimentation layer in this area, making the undulation of the depth in a formation very extreme. The fault is likely to be the main factor responsible for the deviation of the horizon depth. The tectonic activity causing the faults are likely to create a high degree of compression in the syncline area. Thus, the density of rocks above Tampur Formation may not be too contrasted with Tampur Formation causing the mapping to shift.

The economic basement mapped using the ESA–MWT technique has been justified by the pattern of the estimated horizon from the seismic data. The three deepest horizons (figures 6a, 6b, 6c) have extreme depth undulation which is typical of the formation in the field "X" petroleum system. This is due to the presence of many reverse faults (figure 5), the undulation of a formation varies at around 2-3 km.

IOP Conf. Series: Materials Science and Engineering **854** (2020) 012067 doi:10.1088/1757-899X/854/1/012067

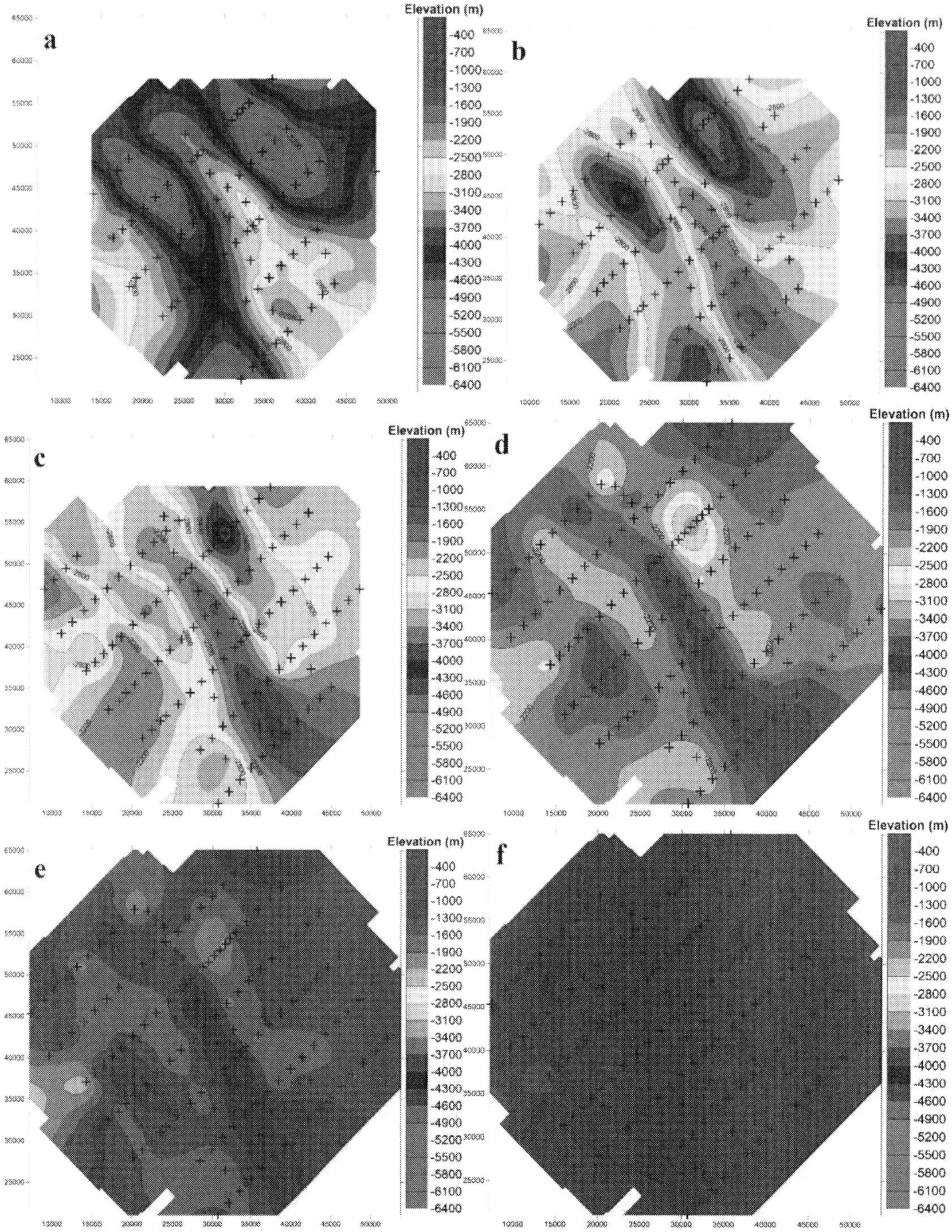

Figure 6. Mapping horizon results: (a) Horizon 1 as the top Tampur Formation (economic basement); (b) Horizon 2 as the top Belumai Formation; (c) Horizon 3 as the top Intra-Sediment 1; (d) Horizon 4 as the top Intra-Sediment 2; (e) Horizon 5 as the top Intra-Sediment 3; (f) Horizon 6 as the top Intra-Sediment 4; cross represents the test points.

The undulation is also shown on the gravity profile as the high rate of gravity change in the SW-NE direction along the cross-section in figure 5, where there is a high anomaly in Southwest whose value drops significantly to the Northeast. Sedimentation depocenter as the low anomaly area separated by a reverse fault group as the group of faults on the left side in figure 5. These reverse faults are a major fault in this study area. Based on geological data (figure 1) these faults appears as a separator of the shallow basement area in the Southwest and the basin depocenter in the Northeast. This field has a high level of structural complexity, which is very potential for the petroleum system with the type of trap as a structural and stratigraphy trap.

The mapping results at 2 shallower horizons (figures 6e and 6f) have no extreme depth. The gravity anomaly data is dominated by the regional anomaly that results in a syncline structure on the shallower horizon that is not well mapped. So, the results of the horizon mapping are flatter than the deeper horizon (see figure 7).

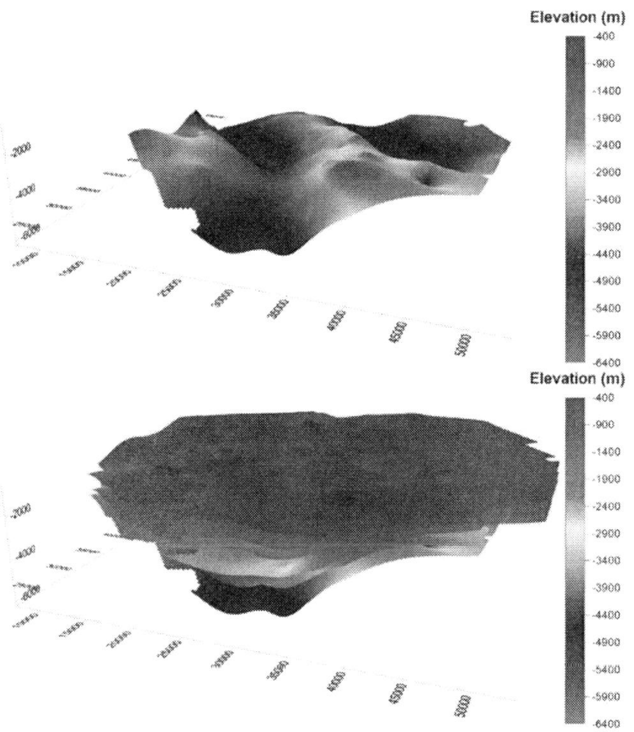

Figure 7. 3D surface view from Tampur economic basement (above) and the whole of mapping horizon result (bottom).

6. Conclusions

We can conclude that in the field "X" petroleum system the mapping of density interfaces using ESA–MWT technique gravity data can provide good estimates of depths of economic basement and the deeper other sedimentary interfaces as the regional anomaly domination. However, there is some deviation in estimating depths of the seismic horizon caused by structures complexity. There are 6 horizons identified, the Horizon 1 as top Tampur Formation (as an economic basement), Horizon 2 as top Belumai Formation, Horizon 3 as top Intra-Sediment 1, Horizon 4 as top Intra-Sediment 2, Horizon 5 as top Intra-Sediment 3, Horizon 6 as top Intra-Sediment 4.

Acknowledgments
We would like to thank the management of PT Pertamina EP for using the data. I also wish to thank the DRPM Universitas Indonesia for financial support in publishing the paper through PITTA 2019 grant.

References

[1] Bhattacharyya, B K. 1966. Continuous Spectrum of the Total-Magnetic-Field Anomaly due to a Rectangular Prismatic Body. *Geophysics,* **31**, 97–121.
[2] Doust, H and Noble, R A. 2008. Petroleum Systems of Indonesia. *Marine and Petroleum Geology*, **25**, pp.103–129.
[3] Kivior, I, Chandola, S K, Markham, S, Chong, L B, Keong, E B, Nordin, A and Hagos, F. 2012. Regional Mapping of Basement and Sedimentary Interfaces in the Andaman Sea Basin Using Marine Magnetic and Gravity Data, *AAPG-ICE 2012 Int. Conf. & Exhibition Singapore* (Singapore: AAPG).
[4] Kivior, I, Markham, S, Hagos, F, Baigent, M, Rudge, T, and Devereux, M. 2018. Improved Imaging of the Subsurface Geology in the Mowla Terrace, Canning Basin Using Gravity Gradiometry Data. *ASEG 2018,* **1**, pp.1-10.
[5] Markham, S, Damte, S, Kivior, I, Randle, S, Shimada, M, Jong, J, Kusaka, H and Tan, T Q. 2011. Mapping Regional Sedimentary Horizons in the Onshore Baram Delta, Sarawak, from Magnetic and Gravity Data Using Energy Spectral Analysis, *The 12th Int. Congress of the Brazilian Geophysical Society,* (Rio de Janeiro: SBGf).
[6] Spector, A and Grant, F S. 1970. Statistical Models for Interpreting Aeromagnetic Data, *Geophysics*, **35**, 293–302.

Interaction of Enzyme-Substrate from Indigenous Cellulolytic Bacteria by Bioinformatics

Yanty Maryanty, Tri adryati, Suharjono, Sutiman Bambang Sumitro

Department of Industrial Chemical Industry, State Polytechnic of Malang, Jl. Soekarno Hatta No.9, Jatimulyo, Kec. Lowokwaru, Kota Malang, Jawa Timur 65141, Indonesia

Email: yanty.maryanty@polinema.ac.id

Abstract-This study aims to predict the degradation mechanism of cellulose substrate in silico from the results of isolation and test the potential of cellulolytic bacteria from rice fields in Greater Malang in producing cellulolytic enzymes, among others: β-1, 4 exoglunase, β-1, 4 endoglucase and β - glucosidase, identified isolates that have high potential in producing cellulolytic enzymes based on 16S rDNA ie isolates A, B, D and F. Only Isolate B was successfully predicted by its substrate enzyme interaction both homologically and based on the results of the isolation of the cellulite gene. Homology B isolate analysis results showed that the hydrogen bonds that occur in glutamic acid GLU257, tryptophan 207, serine 264 and glutamic acid 169, while hydrophobic interactions occur in tryptophan 207 bonds. While the results of the analysis based on isolation of cellulase encoding genes from isolate B were predicted in-silico showed the interaction between hydrogen bonds in tyrosin (TRY 299), glutamine (GLU201 and GLU 342.72), Asparagine (ASN 200) and glycine (GLY 384). This interaction is slightly different from the insilico results obtained homologically i.e there is no interaction with serine and glutamate acid, but with asparagine. Isolate B predicted that its homology in silico can degrade cellulose substrate with the binding affinity of -7.0 Kcal/mol while based on the results of isolation of cellulase encoding genes shows the degradation ability of cellulose substrate with the binding affinity of -6.5 Kcal/mol.

Keywords: Cellulase, in silico, hydrophobic interaction, the interaction of hydrogen bonds, binding affinity

1. Introduction

Cellulose biomass can be degraded by cellulase enzymes produced by cellulolytic bacteria. These enzymes have a variety of industrial applications and are currently referred to as the main group in the enzyme industry. Cellulase contributes eight percent of the needs of the enzyme industry worldwide (Alam et al., 2004; Meryandini et al., 2009; Sadhu & Maiti, 2013). In the past decade, cellulose-degrading bacteria have been isolated from various sources such as soil, weathered plants, hot springs, organic sources, ruminant manure, compost, and rice straw to obtain a more effective source of cellulase enzymes. Enzymes derived from cellulolytic bacteria are generally more resistant to alkali and thermophilic conditions with high cellulase activity (Immanuel et al., 2006; Doi, 2008; Irfan et al., 2012; Sadhu and Maiti, 2013; Feng et al., 2013; Abdel-Rahman, 2016). Enzymatic hydrolysis of cellulose depends on many factors including physical properties of the substrate (composition, crystallinity, and degree of polymerization), enzyme synergy (origin, composition), mass transfer (adsorption of the substrate, bulk and pore diffusion) and intrinsic kinetics. Previous research has succeeded in isolating some cellulolytic bacteria from agricultural land in the Greater Malang area. Interaction of substrate enzymes from cellulolytic bacteria can be predicted by bioinformatics.

Bioinformatics is a study of science that combines the disciplines of molecular biology, mathematics and information technology. Bioinformatics can be defined as the application or use of tools and computational methods (in silico methods) that aim to handle molecular biology data or information. (Thompson et al, 2000). Molecular docking is a computerized method used to describe interactions between a molecule as a ligand and a receptor or protein. Receptors or targets in the docking process can be obtained from experimental results or modelling homology results. Molecular docking as a methodology in structure-based virtual screening began in the early 1980s. In general, the purpose of docking studies is modelling accurate structures and predictions of appropriate activities (Thompson et al., 2000). In the docking process, the molecules involved are considered as atoms, surfaces or grids that represent the molecules. The docking process consists of several complex stages. This process begins with the application of a docking algorithm that positions the ligand on the active site with certain conformations and specific conformational search sequences, then the scoring function that completes the docking algorithm will evaluate the conformation by calculating based on physicochemical properties to obtain optimal molecular structure (Kitchen et al., 2004). Based on this process, a molecular docking program is a combination of algorithm docking and scoring functions (Thompson et al., 2000). Utilization and understanding of the interaction of substrate enzymes in bioinformatics in silico are carried out to help identify and measure the factors that influence the system for the hydrolysis process. This study aims to predict the mechanism of degradation of cellulolytic enzymes from isolates resulting from the isolation of Malang Raya rice fields in degrading cellulose substrate to cellulose substrate in silico.

2. Methodology

2.1. In Silico Analysis of Cellulase and Substrate Interactions

This study predicts the interaction of substrate enzymes from the isolation of cellulolytic bacteria from agricultural land in the Greater Malang area. The selected isolates were isolates A, B, D and F. These four bacteria were tested for cellulolytic activity and showed the highest activity compared to other isolates that had been successfully isolated. The DNA sequences of the four isolates from the sequenced cellulase gene sequences were validated and translated and then modelled using a homology modelling approach with the SWISS-MODEL webservice. The results of modelling simulated its interaction with the glucose substrate. 3D models of glucose substrate were obtained from the PubChem database. Structures in the SDF format are saved and used for further analysis. To find out the binding position of the enzyme and substrate so that the mechanism of action is known, a molecular docking study is performed. The interaction simulation between enzymes and substrate was studied using the Molecular Docking PyRx 0.8 program. The docking process uses autodock vina and is carried out specifically on the active site of the target protein.

Molecular docking can predict the binding affinity of a compound to a specific target protein (Trott & Olson, 2010). The higher the binding affinity of the substrate, the more accurate the interaction prediction will be. The molecular docking process is also based on control of the ligands that have previously interacted. Interactions between substrates and enzymes were analyzed with the Discovery Studio V.4.0 program. The purpose of this analysis is to understand the types of amino acids involved in interactions, so it can be concluded the active side and mechanism of action of the enzyme (Trott & Olson, 2010).

3. Results and Discussions

3.1. In silico analysis of cellulase and substrate interactions

The purification results of the selected cellulase encoding genes A, B, D, and F have been sequenced to ensure that the gene is well amplified. The results of this sequencing are useful for knowing the sites of cutting restriction enzymes contained in cellulase genes. These restriction enzyme sites are very useful for genetic engineering.

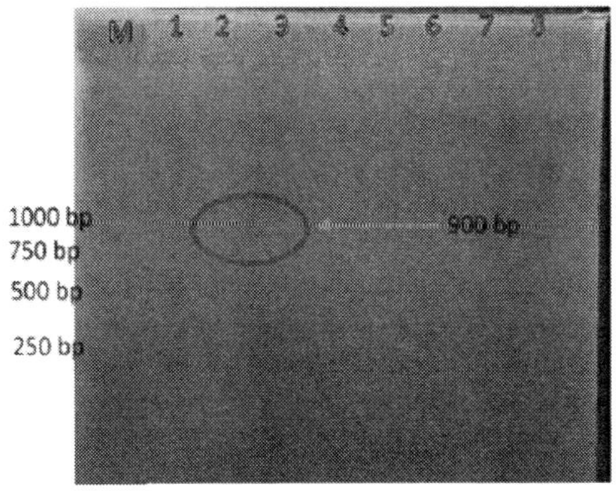

Figure 1: Electrophoresis results of cellulase-encoding genes selected isolates A, B, D, and F, along with Bacillus subtilis isolates; 1. Bacillus subtilis, 2 & 3 isolates B, 4 & 5 isolates D, 6 isolates D (with endoglucanase genes), 7 isolates A and 8 isolates F

The results of the sequencing were BLAST to analyze the level of similarity with cellulase-encoding nucleotides from other bacteria. The results of this study only succeeded in isolating endoglucanase-coding genes from isolate B. The results showed that the sequence of nucleotide sequences from isolate B had a high degree of similarity with *Bacillus subtilis* and *Bacillus amyloliquefaciens* bacteria by 99%. This shows that the amplified genes are cellulase genes and primers that are designed can be used to isolate and classify these cellulase genes. The sequence of cellulase genes can also be used to predict amino acid sequences and tertiary structure of cellulase isolate B. Prediction of the structure of tertiary cellulase isolate B is carried out by the comparative protein modelling method that is homology modeling. In this method, the structure of the target protein is determined based on other proteins namely the insolate protein that has been recognized by its tertiary structure. The protein template chosen is the one that has the greatest similarity to the target protein sequence. The template used is cellulase from *Bacillus amyloliquefaciens* (KF240848.1. Protein ID: AGW99978.1) because the sequence has the highest alignment score compared to cellulase for other bacteria in the Protein Data Bank (http://www.rcsb.org/pdb/ home / home.do) The modeling process is carried out using the help of a program on the site http://swissmodel.expasy.org The protein model selected has the best QMEAN Z value or quality value. The modeling results are downloaded in the form of a file with PDB type so Structural analysis can be done Analytical structure of cellulase enzymes is carried out with the VMD program.

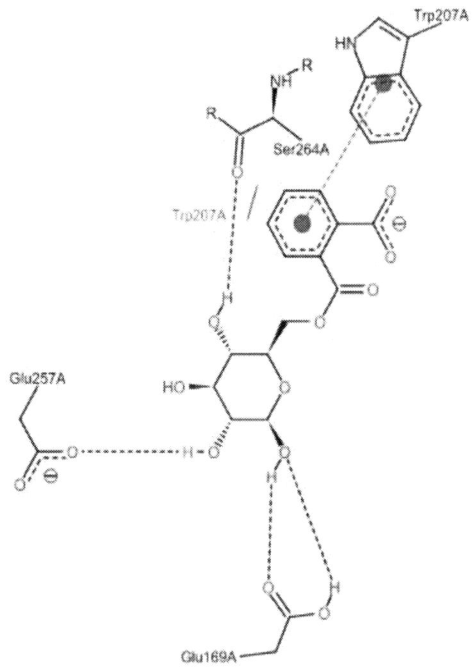

Figure 2: Interaction between cellulose substrate and endoglucanase enzyme isolate *Bacillus amyloliquefaciens* homology in silico

Determination of docking enzyme-ligation complex conformation by selecting one ligation conformation with the best or smallest value of Go binding affinity (kcal/mol). The tendency or strength of a bond or can be called the affinity of a ligand for a receptor or enzyme can be determined by looking at the value of binding affinity (kcal/mol) produced when the ligand enzyme complex is formed. The highest affinity of a ligand for enzymes results from the large intermolecular force between the ligand and the enzyme. The higher the affinity of a ligand for enzymes, the value of binding affinity (kcal/mol) is getting smaller, conversely, if the affinity is lower, the value of ΔG^o binding affinity (kcal/mol) will be greater.

The results of interactions between the substrate cellulose and enzymes are predicted using molecular docking. The bond between the ligand with the receptor is said to be good if the ligand is chemically bound to amino acids in the binding region of the receptor, such as hydrogen bonds, van der Waals, hydrophobic region, and so on. The ligand conformation can affect the affinity of the ligand with the receptor. The analysis showed that the hydrogen bonds that occur in glutamate acid GLU257, tryptophan 207, serine 264 and glutamic acid 169, while hydrophobic interactions occur in tryptophan 207 bonds. The binding affinity of -7.0 Kcal/mol. A further process is to isolate the cellulase-encoding gene from isolate B and predict in silico the possible interactions of the substrate and enzymes produced by the selected isolate and shown in Figure 2.

These results indicate that there are interactions of hydrogen bonds in tyrosin (TRY 299), glutamine (GLU201 and GLU 342.72), Asparagine (ASN 200) and glycine (GLY 384). This interaction is slightly different from the in-silico results obtained homologically ie there is no interaction with serine and glutamate acid, but with asparagine. The isolation of *Bacillus amyloliquefaciens* showed the degradation ability of cellulose substrate with a binding affinity of -6.5 Kcal/mol.

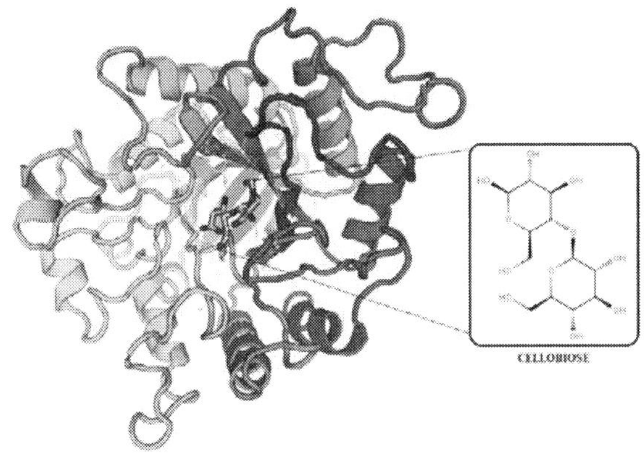

Figure 3: Interaction between the cellobiose substrate and the endoglucanase enzyme in silico

Compared with the results of research conducted by Manhar 2016 that docking CMC bonds with *Bacillus amyloliquefaciens* AMS1 cellulase (endoglucanase). Hydrophobic interactions that occur in Asn134, Lys33, His131, Gln297, Lys296 and Trp69 and the hydrogen concentration on Ala98, Asp 99, Thr97, Ala36, His65 with an affinity binding of - 7.97648 Kcal/mol. Studies show that the interaction of AMS1 with carbohydrate-binding surfaces is mainly by charged and polar residues such as aspartate, lysine, glutamine, threonine, histidine, asparagine and tryptophan. The only aromatic residue that interacts with CMC is Trp 69. Polar residues are not only responsible for changes in carbohydrate conformation, but can exclusively replace inter-chain hydrogen bonds which in turn damage the complex cellulose structure. Threonine and aspartate residues are capable of protonating histidine, which allows AMS1 to break down glycosyl hydrolase according to the donated proton histidine donor.

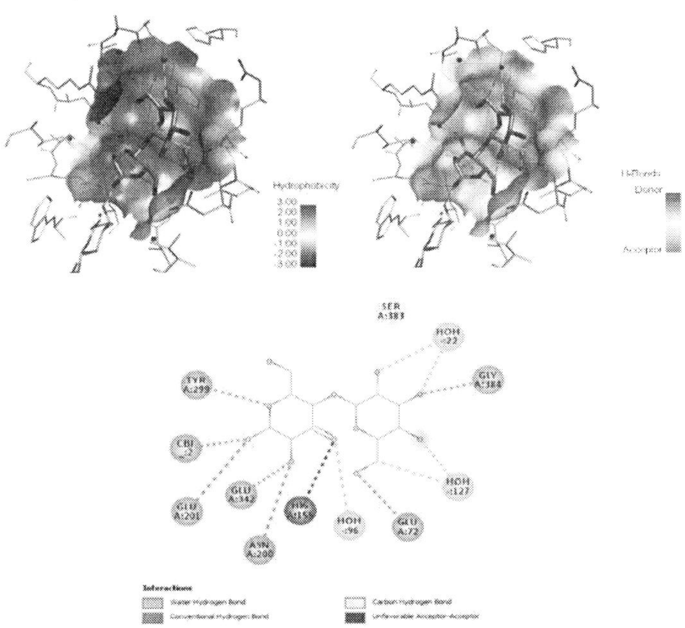

Figure 4: Interaction between cellulose substrate with selected isolate endoglucanase enzyme in silico

For isolate D which was predicted as *Acinetobacter baumannii* was also analyzed as isolate B. Prediction of tertiary cellulase D isolate structure was carried out by the comparative protein modeling method, namely homology modeling. The template used is cellulase from *Acinetobacter baumanni* (Cellulase Synthase (predicted) (JEXJ01000380.1, Protein ID: A0A009RP65 Whole-genome Seq).

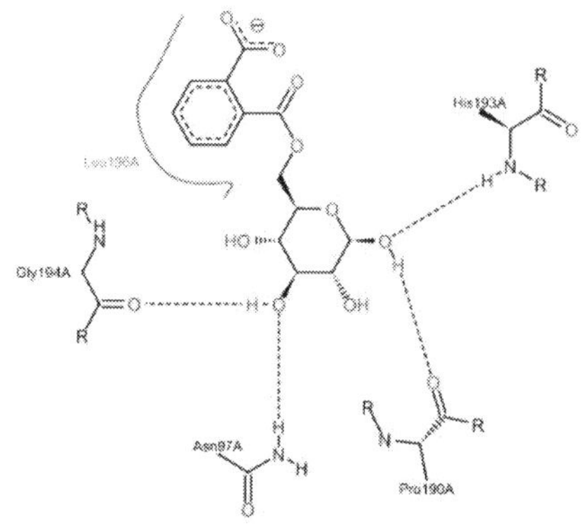

Figure 5: Interaction between cellulose substrate and Cellulase Synthase (predicted) isolate of *Acinetobacter baumanni* homology in silico

The analysis shows that the hydrogen bonds that occur in asparagine ASN97, proline PRO190, Histidine HIS193, and Glycine GLY194, while hydrophobic interactions occur in Leucine LEU196. The binding affinity of -5.7 Kcal/mol. In isolate, D was unable to isolate the encoding gene after isolation was carried out using primary primers based on homology prediction in silico. This is possible because in the isolate D the gene reference used is Cellulase Synthase (predicted) so that when sequencing is not successful.

4. CONCLUSION

The isolates of B and D are predicted to have homologous degradation ability in cellulose substrate with the binding affinity of -7.0 Kcal/mol for isolate B, namely *Bacillus amyloliquefaciens* and isolate D, namely *Acinetobacter baumanni* of -5.7 Kcal/mol. The isolation of *Bacillus amyloliquefaciens* showed the degradation ability of cellulose substrate with a binding affinity of -6.5 Kcal/mol.

ACKNOWLEDGMENT

The research was funded by the Ministry of Research, Technology and Higher Education of the Republic of Indonesia through the BPDN scholarship.

IOP Conf. Series: Materials Science and Engineering **854** (2020) 012068 doi:10.1088/1757-899X/854/1/012068

REFFERENCE

[1] Abdel-Rahman, M.A. 2016. Establishment of efficient cellulolytic bacterial consortium potential for designed composting of rice straw. *International Journal of Advanced Research in Biological Sciences*. 3(4):211-228.

[2] Alam, M.Z., M.A. Manchur & M.N. Anwar. 2004. Isolation, purification, characterization of cellulolytic enzymes produced by the isolat *Streptomyces omiyaensis*. *Journal Bioscience*. 7: 1647-1653.

[3] Feng, H.W., Y.E. Zhi., W.W. Shi., L. Mao & P. Zhou. 2013. Isolation, identification and characterization of a straw degrading *Streptomyces griseorubens* JSD-1. *African Journal of Microbiology Research*. 7(22) : 2730-2735.

[4] Immanuel, G., Dhanusha ., R. Premap & A. Palavesam. 2006. Effect of different growth parameters on endoglucanase activity by bacteria isolated from coir retting effluents of estuarine envirenment. *International journal environmental science technology*. 3:25-34.

[5] Irfan, M., S. Asma., S. Quratulain & N. Muhammad. 2012. Isolation and screening of cellulolytic bacteria from soil and optimization of cellulase production and activity. *Turkish Journal of Biochemistry*. 37(3):287-293.

[6] Kitchen, D.B., H. Decornez., J.R. Furr & J. Bajorath. 2004. Docking and scoring in virtual screening for drug discovery: methods and applications. *Drug Discovery*. 3: 935-949.

[7] Meryandini, A., W. Wahyu., M. Besty., C . Titi., Sunarti,. R. Nisa & S. Hasrul. 2009. Isolasi bakteri selulolitik dan karakterisasi enzimnya. *Makara Sains*.13(1):33-38.

[8] Sadhu, S. & M.T. Maiti. 2013. Cellulase production by bacteria; a riview. *British Microbiology Research Journal*. 3(3):235-258.

[9] Thompson, J.D., F. Plewniak., J.C. Thierry & O. Poch. 2000. DbClustal: rapid and realiable global multiple alignments of protein sequences detected by database searches. *Nucleic Acids Research*. 28(15):2919-2926.

[10] Trott, O. & A.J. Olson. 2010. Auto Dock Vina: improving the speed and accuracy of docking with a new scoring function, efficient optimization and multithreading, *Journal of Computational Chemistry*. 31:455-461.

IOP Conf. Series: Materials Science and Engineering **854** (2020) 012069 doi:10.1088/1757-899X/854/1/012069

The Effectiveness of Coffee Bean Processing Machines in the Rezeki Farmers Group in Takengon, Central Aceh

Amri , Salahuddin, Muhammad Rizka

Department of Information Technology and Computer, Polytechnic Lhokseumawe State, Jl. Banda Aceh-Medan Km. 280,3 Buketrata, Lhokseumawe, 24301, Indonesia

E-mail: amri@pnl.ac.id

Abstract. The appropriate technology program in the Community Service Scheme of Science and Technology for the community entitled "Effectiveness of Coffee Bean Processing Machines in the Rezeki Farmers Group in Takengon, Central Aceh" is to assist partner communities in efforts to improve UD Rezeki Group productivity through the application of appropriate technology and training in product diversification skills. coffee preparations. The method in implementing the Community Service Program consists of three activities, namely (1) making Coffee Roasting machines, flour and packaging, (2) training in processing skills for types of coffee beans, Arabica and Robusta and (3) Coffee Packaging Systems. The purpose of this Coffee Grinder Machine Design activity can facilitate the work in large-scale coffee milling. The methods taken in this design include: the survey stage, the planning stage, the implementation phase, post-construction and report making. Through engine performance test results obtained as follows: 70 watt wiper motor capable of rotating a load of 10 kg with a 72 rpm rotational speed. Gear transmission can reduce rotation from 180 rpm to 72 rpm. The temperature control unit can regulate temperatures from 500 C to 2500 C, while the rotation speed control unit can set the speed from 10 rpm to 72 rpm.

1. Introduction

The current economic condition is in a state of economic crisis, the current economic crisis is very influential in terms of the economic life of the community, so that people are very difficult to meet their needs, resulting in low economic life of the community and it is very alarming. In addition, where the production of large companies also declined, many companies also stopped producing because the company could no longer produce large quantities and many companies went bankrupt / bankruptcy resulting in the termination of employees / commonly referred to as layoffs (employment terminations) thus increasing unemployment everywhere and especially in my own area due to the difficulty of finding work again. Given the importance of economic changes in society today, the community is demanded to be more enthusiastic and try again to achieve a job and experience better changes to meet their own lives, family, and to help others who are experiencing difficulties. Not only these factors but also the people today must be more creative in finding work.

Takengon Regency is an agrarian area, there is still a lot of agricultural land and the agricultural land is widely used to grow coffee trees because the land conditions are very suitable for planting coffee trees, but coffee is not a basic need in my area so when they harvest they are confused about selling coffee where this is, I am increasingly concerned about such conditions, because it is the only agricultural land that they can make as their livelihood. By looking at such conditions I will make an effort that can certainly help the difficulties that are happening in Takengon, especially in Blang Kolak

1 village, namely by helping people who have coffee bean processing and also helping people who do not have jobs. I will make a way of processing coffee beans and packaging them, with that way coffee traders will no longer be confused to sell to supermarkets, supermarkets nearby, even selling to the national and international level by making this packaging, I can also attract workers so that unemployment can be reduced. At this time many people who want to be practical in all drinks. Therefore, I will make this coffee packaging with a different making so that people who like to drink this coffee do not feel bored because many people who like drinks, especially this coffee, then I will try to provide good quality and the price is relatively cheap for all society.

2. Research methodology

To help overcome the problems faced by the small household industry "UD REZEKI" in the effort to produce coffee as a superior product for partners in the Applying Appropriate Technology to the community program, several problem-solving methods were chosen, as follows.

2.1. Machine Design and Manufacturing Methods

The design process is the first step in the work of making roasted coffee machines for the application of Appropriate Technology to the community (PPTTG). Based on the results of the design, it can be known the dimensions of the machine, the amount of material requirements, the engine power plan, the machine productivity plan, and other problems related to the engine manufacturing plant.

Based on the results of the design drawings, continued with the procurement of equipment (coffee roasting machines, pulverizing and packaging) in accordance with the absolute design must be carried out to support the success of all plans for the application of Appropriate Technology Application to the community (PPTTG). In addition, the procurement of equipment in the form of a sangria coffee machine, powder and packaging will gradually help greatly in solving the problems faced by the small household industry "UD REZEKI" in Kampung Ulu Nuwih Village, Bebesen District, Central Aceh District, Aceh Province in order to develop its business.

2.2. Theory and Lecture Methods

The theoretical method and lecture were chosen to convey several supporting theories which are closely related to the problem of the use of the Roasted Coffee machine, the sculpting and packaging and the roasting process. Problems presented in this method, such as:

(1) determine the size of the sangria coffee machine that is suitable for small home industries,
(2) the technique of operating a roasted machine, and
(3) how to maintain occupational safety and health (K3) in using the freying vacuum frying machine.

2.3. Method of Exercise / Practice

This method aims to equip the skills of training and home industry operators in "UD REZEKI" as a small-scale industry partner (partner) in the implementation of the Program for Applying Appropriate Technology to the community (PPTTG). The training and practice in this program are specific to the techniques of operating a coffee sangria and flour machine, how to maintain occupational safety and health in using the machine, and the roasting technique.

Figure 1 shows Engine Component Design Design Before making a semi-automatic coffee roasting machine in a workshop, we need to design the machine components. Such as buffer components, driving components, production units and control components. The component design was created using Inventor 2014 software. The image below is a design of several components in a semi-automatic coffee roasting machine.

The design of this semi-automatic coffee roasting machine will work when the motor is electrified so the motor will rotate the existing small gears, then will move the large gears on the shaft of the production unit to rotate the production unit so that it can roast coffee evenly and the temperature control system on roasting machine will regulate the temperature so that the coffee stays cooked evenly. After passing the roasting stage, the coffee will fall into the drainage section.

This drive component functions to rotate and start the roasting machine, the drive component consists of drive. Transmission and production units.

The production unit is a coffee roasting area, in the roasting tube, there is a stirrer which functions to flatten the coffee maturity during the roasting process.

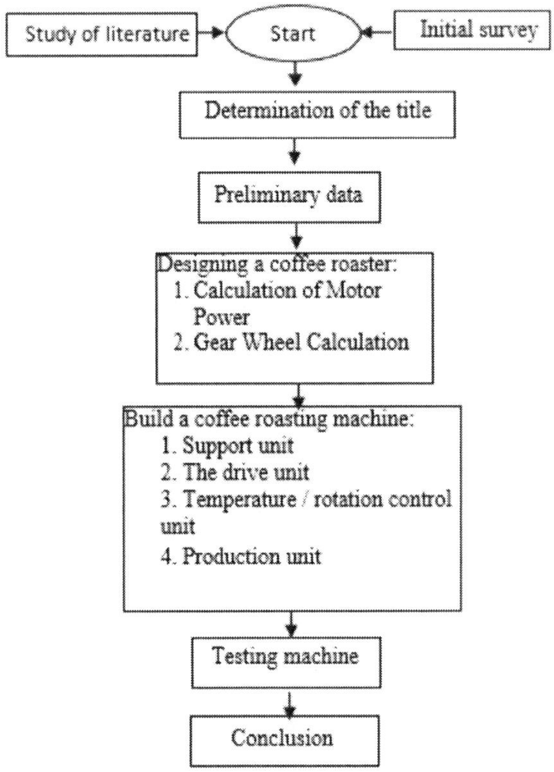

Figure 1. Design of the test process

2.4. Preparation and Manufacturing Equipment
In the manufacturing process required preparation of manufacturing equipment for assemblers of semi-automatic coffee roasting machines to match their functions, for the preparation of equipment adapted to the type of production of each component.

3. Results and discussions

3.1. Torque Calculation
Based on preliminary data obtained, the calculation of torque and engine speed will be planned.
Known:

F = 16 kg
L = 6 cm
$T = ?$
T = F x L
T = 16 kg x 6 cm
T = 96 kg cm

IOP Conf. Series: Materials Science and Engineering **854** (2020) 012069 doi:10.1088/1757-899X/854/1/012069

3.2. Calculation of Power and Motor Selection

Researchers design the rotational speed of a semi-automatic coffee roaster machine is 60 rpm.
Known:

$N = 60$ (rpm)

$T = 96$ (kg.cm) $= 6.96$ *(lb.ft)*

$P = ?$

$$P(hp) = \frac{N(rpm).T(lbf.ft)}{5252}$$

$$P(hp) = \frac{60(rpm)\ 6.96(lbf.ft)}{5252}$$

$P(hp) = 0.079\ (hp)$

$P(hp) = 0.079(hp) \times 745.7(watt\ /\ hp) = 58{,}91\ (watt)$

3.3. After transmission efficiency is known furthermore

calculate the engine power benchmark.

Known :

= 1.0 (because looking for maximum power)

P = 58.91 (watts) = 0.059 (kw)

Pd =?

Pd =. P (3)

= 1.0 x 0.059

= 0.059 (kw) = 0.08 (hp)

3.4. Torque Check

If the torque counts at 6.96 *lb.ft* and the installed power is 0.09 *hp*.

$$T(lbf.ft) = \frac{P(hp)\ x\ 5252}{N(rpm)}$$

$$T(lbf.ft) = \frac{0.09(hp)\ x\ 5252}{60(rpm)}$$

$T(lbf.ft) = 7.878(lb.ft)$

The community service activities through PPTG generally aim to create new jobs, especially for new business groups in Blang Kolak 1 village, Bebesen Sub-district, Central Aceh District, so that the income of poor households increases and can even be separated from poverty that has been shackling them. The target of science and technology activities for the community are as follows:

(1) To increase the skills of coffee traders in making various types of processed coffee that are fast, easy, tasty and inexpensive with attractive and hygienic packaging. The processed coffee products that will be produced are the types of Arabica coffee that is packaged and Robusta coffee.

(2) To add insight into coffee powder sellers in applying marketing strategies through product strategies, pricing strategies, distribution/place strategies and promotion strategies for various coffee beverage products (mainly for arabica coffee).

(3) To assist sales in the business group in preparing a business plan for coffee drinks (and for Robusta coffee),
(4) To assist coffee farmers who are members of a business group in expanding the marketing of various types of coffee made in Blang Kolak Village 1, Bebesen District, Central Aceh Regency,
(5) The compilation of business management and marketing strategies for various types of processed coffee in Blang Kolak Village 1, Bebesen District, Central Aceh Regency.
(6) The social and economic activities of Blang Kolak 1 Village, Bebesen Subdistrict, Central Aceh District, especially coffee entrepreneurs are becoming more qualified.

3.5. Roasted Process

The degree of maturity of roasted coffee is generally divided into three levels, i.e. In the process of roasting coffee occurs in two sequential stages. This fluidizing model roaster machine is designed when it has reached the time limit of the roaster, will turn off the heat source and automatically the air that blows to roast the coffee. The temperature profile readings during the roasting and tempering process can be seen in Figures 2, and 3. The effectiveness of this roasted machine performance, with the types of Arabica and Robusta coffee at the same temperature profile, only differ according to the degree of maturity determined by the process temperature.

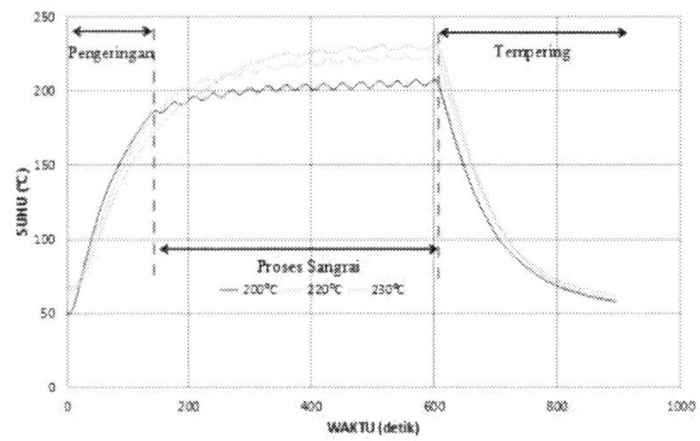

Figure 2. Process temperature type of Arabica coffee with light degree maturity (200 ° C),

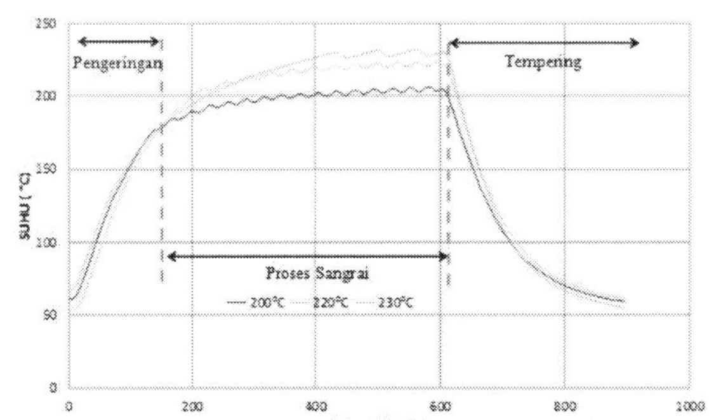

Figure 3. Process temperature type of Robusta coffee with light degree maturity (200 ° C), medium (220 ° C) and dark (230 ° C)

4. Conclusion

The 10 kg coffee roasting machine test showed that the machine functioned well, the temperature measurement accuracy was quite good, the roasting results were evenly demonstrated by the consistency of roasting process from 2 types of Arabica and Robusta coffee, and 3 degrees of Light, Medium and Dark maturity. All of them need the same 10 minutes of baking and 5 minutes of processing. Coffee roast machine test results produce activity above 80%. This shows that the processing process and post-harvest handling methods affect the activity of roasted coffee types. Roasted machine test results at the degree of light maturity produce taste test values above 80 for all types of coffee both arabica and robusta. But for Arabica coffee at all degrees of maturity it has a value above 80. In comparison to the type of roasted machine,

References

[1] Lumbantoruan, Hardono Christanto. 2014 (April). "Decision Support System in Determining Export Quality Coffee Beans with Simple Additive Weighting method (SAW) (Case Study: PT Volkopi Indonesia Lintongnihuta Humbang Hasundutan)," Journal of Pelita Informatics Budi Darma. VI. http://pelita-informatika.com/berkas/jurnal/14.%20Hardono.pdf.Retrieved 3 December 2015

[2] Mulato, FebryYuni. 2015. "Maturity Classification of Red Guava (Psidium Guajava) Using Fuzzy Model," (Student's Undergraduate Thesis of Mathematics Program, Faculty of Mathematics and Natural Sciences of Yogyakarta), Yogyakarta. http://eprints.uny.ac.id/12515/1/Skripsi_Febry_(10305144002).pdf . Retrieved 3 December 2015

[3] Munarto, Ri., Endi Permata, and Rensi Salsabilla. 2014. "Classification of Sweet Corn Seed Quality Based on Color Features Using Fuzzy Logic," Journal of National Symposium RAPI, XIII.https://publikasiilmiah.ums.ac.id/bitstream/handle/11617/5480/2.Ri %20Munarto.pdf?sequence=1 . Retrieved 11 April 2016

[4] Yazdi, Mohammad and Gunawan Ferry Handono. 2013. "Fuzzy Expert System in Determining Cocoa Quality," National Journal of Information Engineering Education (JANAPATI). I. http://pti.undiksha.ac.id/janapati/vol2no2/6.pdf.Accessed 20 April 2016

[5] Maryaningsih. 2013. "Tsukamoto Fuzzy Logic Method in Decision Making System of Scholarship Acceptance," Journal of Media Infotama .9. http://eprints.dinus.ac.id/17029/1/jurnal_16323.pdf . Retrieved 30 April 2016

[6] Sutojo, T., Edy Mulyanto, Vincent Suhartono. 2011. Artificial Intelligence. Semarang: Andi

[7] Agastya,D.G.A.P. 2017. Mesin Roasting Biji Kopi Portable Berbasis Mikrokontroler. Universitas Senata Dharma. Yogyakarta.

[8] Mita, A,A Ali Imron, Sryang Tera Sarena. 2016. Rancang Bangun Alat Penyangrai.(Roaster) Kopi dan Penggiling (Grinder) KopiiOtomatis BerbasissMikrokontroler. Jurusan Teknik Permesinan Kapal, Politeknik Perkapalan Negeri Surabaya, Surabaya

IOP Conf. Series: Materials Science and Engineering **854** (2020) 012070 doi:10.1088/1757-899X/854/1/012070

Smart Power Management System for Ambulance

Fauziah Che Mat[1], Shah Erawati Binti Mokhtar[2], Mohamad Badrol Hisyam[3] and Sulaiman Wadi Harun[3]

[1] Jabatan Kejuruteraan Elektrik, Politeknik Port Dickson, Km 14 Jalan Pantai 71050 Si Rusa, Negeri Sembilan
[2] Politeknik Sultan Azlan Shah, Behrang Stesyen,35950 Behrang, Perak
[3] Jabatan Kejuruteraan Elektrik, Fakulti Kejuruteraan, Universiti Malaya, 50603, Kuala Lumpur

Email: mbhisyam@um.edu.my

Abstract. Ambulance is one the most critical component in healthcare where it must have highest reliability and mobility to ensure minimum travel time of patient to the hospital. In Malaysia, there are significant number of cases where ambulance tend to have breakdown or caught on fire due to electrical current overload which caused by unmanaged electrical usage by medical equipment. To overcome this problem, a smart power management system has been developed by employing Hall-Effect based current sensor replacing conventional fuse. The sensor reading is actively analysed by a microcontroller which will cut off electrical power from over-current equipment. The system also can prioritize some critical equipment to be powered when there are limited current available.

1. Introduction

In this modern era, medical service provider get a lot of technological improvement in order to provide better healthcare service to the society. Despite of this, most of the ambulance technology is still left behind compared to other medical equipment such as MRI, X-Ray, CT Scan, Mammography and Ultrasound. This is due to the lack of dedicated industry for developing comprehensive ambulance platform.

Ambulance normally require combination of automotive and medical industry, which most of the work is carried out by vehicle integrator company that specialized in making special purpose vehicle [1]. However, since it involves with two different engineering disciples, the integration process is not properly done, inducing several problems that reducing the ambulance reliability and effectiveness [2]. There are much improvement have been made including developing more efficient lighting and siren [3-5]. To reduce the mechanical stress on the vehicle alternator, alternate power source have been develop to support the increasing electric load of the ambulance [6,7]. With the emergence of Industrial Revolution 4.0, there are number of study that uses Internet-of-Things (IOT) to provide interactive control and monitoring of the ambulance [8]. However, there are rare study on providing better power management system, primarily focusing on fuse-less approach and priority-based equipment power channelling.

2. Problems

Each year, additional brand-new equipment being put on-board the ambulance to comply with user requirement. The use of these equipment dramatically increases the power consumption that often higher than those provided by the vehicle. In this study, we made an assumption of possible ambulance equipment list that will have power consumption as below.

Table 1. Power consumption calculation of ambulance system

Item	Description	Qty	Current (Amp)		Power (Watt)	
			Unit	Total	Unit	Total
A	Chassis					
1	Electrical Fuel Injection	1	3.00	3.00	36	36
2	Headlamps	2	3.00	6.00	36	72
3	Rear lamp	4	0.50	2.00	6	24
4	Position lamp, Front	2	0.25	0.50	3	6
5	End-outline marker lamp, rear	2	0.25	0.50	3	6
6	Instrument Panel Lighting	30	0.05	1.50	0.6	18
7	Fan Motor	1	5.00	5.00	60	60
8	Windscreen Wiper Motor	1	4.00	4.00	48	48
9	Radio	1	2.00	2.00	24	24
	Sub Total			**24.5**		**294**
B	Accessories					
1	Siren	1	8.30	8.30	100	100
2	Scene light (White)	2	0.80	1.60	9.6	19.2
3	Scene light (Red)	6	0.80	4.80	9.6	57.6
4	Scene light (Front)	2	0.40	0.80	4.8	9.6
5	Cabin light	5	1.33	6.67	16	80
6	Light bar	1	4.00	4.00	48	48
7	Beacon light	2	1.20	2.40	14.4	28.8
8	Spotlight	1	1.00	1.00	12	12
9	Inverter	1	1.00	80.00	1000	1000
10	Fan	2	0.50	1.00	6	12
11	Girn	1	1.50	1.50	18	18
	Medical Equipment					
12	Defibrillator and Monitor	1	9.00	9.00	108	108
13	Ventilator	1	8.30	8.30	100	100
14	Suction Unit	1	2.75	2.75	33	33
15	Syringe Pumps	2	0.83	1.67	10	20
	Sub Total			**133.8**		**1646**
	Total (Chassis + Accessories)			**158.3**		**1940**
	Power Available from Alternator 130A			**130**		**1560**
	Power Available after (chassis + Accessories) consumption			**-28.3**		**-380**
	% Power Available			**-22%**		**-24%**

As shown in the table above, the total power consumption is about 1940 W, while the supplied power from the vehicle is only 1560 W thus lead to power shortage of about 24%. This situation can lead to severe alternator damage and can bring down the whole vehicle system including the engine which will caused the vehicle to be at full stop. This is the worst condition that can happened to an ambulance,

IOP Conf. Series: Materials Science and Engineering **854** (2020) 012070 doi:10.1088/1757-899X/854/1/012070

which can lead to fatality of patient on-board. In some cases, the overcurrent event will lead to slow-burning in the compartment engine, which will spread to the whole vehicle if being left uncheck.

Figure 1. Ambulance caught on fire due to electrical problem

3. System Design

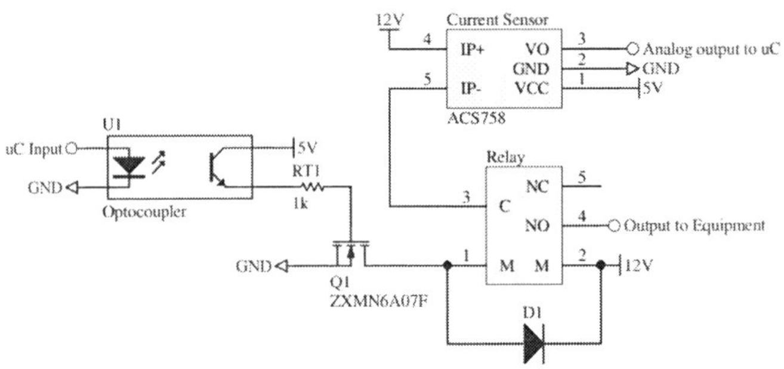

Figure 2. Current sensing mechanism

Figure 2 shows the current sensing circuit used to monitor the current consumption in each power output channel in realtime. An ACS758 Hall-Effect current sensor is used to measure the current where it is propotionally converted into 0-5 analogue output corresponding to 0 – 50 A load current. The analogue output is received by a high-speed microcontroller for current analysis. The current value is compared to allowed threshold current value where any value above this threshold will force microcontroller to cut off power source to that channel by powering down the Optocoupler which lead to powering down MOSFET transistor ZXMN6A07F followed by demagnetization of Relay. Optocoupler is being used to prevent high-noise signal generated by high-current channel from interrupting low current microcontroller signal.

IOP Conf. Series: Materials Science and Engineering **854** (2020) 012070 doi:10.1088/1757-899X/854/1/012070

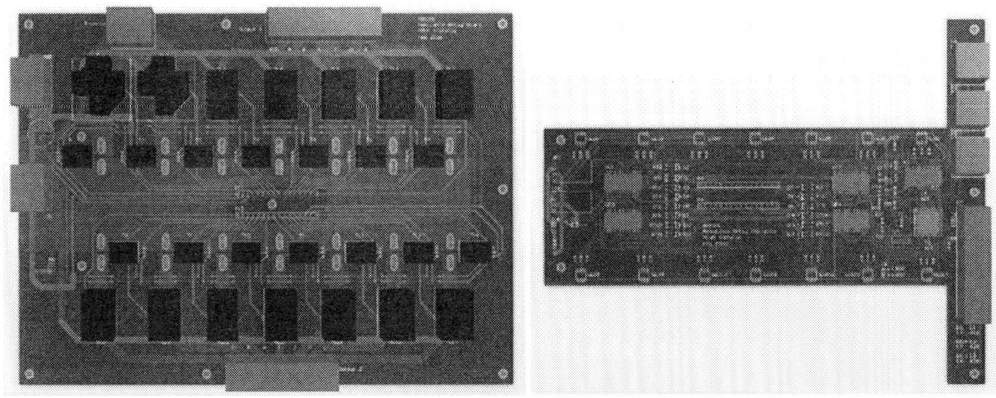

Figure 3. PCB Layout for Power Management Board, Left: High Current, Right: Signal Controller

Figure 3 shows the PCB layout of a Power Management Board, capable of handling up to 15 power output channel. The first 2 relays are able to traverse up to 40A each that normally used to power up a 1000W DC-AC Inverter which used to supply power to a standard 240VAC powered medical equipment. The rest 13 relays are able to traverse up to 30A each. To prevent the low current and sensitive microcontroller from easily get damaged, the signal controller side is separated from high current side as shown in picture. To minimize the usage of microcontroller pin, an MCP3221 ADC is used where current reading from 8 channel can be simultaneously transmitted via only 2 wire using I2C protocol.

Apart from current reading, the system is also employing RGB LED on each channel to indicate availability (Blue), turned-on (Green) and overcurrent (Red) to easily identify each channel status. The total of 45 statuses are triggered by a MCP23018 I/O expander, which also use I2C protocol to reduce number of wire needed. The system also able to receive special trigger such as door switches and manual switch panel trigger. To enhance the connectivity to other system, 3 units of RS-232 serial communication interface are employed which can be use for interconnection to IOT hub, driver switch panel and cabin switch panel.

4. Implementation

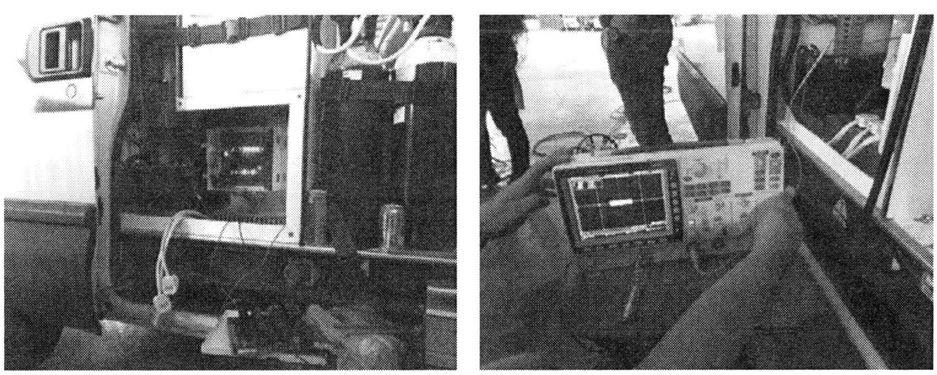

Figure 4. System deployment and test on actual ambulance.

The system have been successfully installed on an actual ambulance by replacing conventional power management board. To emulate the input switch, a computer is connected to the system using RS-232 interface where a simplified software is being used to interact with the system.

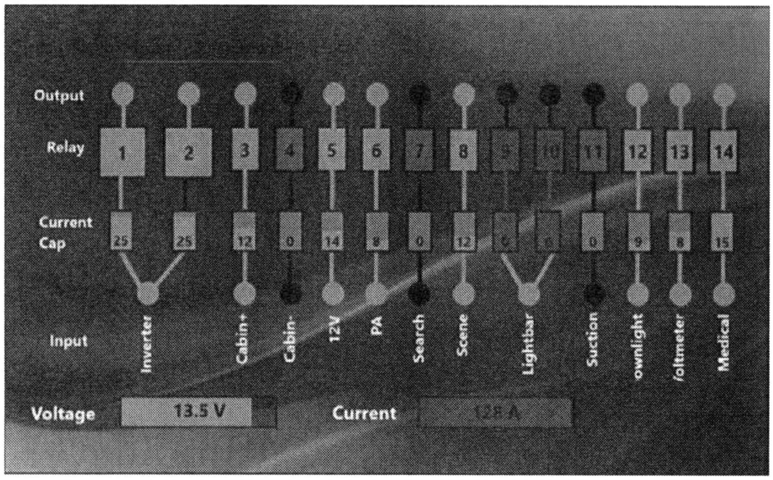

Figure 5. Current consumption monitoring result

The monitoring software is shown in Figure 5 where all current consumption for each channel is been monitored in real-time. In the result shown above, the total current consumption for the vehicle is 128 A, which is nearly reach the maximum value of 130A. To avoid damage to the vehicle AC-DC alternator, the system automatically take preventive action by disabling relay 9 and 10, which corresponding to Lightbar output. In the system priority management, Lightbar is one of the non-critical component while Medical output is the most critical component. In the conventional system, this feature is not available therefore incur high risk to alternator damage.

5. Conclusion

The ambulance smart power management system presented in this study has proven to provide much benefit to the healthcare sector by potentially reducing risk of damage due to electrical current overload. Several upgraded version has been planned including interconnection with GSM/Long Range Wireless communication and adding monitoring capability of vehicle chassis status such as engine condition, fuel level, odometer etc.

Acknowledgements
The research was supported by the grant of TVET Applied Research Grant Scheme (T-ARGS) of Kementerian Pendidikan Malaysia (KPM).

References

[1] Zhou, G. H., & Xie, P. (2013). Study on Inversion Power Supply of Jiangling Ford Transit Ambulance. *Chinese Medical Equipment Journal,* (3), 14.

[2] Jennings, B. C., Effraimidis, D., Perruccio, M. J., & Fournier, N. S. (2013). Improving Ambulance Power Systems.

[3] Owusu-Ansah, B. K., Alberti Arroyo, H. E., Smalcuga, P. K., & Lushllari, S. (2013). Improving Ambulance Compartment Insulation, Energy and Lighting Design.

[4] Catchpole, K., & McKeown, D. (2007). A framework for the design of ambulance sirens. *Ergonomics*, *50*(8), 1287-1301.

IOP Conf. Series: Materials Science and Engineering **854** (2020) 012070 doi:10.1088/1757-899X/854/1/012070

[5] O'Brien, D. J., Price, T. G., & Adams, P. (1999). The effectiveness of lights and siren use during ambulance transport by paramedics. *Prehospital Emergency Care*, *3*(2), 127-130.

[6] Apodaca-Madrid, J. R., & Newman, K. (2010, April). Design and evaluation of a Green Ambulance. In *2010 IEEE Green Technologies Conference* (pp. 1-5). IEEE.

[7] Baker, W., Mansfield, N., Storer, I., & Hignett, S. (2013). Future electric vehicles for ambulances (FEVA). *Journal of Paramedic Practice*, *5*(2), 77-82.

[8] Venkatesh, H., Perur, S. D., & Jagadish, M. C. (2015). An approach to make way for intelligent ambulance using IoT. *International Journal of Electrical and Electronics Research*, *3*(1), 218-223.

IOP Conf. Series: Materials Science and Engineering **854** (2020) 012071 doi:10.1088/1757-899X/854/1/012071

Fibre-based Saturable Absorbers for Pulsed Generations in the 1-micron Region

M F A Rahman[1,2], M T Ahmad[2,3], M F M Rusdi [2,3], A A Latiff [2,4], S W Harun[3]

[1] Fakulti Teknologi Kejuruteraan Elektrik dan Elektronik, Universiti Teknikal Malaysia Melaka, 76100 Hang Tuah Jaya, Melaka, Malaysia.
[2] Photonics Research Group, Centre for Telecommunication Research & Innovation, Universiti Teknikal Malaysia Melaka, 76100 Hang Tuah Jaya, Melaka, Malaysia
[3] Photonics Engineering Laboratory, Department of Electrical Engineering, Faculty of Engineering, University of Malaya, 50603 Kuala Lumpur, Malaysia
[4] Faculty of Electronic and Computer Engineering, Universiti Teknikal Malaysia Melaka, 76100 Hang Tuah Jaya, Melaka, Malaysia

Email: mfauziar@utem.edu.my

Abstract. In this paper, we demonstrated and compared several fibre-based saturable absorbers (SAs) used in the generation of stable Q-switched fibre lasers in the 1-micron region, using 1.5 m long Ytterbium-doped fibre (YDF) as the gain medium. The tested fibre SAs consist of four segments; (1) 8 cm long Hafnium bismuth erbium-doped fibre (HBEDF), (2) 12 cm long Ytterbium-doped fibre (YDF), (3) 20 cm long Titanium dioxide-doped fibre (TDDF) and (4) 11 cm long Thulium-doped fibre (TDF). The fibre segments were joint with a single-mode fibre (SMF28) of approximately 10 cm at both of their ends via fusion splicing to become useful fibre-based SAs. The Q-switched fibre lasers stably operated at ~1060 nm wavelength. Among these fibre SAs, the highest pulse energy (191 nJ) was achieved through the TDDF SA, while the shortest pulse width (1.97 μs) and the highest pulse frequency (74.2 kHz) were obtained via the YDF SA. These fibre SAs provide simple operation, high stability and high thermal damage threshold which are desirable for high power laser operations.

Keyword: Fibre-based SA, passive Q-switching, Ytterbium-doped fibre laser

1. Introduction

Q-switched fibre lasers have many benefits, such as compactness, possesses high temperature and vibrational stability, high-quality beam, and low maintenance cost, which suits many applications, including material processing, medicine and remote sensing. The Q-switched fibre lasers can be generated via an active or a passive method. The passive method is much simple in preparation and handling, and is much cheaper and reliable in comparison to the active method. It uses saturable absorber (SA), such as graphene [1-3], carbon nanotubes (CNTs) [4, 5], topological insulators (TIs) [6, 7], transition metal dichalcogenides (TMDs) [8-11], and metal oxides [12-15] to modulate intra-cavity losses and Q-factor in the cavity.

Many of these SAs present in the form of exfoliated bulk materials or thin films and are integrated in the cavity by attaching them between two fibre ferrules tightened with fibre adapters. Bulk material SA, such as graphene, tends to be oxidised when contacted with the air and thus is not so reliable. In contrary, the thin film SAs, have low melting points of ~230°C, and may deform when subjected to high pump power. Another technique is by using a side-polished fibre deposited with SA materials. The side-polished fibres deposited SA, although have an excellent thermal damage threshold, however, are fragile (particularly at the polishing area), and produce high power losses. Thus, to overcome this problem, fibre-based SAs, which doped with certain materials (that have sufficient linear absorption at the specified operating bandwidth) are proposed for alternative solutions. The present of the linear absorption attribute enables losses modulation occurs in the cavity, particularly at the Q-switching operating wavelength. These fibre SAs are renowned for excellent thermal handling with a softening temperature of around 1200℃, besides, are suitable for applications that requires high flexibility, maintenance-free and simple operation. Several fibre-based SAs with stable pulse performances have also been demonstrated in literature in the past few fears [16-18].

In this paper, we demonstrated and proposed several fibre-based SAs, consisting of newly developed fibres and commercial fibres; (1) 8 cm of Hafnium bismuth erbium-doped fibre (HBEDF), (2) 12 cm of Ytterbium-doped fibre (YDF), (3) 20 cm of Titanium dioxide-doped fibre (TDDF) and (4) 11 cm of Thulium-doped fibre (TDF) for the Q-switching in the 1-micron region. Theses fibre SAs were integrated in the YDFL ring cavity, with the help of fibre connectors and adapters. Several important Q-switched laser characteristics were investigated and reported such as the pulse width, pulse period, pulse frequency, and pulse energy. These parameters are summarised in the chapters to follow.

2. Method

2.1. Fibre SA preparation and characterisation

The selected fibre was cut into segment and then connected with a single-mode fibre (SMF28, length ~10 cm) at both of its ends through fusion splicing. The simplified structure of the fibre SA is shown in Figure 1. However, in this paper, the additional length of SMF28 will not be addressed, and only the fibre SA's length will be reported for simplicity. The fibre SAs' characterisation details could also be found in the previous reports [19-23]. The appropriate length for the fibre segment was determined from the linear absorption profile, measured at the respective Q-switched operating wavelength. The linear absorption profile was investigated by propagating low density and wideband white light source into one end of the fibre-based SAs. The output was then collected from another end of the fibre SA, and then channelled into an Optical spectrum analyser (OSA). The fibres segment length with their respective measured linear absorption value are summarized in Table 1.

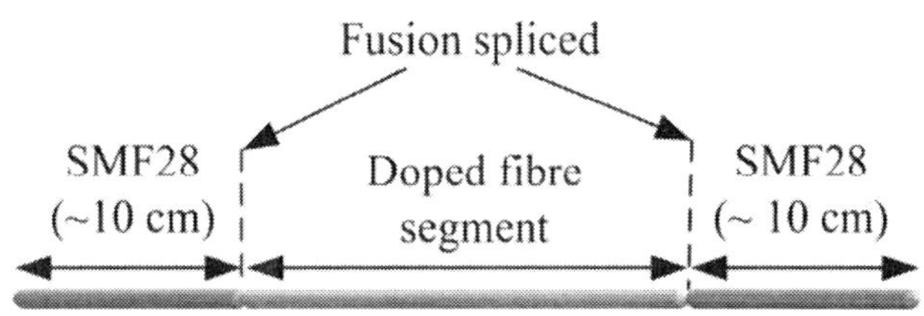

Figure 1. Fibre-based SA structure, consisting of single-mode fibre, SMF28 (~ 10 cm) fusion spliced at both of the fibre segment' ends.

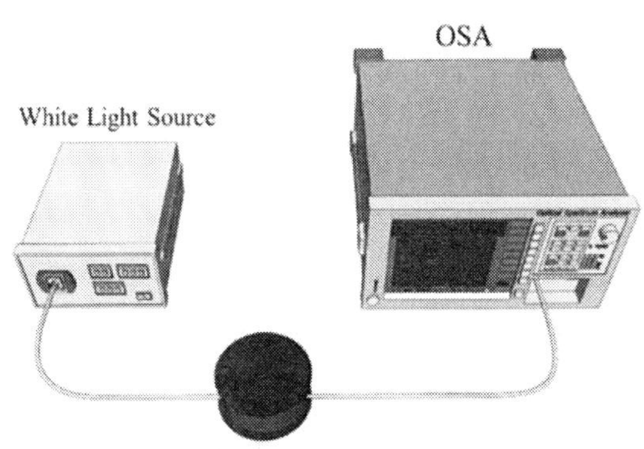

Figure 2. Fibre-based SA linear absorption measurement set-up

IOP Conf. Series: Materials Science and Engineering **854** (2020) 012071 doi:10.1088/1757-899X/854/1/012071

Table 1. Fibre SA's segment length and its linear absorption profile

Teste fibre SA	Length (cm)	Linear absorption profile (dB)
HBEDF	8	3
YDF	12	2.44
TDDF	20	3.24
TDF	11	1.35

2.2. YDFL cavity configuration

The experimental design used for the YDFL Q-switching operation is illustrated in Figure 3. The cavity is pumped by 980 nm LUMICS laser diode (LD) through a wavelength division multiplexer (980/1060 WDM). The light is then propagated into YDF gain medium (1.5 m), 3 dB coupler 1, fibre-based SA (8 cm HBEDF/ 12cm YDF/ 20 cm TDDF/ 11 cm TDF) and finally into an isolator, before reaching back to the WDM through a 1060 nm port, for a complete light propagation. The YDF gain medium (LIEKKI, YB 1200-4125 from the THORLABS) used in the set-up has a core and cladding diameter of around 4.4 µm and 125 µm, respectively. It has a core absorption of 1200 dB/m at 975 nm wavelength. The 3 dB coupler 1 splits the light into 50:50 separations. 50% of the output is channelled into 3 dB coupler 2, while half of the quantity is left circulated in the cavity. The extended 3 dB coupler 2 further splits the output into halves, to ease the process of monitoring dual optical quantities, using optical equipment (OSA, radio frequency spectrum analyser (RFSA), oscilloscope) simultaneously. The fibre SA is integrated in the cavity to modulate losses, as well as Q-factor in the cavity. The isolator ensures unidirectional light propagation. The OSA measures the optical spectrum of the laser, while the RFSA and the oscilloscope displays the Q-switching characteristic in the frequency domain and time domain, respectively.

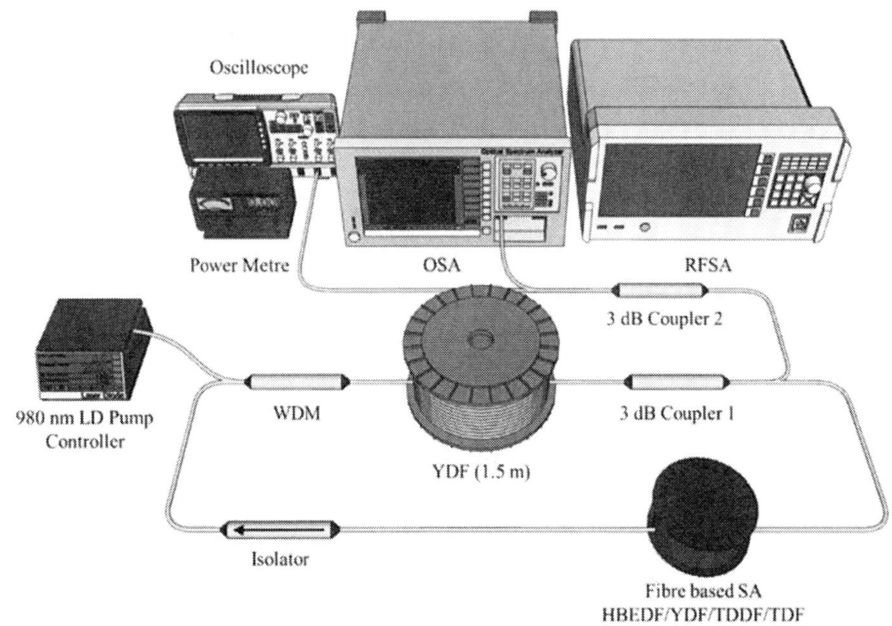

Figure 3. YDFL cavity design for the Q-switching operation

3. Result and Discussion

3.1. Q-switching performance via HBEDF SA

The Q-switched YDFL using HBEDF SA stably emerged at a threshold pump power of 106 mW and steadily appeared until 175 mW. The pulse frequency could be varied from 41 kHz to 67 kHz. Figure 4(a) shows the YDFL Q-switched output spectrum, which is centred at 1069 nm. Figure 4(b) shows the temporal performance of the Q-switched YDFL at the maximum pump power of 175 mW. As illustrated, the peak to peak separation time (pulse period, T) is measured as 14.9 μs (67 kHz), while the pulse width, τ is obtained as 3.48 μs. The fundamental frequency has a signal to noise ratio (SNR) of 48 dB, and the maximum pulse energy was around 70.2 nJ.

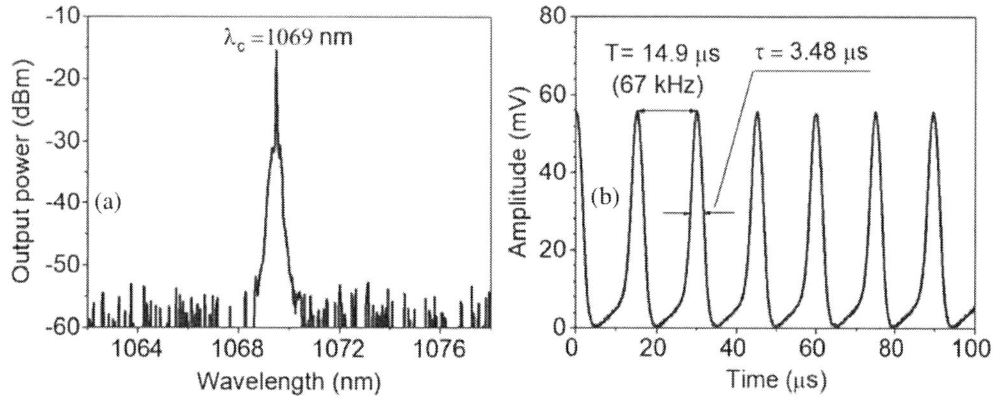

Figure 4. YDFL via HBEDF SA Q-switching performance (a) optical spectrum, and (b) time domain

3.2. Q-switching performance via YDF SA

The Q-switched YDFL also could be obtained by using a short length of YDF (12 cm). With the YDF SA in the cavity, a stable Q-switched laser with a peak wavelength of 1068 nm was observed. The related optical spectrum is shown in Figure 5(a). The Q-switched YDFL stably arose at a threshold pump power 151 mW and diminished into a continuous wave, CW beyond the 233 mW. The pulse frequency has an initial value of 63.2 kHz which could be further lifted up to 74.2 kHz, with the increase of pump power. Figure 5(b) shows the Q-switching temporal performance at the maximum pump power (233 mW). As illustrated, the pulse period is measured as 13.4 μs (74.2 kHz), while the pulse width is recorded as 1.97 μs. The fundamental frequency had an SNR of 45 dB, while the maximum pulse was obtained as 49.9 nJ.

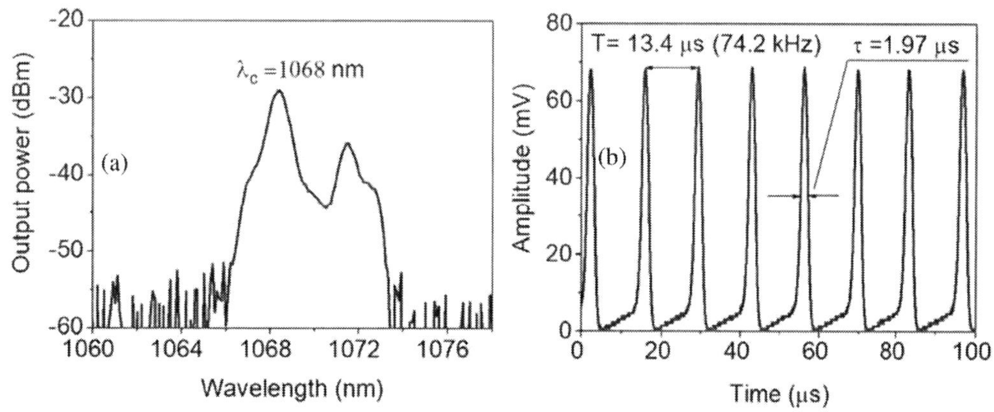

Figure 5. YDFL via YDF SA, Q-switching performance (a) optical spectrum, and (b) time domain

3.3. Q-switching performance via TDDF SA

The TDDF SA was integrated in the same YDFL cavity configuration (Figure 2). The Q-switched laser stably emerged within a pump power range of 109-233 mW. The initial frequency was recorded as 35 kHz. Figure 6(a) shows the optical spectrum of the Q-switched laser, with an output spectrum centred at 1062 nm wavelength. Meanwhile, Figure 6(b) illustrates the Q-switching in the time domain, observed at the maximum pump power of 233 mW. As shown, the pulse train is quite uniform in shape, suggesting excellent pulsing stability. The pulse has a peak to peak separation of 18 μs, that corresponds to a pulse frequency of 53 kHz, and a pulse width of 2.55 μs. The first harmonic has an SNR of 47 dB, while the most attainable pulse energy was measured to be around 191 nJ.

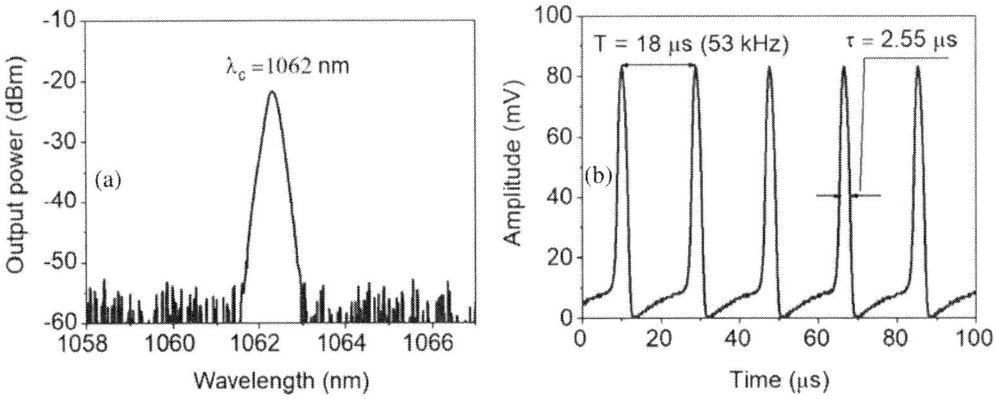

Figure 6. YDFL via TDDF SA, Q-switching performance (a) optical spectrum, and (b) time domain

3.4. Q-switching performance via TDF SA

A reliable YDFL Q-switched operation with a starting frequency of 40 kHz until 60.2 kHz emerged stably as the pump power rose from 109 mW to 206 mW. The Q-switched laser has an output spectrum centred at 1069 nm wavelength, as illustrated in Figure 7(a). Figure 7(b) shows the Q-switched pulse train, traced by the oscilloscope at the maximum pump power (206 mW). As shown, the pulse signal has a pulse width of 2.87 μs and a pulse period of 16.6 μs (60.2 kHz). The maximum pulse energy was 80.7 nJ, and the Q-switched output has a fundamental frequency with an SNR of 47 dB.

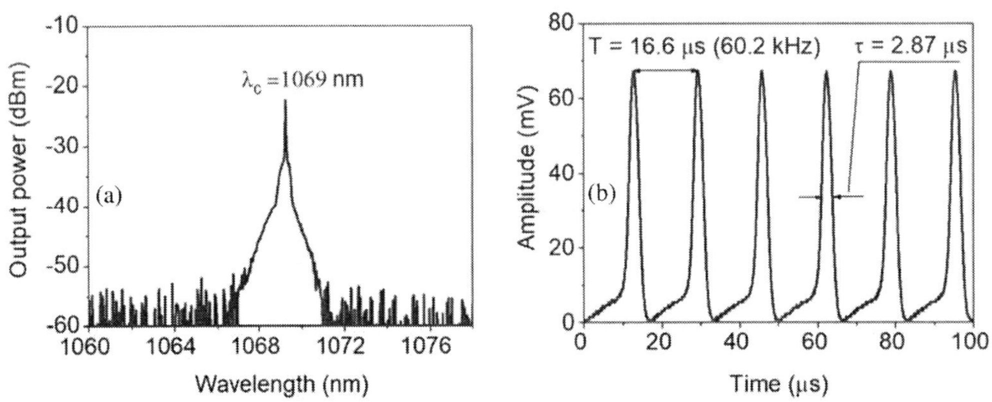

Figure 7. YDFL via TDF SA, Q-switching performance (a) optical spectrum, and (b) time domain

3.5. Q-switching overall performance

The overall performance of the Q-switched lasers using the proposed fibre-based SAs are summarized in Table 1. Within these fibre-based SAs, the highest pulse energy of 191 nJ was obtained from the TDDF SA, while the narrowest pulse width of 1.97 μs was attained via the YDF SA. These fibre SAs shows quite good maximum pulse energy, minimum pulse width and the tunable range of pulse frequency, which are comparable to other SAs demonstrated in the literature.

To test the Q-switching stability, the fibre SAs were integrated in the laser cavity for few hours, at the maximum Q-switching pump power. Based on the experimental results, the Q-switching pulse remained almost unchanged (during the test period), indicating that the fibre SAs were in still excellent condition and performed under their thermal damage threshold.

Table 2. YDFL Q-switching performance using fibre-based SAs (our work) in comparison with other SAs in the literature.

Saturable absorber (SA)	Maximum pulse energy (nJ)	Minimum pulse width, τ (μs)	Pulse frequency, f (kHz)	Central wavelength, λ_c (nm)	Reference
HBEDF (8.cm)	70.2	3.48	41-67	1069	This work
YDF (12.cm)	49.9	1.97	63.2-74.2	1068	This work
TDDF (20.cm)	191	2.55	35-53	1062	This work
TDF (11.cm)	80.7	2.87	40-60.2	1069	This work
CNTs	143.5	12.18	7.9-24.27	1060.2	[4]
CNTs	18.4	1	30-50	1061	[5]
Graphene	46	~70 (ns)	140- 257	1064.2	[2, 3]
Graphene	141.8	1.3	28.9- 110	1027	[3]
MoS_2	32.6	5.8	6.4-28.9	1066.5	[10]
$MoSe_2$	116	2.85	60-74.9	1060	[11]
BP	2.09	1.16	52.52-58.73	1038.8 & 1042.1	[24]
BP	7.1	4	6-44.8	1056.6-1083.3	[25]
Bi_2Se_3	6.2	2.1	14.9-62.5	1050.4	[26]
Bi_2Te_3	38.3	1	35-77	1056	[7]

4. Conclusion

We have demonstrated and compared several fibre-based SAs; (1) 8 cm of HBEDF, (2) 12 cm of YDF (3) 20 cm of TDDF and (4) 11 cm of TDF in the 1-micron region, using 1.5 m long YDF as the gain medium. All these fibres showed quit excellent Q-switching stability as well as comparable laser performances to several of other SAs in the literature. The Q-switched fibre lasers operated near the 1060 nm wavelength. The highest pulse energy (191 nJ) was achieved via the TDDF SA, while the shortest pulse width (1.97 μs) and the highest pulse frequency (74.2 kHz) were obtained via the YDF SA. Incorporating of these fibre-based SAs, provide better flexibility and high thermal damage threshold, which significantly desired for applications such as in medical and material processing.

Acknowledgement
This work is supported by the Universiti Teknikal Malaysia Melaka, under research grant JURNAL/2019/FTKEE/Q00021.

Reference

[1] Sobon G and Sotor J 2016 Recent Advances in Ultrafast Fiber Lasers Mode-locked with Graphenebased Saturable Absorbers *Current Nanoscience* **12** 291-8
[2] Liu J, Wu S, Yang Q-H and Wang P 2011 Stable nanosecond pulse generation from a graphene-based passively Q-switched Yb-doped fiber laser *Optics letters* **36** 4008-10

[3] Zhang L, Fan J, Wang J, Hu J, Lotya M, Wang G, Li R, Blau W, Coleman J and Wang J 2012 Graphene incorporated Q-switching of a polarization-maintaining Yb-doped fiber laser *Laser Physics Letters* **9** 888

[4] Kasim N, Al-Masoodi A, Ahmad F, Munajat Y, Ahmad H and Harun S 2014 Q-switched ytterbium doped fiber laser using multi-walled carbon nanotubes saturable absorber *Chinese optics letters* **12** 031403

[5] Li X, Wang Y, Wang Y, Wang Q, Zhao W, Zhang Y, Ya X and Zhang Y 2013 A single-walled carbon nanotube wall paper as an absorber for simultaneously achieving passively mode-locked and Q-switched Yb-doped fiber lasers. In: *Nanoelectronics Conference (INEC), 2013 IEEE 5th International*: IEEE) pp 139-42

[6] Haris H, Arof H, Muhammad A R, Anyi C L, Tan S J, Kasim N and Harun S W 2019 Passively Q-switched and mode-locked Erbium-doped fiber laser with topological insulator Bismuth Selenide (Bi_2Se_3) as saturable absorber at C-band region *Optical Fiber Technology* **48** 117-22

[7] Lee J, Koo J, Chi C and Lee J H 2014 All-fiberized, passively Q-switched 1.06 μm laser using a bulk-structured Bi_2Te_3 topological insulator *Journal of Optics* **16** 085203

[8] Mao D, Cui X, Gan X, Li M, Zhang W, Lu H and Zhao J 2017 Passively Q-switched and mode-locked fiber laser based on an ReS_2 saturable absorber *IEEE Journal of Selected Topics in Quantum Electronics* **24** 1-6

[9] Woodward R, Kelleher E, Howe R, Hu G, Torrisi F, Hasan T, Popov S and Taylor J 2014 Tunable Q-switched fiber laser based on saturable edge-state absorption in few-layer molybdenum disulfide (MoS_2) *Optics express* **22** 31113-22

[10] Luo Z, Huang Y, Zhong M, Li Y, Wu J, Xu B, Xu H, Cai Z, Peng J and Weng J 2014 1-, 1.5-, and 2-μm fiber lasers Q-switched by a broadband few-layer MoS_2 saturable absorber *Journal of Lightwave Technology* **32** 4077-84

[11] [11] Woodward R, Howe R, Runcorn T, Hu G, Torrisi F, Kelleher E and Hasan T 2015 Wideband saturable absorption in few-layer molybdenum diselenide ($MoSe_2$) for Q-switching Yb-, Er-and Tm-doped fiber lasers *Optics Express* **23** 20051-61

[12] Al-Hiti A S, Rahman M, Harun S, Yupapin P and Yasin M 2019 Holmium oxide thin film as a saturable absorber for generating Q-switched and mode-locked erbium-doped fiber lasers *Optical Fiber Technology* **52** 101996

[13] Baharom M, Rahman M, Latiff A, Wang P, Arof H and Harun S 2019 Lutetium oxide film as a passive saturable absorber for generating Q-switched fiber laser at 1570 nm wavelength *Optical Fiber Technology* **50** 82-6

[14] Rahman M, Rusdi M, Harun S and Yasin M 2018 Q-Switched Thulium-Doped Fiber Laser Using Vanadium Oxide Saturable Absorber *Nonlinear Optics, Quantum Optics: Concepts in Modern Optics* **48**

[15] Rusdi M, Rosol A, Rahman M, Mahyuddin M, Latiff A, Ahmad H, Harun S and Yasin M 2019 Q-switched and mode-locked thulium doped fiber lasers with nickel oxide film saturable absorber *Optics Communications* **447** 6-12

[16] Kurkov A, Sadovnikova Y E, Marakulin A and Sholokhov E 2010 All fiber Er-Tm Q-switched laser *Laser Physics Letters* **7** 795

[17] Kurkov A, Sholokhov E and Medvedkov O 2009 All fiber Yb-Ho pulsed laser *Laser Physics Letters* **6** 135-8

[18] Rahman M F A, Dhar A, Das S, Dutta D, Paul M C, Rusdi M F M, Latiff A A, Dimyati K and Harun S W 2018 An 8 cm long holmium-doped fiber saturable absorber for Q-switched fiber laser generation at 2-μm region *Optical Fiber Technology* **43** 67-71

[19] Rahman M, Latiff A, Reddy P, Das S, Dhar A, Paul M and Harun S 2019 Passively Q-switched fiber laser utilizing new hafnium–bismuth–erbium co-doped fiber as saturable absorber *Indian Journal of Physics* 1-5

[20] Rahman M F, Latiff A A, Rusdi M F, Dimyati K and Harun S W 2018 Q-switched ytterbium-doped fiber laser via a thulium-doped fiber saturable absorber *Applied optics* **57** 6510-5

[21] Reddy P H, Rahman M F A, Paul M C, Latiff A A, Rosol A H A, Das S, Dhar A, Bhadra S K, Dimyati K and Harun S W 2018 Titanium dioxide doped fiber as a new saturable absorber for generating mode-locked erbium doped fiber laser *Optik* **158** 1327-33

[22] Rahman M, Reddy P, Paul M, Das S, Dhar A, Rusdi M, Latiff A, Dimyati K and Harun S 2019

Titanium dioxide-based picoseconds pulsed fiber laser performances comparison in the 1.5-micron region. In: *Journal of Physics: Conference Series*: IOP Publishing) p 012023
[23] Rahman M, Rusdi M, Muhammad A, Latiff A, Ahmad A, Dimyati K and Harun S 2019 Ytterbium doped fiber saturable absorber for a stable passively Q-switched fiber laser in the 1-micron region. In: *Journal of Physics: Conference Series*: IOP Publishing) p 012008
[24] Rashid F, Azzuhri S R, Salim M, Shaharuddin R, Ismail M, Ismail M, Razak M and Ahmad H 2016 Using a black phosphorus saturable absorber to generate dual wavelengths in a Q-switched ytterbium-doped fiber laser *Laser Physics Letters* **13** 085102
[25] Ahmad H, Salim M A M, Thambiratnam K, Norizan S F and Harun S 2016 A black phosphorus-based tunable Q-switched ytterbium fiber laser *Laser Physics Letters* **13** 095103
[26] Haris H, Harun S, Muhammad A, Anyi C, Tan S, Ahmad F, Nor R, Zulkepely N and Arof H 2017 Passively Q-switched Erbium-doped and Ytterbium-doped fibre lasers with topological insulator bismuth selenide (Bi_2Se_3) as saturable absorber *Optics & Laser Technology* **88** 121-7

IOP Conf. Series: Materials Science and Engineering **854** (2020) 012072 doi:10.1088/1757-899X/854/1/012072

Different Techniques of Rooftile BIPV System Implementation: Materials and Constructions

Nelly Safitri[1]*, Rizal Syahyadi[2], Faisal Rizal[2], Yassir[1]

[1] Electrical Engineering Department, Politeknik Negeri Lhokseumawe
[2] Civil Engineering Department, Politeknik Negeri Lhokseumawe

*Email: nellysafitri@pnl.ac.id

Abstract. Roof integrated systems which are integrating photovoltaic (PV) modules into rooftiles now become a trend of building integrated photovoltaic (BIPV) technology. BIPV refers to PV that are integrated into the building envelope (such as facade or roof) to generate clean and environmental-friendly electricity from the sun radiation. The architectural integration of PV collector modules in new construction makes possible to create glazed surfaces which, besides being an aesthetical and functional innovation, generate electricity, that is allowing the realization of solar control and electric range with consequent energy savings. However, up until now, there have been limited studies that analyzed BIPV from the materials and constructions point of views. This paper aims to expose rooftop BIPV materials that further might be used to replace conventional rooftiles such as terracotta or slate, concrete, plastic, and clay tiles. The flat rooftiles of monocrystalline- and polycrystalline-based BIPV cells are being considered in this paper. Their benefits and drawbacks are also put into account. The parameters that were compared are materials, physical constructions, installations and performances at all weather conditions. All the comparisons of those rooftiles that have been made turned out to meet the modern households' aesthetical, energy and costs saving.

1. Introduction

As the renewable energy used become trending topic for over two decades, and the non-polluting energy, together with the energy efficiency, are increasing, also zero energy and zero emission buildings are rapidly drawing attention. Due to these circumstances, several buildings in the city are now design to harvest energy from its surroundings. Sun radiation as a clean and friendly-environment energy source is one of the best options. Building integrated photovoltaic (BIPV) systems, where solar cells are integrated within the climate envelopes of buildings and utilizing solar radiation to produce electricity, may represent a powerful and versatile tool for reaching these goals with respect to both aesthetical, economic and technical solutions [1].

According to [2], PV cells may be carried out on sloped roofs, flat roofs, facades and solar shading systems. PV cells could be mounted above or onto the existing or traditional roofing. BIPV systems replace the outer building envelope skin, thus it is also serving simultaneously as both a climate screen and a power source generating electricity [1]. Both studies by Jelle, [1] and [2] reviewed the considered and evaluated aspects of the integration of PV cells into the outer building envelope skin are i) ensuring the air gap underneath the solar cells due to provide air flow; ii) reducing the temperature of the solar cells; iii) inclining the BIPVs, both with respect to existing and new buildings, as the solar cells necessarily need to follow the roof inclination to be integrated solutions; and iv) geographical position orientation towards the sun and area coverage; and v) building physical issues, for examples heat and moisture that flow in the building envelope.

Reference [3] reported that BIPV can also bring some other advantages to a building seasonally. For example, in summer, it can help to reduce the heat gain by preventing the envelopes from being directly exposed to solar radiation, thus effectively reducing the indoor cooling load. Additionally, by reserving an air channel between the PV modules and the external envelopes of a building could bring benefit from the air circulation in the channel by decreasing the operating temperature of the PV modules, which provides an effective method to increase the energy efficiency of PV module.

Reference [4] reported that the fundamental raw material that still rules the PV market is silicon. It is the second most common chemical element on earth after oxygen. It is a very hard, grey material that glitters like metal. Naturally it can be found in oxidized form for instance in gravel, sand, clay or quartz. The use of silicon came into focus due to its semiconductor properties.

Rooftile integrated PV module is a trend in BIPV technology. The integration of PV collector modules in new construction, makes possible to create glazed surfaces. This integrated PV into rooftile replaces the conventional rooftiles such as terracotta or slate, concrete, plastic and clay tiles. This paper reviews the silicon-based, namely, monocrystalline and polycrystalline silicon, respectively. Then the model of solar cell and the test to get its energy efficiency. The benefits and drawbacks of those two silicon-based BIPV rooftile solar collector module are also discussed in briefly.

2. Silicon-based BIPV

Silicon is the second most plentiful element in the shell of the earth, right after oxygen. It appears in sand, quartz, stone, rocks, gravel, even in seawater silicon appears in gigantic quantities. The extraction of silicon is less harm to the environment and our planet at all, if the manufacture considered the environmentally friendly processes. Since silicon does not rot, by manufacturing it in a good quality process, this also applies to the other components, such. copper, tin and the high-quality glass. For this reason, the flat-plates of high-purity silicon can generate clean electricity [5].

The discussion of generating electricity by utilizing sun radiation, the distance from the sun to the earth is between 147-152 million kilometres and this leads to oscillations in the solar constant between 1325-1420 watts/m^2 (in the long term, if the shape of the earth's orbit changes - sometimes it deviates more from the circular path, sometimes less). The sun radiates its energy equally strong in all directions. If the average distance from sun to earth is 1367 watts/m^2, then this also applies to every other square meter that is located on a gigantic sphere with the radius of the earth 's orbit. Then by multiplying the area of this sphere by the sun constant, the sphere has a radius of 150 million kilometers. Additionally, in every square kilometer it provides 1367 watts/m^2. Overall, this results in a value of gigantic 386 quadrillion watts for the radiance of the sun every second. This amount is a massive power that a power station can possibly generates [5].

Silicon tiles may be applied to make a BIPV roof look very much like a standard tiled roof, while semi-transparent modules may be applied in glass ceilings to create different visual effects [2]. 1367 watts or 1.367 kW/m^2 are the basis of calculation. An average per year. If a roof with an area of 100m^2, 13,67kWh of energy per day fall on every m^2 with 10 hours of sunshine, then in total, this roof has a total energy of 1,367 kWh per day, in the annual average. This is an overall solar rooftile concept that has been developed by a solar rooftile BIPV company in Germany [5].

The crystalline silicon ingots that commonly used to create solar cells from silicon get solar cells and may result in cells with specific thicknesses, sizes and shapes [2] and [4]. Silicon-based PV material technologies can be grouped into three types: monocrystalline silicon (mono-Si), (poly-Si), and thin-film amorphous silicon (a-Si). Table 1 describes the differences of those three silicon-based PV cells.

Utilizing electrical energy without emissions and pollutants, and even completely for free is the aim of every country due to save the environment. This issue can be implemented for the purpose of households 'needs whereas each house might cover 100% of its own energy needs and also generate energy surplus, which is then fed into the grid.

IOP Conf. Series: Materials Science and Engineering **854** (2020) 012072 doi:10.1088/1757-899X/854/1/012072

Figure 1. Silicon materials [5]

This section discusses the silicon-based rooftile solar BIPV module, mono-Si and poly-Si, respectively. The discussion details in their materials, configurations and constructions.

Table 1. silicon-based solar PV [4] and [6]

Type of Cells and their criteria	Silicon-based semiconductor materials		
	Monocrystalline (Mono-Si)	**Polycrystalline (Poly-Si)**	**Thin-film amorphous silicon (a-Si)**
Physical photo		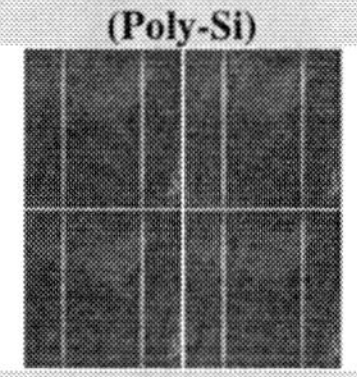	
Cell efficiency (at 25 °C, light intensity of 1000W/m^2, air mass = 1.5)	16-17%	16-17%	16-17%
Module efficiency (%)	15-20	11-15	5-7
Temperature coefficient P_{max} [%/°C] (±0.03)	-0.45	-0.45	-0.21
Characteristic cells	Shape, size, texture and color (antireflective layer) of the cells	Shape, size, texture and color (antireflective layer) of the cells	Homogeneous surface
Characteristic panels	Arrangement of cells (pattern and translucency), Colors and patterns (background), Colors and types (frame) Glass coating (optical reflection)	Arrangement of cells (pattern and translucency), Colors and patterns (background), Colors and types (frame) Glass coating (optical reflection)	Flexibility: different substrates of the building skin Optical reflection

2.1. Monocrystalline silicon BIPV

Materials, configurations and construction

The mono-Si modules are normally of a black or grey colour and have the highest efficiency, but also high prices [7].

The considered mono-Si PV in this study is solar tiles type BIPV 36 cells 100w [8]. This solar roof tiles that has high-efficiency to transform the energy of sunlight into electric energy. Each cell is electrically rated to optimize the behavior of the module. The physical sides (Figure 2) has front-, cell circuit and back-sides. The front-side of the solar tile contains a tempered solar glass with high transmissivity, low reflectivity and low iron content. The cell circuit is laminated using polyvinyl butyral (PVB) as an encapsulant in combination with a tempered glass on both its front and back which provides complete protection and seals against environmental agents and electrical insulation. The back-side of the tile contains tempered glass with low iron content.

The junction boxes with IP65, are made from high temperature resistant plastics and containing terminals, connection terminals and protection diodes (by-pass). These tiles are supplied with symmetric lengths of cable, with a diameter of copper section of 4 mm and an extremely low contact resistance, all designed to achieve the minimum voltage drop losses.

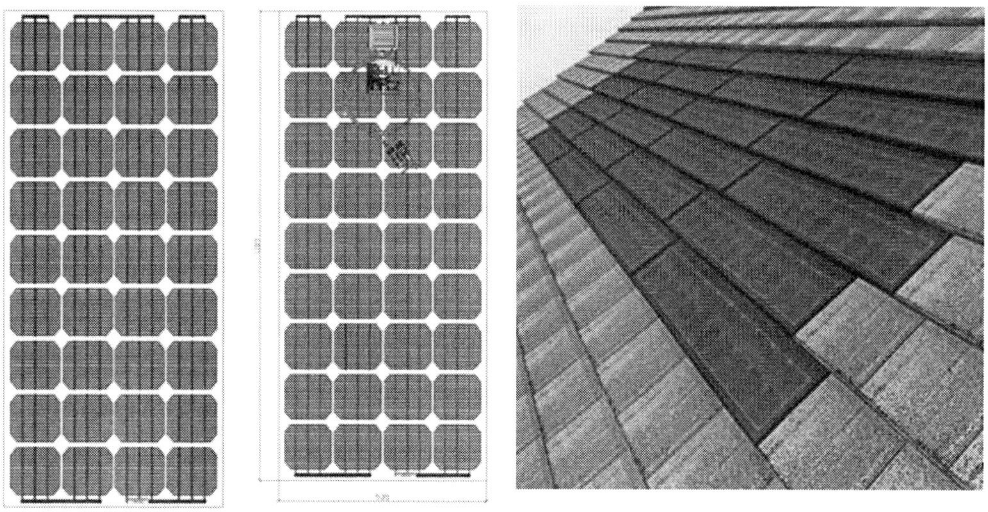

(a) front (b) back (c) physically mounted

Figure 2. Mono-Si solar rooftile BIPV 36 cells 40W module [8].

2.2. Polycrystalline silicon BIPV

Materials, configurations and construction

The poly-Si modules have a blue colour deriving from small crystals and are cheaper but yield lower efficiencies [7].

The considered poly-Si PV in this study is solar tiles type BIPV 36 cells 100w modules that cells are made of several crystals of high purity silicon to transform the energy of sunlight into electric energy. Each cell is electrically rated to optimize the behaviour of the module. Figure 3 illustrates the front-, cells, and back-side of the module. The front of the module contains a tempered solar glass with high transmissivity, low reflectivity and low iron content. The cell circuit is laminated using EVA (Ethylene-Vinyl Acetate) as an encapsulant in combination with a tempered glass on its front and a tempered glass with low iron content on the backside.

The junction boxes with IP67, are made from high-temperature resistant plastics and containing terminals, connection terminals and protection diodes (by-pass). These modules are supplied with symmetric lengths of cable, with a diameter of copper section of 4 mm and an extremely low contact resistance, all designed to achieve the minimum voltage drop losses.

IOP Conf. Series: Materials Science and Engineering **854** (2020) 012072 doi:10.1088/1757-899X/854/1/012072

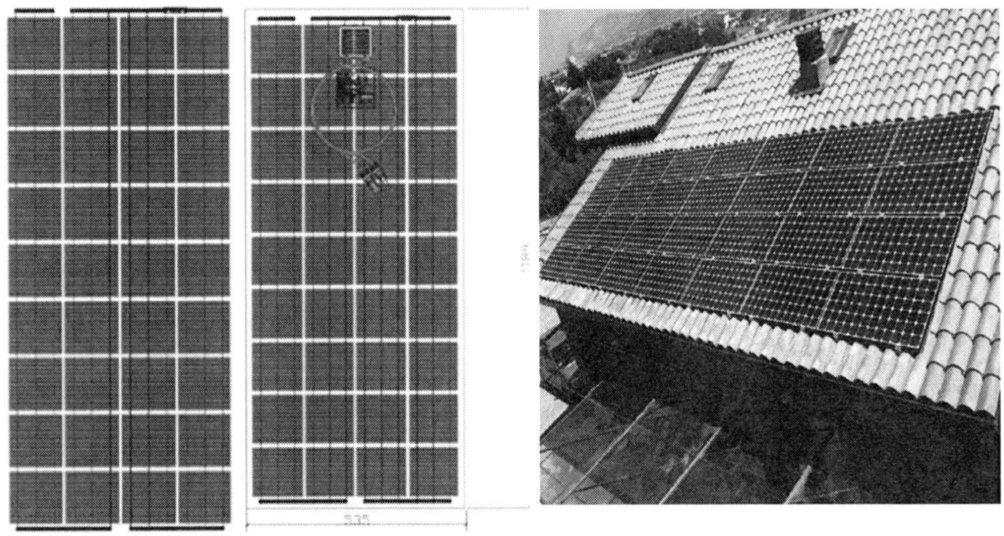

(a) front (b) back (c) physically mounted

Figure 3. Poly-Si solar rooftile BIPV 36 cells 100W module [8].

3. Test method

According to [9], the evaluation of BIPV involve several properties, for example solar cell current and voltage (1), maximum power (2), efficiency (3) and the effect of temperature on cell efficiency are as follow.

$$I = I_p - I_0\left[\exp\left(\frac{q(V + IR_{se})}{AkT}\right) - 1\right] - \left(\frac{V + IR_{se}}{R_{sh}}\right) \quad (1)$$

where I_p is the photocurrent, R_{se} is the series resistance, R_{sh} is the shunt resistance, A is the diode ideality factor, and I_0 is the diode saturation current.

$$P_m = V_m I_m = (FF) V_{oc} I_{sc} \quad (2)$$

where FF is the fill factor, m refers to maximum power point of the I-V curve, V_{oc} is the open circuit voltage, and I_{sc} is the short circuit current.

$$\eta = \frac{V_m I_m}{AG} \quad (3)$$

where A is PV module area, and G is the irradiance.
The effect of temperature on cell efficiency can be deduced from the relation of (4).

$$\eta c = \eta_{Tref}[1 - \beta_{ref}(Tc - Tref)] \quad (4)$$

where η_{Tref} is the module efficiency at reference conditions, β_{ref} is the temperature coefficient, T_{ref} is the reference temperature, and Tc is the cell/module temperature.

The values reported by solar cell manufacturers are mainly obtained according to standard test conditions (STC, irradiance 1000 W/m^2, temperature of PV cell 25°C, solar radiation distribution AM

1.5) or nominal operating cell temperature (NOCT, irradiance 800 W/m^2, ambient air temperature 20°C) [2]. By using the equations (1)-(3), and modeling them using MATLAB platform, then the output power (Wp/m^2), and transparency of mono-Si, poly-Si rooftile and a- Si solar BIPV, respectively are illustrated in Figure 4.

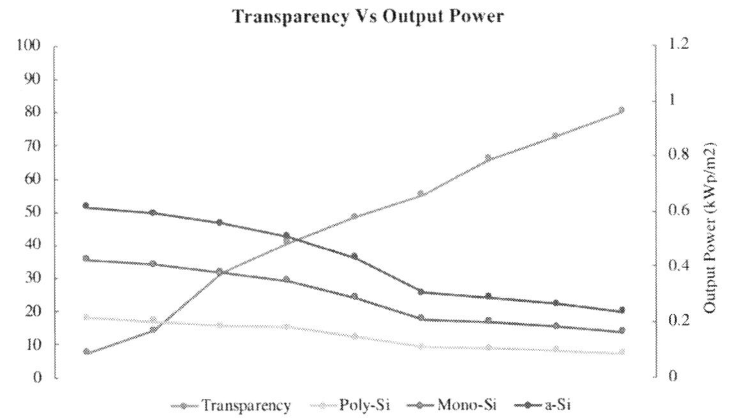

Figure 4. Mono-Si and poly-Si output power versus their module transparency.

Thus, it can be stated that the more transparent the mono-Si, poly-Si and a-Si modules, the more the output power increased, since the more sun radiation has been absorbed through the modules.

4. Conclusion

The persistent research and development of both PV and BIPV materials such as mono-Si, poly-Si, and a-Si on their construction technologies perspective then generate better sun radiation and result clean and environmentally friendly electricity. Additionally, better BIPV solutions with respect to increased solar cell efficiency, reduced production costs improved building integration problems.

References

[1] B. P. Jellea and C. Breivikb, "State-of-the-art building integrated photovoltaics.', Energy Prodia, Vol. 20, pp. 68-77, 2012.

[2] B. P. Jellea, "Building Integrated Photovoltaics: A Concise Description of the Current State of the Art and Possible Research Pathways.", Energy Journal, Vol. 9, doi:10.3390/en9010021, 2016.

[3] T. Zhang, M. Wang and H. Yang, "A Review of the Energy Performance and Life-Cycle Assessment of Building-Integrated Photovoltaic (BIPV) Systems.", Energy Journal, Vol. 11, pp. 31-57, doi:10.3390/en11113157, 2018.

[4] K. Farkas, I. Andersen and A. G. Hestnest, "Architectural Integration Of Photovoltaic Cells Overview Of Materials And Products From An Architectural Point Of View.", PhD thesis report, Department of Architectural Design, History and Technology, Norwegian University of Science and Technology, Trondheim, Norway, 2009.

[5] Solteq, "The Energy source of the Future: Your Roof - Clean and free Energy for Electricity and Heating.", www.SOLTEQ.uk + www.TheSolarroof.uk, 29/05/2019.

[6] P. Heinstein, C. Ballif and L. Perret-Aebi, "Building Integrated Photovoltaics (BIPV): Review, Potentials, Barriers and Myths.", De Gruyter Journal, DOI 10.1515/green-2013-0020Green, 2013.

[7] T. Salema, E. Kinabb, "Analysis of Building-Integrated Photovoltaic Systems: A Case Study of Commercial Buildings under Mediterranean Climate.", International Conference on Sustainable Design, Engineering and Construction, Procedia Engineering Vol.118, pp. 538 – 545, 2015.

[8] Solarinnova, "Looking for the future energy in your house.", www.solarinnova.net , 5/8/2014.

[9] S. Ziuku and E. L Meyer, "Implementing building integrated photovoltaics in the housing sector in South Africa.", Journal of Energy in Southern Africa, Vol 24 No 2, 2013.

Soot Formation Model Performance in Turbulent Non-Premix Ethylene Flame: A Comparison Study

E Elwina[1], N Sylvia[2], W Wusnah[2], Y Yunardi[3], Y Bindar[4]

[1]Chemical Engineering Department, Politeknik Negeri Lhokseumawe, Aceh, Indonesia
[2]Chemical Engineering Dept., University of Malikussaleh, Lhokseumawe, Indonesia
[3]Chemical Engineering Dept., Syiah Kuala University, Banda Aceh, Indonesia
[4]Energy and of Processing System Chemical Engineering Dept., Faculty of Industrial Technology, Bandung Institute of Technology, Bandung, Indonesia

Email: elwina@pnl.ac.id

Abstract. This paper presents results obtained from the application of a computational fluid dynamics (CFD) approach to modelling of non-premixed turbulent ethylene sooting flame. The study focuses on comparing the two soot models available in the Fluent in predicting the soot level in the turbulent non-premixed ethylene flame. A standard k-ε model and Eddy Dissipation model are utilized for the representation of flow field and combustion of the flame being investigated. For performance comparison study, a single step soot model of Khan and Greeves and two-step soot model proposed by Tesner are tested. The results of calculations are compared with experimental data for a turbulent sooting flame taken from literature. The results of the study show that a combination of the standard k-ε turbulence model and eddy dissipation model is capable of producing reasonable predictions of temperature both in axial and radial profiles; although further downstream of the flame over-predicted temperatures are evidence. With regard to soot model performance study, it shows that the two-step model clearly performed far better than the single-step model in predicting the soot level in ethylene flame at both axial and radial profiles.

Keywords: soot, single-step model, two-step model, turbulent, ethylene

1. Introduction

Modelling the soot formation and destruction does not only play an important role in the design, but also essential in the operation of in a turbulent combustion process. In order to accurately predict the production and destruction of soot in such systems, an integrated model that at least represents turbulence, combustion process, soot particle production and removal, and radiation heat losses is required. The ability to apply such an integrated model, of demonstrated accuracy, to minimise soot production and emission in relation to safety and environmental considerations, and to maximize production in systems where this would be beneficial for heat transfer, would represent a major step-forward in our ability to design and manage combustion processes.

A number of soot models, ranging from empirical to detailed approaches, have been proposed in the literature, and research to develop new models and improve the existing ones is ongoing. Those approaches that have been proposed for modelling soot formation in turbulent flames have their own advantages and disadvantages. The empirical soot models as reviewed in Kennedy [1] are mainly based on model parameters which are different for different fuels. The most sophisticated soot models today, such as that proposed by Frenklach and Wang [2] employ detailed chemical kinetic and physical models to describe each sub-process that occurs in the gas phase, solid phase, and on the surface of soot particles. Although such models are applicable over a wide range of combustion conditions, their application at present has been impaired by the excessive requirement for computer resources even for simple flames and the poor representation of soot inception chemistry, with some of the relevant reaction rates purely estimations [3]. Thus, for predictions of soot in practical engineering devices, it is often necessary to use simplified models to keep the computational cost at an acceptable level without losing an acceptable degree of accuracy.

In this paper, the results of the application of a computational fluid dynamics (CFD) code Fluent 6.3 to modelling of ethylene non-premixed sooting flames are presented. The objective of the present work is the assessment of the two soot models available directly in the code for the prediction of soot levels in the flame. Soot formation in the flame is calculated using a single-step and tow-step models, respectively, and predicted results are compared each other. Validation is achieved by comparing the computed prediction by each soot model against available experimental data of ethylene non-premixed sooting flames [4].

2. Mathematical formulation and metode

2.1. Formatting the title

The evolution of soot in a flame can be described by two competing processes, soot formation and destruction. As a consequent, almost all soot models include these two phenomena in their mathematical development. In this study, a single-step soot model proposed by Khan and Greeves [5] and two-step soot model proposed by Tesner et al [6] will be implemented. The former model has been applied with some success in calculations of soot emissions from diesel engines [7] and gas turbines [8]. With respect to the latter model, Magnussen and Hjertager [9] were one of the first to apply the Tesner model coupled with the eddy dissipation combustion model (EDM) to simulate soot formation in flames.

The calculation of Khan and Greeves model requires a solution of a single transport equation, presented in Eq.1, in which the source term is given by Eq.2.

$$\frac{\partial}{\partial t}(\rho Y_{soot}) + \nabla.(\rho \vec{v} Y_{soot}) = \nabla.\left(\frac{\mu_t}{\sigma_{soot}} \nabla Y_{soot}\right) + R_{soot} \quad (1)$$

$$R_{soot} = R_{soot;form} - R_{soot;comb} \quad (2)$$

where R_{soot} = net rate of soot generation (kg/m^3.s); $R_{soot;form}$ = rate of soot formation (kg/m^3.s) and $R_{soot;comb}$ = rate of soot destruction (kg/m^3.s). The rate of soot formation is given by a simple empirical rate expression as in Eq. 3.

$$R_{soot;form} = C_s p_{fuel} \Phi^r e^{-|E/RT} \quad (3)$$

where C_s = soot formation constant (kg/N-m-s); p_{fuel} = fuel partial pressure (Pa); Φ= equivalence ratio; r = equivalence ratio exponent; and E/R = activation temperature (K). The rate of soot combustion is the minimum of two rate expressions of Eq.4

$$R_{soot;comb} = \min\left[R_1; R_2\right] \quad (4)$$

The two rates are calculated in accordance to Equations 5 and 6.

$$R_1 = A\rho Y_{soot} \frac{\varepsilon}{k} \quad (5)$$

$$R_2 = A\rho\left(\frac{Y_{ox}}{v_{soot}}\right)\left(\frac{Y_{soot}v_{soot}}{Y_{soot}v_{soot}+Y_{fuel}v_{fuel}}\right)\frac{\varepsilon}{k} \tag{6}$$

where A = Constant in Magnussen model; Y_{ox}, Y_{fuel} = mass fractions of oxidizer and fuel; v_{soot}, v_{fuel} = mass stoichiometric for soot and fuel combustion; ε= energy dissipation; and k = kinetic energy

In addition to the solution of transport equation for soot mass fraction, as given in Eq. 1, Tesner model requires the solution for another transport equation for radical nuclei concentration, as shown in Eq. 7

$$\frac{\partial}{\partial t}\left(\rho b^*_{nuc}\right)+\nabla.\left(\rho\vec{v}b^*_{nuc}\right)=\nabla.\left(\frac{\mu_t}{\sigma_{nuc}}\nabla Y_{nuc}\right)+R^*_{nuc} \tag{7}$$

where b^*_{nuc} = radical nuclei concentration (particles x 10^{-15}/kg); σ_{nuc} = turbulent Prandtl number for nuclei transport; R^*_{nuc} = net rate of nuclei generation (particles x10^{-15}/m^3.s). The rate of soot combustion, $R_{soot;comb}$, is computed in the same way as for the single-step following Eq. 4. However, the rate of soot formation, $R_{soot;form}$, depends on the concentration of radical nuclei, as presented by Eq. 8.

$$R_{soot;form} = m_p\ \left(\alpha-\beta N_{soot}\right)C_{nuc} \tag{8}$$

where m_p= mean mass of soot particle (kg/particle); N_{soot}= concentration of soot particles (particles/m^3); C_{nuc}= radical nuclei concentration (particles/m^3); α= empirical constant (s^{-1}); and β = empirical constant (m^3/particle-s).

The ethylene-air jet turbulent non-premixed sooting flames measured by Kent and Honnery [4] is investigated for comparing the performance of Khan and Greeves' and Terner's models. The related reference may be consulted for important characteristics, including the flame geometry, methods of data collection, and processing. The calculation of flow and mixing fields was achieved by solving the partial differential equations which describe conservation of mass and momentum. A standard k-ε turbulence model was used to close the above equation set, with an adjustment was made to the value of $C_{\varepsilon 1}$. The combustion calculation which solved the energy and species concentration was performed by utilizing eddy dissipation model (EDM). All the above calculations were simultaneously performed using commercial CFD software FLUENT ver. 6.3 which functions as the processor as well as post processor. The soot models calculations were implemented after the calculations of turbulence and combustion.

3. Result and discussion

3.1. *Temperature Profile and contour*

Figure 1 presents a comparison between the centerline and radial temperature predictions in the 1 atm ethylene flame and the experimental data. The solid line represents the temperature predictions calculated using a combination of a standard k-ε turbulence model and eddy dissipation combustion model, while the symbol of small circle represents the experimental measurements. The axial temperature predictions generally display qualitatively good results in comparison to the experimental data. The evolution of the computed axial temperature follows the trend of the axial temperature measurements. However, from quantitative point of view, from a region between 150 and 350 mm (fig. 1a) above the nozzle, the temperature is over-predicted. Such over-predictions might be due to simple representation of combustion chemistry by eddy dissipation combustion model and simple radiation model selected during the combustion calculation. It is to be noted that the EDM assumes the fuel is fast burning, and the overall rate of reaction is controlled by turbulent mixing. As a consequent, the chemical kinetic can be neglected, which is not the case for other combustion models where the detailed kinetic mechanism can be included in their calculation. In addition, Fluent code provides various radiation models, ranging from simple Roseland model up to more complex models, such as Discrete Order (DO) radiation model. For the sake of simplicity in the calculation, a radiation model

P1 was selected. Nonetheless, the value and location of peak temperature are well predicted by the model. With respect to the radial temperature profile, it is evidence that predictions are in reasonable agreement with the experimental as also shown in the same figure. The temperature profiles in the radial fuel-lean and fuel-rich regions are captured reasonably well.

Figure 2 shows the contours of temperature along the flame using the EDM model. From that, can see the colour differences to illustrate the value of the temperature, more red the contour colour, higher the temperature. The blue colour indicates the air temperature. From that picture shows that the EDM model are able to describe the contours of the flame as the actual situation, although not yet providing high accuracy, as compared with temperature data from previous experiment. But it is clear that in the middle zone the flame of temperature is higher than the rich areas of oxygen. So that the combustion phenomena can described by this model.

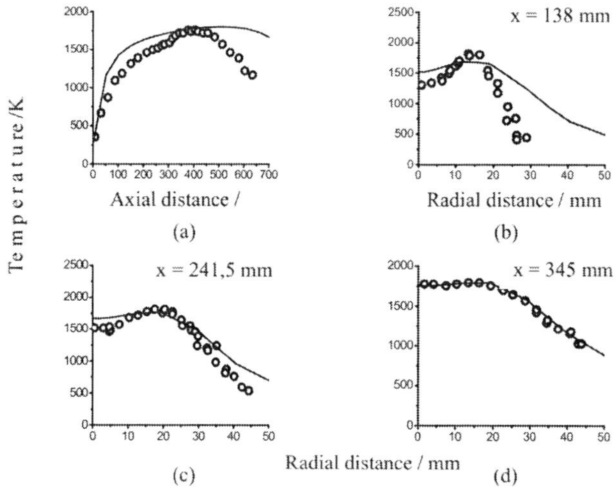

Figure 1. Axial and radial temperatures for the 1 atm ethylene flame (symbol circle O measured, solid line – predictions)

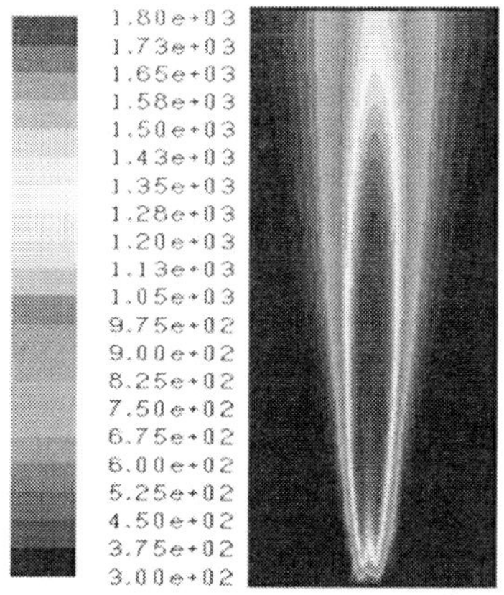

Figure 2. The Contour of the flame temperature of ethylene

3.2. Soot Profile and Contour

Figure 3 presents a comparison of axial and radial predictions of soot volume fraction for the ethylene flame with experimental data. The solid line represents the simulations resulting from the use of Tesner's model, and the dashed line the simulations resulted from employing Khan and Greeves' model. It is clearly seen that the centreline soot volume fraction profile is very well represented by Tesner's model, when the empirical constant, α in Eq. 8 was adjusted around 3 times of the default value in the code. Although the same adjustment was made to soot formation constant, C_s in the Khan and Greeves' model, the soot predictions yielded by this model are very unrealistically low than measurements which is due to the strong role of fuel concentration in the model [10]. With respect to radial soot profile of 1 atm flame, although quantitatively the predictions by Tesner's model are slightly under-predicted at all axial locations, qualitatively the predicted trend is agreement with the measurements, in which Khan and Greeves' model failed to produce. With a slight adjustment in the constant α, the current results are comparable to those produced using other semi-empirical models [11,12,13].

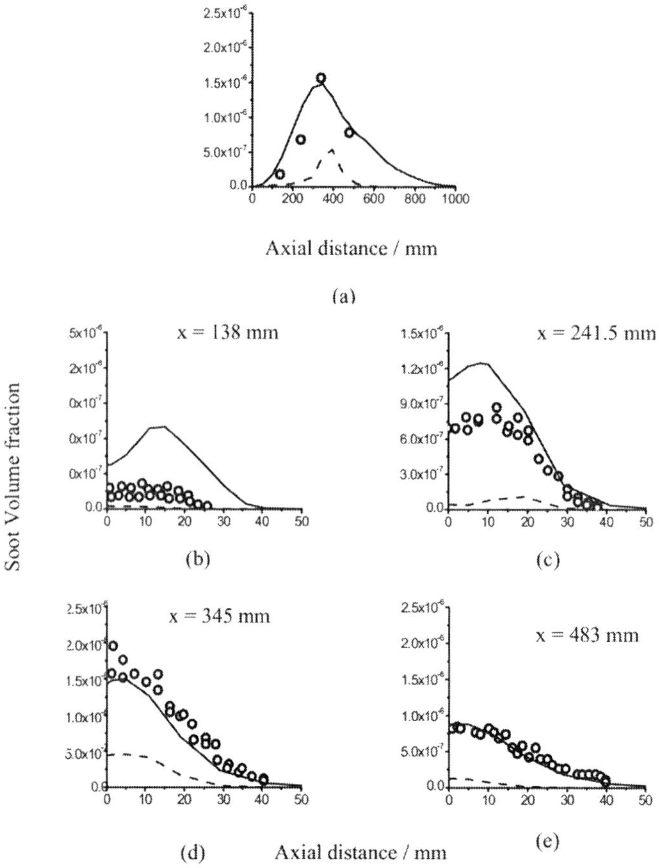

Figure 3. The Comparison of experimental results and predictions for soot volume fraction in the axial and radial position in the ethylene flame KH (O experiments symbol, — two-step model prediction, — — one-step model prediction).

In the axial profile (Fig. 3a), shows that the two-step model is able to produced satisfactory predictions. It is characterized by the conformity of prediction to the experimental data on soot growth phase and the destruction phase. The highest value of soot position can also be predicted well by two-steps model. However, it should be noted, that the accuracy is achieved by modifying the constant alpha value. This modification is performed randomly to get the results close to experimental data, considering there are no experimental data for alpha constanta value for ethylene flame. For default alpha value represents the value for the acetylene flame. Such modifications are also mostly done by other researchers [11].

In the radial profiles, the prediction of soot in the region close to the nozzle is less satisfactory. Tesner model provides the predictions of soot which exceeds the experimental value, while the one-step model provides a prediction that is very low compared to experimental data. The Predictions which is exceed in the experimental value at position 138 mm (Figure 3b) from the surface of the nozzle is also experienced by Pitsch [14]. But, two-step model could improve the predictions of soot in radial profiles at height of 345 and 483 mm (Figure 3d and e) from the surface of the nozzle. This is possible because the prediction of temperature is more accurate in both these altitudes. The Prediction at altitude of 345 and 483 mm using by two-step soot model is better obtained by Pitsch [14]. But not so with one-step method. Although the accuracy assumption of temperature is sufficient, the one-step method predicting the soot is quite low at these two heights.

Furthermore, for two-steps model at height of 138 and 241 mm (Figure 3b and c), the deviates from the results is quite large, especially at the height of 138 mm, the simulation results none could capture the appropriate experimental data, except for one-step model obtained the simulation results close to the experimental data. In contrast to the positions 345 and 483 mm, where the two-steps model simulation is capable to capturing accordance with the experimental data, especially at a position 483 mm which is in accordance with experimental results. The Deviations that occur cannot be separated from the minimum data obtain for the kinetic reaction mechanism. However, for one-step model prediction, the results are still very far from the experimental data.

Figure 4 shows the contour of soot mass fraction which produced using two-step soot model [6] and one-step soot model [5]. The blue color indicates the minimum soot mass fraction (close to zero, or no soot), while red indicates the maximum soot mass fraction. From the pictures can be seen that the two-step model of soot begins to form in areas close to the nozzle. This is understandable because in this region rich in fuel and poor air, so that resulting a tendency to form the soot. Soot maximum is reached still in the regions rich of fuels and along with the movement direction of soot with the flow flame, the diffusion of air more higher causes the oxidation of soot comes into play. Consequently, after reaching the maximum indication, soot mass fraction decreases due to oxidation and turns into a gas phase.

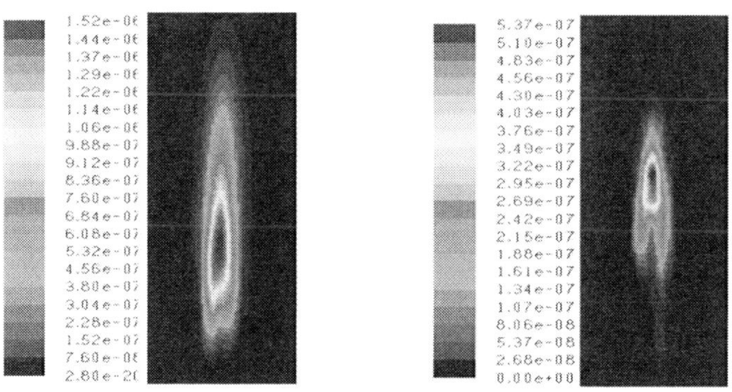

Figure. 4. The Contours of soot mass fraction in the flame of ethylene (a) with a two-step model and (b) with one step model

Different with two-step model, one-step model cannot predict the formation of soot in the region close to the nozzle. The soot begins to formed in areas away from the nozzle. In this area, the air is rapidly diffused into the flame. As a result, the soot is oxidized too quickly. This is evident in the mass fraction contours generated by one step model compared with the soot mass fraction contours generated by two-steps model.

The constant in both models also have an influence on outcome prediction. In the two-step model, the constanta that holds important role is an alpha value, whereas in the one-step model is the constanta of soot formation. From Figure 4 can also see the contours of the soot in one-step model, where the rate of the soot formed is slower than the two-step model, so that the peak area on one step model are shifting. As mentioned on the theory that the one step model is a simple empirical model that the application is a little difficult to be applied to other hydrocarbons, the constanta used should be arranged so that with a certain constanta value through the sensitivity test will be obtained the accordance results with the experiments. However, the results obtained have not been close to the experimental data because in this model the soot calculation just look at the soot formation and oxidation without other processes such as the formation of soot nuclei as in the two-step model. But overall the predicted results from Tesnet et al two-step model so far has given satisfactory results that can be applied in industry because of the predictions results are not difference too much with the predicted results which is obtained by using semi-empirical soot model which more complex [14].

Pitsch [14] also conducted same simulation of the ethylene flame. The Axial temperature predictions made by Pitsch [14] reviewed on the basis of taking into account the soot and gas radiation, without the soot and gas radiation, and without soot radiation. It should be noted that Pitsch [14] using Unsteady Laminar Flamelet model with kinetics through detailed combustion. Seen that the predictions are very consistent with the experimental data when the predictions considering the soot and gas radiation. If seen the comparison results of the predicted results in this study, for height at 450 mm in terms of prediction quality in this study is slightly below with Pitsch [14] prediction results. Above 450 mm Pitsch [14] prediction is much better. From this, it can be presumed that the kinetics of combustion reaction and radiation models play an important role in generating the proper temperature predictions.

The calculated axial and radial soot volume fractions again show that better predictions can be achieved by using Tesner's soot model. Soot volume fraction distribution in axial profile calculated using Tesner's model is qualitatively and quantitatively in excellent agreement with the measurements, both in the soot formation and oxidation zones. It is important to note here that the Tesner's model is capable of capturing the increase of soot level due to the increase of flame pressure which on the contrary Khan and Greeves' model failed to produce so.

4. Conclusion

A numerical simulation of soot formation and destruction has been adopted to study the performance of two soot models for prediction of soot levels in turbulent non-premixed flames. Soot formation is modelled by a single-step model and a two-step model as the results are compared with each other and with experimental data. The two-step soot model proposed by Tesner et al [6], with a light adjustment in the modeling constant, is capable of producing the predictions closer to the experimental in ethylene flames. On the contrary, the single-soot model proposed by Khan and Greeves [5] produced very poor results, leading to a significant under-prediction of soot levels in both flames. Although the Tesner's soot model is simple in its mathematical formulation, this model is still capable of providing reasonable agreement with experimental data, allowing its application for the purpose of design and operation of an industrial combustion system.

References

[1] I.M. Kennedy: Prog. Energ. Combust. Vol. 23 (1997), p. 95
[2] M. Frenklach and H. Wang, in: Soot Formation in Combustion edited by H. Bockhorn, Springer-Verlag, Berlin.(1994).
[3] J.Z. Wen, M.Thomson, S. Park, S. Rogak, and M. Lightstone: P. Combust. Inst. Vol. 30 (2005), p. 1477
[4] Kent, J.H and Honnery, D., Soot and mixture fraction in turbulent diffusion flames. *Combustion Science and Technol*ogy (1987), 54: 383-397.
[5] I.M. Khan and G. Greeves, in: Heat Transfer in Flames edited by N.H. Afgan and J.M. Beer, Chapter 25, Scripta, Washington, USA (1974).
[6] P.A. Tesner, T.D. Snegriova and V.G. Knorre: Combust. Flame Vol. 17 (1971), p.253
[7] P.S. Mehta and S. Das: Fuel Vol. 71 (1992), p. 689
[8] A. Lefebvre: Int. J. Heat Mass Tran. Vol. 27, p. 1493

IOP Conf. Series: Materials Science and Engineering **854** (2020) 012074 doi:10.1088/1757-899X/854/1/012074

Effect of the Seeding Thickness on the Growth of ZnO Nanorods prepared by CBD

N Rosli[1], M M Halim[1], M R Hashim[1], W Maryam[1], M F M Rusdi[2], A R Muhammad[2]

[1]School of Physics, Universiti Sains Malaysia, 11800 USM, Penang, Malaysia.
[2]Department of Electrical Engineering, University of Malaya, Kuala Lumpur, Malaysia.

E-mail: nurizatirosli@student.usm.my

Abstract. Zinc Oxide (ZnO) nanostructure (NS), with nanorods (NRs) forward-facing alignment were deposited by chemical bath deposition (CBD) method on seeded glass substrates. Two different ZnO seed layer thickness, 100 nm and 200 nm, prepared by RF sputtering were employed to investigate the effect of seed layer thickness on the characteristics of the grown ZnO NRs. The investigations on physical, structural, and optical properties were carried out through field emission scanning electron microscope (FESEM) with integrated energy dispersive X-ray (EDX), X-Ray Diffraction (XRD), and UV and visible (UV-Vis) analyses. The morphological results reveal that with a thicker seed layer, the grown NRs diameter increased double while the NRs height was maintained at 1.5 μm. From EDX, the stoichiometric ratios of Zn:O were mostly towards 1:1. While the XRD evaluates that the preferred structure was wurtzite hexagonal with c-axis orientation along (002) plane. The UV-Vis analysis showed the increased in optical band gap energy closes to the bulk ZnO at 3.37 eV, for the ZnO NRs grown on the thicker seed layer. The obtained results illustrated high quality single-crystal ZnO NRs with different diameters were realized by only adjusting the seed layer thickness.

1. Introduction

ZnO nanorods enticed much attention as metal oxide nanoparticles with a wide direct bandgap, 3.37 eV, large exciton binding energy, 60 meV, inherent toxicity, as well as peculiar physical, chemical and optical properties. With this high specification ability, ZnO is considered as one of the preferred materials in optoelectronic, photonics, electronics and biomedical field [1-3].

There are numerous processes to fabricate ZnO materials. Deposition on a blank substrate such as glass without any pre-treatment leads to the formation of the lower density of ZnO NS and the non-supportive seeding leads to the random growth of NRs. Several studies presented the useful of the seed layer such as it plays the role as a template for the NRs [4], intended for growth parameter alignment [5], aiming for growth quality [6], decrease the nucleation edges by reducing the interface energy [7] and so on. Numerous works verify the seeding and the growth techniques via variety method had been published such as sol-gel with hydrothermal [8], spin-coating with CBD [9], spin-coating and hydrothermal [10], RF sputtering and hydrothermal [4,11,12], RF sputtering and thermal evaporation [13] and many more. The seed layer acts as a nucleation site to assure the formation of a high density of ZnO NS, assent for preferred or self-organized growth orientation increase and attend on well structural as well as optical properties likewise the bulk ZnO. RF sputtering was chosen as it promotes a large deposition area with easy proceeding and low-time consuming.

IOP Conf. Series: Materials Science and Engineering **854** (2020) 012074 doi:10.1088/1757-899X/854/1/012074

Thus, this work presents the growth control techniques employed by RF sputtering for variation of seed layer thickness, while CBD methods intended for well-aligned growth and high-quality growth of ZnO nanostructures.

2. Methodology

2.1. Preparation of seed layer
RF sputtering process was carried using Auto HHV 500 Sputter Coater with RF power of 150 W, using the ZnO target, at the base pressure of 5.5 x 10-5 mbar and coating rate of 0.6 Å/s, as shown in **Fig. 1a** and large deposition area of sputtering coating plate as in Fig. 1b. Two samples with different seed layer thickness at 100 nm and 200 nm were prepared on glass substrates. Then the samples were annealed in a furnace at 400°C for 1 hour.

2.2. Growth of ZnO nanorods
The annealed seed layer was then pursued for nanostructures growth via CBD at temperature 96 oC for 3 hours. The concentration of both CBD reactants, HMTA and Zn(NO3)2 was fixed to 0.10 M as 1:1 ratio. Then the samples were annealed in the furnace at 400°C for 1 hour.

2.3. Characterization of ZnO Nanorods
The morphology, growth rate, structural, optical, properties were analysed thru FESEM (Nova Nano SEM 450, FEI, Japan with EDX), XRD (PANalytical X'pert, PRO MRD PW3040, Netherlands), and UV-Vis (Cary 5000, Agilent, US). Image processing software, ImageJ was employed to analyse the acquired images, while some equations were utilized to reveal the NS properties.

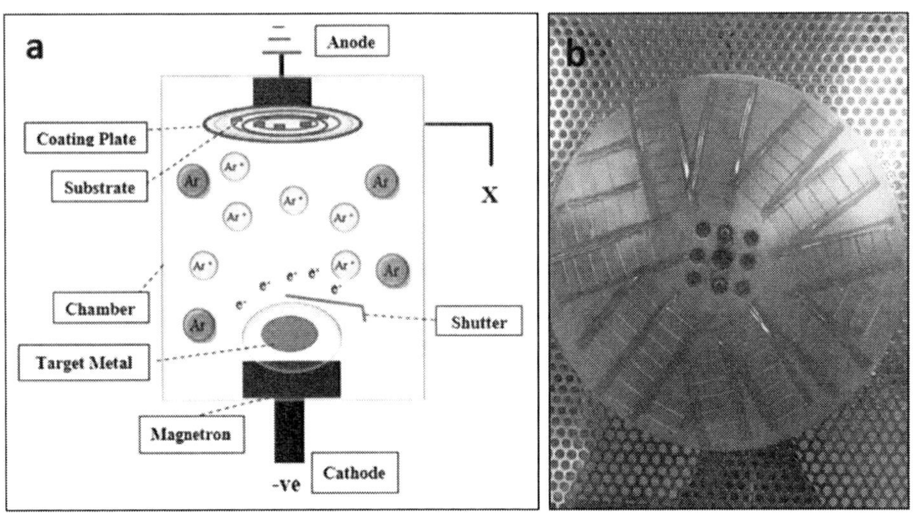

Fig. 1: (a) Schematic diagram of Auto HHV 500 Sputter Coater with RF power, and (b) sputtering coating plate with large deposition area for glass substrates.

3. Results and Discussion

3.1. Morphology by FESEM
As shown in **Fig. 2a** and **Fig. 2b**, FESEM images reveal the top morphology of the grown ZnO 1-D nanostructures. It shown that there was a significant change in size when the ZnO NRs were grown on 200 nm seed layer as compared to 100 nm. With the former seed layer twice size thicker, compared to

IOP Conf. Series: Materials Science and Engineering **854** (2020) 012074 doi:10.1088/1757-899X/854/1/012074

the later, as can be seen the resulting NRs were almost double in diameter size. However, the length of the NRs for both samples was maintained closes to 1.5 λ m on average. This was due to the nature of the chemical synthesis of CBD process as it reached the saturated condition, the maximum height is achieved. Elemental analysis from the EDX integrated with the FESEM as tabulated in **Table 1**, reveals the atomic percentage for both samples was quite close to 1:1 ratio with 56% of Zn and 44% for O, in overall.

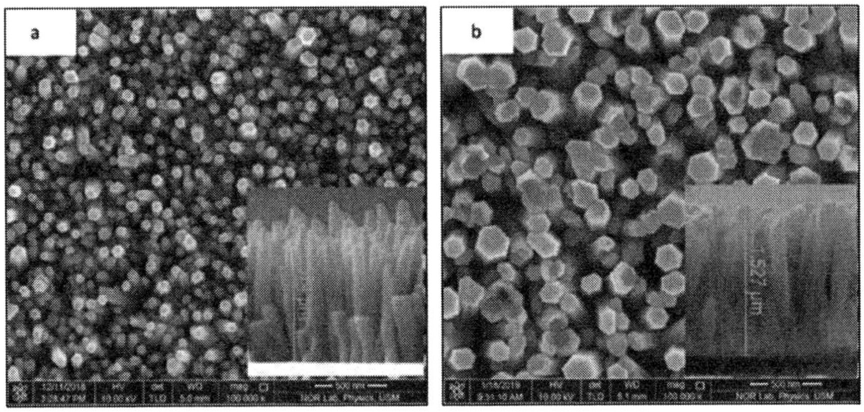

Fig. 2 Top-down morphology and its side profile of the ZnO NRs grow on top of the seed layer with thickness of (a) 100 nm, and (b) 200 nm.

3.2. Density Distribution by ImageJ

ImageJ software was used to analyse and to identify the growth density of the NRs from the morphology of the FESEM images. As can be seen from **Fig. 3**, the density distribution of NRs grown on 100 nm seed layer is given by 113/μm^2 (**Fig. 3a**), and the number is much less on 200 nm seed layer at 24/μm^2 (**Fig. 3b**). It was well understood that larger sizes of ZnO NRs on 200 nm seed layer encourage faster nucleation process. The increase in crystallite size, effectively and promoting the growth of larger ZnO NRs, while reducing the numbers of NRs on a surface area of 1 μm x 1μm [14].

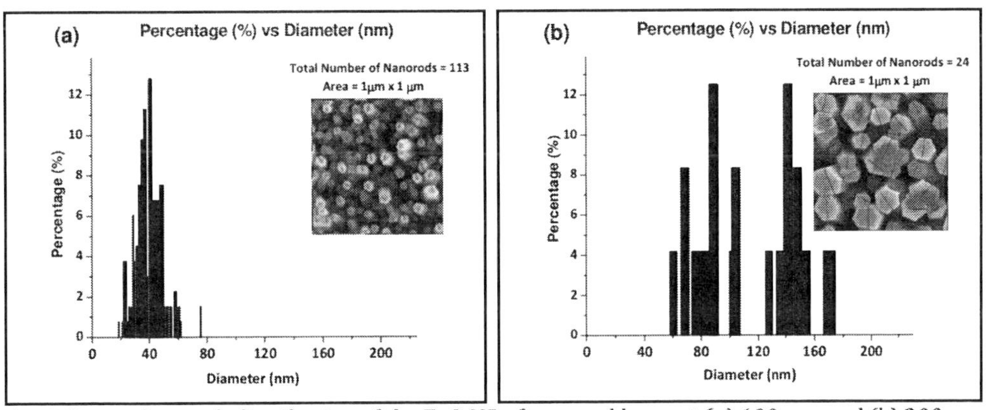

Fig. 3 *Size and growth distribution of the ZnO NRs from seed layer at (a) 100 nm, and (b) 200 nm.*

3.3. Structural properties by XRD

The XRD analysis provides the structural properties of the grown ZnO NRs for diffraction angle 2θ = 20° – 80°, with ZnO reference code of 01-080-0075 as the indicator. As shown in **Fig. 4**, the dominant

IOP Conf. Series: Materials Science and Engineering **854** (2020) 012074 doi:10.1088/1757-899X/854/1/012074

peaks were at 34.425° and 34.375°, very close to the standard lattice constant for unstrained ZnO, *hkl* position of 002 at 34.40°. As can be seen, ZnO NRs grown on 200 nm seed layer reveal a sharper peak, thus smaller full width half maximum (FWHM), which commonly correlated to the improvement of the ZnO crystal quality and related to the enlargement of the grain size [15,16]. Details on the data analysis has been tabulated in **Table 1**. From the data, both samples show that the NRs grown preferentially along *c*-axis, perpendicular to the substrate surface, which equivalent to (002) hexagonal close-packed (hcp) character, and this agrees well with the FESEM results. There is also a small shift, considerably affect the strain percentage implies to value of $\pm$0.07% for both samples and these was caused by tensile strain and compressive strain in the ZnO NRs.

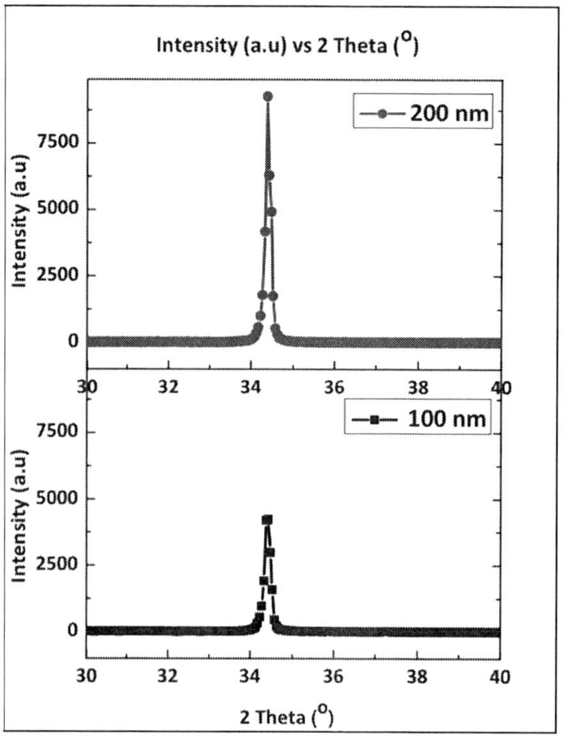

Fig 4. XRD plot of the grown ZnO NRs on the seed layer with a thickness of 100 nm (bottom), and 200 nm (top).

3.4. Diffused Reflectance UV-Vis Spectra and Kubelka-Munk Function

The optical properties of ZnO NRs were characterized using diffuse reflectance UV-Vis spectra in the range from 200 nm to 600 nm for both samples. As depicted in **Fig. 5a**, the grown ZnO NRs reflectance peaks show a blue-shifted characteristic for sample grown on 100 nm seed layer to 200 nm seed layer, with their peak shifted from 383 nm to 377 nm, respectively. This progressive observation was attributed to the increase in particle size [17]. Due to the larger particle size, the arrangement of atoms for the NRs follows the long range order of ZnO thus absorb towards smaller wavelength (higher energy) as the properties of bulk ZnO [18]. These data were pursued for Kubelka-Munk function analysis to determine the optical energy bandgap (E_g) of the ZnO NRs by extrapolating the plot to the *x*-axis as in **Fig. 5b**. The acquired values of the E_g were found to be, respectively at 3.10 eV and 3.16 eV, and this was towards the E_g of bulk ZnO. As listed in **Table 1**, the increase in the E_g value was observed due to the smaller crystallites size of the former sample having a smaller energy gap, as compared to the to the

IOP Conf. Series: Materials Science and Engineering **854** (2020) 012074 doi:10.1088/1757-899X/854/1/012074

larger crystallites size of the later sample, having a larger energy gap. This observation were normally found elsewhere when a small crystallite agglomerates, and formed a bigger crystallite [19].

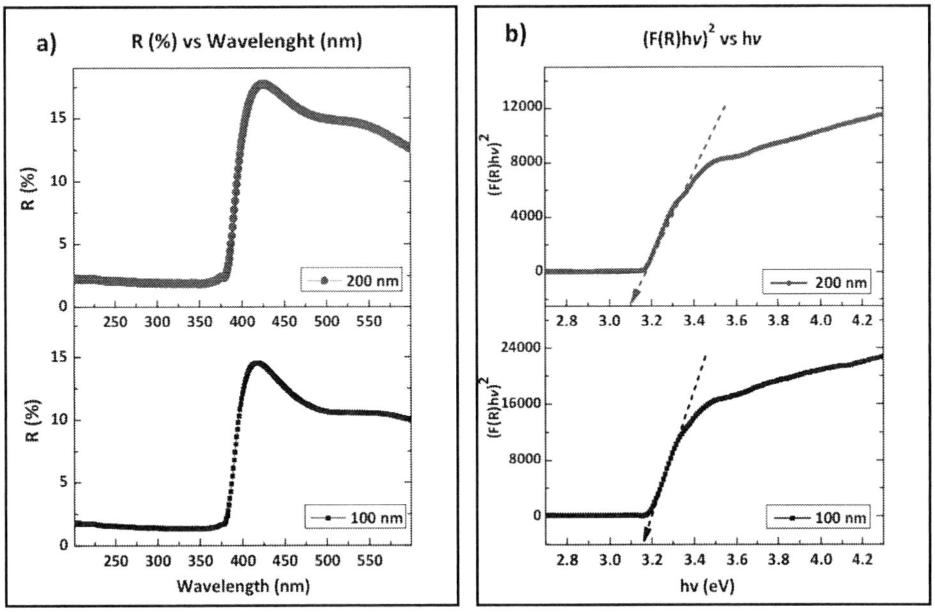

Fig. 5: (a) Diffused reflectance UV-Vis spectra of the grown ZnO NRs, and (b) plot of K-M function.

Table 1. Overall analysed data from FESEM, EDX, XRD, and UV-Vis from the ZnO NRs.

Result by FESEM and EDX				
NRs grown on seed layer thickness, (nm)	Average diameter, (nm)	Thickness CS (μm)	% Atomic, (%)	% Weight, (%)
100 nm	48.70	1.574	O - 43.51 Z - 56.49	O - 15.86 Z - 84.14
200 nm	90.86	1.542	O - 43.57 Z - 56.25	O - 15.89 Z - 84.11
Result by XRD				
NRs grown on seed layer thickness, (nm)	Peak position, 2θ, (°)	FWHM	D, (nm)	Lattice strain, (%)
100 nm	34.425	0.13° 0.0023 rad	41.91	0.07
200 nm	34.375	0.16° 0.0028 rad	44.60	-0.07
Result by UV-Vis spectra and K-M plot				
NRs grown on seed layer thickness, (nm)	Energy band gap, (eV)			
100 nm	3.10			
200 nm	3.16			

4. Conclusion

This work revealed the thicker seeding thickness from 100 nm to 200 nm enhances the structural and optical properties of the growth of ZnO NRs by CBD. FESEM analysis shows the well-aligned, same-height NRs morphology for both samples with the diameter size varies from 40.7 nm to 90.86 nm. XRD analyses revealed the intense and dominant peak at 2θ = 34.375° for ZnO NRs on 100 nm seed layer, compared to 2θ = 34.425° for ZnO NRs on 200 nm seed layer. It proved the ZnO NRs high quality of crystallinity through the sharp peaks. From UV-Vis and K-M analysis, the E_g was found to increase from 3.10 eV to 3.16 eV as the seed layer thicker. These values were towards the E_g of bulk ZnO at 3.37 eV. The seed layer introduced by RF sputtering in this work is found to be incorporated well with the CBD method, yields to good structural and optical quality of the ZnO NRs. The outcome of this result potentially to be utilized to study the dependence of ZnO NRs geometry in optoelectronic device performance.

Acknowledgement
The authors acknowledged contribution from RUI grant under Universiti Sains Malaysia, 1001/PFIZIK/8011092, and extended thanks to staff from School of Physics, USM. Appreciation goes to all the committees of International Technical Postgraduate Conference for this great opportunity.

References
[1] J. Kong, "Development of ZnO based light emitting diodes and laser diodes," 2012.
[2] J. Jiang, J. Pi, and J. Cai, "The advancing of zinc oxide nanoparticles for biomedical applications," *Bioinorg. Chem. Appl.*, vol. 1155, p. 1019, 2018.
[3] O. F. Farhat, M. M. Halim, N. M. Ahmed, and M. A. Qaeed, "ZnO nanofiber (NFs) growth from ZnO nanowires (NWs) by controlling growth temperature on flexible Teflon substrate by CBD technique for UV photodetector," *Superlattices Microstruct.*, vol. 100, pp. 1120–1127, 2016.
[4] H. Ghayour, H. R. Rezaie, S. Mirdamadi, and A. A. Nourbakhsh, "The effect of seed layer thickness on alignment and morphology of ZnO nanorods," *Vaccum*, vol. 86, no. 1, pp. 101–105, 2011.
[5] V. Manthina, T. Patel, and A. G. Agrios, "Number density and diameter control of chemical bath deposition of ZnO nanorods on FTO by forced hydrolysis of seed crystals," *J. Am. Ceram. Soc.*, vol. 97, no. 4, pp. 1028–1034, 2014.
[6] Y.-Z. Gu, H.-L. Lu, Y. Zhang, P.-F. Wang, S.-J. Ding, and D. W. Zhang, "Effects of ZnO seed layer annealing temperature on the properties of n-ZnO NWs/Al2O3/p-Si heterojunction," *Opt. Express*, vol. 23, no. 19, p. 24456, 2015.
[7] O. F. Farhat, M. M. Halim, N. M. Ahmed, A. A. Oglat, A. A. Abuelsamen, "A study of the effects of aligned vertically growth time on ZnO nanorods deposited for the first time on Teflon substrate," *Appl. Surf. Sci.*, vol. 426, pp. 906–912, 2017.
[8] H. Jin, S. Lee, Y. Yu, S. Min, H. Chul, and M. Yong, "Low-temperature hydrothermal growth of ZnO nanorods on sol – gel prepared ZnO seed layers : Optimal growth conditions," *Thin Solid Films*, vol. 524, no. December 2017, pp. 144–150, 2012.
[9] L. L. Yang, Q. X. Zhao, M. Willander, and J. H. Yang, "Effective way to control the size of well-aligned ZnO nanorod arrays with two-step chemical bath deposition," *J. Cryst. Growth*, vol. 311, pp. 1046–1050, 2009.
[10] K. H. Kim, K. Utashiro, Y. Abe, and M. Kawamura, "Growth of zinc oxide nanorods using various seed layer annealing temperatures and substrate materials," vol. 9, pp. 2080–2089, 2014.
[11] A. A. Semenova, N. A.Lashkova, A. I.Maximov, V. A.Moshnikov, D. A. Kudryashov, A. M.Mozharov, V. N. Verbitsky, P. A. Somov, "Formation of ZnO nanorods on seed layers for piezoelectric nanogenerators," *J. Phys. Conf. Ser. Pap.*, vol. 917, no. 032022, pp. 1–7, 2017.
[12] T. S. Tlemcani, C. Justeau, K. Nadaud, G. Poulin-vittrant, and D. Alquier, "Deposition time and annealing effects of ZnO seed layer on enhancing vertical alignment of piezoelectric ZnO nanowires," *Chemosensors*, vol. 7, no. 7, pp. 1–13, 2019.
[13] F. H. Alsultany, Z. Hassan, N. M. Ahmed, N. G. Elafadill, and H. R. Abd, "Effects of ZnO seed layer thickness on catalyst-free growth of ZnO nanostructures for enhanced UV photoresponse," *Opt. Laser Technol.*, vol. 98, pp. 344–353, 2018.
[14] A. Takeyama, S. Yamamoto, K. Kitamura, T. Yatsui, M. Ohtsu, S. H. Baek, I. K. Park,

"Formation of ZnO nanorods on seed layers for piezoelectric nanogenerators," *J. Phys. Conf. Ser. Pap.*, vol. 917, no. 03202, pp. 1742–6596, 2017.

[15] P. S. Shinde, H. Divi, C. Bhosale, and K. Rajpure, "Photoelectrocatalytic degradation of oxalic acid by spray deposited nanocrystalline zinc oxide thin films," *J. Alloys Compd.*, no. July, 2012.

[16] W. Maryam, N. M. Subhi, S. M. Saad, N. Fazrina, M. Halim, and M. Hashim, "Annealing effects on photonic band gap of ZnO nanorod grown by chemical bath deposition," *Bul. Opt.*, vol. 2016, no. Cvd, pp. 22–27, 2016.

[17] R. Raji and K. G. Gopchandran, "ZnO nanostructures with tunable visible luminescence: Effects of kinetics of chemical reduction and annealing," *J. Sci. Adv. Mater. Devices*, vol. 2, no. 002, pp. 1–35, 2017.

[18] X. Li, X. Chen, Z. Yi, Z. Zhou, Y. Tang, and Y. Yi, "Fabriction of ZnO nanorods with strong UV absorption and different hydrophobicity on foamed nickel under different hydrothermal conditions," *Micromachines*, vol. 10, no. 164, pp. 1–13, 2019.

[19] H. Sarma, D. Chakrabortty, and K. C. Sarma, "Structural and optical properties of ZnO nano particles," *J. Appl. Phys.*, vol. 6, no. 4, pp. 8–12, 2014.

IOP Conf. Series: Materials Science and Engineering **854** (2020) 012075 doi:10.1088/1757-899X/854/1/012075

Microbottle-Resonator Ethanol Liquid Sensor

M A M Johari[1,2], A H Rosol[1], N A Baharuddin[1], M I M A Khudus[3], M H Jali[1,4], Maslinda M S[1], S S Jaapar[5,6], S W Harun[1].

[1] Department of Electrical Engineering, University of Malaya, 50603 Kuala Lumpur, Malaysia.
[2] Faculty of Engineering Technology, Universiti Teknikal Malaysia Melaka, 76100 Melaka, Malaysia
[3] Department of Physics, Faculty of Science, University of Malaya, 50603 Kuala Lumpur, Malaysia
[4] Faculty of Electrical Engineering, Universiti Teknikal Malaysia Melaka, 76100 Melaka, Malaysia
[5] School of Data Sciences, Perdana University
[6] Adnuri SMA Research Center Sdn.Bhd

Email: swharun@um.edu.my

Abstract. This investigation studies the use microbottle resonator (MBR) as ethanol liquid sensor. This resonator made-up from silica fiber SMF28 using process recognized as "soften-and-compress" to form bottle structure with a diaphragm diameter $D_b = 170\mu m$, stem diameter of $D_s = 125\mu m$ and bottle length $L_b = 180\mu m$. The MBR characterized by via 3µm tapered microfiber and bright to have >10^4 for the quality factor value (Q-factor). The MBR was formerly used as ethanol liquid sensor for the range between 10% to 100% and compared with non-MBR microfiber for wide-ranging performance. The MBR sensor was initiate to have a sensitivity 0.1756 dB/%ppm with linearity 99.28% and standard deviation of 3.355 dB, which was exceptional compare to the non-MBR microfiber in all limitation measured.

1. Introduction

Optical microresonator (OMR) has established significant reputation due to widespread ability in sensors, laser and also for some plasmonic devices [1-9]. The OMR has been designed in perceptible structure such as micropillar, microring also microdisc and it was expressively work at some definite wavelength by internal reflection in continuous resonant [10-18]. The concept of this reflection was primarily exposed by Lord Rayleigh in the dome of St. Paul's cathedral in London which is acknowledged as whispering gallery modes (WGM) [19]. Lord Rayleigh revealed that along the dome length at some specific point, the sound could be hear obviously. The sound wave was then so-called as whispering gallery modes. The idea of the sound along the dome dimension was then interpreted in the optical micro resonator with the alike construction, which produce alike resonance structures for WGM. The WGM which functioned on resonator, was then revealed having numerous advantages including in fabrication, minimal losses and able to produce high quality factor (Q-factor) based on resonator classification [20, 21]. One of WGM resonator geometry apprehended interest recently due to competence on having high Q-factor. It was known as micro-bottle resonator (MBR); designed in micro size from silica-based fiber [21, 22]. Due to these advantages including high sensitivity, high accuracy lesser losses, the MBR has been use in variety sensing purpose, including humidity and formaldehyde sensing [23-27].

In this paper we explore the capability of MBR for ethanol liquid sensor. The performance of MBR as ethanol liquid sensor been equated with non-MBR microfiber. The MBR was moulded from standard silica fiber SMF28 using technique so-called "soften-and-compress". For characterization of the MBR, 3μm microfiber is used together with optical power meter and tuneable laser source. The MBR then employed for ethanol liquid in range of 10% - 100% ppm and compared with non-MBR microfiber for complete sensing performance.

2. Micro-bottle Resonator Fabrication and Characterization

The optical micro-resonator was form using silica fiber SFM28 using method recognized as "soften-and-compress" [28]. A manual splicing machine (Furukawa Electric Fitel S178A) is used to fixed silica fiber position. The heat from the splicing machine is applied on middle area and the fiber is compressed inwards from both sides. This procedure allowed to create bugle on silica fiber and the bugle size determined by numbers of arc employed [22]. The resonator in bottle structure define in three parameters as the bottle diameter D_b, the stem diameter D_s and the bottle length L_b, as show in **Figure 1**. The bare microfiber with a diameter waist of 3μm is made-up from SMF28 silica fiber using "flame brushing" procedure [29].

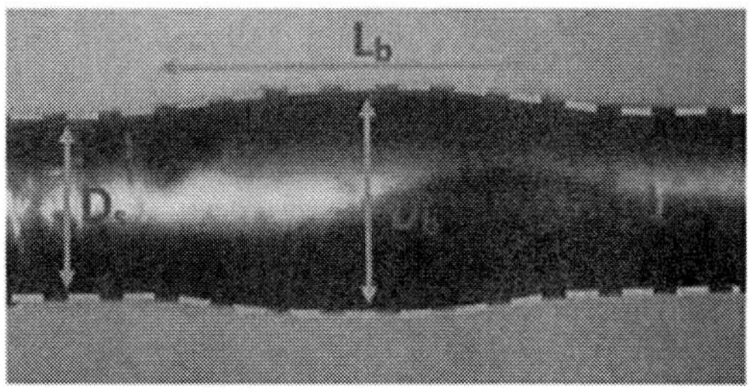

Figure 1. The MBR fabricated from SMF28 with Lb = 180 μm, Db = 170 μm and Ds = 125 μm

For the characterization of the MBR, tuneable laser source (ANDO AQ4321D) is used to supply range of wavelength and optical power meter (THORLABS S145C) for collector to record final power output. This tuneable laser source (TLS) able to supply wavelength in range between 1520 nm to 1620 nm, which is fit for the MBR characterization. The wavelength from the TLS was adjusted from 1550 nm to 1575 nm (with 0.001nm interval) and launched into the MBR over tapered microfiber with 3μm waist diameter. **Figure 2** denoted the transmission spectral of the MBR with numerous sharp resonant peak [30]. The graph displayed the insertion loss roughly at -17dBm and it possibly enhanced by monitoring a gap of the coupling between the MBR and tapered microfiber [31]. The Q-factor used to determine the quality of resonant peak of the MBR and define as $\Delta\lambda/\lambda$ where λ is the resonant wavelength and is recorded to be 1.534 x 10^5 which is lesser than the earlier work [32]. The is owed to the tapered microfiber size and the size of the MBR and give a huge influence in Q-factor.

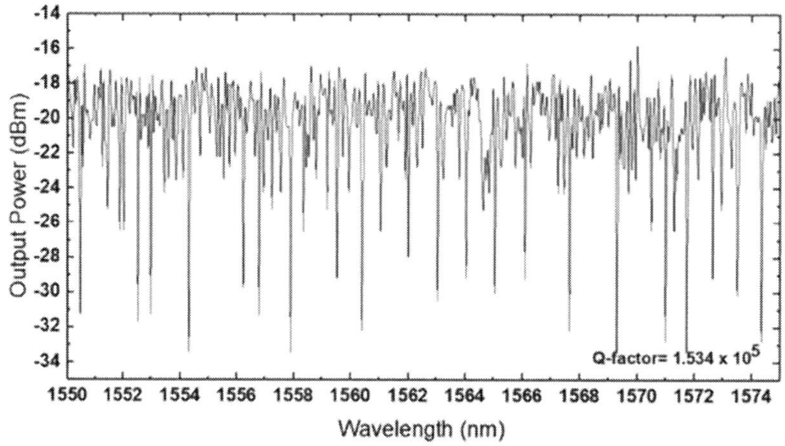

Figure 2. The transmission spectral of the MBR with several resonant peaks while couple with tapered microfiber 3μm waist diameter.

3. Ethanol Liquid with Different Percentage of Concentration Preparation Process

Ethanol is a simple alcohol which can be produced synthetically from petroleum by-product or naturally from plant fermentation. It appears as colourless liquid organic solvent and possess hydroxyl (OH) group moieties, two single bonded carbon atoms with five hydrogen atoms in its molecular structure [33, 34]. Ethanol solvent is particularly important chemicals in research, medicine, industrial as well as the most critical ingredient in food and beverage in our daily lives [33, 35, 36] highlights the considerable amount of ethanol percentage should be presence in the food and beverage industry as it is crucial to maintain the quality and safety of products. This is due to the fact that ethanol is suffered from alcohol intoxication which may cause anxiolytic effect, short term memory loss, locomotor suppression, motor incoordination and the loss of righting reflex or narcosis to the consumer [33]. On the other hand, in ASEAN country such as Malaysia, Thailand and Singapore the halal products are allow to have a little of ethanol content (0.5-1%), while Brunei, United Kingdom and Canada ethanol is not allowed [37]. Due to these phenomena, variable methods have been used in detecting ethanol such as Fourier Transform Infrared Spectrometer (FTIR), chromatography and raman spectrophotometer. Unfortunately, the drawback of this method is time consuming and require large sample [38]. To overcome this limitation, researchers have developed optical ethanol sensor which is able to detect the concentration of ethanol in rapid and accurate method for the on-site ethanol detection [36, 38].

4. Experiment Setup for The MBR Ethanol Liquid Sensing

An experiment setup of the MBR ethanol liquid sensor shown in **Figure 3**. The microfiber and the MBR are positioned on the stage with an ethanol liquid located under the stage. The MBR is situated between the microfiber and ethanol liquid. This is to certify the MBR able to couple with the microfiber and touch the ethanol in the same time. The setup was seal inside the compartment at ambient room temperature of 250C at atmospheric pressure 1.0 atm. One end of fiber connected to TLS while another end is connected to OPM. The ethanol liquid was then changeable from 10% to 100% concentrations. The TLS wavelength is 1560 nm and the output power recorded at different ethanol concentrations. Both performances are compared to define the best ethanol liquid sensing.

IOP Conf. Series: Materials Science and Engineering **854** (2020) 012075 doi:10.1088/1757-899X/854/1/012075

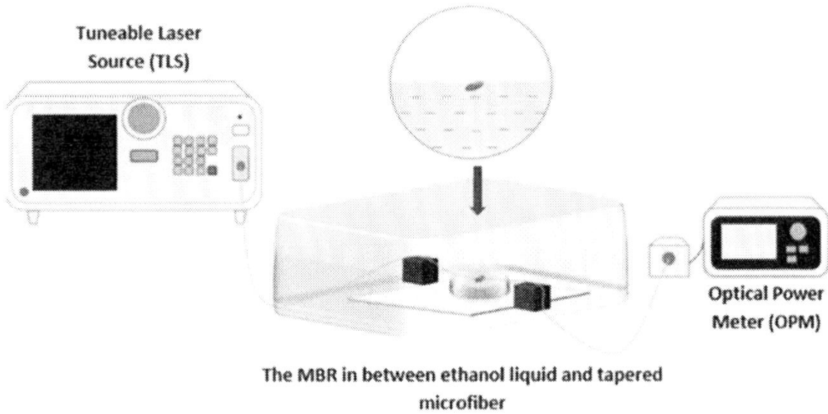

Figure 3. The MBR in between ethanol liquid and tapered microfiber of 3μm. The tuneable laser source and optical power meter connected to the end of fiber.

5. The MBR and Bare Microfiber Performance as Ethanol Liquid Sensor

The averaged output power (dBm) of the MBR and non-MBR microfiber at different concentrations of ethanol display in **Figure 4**. Usually, the output power increased once the level of concentration rises. The outcomes of the sensor performance showed in **Table 1** wherever sensitivity, linearity and standard deviation used as parameters. The performance of the MBR is so much surpass than non-MBR microfiber for all parameter. Truly, the sensitivity of the MBR is 0.07 dB/%ppm then non-MBR microfiber (0.1756 dB/% as compare to 0.1049 dB/%). The linearity of the MBR is 99.28%, better than non-MBR microfiber where is only 83.01%. All these outcomes showed the MBR is outperforms the non-MBR microfiber as per ethanol liquid sensor. Throughout spectral transmission, the light experienced multiple rotation on the MBR where is each circular were increase the adsorption, consequently increase the sensitivity [39, 40].

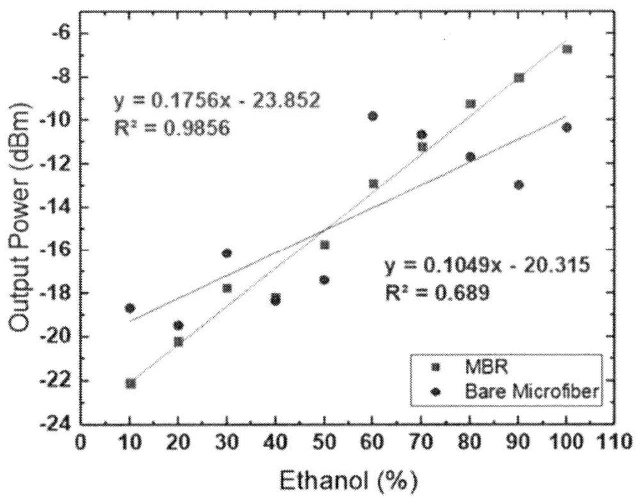

Figure 4. The output power value for MBR and bare microfiber varies with percentage of ethanol concentrations.

Table 1. Analysis of MBR and bare microfiber performance in ethanol liquid sensing activity.

Parameters	Non-MBR Microfiber	With MBR
Linearity (%)	83.01%	99.28%
Sensitivity (dB/%ppm)	0.1049	0.1756
Standard deviation (dBm)	5.827	3.355
Linear Range (% ppm)	10 - 100	10 - 100

6. Conclusion

We have examined the performance of MBR ethanol liquid sensors. The "soften-and-compress" procedure was used to structure the MBR with bottle diameter of D_b=170 μm, stem diameter of D_s=125 μm and bottle length of L_b=180 μm. A silica microfiber by a waist diameter of 3 μm was applied to launch a light from a TLS to the MBR. The TLS wavelength range remained set to 1550.00 nm and 1575.00 nm with an intermission of 0.001 nm for the measurement. A Q-factor of >105 was documented for the MBR. Following, the performance of the MBR ethanol liquid sensor were examine and compare with non-MBR microfiber. The sensitivity, linearity and standard deviation would be being documented to evaluate the performance of the MBR sensor. The linearity of the MBR are >90% which is show that the MBR are performed so well than non-MBR microfiber is only <90%. However, the MBR showed batter sensitivity as compared to non-MBR microfiber, whose highest sensitivity 0.1756dB/%RH. standard deviation for non-MBR microfiber is much higher then the MBR which less accuracy for non-MBR microfiber as ethanol liquid sensor.

Acknowledgement

The authors would like to acknowledge University of Malaya and Ministry of Education, Malaysia, Fakulti Teknologi Kejuruteraan Elektrik dan Electronik, Universiti Teknikal Malaysia Melaka and the University of Malaya for their financial support.

References

[1] Chiasera A, Dumeige Y, Feron P, Ferrari M, Jestin Y, Nunzi Conti G, Pelli S, Soria S and Righini G C 2010 Spherical whispering-gallery-mode microresonators *Laser & Photonics Reviews* **4** 457-82

[2] Matsko A, Savchenkov A, Strekalov D, Ilchenko V and Maleki L 2005 Review of applications of whispering-gallery mode resonators in photonics and nonlinear optics *IPN Progress Report* **42** 1-51

[3] Berini P 2009 Long-range surface plasmon polaritons *Advances in optics and photonics* **1** 484-588

[4] Berini P, Charbonneau R and Lahoud N 2007 Long-range surface plasmons on ultrathin membranes *Nano letters* **7** 1376-80

[5] Bykov V and Silichev O 1995 Laser Resonators, Cambridge Intl *Science Publ*

[6] Rosol A H A, Rahman H A, Ismail E I, Irawati N, Jusoh Z, Latiff A A and Harun S W 2017 Cadmium Selenide Polymer Microfiber Saturable Absorber for Q-Switched Fiber Laser Applications *Chinese Phys. Lett* **34** 094202

[7] Rosol A, Latiff A, Khudus M A and Harun S J I J o P 2019 Nanosecond pulses generation with rose gold nanoparticles saturable absorber 1-5

[8] Rosol A H A, Rahman H A, Ismail E I, Jusoh Z, Latiff A A and Harun S W J C O L 2017 Mode-locked fiber laser with a manganese-doped cadmium selenide saturable absorber **15** 071405

[9] Rosol A H A, Rahman H A, Ismail E I, Irawati N, Jusoh Z, Latiff A A and Harun S W J C P L 2017 Cadmium selenide polymer microfiber saturable absorber for Q-switched fiber laser applications **34** 094202

[10] Little B E, Chu S T, Haus H A, Foresi J and Laine J-P 1997 Microring resonator channel dropping filters *Journal of lightwave technology* **15** 998-1005

[11] Absil P P 2000 Microring resonators for wavelength division multiplexing and integrated photonics applications

[12] Kokubun Y 2005 Wavelength selective integrated device by vertically coupled microring resonator filter *Photonics Based on Wavelength Integration and Manipulation IPAP Books* **2** 303-16

[13] Armani D, Kippenberg T, Spillane S and Vahala K 2003 Ultra-high-Q toroid microcavity on a chip *Nature* **421** 925

[14] Armani D, Min B, Martin A and Vahala K J 2004 Electrical thermo-optic tuning of ultrahigh-Q microtoroid resonators *Applied physics letters* **85** 5439-41

[15] Gérard J, Sermage B, Gayral B, Legrand B, Costard E and Thierry-Mieg V 1998 Enhanced spontaneous emission by quantum boxes in a monolithic optical microcavity *Physical review letters* **81** 1110

[16] Sarma J and Shore K 1985 Electromagnetic theory for optical disc resonators *IEE Proceedings J (Optoelectronics)* **132** 325-30

[17] Srinivasan K, Barclay P E, Painter O, Chen J, Cho A Y and Gmachl C 2003 Experimental demonstration of a high quality factor photonic crystal microcavity *Applied Physics Letters* **83** 1915-7

[18] Lee P-T, Lu T-W, Tsai F-M and Lu T-C 2006 Investigation of whispering gallery mode dependence on cavity geometry of quasiperiodic photonic crystal microcavity lasers *Applied physics letters* **89** 231111

[19] Rayleigh L 1910 CXII. The problem of the whispering gallery *The London, Edinburgh, and Dublin Philosophical Magazine and Journal of Science* **20** 1001-4

[20] Lee P-T, Lu T-W, Fan J-H and Tsai F-M 2007 High quality factor microcavity lasers realized by circular photonic crystal with isotropic photonic band gap effect *Applied physics letters* **90** 151125

[21] Bianucci P 2016 Optical microbottle resonators for sensing *Sensors* **16** 1841

[22] Murugan G S, Wilkinson J S and Zervas M N 2009 Selective excitation of whispering gallery modes in a novel bottle microresonator *Optics express* **17** 11916-25

[23] Johari M A M, Khudus M I M A, Jali M H B, Al Noman A and Harun S W 2019 WHISPERING GALLERY MODES ON OPTICAL MICRO-BOTTLE RESONATOR FOR HUMIDITY SENSOR APPLICATION *Optik*

[24] Johari M, Khudus M A, Al Noman A, Jali M, Yusof H, Harun S and Yasin M 2019 Microbottle resonator formaldehyde sensor. In: *Journal of Physics: Conference Series*: IOP Publishing) p 012021

[25] Johari M, Al Noman A, Khudus M A, Jali M, Yusof H, Harun S and Yasin M 2018 Microbottle resonator for formaldehyde liquid sensing *Optik* **173** 180-4

[26] Johari M A M, Khudus M I M A, Jali M H B, Al Noman A and Harun S W 2018 EFFECT OF SIZE ON SINGLE AND DOUBLE OPTICAL MICROBOTTLE RESONATOR HUMIDITY SENSORS *Sensors and Actuators A: Physical*

[27] Matsko A B and Ilchenko V S 2006 Optical resonators with whispering gallery modes I: basics *IEEE J. Sel. Top. Quantum Electron* **12** 3

[28] Murugan G S, Petrovich M, Jung Y, Wilkinson J and Zervas M 2011 Hollow-bottle optical microresonators *Optics express* **19** 20773-84

[29] Lim K, Harun S, Arof H and Ahmad H 2012 *Selected Topics on Optical Fiber Technology*:

InTech)

[30] Nasir M N M, Murugan G S and Zervas M N 2016 Broadly tunable solid microbottle resonator. In: *Photonics Conference (IPC), 2016 IEEE*: IEEE) pp 759-60

[31] Cai M, Painter O and Vahala K J 2000 Observation of critical coupling in a fiber taper to a silica-microsphere whispering-gallery mode system *Physical review letters* **85** 74

[32] Nasir M N M, Murugan G S and Zervas M N 2016 Spectral cleaning and output modal transformations in whispering-gallery-mode microresonators *JOSA B* **33** 1963-70

[33] Pauzi N, Man S, Nawawi M S A M and Abu-Hussin M F B 2019 Ethanol standard in halal dietary product among Southeast Asian halal governing bodies *Trends in Food Science & Technology*

[34] Alzeer J and Hadeed K A 2016 Ethanol and its Halal status in food industries *Trends in Food Science & Technology* **58** 14-20

[35] Gorgus E, Hittinger M and Schrenk D 2016 Estimates of ethanol exposure in children from food not labeled as alcohol-containing *Journal of analytical toxicology* **40** 537-42

[36] Nguyen N L T, Baek S H, Akbar Z A, Jang S-Y, Ha S, Park J P and Park T J 2018 Rapid Determination of Ethyl Alcohol in Alcoholic Beverages Using a Fluorescent Nanofiber Film *BioChip Journal* **12** 240-8

[37] Jamaludin M, Hashim D, Rahman R, Ramli M, Majid M, Othman R and Amin A 2016 Determination of permissible alcohol and vinegar in Shariah and scientific perspectives *International Food Research Journal* **23**

[38] Mulijani S, Iswantini D, Wicaksono R and Notriawan D 2018 Optical Sensor based Chemical Modification as a Porous Cellulose Acetate Film and Its Application for Ethanol Sensor. In: *IOP Conference Series: Materials Science and Engineering*: IOP Publishing) p 012014

[39] Arregui F J, Liu Y, Matias I R and Claus R O 1999 Optical fiber humidity sensor using a nano Fabry–Perot cavity formed by the ionic self-assembly method *Sensors and Actuators B: Chemical* **59** 54-9

[40] Batumalay M, Harun S W, Irawati N, Ahmad H and Arof H 2015 A study of relative humidity fiber-optic sensors *IEEE Sensors Journal* **15** 1945-50

IOP Conf. Series: Materials Science and Engineering **854** (2020) 012076 doi:10.1088/1757-899X/854/1/012076

Formulating an Anorganic Membrane Using Clay, Activated Carbon and Micro Zeolite as Filter Media for Peat Water Purification

N Nahar, S Saifuddin, Muhammad Sami

Chemical Engineering Department, Politeknik Negeri Lhokseumawe
Jl. Medan-Banda Aceh, Buketrata, Lhokseumawe

E-mail: nahar@gmail.com

Abstract. Peat water is surface water in swampy or lowland areas that has a high color intensity, low pH (3-5) and strongly metals' bonded. In proper treatment, the existence of peat water can meet the needs of the surrounding community. Additionally, keeping clean water or drinking water can be helped by the alternative treatment of peat water. Peat inorganic ceramic membrane which can be in accordance with water quality based on the Regulation of the Minister of Health of the Republic of Indonesia Number: 492 / Menkes / Per / IV / 2010. Ceramic membrane made from clay and modified with activated carbon and zeolite in such way turned out to be able to produce drinking water in accordance with drinkable water standards. This study aims to determine the membrane's ability to treat peat water and determine the magnitude of turbidity reduction, total dissolved solid (TDS), pH, Fe and Mn and E. coli bacteria. This research was carried out by filtering peat water through the membrane. Membranes was made of tubes with specifications used are 7 cm high and 6.8 cm in diameter and 0.8 cm wide with various mixture of clays, activated carbon and zeolites. The compositions are 80; 10; 10; 70; 20; 10; 60; 30; 10, 50; 40; 10, respectively, and 140 mesh active carbon size variations. After the investigation, it turns out that all of the composition 80; 10; 10 (clay: activated carbon: zeolite) with a combustion temperature of 800°C is the best membrane removal result that produces standards drinkable water content with clear color intensity.

1. Introduction

Well water and river water are peat water sources in Indonesia that can be processed into water consumption. Since these waters generally do not meet the requirements of clean water so further processing is needed.

The development of clean water treatment technology with membrane technology is a promising treatment process with good quality and also suitable for drinking water treatment in developing countries because membranes have many advantages. The common filter used is the ceramic membrane as one of the appropriate technologies. Among the advantages of membrane technology are producing quality water, using less chemicals, ease of operation (automatic), being able to set aside pollutants in a large range so that it does not require a large place and a ceramic water filter then produce water that is free from dirt or microorganisms.

In reference [1], the authors have made ceramic membranes to improve the quality of swamp water. The results of this study show that the performance of ceramic filters made from a mixture of clay, coal fly ash and iron powder is relatively effective in producing good quality permeate. This is reflected in

IOP Conf. Series: Materials Science and Engineering **854** (2020) 012076 doi:10.1088/1757-899X/854/1/012076

the reduction in the content of heavy metal iron (Fe) ions in swamp water to reach 91.54% and the content of organic matter (KmnO4) reached 84.83% followed by a reduction in total dissolved solid (TDS) approximately 60.2% [2] and [3].

Porous carbon, or better known as activated carbon, is used as an adsorbent for color removal, waste treatment, water purification. Activated carbon is charcoal that has changed its physical and chemical properties due to the activation treatment carried out with activators of chemicals or by heating at high temperatures. Activated carbon then forms amorphous which consists mainly of free carbon and has a hollow inner surface, black color, odorless, tasteless, and has a far greater absorption compared to carbon that has not undergone the activation process [4] and [5].

Zeolites are a group of non-metal natural minerals which have a chemical structure of aluminum-silicate framework found in nature with high cation exchange capacity, high adsorption and hydration-dehydration properties. The color of the zeolite itself is white-grayish, white-greenish, and yellowish-white. The size of the zeolite ranges from no more than 10-15 microns.

In this research, a ceramic membrane was made using clay, activated carbon nanoparticles and zeolite nanoparticles as drinking water filter media. Nanoparticles are currently used for the developments in science and technology. Nano-scale material that can enhance the physical, mechanical and chemical properties of a material without having to damage the atomic structure.

Additionally, this research aims to find alternative water treatment. Basically, ceramic material has sufficient porosity to separate impurities, is relatively durable and has sufficient mechanical strength and is not reactive to water.

2. Method

2.1. Materials and Tools

The tools used are Digital Scales, Turbidimeters, TDS meters, pH meters, Sieve size 140, planetary ball milling Fritsch pulverisette, PSA and AAS. Materials used are Clay, Zeolite, Activated Carbon, HCl 2M and Aquadest.

2.2. Substance Activation

The used chemical substances such as zeolite, activated carbon and zeolite particles must activate in a chemical process before being used as filter media for the peat water treatment. The making of ceramic membrane as the mixture of clays and those mentioned substances also put into account.

2.2.1. Zeolite Activation. This activation is proceeded by grinding zeolite as much as 2 kg then soaked with 2.2 litters of 2M HCl solution for 1 hour, filtered, washed with distilled water until the pH of the zeolite is neutral and roasted at 130°C for 4 hours.

2.2.2. The making of Activated Carbon Particles and Zeolite Particles. Zeolites are sieved with 140 mesh sieves. The making of zeolite particles and activated carbon particles was carried out using the top down method using the planetary ball milling Fritsch pulverisette by grinding the starting material (zeolite and activated carbon) into the milling device. Every milling time, 80 grams of zeolite is utilized by using a ball mill with a speed of 350 rpm lasts for 6 hours.

2.2.3. The making of Ceramic Membranes. Mixing clay with activated carbon and zeolite particles in a ratio 80%; 10%; 10%, 70%; 20%; 10%, 60%; 30%; 10% and 50%; 40%; 10%. Then added water and stirred evenly. After that the material is printed manually shaped tube. The printed material is dried at room temperature for 4 days. The printed material is burned with a combustion temperature of 600°C, 700°C and 800°C so that a complete dehydration (drying) and vitrification (change of chemical elements from clay to ceramic) forms a ceramic filter.

3. Results and Discussion

Based on the peat water test results obtained from the analysis of the initial characteristics of several parameters listed in table 1.

In reference to the standard of drinking water quality in the Republic of Indonesia Ministry of Public Health Regulation No. 492 / MECHANISM / PER / IV / 2010 [6] and thus the peat water has exceeded the threshold for use as drinking water. Therefore, peat water should be treated first before being used as drinking water.

Table 1. Considered parameters

Parameter	Units	Value
TDS	mg/L	550
Turbidity	NTU	2.99
pH	-	5.5
Fe	mg/L	1.2143
Mn	mg/L	7.1294

However, in this paper the considered parameter that put into the discussion is turbidity. As illustrated in Figure 1.

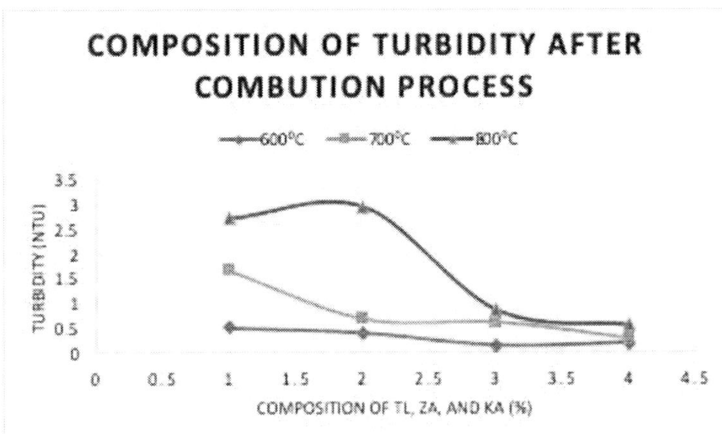

Figure 1. Effect of burning temperature of membranes on the efficiency of allowance for turbidity content

The highest efficiency of turbidity reduction is a membrane with a composition of 60%: 30%: 10% at a temperature of 600°C of 94.64%. in addition, the lowest turbidity reduction efficiency is membrane with a composition of 80%: 10%: 10% at a temperature of 800°C that is equal to 38.46%. The amount of turbidity removal efficiency in peat water is influenced by the combustion temperature that is not too high as well as the large amount of activated carbon that varies with the clay and zeolites in the membrane. While the value of turbidity removal efficiency is small which is influenced by high combustion temperatures, causing large pores and cracks in the membrane that make the turbidity in peat water not filtered properly.

4. Conclusion

Based on the results, it can be concluded that clay-based earthenware filters with a mixture of activated carbon and zeolite can reduce turbidity. The best decrease in turbidity is at a temperature of 600°C with a composition of clay: active carbon: zeolite (50%: 40%: 10%). The more activated carbon used in the making of membranes, the more pores are formed and also form a larger surface area so that more particles are retained and conversely the less active carbon is used the smaller the surface area so that the particles are retained less.

References

[1] Agmalini, S., Lingga, N.N and Nasir, S. 2013. Improving Swamp Water Quality Using Ceramic Membranes Made from Clay and Coal Fly Ash. Journal of Chemical Engineering. Vol.19 (2), p. 59-68.

[2] Anggriawan, A., Saputra, E and Olivia, M. 2015. Allowance for Fe and Mn Metal Content in Peat Water by the Utilization of Kaolin Geopolymers as Adsorbents. Jom FTEKNIK. 2 (1), p. 1-6.

[3] Baker, W.R. 2004. Membrane technology and Applications., 2nd edition., California: Jon Willey & Sons.

[4] Asip, F and Okta, T. 2013. H2s Adsorption on Natural Gas Using Ceramic Membranes with Iodometric Titration Method. Journal of Chemical Engineering. 19 (4), p. 22-28.

[5] Nurhayati, C and Susanto, T. 2015. Utilization of Fly Ash Coal as a Ceramic Membrane Material in the Peat Water Treatment Unit. Journal of Industrial Research Dynamics. 26 (2), p. 95-105.

[6] Indonesia Ministry of Public Health, 2010. Regulation of the Minister of Health of the Republic of Indonesia. No. 492 / MENKES / PER / IV / 2010 concerning Drinking Water Quality Requirements. Jakarta: Ministry of Health.

AUTHOR INDEX